Bacterial Pathogen and their Resistance Mechanisms

Bacterial Pathogen and their Resistance Mechanisms

Editor: Adrianna Butcher

AMERICAN
MEDICAL PUBLISHERS
www.americanmedicalpublishers.com

AMERICAN
MEDICAL PUBLISHERS
www.americanmedicalpublishers.com

Cataloging-in-Publication Data

Bacterial pathogen and their resistance mechanisms / edited by Adrianna Butcher.
p. cm.
Includes bibliographical references and index.
ISBN 978-1-63927-908-1
1. Pathogenic bacteria. 2. Pathogenic microorganisms. 3. Bacterial diseases.
4. Medical bacteriology. 5. Host-parasite relationships. 6. Drug resistance in microorganisms.
I. Butcher, Adrianna.
QR201 .B33 2023
616.014--dc23

American Medical Publishers,
41 Flatbush Avenue,
1st Floor, New York,
NY 11217, USA

ISBN 978-1-63927-908-1 (Hardback)

Contents

Preface

It is often said that books are a boon to mankind. They document every progress and pass on the knowledge from one generation to the other. They play a crucial role in our lives. Thus I was both excited and nervous while editing this book. I was pleased by the thought of being able to make a mark but I was also nervous to do it right because the future of students depends upon it. Hence, I took a few months to research further into the discipline, revise my knowledge and also explore some more aspects. Post this process, I begun with the editing of this book.

Bacteria are microscopic, single-celled organisms. They can be found in nearly all places on Earth and are essential to the planet's ecosystems. The majority of bacteria in the body are harmless and even beneficial in certain cases, while a small number of species cause disease. Pathogenic bacteria are a group of disease causing bacteria. They are uniquely endowed and adapted with mechanisms for overcoming the body's natural defenses. They can also attack parts of the body where bacteria are not found often like the blood. There are many different mechanisms through which bacteria develop antimicrobial resistance, but the main cause of developing resistance is the abuse and overuse of antibiotics. The majority of pathogenic bacteria can be developed in cultures and are identified using Gram stain as well as other techniques. This book includes some of the vital pieces of work being conducted across the world, on various topics related to pathogenic bacteria. Students and professionals actively engaged in the study of resistance mechanisms of bacteria will find it full of crucial and unexplored concepts.

I thank my publisher with all my heart for considering me worthy of this unparalleled opportunity and for showing unwavering faith in my skills. I would also like to thank the editorial team who worked closely with me at every step and contributed immensely towards the successful completion of this book. Last but not the least, I wish to thank my friends and colleagues for their support.

Editor

1

Increased Prevalence of Carbapenem Resistant *Enterobacteriaceae* in Hospital Setting Due to Cross-Species Transmission of the *bla*$_{NDM-1}$ Element and Clonal Spread of Progenitor Resistant Strains

Xuan Wang[1], *Gongxiang Chen*[1], *Xiaoyan Wu*[2], *Liangping Wang*[3], *Jiachang Cai*[1], *Edward W. Chan*[4,5], *Sheng Chen*[4,5]* *and Rong Zhang*[1]*

[1] *Second Affiliated Hospital of Zhejiang University, Zhejiang University, Hangzhou, China,* [2] *Second People's Hospital of Jiaxing, Jiaxing, China,* [3] *People's Hospital of Pinghu, Jiaxing, China,* [4] *Shenzhen Key lab for Food Biological Safety Control, Food Safety and Technology Research Center, Hong Kong PolyU Shen Zhen Research Institute, Shenzhen, China,* [5] *State Key Lab of Chirosciences, Department of Applied Biology and Chemical Technology, The Hong Kong Polytechnic University, Kowloon, Hong Kong*

***Correspondence:**
Sheng Chen,
State Key Lab of Chirosciences, Department of Applied Biology and Chemical Technology, The Hong Kong Polytechnic University, 11 Yuk Choi Rd, Hung Hom, Kowloon, Hong Kong, China
sheng.chen@polyu.edu.hk;
Rong Zhang,
Second Affiliated Hospital of Zhejiang University, Zhejiang University, 88 Jiefang Rd, Shangcheng, 310009 Hangzhou, China
brigitte_zx@163.com

This study investigated the transmission characteristics of carbapenem-resistant *Enterobacteriaceae* (CRE) strains collected from a hospital setting in China, in which consistent emergence of CRE strains were observable during the period of May 2013 to February 2014. Among the 45 CRE isolates tested, 21 (47%) strains were found to harbor the *bla*$_{NDM-1}$ element, and the rest of 24 CRE strains were all positive for *bla*$_{KPC-2}$. The 21 *bla*$_{NDM-1}$—borne strains were found to comprise multiple *Enterobacteriaceae* species including nine *Enterobacter cloacae*, three *Escherichia coli*, three *Citrobacter freundii*, two *Klebsiella pneumoniae*, two *Klebsiella oxytoca,* and two *Morganella morganii* strains, indicating that cross-species transmission of *bla*$_{NDM-1}$ is a common event. Genetic analyses by PFGE and MLST showed that, with the exception of *E. coli* and *E. cloacae*, strains belonging to the same species were often genetically unrelated. In addition to *bla*$_{NDM-1}$, several CRE strains were also found to harbor the *bla*$_{KPC-2}$, *bla*$_{VIM-1}$,and *bla*$_{IMP-4}$ elements. Conjugations experiments confirmed that the majority of carbapenem resistance determinants were transferable. Taken together, our findings suggest that transmission of mobile resistance elements among members of *Enterobacteriaceae* and clonal spread of CRE strains may contribute synergistically to a rapid increase in the population of CRE in clinical settings, prompting a need to implement more rigorous infection control measures to arrest such vicious transmission cycle in CRE-prevalent areas.

Keywords: carbapenem-resistant *Enterobacteriaceae*, NDM-1, clonal spread, mobile element

Introduction

β-Lactams have been a cornerstone in treatment of infections caused by Gram-negative bacterial pathogens due to their high efficacy and low toxicity to humans, among which carbapenems are considered agents of the last resort, especially in cases where extended-spectrum β-Lactamase (ESBL) producing organisms were involved (Dalhoff and Thomson, 2003). In the past two decades, usage of carbapenems such as imipenem and meropenem has been substantially increased due to the emergence of multidrug-resistant organisms (Goel et al., 2011; Zilberberg and Shorr, 2013). However, increased carbapenem consumption in turn initiated a vicious cycle in which carbapenem-resistant Gram-negative pathogens (CRGNP), which often cause untreatable hospital infections (Livermore, 2004, 2009; Karaiskos and Giamarellou, 2014), further gained selection advantage. Species belonging to the family *Enterobacteriaceae* are common human pathogens which can cause a wide range of community-acquired and nosocomial infections (Stock, 2014). The emergence of carbapenem-resistant *Enterobacteriaceae* (CRE) has posed a huge challenge to clinical infection control. Carbapenem resistance in CRE was mainly mediated by the production of carbapenemases, among which KPC, Metallo-β-lactamases (VIM, IMP, NDM) and OXA-48 type of enzymes were the most common (Nordmann et al., 2011). New Delhi Metallo-β-lactamase-1 (NDM-1) was one of the most important carbapenemases of CRE. Since its first discovery in 2008 in a *Klebsiella pneumoniae* isolate recovered from a patient at a hospital in New Delhi, India, it has been transmitted to many species of *Enterobacteriaceae* in various countries (Yong et al., 2009). NDM-1 is most frequently identified in the Indian subcontinent, followed by the Balkans region and the Middle East, and is mainly associated with community-acquired infections.

In China, the first clinical report of $bla_{\text{NDM}-1}$ involved carbapenem-resistant *Acinetobacter baumannii* strains detectable in four patients indifferent provinces in 2011 (Chen et al., 2011). Since then it has been recoverable in most species of *Enterobacteriaceae*, including *K. pneumoniae, Klebsiella oxytoca, Escherichia coli, Enterobacter cloacae, Enterobacter aerogenes,* and *Citrobacter freundii.* To date, NDM-1-producing isolates have been reported in various cities in China including Beijing, Changsha, Chongqing, Fuzhou, Guangzhou, Hangzhou, Hebei, Hong Kong, and Zhengzhou (Berrazeg et al., 2014; Qin et al., 2014). Although various $bla_{\text{NDM}-1}$-carrying CRE strains have been sporadically identified, few outbreaks of CRE carrying the $bla_{\text{NDM}-1}$ element have been reported in China, suggesting that the transmission of NDM-1 is mainly mediated by conjugative plasmids, which is consistent with the features of NDM-1 transmission observable in other parts of the world (Hu et al., 2014). In this study, we report an increasing prevalence of $bla_{\text{NDM}-1}$-positive *Enterobacteriaceae* in a Jiaxing hospital, which is located in Zhejiang Province, China. Through time-series analysis of the molecular features and epidemiological linkage of CRE recovered within the hospital, we demonstrated that emergence of new CRE strains was due to a combination of clonal spread of existing $bla_{\text{NDM}-1}$-carrying strains and efficient horizontal transfer of the $bla_{\text{NDM}-1}$ elements from such strains to other drug sensitive organisms. This combinatorial mode of transmission of both CRE organisms and the resistance elements that they harbor can theoretically result in an exponentially increasing rate of spread of NDM positive CRE in clinical settings if proper infection control measures are not implemented to disrupt the transmission routes.

Materials and Methods

Bacterial Strains and Species Identification

From May 2013 to February 2014, a total of 6598 clinical *Enteroabcteriaceae* strains were isolated from different specimens (urine, feces, and sputum) collected from patients in Second People's Hospital of Jiaxingin Zhejiang Province, China. All isolates were identified using the Vitek 2 system (bioMérieux, Marcy-l'E' toile, France), and confirmed by the MALDI-TOF MS apparatus (Bruker Microflex LT, Bruker Daltonik GmbH, Bremen, Germany). These isolates were screened for their ability to produce carbapenemases by a disc diffusion test, in which 10-mg imipenem discs were used (Oxoid, Basingstoke, UK) (Zhou et al., 2015). A total of 45 CRE isolates were recovered from these *Enterobacteriaceae* strains.

Molecular Detection of Resistance Genes

PCR and nucleotide sequencing were employed to screen for the presence of carbapenemase-encoding genes, including bla_{VIM}, bla_{IMP}, bla_{KPC}, $bla_{\text{OXA}-48}$ and $bla_{\text{NDM}-1}$, as well as ESBL genes, including $bla_{\text{CTX}-\text{M}}$, bla_{TEM} and bla_{SHV}, as described previously (Dallenne et al., 2010). An imipenem-EDTA double-disc synergy test and a modified Hodge test were used to screen for the presence of Metallo-β-lactamases(MBLs) and carbapenemases, respectively, and were analyzed according to CLSI guidelines (Zhou et al., 2015).

Antimicrobial Susceptibility Testing

The MICs of 10 antibiotics, including imipenem, meropenem, ceftazidime, cefotaxime, aztreonam, piperacillin-tazobactam, fosfomycin, amoxicillin-clavulanic acid, amikacin, tigecycline, were determined using the agar dilution method, and the results were analyzed according to the CLSI criteria of 2014 (Zhou et al., 2015). The 2014 EUCAST breakpoints were used (available at http://www.eucast.org/clinical_breakpoints/) for tigecycline.

PFGE and MLST

Clonal relationships between $bla_{\text{NDM}-1}$-positive isolates were investigated by PFGE of *Xba*I-digested genomic DNA using a Rotaphor System 6.0 instrument (Whatman Biometra, Goettingen, Germany), with a running time of 24 h and pulse times of 3–40 s. *Salmonella* strain H9812 was used as the control strain. Bands were stained with ethidium bromide (0.5 μg/mL) prior to visualization under UV light. A dendrogram depicting the genetic relatedness of the test strains was generated from the homology matrix with a 0.2% coefficient by the unweighted pair-group method, and by using arithmetic averages (UPGMA), to describe the relationships of the PFGE profiles. Isolates were allocated to the same PFGE group if their dice similarity index was ≥85%.

MLST was performed using seven housekeeping genes inbla_{NDM-1}-producing *E. coli*, *K. pneumonia* and *E. cloacae* isolates, which were amplified using primers listed in the online databases (http://pubmlst.org/ecloacae/ for *E. cloacae*, http://bigsdb.web.pasteur.fr/klebsiella/klebsiella.html for *K. pneumoniae* and http://mlst.warwick.ac.uk/mlst/dbs/Ecoli for *E. coli*). The resultant PCR products were purified and sequenced. Sequence types (STs) were assigned using online database tools.

Conjugation Experiments

The conjugation experiment was carried out using the mixed broth method as previously described (Borgia et al., 2012). Both the donor (bla_{NDM-1}-positive *Enterobacteriaceae*) and the recipient strains (sodium azide-resistant *E. coli* J53) were mixed on Luria-Bertani agar at a ratio of 1:1, and the mixtures were incubated for 24 h at 35°C. Transconjugants were selected on LB agar supplemented with sodium azide (100 mg/L) and meropenem (0.3 mg/L). Colonies that grew on the selective medium were picked for identification by the Vitek MS system. Transformants that harbored bla_{NDM-1} and exhibited resistance to carbapenems and cephalosporins were defined as transconjugants.

Results

The study period and venue were May 2013 to February 2014 and the Second Hospital of Jiaxing, Zhejian Province, China, respectively. During this period, a total of 45 CRE isolates were recovered from 6598 clinical specimens (urine, feces and sputum), among which 21 (47%) were found to harbor the bla_{NDM-1} elements; the rest of 24 CRE strains were all positive for bla_{KPC-2}. The objective of the study was to investigate the molecular events that lead to emergence of CRE in the hospital, with a focus on understanding the molecular and epidemiological features of transmission of both bla_{NDM-1}-borne strains and the bla_{NDM-1} element itself. The 21bla_{NDM-1}–borne CRE strains were found to comprise a variety of different species of *Enterobacteriaceae* including nine *E. cloacae*, three *E. coli*, three *C. freundii*, two *K. pneumoniae*, two *K. oxytoca* and two *M. morganii* strains, indicating that bla_{NDM-1}can be efficiently acquired by various *Enterobacteriaceae* species. Most of these bla_{NDM-1}–borne CRE were recovered from patients in the Neurosurgery Department (43%),who were subjected to various neuronal system surgical procedures, followed by patients from the Respiratory Department (19%) suffering mainly from pulmonary infection, patients from the Pediatric Department and Recovery department (14% each), and one patient each from the Cardio-Thoracic Surgery and Breast Department (**Table 1**). These bla_{NDM-1} borne CRE strains were isolated from urine (85%), sputum (10%), and breast secretion (5%), respectively.

Clinical records showed that most of patients from whom bla_{NDM-1}–borne CRE were recovered have been subjected to different types of antimicrobial treatment including the use of cephalosporins (ceftazidime, cefoperazone/sulbactam, cefuroxime, piperacillin/tazobactam), amikacin, levofloxacin and tigecycline, and the vast majority of patients recovered and were discharged from the hospital. Our record also showed that although bla_{NDM-1}-borne CRE were resistant to all cephalosporins, clinical treatment with cephalosporins for non-blood infections caused by bla_{NDM-1}borne CRE remained effective (**Table 1**). In addition, levofloxacin could be a choice for treatment of infections caused by fluoroquinolone-susceptible CREs. Likewise, amikacin was also a good choice since most of the CRE were susceptible to this antibiotic.

The first case of CRE infection occurred in the Neurosurgery Department on June 24, 2013, with the causative organism being identified as *K. oxytoca*, recurrent infection due to the same clone with identical PFGE pattern occurred on August 15, 2013, suggesting the long-term persistence of similar clones in a hospital. Interestingly, non-clonal spread was seen in specific species of CRE such as *C. freundii* and *K. pneumoniae.* For instance, the first case of NDM-producing *C. freundii* infection was recorded on June 26, 2014 in the Neurosurgery Department, followed by the second case in the Respiratory Department on August 17, 2013; the third case was recorded back in Neurosurgery Department on February 20, 2014, however, pneumonia-associated *C. freundii* strains recoverable form each of these three cases were found to be genetically non-identical (**Figure 1**, **Table 1**). Likewise, two genetically un-related strains of *K. pneumoniae* were found to cause infections in different departments at different dates (**Figure 1,Table 1**). On the other hand, both non-clonal and clonal spread could be seen upon analysis of infection caused by *E. coli* and *E. cloacae.* The first case of NDM-producing *E. coli* was observable in the Pediatric Department on September 26, 2013, whereas a different strain was found to cause infections in the Recovery Department in 2014. Although the NDM-1 producing *E. coli* strains recovered belonged to two different clones, three of the strains were found to belong to ST167 (**Figure 1**, **Table 1**). *E. cloacae* were the most common CRE species in this hospital. The first case of *E. cloacae* was reported on June 29, 2013 in the Pediatric Department. The same clone, which belonged to ST114, was causing an outbreak in the Neurosurgery Department, in which four infections recorded during the period of August to October, 2013. In August 2013, infections caused by a different clone also occurred in the Pediatric Department, and other clones were also found to cause infections in different departments at different times. These clones belonged to different ST types such as ST93, ST190, and ST66 (**Figure 1**, **Table 1**). Finally, NDM-1 producing *Morganella morganii* was reported for the first time in this hospital. Two very similar clones were found to cause infections in different departments and dates (**Figure 1**, **Table 1**).

Antimicrobial susceptibilities were performed on all CRE strains. All 21bla_{NDM-1}-positive isolates were resistant to most β-lactam antibiotics, including expanded-spectrum cephalosporins and the carbapenems. Moreover, all isolates were resistant to fosfomycin (≥256 mg/L) and amoxicillin-clavulanic acid (**Table 2**). In addition, 10 isolates were resistant to amikacin (≥64 mg/L). Only two isolates were sensitive to aztreonam (≤4 mg/L). Sixteen isolates exhibited intermediate susceptibility to tigecycline (MICs ≤ 1 mg/L), and 11 were tigecycline-resistant (MICs > 2 mg/L). The high frequency of resistance observed to

TABLE 1 | Origin and molecular features of *Enterobacteriaceae* isolates harboring bla_{NDM-1}, and clinical outcome of diseases that they caused.

Isolate	Isolation date	Location of acquisition	Culture type	Diseases	Treatment	Outcome	PFGE	MLST
EC-06	26/9/2013	PD	Urine	Neonatal pneumonia	Ceftazidime	Recovered	Ec1	ST167
EC-33	18/01/2014	ECD	Urine	Myelitis	NT	Uracratia	Ec2	ST167
EC-34	12/02/2014	RCD	Urine	Myelitis, Urinary tract infection	NR	NR	Ec2	ST167
KO-03	24/06/2013	ND	Urine	Subarachnoid hemorrhage,Oculomotor paralysis, Pneumonia	NR	Recovered	Ko1	ND
KO-16	15/08/2013	ND	Urine	Cerebroma recurrence	NT	Discharged	Ko1	ND
CF-05	26/06/2013	ND	Urine	Traumatic subarachnoid hemorrhage	Levofloxacin, Amikacin	Recovered	Cf1	ND
CF-17	20/08/2013	RD	Sputum	Pulmonary infection	Tigecycline	Recovered	Cf2	ND
CF-35	20/02/2013	ND	Urine	Cerebral contusion	Levofloxacin	Recovered	Cf3	ND
ECL-07	29/06/2013	PD	Urine	Neonatal pneumonia	Ceftazidime	Recovered	Ecl1	ST114
ECL-08	05/08/2013	ND	Urine	Cerebral hemorrhage	Levofloxacin	Transferred	Ecl1	ST114
ECL-09	07/08/2013	ND	Urine	Intracranial aneurysm	Piperacillin/Tazobactam	Discharged	Ecl2a	ST93
ECL-10	09/08/2013	ND	Urine	Traumatic subdural hemorrhage	Levofloxacin	Recovered	Ecl1	ST114
ECL-18	28/08/2013	PD	Urine	Neonatal infections	Ceftazidime	Recovered	Ecl3	ST190
ECL-19	28/08/2013	RD	Urine	Pulmonary infection	Mupirocin ointment		Ecl1	ST114
ECL-22	02/10/2013	ND	Urine	Subarachnoid hemorrhage	Levofloxacin	Recovered	Ecl1	ST114
ECL-27	11/11/2013	CTS	Sputum	Abdominal aortic aneurysm	Cefuroxime	Recovered	Ecl4	ST66
ECL-28	27/11/2013	RD	Urine	Chronic obstructive bronchitis	Cefoperazone/Sulbactam		Ecl2b	ST93
KP-04	26/06/2013	RD	Urine	Bladder neoplasmpulmonary infection	Ceftazidime, Piperacillin/Tazobactam	Discharged	Kp1	ST147
KP-14	08/08/2013	ND	Urine	Cerebral hernia	Levofloxacin	Discharged	Kp2	ST1724
MM-23	04/10/2013	BD	Secretion	Cholangitis	Amikacin, Metronidazole	Recovered	Mm1a	ND
MM-26	29/10/2013	RCD	Urine	Spinal cord injury	Amikacin	Recovered	Mm1b	ND

EC, Escherichia coli; KO, Klebsiella oxytoca; CF, Citrobacter freundii; ECL, Enterobacter cloacae; KP, Klebsiella pneumoniae; MM, Morganella morganii. JH, the Second People's Hospital of JiaXing; NT, not treated; NR, no record; ND, not determined. PD, Pediatric Department; RCD, Recovery Department; ND, Neurosurgery Department; CTS, Cardio-Thoracic Surgery; RD, Respiratory Department; BD, Breast Department.

alternative therapeutic antibiotics is a great concern for clinicians in charge of treating infections caused by these CREs.

Transferability of resistance determinants in these CRE strains, in particular those producing NDM-1, was investigated. Conjugation experiments were successful for all isolates except for two (ECL-8 and CF-17) bla_{NDM-1}-positive isolates. The resulting 19 transconjugants all exhibited resistance to carbapenems and cephalosporins. Importantly, fosfomycin and amikacin resistance determinants could also be transferred to the recipient *E. coli* strain (**Table 2**). Phenotypic resistance to all β-lactams in these isolates suggested that they might express additional β-lactamases since NDM-1 would not mediate resistance to aztreonam. In view of the discrepancy between the existence of known carbapenemase genes and the drug susceptibility profiles of the test strains, all CRE strains and their corresponding transconjugants were screened for the presence of additional β-lactamase genes. Our data revealed that in addition to bla_{NDM-1}, strains CF35, ECL19, and KP14 also harbored bla_{KPC-2}, strain KP04 was found to harbor bla_{VIM-1}, and bla_{IMP-4} was detectable in ECL18. To the best of our knowledge, this is the first case of *K. pneumoniae* containing both bla_{NDM-1} and bla_{VIM-1}, hence further works are required to elucidate the mechanisms governing the uptake of multiple carbapenemase genes in a single organism. In addition, 18 out of the 21 (86%) bla_{NDM-1}-positive isolates were found to harbor ESBL genes in various combinations (**Table 2**). It should also be noted that the additional resistance genes detectable in several isolates, including bla_{SHV-12}, $bla_{CTX-M-3/14}$, bla_{TEM-1} and bla_{KPC-2}, could be co-transferred to the *E. coli* recipient strain J53, along with the bla_{NDM-1} element (**Table 2**).

Discussion

Since the first report of their emergence, NDM-1-producing *Enterobacteriaceae* have become a worldwide public health concern (Patel and Bonomo, 2013). In China, currently available data tend to suggest that bla_{NDM-1} is only present at a relatively low frequency and spreading sporadically amongst

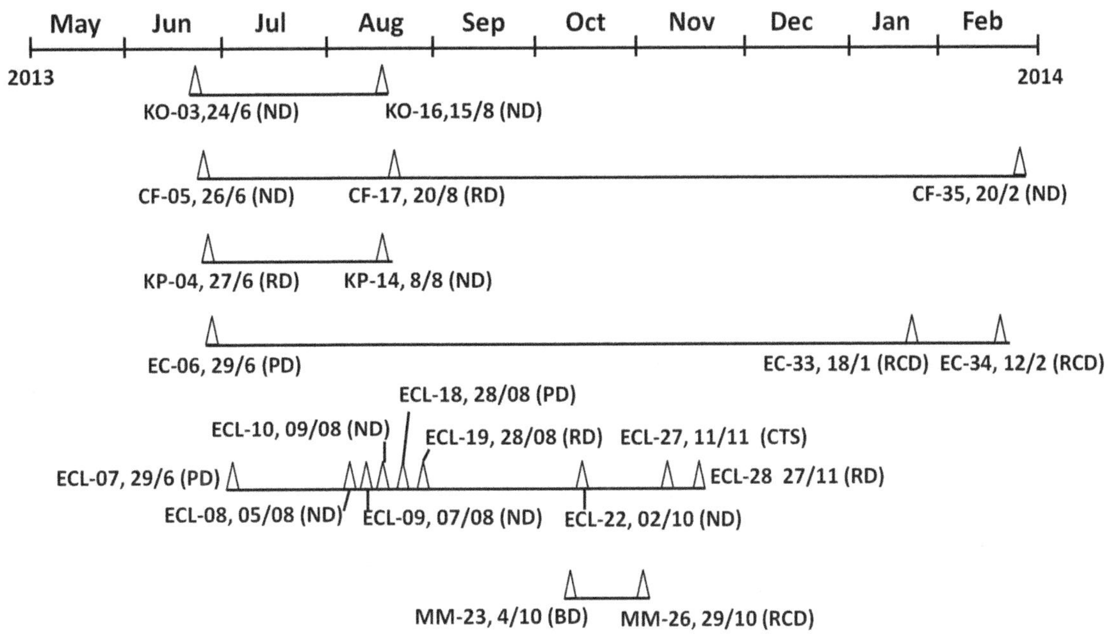

FIGURE 1 | Timeline of events in which epidemiologically linked NDM-1–producing *Enterobacteriaceae* strains were recovered. Test strains were isolated from May2013 to February 2014 in the Second People's Hospital of Jiaxing. Dates are shown as date/month. Abbreviations: Δ, NDM-1 positive isolate; EC, *Escherichia coli*; KO, *Klebsiella oxytoca*; CF, *Citrobacter freundii*; ECL, *Enterobacter cloacae*; KP, *Klebsiella pneumoniae* MM, *Morganella morganii*. PD, Pediatric Department; RCD, Recovery Department; ND, Neurosurgery Department; CTS, Cardio-Thoracic Surgery; RD, Respiratory Department; BD, Breast Department.

Enterobacteriaceae (Wang et al., 2013; Hu et al., 2014). A recent study reported a high rate [33.3% (16/48)] of NDM-1 positive CRE organisms in a hospital in Henan Province, China, most of which were due to plasmid mediated transmission of the bla_{NDM-1} elements among different members of *Enterobacteriaceae* (Qin et al., 2014). Although the prevalence of bla_{NDM-1}-positive CRE has been increasing over the past several years in China, very few epidemiological data are available to elucidate the underlying mechanisms that mediate the increased rate of transmission of NDM-1 positive CRE strains in hospitals. In this study, we identified 21 NDM-1-producing strains of *Enterobacteriaceae* in a hospital over a short period of time, from May 2013 to February 2014, with the majority of the strains (18/21) being collected during the period of June–November 2013. Our data suggest that a combination of outbreak of CRE infections and sporadic emergence of genetically unrelated resistant organisms contributed to the dramatic increase of CRE infections in this hospital during this period, prompting a need to investigate the molecular basis of these events. It should be noted that the series of CRE infections investigated in this study represents the most serious of its kind in China to date (Berrazeg et al., 2014; Zhou et al., 2014).

MLST-diverse NDM-1-producing *E. coli* (ST410, ST131, ST684, and ST101) and *K. pneumoniae* strains (ST147, ST14, ST11, and ST340) have been identified worldwide (Poirel et al., 2011). In this work, the NDM-1 positive *E. coli* strains tested were found to belong to ST167 with two distinct PFGE types. NDM-1 positive ST167 *E. coli* were previously shown to be an animal-associated clone recoverable in both France and China (Cuzon et al., 2013; Zhang et al., 2013; Yang et al., 2014). Repeated emergence of these NDM-1 positive *E. coli* strains in China represents a significant clinical and public health concern. The two *K. pneumoniae* isolates reported in the current study belonged to ST147 and a novel type, ST1724. Clinical NDM-1-producing *K. pneumoniae* ST147 strains are frequently detected worldwide, suggesting that they play an important role in the dissemination of the bla_{NDM-1} elements to other *K. pneumoniae* strains (Giske et al., 2012; Peirano et al., 2014). Identification of carbapenem-resistant *K. pneumoniae* strains belonging to the novel STs (ST359 and ST1724) in this study infers that the size of the existing pool of NDM-1-producing strains has been further expanded.

In contrast to *E. coli* and *K. pneumoniae*, MLST type of NDM-1-producing *E. cloacae* strains in this study included both epidemic ST types and new ST types. A carbapenem-resistant *E. cloacae* strain belonging to ST89 and producing the OXA-48 carbapenemase was isolated in Poland in 2014 (Majewski et al., 2014). Izdebski et al. reported that the ST66, ST78, and ST114 types have spread worldwide and were commonly associated with production of ESBLs and carbapenemases in many countries (Izdebski et al., 2014). NDM positive *E. cloacae* strains are among the most frequently reported CRE species and are known to cause occasional hospital outbreaks and community-acquired infections (Rozales et al., 2014; Yanik et al., 2014; Stoesser et al., 2015). In this study, we reported a similar situation in which *E. cloacae* belonging to the ST114 type was linked to five infection

TABLE 2 | Profiles of antimicrobial drug susceptibility and detectable-lactamase determinants of NDM-1 positive CRE strains and the corresponding transconjugants.

Strains	MICs (mg/L)												Additional β lactamase determinants
	IPM	MEM	CAZ	CTX	ATM	TZP	SCF	AMC	LEV	FOS	AK	TIG	
EC-06	16	32	>256	>256	>256	>256	>256	128	8	32	4	2	CTX-M-3, CTX-M-14,SHV-12
T-EC06	16	32	>256	>256	2	>256	>256	128	<0.125		1	1	–
EC-33	4	4	>256	>256	>256	>256	>256	64	32	32	2	1	CTX-M-15
T-EC33	4	4	>256	>256	>256	>256	>256	64	<0.125		1	1	CTX-M-15
EC-34	4	8	>256	>256	>256	>256	>256	128	32	32	2	2	CTX-M-15, TEM-1,SHV-12
T-EC34	4	4	>256	>256	>256	>256	>256	64	<0.125		1	1	TEM-1,SHV-12
KO-03	8	16	>256	>256	>256	>256	>256	128	16	16	1	4	SHV-12
T-KO03	16	32	>256	>256	2	>256	>256	128	<0.125		1	1	–
KO-16	8	16	256	>256	>256	>256	>256	128	16	16	1	4	SHV-12
T-KO16	16	32	>256	>256	>256	>256	>256	128	<0.125		1	1	SHV-12
CF-05	16	32	>256	>256	>256	>256	>256	256	4	8	1	2	SHV-12
T-CF05	16	32	>256	>256	>256	>256	>256	128	<0.125		1	1	SHV-12
CF-17	8	8	>256	>256	256	>256	>256	256	8	32	16	2	SHV-12
CF-35	8	32	>256	>256	>256	>256	>256	>256	8	16	>64	1	CTX-M-3, KPC-2,SHV-12
T-CF35	2	1	16	8	128	128	32	128	<0.125	0.25	>64	0.5	SHV-12
ECL-07	4	16	>256	>256	>256	>256	>256	256	16	64	>64	2	CTX-M-14, TEM-1,SHV-12
T-ECL07	4	2	>256	128	64	256	128	256	<0.125	0.5	1	1	TEM-1
ECL-08	4	4	>256	>256	>256	>256	>256	128	16	64	>64	4	CTX-M-14,TEM-1,SHV-12
ECL-09	4	16	>256	>256	>256	>256	>256	256	32	32	>64	4	CTX-M-3, CTX-M-14, TEM-1,SHV-12
T-ECL09	4	1	>256	256	32	256	256	256	0.25	0.5	1	1	CTX-M-14,TEM-1
ECL-10	8	16	>256	>256	>256	>256	>256	256	16	>64	>64	4	CTX-M-14, TEM-1,SHV-12
T-ECL10	8	8	>256	128	128	256	256	256	<0.125	<0.125	1	1	–
ECL-18	4	32	256	>256	256	256	>256	256	0.5	1	>64	2	IMP-4,TEM-1
T-ECL18	4	8	>256	128	16	256	256	256	<0.125	<0.125	>64	1	TEM-1
ECL-19	8	8	>256	>256	>256	>256	>256	256	16	64	>64	2	CTX-M-14,KPC-2
T-ECL19	8	4	>256	128	64	256	256	256	<0.125	<0.125	1	1	–
ECL-22	8	4	>256	>256	>256	>256	>256	256	32	32	>64	4	CTX-M-14, TEM-1,SHV-12
T-ECL22	8	8	>256	128	128	>256	256	256	0.25	0.5	1	1	TEM-1,SHV-12
ECL-27	4	4	256	>256	256	>256	>256	128	0.5	2	0.5	2	CTX-M-3, CTX-M-14, TEM-1,SHV-12
T-ECL27	8	8	>256	128	128	>256	256	256	<0.125	0.5	1	1	TEM-1
ECL-28	4	4	>256	>256	256	>256	>256	256	64	>64	4	4	TEM-1,SHV-12
T-ECL28	4	8	>256	>256	256	>256	256	256	0.25	1	2	1	TEM-1,SHV-12
KP-04	8	16	>256	>256	>256	>256	>256	128	2	4	1	2	CTX-M-14, VIM-1,TEM-1,SHV-12
T-KP04	8	16	>256	>256	>256	>256	>256	128	<0.125	0.5	1	1	SHV-12
KP-14	16	16	256	>256	>256	>256	>256	128	0.15	0.5	1	2	CTX-M-14, KPC-2,SHV-12
T-KP14	16	16	256	>256	>256	>256	>256	128	0.15	0.5	1	2	SHV-12
MM-23	8	4	128	>256	2	256	>256	>256	1	4	4	2	–
T-MM23	4	4	>256	128	1	>256	256	256	2	4	1	2	–
MM-26	8	4	64	>256	2	64	>256	>256	16	16	4	4	–
T-MM26	4	8	>256	128	1	>256	256	128	1	2	1	2	–

T denotes transconjugant. IPM, imipenem; MEM, meropenem; CAZ, ceftazidime; CTX, cefotaxime; ATM, aztreonam; TZP, piperacillin-tazobactam; CPS, cefoperazone/sulbactam; AMC, amoxicillin-clavulanic acid; LEV, levofloxacin, FOS, fosfomycin; AK, amikacin; TIG, tigecycline.

cases, organisms belonging to the ST93 type were linked to two cases of infections, and two new ST types, ST66 and ST190 were responsible for one infection each. Since the five ST114 isolates were collected from different wards at the Second People's Hospital of Jiaxing, we speculated that NDM-1-producing *E. cloacae* strain ST114 may have already spread in the hospital.

In the United Kingdom and the Indian subcontinent, $bla_{\mathrm{NDM-1}}$ has been found to be located in plasmids of various sizes (ca. 50–300 kb), which generally belong to at least four different Inc groups, including A/C,L/M, FI/FII and an undefined type (Kumarasamy et al., 2010; Poirel et al., 2011). In our study, most of $bla_{\mathrm{NDM-1}}$ elements recoverable from CRE isolates were harbored by self-transmissible plasmids, which also encoded multiple β-lactamases and other determinants of amikacin and fosfomycin resistance. This finding is consistent with currently available data in China and various other countries. Finally, our finding that transmission of conjugative plasmids encoding various carbapenemases and clonal spread of strains containing such plasmids were both responsible for a significant increase in the number of NDM-1 positive CRE infections in a hospital in China raised an alarming possibility that the prevalence of CRE can increase rapidly in a hospital setting within a short period. Whether the rate of infections due to such strains increases at a similar rate in the future depends on numerous factors including the effectiveness of infection control measures of the hospital concerned and the immune status of the infected patients. More works are urgently needed to investigate factors that determine the rate of transmission of CRE and the mobile resistance elements in order to help design appropriate intervention strategies that pinpoint the core of the problem, that is, to target the pool of existing CRE and the resistance elements that they harbor.

Funding

This work was supported by the Chinese National Key Basic Research and Development Program (2013CB127200) and Natural Science Foundation of China (Grant No. 81371871).

Acknowledgments

We thank the team of curators of the Institute Pasteur MLST and whole genome MLST databases for curating the data and making them publicly available at http://bigsdb.web.pasteur.fr. We extend special thanks to hospitals which kindly provided the reference strains.

References

Berrazeg, M., Diene, S., Medjahed, L., Parola, P., Drissi, M., Raoult, D., et al. (2014). New Delhi Metallo-beta-lactamase around the world: an eReview using Google Maps. *Euro Surveill.* 19:20809. doi: 10.2807/1560-7917.es2014.19.20.20809

Borgia, S., Lastovetska, O., Richardson, D., Eshaghi, A., Xiong, J., Chung, C., et al. (2012). Outbreak of carbapenem-resistant enterobacteriaceae containing blaNDM-1, Ontario, Canada. *Clin. Infect. Dis.* 55, e109–e117. doi: 10.1093/cid/cis737

Chen, Y., Zhou, Z., Jiang, Y., and Yu, Y. (2011). Emergence of NDM-1-producing *Acinetobacter baumannii* in China. *J. Antimicrob. Chemother.* 66, 1255–1259. doi: 10.1093/jac/dkr082

Cuzon, G., Bonnin, R. A., and Nordmann, P. (2013). First identification of novel NDM carbapenemase, NDM-7, in *Escherichia coli* in France. *PLoS ONE* 8:e61322. doi: 10.1371/journal.pone.0061322

Dalhoff, A., and Thomson, C. J. (2003). The art of fusion: from penams and cephems to penems. *Chemotherapy* 49, 105–120. doi: 10.1159/000070616

Dallenne, C., Da Costa, A., Decre, D., Favier, C., and Arlet, G. (2010). Development of a set of multiplex PCR assays for the detection of genes encoding important beta-lactamases in Enterobacteriaceae. *J. Antimicrob. Chemother.* 65, 490–495. doi: 10.1093/jac/dkp498

Giske, C. G., Froding, I., Hasan, C. M., Turlej-Rogacka, A., Toleman, M., Livermore, D., et al. (2012). Diverse sequence types of *Klebsiella pneumoniae* contribute to the dissemination of blaNDM-1 in India, Sweden, and the United Kingdom. *Antimicrob. Agents Chemother.* 56, 2735–2738. doi: 10.1128/AAC.06142-11

Goel, N., Wattal, C., Oberoi, J. K., Raveendran, R., Datta, S., and Prasad, K. J. (2011). Trend analysis of antimicrobial consumption and development of resistance in non-fermenters in a tertiary care hospital in Delhi, India. *J. Antimicrob. Chemother.* 66, 1625–1630. doi: 10.1093/jac/dkr167

Hu, L., Zhong, Q., Shang, Y., Wang, H., Ning, C., Li, Y., et al. (2014). The prevalence of carbapenemase genes and plasmid-mediated quinolone resistance determinants in carbapenem-resistant Enterobacteriaceae from five teaching hospitals in central China. *Epidemiol. Infect.* 142, 1972–1977. doi: 10.1017/S0950268813002975

Izdebski, R., Baraniak, A., Herda, M., Fiett, J., Bonten, M. J., Carmeli, Y., et al. (2014). MLST reveals potentially high-risk international clones of *Enterobacter cloacae*. *J. Antimicrob. chemother.* 70, 48–56. doi: 10.1093/jac/dku359

Karaiskos, I., and Giamarellou, H. (2014). Multidrug-resistant and extensively drug-resistant Gram-negative pathogens: current and emerging therapeutic approaches. *Expert Opin. Pharmacother.* 15, 1351–1370. doi: 10.1517/14656566.2014.914172

Kumarasamy, K. K., Toleman, M. A., Walsh, T. R., Bagaria, J., Butt, F., Balakrishnan, R., et al. (2010). Emergence of a new antibiotic resistance mechanism in India, Pakistan, and the UK: a molecular, biological, and epidemiological study. *Lancet Infect. Dis.* 10, 597–602. doi: 10.1016/S1473-3099(10)70143-2

Livermore, D. M. (2004). The need for new antibiotics. *Clin. Microbiol. Infect.* 10(Suppl. 4), 1–9. doi: 10.1111/j.1465-0691.2004.1004.x

Livermore, D. M. (2009). Has the era of untreatable infections arrived? *J. Antimicrob. Chemother.* 64(Suppl. 1), i29–i36. doi: 10.1093/jac/dkp255

Majewski, P., Wieczorek, P., Sacha, P. T., Frank, M., Juszczyk, G., Ojdana, D., et al. (2014). Emergence of OXA-48 carbapenemase-producing *Enterobacter cloacae* ST89 infection in Poland. *Int. J. Infect. Dis.* 25, 107–109. doi: 10.1016/j.ijid.2014.02.024

Nordmann, P., Naas, T., and Poirel, L. (2011). Global spread of Carbapenemase-producing Enterobacteriaceae. *Emerging Infect. Dis.* 17, 1791–1798. doi: 10.3201/eid1710.110655

Patel, G., and Bonomo, R. A. (2013). "Stormy waters ahead": global emergence of carbapenemases. *Front. Microbiol.* 4:48. doi: 10.3389/fmicb.2013.00048

Peirano, G., Ahmed-Bentley, J., Fuller, J., Rubin, J. E., and Pitout, J. D. (2014). Travel-related carbapenemase-producing Gram-negative bacteria in Alberta, Canada: the first 3 years. *J. Clin. Microbiol.* 52, 1575–1581. doi: 10.1128/JCM.00162-14

Poirel, L., Dortet, L., Bernabeu, S., and Nordmann, P. (2011). Genetic features of blaNDM-1-positive Enterobacteriaceae. *Antimicrob. Agents Chemother.* 55, 5403–5407. doi: 10.1128/AAC.00585-11

Qin, S., Fu, Y., Zhang, Q., Qi, H., Wen, J. G., Xu, H., et al. (2014). High incidence and endemic spread of NDM-1-positive Enterobacteriaceae in Henan Province, China. *Antimicrob. Agents Chemother.* 58, 4275–4282. doi: 10.1128/AAC.02813-13

Rozales, F. P., Ribeiro, V. B., Magagnin, C. M., Pagano, M., Lutz, L., Falci, D. R., et al. (2014). Emergence of NDM-1-producing Enterobacteriaceae in Porto Alegre, Brazil. *Int. J. Infect. Dis.* 25, 79–81. doi: 10.1016/j.ijid.2014.01.005

Stock, I. (2014). [Infectious diseases caused by carbapenemase-producing Enterobacteriaceae–a particular challenge for antibacterial therapy]. *Med. Monatsschr. Pharm.* 37, 162–172; quiz 73–74.

Stoesser, N., Sheppard, A. E., Shakya, M., Sthapit, B., Thorson, S., Giess, A., et al. (2015). Dynamics of MDR *Enterobacter cloacae* outbreaks in a neonatal unit in Nepal: insights using wider sampling frames and next-generation sequencing. *J. Antimicrob. Chemother.* 70, 1008–1015. doi: 10.1093/jac/dku521

Wang, X., Liu, W., Zou, D., Li, X., Wei, X., Shang, W., et al. (2013). High rate of New Delhi metallo-beta-lactamase 1-producing bacterial infection in China. *Clin. Infect. Dis.* 56, 161–162. doi: 10.1093/cid/cis782

Yang, Z., Liu, W., Cui, Q., Niu, W., Li, H., Zhao, X., et al. (2014). Prevalence and detection of *Stenotrophomonas maltophilia* carrying metallo-beta-lactamase blaL1 in Beijing, China. *Front. Microbiol.* 5:692. doi: 10.3389/fmicb.2014.00692

Yanik, K., Guney, A. K., Karadag, A., and Eroglu, C. (2014). [NDM-1 cases and outbreaks in Turkey]. *Mikrobiyol. Bul.* 48, 709–710. doi: 10.5578/mb.8334

Yong, D., Toleman, M. A., Giske, C. G., Cho, H. S., Sundman, K., Lee, K., et al. (2009). Characterization of a new metallo-beta-lactamase gene, bla(NDM-1), and a novel erythromycin esterase gene carried on a unique genetic structure in *Klebsiella pneumoniae* sequence type 14 from India. *Antimicrob. Agents Chemother.* 53, 5046–5054. doi: 10.1128/AAC.00774-09

Zhang, X., Lou, D., Xu, Y., Shang, Y., Li, D., Huang, X., et al. (2013). First identification of coexistence of blaNDM-1 and blaCMY-42 among *Escherichia coli* ST167 clinical isolates. *BMC Microbiol.* 13:282. doi: 10.1186/1471-2180-13-282

Zhou, G., Guo, S., Luo, Y., Ye, L., Song, Y., Sun, G., et al. (2014). NDM-1-producing strains, family Enterobacteriaceae, in hospital, Beijing, China. *Emerging Infect. Dis.* 20, 340–342. doi: 10.3201/eid2002.121263

Zhou, S., Chen, X., Meng, X., Zhang, G., Wang, J., Zhou, D., et al. (2015). "Roar" of blaNDM-1 and "silence" of blaOXA-58 co-exist in *Acinetobacter pittii. Sci. Rep.* 5:8976. doi: 10.1038/srep08976

Zilberberg, M. D., and Shorr, A. F. (2013). Secular trends in gram-negative resistance among urinary tract infection hospitalizations in the United States, 2000-2009. *Infect. Control Hosp. Epidemiol.* 34, 940–946. doi: 10.1086/671740

2

MRSA Prevalence and Associated Risk Factors among Health-Care Workers in Non-outbreak Situations in the Dutch-German EUREGIO

Ricarda Sassmannshausen[1†], *Ruud H. Deurenberg*[2†], *Robin Köck*[1,3], *Ron Hendrix*[2,4], *Annette Jurke*[5], *John W. A. Rossen*[2] *and Alexander W. Friedrich*[2*]

[1] *Institute of Hygiene, University Hospital Münster, Münster, Germany,* [2] *Department of Medical Microbiology, University of Groningen, University Medical Center Groningen, Groningen, Netherlands,* [3] *Institute of Medical Microbiology, University Hospital Münster, Münster, Germany,* [4] *CERTE-Lvl, Groningen, Netherlands,* [5] *Department of Infectiology and Hygiene, Centre for Health North Rhine-Westphalia, Münster, Germany*

***Correspondence:**
Alexander W. Friedrich
alex.friedrich@umcg.nl

[†] *These authors have contributed equally to this work.*

Preventing the spread of methicillin-resistant *Staphylococcus aureus* (MRSA) in healthcare facilities is a major infection control target. However, only a few studies have assessed the potential role of healthcare workers (HCWs) for MRSA dissemination. To investigate the MRSA prevalence and the risk factors for MRSA colonization among HCWs, nasopharyngeal swabs were taken between June 2010 and January 2011 from 726 employees from nine acute care hospitals with different care levels within the German part of a Dutch-German border region (EUREGIO). The isolated MRSA strains were investigated using *spa* typing. The overall MRSA prevalence among HCWs in a non-outbreak situation was 4.6% (33 of 726), and was higher in nurses (5.6%, 29 of 514) than in physicians (1.2%, 1 of 83). Possible risk factors associated with MRSA colonization were a known history of MRSA carriage and the presence of acne. Intensive contact with patients may facilitate MRSA transmission between patients and HCWs. Furthermore, an accumulation of risk factors was accompanied by an increased MRSA prevalence in HCW.

Keywords: MRSA, decolonization, MRSA risk factors, personnel, staff, EurSafety health-net

INTRODUCTION

Methicillin-resistant *Staphylococcus aureus* (MRSA) is one of the most important hospital-associated (HA) pathogens (Voss and Doebbeling, 1995). Although its prevalence among patients decreased in several European countries (e.g., UK, France, and Germany) during the last years, the MRSA prevalence increased in other countries (e.g., Norway and Poland; http://ecdc.europa.eu). In healthcare institutions, MRSA can be transmitted between patients, or through the hands, clothes, or equipment of healthcare workers (HCWs), and the environment (Haley et al., 1982; Hardy et al., 2006; Henderson, 2006). Furthermore, it has been reported that HCWs have been the source of MRSA outbreaks in several cases (Vonberg et al., 2006). In a systematic review of Albricht et al., in which 127 studies and outbreak reports published between January 1980 and March 2006 were reviewed, transmission of MRSA from HCWs to patients was likely in 63 of 68 (93%) studies as shown by genotyping (Albrich and Harbarth, 2008). The average MRSA prevalence among HCWs in these 127 studies was 4.6%, with a broad range from 0 to 59% (95% CI 1.0–8.2%) between countries and institutions. Notably, the MRSA prevalence rates were found to be higher in endemic situations (8.1%) compared to outbreak situations (3.9%). However, it has to be stated that different designs of the studies included limit the informative value of this review (Albrich and Harbarth, 2008).

The implementation of MRSA screening among HCWs has been shown to have a positive impact on outbreak management (Peters et al., 1999; Blok et al., 2003; Ben-David et al., 2008). Hence, infection control guidelines from many countries, including Germany, recommend a systematic screening of HCWs in the case of an acute outbreak of MRSA (Peters et al., 1999, 2014). In the Netherlands, having one of the lowest MRSA prevalence rates in Europe (http://ecdc.europa.eu), MRSA screening of HCWs is not conducted regularly, but in addition to outbreak cases, the national "Search-and-Destroy" strategy (www.wip.nl) recommends screening for further risk groups among personnel, such as staff returning from work in healthcare institutions abroad (Blok et al., 2003).

However, the "Search-and-Destroy" strategy remains subject of discussion (Hawkins et al., 2011; Hill, 2011) due to the lack of reliable prevalence data and an insufficient differentiation between stable and transient MRSA carriage. To the best of our knowledge, only a few studies have focused on the prevalence of MRSA colonization among HCWs in non-outbreak situations and the risk factors for stable and transient MRSA carriage in Germany. Furthermore, none of these studies involved HCWs from more than one hospital (Witte et al., 2005; Kaminski et al., 2007; Reich-Schupke et al., 2010). Therefore, the aim of the present study was to investigate the MRSA prevalence among HCWs in several hospitals in non-outbreak situations in the German part of a Dutch-German border region ("EUREGIO") within the Dutch-German prevention network EurSafety Health-net (www.eursafety.eu) and assess data on (including non-occupational-related) risk factors for MRSA colonization among hospital staff.

MATERIALS AND METHODS

Participants

All hospitals in the German part of the EUREGIO were asked to participate in this study, and representatives of them were informed at the annual meeting of the German EUREGIO within the Dutch-German prevention network EurSafety Health-net. They discussed the study in their, respectively, hospital committees before the hospitals decided to participate in the study. Participation was voluntary for all HCWs and the directorate had to accept participation.

Between June 2010 and January 2011, nine acute care hospitals with different care levels within the German part of a Dutch-German EUREGIO participated in this study; two hospitals of basic care (No. 5 and 8), four hospitals of secondary care (No. 1, 4, 6, and 7) and three specialized clinics (No. 2, 3, and 9). The number of staff members within the hospitals ranged from 33 to 1640 and their capacities from 20 to 653 beds (**Table 1**). The participating hospitals had a total capacity of 2249 beds and employed 5264 staff members, including 667 physicians and 2325 nurses. In this study, both the medical staff and employees working outside direct patient-care, such as administration, maintenance, and cleaning, were encouraged to participate, and a questionnaire was used to obtain the needed information from the participants. Overall, 726 employees participated in this study (83 physicians, 514 nurses, 109 other staff members, and 20 participants of unknown professional group), covering 12% of the hospitals physicians and 22% of the nurses, respectively. As shown in **Table 1**, the study's coverage of medical staff varies between 9 and 91% between the different hospitals.

Survey of Risk Factors

The participants' risk factors were collected using a standardized paper-based questionnaire (Supplementary Material). The following potential risk-factors for MRSA colonization were surveyed: history of MRSA carriage, profession, contact with MRSA carriers in a professional or private setting (with and without protective clothing), involvement in home-care of relatives, former professional occupation in a country known to be endemic for community-associated MRSA (CA-MRSA), dermatosis/diseases of the skin, chronic inflammatory bowel diseases (IBD), diseases of the upper respiratory tract, antibiotic therapy within the last 6 months, acute diseases (at the time of risk factor assessment), inpatient treatment within the last 6 or 12 months, contact with domestic or farm animals, and consumption of raw meat within the last 12 h. All questionnaires were processed using random pseudonyms that could only be decoded by the duty hygiene officer or the company medical officer.

Nasopharyngeal Swabs

HCWs can be MRSA positive after a working shift due to short-term contact with patients (transient MRSA carriers), while other HCWs are MRSA carriers for longer than 24 h (stable MRSA carriers) (Cookson et al., 1989; van Cleef et al., 2011). To differentiate between transient and stable MRSA carriers, swabs were taken twice on different days from all HCWs. Nasopharyngeal swabs were collected by trained personnel on Monday after an off-work weekend and subsequently on Wednesday before starting to work. To minimize the risk of considering a stable carrier as transient due to false-negative swabs, a third validation swab was collected in cases of discrepant results between the first and second sample. If the validation swab was positive, the HCW was considered to be a stable MRSA carrier, and if the validation swab was negative, the participant was regarded as a transient carrier. In addition, as skin colonization influenced decolonization-strategy, skin swabs (from the axilla and the groin) were taken from stable carriers immediately after being identified as such.

MRSA Isolation and *spa* Typing

Swabs were applied to chromogenic media (BioMérieux, Nürtingen, Germany) the same day as the swab was collected and incubated for 48 h at 37°C. *S. aureus* colonies were tested for antibiotic resistance using the Vitek 2 (BioMérieux, Nürtingen, Germany) and the presence of the MRSA-specific penicillin binding protein 2a was tested by

TABLE 1 | Overview of the participating hospitals and its personnel coverage shown as different professional groups.

Hospital	Beds (n)		Physicians	Nurses	Total medical staff	Others	Unknown	Total staff
1	282	Total (n)	72	347	419	346		765
		Study (n%)	15 (21)	54 (16)	69 (17)	31 (9)	2	102 (13)
2	20	Total (n)	11	20	31	3		34
		Study (n%)	11 (100)	14 (70)	25 (81)	2 (67)	5	32 (94)
3	20	Total (n)	10	33	43	4		47
		Study (n%)	6 (60)	33 (100)	39 (91)	4 (100)	1	43 (92)
4	582	Total (n)	231	653	884	756		1640
		Study (n%)	5 (2)	81 (12)	86 (10)	5 (1)	0	91 (6)
5	271	Total (n)	63	210	273	212		485
		Study (n%)	29 (46)	183 (87)	212 (78)	63 (30)	1	276 (57)
6	405	Total (n)	135	484	619	469		1088
		Study (n%)	2 (2)	52 (11)	54 (9)	3 (1)	0	57 (5)
7	361	Total (n)	95	387	482	378		860
		Study (n%)	2 (2)	58 (15)	60 (12)	0 (0)	0	60 (7)
8	185	Total (n)	37	120	157	37		194
		Study (n%)	6 (16)	19 (16)	25 (16)	0 (0)	6	31 (16)
9	123	Total (n)	13	84	97	68		165
		Study (n%)	7 (54)	20 (24)	27 (28)	1 (2)	5	33 (20)

a latex-agglutination-test (Oxoid, Cambridge, UK). Every first MRSA isolate was further characterized using *S. aureus* protein A gene (*spa*) typing as described by Mellmann et al. (2008). Analysis was performed using the Staph Type™ software (Ridom GmbH, Münster, Germany) (Mellmann et al., 2007).

Decolonization

If HCWs were identified as stable MRSA carriers, a decolonization protocol was started. They were asked to use mupirocin ointment intranasally and to gurgle with an octenidine-based solution, both thrice a day during a period of 6 days (Friday to Wednesday or Wednesday to Monday) and to attend a healthcare professional during the decolonization period. Moreover, all MRSA carriers were advised to daily change and wash clothes and bed-linen during the decolonization therapy. Those participants with skin colonization (additionally to nasopharyngeal colonization), as indicated by positive inguinal or axillary swabs, were also asked to daily wash themselves with octenidine-based soap.

Statistical Analysis

Data processing was performed using Microsoft Excel (Microsoft Corp., Redmont, USA) and statistical analyses were conducted with PASW Statistics (IBM Corp., Amonk, USA). Due to the small case numbers, statistical significance of risk factors was calculated in a univariate analysis with Fisher's exact test and $p < 0.05$ were regarded to be significant. To identify independent risk-factors, a multivariate analysis was performed using multiple logistic regression for all variable with $p < 0.2$ in univariate analysis and $p < 0.05$ were considered to be significant.

Ethics

This study of screening HCW was approved by the Medical Chamber's of Westfalen-Lippe and Medical Faculty of the University of Münster's research ethic committee (2006-268-f-S). In all participating hospitals, the responsible member of the board of directors agreed to conduct the study. The employees participated in this study on a voluntary basis and written informed consent was obtained from all participants in accordance with the Declaration of Helsinki after the study's principal investigator and local infection control manager held meetings to inform the participant about the study.

RESULTS

Swabs

At least two nasopharyngeal swabs from each of the 726 HCWs were collected. In 21 HCWs, both swabs were MRSA positive, and in 12 HCWs only one swab was positive for MRSA. A third "validation"-swab was required for these 12 HCWs. As five participants refused to give permission to take a third swab, and started already with decolonization, persistence of colonization could not be determined for these HCWs (**Table 2**).

TABLE 2 | MRSA colonization among healthcare-workers in the EUREGIO screened twice for nasopharyngeal carriage.

	Initial swab		Validation swab*		
Result**	Number	Percentage	Participation	Positive swab	Negative swab***
(−/−)	693	95.45	NA	NA	NA
(+/+)	21	2.89	NA	NA	NA
(−/+)	4	0.55	2 of 4	0	2
(+/−)	8	1.11	5 of 8	2	3
Total	726	–	–	–	–

*Validation swab to confirm or exclude stable MRSA colonization among HCWs with discordant results of initial two nasopharyngeal swabs; **MRSA test result (−/−), negative in two nasopharyngeal swabs; (+/+), MRSA detected in two nasopharyngeal swabs; (−/+) and (+/−), MRSA detected in one of two nasopharyngeal swabs, a validation swab was taken in these cases; ***NA, not applicable.*

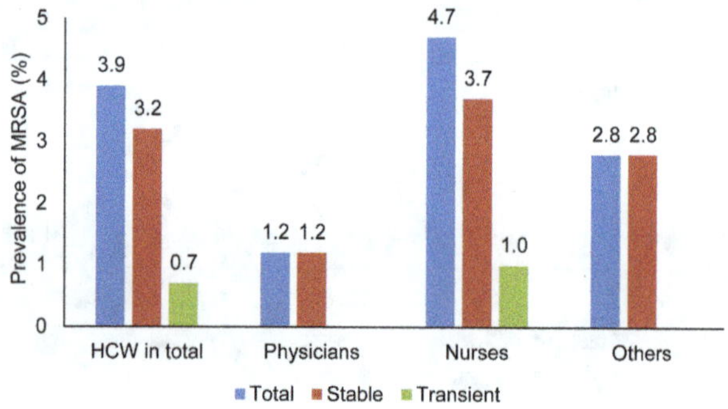

FIGURE 1 | MRSA prevalence among HCWs in different professions and the differentiation between stable and transient MRSA carriers.

MRSA Prevalence

In 33 of the 726 HCWs at least one swab was MRSA positive. Consequently, the MRSA prevalence was 4.6%. Two of the seven participants with discrepant swab results had a positive validation swab. Thus, 23 HCWs appeared to be stable carriers, whereas 5 HCWs were transient carriers resulting in a prevalence of 3.2% for stable carriers and 0.7% for transient carriers. In total, 17.9% of the detected cases of MRSA colonization were transient according to the applied definition.

The prevalence of MRSA differed for the type of healthcare profession. Among physicians, the MRSA prevalence was lower (1.2%, 1 of 83) compared to nurses (5.6%, 29 of 514). In nurses, stable MRSA carriage was found in 3.7% (19 of 514) of the cases, while transient MRSA carriage was observed in 1.0% (5 of 514) (**Figure 1**). Interestingly, stable MRSA positivity was observed in 2.8% (3 of 109) of other staff members who had no direct contact with patients.

spa Typing

Nine different *spa* types were observed among the 34 MRSA isolates [t003 ($n = 13$), t032 ($n = 11$), t151 ($n = 3$), t004 ($n = 1$), t020 ($n = 1$), t022 ($n = 2$), t179 ($n = 1$), t2261 ($n = 1$), and t6015 ($n = 1$)]. One HCW was colonized with two genetically non-related MRSA strains (*spa* type t004 and t032). Most frequently, typical HA-MRSA strains with *spa* type t003 (38%) and t032 (32%) were observed. The observation of these *spa* types is in accordance with the most prevalent *spa* types found among patients in the study region (Köck et al., 2009).

Risk Factors for MRSA Colonization

For analysis of the risk factors for MRSA colonization in HCWs, all participants with at least one positive swab were included. The statistical analysis of the risk factors is shown in **Table 3**. Previous MRSA colonization ($p = 0.01$, CI = 2.2–24.8), acne ($p = 0.049$, CI = 1.008–19.07) and chronic IBD ($p = 0.005$, CI = 3.66–1457.33) were significantly associated with MRSA carriage among HCWs in the multivariate analysis.

For the evaluation of risk factor groups, all risk factors with an odds ratio >1 in the univariate analysis were taken into account. In order to analyze a possible impact of the accumulation of risk factors, all participants were grouped in three categories: 0, 1, to 3, or >3 risk factors. In total, 12.4% of the staff members were free of all risk factors analyzed, for the majority of participants (78.4%) 1–3 risk factors were applicable, and 9.2% of the HCWs were associated with more than three risk factors. The differences in MRSA prevalence according to the risk factor group are shown in **Figure 2**.

Decolonization

Only stable MRSA carriers were decolonized. Five participants who did not participate in the validation screening were excluded from this analysis, and two HCWs wished to be treated outside the study. So, 21 HCWs participated in the controlled decolonization. Two were colonized with MRSA on the skin and underwent the entire decolonization measures including skin washes. Nineteen HCWs (90%) underwent nasopharyngeal decolonization only. All control swabs after the decolonization attempt were negative. Six of the 21 HCWs (29%), among those the HCWs with skin-colonization, agreed on re-swabbing after 1 month, while nine HCWs (39%) were re-swabbed after 6 months. All of these control swabs were negative and confirmed the initial test result.

DISCUSSION

A 4.6% MRSA prevalence was observed among HCWs in the present study, which was lower than described by Albrich et al. (8.1% outside outbreak situations in several different countries), but similar to the results of Kaminski et al. (4.6% inside outbreak situations) in Germany in 2001 and 2002 (Kaminski et al., 2007; Albrich and Harbarth, 2008). Intensive training of healthcare personnel and an increased public awareness toward MRSA in the framework of the EUREGIO MRSA-net (Friedrich et al., 2008) may have contributed to the successful prevention of MRSA transmission in- and out-side outbreak situations.

The MRSA prevalence among HCWs in the EUREGIO is high compared to the MRSA prevalence observed among patients upon hospital admission. As Köck et al. have shown, MRSA admission prevalence rates in patients are approximately 2.5 times lower than in HCWs (Köck et al., 2010). In consequence, HCWs are at higher risk for MRSA colonization than the general

TABLE 3 | Risk-factor analysis for MRSA colonization in HCWs.

Risk factor	MRSA negative ($n = 693$)	MRSA positive ($n = 33$)	Fisher's exact test			Logistic regression*		
			Odds ratio	p (FE)*	95%CI (FE)	Odds ratio	p (FE)	95%CI (FE)
Physicians	82	1	0.23	0.162	0.31–1.72	0.7	0.7	0.63–7.79
Nurses	485	29	3.11	0.030	1.08–8.96	2.81	0.15	0.69–11.43
Staff without direct contact with patients	109	3	0.55	0.460	0.17–1.85	–	–	–
Unknown profession	20	0	0.97	1.000	0.96–0.98	–	–	–
Previous MRSA carrier	16	5	7.56	0.002	2.58–22.09	7.44	0.01	2.2–24.8
Occupation on ICU	102	9	2.17	0.077	0.98–4.81	2.05	0.12	0.84–4.96
Occupation with new-born	16	1	1.32	0.551	0.17–10.28	–	–	–
Occupation in foreign countries	3	0	1.00	1.000	1–1	–	–	–
Contact with MRSA carriers without protective clothing	68	4	1.31	0.549	0.45–3.84	–	–	–
Contact with MRSA carriers with protective clothing	227	16	1.92	0.088	0.95–8.87	1.04	0.94	0.39–2.7
Home-care of relatives	14	2	3.13	0.162	0.68–14.37	4.28	0.097	0.77–23.76
Contact with MRSA carriers at home	7	2	6.32	0.059	1.26–31.7	6.63	0.063	0.9–48.91
Atopic dermatitis	33	3	2.00	0.221	0.58–6.90	–	–	–
Paronchyia	1	0	1.00	1.000	1–1	–	–	–
Acne	13	3	5.23	0.032	1.42–19.34	4.38	0.049	1.008–19.07
Otitis	4	1	5.38	0.208	0.59–49.55	–	–	–
Open wounds	5	0	1.00	1.000	1–1	–	–	–
Chronic inflammatory bowel disease	1	1	21.63	0.089	1.32–353.62	73.12	0.005	3.66–1457.33
Sinusitis	23	0	0.97	0.619	0.96–0.98	–	–	–
Diabetes mellitus	8	1	2.68	0.344	0.325–22.05	–	–	–
Rhinitis	25	2	1.72	0.350	0.39–7.61	–	–	–
Other chronic diseases	64	3	0.98	1.000	0.29–3.3	–	–	–
Antibiotics	81	4	1.04	1.000	0.36–3.04	–	–	–
Hospitalization for >24 h in the last 6 months	23	1	0.91	1.000	0.12–6.95	–	–	–
Hospitalization for >24 h in the last 6–12 months	24	0	0.97	0.620	0.95–0.98	–	–	–
Hospitalization for >24 h in a foreign country	0	1	0.45	0.715	0.6–3.37	–	–	–
Contact with farm animals	27	2	1.59	0.384	0.36–7	–	–	–
Thereof contact with pigs	17	1	1.24	0.570	0.16–9.63	–	–	–
Contact with domestic animals	265	16	1.52	0.270	0.76–3.06	–	–	–
Contact with raw meat within last 12 h	62	5	1.82	0.218	0.678–4.88	–	–	–
Acute illness	32	3	2.07	0.209	0.6–7.13	–	–	–

**Risk factors with $p < 0.2$ in univariate analysis (Fisher's exact test) were included in logistic regression.*

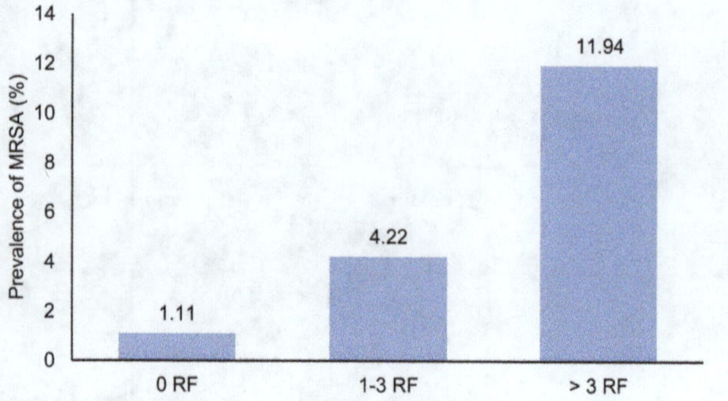

FIGURE 2 | Prevalence of MRSA in HCWs grouped by the number of the participants' overall risk factors (RF).

population due to the constant contact exposure to known or unknown MRSA-positive patients. This is also underpinned by the typing results of MRSA isolated from HCWs. In this study, mainly *spa* types t003 and t032, associated with the classical German healthcare-associated MRSA lineages CC5 and CC22, respectively, were observed. Transmission between patients and HCWs is likely, as these *spa* types also predominated among regional patients at the same time (Köck et al., 2013). By contrast, livestock-associated MRSA (LA-MRSA), e.g., *spa* type t011 and t034, which are the third and fourth most prevalent *spa* types among patients in the German part of the EUREGIO (Köck et al., 2011, 2013), could not be detected at all in HCWs. Since all MRSA isolates from patients in the EUREGIO are characterized, *spa* types observed in HCWs and patients in the corresponding hospital could be compared. The most prevalent *spa* types observed among HCWs were: t003 (38.2%), t032 (32.4%), t151 (8.8%), t022 (5.9%), t004, t020, t179, t2261, and t6015 (each 2.9%), while *spa* types t003 (24.7%), t032 (19.1%), t011 (13.1%), t034 (9.2%), and t020 (2.5%) were most prevalent in patients in the EUREGIO. This discrepancy of the distribution of *spa* types in patients and HCWs can be explained by the observations that the risk for LA-MRSA-colonization is 168 times higher for people who work daily with pigs than for family members living on a farm and that transmission of LA-MRSA between humans outside farms is limited (Cuny et al., 2009). The reason therefore is still unknown. Among the HCWs screened in this study, only 4% had contact with livestock.

In the different professional groups, our observations confirm the results found in previous studies, which showed a higher MRSA prevalence in nurses compared to medical doctors (Kaminski et al., 2007; Albrich and Harbarth, 2008). HCWs are likely to act more frequently as vectors, rather than being the main source of MRSA transmission. Furthermore, MRSA can be transmitted both from nurses to patients and from patients to nurses. Although colonized HCWs appeared to be most often transiently carriers, they may become stable carriers if they have sinusitis, or chronic dermatitis, leading to MRSA transmission over a longer time period (Dulon et al., 2014). The discussion on who is transmitting MRSA to whom has also been discussed in relation with MRSA observed in (pet) animals and possible transmission to humans. A Dutch study showed that a dog living in close contact with a nurse became colonized with MRSA resulting in recurrent MRSA colonization of the nurse (van Duijkeren et al., 2004). In addition, a narrative review concluded that both dogs and cats can serve as vectors for MRSA transmission (Bramble et al., 2011). Consequently, it can be hypothesized that close physical contact with patients, e.g., washing and dressing, contributes to a higher MRSA transmission rate between patients and HCWs. The high MRSA prevalence among employees working in non-patient related activities (2.8%), also higher than in medical doctors (1.2%), seems to contradict this assumption. A possible explanation for this could be that non-patient related activities, e.g., cleaning personnel, is exposed to patients and especially patient fluids and materials in toilets and bathrooms, but may be less informed about the protocols to prevent indirect MRSA transmission. Furthermore, two of the three MRSA-positive samples from people with non-patient related activities in this study were either in contact with MRSA-positive relatives, or cared for relatives at home. Both these risk factors had a high odds ratio in the univariate analysis, but were not statistically significant risk factors for MRSA colonization.

It would be interesting to know the MRSA prevalence among surgeons and the staff from surgical wards. Unfortunately, the present study was underpowered to detect such possible differences. This was due to the fact that collection of data on the profession of the participants was not mandatory. From 85 participants, it is known that they were medical doctors. Of these 85 medical doctors, it was possible in only 39 cases to distinguish between surgeons and non-surgeons, and analyses resulted in 15 surgeons and 24 non-surgeons. Only one of these 15 surgeons was MRSA positive.

S. aureus colonization may be dependent on both host and bacterial factors. Furthermore, it has been reported that after decolonization, stable carriers often become re-colonized with the same *S. aureus* strain, whereas non-carriers resist experimental colonization. Long-term carriers of *S. aureus* were reported to carry *S. aureus* between 70 days and 8 years (Brown et al., 2014). Two MRSA swabs were taken from HCWs in order to differentiate between transient and stable carriage. Previously, it has been shown that transient carriage is especially found when HCWs are swabbed during or after their shifts (Cookson et al., 1989). Even though HCWs were sampled before starting their shifts, we observed heterogeneous swab results, indicating transient carriage among 5 of the 28 MRSA-positive HCWs. Besides transient carriage, heterogeneous results may also be the consequence of false-negative results in one swab of an actual stable carrier due to a sampling error, the less than 100% sensitivity of the nasopharyngeal swabs (Kunori et al., 2002), or the lack of the use of a semi-selective broth before application of the swabs on chromogenic media (Bocher et al., 2010) as used in the present study. To exclude false-negative results, a third validation swab was taken. This validation swab confirmed negative swab results, and thereby a stable carriage rate of 69.7%. So, 17.9% of all MRSA carriers proved to be transient carriers, and we conclude that differentiation between transient and stable carriers seems appropriate and reduces the number of HCWs who needed to be included in decolonization therapies.

By contrast, abdication of differentiation might overestimate actual MRSA prevalence in HCWs and lead to unnecessary decolonization.

According to the multivariate analysis, a positive anamnesis of MRSA carriage, acne and chronic IBD were possible risk factors for MRSA colonization in HCWs. As the latter only refers to one MRSA positive HCW and one MRSA negative HCW, its risk impact is not clear and should be confirmed using a larger cohort of individuals suffering from IBD. The other two possible risk factors confirmed observations in previous studies (Albrich and Harbarth, 2008). The fact that a positive anamnesis is a highly significant ($p = 0.01$) risk factor, encourages attempts for long-term control of MRSA-positive HCWs. Although we have shown that all HCWs could be decolonized initially, a positive MRSA anamnesis still increases the risk for re-colonization or resurge of suppressed colonization. The reason for re-colonization/resurge is unclear, but a possible vulnerability for MRSA colonization and frequent long-term failure of topical decolonization therapies might contribute to it (Ammerlaan et al., 2009). As shown before, chronic skin diseases, such as acne in the present study, are known risk factors for MRSA colonization (Berthelot et al., 2003). As the number of individuals included in our study is rather small, the factors identified as possible risk factors should be confirmed in a larger group.

The drawback of this study was the great variation in the participation rate of HCWs among the hospitals, e.g., in hospital no. 6 only 9% of the HCWs participated, whereas in hospital no. 3 the participation rate was 91%. This variation in the participation rate of HCWs was due to different factors. Our aim was to screen approximately 100 persons per hospital, or less if the hospital was smaller. We argued with hospital directorates that not less than 100 persons per hospital should participate in order to enlarge the acceptability of the study. The directorate was concerned that the hospital would be understaffed because of the expected high rates of MRSA-positive employees and the fact that MRSA-positive employees had to be decolonized and could not work for a few days. In the future, well-organized campaigns focusing on the benefit for the hospital to participate in studies, such as ours, may take away their concern. Furthermore, the number of participants was also dependent on the commitment and the networking of the representatives of the hospitals. If these representatives were committed and networked well, which is much easier in smaller hospitals, more employees could be screened. To the best of our knowledge, there were no studies in Germany in 2012 in which more physicians were included for MRSA screening.

In this study, the MRSA prevalence among HCWs in a non-outbreak situation is not as high as observed in other studies, but still higher compared to the MRSA prevalence in patients at hospital admission. We showed that 70% of HCWs were stable MRSA carriers, whereas 30% "lost" MRSA in control swabs taken without decolonization attempts and indicating transient colonization/contamination. Since a higher MRSA prevalence in nurses compared to medical doctors, it can be suggested that close physical contact with patients contributes to a higher MRSA transmission rate between patients and HCWs.

AUTHOR CONTRIBUTIONS

RS, RD, RK, RH, AJ, JR, and AF planned the study, collected and analyzed the data and wrote the manuscript.

FUNDING

This study was supported by the Interreg IVa-funded projects EurSafety Heath-net (III-1-02 = 73) and SafeGuard (III-2-03 = 025), part of a Dutch-German cross-border network supported by the European Commission, the German Federal States of Nordrhein-Westfalen and Niedersachsen, and the Dutch provinces of Overijssel, Gelderland, and Limburg and by the German Ministry of Education and Research (MedVet-Staph No. 01KI1014A).

REFERENCES

Albrich, W. C., and Harbarth, S. (2008). Health-care workers: source, vector, or victim of MRSA? *Lancet Infect. Dis.* 8, 289–301. doi: 10.1016/S1473-3099(08)70097-5

Ammerlaan, H. S., Kluytmans, J. A., Wertheim, H. F., Nouwen, J. L., and Bonten, M. J. (2009). Eradication of methicillin-resistant *Staphylococcus aureus* carriage: a systematic review. *Clin. Infect. Dis.* 48, 922–930. doi: 10.1086/597291

Ben-David, D., Mermel, L. A., and Parenteau, S. (2008). Methicillin-resistant *Staphylococcus aureus* transmission: the possible importance of unrecognized health care worker carriage. *Am. J. Infect. Control* 36, 93–97. doi: 10.1016/j.ajic.2007.05.013

Berthelot, P., Grattard, F., Fascia, P., Fichtner, C., Moulin, M., Lavocat, M. P., et al. (2003). Implication of a healthcare worker with chronic skin disease in the transmission of an epidemic strain of methicillin-resistant *Staphylococcus aureus* in a pediatric intensive care unit. *Infect. Control Hosp. Epidemiol.* 24, 299–300. doi: 10.1086/502208

Blok, H. E., Troelstra, A., Kamp-Hopmans, T. E., Gigengack-Baars, A. C., Vandenbroucke-Grauls, C. M., Weersink, A. J., et al. (2003). Role of healthcare workers in outbreaks of methicillin-resistant *Staphylococcus aureus*: a 10-year evaluation from a Dutch university hospital. *Infect. Control Hosp. Epidemiol.* 24, 679–685. doi: 10.1086/502275

Böcher, S., Middendorf, B., Westh, H., Mellmann, A., Becker, K., Skov, R., et al. (2010). Semi-selective broth improves screening for methicillin-resistant *Staphylococcus aureus*. *J. Antimicrob. Chemother.* 65, 717–720. doi: 10.1093/jac/dkq001

Bramble, M., Morris, D., Tolomeo, P., and Lautenbach, E. (2011). Potential role of pet animals in household transmission of methicillin-resistant *Staphylococcus aureus*: a narrative review. *Vector Borne Zoonotic Dis.* 11, 617–620. doi: 10.1089/vbz.2010.0025

Brown, A. F., Leech, J. M., Rogers, T. R., and McLoughlin, R. M. (2014). *Staphylococcus aureus* colonization: modulation of host immune response and impact on human vaccine design. *Front. Immunol.* 4:507. doi: 10.3389/fimmu.2013.00507

Cookson, B., Peters, B., Webster, M., Phillips, I., Rahman, M., and Noble, W. (1989). Staff carriage of epidemic methicillin-resistant *Staphylococcus aureus*. *J. Clin. Microbiol.* 27, 1471–1476.

Cuny, C., Nathaus, R., Layer, F., Strommenger, B., Altmann, D., and Witte, W. (2009). Nasal colonization of humans with methicillin-resistant *Staphylococcus aureus* (MRSA) CC398 with and without exposure to pigs. *PLoS ONE* 4:e6800. doi: 10.1371/journal.pone.0006800

Dulon, M., Peters, C., Schablon, A., and Nienhaus, A. (2014). MRSA carriage among healthcare workers in non-outbreak settings in Europe and the United States: a systematic review. *BMC Infect. Dis.* 14:363. doi: 10.1186/1471-2334-14-363

Friedrich, A. W., Daniels-Haardt, I., Köck, R., Verhoeven, F., Mellmann, A., Harmsen, D., et al. (2008). EUREGIO MRSA-net Twente/Munsterland - a Dutch-German cross-border network for the prevention and control of infections caused by methicillin-resistant *Staphylococcus aureus*. *Euro Surveill.* 13, 1–5.

Haley, R. W., Hightower, A. W., Khabbaz, R. F., Thornsberry, C., Martone, W. J., Allen, J. R., et al. (1982). The emergence of methicillin-resistant *Staphylococcus aureus* infections in United States hospitals. Possible role of the house staff-patient transfer circuit. *Ann. Intern. Med.* 97, 297–308.

Hardy, K. J., Oppenheim, B. A., Gossain, S., Gao, F., and Hawkey, P. M. (2006). A study of the relationship between environmental contamination with methicillin-resistant *Staphylococcus aureus* (MRSA) and patients' acquisition of MRSA. *Infect. Control Hosp. Epidemiol.* 27, 127–132. doi: 10.1086/500622

Hawkins, G., Stewart, S., Blatchford, O., and Reilly, J. (2011). Should healthcare workers be screened routinely for meticillin-resistant *Staphylococcus aureus*? A review of the evidence. *J. Hosp. Infect.* 77, 285–289. doi: 10.1016/j.jhin.2010.09.038

Henderson, D. K. (2006). Managing methicillin-resistant Staphylococci: a paradigm for preventing nosocomial transmission of resistant organisms. *Am. J. Med.* 119(6 Suppl. 1), S45–S52. discussion: S62–S70. doi: 10.1016/j.amjmed.2006.04.002

Hill, S. F. (2011). Should healthcare workers be screened routinely for meticillin-resistant *Staphylococcus aureus*? *J. Hosp. Infect.* 79, 275. doi: 10.1016/j.jhin.2011.05.019

Kaminski, A., Kammler, J., Wick, M., Muhr, G., and Kutscha-Lissberg, F. (2007). Transmission of methicillin-resistant *Staphylococcus aureus* among hospital staff in a German trauma centre: a problem without a current solution? *J. Bone Joint Surg. Br.* 89, 642–645. doi: 10.1302/0301-620X.89B5.18756

Köck, R., Becker, K., Cookson, B., van Gemert-Pijnen, J. E., Harbarth, S., Kluytmans, J., et al. (2010). Methicillin-resistant *Staphylococcus aureus* (MRSA): burden of disease and control challenges in Europe. *Euro Surveill.* 15, 19688.

Köck, R., Brakensiek, L., Mellmann, A., Kipp, F., Henderikx, M., Harmsen, D., et al. (2009). Cross-border comparison of the admission prevalence and clonal structure of meticillin-resistant *Staphylococcus aureus*. *J. Hosp. Infect.* 71, 320–326. doi: 10.1016/j.jhin.2008.12.001

Köck, R., Schaumburg, F., Mellmann, A., Köksal, M., Jurke, A., Becker, K., et al. (2013). Livestock-associated methicillin-resistant *Staphylococcus aureus* (MRSA) as causes of human infection and colonization in Germany. *PLoS ONE* 8:e55040. doi: 10.1371/journal.pone.0055040

Köck, R., Siam, K., Al-Malat, S., Christmann, J., Schaumburg, F., Becker, K., et al. (2011). Characteristics of hospital patients colonized with livestock-associated meticillin-resistant *Staphylococcus aureus* (MRSA) CC398 versus other MRSA clones. *J. Hosp. Infect.* 79, 292–296. doi: 10.1016/j.jhin.2011.08.011

Kunori, T., Cookson, B., Roberts, J. A., Stone, S., and Kibbler, C. (2002). Cost-effectiveness of different MRSA screening methods. *J. Hosp. Infect.* 51, 189–200. doi: 10.1053/jhin.2002.1247

Mellmann, A., Weniger, T., Berssenbrügge, C., Keckevoet, U., Friedrich, A. W., Harmsen, D., et al. (2008). Characterization of clonal relatedness among the natural population of *Staphylococcus aureus* strains by using *spa* sequence typing and the BURP (Based Upon Repeat Patterns) algorithm. *J. Clin. Microbiol.* 46, 2805–2808. doi: 10.1128/JCM.00071-08

Mellmann, A., Weniger, T., Berssenbrügge, C., Rothganger, J., Sammeth, M., Stoye, J., et al. (2007). Based Upon Repeat Pattern (BURP): an algorithm to characterize the long-term evolution of *Staphylococcus aureus* populations based on *spa* polymorphisms. *BMC Microbiol.* 7:98. doi: 10.1186/1471-2180-7-98

Peters, G., Becker, K., Briesch, H., Hergenröder, H., Heudorf, U., Just, H. M., et al. (2014). Empfehlungen zur Prävention und Kontrolle von Methicillin-resistenten *Staphylococcus aureus*-Stämmen (MRSA) in medizinischen und pflegerischen Einrichtungen. *Bundesgesundheitsblatt* 57, 696–732. doi: 10.1007/s00103-014-1980-x

Peters, G., Becker, K., Kipp, F., Heuck, D., Nassauer, A., Unger, G., et al. (1999). Empfehlung zur Prävention und Kontrolle von Methicillin-resistenten *Staphylococcus aureus*-Stämmen (MRSA) in Krankenhäusern und anderen medizinischen Einrichtungen Mitteilung der Kommission für Krankenhaushygiene und Infektionsprävention am RKI. *Bundesgesundheitsblatt* 42, 954–958.

Reich-Schupke, S., Geis, G., Reising, M., Altmeyer, P., and Stücker, M. (2010). MRSA in dermatology - prospective epidemiological study in employees and patients of a dermatological department of a university hospital. *J. Dtsch. Dermatol. Ges.* 8, 607–613. doi: 10.1111/j.1610-0387.2010.07381.x

van Cleef, B. A., Graveland, H., Haenen, A. P., van de Giessen, A. W., Heederik, D., Wagenaar, J. A., et al. (2011). Persistence of livestock-associated methicillin-resistant *Staphylococcus aureus* in field workers after short-term occupational exposure to pigs and veal calves. *J. Clin. Microbiol.* 49, 1030–1033. doi: 10.1128/JCM.00493-10

van Duijkeren, E., Wolfhagen M. J., Box A. T., Heck M. E., Wannet W. J., and Fluit A. C. (2004). Human-to-dog transmission of methicillin-resistant *Staphylococcus aureus*. *Emerg. Infect. Dis.* 10, 2235–2237. doi: 10.3201/eid1012.040387

Vonberg, R. P., Stamm-Balderjahn, S., Hansen, S., Zuschneid, I., Ruden, H., Behnke, M., et al. (2006). How often do asymptomatic healthcare workers cause methicillin-resistant *Staphylococcus aureus* outbreaks? A systematic evaluation. *Infect. Control. Hosp. Epidemiol.* 27, 1123–1127. doi: 10.1086/507922

Voss, A., and Doebbeling, B. N. (1995). The worldwide prevalence of methicillin-resistant *Staphylococcus aureus*. *Int. J. Antimicrob. Agents* 5, 101–106.

Witte, W., Mielke, M., Ammon, A., Nassauer, A., and Wischnewski, N. (2005). Fachtagung der AG Nosokomiale Infektionen am RKI zur Intensivierung der Umsetzung von Präventionsstrategien bei MRSA. *Epid. Bull.* 5, 31–38.

3

Antibiotic Resistance Patterns of *Pseudomonas* spp. Isolated from the River Danube

*Clemens Kittinger[1], Michaela Lipp[1], Rita Baumert[1], Bettina Folli[1], Günther Koraimann[2], Daniela Toplitsch[1,2], Astrid Liebmann[1,2], Andrea J. Grisold[1], Andreas H. Farnleitner[3,4], Alexander Kirschner[3,5] and Gernot Zarfel[1]**

[1] Institute of Hygiene, Microbiology and Environmental Medicine, Medical University Graz, Graz, Austria, [2] Institute of Molecular Biosciences, University of Graz, Graz, Austria, [3] Interuniversity Cooperation Centre for Water and Health, Vienna, Austria, [4] Research Group Environmental Microbiology and Molecular Ecology, Institute of Chemical Engineering, Vienna University of Technology, Vienna, Austria, [5] Institute for Hygiene and Applied Immunology, Medical University of Vienna, Vienna, Austria

***Correspondence:**
Gernot Zarfel
gernot.zarfel.@medunigraz.at

Spread and persistence of antibiotic resistance pose a severe threat to human health, yet there is still lack of knowledge about reservoirs of antibiotic resistant bacteria in the environment. We took the opportunity of the Joint Danube Survey 3 (JDS3), the world's biggest river research expedition of its kind in 2013, to analyse samples originating from different sampling points along the whole length of the river. Due to its high clinical relevance, we concentrated on the characterization of *Pseudomonas* spp. and evaluated the resistance profiles of *Pseudomonas* spp. which were isolated from eight sampling points. In total, 520 *Pseudomonas* isolates were found, 344 (66.0%) isolates were identified as *Pseudomonas putida*, and 141 (27.1%) as *Pseudomonas fluorescens*, all other *Pseudomonas* species were represented by less than five isolates, among those two *P. aeruginosa* isolates. Thirty seven percent (37%) of all isolated *Pseudomonas* species showed resistance to at least one out of 10 tested antibiotics. The most common resistance was against meropenem (30.4%/158 isolates) piperacillin/tazobactam (10.6%/55 isolates) and ceftazidime (4.2%/22 isolates). 16 isolates (3.1%/16 isolates) were multi-resistant. For each tested antibiotic at least one resistant isolate could be detected. Sampling points from the upper stretch of the River Danube showed more resistant isolates than downriver. Our results suggest that antibiotic resistance can be acquired by and persists even in *Pseudomonas* species that are normally not in direct contact with humans. A possible scenario is that these bacteria provide a reservoir of antibiotic resistance genes that can spread to related human pathogens by horizontal gene transfer.

Keywords: JDS3, *Pseudomonas*, antibiotic resistance, water, Danube

INTRODUCTION

Multiresistant bacteria are present in many surface waters (Girlich et al., 2011; Czekalski et al., 2012; Tissera and Lee, 2013; Blaak et al., 2015; Maravic et al., 2015). Typically, evidence is provided through fecal indicators, and mostly relates to short river sections or sampling at individual points. Thus, investigations of whole water systems are rare, especially if the river passes through 10 riparian countries, like the River Danube. Water samples from the third Joint Danube Survey (JDS3), the world's biggest river research expedition of its kind, offered a chance for evaluating resistance of bacteria over a whole river system. Based on these samples, a resistance profile of *Pseudomonas* spp. over the course of the multinational River Danube was created in our study.

Pseudomonas species can be naturally found in all surface waters, lakes and rivers, but they are rarely found in drinking water. *Pseudomonas* spp. can survive in both low and high nutrition environments (Mena and Gerba, 2009) or even in double distilled water and, in addition, can help Salmonellae survive in this environment (Warburton et al., 1994). The whole group of non-fermenting Gram negative bacilli is suspected of establishing the basis for multiresistance in Gram negative bacteria, as the members of this group carry multiple intrinsic resistances and have the ability to acquire and evolve additional resistances (Farinas and Martinez-Martinez, 2013). *Pseudomonas* species are known to harbor multiple intrinsic and acquired resistance genes, host several mobile genetic elements, and also exchange them with other families of Gram negative bacilli like *Enterobacteriaceae* (Juan Nicolau and Oliver, 2010; Pfeifer et al., 2010). Hence *Pseudomonas* are known starting points of several important carbapenemases families (Pfeifer et al., 2010). The occurrence and spread of carbapenemases have become a substantial global health problem, as they inactivate a substantial antibiotic class.

The most common pathogen in this genus is *Pseudomonas aeruginosa*. It causes a variety of different infections, from easy-to-cure ear infections, serious infections of burn patients, to severe lung infections which lead to major complications in cystic fibrosis patients (Barbier and Wolff, 2010; Azzopardi et al., 2014). Besides *Pseudomonas aeruginosa*, other species e.g., *Pseudomonas putida* or *Pseudomonas fluorescens* are also a cause for infections in clinical settings (Gilarranz et al., 2013; Erol et al., 2014; Bhattacharya et al., 2015; Mazurier et al., 2015).

The aim of the study was to evaluate the resistance profiles of *Pseudomonas* spp. isolated at selected sites along the whole course of the River Danube. *Pseudomonas* spp. were chosen for various reasons: they belong to the native bacterial community in surface waters, they are clinically relevant, and changes in their natural resistance profiles indicate anthropogenic influence. This study, therefore, aims at monitoring the presence of resistances of *Pseudomonas* spp. to clinically important antibiotics along the river course. Doing so, changes in the resistance profiles were to be detected, if possible.

MATERIALS AND METHODS

Sample Collection

Samples were collected during the JDS 3, which was organized by the International Commission for the Protection of the Danube River (ICPDR).The ICPDR is a transnational body, which has been established to implement the Danube River Protection Convention. All Danube countries are member states of the ICPDR on the base of the "Convention on Cooperation for the Protection and Sustainable use of the Danube River" (Danube River Protection Convention).

Between Aug. 12 and Sep. 26, 2013, surface water samples for microbiological investigations were collected from 68 sampling sites along the river Danube (JDS 3, 2015).

For each sampling site, water samples were taken at three sampling points, on the left, in the middle and on the right side of the River Danube. Samples were collected in sterile 1-L glass flasks, from 30 cm below the river surface (**Figure 1**, pink and violet dots, high resolution map is added as **Supplementary Figure S1**). Duplicate volumes of the samples (45 ml) were filled into sterile non-toxic 50 ml plastic vials (Techno Plastic Products AG, TPP, Switzerland), containing 5 ml glycerol (final conc. 10% v/v). The vials were shaken and turned around to homogenize glycerol and water and after that immediately stored at −20°C on board of the JDS3 research ship. After transfer to the home laboratory (beginning of October 2013) the samples were stored at −80°C. Out of the 68 sampling sites the four sites directly downstream from the cities Vienna, Budapest, Novi Sad, and Bucharest were chosen for investigation. In addition four non-city related sampling sites were chosen (including the delta and near the starting point of the JDS3; **Table 1**).

Isolation of Bacteria

The frozen samples were thawed and 15 ml (left, middle, right 5 ml each) were plated in 0.5 ml portions on different selective agars: Endo Agar, Xylose Lysine Deoxycholat Agar (XLD agar), and Chromocult Coliform Agar (CCA), (all Merck, Austria). Growth conditions were 37 ± 1°C for 18–24 h. Identification of Pseudomonas by MALDI-TOF-MS (matrix-assisted laser desorption ionization time-of-flight mass spectrometry) was per-formed as described previously (Jamal et al., 2014). A single bacterial colony was deposited on the target slide, followed by the addition of the matrix (VITEK MS-CHCA) and air drying. Samples were processed in the MALDI-TOF-MS spectrometer VITEK® MS (Biomerieux, Austria). Microbial identification was achieved by obtaining the spectra using MALDI-TOF technology and analyzing the spectra with the VITEK MS database. The peaks from these spectra were compared with the characteristic pattern for the species, genus or family of the microorganism, leading to identification of the organism.

Antibiotic Susceptibility Testing

For all identified *Pseudomonas* spp., antibiotic susceptibility testing was performed as recommended by the European Committee on Antimicrobial Susceptibility Testing (EUCAST) including recommended controls. Inhibition zone diameters were interpreted according to EUCAST guidelines (http://www.eucast.org/clinical_breakpoints/) (EUCAST, 2013; Matuschek et al., 2014). Classification of multiresistance of *Pseudomonas* spp. was evaluated according to the Robert Koch Institut (RKI, Germany, http://www.rki.de/EN/Home/homepage_node.html). The suspension for inoculation was prepared from an over-night pure culture on a blood agar (non-selective medium). Colonies were picked with

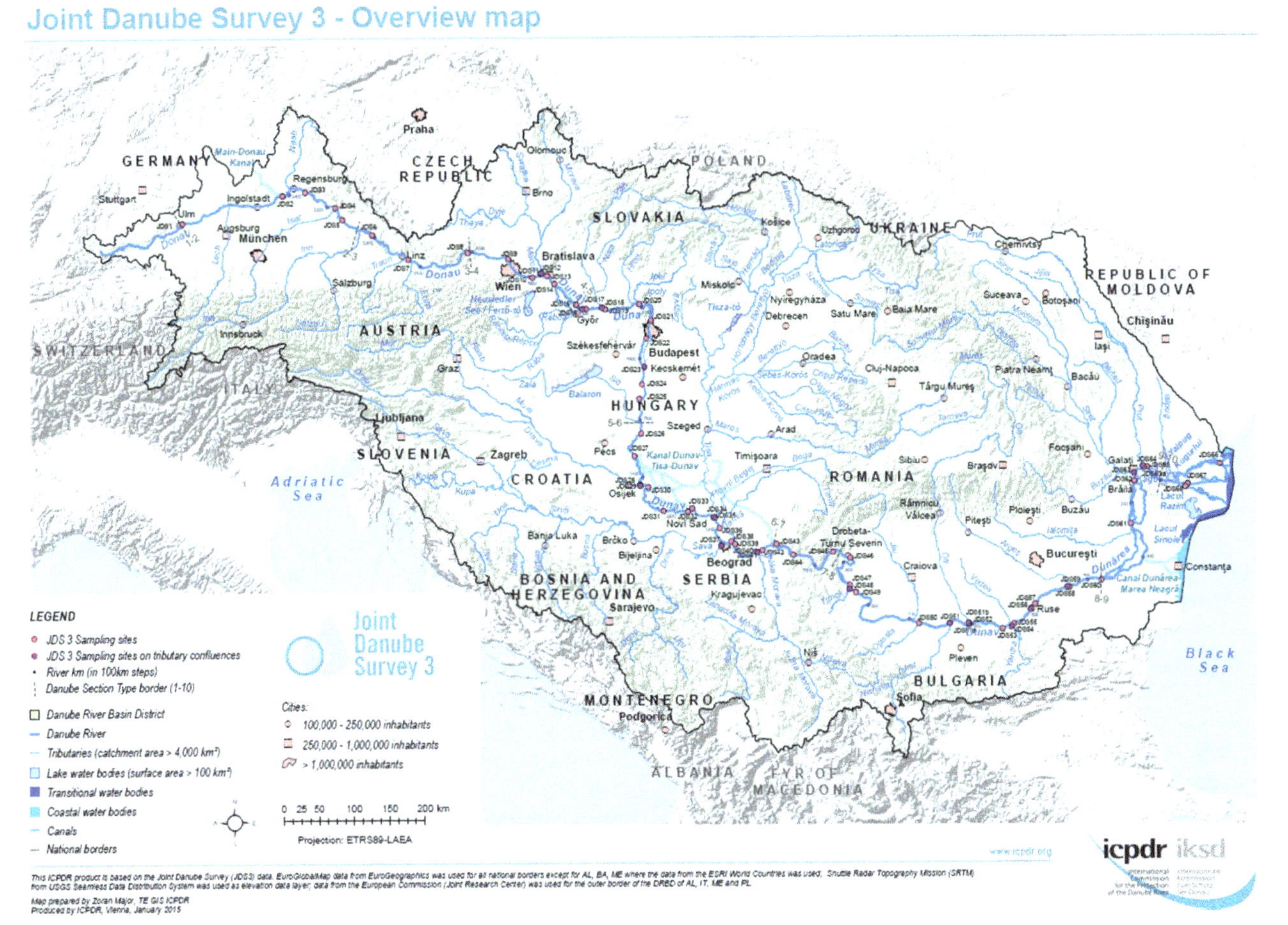

FIGURE 1 | Overview of the JDS3 sampling points along the river Danube. The map was taken with kind permission of the ICPDR. (http://www.danubesurvey.org/results).

TABLE 1 | Investigated sampling sites names and numbers according to JDS3, country and detailed location (ds = downstream, us = upstream).

					E. coli [MPN/100ml]		
Site	**Name**	**Country**	**Location**	**rkm**	**Left**	**Middle**	**Right**
JDS03	Geisling	Germany	ds Regensburg	2354	1728	1739	1304
JDS10	Wildungsmauer	Austria	ds Vienna	1895	1044	917	1739
JDS22	DS Budapest	Hungary	ds Budapest	1632	4320	6792	6310
JDS28	US Drava	Croatia/Serbia	us tributary Drava	1384	2445	1998	2880
JDS36	DS Tisa, Serbia	Serbia	ds Novi Sad	1200	11060	9900	4960
JDS59	DS Arges	Romania/Bulgary	ds tributary Arges*	429	288000	720	1983
JDS63	Siret	Romania	Galati	154	7565	11851	7120
JDS68	St. Gheorge arm	Romania	river delta	104	4880	2424	2431

Basic microbiological parameters. MPN, most probable numbers; rkm, river kilometer. E. coli concentrations were determined according to ISO 9308-2 (JDS 3, 2015).
**Arges is the river flowing through Bucharest.*

a sterile loop and suspended in sterile saline (0.85% NaCl w/v in water) to the density of a McFarland 0.5 standard (DensiCheck, Biomerieux, Austria). The suspension was then plated on Mueller-Hinton Agar by using an automatic plate rotator (Retro C80, Biomerieux, Austria). Antibiotic test disks were applied firmly on the agar surface within 15 min of inoculation of the plates. Plates were incubated at 36°C for 16–20 h. After incubation, inhibition zones were measured.

TABLE 2 | EUCAST Clinical Breakpoints for *Pseudomonas* spp.

Antibiotic	Antibiotic centration on test disk (μg)	Susceptibly inhibition zone (mm)	Resistant (including intermediate) inhibition zone (mm)
Piperacilin/tazobactam	30/6	≥18	<18
Ceftazidime	10	≥16	<16
Cefepime	30	≥19	<19
Meropenem	10	≥18	<18
Imipenem	10	≥20	<20
Amikacin	30	≥24	<24
Gentamicin	10	≥15	<15
Tobramycin	10	≥16	<16
Ciprofloxacin	5	≥25	<25
Levofloxacin	5	≥20	<20

The following antibiotics were tested:

Piperacilin/tazobactam (TZP), ceftazidime (CAZ), cefepime (FEP), meropenem (MEM), imipenem (IPM), amikacin (AN), gentamicin (GM), tobramycin (NN), ciprofloxacin (CIP), levofloxacin (LEV), and sulfamethoxazole/trimethoprim (SXT) (all Becton Dickinson, Schwechat, Austria; **Table 2**).

SXT was evaluated because sulfamethoxazole was part of the chemical analysis of the River Danube water. There are no sulfamethoxazole/trimethoprim breakpoints according to EUCAST for *Pseudomonas* spp. To include the sensitivity of *Pseudomonas* spp. to sulfamethoxazole/trimethoprim, diameters of inhibition areas were evaluated and compared (SXT test discs specification: sulfamethoxazole/trimethoprim: 1.25/23.75 μg).

Modified Hodge Test

To estimate the presence of carbapenemases, a modified Hodge test was performed with all isolates resistant to at least one carbapenem. In brief: after plating of a carbapenem sensitive *Klebsiella pneuomoniae* (ATCC 700603) a 10 μg imipenem or meropenem disc was placed in the center, and each test isolate was streaked from the disk to the edge of the plate. After incubation (37 ± 1°C for 18–24 h) the plates were checked for showing a "cloverleaf shaped" inhibition zone. Isolates that produced carbapenemases enabled growth of the sensitive Klebsiella closer to the antibiotic disk (Bennett et al., 2009).

Data Analyses

Statistical analyses were calculated with GraphPadPrism™ 5.01 for Windows, GraphPad Software, San Diego California USA, www.graphpad.com.

RESULTS

Species Composition of Isolates

In total, 520 *Pseudomonas* spp. were isolated, the fewest isolates were obtained from JDS68 (32 isolates), and the highest number could be isolated from JDS28 sample with 117 isolates (**Table 3**). The most abundant *Pseudomonas* species were *Pseudomonas putida* (66.0%/344 isolates) and *Pseudomonas fluorescens* (27.1%/141 isolates). Each of the other detected species represented less than 1% of all isolates (five or fewer isolates). Only two *Pseudomonas aeruginosa* were isolated, both from JDS28.

Antibiotic Resistances

Wild type *Pseudomonas* species are susceptible to all tested antibiotics except SXT (EUCAST, 2013). The highest number of resistances was the one against meropenem with 158 resistant isolates (30.4%), 55 isolates (10.6%) were resistant to piperacilin/tazobactam and 22 isolates (4.2%) resistant to ceftazidime, 11 (2.1%) to imipenem and four (0.8%) to cefepime (**Table 3**). In the fluoroquinolone group only six (1.2%) of the isolates showed levofloxacin resistance, whereas 18 (3.4%) isolates were resistant against ciprofloxacin. Resistance to ciprofloxacin turned out to be the most frequent resistance of all tested non beta-lactam antibiotics in this study. One isolate which was resistant to levofloxacin was still sensitive to ciprofloxacin. Resistance to aminoglycosides was very rare, two isolates were resistant to amikacin and gentamicin respectively and only one isolate showed no susceptibility to tobramycin (**Table 3**). The 11 isolates resistant to imipenem were also resistant to meropenem and were positive in the modified Hodge test, indicating carbapenemase activity. 327 (62.9%) isolates were susceptible to all tested antibiotics with EUCAST breakpoints (clinical resistance wild type), 128 (24.6%) isolates showed resistance to one, and 49 (9.4%) isolates were resistant to two tested antibiotics. The most common combination of resistances was to meropenem and piperacillin/tazobactam. Sixteen isolates revealed resistance to 3 or more antibiotics, including 12 (2.3%) isolates with three resistances, three (0.6%) isolates with four and one (0.2%) with five. Eight of them were classified as multi-resistant as they were resistant to three or four different antibiotic classes (**Table 4**).

The 12 isolates with three antibiotic resistances split in eight different resistance patterns, the isolates with four resistances all displayed the same pattern. The two *Pseudomonas aeruginosa* were susceptible to all tested antibiotics. Out of all other isolated species in this study at least one showed the EUCAST defined antibiotic susceptibility wild type, susceptible to all tested antibiotics (except SXT; **Supplementary Table S1**).

Four sampling points had less than 50% isolates with clinical resistance wild type, JDS03, JDS10, JDS22, and JDS59. Three sampling points revealed no isolate that was non-susceptible to three or more tested antibiotics, JDS10, JDS63 and JDS68 (**Table 4**).

Surprisingly, the upstream sampling points (JDS03, JDS10, JDS22, and JDS28) revealed higher proportions of resistant bacteria than the downstream ones (**Figure 2**). Sampling point JDS03 revealed the highest proportions of resistance to ceftazidime (28.6% of JDS03 isolates) and ciprofloxacin (25.7% of JDS03 isolates). Sampling point JDS22 showed high rates for piperacillin/tazobactam (32.6% of the JDS22 isolates), meropenem (69.6% of the JDS22 isolates), and imipenem (13.0% of the JDS22 isolates).

SXT was chosen for testing as sulfamethoxazole was measured directly during JDS3. Chemical analysis revealed a sampling

TABLE 3 | Number of isolated *Pseudomonas* spp. at the investigated sampling points (SP) and the number of resistant isolates to the testes antibioitic.

SP	TZP	CAZ	FEP	MEM	IPM	AN	GM	NN	CIP	LEV	No. Isolates
JDS03	3	10	1	15	0	0	1	0	9	1	35
JDS10	2	2	0	16	2	0	0	0	1	0	33
JDS22	15	5	0	32	6	1	0	0	2	0	46
JDS28	19	4	2	35	3	0	0	0	2	3	117
JDS36	6	0	0	10	0	1	1	1	1	1	109
JDS59	6	1	1	21	0	0	0	0	0	0	46
JDS63	3	0	0	21	0	0	0	0	2	1	102
JDS68	1	0	0	8	0	0	0	0	1	0	32
Sum	55	22	4	158	11	2	2	1	18	6	520

Tested antibiotics and their abbreviation: (TZP), piperacilin/tazobactam; (CAZ), ceftazidime; (FEP), cefepime; (MEM), meropenem; (IPM), imipenem; (AN), amikacin; (GM), gentamicin; (NN), tobramycin; (CIP), ciprofloxacin; (LEV), levofloxacin.

TABLE 4 | Resistance pattern of isolates showing resistances to three or more of the tested antibiotics.

Isolate	Species	Resistance Pattern	Multidrug Resistance (MDR)
JDS03PS007	*Pseudomonas fluorescens*	CAZ, MEM, CIP	MDR 3
JDS03PS016	*Pseudomonas fluorescens*	TZP, CAZ, GM, MEM, CIP	MDR 4
JDS03PS019	*Pseudomonas putida*	TZP, MEM, CIP	MDR 3
JDS03PS020	*Pseudomonas fluorescens*	CAZ, MEM, CIP	MDR 3
JDS03PS032	*Pseudomonas fluorescens*	CAZ, CIP, LEV	
JDS22PS016	*Pseudomonas putida*	TZP, CAZ, MEM	MDR 3
JDS22PS018	*Pseudomonas fluorescens*	CAZ, IMP, MEM	
JDS22PS032	*Pseudomonas putida*	TZP, IMP, MEM	
JDS22PS035	*Pseudomonas putida*	TZP, IMP, MEM	
JDS22PS043	*Pseudomonas putida*	TZP, MEM, CIP	MDR 3
JDS28PS083	*Pseudomonas fluorescens*	CAZ, FEP, IMP, MEM	
JDS28PS113	*Pseudomonas fluorescens*	CAZ, IMP, MEM	
JDS28PS115	*Pseudomonas putida*	TZP, MEM, CIP, LEV	MDR 3
JDS28PS117	*Pseudomonas putida*	TZP, MEM, LEV	MDR 3
JDS36PS036	*Pseudomonas putida*	TZP, AN, GM, NN	
JDS59PS020	*Pseudomonas fluorescens*	TZP, CAZ, FEP	

Multidrug resistance was assigned to an isolate if it revealed a resistance to three (MDR 3) or four (MDR 4) antibiotic classes. Antibiotic classes: acylureidopenicillins: (TZP) piperacilin/tazobactam; cephalosporins: (CAZ) ceftazidime, (FEP) cefepime; carbapenems: (MEM) meropenem, (IPM) imipenem; aminoglycosides: (AN) amikacin, (GM) gentamicin, (NN) tobramycin; fluoroquinolones: (CIP) ciprofloxacin, (LEV) levofloxacin.

site (JDS58) with elevated levels for sulfamethoxazole (Arges, tributary) (JDS 3, 2015). The subsequent River Danube sampling site JDS59 did not show significantly reduced diameters ($p = 0.68$) (**Figure 3**). JDS03 and JDS68 showed both elevated diameters (6.59 mm JDS03, 8.13 mm JDS68) but only JDS68 differed significantly from the other sampling points ($p < 0.05$).

DISCUSSION

The spread of antibiotic resistant bacteria, their distribution, and their reservoirs in the environment are important issues. Within the last years many different possible sources have been intensively investigated to shed light on the spread of antibiotic resistant bacteria. Farming and the spread of liquid manure are known to contribute to the spread of multiresistant bacteria (Sengupta et al., 2011; Friese et al., 2013). Recently, the focus has been put on waste water, as bacteria of all kind and with all possible genetic features are mixed up there. And a very critical feature in waste water is the possibility that bacteria harboring resistance exchange their resistance determinants with other bacteria (Korzeniewska and Harnisz, 2013; Reinthaler et al., 2013; Amador et al., 2015). Microorganisms from these and other sources can be relatively easily flushed into surface waters (Czekalski et al., 2012; Zurfluh et al., 2013; Hess and Gallert, 2014), but except for a few studies on relatively small rivers that deal with this topic, the fate of deposited bacteria is quite unclear.

The distribution of susceptible and resistant *Pseudomonas* spp. at the investigated JDS3 sampling points showed site-specific differences. At the upstream sampling points, there was a trend to more resistant bacteria (JDS03, JDS10, JDS22, JDS28), and multiresistant *Pseudomonas* spp. could only be detected in this part of the river. One reason for this finding

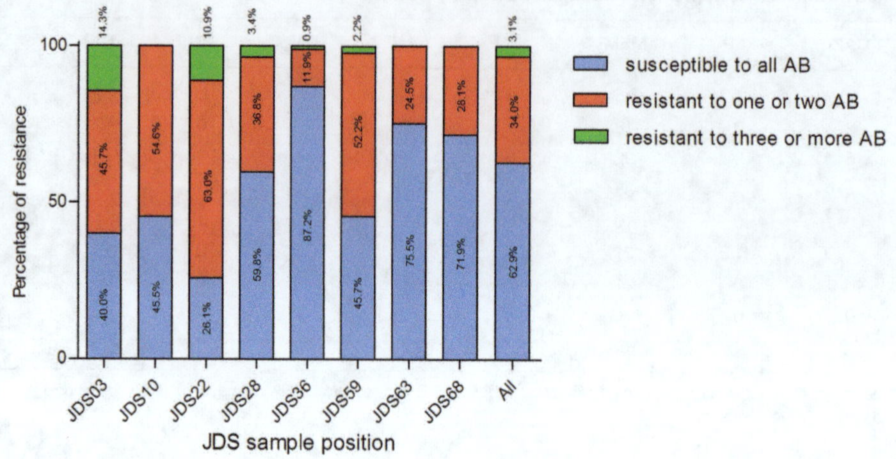

FIGURE 2 | Proportion of isolates susceptible to all tested antibiotics (blue bars), resistant to one or two tested antibiotics (red bars) and resistant to three or more tested antibiotics (green bars), at different sampling points. JDS03, JDS28, JDS63, and JDS68 are non-urban sampling sites, whereas JDS10 is ds Vienna, JDS22 ds Budapest, JDS36 ds Novi Sad and JDS59 is ds Bucharest. Kruskal-Wallis test revealed a non-Gaussian distribution with a $p < 0.0001$.

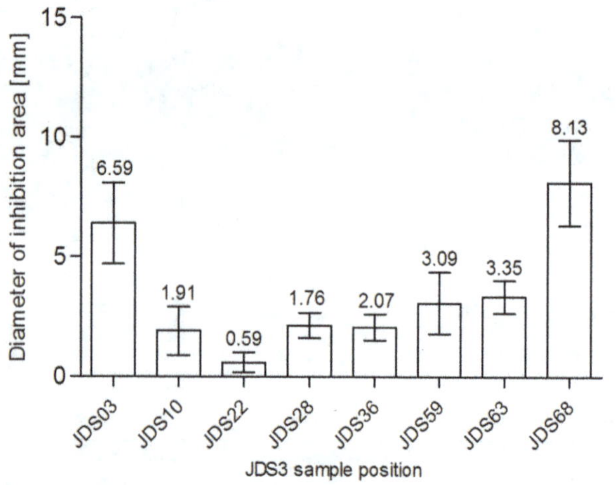

FIGURE 3 | Zone of inhibition diameters (mm) for SXT at the investigated sampling sites. Bars show mean values (values given on top of the bars, mm) and standard deviation errors bars (SD error bars) for each sampling site.

could be the lower river water volume in the upstream parts of the River Danube, which might result in less dilution of the resistant microorganisms. Downstream from the cities of Vienna, Budapest, and Bucharest (JDS10, JDS22, and JDS59 via Argeş; **Figure 3**), the occurrence of resistant bacterial isolates was also elevated, with anthropogenic influence very likely being the reason for that.

Under similar non-selective isolation conditions Suzuki et al. found no resistance to meropenem, gentamicin, amikacin, or ciprofloxacin (Suzuki et al., 2013). A presence of only 8 (1.5%) multiresistant *Pseudomonas* spp. present in three of the eight sampling points seems to be a low number. But still, if we extrapolate the number of 8 multiresistant *Pseudomonaceae* in a collected volume of 75 ml at the 5 sampling points (5 × 15 ml) to 1 liter, we might estimate over 100 multiresistant bacteria of the *Pseudomonas* group in one liter of Danube water.

The isolates all over the course of the river Danube showed high resistance rates against meropenem (9.2–69.6%). The resistance against carbapenems in *Pseudomonas* spp. is mostly mediated via efflux pumps (intrinsic resistance), especially in water environment (Tacao et al., 2015). However, carbapenem resistance poses a challenge for therapy, regardless of the underlying mechanism. For example, *Pseudomonas putida* (66% of all isolates in this study) is increasingly involved in hospital infections (Kim et al., 2012; Molina et al., 2014). These infections come up with severe complications and high mortality rates (up to 40%). In most of these cases multiresistant *Pseudomonas putida* was the reason for the infection or the nosocomial outbreak (Kim et al., 2012).

This study is the first study investigating bacterial resistance in a transnational river survey (2500 rkm). Although it was limited by a small sample volume and a fixed time course, the results of this study substantiate the occurrence of waterborne *Pseudomonas* spp. with non-wild type resistance pattern in the whole River Danube. Their presence and their distribution suggest human influence.

AUTHOR CONTRIBUTIONS

CK study design, laboratory work, manuscript preparation, data analysis, ML laboratory work, data conservation. RB laboratory work. BF laboratory work. GK data analysis, manuscript preparartion. DT laboratory work, data analysis. AL laboratory work. AG manuscript preparation. AF manuscript preparation. AK sample collection, manuscript preparation. GZ study design, laboratory work, data analysis, manuscript preparation.

FUNDING

The Joint Danube Survey was organized by the International Commission for the Protection of the Danube River (ICPDR). The study was supported by the Austrian Science Fund (FWF), project nr P25817-B22.

ACKNOWLEDGMENTS

We would like to thank Georg Reischer, Stefan Jakwerth, and Stoimir Kolarevic for their help in sampling.

REFERENCES

Amador, P. P., Fernandes, R. M., Prudencio, M. C., Barreto, M. P., and Duarte, I. M. (2015). Antibiotic resistance in wastewater: occurrence and fate of Enterobacteriaceae producers of class A and class C beta-lactamases. *J. Environ. Sci. Health. A. Tox. Hazard. Subst. Environ. Eng.* 50, 26–39. doi: 10.1080/10934529.2015.964602

Azzopardi, E. A., Azzopardi, E., Camilleri, L., Villapalos, J., Boyce, D. E., Dziewulski, P., et al. (2014). Gram negative wound infection in hospitalised adult burn patients-systematic review and metanalysis-. *PLoS ONE* 9:e95042. doi: 10.1371/journal.pone.0095042

Barbier, F., and Wolff, M. (2010). Multi-drug resistant *Pseudomonas aeruginosa*: towards a therapeutic dead end? *Med. Sci. (Paris)* 26, 960–968. doi: 10.1051/medsci/20102611960

Bennett, J. W., Herrera, M. L., Lewis, J. S. II., Wickes, B. W., and Jorgensen, J. H. (2009). KPC-2-producing Enterobacter cloacae and pseudomonas putida coinfection in a liver transplant recipient. *Antimicrob. Agents Chemother.* 53, 292–294. doi: 10.1128/AAC.00931-08

Bhattacharya, D., Dey, S., Kadam, S., Kalal, S., Jali, S., Koley, H., et al. (2015). Isolation of NDM-1-producing multidrug-resistant Pseudomonas putida from a paediatric case of acute gastroenteritis, India. *New Microbes New Infect.* 5, 5–9. doi: 10.1016/j.nmni.2015.02.002

Blaak, H., Lynch, G., Italiaander, R., Hamidjaja, R. A., Schets, F. M., and de Roda Husman, A. M. (2015). Multidrug-resistant and extended spectrum beta-lactamase-producing *Escherichia coli* in Dutch Surface Water and Wastewater. *PLoS ONE* 10:e0127752. doi: 10.1371/journal.pone.0127752

Czekalski, N., Berthold, T., Caucci, S., Egli, A., and Burgmann, H. (2012). Increased levels of multiresistant bacteria and resistance genes after wastewater treatment and their dissemination into lake geneva, Switzerland. *Front. Microbiol.* 3:106. doi: 10.3389/fmicb.2012.00106

Erol, S., Zenciroglu, A., Dilli, D., Okumus, N., Aydin, M., Gol, N., et al. (2014). Evaluation of nosocomial blood stream infections caused by Pseudomonas species in newborns. *Clin. Lab.* 60, 615–620. doi: 10.7754/Clin.Lab.2013.130325

EUCAST (2013). *European Committee on Antimicrobial Susceptibility Testing (EUCAST).* Växjö.

Farinas, M. C., and Martinez-Martinez, L. (2013). Multiresistant Gram-negative bacterial infections: Enterobacteria, *Pseudomonas aeruginosa, Acinetobacter baumannii* and other non-fermenting Gram-negative bacilli. *Enferm. Infecc. Microbiol. Clin.* 31, 402–409. doi: 10.1016/j.eimc.2013.03.016

Friese, A., Schulz, J., Laube, H., von Salviati, C., Hartung, J., and Roesler, U. (2013). Faecal occurrence and emissions of livestock-associated methicillin-resistant Staphylococcus aureus (laMRSA) and ESbl/AmpC-producing E. coli from animal farms in Germany. *Berl. Munch. Tierarztl. Wochenschr.* 126, 175–180. doi: 10.2376/0005-9366-126-175

Gilarranz, R., Juan, C., Castillo-Vera, J., Chamizo, F. J., Artiles, F., Alamo, I., et al. (2013). First detection in Europe of the metallo-beta-lactamase IMP-15 in clinical strains of *Pseudomonas putida* and *Pseudomonas aeruginosa. Clin. Microbiol. Infect.* 19, E424–E427. doi: 10.1111/1469-0691.12248

Girlich, D., Poirel, L., and Nordmann, P. (2011). Diversity of clavulanic acid-inhibited extended-spectrum beta-lactamases in Aeromonas spp. from the Seine River, Paris, France. *Antimicrob. Agents Chemother.* 55, 1256–1261. doi: 10.1128/AAC.00921-10

SUPPLEMENTARY MATERIAL

Supplementary Figure S1 | Figure of the JDS3 River Danube Survey (high resolution.pdf).

Supplementary Table S1 | Table of all *Pseudomonas* spp. isolates with resistance pattern. (TZP), Piperacilin/tazobactam; (CAZ), ceftazidime; (FEP), cefepime; (MEM), meropenem; (IPM), imipenem; (AN), amikacin; (GM), gentamicin; (NN), tobramycin; (CIP), ciprofloxacin; (LEV), levofloxacin; and (SXT), sulfamethoxazole/trimethoprim.

Hess, S., and Gallert, C. (2014). Demonstration of staphylococci with inducible macrolide-lincosamide-streptogramin B (MLSB) resistance in sewage and river water and of the capacity of anhydroerythromycin to induce MLSB. *FEMS Microbiol. Ecol.* 88, 48–59. doi: 10.1111/1574-6941.12268

Jamal, W., Albert, M. J., and Rotimi, V. O. (2014). Real-time comparative evaluation of bioMerieux VITEK MS versus Bruker Microflex MS, two matrix-assisted laser desorption-ionization time-of-flight mass spectrometry systems, for identification of clinically significant bacteria. *BMC Microbiol.* 14:289-014-0289-0. doi: 10.1186/s12866-014-0289-0

JDS 3, I. (2015). *Joint Danube Survey 3. A Comprehensive Analysis of Danube Water Quality.* Vienna: ICPDR – International Commission for the Protection of the Danube River.

Juan Nicolau, C., and Oliver, A. (2010). Carbapenemases in Pseudomonas spp. *Enferm. Infecc. Microbiol. Clin.* 28 (Suppl. 1), 19–28. doi: 10.1016/S0213-005X(10)70004-5

Kim, S. E., Park, S. H., Park, H. B., Park, K. H., Kim, S. H., Jung, S. I., et al. (2012). Nosocomial Pseudomonas putida Bacteremia: high rates of carbapenem resistance and mortality. *Chonnam Med. J.* 48, 91–95. doi: 10.4068/cmj.2012.48.2.91

Korzeniewska, E., and Harnisz, M. (2013). Extended-spectrum beta-lactamase (ESBL)-positive Enterobacteriaceae in municipal sewage and their emission to the environment. *J. Environ. Manage.* 128, 904–911. doi: 10.1016/j.jenvman.2013.06.051

Maravic, A., Skocibusic, M., Cvjetan, S., Samanic, I., Fredotovic, Z., and Puizina, J. (2015). Prevalence and diversity of extended-spectrum-beta-lactamase-producing Enterobacteriaceae from marine beach waters. *Mar. Pollut. Bull.* 90, 60–67. doi: 10.1016/j.marpolbul.2014.11.021

Matuschek, E., Brown, D. F., and Kahlmeter, G. (2014). Development of the EUCAST disk diffusion antimicrobial susceptibility testing method and its implementation in routine microbiology laboratories. *Clin. Microbiol. Infect.* 20, O255–O266. doi: 10.1111/1469-0691.12373

Mazurier, S., Merieau, A., Bergeau, D., Decoin, V., Sperandio, D., Crepin, A., Barbey, C., et al. (2015). Type III secretion system and virulence markers highlight similarities and differences between human- and plant-associated pseudomonads related to *Pseudomonas fluorescens* and *P. putida. Appl. Environ. Microbiol.* 81, 2579–2590. doi: 10.1128/aem.04160-14

Mena, K. D., and Gerba, C. P. (2009). Risk assessment of *Pseudomonas aeruginosa* in water. *Rev. Environ. Contam. Toxicol.* 201, 71–115. doi: 10.1007/978-1-4419-0032-6_3

Molina, L., Udaondo, Z., Duque, E., Fernàndez, M., Molina-Santiago, C., Roca, A., Porcel, M., et al. (2014). Antibiotic resistance determinants in a *Pseudomonas putida* strain isolated from a hospital. *PLoS ONE* 9:e81604. doi: 10.1371/journal.pone.0081604

Pfeifer, Y., Cullik, A., and Witte, W. (2010). Resistance to cephalosporins and carbapenems in Gram-negative bacterial pathogens. *Int. J. Med. Microbiol.* 300, 371–379. doi: 10.1016/j.ijmm.2010.04.005

Reinthaler, F. F., Galler, H., Feierl, G., Haas, D., Leitner, E., Mascher, F., et al. (2013). Resistance patterns of *Escherichia coli* isolated from sewage sludge in comparison with those isolated from human patients in 2000 and 2009. *J. Water. Health.* 11, 13–20. doi: 10.2166/wh.2012.207

Sengupta, N., Alam, S. I., Kumar, R. B., and Singh, L. (2011). Diversity and antibiotic susceptibility pattern of cultivable anaerobic bacteria from soil and sewage samples of India. *Infect. Genet. Evol.* 11, 64–77. doi: 10.1016/j.meegid.2010.10.009

Suzuki, Y., Kajii, S., Nishiyama, M., and Iguchi, A. (2013). Susceptibility of *Pseudomonas aeruginosa* isolates collected from river water in Japan to antipseudomonal agents. *Sci. Total Environ.* 450-451, 148–154. doi: 10.1016/j.scitotenv.2013.02.011

Tacao, M., Correia, A., and Henriques, I. S. (2015). Low prevalence of carbapenem-resistant Bacteria in River Water: resistance is mostly related to intrinsic mechanisms. *Microb. Drug Resist.* 21, 497–506. doi: 10.1089/mdr.2015.0072

Tissera, S., and Lee, S. M. (2013). Isolation of Extended Spectrum beta-lactamase (ESBL) Producing Bacteria from Urban Surface Waters in Malaysia. *Malays. J. Med. Sci.* 20, 14–22.

Warburton, D. W., Bowen, B., and Konkle, A. (1994). The survival and recovery of *Pseudomonas aeruginosa* and its effect upon salmonellae in water: methodology to test bottled water in Canada. *Can. J. Microbiol.* 40, 987–992. doi: 10.1139/m94-158

Zurfluh, K., Hachler, H., Nuesch-Inderbinen, M., and Stephan, R. (2013). Characteristics of extended-spectrum beta-lactamase- and carbapenemase-producing Enterobacteriaceae Isolates from rivers and lakes in Switzerland. *Appl. Environ. Microbiol.* 79, 3021–3026. doi: 10.1128/AEM.00054-13

4

Co-diversification of *Enterococcus faecium* Core Genomes and PBP5: Evidences of *pbp5* Horizontal Transfer

*Carla Novais[1], Ana P. Tedim[2,3], Val F. Lanza[2,3], Ana R. Freitas[1,2], Eduarda Silveira[1], Ricardo Escada[1,4], Adam P. Roberts[5], Mohammed Al-Haroni[5], Fernando Baquero[2,3,6], Luísa Peixe[1] and Teresa M. Coque[2,3,6]**

[1] UCIBIO/REQUIMTE, Laboratório de Microbiologia, Departamento de Ciências Biológicas, Faculdade Farmácia, Universidade do Porto, Porto, Portugal, [2] Servicio de Microbiología, Instituto Ramón y Cajal de Investigación Sanitaria, Madrid, Spain, [3] Consorcio de Investigación Biomédica en Red de Epidemiología y Salud Pública, Barcelona, Spain, [4] Faculdade de Ciências da Saúde, Universidade Fernando Pessoa, Porto, Portugal, [5] Division of Microbial Diseases, UCL Eastman Dental Institute, University College London, London, UK, [6] Unidad de Resistencia a Antibióticos y Virulencia Bacteriana (RYC-CSIC), Madrid, Spain

***Correspondence:**
Teresa M. Coque
mariateresa.coque@salud.madrid.org

Ampicillin resistance has greatly contributed to the recent dramatic increase of a cluster of human adapted *Enterococcus faecium* lineages (ST17, ST18, and ST78) in hospital-based infections. Changes in the chromosomal *pbp5* gene have been associated with different levels of ampicillin susceptibility, leading to protein variants (designated as PBP5 C-types to keep the nomenclature used in previous works) with diverse degrees of reduction in penicillin affinity. Our goal was to use a comparative genomics approach to evaluate the relationship between the diversity of PBP5 among *E. faecium* isolates of different phylogenomic groups as well as to assess the *pbp5* transferability among isolates of disparate clonal lineages. The analyses of 78 selected *E. faecium* strains as well as published *E. faecium* genomes, suggested that the diversity of *pbp5* mirrors the phylogenomic diversification of *E. faecium*. The presence of identical PBP5 C-types as well as similar *pbp5* genetic environments in different *E. faecium* lineages and clones from quite different geographical and environmental origin was also documented and would indicate their horizontal gene transfer among *E. faecium* populations. This was supported by experimental assays showing transfer of large (≈180–280 kb) chromosomal genetic platforms containing *pbp5* alleles, *ponA* (transglycosilase) and other metabolic and adaptive features, from *E. faecium* donor isolates to suitable *E. faecium* recipient strains. Mutation profile analysis of PBP5 from available genomes and strains from this study suggests that the spread of PBP5 C-types might have occurred even in the absence of a significant ampicillin resistance phenotype. In summary, genetic platforms containing *pbp5* sequences were stably maintained in particular *E. faecium* lineages, but were also able to be transferred among *E. faecium* clones of different origins, emphasizing the growing risk of further spread of ampicillin resistance in this nosocomial pathogen.

Keywords: ampicillin resistance, PBP5 mutations, PonA, chromosomal transfer, phylogenomic diversification

INTRODUCTION

Infections caused by *Enterococcus faecium* have increasingly been reported since the early 1980s', as have the number of antibiotic resistant isolates of this species (Top et al., 2007). Currently, most clinical strains of *E. faecium* are ampicillin resistant (AmpR), which often also acquire transferable mobile genetic elements, some encoding vancomycin resistance (Arias and Murray, 2012; Wagenvoort et al., 2015). Multidrug resistant *E. faecium* isolates are responsible for infections associated with treatment failures and high mortality rates (Arias et al., 2010; Centers for Disease Control and Prevention, 2013). However, the influence of ampicillin resistance in the population structure of *E. faecium* remains largely unexplored.

Enterococcus faecium is intrinsically resistant to cephalosporins and exhibit natural reduced susceptibility to penicillins. This species has six penicillin binding proteins (PBPs) belonging either to class B (monofunctional D,D-transpeptidases), or class A (bifunctional enzymes with glycosyltransferase and D,D-transpeptidase activity), some of them being associated with resistance to β-lactams (Zorzi et al., 1996; Rice et al., 2009). In *E. faecium* resistance to β-lactam antibiotics is conferred by the low-affinity class B PBP5, that requires the participation of class A PBPs (PonA, PbpF) to synthesize the cell wall in the presence of cephalosporins (Williamson et al., 1983; Rice et al., 2009). Resistance to high ampicillin concentrations in *E. faecium* was initially explained by either the enhanced production of PBP5, and/or by polymorphisms in the beta subunit of this protein (Fontana et al., 1996). Soon after, it was demonstrated that such changes were frequently strain specific and did not necessarily correlate with differences in the ampicillin MIC values (Fontana et al., 1996; Zorzi et al., 1996; Rybkine et al., 1998; Rice et al., 2001, 2004; Sifaoui et al., 2001; Belhaj et al., 2016). Further analysis of *E. faecium* strains with different levels of ampicillin susceptibility (0,5- > 128 mg/L) revealed that the variability of PBP5 sequences is mostly due to changes in 21 specific positions of the protein, suggesting that a sequential acquisition of mutations could have contributed to the progressive resistance to ampicillin from the early 1980s (Galloway-Peña et al., 2011; Pietta et al., 2014). Mutations in genes encoding other species-specific proteins that participate in the cell wall synthesis, such as D,D-carboxypeptidases (Ddcp and DdcY), L,D-transpeptidases (Ldt_{fm}), glycosyltransferases (PgtA) and acetylmuramoyl-L-alanine amidase (LytG) may also slightly increase the MIC values (Rice et al., 2007; Zhang et al., 2012; Kristich et al., 2014), even in the absence of PBP5 (Sacco et al., 2014). Although the occurrence of β-lactamases has been documented in *E. faecium*, β-lactamases producing *E. faecium* remain rare (Coudron et al., 1992; Klare et al., 1992; Zhang et al., 2012; Hendrickx et al., 2013).

Enterococcus faecium population biology is dominated by two main phylogenomic groups, clade A and clade B. Most AmpR isolates belong to clade A or "hospital-associated clade" mainly comprising *E. faecium* from hospitalized patients (Galloway-Peña et al., 2012; Palmer et al., 2012; Lebreton et al., 2013). A subgroup within the clade A, clade A1, is enriched in mobile genetic elements and have enhanced ability to colonize and persist in human hosts due to the presence of adhesins and specific metabolic traits. In contrast, the clade B, or "community-associated *E. faecium*," mostly comprises ampicillin susceptible (AmpS) isolates from healthy, non-hospitalized individuals (Willems et al., 2005; Lebreton et al., 2013; Tedim et al., 2015; Freitas et al., 2016; Guzman Prieto et al., 2016). The increasing detection of AmpR among isolates from hosts not associated with the hospital setting (Gonçalves et al., 2011; de Regt et al., 2012; Novais et al., 2013; Santos et al., 2013; Tremblay et al., 2013) is of concern, as the acquisition of transferable genes encoding AmpR might facilitate its further spread into AmpS *E. faecium* populations or to other less frequently recovered enterococcal species for which AmpR has been rarely documented (Raze et al., 1998). A relevant observation was the identification of a transferable chromosomal region of 60 kb comprising *pbp5* and a transposon that confers resistance to glycopeptides (*vanB2*-CTn*5386*) (Carias et al., 1998; Rice et al., 2005b). Homologous recombination between similar chromosomal regions carrying *pbp5* of different strains potentially associated with a conjugation mechanism has been recently suggested (García-Solache et al., 2016).

The known diversity of genotypes of AmpR *E. faecium* is mainly focused on the PBP5 polymorphisms analyzed in a limited number of strains (Poeta et al., 2007; Klibi et al., 2008; Galloway-Peña et al., 2011). Similarly, knowledge related to the *pbp5* transferability is limited to the characterization of conjugative Tn*916*-like elements (CTn*5382* or the interaction of CTn*916* and CTn*5386*) in a few *E. faecium* strains (Carias et al., 1998; Rice et al., 2005a,b; García-Solache et al., 2016). Here, we extend previous knowledge on these issues by using a comparative genomics approach of the chromosomal regions containing *pbp5*, that demonstrate a relationship between the diversity of PBP5 and *E. faecium* phylogenomic groups. We also assess the *pbp5* transferability among isolates from different clonal lineages.

MATERIALS AND METHODS

Bacterial Strains

Seventy-eight isolates from a collection of 205 AmpR *E. faecium* (MIC ≥ 16 mg/L) recovered in Portugal from the last decades were analyzed (Novais et al., 2005a,b,c, 2006, 2013). The 78 *E. faecium* were selected in order to include isolates with different antibiotic resistant phenotypes, from diverse origins (39 from different patients admitted in five hospitals from different cities, 18 from swine feces and piggeries, 4 from retail poultry carcasses, 2 from healthy human feces and 15 from hospital wastewater), isolation date and, whenever possible, clonal relatedness. Susceptibility to ampicillin and 10 other antibiotics of different classes was evaluated by disk diffusion and/or agar dilution method (Clinical and Laboratory Standards Institute, 2012). β-lactamase production was tested in AmpR *E. faecium* isolates by the nitrocefin test (5 μl were directly placed in bacteria growing around the ampicillin disk) and PCR amplification of *blaZ* using primers based on the GenBank sequence no. M25257.1 (blaZF-3′-TTGCCTATGCTTCGACTTCA-5′, blaZR-3′-AGTGAAACCGCCAAGAGTGT-5′). Clonal relatedness was established by Pulsed-Field Gel Electrophoresis (PFGE) and analysis of Multilocus Sequence Typing (MLST) data (Tenover et al., 1995; Homan et al., 2002; Freitas et al., 2013)[1] using the Bayesian Analysis of Population Structure (BAPS) algorithm (Willems et al., 2012; Tedim et al., 2015).

Transferability of Ampicillin Resistance

Filter-mating assays were performed in plain Brain Heart Infusion (BHI) not supplemented with antibiotics at 37°C overnight, using a donor/recipient ratio of 1:1 (1-3 experiments per isolate) and *E. faecium* GE1 as the recipient strain. Further filter mating assays under the same experimental conditions but using *E. faecium* BM4105RF and *E. faecium* 64/3 strains as recipients, were performed for those field isolates able to transfer *pbp5* to *E. faecium* GE1. The three laboratory recipient strains used differed in the susceptibility against ampicillin and the presence of *pbp5*: GE1 [Δ*pbp5*; $MIC_{Amp} < 0.016$mg/L; tetracycline (Tet), rifampicin (Rif) and fusidic acid (Fus) resistant (^{R}); ST515/BAPS 2.3b], BM4105RF (*pbp5*; Rif^R, Fus^R; $MIC_{Amp} = 0.5$mg/L; ST172/BAPS 1.3) and 64/3 (*pbp5*; Rif^R, Fus^R; $MIC_{Amp} = 1$mg/L; ST21/BAPS 2.3a). Recipients acquiring *pbp5* are from this point named as transconjugants, based on the most likely transfer mechanism related to *pbp5* (conjugation) published during the revision of this article (García-Solache et al., 2016). Transconjugants were selected on BHI agar supplemented with antibiotics (ampicillin-10 mg/L, fusidic acid-25 mg/L; rifampicin-30 mg/L) and incubated for 24 up to 96 h (37°C) to recover potential transconjugants growing at different times. Transconjugants were confirmed by comparison of their antibiotic resistance phenotype and PFGE profile with those of the donors and the recipients. Stability of acquired *pbp5* chromosomal regions (from now on designated *pbp5* genetic platforms) after serial daily passages (x30) on antibiotic free BHI agar was evaluated in both donor strains and transconjugants. Colonies from each passage were inoculated on plates containing the same agar medium and tested for ampicillin susceptibility by disk diffusion (Clinical and Laboratory Standards Institute, 2012).

Characterization of the Region Conferring AmpR

Total DNA from a donor strain (AmpR *E. faecium* clinical isolate HPH2), a recipient strain (AmpS *E. faecium* GE1) and the resulting GE1 transconjugant (AmpR *E. faecium* TCGEHPH2.1) was extracted with PureElute™ Bacterial Genomic Kit (EdgeBio, Gaithersburg, MD, USA). DNA concentration was measured with Nanodrop 2000 (Thermo Scientific, Waltham, MA, USA) and Qubit 2.0 fluorometer (Life Technologies). Genomic DNA (250–350 ng) was sequenced on the Illumina MiSeq system using the MiSeq reagent kit v3 and a read length configuration of 2 × 300 bp (Illumina, San Diego, CA, USA). Sequencing was carried out using a standard 2 × 71 base protocol (300–400 bp insert size) in a Genome Analyzer IIx (Illumina, San Diego, CA, USA) at the sequencing facility of the University of Newcastle (United Kingdom). The main statistics for the three sequence datasets (number of reads and coverage) analyzed are shown in Supplementary Table S1. Assembly of sequence data was done using Newbler software (454 life sciences, Roche, Branford, CT, USA). Supplementary Table S2 shows the final assembly results.

Gene prediction was performed using GeneMark.hmm 3.05 (Besemer et al., 2001). Similarity searches for potential protein-coding regions were carried out against a UniRef100 database[2] using Best Blast Hit approach (Supplementary Table S3). The sequence of the transferred region containing the *pbp5* gene was predicted by a gene-by-gene comparison strategy using *blastn* (genes present in transconjugant strain TCGEHPH2.1 and donor strain HPH2 but not in recipient strain GE1) (Supplementary Table S4). Sequence of the putative "transferred region" was mapped against the genome of the *E. faecium* strain DO (RefSeq Accession: NC_017960) using Easyfig 1.2 software (Sullivan et al., 2011). In addition to *E. faecium* DO, the "transferred region" was also aligned against fully sequenced and closed genomes of *E. faecium* Aus0004, *E. faecium* NRRL B-2354 and *E. faecium* Aus0085 strains (RefSeq accessions numbers: NC_017022, NC_020207 and NC_021994, respectively) with Mauve (Darling et al., 2004). Functional annotation of the query sequences was performed using Blast2GO and Uniprot and KEGG databases[3].

Analysis of the *pbp5* Genetic Environment

An 8–10 kb DNA fragment comprising the *pbp5* gene and its boundaries was characterized by PCR assays in 15 transconjugants and 21 field isolates (**Table 1**). PCR conditions were adapted according to the expected amplicon size (<3 kb/>3 kb): 0.5/1 mM of each primer, 2/2.5 mM $MgCl_2$, 1x of reaction buffer, 0.2/0.4 mM of each deoxynucleoside triphosphate, 1,25U GoTaq® Flexi DNA Polymerase [Promega Corporation, Madison, WI, USA]/2.5U Takara LA Taq polymerase (Takara™ Bio Inc., Shiga, Japan). The amplification program was 25 cycles of 30 s at 94°C, 30 s at 55°C, 30 s at 72°C; 1 cycle 10 min at 72°C (for fragments <3 kb) or 35 cycles of 30 s at 96°C, 1 min at 55°C, 7 min at 72°C; 1 cycle 10 min at 72°C (for fragments >3 kb). Amplicons were further discriminated by comparing the corresponding RFLP patterns obtained after digestion with *Dde*I or *Apa*I and sequences. Genomic location of *pbp5* was identified by hybridization of I-*Ceu*I and *Sma*I-digested genomic DNA using 23S rDNA and/or *pbp5* probes (Liu et al., 1993; Freitas et al., 2013). Labeling and detection were carried out with Gene Images Alkphos direct labeling system kit following the manufacturer's instructions (Amersham GB/GE Healthcare Life Sciences UK limited). The presence of CTn*916*, CTn*5386* and CTn*5382*, previously associated with transferable AmpR, was determined by analyzing the presence of specific sequences (integrases, excisionases, non-integrase left region) (Carias et al., 1998; León-Sampedro et al., 2016). The transferred 8–10 kb genetic regions carrying *pbp5* characterized in this study were compared with those of all *E. faecium* available genomes at the National Center for

[1] http://pubmlst.org/

[2] http://www.uniprot.org/uniref/

[3] http://www.genome.jp/kegg/genome.html

TABLE 1 | Epidemiological background, *pbp5* gene characterization and antibiotic resistance among *E. faecium* strains and transconjugants from this study.

Isolate type (isolate name)	*pbp5* genetic environment type (*pbp5*-fstW) (RFLP_*Dde*I)	Sequencing	PBP5 alleles	PFGE-*Sma*I *pbp5* hybridization (kb)	Epidemiological background ST	BAPS	PFGE[a]	Date	Source/ sample[d]	MIC to Amp (mg/L)	Resistance to other antibiotics
***E. faecium* recipient strains**											
Rc (*Efc* GE1)	–	–	–	–	515	2.3b	A			≤ 0.016	TET, RIF, FA
Rc (*Efm* BM4105RF)	A	I		290	172	1.3	B			0.5	RIF, FA
Rc (*Efc* 64/3)	B	I		250	21	2.3a	C			1	RIF, FA
***E. faecium* donor strains and their corresponding transconjugants**											
WT (H323)	C	I[b]		210	280	3.1	100	2002	H(C)/urine	128	VAN, TEC, ERY, CIP, GEN, STR
TC*Efm*GE1 (GEH323.3)	C	I	C12	210						32	VAN, TEC, TET, ERY, RIF, FA
WT (HPH2)	C	I[b]		210	125	3.1	126	2007	H(A)/urine	128	VAN, TEC, ERY, CIP, NIT
TC*Efm*GE1 (GEHPH2.1)	C	I	C20	210						64	VAN, TEC, TET, ERY, RIF, FA
TC*Efm*64/3 (64HPH2.1)		ND		170 + 250						8	RIF, FA
WT(70411)	D	II[b]		210	670	3.3a	90	1997	H(B)/urine	32	VAN, TEC, TET, ERY, Q/D, STR
TC*Efm*GE1 (GE70411.2)	D	II	C4	210						32	VAN, TEC, TET, ERY, RIF, FA
TC*Efm*BM4105RF (BM70411.5)	D	II	C4	170 + 200						8	VAN, TEC, TET, ERY, RIF, FA
WT (E4)	D	II[b]		210	132	3.3a	Z	2001	HS(E)/water	128	TET, ERY, CIP, GEN, STR
TC*Efm*GE1 (GEE4.1)	D	II	C19	210						16	TET, RIF, FA
WT (28798)	D	III[b]		210	132	3.3a	88	1999	H(B)/blood	64	VAN, TEC, TET, ERY, QD, CIP, GEN, STR
TC*Efm*GE1 (GE28798.1)	D	III	C8	210						32	VAN, TEC, TET, ERY, STR, RIF, FA
WT (E233)	D	III[b]		210	132	3.3a	88.1	2002	HS(E)/water	64	ERY, CIP, GEN, STR
TC*Efm*GE1 (GEE233.1)	D	III		210						32	TET, RIF, FA
WT (E169)	D	III[b]		210	132	3.3a	88.2	2001	HS(C)/water	64	ERY, CIP, GEN, STR

(Continued)

TABLE 1 | Continued

Isolate type (isolate name)	*pbp5* genetic environment type (*pbp5*-fstW) (RFLP_*Dde*I)	*pbp5* genetic environment type (*pbp5*-fstW) Sequencing	PBP5 alleles	PFGE-*Sma*I *pbp5* hybridization (kb)	Epidemiological background ST	Epidemiological background BAPS	Epidemiological background PFGE[a]	Epidemiological background Date	Epidemiological background Source/sample[d]	MIC to Amp (mg/L)	Resistance to other antibiotics
TC*Efm*GE1 (GEE169.3)	D	III		210						32	TET, RIF, FA
TC*Efm*BM4105RF (BME169.3)		ND		200 + 240						8	RIF, FA
WT (E49)	D	III[b]		210	132	3.3a	C3	2001	HS(E)/water	128	VAN, TEC, ERY, CIP, STR
TC*Efm*GE1 (GEE49.1)	D	III	C8	210						32	VAN, TEC, TET, RIF, FA
WT (H207)	ND	ND	C4	210	280	3.1	150	2002	H(C)/exsudate	32	TET, ERY, CIP, GEN, STR
TC*Efm*GE1 (GEH207.1)	ND	Unknown[c]		180						32	TET, RIF, FA
TC*Efm*BM4105RF (BMH207.3)	ND	ND		210						8	RIF, FA
WT (SN71, SN133)	C	I[b]		200	393	2.1a	SN208	2006	PG/n = 4-waste lagoon, animal faeces	64	TET, ERY, STR, NIT
TC*Efm*GE1 (GESN71.1; GESN133.1)	C	I	C9	200						64	TET, RIF, FA
TC*Efm*BM4105RF (BMSN71.1)	C	I		200						8	RIF, FA
TC*Efm*64/3 (64SN71.1)	C	I		210						64	RIF, FA
WT (VD79C1)	D	II[b]		200	18	3.3a	59	2001	HH/faeces	128	VAN, TEC, TET, ERY, CIP, GEN
TC*Efm*GE1 (GEVD79C1.5)	D	II	C7	210						128	VAN, TEC, TET, ERY, GEN, RIF, FA
TC*Efm*BM4105RF (BMVD79C1.6)	D	II	C7	200						128	VAN, TEC, RIF, FA
Representative *E. faecium* without *pbp5* transfer											
WT (529940)	ND	II	C6	ND	16	3.3a	74	2000	H(B)/urine	> 64	VAN, TEC, TET, ERY, STR
WT (H352)	ND	II	C4	ND	280	3.1	71	2000	H(D)/sputum	> 32	VAN, TEC, ERY, CIP, GEN
WT (HPH6)	ND	II	C5	ND	18	3.3a	128	2007	H(A)/pus	> 32	VAN, TEC, ERY, CIP, NIT

(Continued)

TABLE 1 | Continued

Isolate type (isolate name)	*pbp5* genetic environment type (*pbp5-fstW*) (RFLP_*DdeI*)	Sequencing	PBP5 alleles	PFGE-*SmaI pbp5* hybridization (kb)	Epidemiological background: ST	BAPS	PFGE[a]	Date	Source/ sample[d]	MIC to Amp (mg/L)	Resistance to other antibiotics
WT (E197)	ND	III	C8	ND	368	3.3a	H4	2001	HS(C)/water	> 32	VAN, ERY, CIP, GEN, STR
WT (SN446)	ND	III	C8	ND	132	3.3a	H119.5	2007	PG/ water	> 32	VAN, TEC, ERY, GEN
WT (H196)	ND	V	C1	ND	390	3.1	129	2002	H(D)/unknown	> 32	TET, ERY, STR
WT (C373)	ND	I	C11	ND	148	2.1b	36.2	2001	HH/faeces	> 32	TET, ERY, CIP, GEN, STR, NIT
WT (SE97F1)	ND	I	C10	ND	*purK6*	ND	ND	2001	P/skin	> 32	TET, ERY, CIP, STR
WT (164306)	ND	V	C2	ND	190	2.3a	98	1998	H(B)/urine	32	VAN, TEC, TET, ERY, CIP, STR
WT (SN194)	ND	V	C3	ND	264	2.3a	ND	2006	PG/residual water	32	TET, CIP, CHL, STR

[a]Each different PFGE number or letter represent a different clone. In clones 88.1 and 88.2, the number after a dot represents the number of bands differing from the more ancient clone 88. [b]The pbp5 genetic platform type of these wild strains was determined by RFLP, which pattern was the same as the respective transconjugants with pbp5 platform characterized by sequencing. [c]In this strain we could not amplify the region between pbp5 and fstW or pbp5 and N-acylglucosamine-6-phosphate 2-epimerase. Abbreviations: Rc, recipient strain; WT, wild type; TC, transconjugant; Efm, E. faecium; H, hospital; HS, hospital sewage; HH, healthy human; PG, piggery; P, poultry; VAN, vancomycin; TEC, teicoplanin; TET, tetracycline, ERY, erythromycin; Q/D, quinupristin-dalfopristin; CIP, ciprofloxacin; STR, High level resistance (HLR) to streptomycin; GEN, HLR to gentamicin; NIT, nitrofurantoin; RIF, rifampicin; FA, Fusid Acid; ND, not determined. [d]Letters appearing in parenthesis of H (hospitals) and HS (hospital sewage) represent the five hospitals included (A-E).

Biotechnology Information (NCBI) database (Supplementary Table S5).

Phylogenetic Analysis of PBP5

Comparative analysis of all PBP5 protein sequences available in the GenBank database with those identified in this study was performed using the basic local alignment sequence tool (BLAST) (Altschul et al., 1997) and the ClustalW2 program for multiple sequence alignment (Larkin et al., 2007; Goujon et al., 2010), hosted by the National Center for Biotechnology Information (NCBI) and the European Bioinformatic Centre (EBI), respectively. The PBP5 amino acid sequences were designated by a "C" followed by a number in agreement with the nomenclature used in previous works (Supplementary Table S6) (Galloway-Peña et al., 2011). Multiple sequence alignments of all PBP5 sequences were performed through MEGA 7 software[4] and the topology of the phylogenetic tree was inferred by maximum-likelihood algorithm using PhyML (Guindon et al., 2010) with bootstrap analysis based in 1000 permutations and a cut-off of $\geq$70%.

Phylogenetic congruence of core genomes and *pbp5* sequences was established by comparative analysis of their gene tree topologies. Core genomes for the 233 *E. faecium* strains whose draft genome sequences were available at NCBI Trace Database (last updated on February 2014), were reconstructed independently using an all-against-all reciprocal BLAST approach. Briefly, each CDS was used as a BLAST query against a local database of CDS from all genomes in the sample analyzed. Afterward, we re-annotated the chromosome sequences with GeneMarkS v2.8 (Besemer et al., 2001). The core genome was defined as the non-redundant genes present in all strains using CD-HIT (Li and Godzik, 2006) with the 80% coverage, 80% identity cut-offs, and parsed by homemade Perl script. All core genome genes were concatenated and aligned using MAFFT (Katoh and Toh, 2010). The phylogenetic tree was built by FastTree2 (Price et al., 2010). Tree comparison was carried out by *cophyloplot* command of APE (Analyses of Phylogenetics and Evolution) (Paradis et al., 2004) using R software.

RESULTS

Transferability of AmpR (*pbp5*)

Twelve isolates (n = 12/78; 15%; **Table 1**) were able to transfer under our experimental conditions a chromosomal genetic platform containing *pbp5* (associated with AmpR) to *E. faecium* strain GE1. Five of them were also able to transfer *pbp5* to *E. faecium* BM4105RF and two, to *E. faecium* 64/3. All donor strains belonged to major *E. faecium* human lineages mostly associated with clinical isolates causing hospital infections, namely BAPS subgroups 3.3a (ST670, ST132, ST280), 3.1 (ST280, ST125), and 2.1a (ST393). They were obtained from samples of hospitalized humans but also from samples of hospital sewage, healthy humans and piggeries. β-lactamase production was not detected in any of the isolates.

[4]www.megasoftware.net

All transconjugants had similar PFGE profiles to recipient strains (**Figure 1**), were resistant to rifampicin and fusidic acid and exhibited a variable susceptibility to ampicillin (MICs = 8–128 mg/L), which was often lower than that of their corresponding donor isolates (8 mg/L or 16 vs. 32 mg/L to >256 mg/L). Besides the AmpR phenotype, some transconjugants also exhibited resistance to vancomycin ($n = 8$), teicoplanin ($n = 8$), erythromycin ($n = 6$), tetracycline ($n = 1$; in non-GE1 transconjugants; as GE1 strain is tetracycline-resistant), streptomycin ($n = 1$) and/or gentamicin ($n = 1$). Transconjugants carrying the *pbp5* genetic platform either alone or with plasmids harboring genes encoding resistance to different antibiotics were recovered from selection plates supplemented with ampicillin and not with other antibiotics (data not shown).

The *pbp5* Gene of AmpR Isolates Is Located on Large and Transferable Chromosomal Genetic Platforms Containing Metabolic Traits

Hybridization of *pbp5* and 23S rDNA probes with a digested I-*Ceu*I DNA band of high MW revealed a chromosomal location of *pbp5*. Further hybridization of *Sma*I-digested genomic DNA with the *pbp5* probe showed hybridization to fragments of ~210 kb in all but one GE1 transconjugants, for which the *pbp5* probe hybridized to a band of ~180 kb (**Figure 1**). Hybridization of the same probe with fragments of *Sma*I-digested genomic DNA of *E. faecium* BM4105RF and *E. faecium* 64/3 transconjugants of different sizes reflects independent transfer events. Similarity in the PFGE patterns of the transconjugants obtained using different donors indicates that DNA acquisition occurs within a particular region of the genome (**Figure 1**). The donor strains lack integrases/excisionases of Tn*916*-like conjugative transposons (Tn*916*, Tn*5386*, Tn*5382*).

For whole genome sequence a representative donor:recipient pair and its transconjugant were analyzed. The donor was the clinical HPH2 strain; the recipient, *E. faecium* GE1 which did not originally carry *pbp5*, and thus limiting the possibilities of recombination events between the acquired and natural *pbp5* genetic platform in the recipient strain; and the size of the *Sma*I digested genomic DNA fragment, the one that most commonly hybridizing with the *pbp5* probe (~210 kb). Sequencing of the recipient (*E. faecium* GE1), the donor (*E. faecium* HPH2) and the transconjugant (*E. faecium* TCGEHPH2.1) allowed us to identify 7 contigs with genes present in the donor and its transconjugant but absent in the

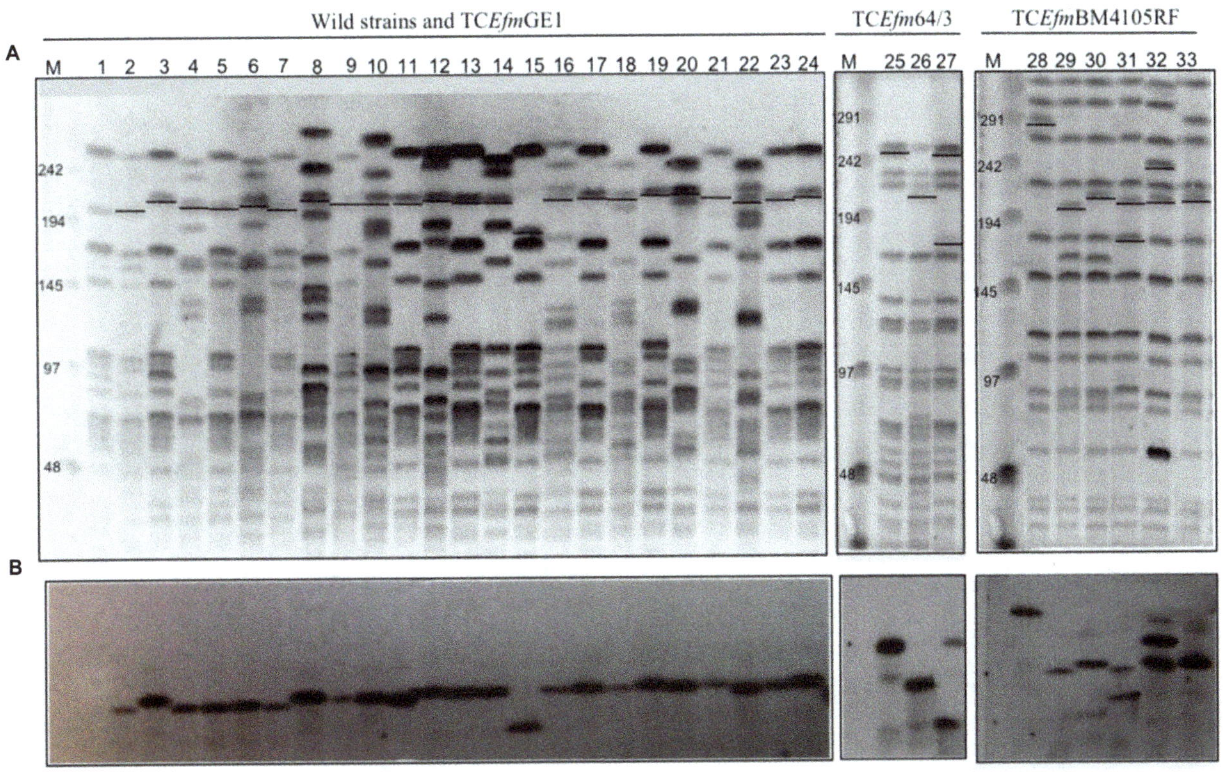

FIGURE 1 | Clonality and hybridization assays with *pbp5* probes of wild type, recipients and transconjugants strains. (A) PFGE SmaI digested DNA of recipient strains (1-*E. faecium* GE1, 25-*E. faecium* 64/3, 28-*E. faecium* BM4105RF), wild type (2-VD79C1, 4-SN71, 6-SN133, 8-HPH2, 10-H323, 12-70411, 14-H207, 16-E49, 18-E233, 20-E169, 22-E4) and transconjugants (3-GEVD79C1.5, 5-GESN71.1, 7-GESN133.1, 9-GEHPH2.1, 11-GEH323.3, 13-GE70411.2, 15-GE207.1, 17-GEE49.1, 19-GEE233.1, 21-GEE169.3, 23-GE28798.1, 24-GEE4.1, 26- 64SN71.1, 27- 64HPH2.1, 29- BMSN71.1, 30- BMH207.3, 31-BM70411.5, 32- BME169.3, 33- BMVD79C1.6). **(B)** Hybridization assays with a *pbp5* probe (primers P1 and P2-**Figure 4**). M- Low Range PFGE Marker, kb (New England, BioLabs). Abbreviations: TC*Efm*- Transconjugant *E. faecium*.

recipient strain. The fragments which comprised the *pbp5* gene, represent a genetic platform of approximately 280 kb that is slightly larger than that inferred from the *Sma*I-PFGE gels. A more detailed analysis of the contigs revealed that two of them (contig00158 and to a lesser extent contig00068) also contained genes present in the recipient and transconjugants but not in the donor strain, which could be explained either by the partial transfer of the *pbp5* genetic platform and/or post-transfer recombination events (**Figure 2**). The comparative analysis of the genomes sequenced in this work with the four closed *E. faecium* genomes available at the NCBI database at the time of writing this manuscript [DO (NC_017960.1), NRRL B-2354 (CP004063.1), Aus0004 (NC_017022.1), Aus0085 (NC_021994.1) strains], revealed a common region of 153kb for our donor/transconjugant strains and the four genomes analyzed, which was absent in the recipient *E. faecium* GE1 genome (**Figure 3**). The remaining sequence of the 280 kb "transferred chromosomal region" was variably present and located in different genomes (**Figure 3**).

Figure 2 shows a detailed characterization of the 280 kb transferred genetic platform, with ORFs of a G + C content ranging from 24 to 41%. Besides the *pbp5*-related resistance to β-lactams, this chromosomal transferable genetic platform contains genes involved in different cellular functions including amino acids and carbohydrate transport, redox processes, survival under stressful conditions in the intestinal environment (e.g., acid and bile tolerance) and also other genes related to β-lactam resistance. Among the last category is a gene (named *ponA*) encoding a bifunctional class A PBP with transpeptidase and transglycosylase activity, which is involved in the synthesis of peptidoglycan, allowing the survival in the presence of cell wall inhibitors (Rice et al., 2009). A copy of the CiaRH operon, coding for a two-component signal-transduction system responsive to cell envelope lysis-stress response and restricted to *Streptococcus* spp. to date, was also detected. The 280 kb genetic platform containing *pbp5* also had genes with presumed influence on microbe-host interactions. These include three phosphotransferase systems (PTS, namely

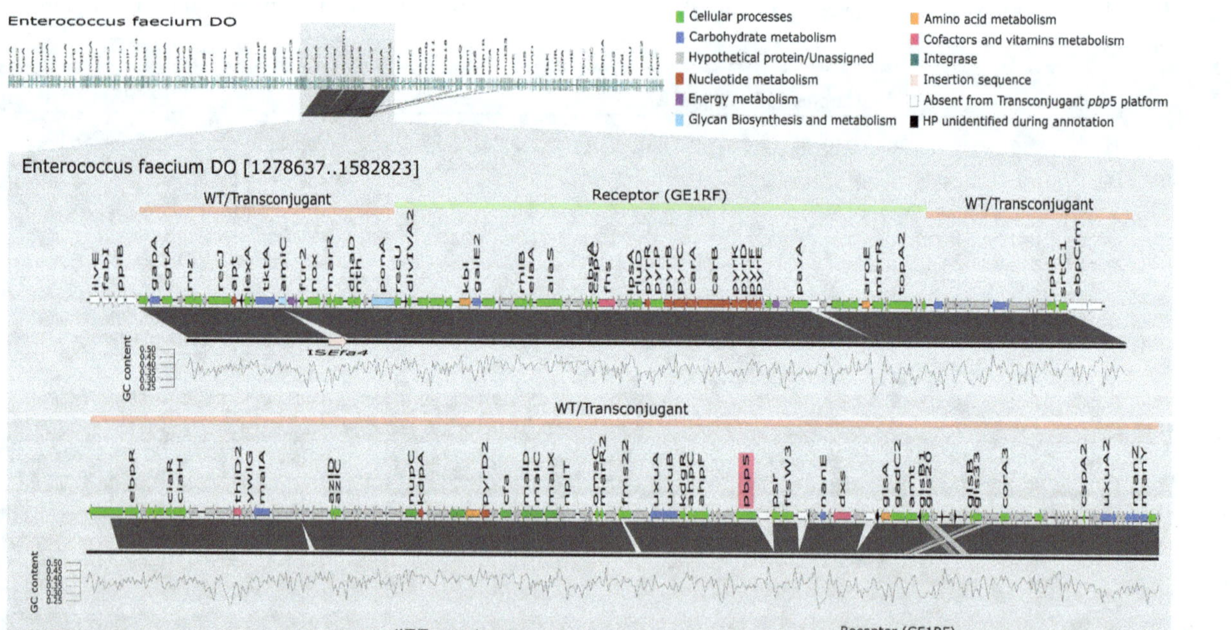

FIGURE 2 | Representation of the transferable chromosomal genetic platform containing *pbp5*. Mapping and annotation (using KEGG database) of transferable *pbp5* genetic platform of transconjugant TCGEHPH2.1 (represented by black line) using *E. faecium* DO as reference genome. Lines above *E. faecium* DO genetic structure represent the transferable *pbp5* genetic platform observed in wild type and transconjugant isolates (pink) and genomic regions of *E. faecium* GE1 recipient and transconjugant (green). GC content was calculated using seqinr in Rstudio. The window used to calculate the GC content was 200 bp, represented in the figure by each vertices of the graph. WT, wild type.

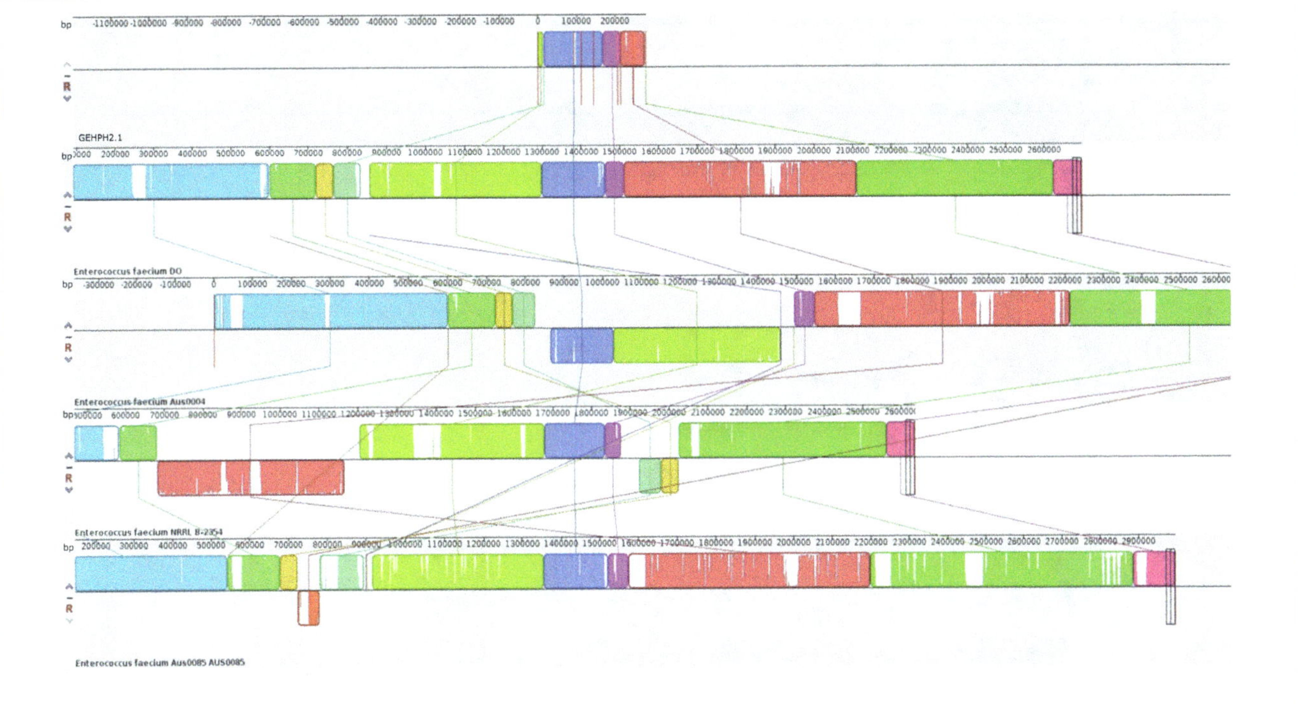

FIGURE 3 | Comparison by MAUVE of the transferable genetic platform containing *pbp5* (TCGEHPH2.1) with the four close *E. faecium* genomes (DO, Aus 0004, Aus0085 and NRRL B-2354), present in GenBank database. Each color block represents a genome region present in at least two of the sequences analyzed. These similar blocks can be identified by their color. Blank regions within the block represent point mutations or even regions that are absent in the same color blocks of the other sequences.

glucitol/sorbitol, L-ascorbate and mannose/fructose/sorbose), the ABC transport and transformation system of maltodextrines (*malACDX*) and the *N*-acetylmannosamine-6-phosphate epimerase (NanE; part of a pathway that allows the usage of sialic acids, major components of glycoproteins, gangliosides, and other sialoglyco-conjugates). Other metabolic genes in the *pbp5* genetic platform could be involved in tolerance to the intestinal acidic environment, including glyceraldehyde-3-phosphate dehydrogenase, ATP synthase subunit α, NADH dehydrogenase, glutaminase and genes encoding enzymes involved in the production of ammonia from glutamine and deamination or transport of branched/nitrogenated amino acids. Finally, five stress response proteins including two belonging to the Csp (cold shock proteins) system (CspA and CspC) and the small chaperone Hsp20, and Gls33 (also present in *E. faecium* genomes available in the GenBank databases) were also identified. They are involved in stress responses to salts, pH and ethanol exposure in *Clostridium*. Hsp20 is a small chaperone protein involved in survival of different abiotic stress conditions including heat (55°C) and salt (5 mM) in *Bifidobacterium longum* (Khaskheli et al., 2015).

To date, only two *E. faecium* strains lack *pbp5*. These are strains GE1 and D344SRF, which are AmpS isolates. The D344SRF strain is susceptible to cephalosporin and ampicillin due to the spontaneous deletion of a 170 kb genome fragment that includes *pbp5* and other *pbp* genes. This deletion occurred by the interaction of CTn*5386* (a 60 kb element that comprises *pbp5* and *vanB2*) with Tn*916* (Rice et al., 2007). The GE1 laboratory strain does only harbor the integrase of Tn*5801*. The causes of the loss of the *pbp5* genetic platform are still unknown.

The Chromosomal Region Harboring *pbp5* Exhibit Hotspots for Insertions

The occurrence of different insertion sequences (ISs), (**Figure 4**) in the boundaries of the *pbp5* gene and the presence of a >3027 bp *psr-pbp5* fragment in *E. hirae* (99.9% identical to that found on *E. faecium*) suggest the presence of hot spots that could facilitate recombination of regions containing the *pbp5*. To test this possibility, we further analyzed the 8–10 Kb genetic environment of *pbp5* from 21 wild type and 15 transconjugant isolates included in this study as well as available enterococcal genomes. Identical RFLP patterns of these 8–10 kb fragment were observed for each pair of donor and its corresponding transconjugants, but were different from those naturally occurring in the *pbp5* carrying recipient strains *E. faecium* BM4105RF and 64/3 (**Figure 4**). Sequencing of fragments representing distinct RFLP patterns and comparative analysis with similar *pbp5* genetic environment in GenBank available genomes, revealed 21 variants of such 8–10 kb chromosomal fragments (designated by roman numerals), which differed in the number and type of ISs (**Figure 4**). Three variants (types I, II, III) corresponded to *pbp5* transferable platforms described in this study (**Figure 4**; **Table 1**).

The predominant 8–10 kb fragment identified in both AmpR and AmpS *E. faecium* analyzed in this study did not contain indels and is considered here as the *pbp5* genetic environment

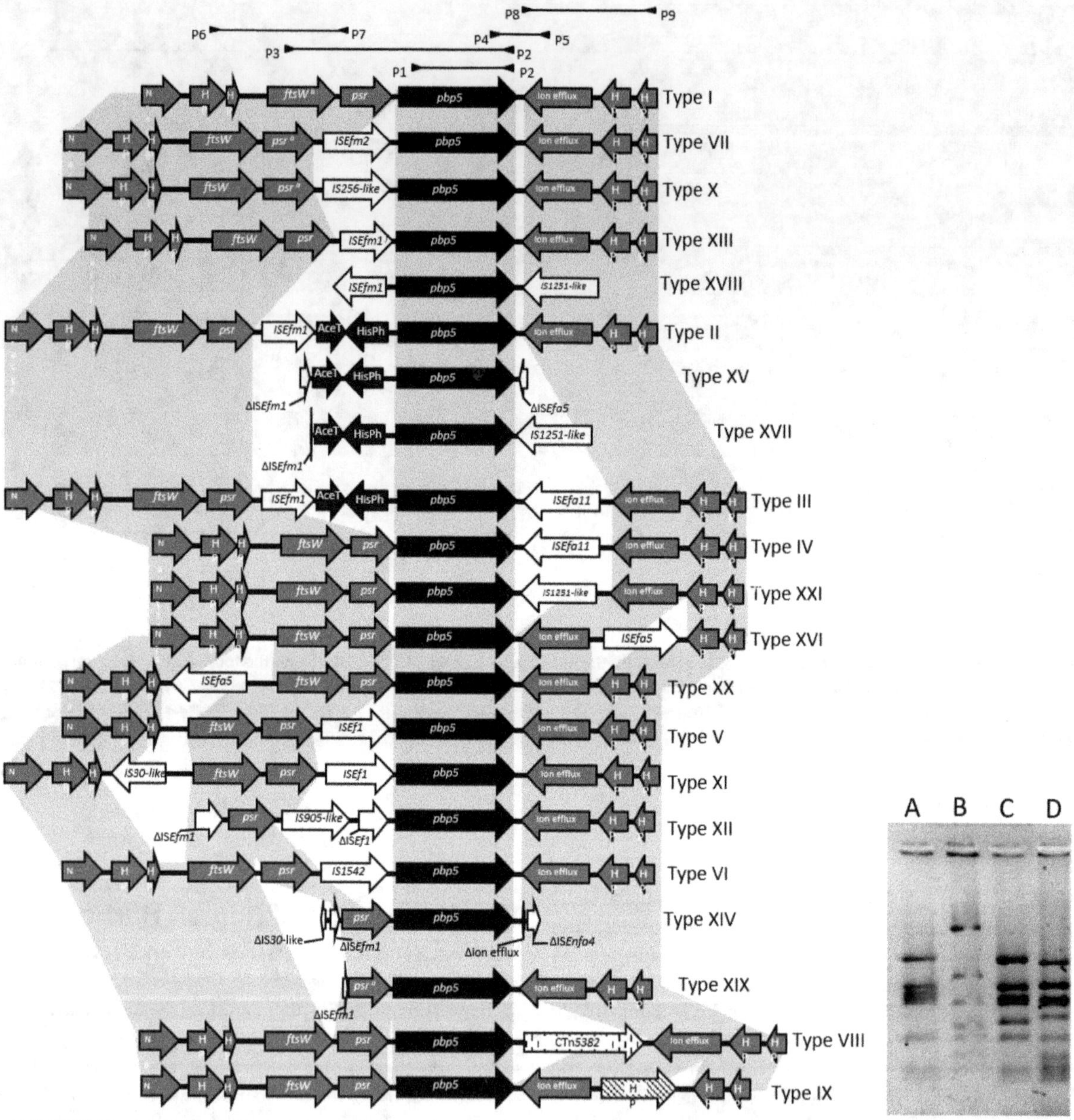

Primer	Sequence (3´-5´)	Amplicon size (bp) considering *pbp5* genetic environment type			
		I	II	III	V
P1:pbp5/1	GCACGGCAAAAATCGAACAGG	1983	1983	1983	1983
P2:pbp5/2 P3:ftsW/1	TGCTCGTTTGGTTGCTGAATCATC AGGCTGTTGCCCAGAAAC	4315	6496	6496	6496
P4:pbp5/2inv1 P5:EFLKRinv/2	GATGATTCAGCAACCAAACGAGCA CGTGGAGAAATGGCTTTGAT	326	326	1948	326
P6:NacylFinv/1 P7:ftsWRinv2	TTCACGAAAGACGCAGACAC TCGGCTTGCAATAAACGAT	1819	1819	1819	1819
P8:EFLKR/1 P9:HPIIEFLK/2	ATCAAAGCCATTTCTCCACG CGGGTTATTCCCAAACAGAA	2030	2030	2030	2030

FIGURE 4 | Continued

FIGURE 4 | Characterization of *E. faecium pbp5* genetic environment by PCR and sequencing. The Roman numbers I, II, III, and V represent the different *pbp5* genetic platforms detected in *E. faecium* from this study. The numbers IV, VI to XXI were detected in available genomes from GenBank database. The different types were named according to diversity of insertions sequences, genomic fragments or conjugative transposons within genes or intergenic regions. Mutations or recombinations within genes or intergenic regions were not considered for type classification. The Table indicates the primers used (designed for this study; P1/P2 described by Dahl et al., 2000) and the size of PCR amplicons from genetic environment of types I–III and V. The A, B, C and D lines of the bottom right side figure represent RFLP patterns of amplicon P3-P2 of mobile platforms I (pattern C) and II/III (pattern D) of isolates included in this study, when digested with DdeI restriction enzyme. The patterns A and B correspond to the amplicons of the recipient strains *E. faecium* BM4105RF and 64/3, respectively. [a] These gene has an extra stop codon within its sequence. Abbreviations: *N*-acyl, (*N*-acyl-glucosamine-6-phosphate-2-epimerase); HP (hypothetical protein); *ftsW* (cell cycle protein); *psr*, (*pbp5* synthesis repressor); *pbp5* (gene encoding penicillin binding protein 5); AceT (acetiltransferase); HisPh (Histidinol Phosphate Phosphatase).

prototype; arbitrarily named Type I (Supplementary Table S5, **Figure 4** and **5**). The *pbp5* or other genes (e.g., *psr*, *ftsw*, ion efflux genes) were flanked by one or two insertions sequences of the IS*256*, IS*L3* or IS*982* families (IS*Ef1*, IS*Efm1*, IS*Efm2*, IS*1542*, IS*256*-like, IS*Efa11*, or IS*1251*-like) in isolates carrying types II to XXI, with the exception of type VIII that had CTn*5382* (*vanB2*) inserted downstream of *pbp5*. The identifiable boundaries of ISs detected within the 8–10 kb boundaries of *pbp5*, as well as the common nucleotide positions at which insertions occurred, suggested recent acquisition events at some hot-spots on an ancestral sequence (**Figure 4**). Despite of predominance of type I, some types seem to be more associated with specific hosts as the case of type V, predominant in strains from pigs and type II/II-like, III or type XIV-like which are frequent among clinical isolates. Epidemiological distribution of isolates appears in Supplementary Table S5.

Diversity of PBP5 Sequences Reflects the Phylogenomic Diversification of *E. faecium*

We identified 75 PBP5 protein variants (Supplementary Tables S5 and S6) corresponding both to AmpR and AmpS strains which comprise 20 of the previously described variants C4, C7, C9, C11, C15-C18, C21, C24, C46-C48, C50, C51, C61, C63, C65, C67, C71 (Rybkine et al., 1998; López et al., 2009; Galloway-Peña et al., 2011). The other 55 variants were firstly detected in this study either in strains from Portugal ($n = 11$) or from available genomes at NCBI database ($n = 44$). The C4, C7, C8, C9, C12, C19, and C20 sequences were identified in isolates able to transfer *pbp5* (**Table 1**, Supplementary Table S6; **Figure 5**), which belonged mostly to ST18, its SLV ST125, ST280 and ST670 (BAPS 3.3a, 3.1 and 2.1a, respectively) (**Table 1**).

Figure 5 shows the phylogenetic tree constructed with all PBP5 protein sequences from this study and those available at GenBank databases. The tree is split into two major clades arbitrarily named B and A, mirroring the clades associated with populations of non-hospitalized persons and hospital isolates respectively, which were previously inferred from phylogenomic studies of *E. faecium* (Lebreton et al., 2013). Clade B comprises PBP5 of AmpS isolates (PBP5-S), which mainly correspond to strains of BAPS subgroups 1.2 and 1.5. Some are similar to the prototype PBP5-S C46 sequence of *E. faecium* BM4107 strain (Sifaoui et al., 2001) but most of them also exhibit mutations at positions T25A, S39T and D644N, which are also common to PBP5 sequences of clade A isolates (Supplementary Table S6). Two strains, isolated in 1964 and 2006, further showed changes at S27G + T324A and S27G respectively, such mutations corresponding to the PBP5-R consensus sequence (Pietta et al., 2014). The type I chromosomal region mentioned above was observed in all available isolates from this group (**Figure 5**; Supplementary Table S5).

Clade A includes PBP5 variants grouped in two main clusters arbitrarily designed here A0 and A1, with most isolates sharing mutations at positions V24A, S27G, E100Q, K144Q, T172A, T324A, N496K, A499T and E525D (7 of the 21 positions used to establish sequence diversity of this protein) (Pietta et al., 2014). The Clade A0, represented by C30, C31, C37, C52, C53, and C54 PBP5 sequences, corresponds to AmpS isolates of different BAPS groups (BAPS 1.2, BAPS 2.3a and BAPS 3.3b), recovered from different hosts (animals and humans), different countries and collected from 1995 to 2001 (**Figure 5**, Supplementary Table S6). Most isolates within clade A0 carried *pbp5* within a type I fragment. Sequences of Clade A1 were classified in three subclusters that parallels *E. faecium* populations of BAPS groups 2 and 3 (Willems et al., 2012; Tedim et al., 2015). They were designated as A1S, comprising a subset of PBP5-AmpS variants (mainly associated with human and animal isolates of BAPS subgroup 2.1b); A1M, comprising a subset of PBP5-AmpS and PBP5-AmpR (mostly associated with animal isolates of diverse BAPS subgroups 2.1b, 2.3a, 2.3b, 3.1, 3.2, and 3.3b); and A1R, including almost all PBP5-AmpR (mostly associated with the clinical setting and BAPS subgroups 3.1, 3.3a and 2.1a, the latter two only observed in this group).

All isolates from Clades A1S, A1M and A1R share mutations R34Q, G66E, L177I and A216S but also present some differences. Mutations S39N and A401S plus A499I were specific for subclades A1S and A1M, respectively. The last two polymorphisms were previously documented in two AmpR strains (Pietta et al., 2014) but this study suggests that they were fixed in certain populations despite having been previously discarded as relevant changes. One of the A1M strains had also variations at specific positions linked to the A1R group (A68T, E85D, M485T, V586L, E629V and P667S). Of note is the PBP5 variant C23 within this cluster, which was overrepresented ($n = 44$).

PBP5 variants within the A1R group exhibited six mutations (A68T, E85D, S204G, 466′S/D, M485A/T, E629V, P667S) predominant in this group, some of them located in the active site of the protein (466′S/466′D, M485A/T) and in the end of a turn between the β1 and β2 strands (E629V; P667S) (Fontana et al., 1996; Rice et al., 2004). It is of note that some AmpS isolates (BAPS subgroups 3.1, 2.3a and 3.2) with PBP5 sequences

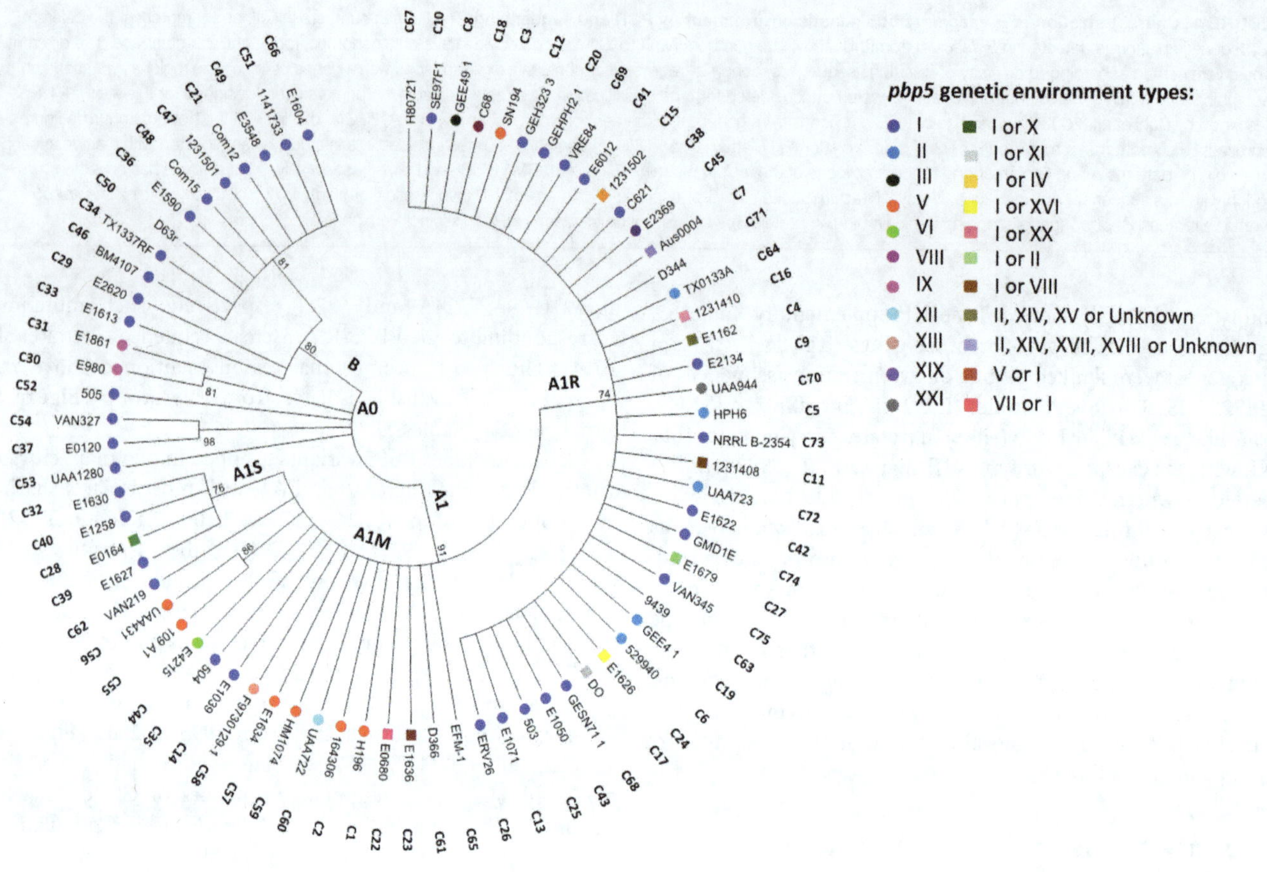

FIGURE 5 | Phylogenetic analysis of PBP5 protein sequences from *E. faecium* isolates of this study and available in GenBank database (last update February 2014). The Maximum Likelihood tree was obtained using Mega 7 and the JTT method. The tree with the highest log likelihood (−2000,3736) is shown. Initial tree(s) for the heuristic search were obtained automatically by applying Neighbor-Join and BioNJ algorithms to a matrix of pairwise distances estimated using a JTT model. Then, the topology with superior log likelihood value was selected. The analysis involved 75 amino acid sequences. All positions containing gaps and missing data were eliminated of the analysis. The final dataset had a total of 445 positions. Evolutionary analyses were conducted in MEGA7. Bootstrap values are indicated and are based in 1000 permutations. The cut-off used was $\geq$70%. "C" followed by a number represent PBP5 amino acid combinations described in Supplementary Table S6. Only one *E. faecium* isolate carrying representative PBP5 sequences (each type of "C") was included in the tree. The *pbp5* genetic environments types described in **Figure 4** are indicated by colored circles or by colored squares when more than one type was detected in isolates carrying the same "C." No symbol reflects that the genetic environment was not elucidated. NA, not applicable.

that cluster in the A1R subgroup (including the PBP5 of the recipient *E. faecium* 64/3) lack the mutation M485A and E629V, which confirmed that such mutations might be necessary for AmpR phenotype, as reported (Rice et al., 2004). Changes A68T, E85D and S204G were shared by AmpS and AmpR isolates of A1R group (Supplementary Table S6). Some strains exhibited the particular pattern of mutations Q408H, A558T, G582S, K632Q and, eventually, V462A, N546T and P642L. They had not previously been associated with AmpR. Distribution of different variants of the 8–10 kb boundaries of *pbp5* gene is shown in **Figure 5** and Supplementary Table S5. Of note, is the detection of type I in isolates of different groups of clade A1 as well as the association of types V and XI with isolates of group A1M and of the similar types II, III, XIV, XV, and XVII with isolates of group A1R.

We also analyzed the diversity of other genes previously associated with AmpR (*ddcP*, *ddcY*, *ldtEfm*, *pgtA*, *lytG*) in available genomes, (protein sequence per UniRef100 available at Uniprot using as reference *E. faecium* strain Aus0004). Single-locus phylogenetic tree of PBP5 was congruent with those of proteins codified by *ldtEfm* and *ddcY* but was non-congruent with those of *ddcP*, *pgtA* and *lytG* (data not shown).

Analysis of gene tree topologies (*E. faecium* core genomes vs. *E. faecium pbp5* genes) revealed discrepancies that indicate transfer of *pbp5* between *E. faecium* populations (**Figure 6**) and explains the occurrence of both the same PBP5 variants in isolates of different clonal lineages, and different PBP5 sequences in strains of the same phylogenomic groups (**Figure 5**, Supplementary Table S5).

Ampicillin Resistance Was Maintained Without Selective Pressure

AmpR was maintained in the wild-type strains and respective *E. faecium* GE1 transconjugants over 200 generations in antibiotic free BHI indicating that the acquired genetic platforms

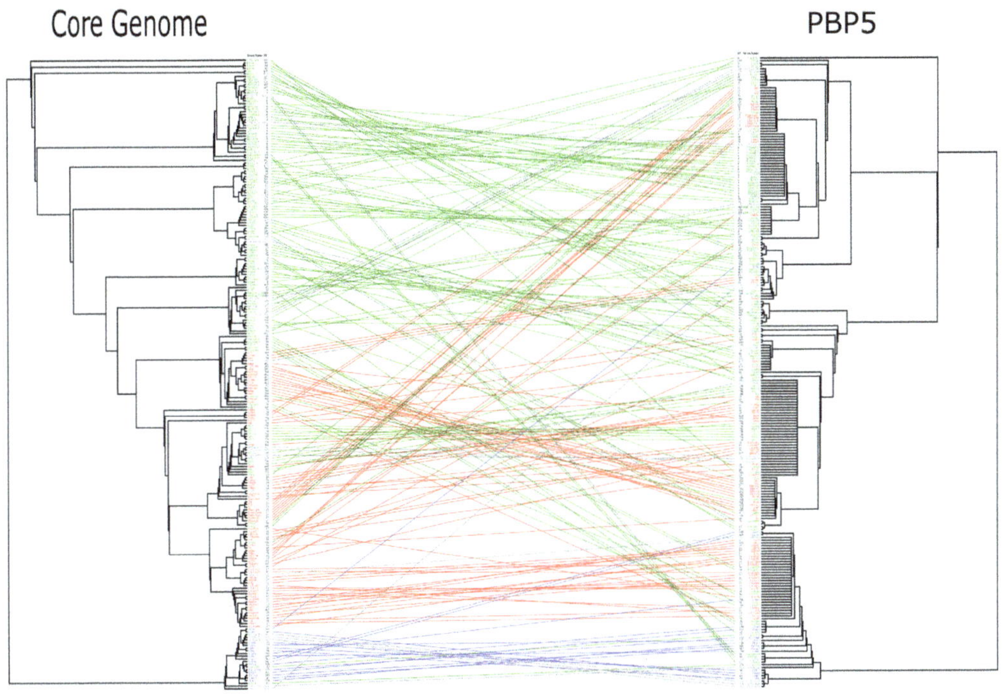

FIGURE 6 | Comparison of gene tree topologies (Core Genome phylogeny vs. *pbp5* phylogeny). Strains of clade B are represented in blue, those of Clade A1 in red and those Clade A2 in green. The edges join the core genome and the corresponding *pbp5* of each strain. Both trees were made by ML using GTR-CAT. Tress and edges were built using APE packages. The results of Mantel test ($r = -0.03$ and significance of 0.761) show the rearrangement of the *pbp5* in comparison with the core genome. The Mantel test (APE package) was made using the distance matrix calculated from nucleotide alignments.

containing *pbp5* may persist in absence of selective pressure in different genetic backgrounds, suggesting no-fitness cost imposed by the acquired *pbp5* platforms. The morphology and growth rate of all evolved transconjugants was similar among each other and to the normal growth of the *E. faecium* GE1 recipient strain.

DISCUSSION

This study documents the occurrence and diversity of a large chromosomal region containing *pbp5* in almost all *E. faecium* genomes and a parallel diversification between the PBP5 and the *E. faecium* core genome. The transferability of this region observed under laboratory conditions using different clonal backgrounds would greatly enhance the possibilities of *E. faecium* adaptation to changing environments.

Polymorphisms in the PBP5 protein sequences allowed grouping of PBP5 variants in clusters that mimic the phylogenomic diversification of *E. faecium* (Galloway-Peña et al., 2012; Lebreton et al., 2013). Lebreton et al. (2013) suggested a model for evolvability of this enterococcal species consisting of a first split of "clade B" and "clade A" coincidental with human and animal co-habitation occurring 30,000 y ago, and a further split of "clade A" in subclades "A1" and "A2" after the introduction of antibiotics in the late 1940s. Two main clusters of PBP5 variants were also identified in the present study, designated as "B" (associated with the *E. faecium* "clade B"), and "A" that further split in subgroups A0 (including only AmpS isolates) and A1 comprising three small groups linked to different *E. faecium* populations differing in the susceptibility to ampicillin: A1S (AmpS from healthy humans of different BAPS groups), A1M (AmpS and AmpR from humans and animals), and A1R (AmpR from clinical isolates). Based on the apparent universal presence of *pbp5* in *E. faecium* populations, we could speculate that *pbp5* might have predated the *E. faecium* evolutionary split among different hosts. Although some authors have suggested a sequential acquisition of amino acid changes in PBP5 sequence (Galloway-Peña et al., 2011; Pietta et al., 2014), such diversification may indicate different evolutionary routes for AmpR in response to distinct selective pressures in different hosts, similarly to what has been observed for different β-lactamase enzymes of Gram negative organisms (Novais et al., 2010).

The full characterization of the transferable 153 kb chromosomal region that contains not only *pbp5* but also genes contributing to survival in the gastrointestinal tract (e.g., resistance to acids and bile) in almost all available *E. faecium* genomes suggest a contribution of this genetic platform to the adaptation to the mammalian intestine and persistence in abiotic environments. Due to the scarcity of *E. faecium* isolates from non-mammals and non-human hosts, it would result highly speculative to hypothesize about the origin of the PBP5 and other genes located in this genetic platform. The PBP5 is a member of the Mec family, which comprises proteins with natural low affinity for β-lactams as PBP2a (*mecA*) of *Staphylococcus aureus* and PBP3r of *Enterococcus hirae* (99% identical to PBP5, suggesting a common origin for PBP5 and PBP3r) (Raze et al., 1998; Hiramatsu et al., 2013). For staphylococci, MecA would have been essential for the survival of ancestral members of staphylococci in the presence of β-lactam antibiotics produced by fungi and Actinobacteria and lost in the Devonian period with the emergence of mammals (more than 400 million of years ago). This would be coincidental with the separation and adaptation of staphylococcal species to different hosts (Hiramatsu et al., 2013), in any case isolating *Staphylococcus* populations from old β-lactam producers. Recent antibiotic pressure would have influenced the transference of PBP2a from environmental to human strains of staphylococci. It is of note that the 153 kb transferred genetic platform characterized in this study has a variable G + C content, reflecting HGT events from different sources and suggesting that a similar evolutionary route as that of PBP2a of staphylococci might have also occurred for *E. faecium* PBP5.

The transfer of *pbp5* in *E. faecium* resulting of the interaction between transposons as Tn*5382* or Tn*916* plus Tn*5386* was previously demonstrated for a few isolates (Rice et al., 2005a). During the revision process of this manuscript, a work by García-Solache et al. (2016) was published. This work suggested that transference of chromosomal regions carrying *pbp5* occurs by homologous recombination between similar DNA regions of donor and recipient strains enriched in AT nucleotides with a possible involvement of plasmids (García-Solache et al., 2016). Tn*916*-like elements were not detected in our strains but the presence of a band of similar size in both donors and recipients and a sequence enriched in AT in the boundaries of the chromosomal transferred region (data not shown) suggested that homologous recombination might occur between *E. faecium* donor and recipient DNA from our study. Although the presence of plasmid genes within this acquired *pbp5* genetic platform was not observed in the transconjugant we sequenced as well as in the case of some transconjugants described in the study of García-Solache et al. (2016), we cannot completely discard this possibility due to the common presence of plasmids of certain families in donors analyzed (data not shown) (Manson et al., 2010; Freitas et al., 2016).

Despite the relevance of elucidation of the mechanism of *pbp5* transfer at molecular level in future studies, it is of note that disparate topologies of the phylogenetic trees of *E. faecium* core genomes and PBP5 sequences obtained in our study reflect how frequent the transfer of *pbp5* genetic platforms under natural circumstances may occur (Tripathi and Sowdhamini, 2008). Similarly, atypical phylogenetic associations have been signaled as evidence of lateral transfer of various types of serine-proteases (PBP5 belongs to this family of enzymes) throughout the prokaryotic world (Tripathi and Sowdhamini, 2008). In fact, we present evidence of inter-chromosomal transfer for different PBP5 C-types. However, in our case, *pbp5* horizontal gene transfer (HGT) events seems to preferentially happen in particular populations, which being more prone to mutation (the subclade A1 that is characterized by a higher mutation rate than clade B) would further facilitate the *pbp5* fixation and evolution followed by their clonal expansion under antibiotic selective pressure, reflecting the "*ex unibus plurum*" evolutionary dynamics (Baquero, 2011). Heterogeneity of populations that colonize humans, designated as "clouds" elsewhere (Stanczak-Mrozek et al., 2015), leads to a possible global adaptive benefit for certain clones and finally for the overall species, that is enhanced by HGT.

The necessary contribution of different genes for the full expression of AmpR phenotype is suggested by the lower MIC values of the transconjugants in comparison with the wild type strains in this and other studies (Rice et al., 2005b; Zhang et al., 2012). The transferable genetic platform identified here contained *pbp5* and *ponA* but also other genes (*ciaRH* operon) previously associated with β-lactam resistance and virulence in *S. pneumoniae*, but still unexplored in enterococci (Guenzi et al., 1994; Jordan et al., 2008; Krawczyk-Balska and Markiewicz, 2016). Outside this *pbp5*, only ldt_{Efm} and *ddcY* showed a similar non-congruent topology with that of *pbp5* indicating that these genes are also under accelerated evolution, which could explain eventual AmpR phenotypes with a lack of correlation with PBP5 sequences (Zhang et al., 2012). The variable ampicillin susceptibility phenotypes observed in different transconjugants even when using the same recipient strain, suggest either a partial transfer of the *pbp5* genetic platform or occurrence of recombination events leading to variation in MICs, which seem to occur frequently in commensal bacteria (Ingle et al., 2016).

CONCLUSION

The arsenal of adaptive traits located on the chromosomal region containing *pbp5* characterized in this study suggests its involvement in the adaptation of *E. faecium* to the gastrointestinal tract of mammals, and its possible contribution to the spread of *pbp5* by horizontal gene transfer, even in the absence of specific β-lactam exposure. However, the apparent frequent transfer events of such genetic platform among "clouds" of closely related *E. faecium* populations, would indicate bacterial shifts in the evolution of pathogenicity (including colonization) and antibiotic resistance that occurred in response to changes in patients' demographics, medical strategies and interventions, within the paradigm of the "Hamiltonian medicine" (Fraser et al., 2005). The data further highlight the increasing need for evolutionary biology to be aligned with medical challenges (Nesse et al., 2010).

NUCLEOTIDE SEQUENCE GENBANK ACCESSION NUMBERS

The sequences corresponding to representative *pbp5* genetic platforms (Types I-IV) were assigned GenBank accession numbers JN208885, JN208888, JN208884 and JN208886, respectively. PBP5 amino acid were analyzed and designated by a "C" followed by a number (Table S6). New sequences correspond to GenBank accession numbers JN208883 (C6 amino acid combination), JN208889 (C19 amino acid combination); JN208887 (C8 amino acid combination); JN208886 (C3 amino acid combination), JN208882 (C2 amino acid combination), KC479673 (C1 amino acid combination), KC479675 (C5 amino acid combination), KC479676 (C10 amino acid combination) and KC479674 (C12 amino acid combination). The Whole Genome Shotgun project has been deposited at DDBJ/ENA/GenBank under the accession MBRG00000000. The version described in this paper is version MBRG01000000.

AUTHOR CONTRIBUTIONS

CN, LP, and TMC designed the study. CN, ARF, APR, MA-H, ES, and RE performed wet lab experiments and participated in the analysis of the data. APT and VFL performed bioinformatic analysis. FB, APR, and LP provided expertise, participated in the analysis of data, and in the revision of the manuscript. CN, APT, and TMC performed the analysis of data and wrote the manuscript. All authors approved the final version of the manuscript.

ACKNOWLEDGMENTS

APT and VFL were funded by the European Union (EvoTAR-282004). ARF and ES were supported by fellowship from FCT [grant SFRH/BPD/96148/2013 through Programa Operacional Capital Humano (POCH) and SFRH/BD/63955/2009 POPH-QREN, respectively]. MA-H was supported by a Marie Curie Intra European Research Fellowship; grant number 272914, Bio-CHARGE; Investigation into the Biological Cost and Adaptation of the Host to Antibiotic Resistance on Mobile Genetic Elements in Enterococcus Species. VFL is further supported by a Research Award Grant 2016 of the European Society for Clinical Microbiology and Infectious Diseases (ESCMID).

FUNDING

This work was supported by Fundação para a Ciência e a Tecnologia (FCT)/Ministério da Educacão e Ciência (MEC) through national funds and co-financed by Fundo Europeu de Desenvolvimento Regional (FEDER) under the Partnership Agreement PT2020 with the reference UID/ MULTI/04378/ 2013–POCI/01/0145/FERDER/007728 to CN and LP, Instituto de Salud Carlos III (Plan Estatal de I+D+i 2013-2016), Grant PI12-01581 and PI15-01307 to TMC and CIBERESP (CB06/02/0053) to FB, European Commission, Seventh Framework Program (EVOTARFP7-HEALTH-282004) to TMC and FB and the Regional Government of Madrid in Spain (PROMPT-S2010/BMD2414) to FB and TMC.

REFERENCES

Altschul, S. F., Madden, T. L., Schäffer, A. A., Zhang, J., Zhang, Z., Miller, W., et al. (1997). Gapped BLAST and PSI-BLAST: a new generation of protein database search programs. *Nucleic Acids Res.* 25, 3389–3402. doi: 10.1093/nar/25.17.3389

Arias, C. A., Contreras, G. A., and Murray, B. E. (2010). Management of multidrug-resistant enterococcal infections. *Clin. Microbiol. Infect.* 16, 555–562. doi: 10.1111/j.1198-743X.2010.03214.x

Arias, C. A., and Murray, B. E. (2012). The rise of the *Enterococcus*: beyond vancomycin resistance. *Nat. Rev. Microbiol.* 10, 266–278. doi: 10.1038/nrmicro2761

Baquero, F. (2011). The 2010 garrod lecture: the dimensions of evolution in antibiotic resistance: ex unibus plurum et ex pluribus unum. *J. Antimicrob. Chemother.* 66, 1659–1672. doi: 10.1093/jac/dkr214

Belhaj, M., Boutiba-Ben Boubaker, I., and Slim, A. (2016). Penicillin-binding protein 5 sequence alteration and levels of plp5 mRNA expression in clinical isolates of *Enterococcus faecium* with different levels of ampicillin resistance. *Microb. Drug Resist.* 22, 202–210. doi: 10.1089/mdr.2015.0211

Besemer, J., Lomsadze, A., and Borodovsky, M. (2001). GeneMarkS: a self-training method for prediction of gene starts in microbial genomes. Implications for finding sequence motifs in regulatory regions. *Nucleic Acids Res.* 29, 2607–2618. doi: 10.1093/nar/29.12.2607

Carias, L. L., Rudin, S. D., Donskey, C. J., and Rice, L. B. (1998). Genetic linkage and cotransfer of a novel, vanB-containing transposon (Tn5382) and a low-affinity penicillin-binding protein 5 gene in a clinical vancomycin-resistant *Enterococcus faecium* isolate. *J. Bacteriol.* 180, 4426–4434.

Centers for Disease Control and Prevention (2013). *Antibiotic Resistance Threats in the United States.* Available at: http://www.cdc.gov/drugresistance/threat-report-2013/

Clinical and Laboratory Standards Institute (2012). *Performance Standards for Antimicrobial Susceptibility Testing; Twenty-Second Informational Supplement M100-S22.* Wayne, PA: CLSI.

Coudron, P. E., Markowitz, S. M., and Wong, E. S. (1992). Isolation of a beta-lactamase-producing, aminoglycoside-resistant strain of *Enterococcus faecium*. *Antimicrob. Agents Chemother.* 36, 1125–1126. doi: 10.1128/AAC.36.5.1125

Dahl, K. H., Lundblad, E. W., Rokenes, T. P., Olsvik, O., and Sundsfjord, A. (2000). Genetic linkage of the *vanB2* gene cluster to Tn*5382* in vancomycin-resistant enterococci and characterization of two novel insertion sequences. *Microbiology* 146, 1469–1479. doi: 10.1099/00221287-146-6-1469

Darling, A. C. E., Mau, B., Blattner, F. R., and Perna, N. T. (2004). Mauve: multiple alignment of conserved genomic sequence with rearrangements. *Genome Res.* 14, 1394–1403. doi: 10.1101/gr.2289704

de Regt, M. J. A., van Schaik, W., van Luit-Asbroek, M., Dekker, H. A. T., van Duijkeren, E., Koning, C. J. M., et al. (2012). Hospital and community ampicillin-resistant *Enterococcus faecium* are evolutionarily closely linked but have diversified through niche adaptation. *PLoS ONE* 7:e30319. doi: 10.1371/journal.pone.0030319

Fontana, R., Ligozzi, M., Pittaluga, F., and Satta, G. (1996). Intrinsic penicillin resistance in enterococci. *Microb. Drug Resist.* 2, 209–213. doi: 10.1089/mdr.1996.2.209

Fraser, C., Hanage, W. P., and Spratt, B. G. (2005). Neutral microepidemic evolution of bacterial pathogens. *Proc. Natl. Acad. Sci. U.S.A.* 102, 1968–1973. doi: 10.1073/pnas.0406993102

Freitas, A. R., Novais, C., Tedim, A. P., Francia, M. V. M. V., Baquero, F., Coque, T. M., et al. (2013). Microevolutionary events involving narrow host plasmid influences local fixation of vancomycin-resistance in *Enterococcus*. *PLoS ONE* 8:e60589. doi: 10.1371/journal.pone.0060589

Freitas, A. R., Tedim, A. P., Francia, M. V., Jensen, L. B., Novais, C., Peixe, L., et al. (2016). Multilevel population genetic analysis of vanA and vanB *Enterococcus faecium* causing nosocomial outbreaks in 27 countries (1986-2012). *J. Antimicrob. Chemother.* doi: 10.1093/jac/dkw312 [Epub ahead of print].

Galloway-Peña, J., Roh, J. H., Latorre, M., Qin, X., and Murray, B. E. (2012). Genomic and SNP analyses demonstrate a distant separation of the hospital and community-associated clades of *Enterococcus faecium*. *PLoS ONE* 7:e30187. doi: 10.1371/journal.pone.0030187

Galloway-Peña, J. R., Rice, L. B., and Murray, B. E. (2011). Analysis of PBP5 of early U.S. isolates of *Enterococcus faecium*: sequence variation alone does not explain increasing ampicillin resistance over time. *Antimicrob. Agents Chemother.* 55, 3272–3277. doi: 10.1128/AAC.00099-11

García-Solache, M., Lebreton, F., McLaughlin, R. E., Whiteaker, J. D., Gilmore, M. S., and Rice, L. B. (2016). Homologous recombination within large chromosomal regions facilitates acquisition of beta-lactam and vancomycin resistance in *Enterococcus faecium*. *Antimicrob. Agents Chemother.* doi: 10.1128/AAC.00488-16 [Epub ahead of print].

Gonçalves, A., Igrejas, G., Radhouani, H., López, M., Guerra, A., Petrucci-Fonseca, F., et al. (2011). Detection of vancomycin-resistant enterococci from faecal samples of Iberian wolf and Iberian lynx, including *Enterococcus faecium* strains of CC17 and the new singleton ST573. *Sci. Total Environ.* 41, 266–268. doi: 10.1016/j.scitotenv.2011.09.074

Goujon, M., McWilliam, H., Li, W., Valentin, F., Squizzato, S., Paern, J., et al. (2010). A new bioinformatics analysis tools framework at EMBL-EBI. *Nucleic Acids Res.* 38, W695–W699. doi: 10.1093/nar/gkq313

Guenzi, E., Gasc, A. M., Sicard, M. A., and Hakenbeck, R. (1994). A two-component signal-transducing system is involved in competence and penicillin susceptibility in laboratory mutants of *Streptococcus pneumoniae*. *Mol. Microbiol.* 12, 505–515. doi: 10.1111/j.1365-2958.1994.tb01038.x

Guindon, S., Dufayard, J.-F., Lefort, V., Anisimova, M., Hordijk, W., and Gascuel, O. (2010). New algorithms and methods to estimate maximum-likelihood phylogenies: assessing the performance of PhyML 3.0. *Syst. Biol.* 59, 307–321. doi: 10.1093/sysbio/syq010

Guzman Prieto, A. M., van Schaik, W., Rogers, M. R. C., Coque, T. M., Baquero, F., Corander, J., et al. (2016). Global emergence and dissemination of enterococci as nosocomial pathogens: attack of the clones? *Front. Microbiol.* 7:788. doi: 10.3389/fmicb.2016.00788

Hendrickx, A. P. A., van Schaik, W., and Willems, R. J. L. (2013). The cell wall architecture of *Enterococcus faecium*: from resistance to pathogenesis. *Future Microbiol.* 8, 993–1010. doi: 10.2217/fmb.13.66

Hiramatsu, K., Ito, T., Tsubakishita, S., Sasaki, T., Takeuchi, F., Morimoto, Y., et al. (2013). Genomic basis for methicillin resistance in *Staphylococcus aureus*. *Infect. Chemother.* 45:117. doi: 10.3947/ic.2013.45.2.117

Homan, W. L., Tribe, D., Poznanski, S., Li, M., Hogg, G., Spalburg, E., et al. (2002). Multilocus sequence typing scheme for *Enterococcus faecium*. *J. Clin. Microbiol.* 40, 1963–1971. doi: 10.1128/JCM.40.9.3548-3548.2002-a

Ingle, D. J., Valcanis, M., Kuzevski, A., Tauschek, M., Inouye, M., Stinear, T., et al. (2016). *In silico* serotyping of *E. coli* from short read data identifies limited novel O-loci but extensive diversity of O:H serotype combinations within and between pathogenic lineages. *Microb. Genom.* 2. doi: 10.1099/mgen.0.000064

Jordan, S., Hutchings, M. I., and Mascher, T. (2008). Cell envelope stress response in Gram-positive bacteria. *FEMS Microbiol. Rev.* 32, 107–146. doi: 10.1111/j.1574-6976.2007.00091.x

Katoh, K., and Toh, H. (2010). Parallelization of the MAFFT multiple sequence alignment program. *Bioinformatics* 26, 1899–1900. doi: 10.1093/bioinformatics/btq224

Khaskheli, G. B., Zuo, F., Yu, R., and Chen, S. (2015). Overexpression of small heat shock protein enhances heat- and salt-stress tolerance of *Bifidobacterium longum* NCC2705. *Curr. Microbiol.* 71, 8–15. doi: 10.1007/s00284-015-0811-0

Klare, I., Rodloff, A. C., Wagner, J., Witte, W., and Hakenbeck, R. (1992). Overproduction of a penicillin-binding protein is not the only mechanism of penicillin resistance in *Enterococcus faecium*. *Antimicrob. Agents Chemother.* 36, 783–787. doi: 10.1128/AAC.36.4.783

Klibi, N., Sáenz, Y., Zarazaga, M., Ben Slama, K., Masmoudi, A., Ruiz-Larrea, F., et al. (2008). Polymorphism in pbp5 gene detected in clinical *Enterococcus faecium* strains with different ampicillin MICs from a Tunisian hospital. *J. Chemother.* 20, 436–440. doi: 10.1179/joc.2008.20.4.436

Krawczyk-Balska, A., and Markiewicz, Z. (2016). The intrinsic cephalosporin resistome of *Listeria monocytogenes* in the context of stress response, gene regulation, pathogenesis and therapeutics. *J. Appl. Microbiol.* 120, 251–265. doi: 10.1111/jam.12989

Kristich, C., Rice, L., and Arias, C. (2014). "Enterococcal infection—treatment and antibiotic resistance," in *Enterococci: From Commensals to Leading Causes of Drug Resistant Infection*, eds M. S. Gilmore, D. B. Clewell, Y. Ike, and N. Shankar (Boston, MA: Massachusetts Eye and Ear Infirmary).

Larkin, M. A., Blackshields, G., Brown, N. P., Chenna, R., McGettigan, P. A., McWilliam, H., et al. (2007). Clustal W and Clustal X version 2.0. *Bioinformatics* 23, 2947–2948. doi: 10.1093/bioinformatics/btm404

Lebreton, F., van Schaik, W., McGuire, A. M., Godfrey, P., Griggs, A., Mazumdar, V., et al. (2013). Emergence of epidemic multidrug-resistant *Enterococcus faecium* from animal and commensal strains. *MBio* 4:e00534-13. doi: 10.1128/mBio.00534-13

León-Sampedro, R., Novais, C., Peixe, L., Baquero, F., and Coque, T. M. (2016). Diversity and evolution of the Tn5801-tet(M)-like integrative and conjugative elements among *Enterococcus, Streptococcus, and Staphylococcus*. *Antimicrob. Agents Chemother.* 60, 1736–1746. doi: 10.1128/AAC.01864-15

Li, W., and Godzik, A. (2006). Cd-hit: a fast program for clustering and comparing large sets of protein or nucleotide sequences. *Bioinformatics* 22, 1658–1659. doi: 10.1093/bioinformatics/btl158

Liu, S. L., Hessel, A., and Sanderson, K. E. (1993). Genomic mapping with I-Ceu I, an intron-encoded endonuclease specific for genes for ribosomal RNA, in *Salmonella* spp., *Escherichia coli*, and other bacteria. *Proc. Natl. Acad. Sci. U.S.A.* 90, 6874–6878. doi: 10.1073/pnas.90.14.6874

López, M., Sáenz, Y., Rojo-Bezares, B., Martínez, S., del Campo, R., Ruiz-Larrea, F., et al. (2009). Detection of vanA and vanB2-containing enterococci from food samples in Spain, including *Enterococcus faecium* strains of CC17 and the new singleton ST425. *Int. J. Food Microbiol.* 133, 172–178. doi: 10.1016/j.ijfoodmicro.2009.05.020

Manson, J. M., Hancock, L. E., and Gilmore, M. S. (2010). Mechanism of chromosomal transfer of *Enterococcus faecalis* pathogenicity island, capsule, antimicrobial resistance, and other traits. *Proc. Natl. Acad. Sci. U.S.A.* 107, 12269–12274. doi: 10.1073/pnas.1000139107

Nesse, R. M., Bergstrom, C. T., Ellison, P. T., Flier, J. S., Gluckman, P., Govindaraju, D. R., et al. (2010). Evolution in health and medicine Sackler colloquium: Making evolutionary biology a basic science for medicine. *Proc. Natl. Acad. Sci. U.S.A.* 107(Suppl. 1), 1800–1807. doi: 10.1073/pnas.0906224106

Novais, A., Comas, I., Baquero, F., Cantón, R., Coque, T. M., Moya, A., et al. (2010). Evolutionary trajectories of beta-lactamase CTX-M-1 cluster enzymes: predicting antibiotic resistance. *PLoS Pathog.* 6:e1000735. doi: 10.1371/journal.ppat.1000735

Novais, C., Coque, T. M., Costa, M. J., Sousa, J. C., Baquero, F., and Peixe, L. V. (2005a). High occurrence and persistence of antibiotic-resistant enterococci in poultry food samples in Portugal. *J. Antimicrob. Chemother.* 56, 1139–1143. doi: 10.1093/jac/dki360

Novais, C., Coque, T. M., Ferreira, H., Sousa, J. C., and Peixe, L. (2005b). Environmental contamination with vancomycin-resistant enterococci from hospital sewage in Portugal. *Appl. Environ. Microbiol.* 71, 3364–3368. doi: 10.1128/AEM.71.6.3364-3368.2005

Novais, C., Coque, T. M., Sousa, J. C., and Peixe, L. V. (2006). Antimicrobial resistance among faecal enterococci from healthy individuals in Portugal. *Clin. Microbiol. Infect.* 12, 1131–1134. doi: 10.1111/j.1469-0691.2006.01542.x

Novais, C., Freitas, A. R., Silveira, E., Antunes, P., Silva, R., Coque, T. M., et al. (2013). Spread of multidrug-resistant *Enterococcus* to animals and humans: an underestimated role for the pig farm environment. *J. Antimicrob. Chemother.* 68, 2746–2754. doi: 10.1093/jac/dkt289

Novais, C., Sousa, J. C., Coque, T. M., and Peixe, L. V. (2005c). Molecular characterization of glycopeptide-resistant *Enterococcus faecium* isolates from Portuguese hospitals. *Antimicrob. Agents Chemother.* 49, 3073–3079. doi: 10.1128/AAC.49.7.3073-3079.2005

Palmer, K. L., Godfrey, P., Griggs, A., Kos, V. N., Zucker, J., Desjardins, C., et al. (2012). Comparative genomics of enterococci: variation in *Enterococcus faecalis*, clade structure in *E. faecium*, and defining characteristics of *E. gallinarum* and *E. casseliflavus*. *MBio* 3:e00318-11. doi: 10.1128/mBio.00318-11

Paradis, E., Claude, J., and Strimmer, K. (2004). APE: analyses of phylogenetics and evolution in R language. *Bioinformatics* 20, 289–290. doi: 10.1093/bioinformatics/btg412

Pietta, E., Montealegre, M. C., Roh, J. H., Cocconcelli, P. S., and Murray, B. E. (2014). *Enterococcus faecium* PBP5-S/R, the missing link between PBP5-S and PBP5-R. *Antimicrob. Agents Chemother.* 58, 6978–6981. doi: 10.1128/AAC.03648-14

Poeta, P., Costa, D., Igrejas, G., Saenz, Y., Zarazaga, M., Rodrigues, J., et al. (2007). Polymorphisms of the pbp5 gene and correlation with ampicillin resistance in *Enterococcus faecium* isolates of animal origin. *J. Med. Microbiol.* 56, 236–240. doi: 10.1099/jmm.0.46778-0

Price, M. N., Dehal, P. S., and Arkin, A. P. (2010). FastTree 2-approximately maximum-likelihood trees for large alignments. *PLoS ONE* 5:e9490. doi: 10.1371/journal.pone.0009490

Raze, D., Dardenne, O., Hallut, S., Martinez-Bueno, M., Coyette, J., and Ghuysen, J. M. (1998). The gene encoding the low-affinity penicillin-binding protein 3r in *Enterococcus hirae* S185R is borne on a plasmid carrying other antibiotic resistance determinants. *Antimicrob. Agents Chemother.* 42, 534–539.

Rice, L. B., Bellais, S., Carias, L. L., Hutton-Thomas, R., Bonomo, R. A., Caspers, P., et al. (2004). Impact of specific pbp5 mutations on expression of beta-lactam resistance in *Enterococcus faecium*. *Antimicrob. Agents Chemother.* 48, 3028–3032. doi: 10.1128/AAC.48.8.3028-3032.2004

Rice, L. B., Carias, L. L., Hutton-Thomas, R., and Rudin, S. (2007). Interaction of related Tn916-like transposons: analysis of excision events promoted by Tn916 and Tn5386 integrases. *J. Bacteriol.* 189, 3909–3917. doi: 10.1128/JB.00859-06

Rice, L. B., Carias, L. L., Hutton-Thomas, R., Sifaoui, F., Gutmann, L., and Rudin, S. D. (2001). Penicillin-binding protein 5 and expression of ampicillin resistance in *Enterococcus faecium*. *Antimicrob. Agents Chemother.* 45, 1480–1486. doi: 10.1128/AAC.45.5.1480-1486.2001

Rice, L. B., Carias, L. L., Marshall, S., Rudin, S. D., and Hutton-Thomas, R. (2005a). Tn5386, a novel Tn916-like mobile element in *Enterococcus faecium* D344R that interacts with Tn916 to yield a large genomic deletion. *J. Bacteriol.* 187, 6668–6677. doi: 10.1128/JB.187.19.6668-6677.2005

Rice, L. B., Carias, L. L., Rudin, S., Hutton, R., Marshall, S., Hassan, M., et al. (2009). Role of class A penicillin-binding proteins in the expression of beta-lactam resistance in *Enterococcus faecium*. *J. Bacteriol.* 191, 3649–3656. doi: 10.1128/JB.01834-08

Rice, L. B., Carias, L. L., Rudin, S., Lakticová, V., Wood, A., and Hutton-Thomas, R. (2005b). *Enterococcus faecium* low-affinity pbp5 is a transferable determinant. *Antimicrob. Agents Chemother.* 49, 5007–5012. doi: 10.1128/AAC.49.12.5007-5012.2005

Rybkine, T., Mainardi, J. L., Sougakoff, W., Collatz, E., and Gutmann, L. (1998). Penicillin-binding protein 5 sequence alterations in clinical isolates of *Enterococcus faecium* with different levels of beta-lactam resistance. *J. Infect. Dis.* 178, 159–163. doi: 10.1086/515605

Sacco, E., Cortes, M., Josseaume, N., Rice, L. B., Mainardi, J.-L., and Arthur, M. (2014). Serine/threonine protein phosphatase-mediated control of the peptidoglycan cross-linking L,D-transpeptidase pathway in *Enterococcus faecium*. *MBio* 5:e01446-14. doi: 10.1128/mBio.01446-14

Santos, T., Silva, N., Igrejas, G., Rodrigues, P., Micael, J., Rodrigues, T., et al. (2013). Dissemination of antibiotic resistant *Enterococcus* spp. and *Escherichia coli* from wild birds of Azores Archipelago. *Anaerobe* 24, 25–31. doi: 10.1016/j.anaerobe.2013.09.004

Sifaoui, F., Arthur, M., Rice, L., and Gutmann, L. (2001). Role of penicillin-binding protein 5 in expression of ampicillin resistance and peptidoglycan structure in *Enterococcus faecium*. *Antimicrob. Agents Chemother.* 45, 2594–2597. doi: 10.1128/AAC.45.9.2594-2597.2001

Stanczak-Mrozek, K. I., Manne, A., Knight, G. M., Gould, K., Witney, A. A., and Lindsay, J. A. (2015). Within-host diversity of MRSA antimicrobial resistances. *J. Antimicrob. Chemother.* 70, 2191–2198. doi: 10.1093/jac/dkv119

Sullivan, M. J., Petty, N. K., and Beatson, S. A. (2011). Easyfig: a genome comparison visualizer. *Bioinformatics* 27, 1009–1010. doi: 10.1093/bioinformatics/btr039

Tedim, A. P., Ruiz-Garbajosa, P., Corander, J., Rodríguez, C. M., Cantón, R., Willems, R. J., et al. (2015). Population biology of intestinal *Enterococcus* isolates from hospitalized and nonhospitalized individuals in different age groups. *Appl. Environ. Microbiol.* 81, 1820–1831. doi: 10.1128/AEM.03661-14

Tenover, F. C., Arbeit, R. D., Goering, R. V., Mickelsen, P. A., Murray, B. E., Persing, D. H., et al. (1995). Interpreting chromosomal DNA restriction patterns produced by pulsed-field gel electrophoresis: criteria for bacterial strain typing. *J. Clin. Microbiol.* 33, 2233–2239.

Top, J., Willems, R., Blok, H., de Regt, M., Jalink, K., Troelstra, A., et al. (2007). Ecological replacement of *Enterococcus faecalis* by multiresistant clonal complex 17 *Enterococcus faecium*. *Clin. Microbiol. Infect.* 13, 316–319. doi: 10.1111/j.1469-0691.2006.01631.x

Tremblay, C.-L., Charlebois, A., Masson, L., and Archambault, M. (2013). Characterization of hospital-associated lineages of ampicillin-resistant *Enterococcus faecium* from clinical cases in dogs and humans. *Front. Microbiol.* 4:245. doi: 10.3389/fmicb.2013.00245

Tripathi, L. P., and Sowdhamini, R. (2008). Genome-wide survey of prokaryotic serine proteases: analysis of distribution and domain architectures of five serine protease families in prokaryotes. *BMC Genomics* 9:549. doi: 10.1186/1471-2164-9-549

Wagenvoort, J. H. T., De Brauwer, E. I. G. B., Penders, R. J. R., van der Linden, C. J., Willems, R. J., Top, J., et al. (2015). Environmental survival of vancomycin-sensitive ampicillin-resistant *Enterococcus faecium* (AREfm). *Eur. J. Clin. Microbiol. Infect. Dis.* 34, 1901–1903. doi: 10.1007/s10096-015-2430-x

Willems, R. J. L., Top, J., van Santen, M., Robinson, D. A., Coque, T. M., Baquero, F., et al. (2005). Global spread of vancomycin-resistant *Enterococcus faecium* from distinct nosocomial genetic complex. *Emerg. Infect. Dis.* 11, 821–828. doi: 10.3201/eid1106.041204

Willems, R. J. L., Top, J., van Schaik, W., Leavis, H., Bonten, M., Sirén, J., et al. (2012). Restricted gene flow among hospital subpopulations of *Enterococcus faecium*. *MBio* 3:e00151-12. doi: 10.1128/mBio.00151-12

Williamson, R., Calderwood, S. B., Moellering, R. C., and Tomasz, A. (1983). Studies on the mechanism of intrinsic resistance to beta-lactam antibiotics in group D streptococci. *J. Gen. Microbiol.* 129, 813–822. doi: 10.1099/00221287-129-3-813

Zhang, X., Paganelli, F. L., Bierschenk, D., Kuipers, A., Bonten, M. J. M., Willems, R. J. L., et al. (2012). Genome-wide identification of ampicillin resistance determinants in *Enterococcus faecium*. *PLoS Genet.* 8:e1002804. doi: 10.1371/journal.pgen.1002804

Zorzi, W., Zhou, X. Y., Dardenne, O., Lamotte, J., Raze, D., Pierre, J., et al. (1996). Structure of the low-affinity penicillin-binding protein 5 PBP5fm in wild-type and highly penicillin-resistant strains of *Enterococcus faecium*. *J. Bacteriol.* 178, 4948–4957.

5

Global Emergence and Dissemination of Enterococci as Nosocomial Pathogens: Attack of the Clones?

*Ana M. Guzman Prieto[1], Willem van Schaik[1], Malbert R. C. Rogers[1], Teresa M. Coque[2,3,4], Fernando Baquero[2,3,4], Jukka Corander[5] and Rob J. L. Willems[1]**

[1] Department of Medical Microbiology, University Medical Center Utrecht, Utrecht, Netherlands, [2] Hospital Universitario Ramon y Cajal, Instituto Ramón y Cajal de Investigación Sanitaria, Madrid, Spain, [3] CIBER Epidemiología y Salud Pública, Madrid, Spain, [4] Unidad de Resistencia a Antibióticos y Virulencia Bacteriana Asociada al Consejo Superior de Investigaciones Científicas, Madrid, Spain, [5] Department of Mathematics and Statistics, University of Helsinki, Helsinki, Finland

***Correspondence:**
Rob J. L. Willems
rwillems@umcutrecht.nl

Enterococci are Gram-positive bacteria that are found in plants, soil and as commensals of the gastrointestinal tract of humans, mammals, and insects. Despite their commensal nature, they have also become globally important nosocomial pathogens. Within the genus *Enterococcus, Enterococcus faecium,* and *Enterococcus faecalis* are clinically most relevant. In this review, we will discuss how *E. faecium* and *E. faecalis* have evolved to become a globally disseminated nosocomial pathogen. *E. faecium* has a defined sub-population that is associated with hospitalized patients and is rarely encountered in community settings. These hospital-associated clones are characterized by the acquisition of adaptive genetic elements, including genes involved in metabolism, biofilm formation, and antibiotic resistance. In contrast to *E. faecium*, clones of *E. faecalis* isolated from hospitalized patients, including strains causing clinical infections, are not exclusively found in hospitals but are also present in healthy individuals and animals. This observation suggests that the division between commensals and hospital-adapted lineages is less clear for *E. faecalis* than for *E. faecium*. In addition, genes that are reported to be associated with virulence of *E. faecalis* are often not unique to clinical isolates, but are also found in strains that originate from commensal niches. As a reflection of more ancient association of *E. faecalis* with different hosts, these determinants Thus, they may not represent genuine virulence genes but may act as host-adaptive functions that are useful in a variety of intestinal environments. The scope of the review is to summarize recent trends in the emergence of antibiotic resistance and explore recent developments in the molecular epidemiology, population structure and mechanisms of adaptation of *E. faecium* and *E. faecalis*.

Keywords: *E. faecium, E. faecalis,* high-risk nosocomial clones, antibiotic resistance, virulence

INTRODUCTION

Enterococci are low-GC Gram-positive ovococci that can form pairs and chains of diverse lengths. Bacteria from the genus *Enterococcus* are facultative anaerobic and grow optimally at 35°C, but can tolerate temperatures ranging from 10°C to 45°C (Byappanahalli et al., 2012). The genus comprises 54 species (Parte, 2014), which are ubiquitously present in nature but the gastrointestinal tract (GIT) of animals, including mammals, reptiles, birds (Mundt, 1963a) and insects (Martin and Mundt, 1972), is thought to be the largest reservoir of enterococci (Gilmore et al., 2013).

In humans, enterococci are common commensals of the GIT. In addition, enterococci have become ever more prominent as a causative agent of nosocomial infections since the 1970s (Arias and Murray, 2012). Two species, *Enterococcus faecalis* and *Enterococcus faecium*, cause the vast majority of hospital-acquired enterococcal infections in humans (Agudelo Higuita and Huycke, 2014). Of these two, *E. faecium* has rapidly acquired resistance to several classes of antibiotics. First, in the 1970s and 1980s, *E. faecium* gained high-level resistance to ampicillin (Grayson et al., 1991; Galloway-Peña et al., 2009) and since the 1980s it acquired resistance to aminoglycosides, fluoroquinolones, and glycopeptides, particularly vancomycin (Leclercq et al., 1988; Uttley et al., 1988; National Nosocomial Infections Surveillance System, 2004). *E. faecalis* has also acquired resistance to aminoglycosides, but resistance to ampicillin and vancomycin is much rarer than in *E. faecium* (Edelsberg et al., 2014). Worryingly, resistance to antibiotics that are used to treat vancomycin resistant enterococci (VRE), like linezolid, tigecycline, and daptomycin, has already been reported (Aksoy and Unal, 2008; Montero et al., 2008; Scheetz et al., 2008; Niebel et al., 2015).

The intrinsic resistance of enterococci to some antibiotics, including aminoglycosides, and the ability to acquire and disseminate antibiotic resistance determinants, like those involved in vancomycin resistance, only partly explain the recent emergence of these organisms as nosocomial pathogens. In addition, the plasticity of the enterococcal genomes allow enterococci to rapidly respond and adapt to selective constraints by acquiring genetic determinants that increase their ability to colonize or infect the host (Hendrickx et al., 2009; Palmer et al., 2012; Lebreton et al., 2013; Van Tyne and Gilmore, 2014). Other host or environmental factors, most notably exposure to antimicrobial agents, may favor an increase in colonization density of enterococci in the GIT of hospitalized patients (Donskey et al., 2000; Ubeda et al., 2010; Ruiz-Garbajosa et al., 2012). Antibiotic therapy that leads to the depletion of Gram-negatives, can reduce production of the antimicrobial peptide REGIIIγ by Paneth cells, and this may promote the outgrowth of VRE (Brandl et al., 2008).

Patients undergoing transplants or with underlying diseases, such as diabetes or renal failure, and patients with long-term catheter usage, are at higher risk of developing infections caused by multi-drug resistant (MDR) enterococci (Arias and Murray, 2012). High-density colonization of the patient GIT facilitates the transmission of MDR enterococci among hospital ward through fecal contamination (Arias and Murray, 2012). Therefore epidemiological surveillance and outbreak investigations, together with infection control polices and interventions, such as the use of protective barriers and proper disinfection, are key for infection control of these organisms in the nosocomial environment (Sydnor and Perl, 2011). High-level enterococcal GIT colonization can also lead to urinary tract infections (UTI) (De Vecchi et al., 2013; Neelakanta et al., 2015), which may progress to bloodstream infections or endocarditis (Patterson et al., 1995; Fernández-Guerrero et al., 2002). Enterococci from high-density intestinal populations may also directly translocate from the GIT into the bloodstream (Kamboj et al., 2014).

Enterococci have become one of the most common causes of health care-associated infections with *E. faecalis* causing approximately 60% of infections and *E. faecium* the remainder (de Kraker et al., 2013; Sievert et al., 2013). This review will focus on *E. faecalis* and *E. faecium* as both species have emerged as important nosocomial pathogens over the last 30 years and represent a major hub for the dissemination of antibiotic resistance genes.

EMERGENCE OF ANTIBIOTIC RESISTANCE IN ENTEROCOCCI

Enterococcus faecalis and *E. faecium* exhibit intrinsic resistance to a broad range of antibiotics. Below we briefly describe the mechanisms that cause resistance to the classes of antibiotics that are currently used in the treatment of enterococcal infections.

Compared to other low-GC Gram-positive cocci, enterococci exhibit decreased susceptibility to β-lactam antibiotics. The β-lactams act through the inactivation of penicillin-binding proteins (PBPs), thereby interfering with synthesis of peptidoglycan. All enterococci display decreased susceptibility to β-lactam antibiotics due to the expression of PBPs with an intrinsic low affinity for this class of antibiotics. Resistance to β-lactams, most notably to ampicillin, is currently far more widespread in *E. faecium* than in *E. faecalis* (Cattoir and Giard, 2014). The most important determinants for β-lactam resistance in *E. faecium* are mutations in the genes encoding the PBP5 (Zorzi et al., 1996; Rice et al., 2001; Zhang et al., 2012). There are clear indications that *E. faecium* progressed toward high-level ampicillin resistance in the 1970s and 1980s through the acquisition of specific mutations in the *pbp5* gene (Grayson et al., 1991; Galloway-Peña et al., 2009, 2011). It is of importance to note that these chromosomally encoded PBPs can be transferable (Dahl et al., 2000; Hanrahan et al., 2000; Rice et al., 2005), which indicates that dissemination of high-level ampicillin resistance can be the result of both clonal spread of strains with mutated *pbp5* genes and horizontal gene transfer. In addition to mutations in *pbp5*, production of β-lactamase has been described in both *E. faecalis* and *E. faecium* (Tomayko et al., 1996; Sarti et al., 2012). In general, the expression level of beta-lactamase in these species is low and impact on ampicillin susceptibility marginal.

Another group of antibiotics to which *E. faecium* and *E. faecalis* exhibit moderate intrinsic and high-level acquired resistance are the aminoglycosides. In *E. faecalis,* intrinsic resistance is thought to be caused by the inability of the antibiotic to enter the cytoplasm and inhibit ribosomal protein synthesis (Aslangul et al., 2006). In *E. faecium,* two chromosomally encoded genes, a 6′-*N*-aminoglycoside acetyltransferase (*aac(6′)-Ii*) (Costa et al., 1993) and an rRNA methyltransferase (*efmM*) (Galimand et al., 2011), have been associated with intrinsic resistance to tobramycin and kanamycin. In addition to intrinsic resistance to aminoglycosides, the therapeutic success of these antibiotics is critically compromised by high-level resistance, due to the gain of aminoglycoside modifying enzymes, such as phosphotransferases, acetyltransferases, and nucleotidyltransferases by *E. faecium* and *E. faecalis* (Chow, 2000; Miller et al., 2014).

The assessment of enterococci, particularly *E. faecium*, as important agents of MDR nosocomial infections was definitively established when they acquired resistance to vancomycin. While vancomycin-resistant enterococci were virtually non-existent in hospitals in the USA before 1990, nowadays 87% of *E. faecium* strains from nosocomial infections are vancomycin-resistant, while this is only 14% for *E. faecalis* (Edelsberg et al., 2014). Vancomycin is a glycopeptide antibiotic, which prevents cross-linking of peptidoglycan by binding to the D-alanine-D-alanine (D-Ala-D-Ala) moiety of the peptide chains that crosslink peptidoglycans. The mechanisms by which enterococci become resistant to vancomycin have been extensively reviewed elsewhere (Courvalin, 2006). In short, enterococci become resistant to vancomycin when the terminal amino acids of peptidoglycan precursors are altered from D-Ala-D-Ala to D-Ala-D-lactate (D-Ala-D-Lac) or to D-Ala-D-Serine (D-Ala-D-Ser), leading to high-level and low-level resistance to vancomycin, respectively. Nine gene clusters are currently known to be involved in vancomycin resistance in enterococci. These vancomycin resistance gene clusters are *vanA, vanB, vanD,* and *vanM*, causing vancomycin resistance through the formation of D-Ala-D-Lac, and *vanC, vanE, vanG, vanL,* and *vanN* which catalyze the formation of D-Ala-D-Ser (Courvalin, 2006; Boyd et al., 2008; Xu et al., 2010; Lebreton et al., 2011). The most frequently found, and thus clinically most relevant, vancomycin resistance determinants are *vanA* and *vanB*, which are both part of larger gene clusters, which encode a two-component regulatory system and enzymes involved in the recycling of D-Ala-D-Ala peptidoglycan precursors to D-Ala-D-Lac. Both *vanA* and *vanB* are located on transposons that contribute to the dissemination of vancomycin resistance among enterococci (Courvalin, 2006). Recently, *vanM* was found to be the most important vancomycin-resistance determinant among different *E. faecium* lineages in hospitals in Shanghai, China (Chen et al., 2015). Differences in the sequence diversity and prevalence of each *van* operon could be due to the different ecological origins of the *van* clusters. While *vanA* seems to have originated from soil organisms, *vanB, vanG,* and *vanD* have been reported in gut commensal microbiota (Guardabassi et al., 2005; Domingo et al., 2007; Howden et al., 2013).

While vancomycin resistance emerged and spread in USA hospitals in the 1990s, carriage of vancomycin-resistant *E. faecium* (VREF) was rare among hospitalized patients in Europe. In contrast, VRE carriage among farm animals, and, to a lesser extent, in healthy humans was higher in Europe than in the USA (Klare et al., 1995; Devriese et al., 1996; Stobberingh et al., 1999). The widespread occurrence of VRE among farm animals was linked to the use of the vancomycin-analog avoparcin as a growth promoter in Europe (Bager et al., 1997). The presence of indistinguishable vancomycin resistance transposons in both animal and human reservoirs provided the first indication that animal-derived enterococci could be a reservoir of antibiotic resistance genes that could be transmitted to humans (Arthur et al., 1993; Jensen et al., 1998; Woodford et al., 1998; Willems et al., 1999). After the ban on the use of avoparcin in 1997, the prevalence of VRE in animal husbandry declined in Europe (Klare et al., 1999; van den Bogaard et al., 2000; Aarestrup et al., 2001). In several European countries, the prevalence of VRE among hospitalized patients increased in the 21st century, but is still lower than the endemic levels reported in US hospitals (Bourdon et al., 2011; Gastmeier et al., 2014; Pinholt et al., 2015; Simner et al., 2015). Levels of VRE among hospitalized patients are also high in Australia, with 37% of *E. faecium* bacteremia isolates exhibiting resistance to vancomycin. Interestingly, vancomycin resistance in Australian *E. faecium* isolates is almost exclusively caused by *vanB*-type transposons, while *vanA* is the major vancomycin-resistance determinant in Europe and the USA (Coombs et al., 2014).

Due to the emergence and rapid spread of VRE, the antibiotics linezolid, daptomycin and tigecycline are increasingly used for the treatment of VRE infections.

Linezolid, the first oxazolidinone antibiotic, was introduced for clinical use in the USA in 2000. It acts on the ribosome by binding to a universally conserved site on 23S rRNA of the large 50S subunit of the ribosome (Colca et al., 2003), thereby inhibiting elongation of the polypeptide chain. The currently described linezolid resistance mechanisms alter the peptidyl transferase center of 23S rRNA, by a mutation in the 23S rRNA gene, with the G2576T mutation being most prominent (Jones et al., 2008), mutations in the genes encoding ribosomal proteins L3 and L4 (Chen et al., 2013) or methylation of the adenine at position 2503, which is catalyzed by the methyltransferase Cfr. The *cfr* gene is encoded on various conjugative and non-conjugative plasmids in enterococci (Diaz et al., 2012; Shen et al., 2013). Recently, the oxazolidinone resistance gene *optrA* has been identified in *E. faecalis* and *E. faecium* isolates of human and animal origin (Cai et al., 2015; Wang et al., 2015). This gene codes for an ATP-binding cassette transporter, which contributes to reduced susceptibility for oxazolidinones (linezolid and tedizolid) and phenicols (chloramphenicol and florfenicol). The *optrA* gene was associated with a conjugative plasmid (Wang et al., 2015). Nevertheless, linezolid resistance is still rare. In a study using data from 19 US hospitals in the period 2007–2010, linezolid resistance was reported in 1.1 and 1.8% of *E. faecium* and *E. faecalis* isolates, respectively (Edelsberg et al., 2014).

Daptomycin is a lipopeptide antibiotic that targets the cell membrane through interactions with phospholipids (Humphries et al., 2013). The genetic basis for resistance in *E. faecalis* and *E. faecium* was first studied through genome sequencing of pairs of strains that developed resistance *in vitro* or *in vivo* (Arias et al., 2011; Palmer et al., 2011; Tran et al., 2013a). Additional genetic and biochemical work has established the role of the three-component regulatory system LiaFSR in contributing to daptomycin resistance in enterococci. The LiaFSR system is conserved in low-GC Gram-positive bacteria and governs the cell envelope stress response (Jordan et al., 2006). Mutations in the *liaF* gene of *E. faecalis* lead to a redistribution of cardiolipin-rich microdomains from the division septum to other regions of the cytoplasmic membrane, which affects the antimicrobial activity of daptomycin. In addition, other mutations in phospholipid biosynthesis genes, most notably in the cardiolipin synthase gene *cls*, are required for full expression of daptomycin resistance in *E. faecalis* (Davlieva et al., 2013; Tran et al., 2013b). Similar mechanisms have been associated with daptomycin resistance in *E. faecium* (Munita et al., 2012, 2014; Davlieva et al., 2013). Notably, emergence of daptomycin resistance in *E. faecium* during daptomycin therapy is not always linked to mutations in *liaFSR* and mutations in *cls* may already be sufficient (Kelesidis et al., 2013; Lellek et al., 2015). Daptomycin resistance is infrequent, but more common in *E. faecium* than in *E. faecalis*. A worldwide study in isolates from hospitalized patients over the period 2005–2012 showed levels of daptomycin resistance of 0.18% for *E. faecium* and 0.02% for *E. faecalis* (Sader et al., 2014). However, in certain settings daptomycin resistance may be considerably more frequent, as illustrated by a study including 4,274 *E. faecium* and 7,007 *E. faecalis* isolates from 19 US hospitals, which found daptomycin resistance among 3.9 and 0.2% of *E. faecium* and *E. faecalis* isolates, respectively (Edelsberg et al., 2014). Similar levels of daptomycin resistance in *E. faecium* were recently reported for a German hospital (Lübbert et al., 2015).

Tigecycline is a semisynthetic derivative of the broad-spectrum tetracycline antibiotic minocycline, which acts on the ribosome by inhibiting its association with aminoacyl-tRNAs. The emergence of resistance during tigecycline therapy was first observed in *E. faecalis* (Werner et al., 2008; Cordina et al., 2012) and has recently been described in *E. faecium* (Niebel et al., 2015). In enterococci, resistance to tigecycline can be mediated through upregulation of tetracycline resistance determinants *tetL* (encoding an efflux pump) and *tetM* (providing ribosomal protection) (Fiedler et al., 2015) and mutations in the ribosomal protein *rpsJ* (Beabout et al., 2015; Cattoir et al., 2015; Niebel et al., 2015). Tigecycline resistance among *E. faecium* and *E. faecalis* is rare at 0.3% for both species (Hoban et al., 2015) and the antibiotic can still be used successfully to treat bacteremias caused by MDR enterococci, especially when it is used in combination with daptomycin (Polidori et al., 2011). The recovery of enterococcal isolates with decreased susceptibility to tigecycline in samples of animal origin is of concern, as extensive use with tetracyclines in the veterinary setting could select for tigecycline tolerant strains (Freitas et al., 2011).

In addition to the antibiotics described above, the antibiotics tedizolid, telavancin, dalbavancin, and oritavancin have recently been approved by the Food and Drug Administration in the USA for the treatment of skin infections by Gram-positive bacteria (Crotty et al., 2016). Tedizolid is an oxazolidinone antibiotic that has improved *in vitro* activity compared to linezolid. Enterococcal strains that have acquired linezolid resistance due to the acquisition of the *cfr* gene or the G2576T mutation in the 23S rRNA gene, may still have relatively low MICs ($\leq$4 μg/ml) for tedizolid (Barber et al., 2016; Silva-Del Toro et al., 2016). Telavancin, dalbavancin and oritavancin are semi-synthetic glycopeptides antibiotics that, similar to vancomycin, affect peptidoglycan stability by binding to the D-Ala-D-Ala moiety of the peptide chains that crosslink peptidoglycan chains. Interestingly, oritavancin is also active against *vanA*- and *vanB*-type vancomycin-resistant enterococci, while telavancin and dalbavancin have limited activity against enterococci carrying the *vanA*-type vancomycin resistance transposon but are active against enterococci with *vanB*-type vancomycin resistance (Jones et al., 2013; Karlowsky et al., 2015). Tedizolid, telavancin, dalbavancin, and oritavancin may be useful as alternatives to linezolid and vancomycin, but they do not have radically different modes of action and they may therefore suffer from the same resistance mechanisms that are threatening the efficacy of linezolid and vancomycin. Therefore, antibiotics that target other structures in the enterococcal cell, including teixobactin (Ling et al., 2015) and the acyldepsipeptides (Brötz-Oesterhelt et al., 2005), hold considerable promise as novel compounds for the treatment of infections with MDR enterococci.

MOLECULAR EPIDEMIOLOGY AND POPULATION STRUCTURE OF *E. faecium* AND *E. faecalis*

E. faecium

As a consequence of the global rise of VREF, a large number of molecular epidemiological studies have been performed to obtain insights into the dissemination of VREF clones in and between hospitals, in farm animals and healthy humans. However, methods like pulse-field gel electrophoresis (PFGE), soon proved to be insufficiently reproducible to study genetic relatedness of isolates (Morrison et al., 1999).

A first insight into the existence of particular ecotypes in *E. faecium*, was obtained by using amplified fragment polymorphism (AFLP) to infer the genetic relatedness of strains from diverse hosts and environments (Willems et al., 2000). This study revealed that strains from hospitalized patients grouped in a specific sub-population that was distinct from groups of strains that were isolated from humans in the community and farm animals.

The use of AFLP for global studies showed the limitations of this technique for comparisons of data obtained from different laboratories. Therefore an alternative method, termed multi locus sequence typing (MSLT), was used in follow-up studies. In MLST, allelic profiles are based on the sequences of fragments of a number of housekeeping genes (seven in the case of the *E. faecium* and *E. faecalis* MLST schemes) (Homan et al., 2002; Ruiz-Garbajosa et al., 2006). Compared to AFLP, MLST has the distinct advantage that data are easily collated and shared through an on-line database. The first analyses of *E. faecium* MLST data were performed with the algorithm eBurst (Feil et al., 2004) and confirmed a distinct clustering of strains derived from the hospital environment. This cluster was named clonal complex 17 (CC17) and soon CC17 was found to be disseminated throughout the world (Willems et al., 2005; Top et al., 2008). The later use of larger datasets and different algorithms for the analysis of MLST data (Didelot and Falush, 2007; Francisco et al., 2009), showed that eBurst can sometimes fail to correctly assign STs to clonal complexes (CCs), particularly when genetic variation in populations is largely driven by recombination rather than mutations, like is the case in *E. faecium* and *E. faecalis* (Turner et al., 2007).

Bayesian analysis of population structure (BAPS) has successfully been used to probabilistically infer the population structure and levels of recombination of several microbial pathogens (Corander et al., 2012; Thomas et al., 2014). When applied to *E. faecium* MLST data, BAPS allowed the partitioning of 519 STs of 1720 *E. faecium* isolates into 13 non-overlapping groups. Of these groups, BAPS 3–3 was significantly associated with isolates from hospitalized patients, while BAPS 2–1 and 2–4 were significantly associated with farm animals. This observation again confirmed that there exists structure in the *E. faecium* population, with a distinct subpopulation of isolates that are almost exclusively found in hospitalized patients (Willems et al., 2012). One of the important nodes in the previously described hospital-associated CC17, ST78, and its descendant STs, grouped in BAPS 2–1 together with farm animal isolates, while two other important CC17 nodes, ST17 and ST18 with their descendant STs, clustered in another BAPS group (BAPS 3–3). These findings indicate that nosocomial *E. faecium* isolates have not evolved from a single ancestor, like previously postulated, but rather the cumulative acquisition of adaptive elements in nosocomial isolates may have occurred multiple times in different genetic backgrounds. Another conclusion that could be drawn from Bayesian modeling of MLST data is that hospital isolates displayed a relative low-level of admixture, despite the high recombination rates in *E. faecium*, suggesting that once strains have adapted to the distinct hospital niche, they become ecologically isolated (including isolation by dominance) and recombination with other populations declines (Willems et al., 2012).

Despite the overt lack of reproducibility, PFGE long remained the "gold standard" for molecular typing of *E. faecium* until the introduction of whole genome sequence (WGS)-based epidemiology. Howden et al. (2013) published a landmark study in which they used whole genome sequencing to track an outbreak of *vanB*-VREF in a large hospital in Australia. Interestingly, detailed phylogenomic analysis and precise mapping of the *vanB* gene revealed that 18 of the 36 *vanB*-VREF had acquired the *vanB* transposon during the outbreak period. This indicates that for *vanB*-VREF, frequent *de novo* generation of VREF through horizontal gene transfer may contribute to the emergence of VREF, in addition to clonal spread. This study was followed by multiple other studies that used whole genome sequencing to investigate the molecular epidemiology of VREF with high resolution and high accuracy (Reuter et al., 2013; Salipante et al., 2015; Brodrick et al., 2016; van Hal et al., 2016). Bender et al. (2016) performed whole genome sequencing of 49 *vanB*-VREF, primarily of ST192 (39/49), from invasive infections from hospitals all across Germany and found that spread of the Tn*1549*-*vanB*-type resistance involved exchange of large chromosomal fragments between *vanB*-positive and *vanB*-negative enterococci rather than independent acquisition events of the *vanB* transposon alone.

WGS-based studies also proposed that the *E. faecium* population was divided into two species-level subdivisions, based on phylogenetic analysis and the determination of average nucleotide identity (ANI) between the two sub-populations. The sub-populations were termed clade A or hospital-associated clade, primarily containing isolates from hospitalized patients and clade B or community-associated clade, mostly containing isolates from healthy, non-hospitalized individuals (Galloway-Peña et al., 2012; Palmer et al., 2012). The high level of diversity between these two clades indicates that the clade A–B split is ancient and precedes the modern antibiotic era. Further work provided further evidence for this split and showed that the population of *E. faecium* in clade A had a second split of more recent date (74 ± 30 years), with clade A1 containing the majority of clinical isolates and clade A2 mostly comprising animal-derived strains (Lebreton et al., 2013). WGS also confirmed that *E. faecium* was subject to high rates of recombination, leading to changes in MLST profile in otherwise closely related strains, which invalidates the use of MLST for tracking transmission events (van Hal et al., 2016). Another finding of van Hal et al. (2016) was that other *E. faecium* strains are the most important donors of imported DNA fragments. Specifically, strains from clade B are an important reservoir for donating foreign DNA to clade A strains (de Been et al., 2013). These findings indicate that hospital-acquired clade A1 *E. faecium* strains have recently emerged from a background of human commensal and animal isolates. Tree-like network representations of *E. faecium* genomic relatedness based on cgMLST data also revealed the distinct clustering of human commensal, animal and human clinical strains (**Figure 1**). Notably, in this analysis the animal isolates do not group in a single clade A2 but form multiple distinct clusters that are located in the network, between the human commensal and clinical isolates, indicating that clade A2 may not be monophyletic, as previously postulated (Lebreton et al., 2013). Interestingly, clade A1 *E. faecium* strains have lower fitness in natural environments, where they are out-competed by other *E. faecium* clones (Leclercq et al., 2013). Similarly, clade B strains outcompete clade A strains in an animal model of gut colonization in the absence of selection by antibiotics (Montealegre et al., 2015). These data highlight the exquisite niche specialization to the hospital environment of clade A1 strains, which may come at a fitness cost in non-hospital environments.

The WGS-based epidemiological studies described above generated phylogenetic trees based on SNPs, by mapping sequencing reads to a reference genome. This approach, although providing high-resolution typing data, complicates comparisons of data between studies and hampers the construction of globally accessible databases. This limitation may be overcome by using a genome-wide gene-by-gene comparison approach, as in classical MLST, but with an important extension of the number of analyzed genes from seven to the entire core genome of the species, i.e., 1423 genes as in the recently published core genome MLST (cgMLST) scheme of *E. faecium* (de Been et al., 2015). The *E. faecium* cgMLST scheme

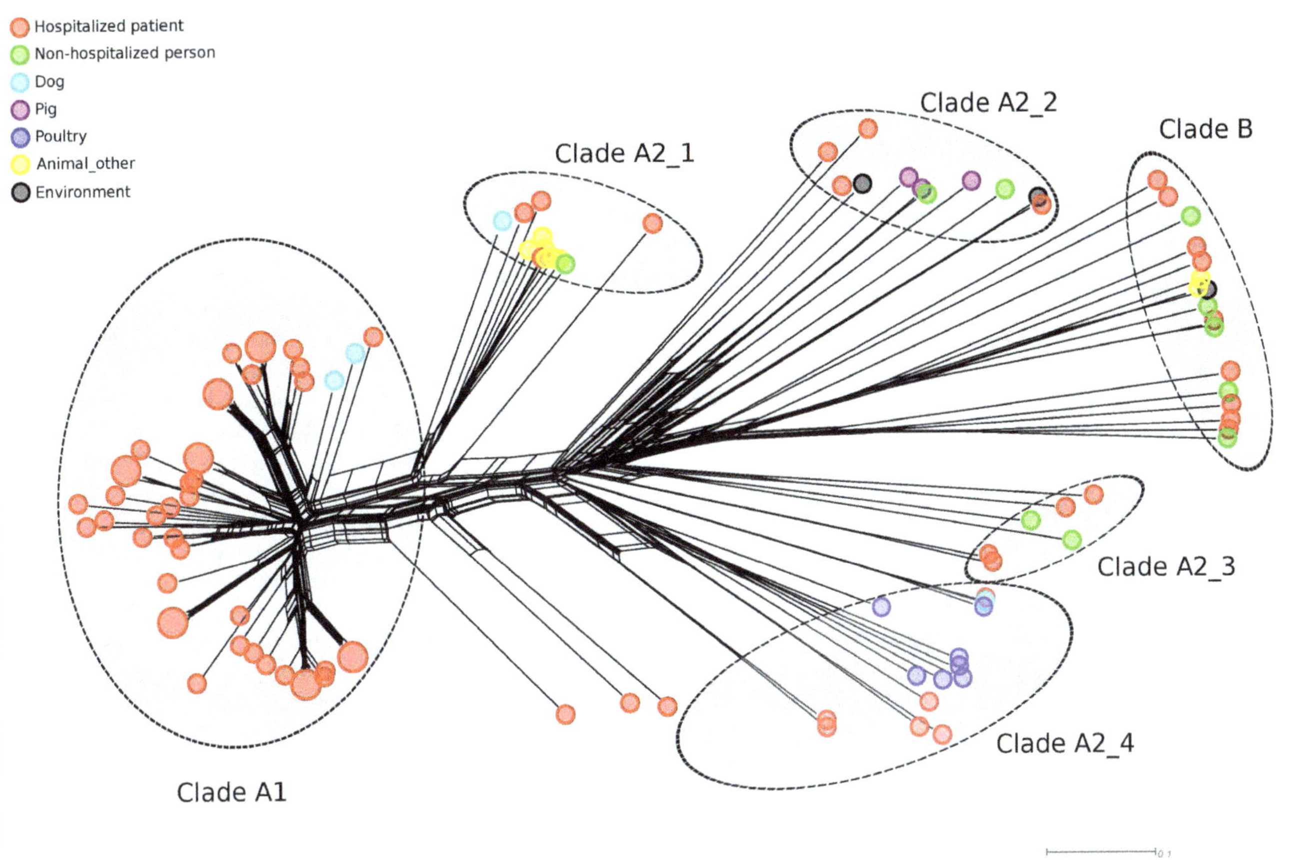

FIGURE 1 | NeighborNet phylogenetic network to visualize the relationships between 196 *E. faecium* isolates. The distance matrix underlying the network was built from all pairwise allelic profile comparisons of 1423 loci of the published *E. faecium* cgMLST scheme (de Been et al., 2015). Allelic profiles were extracted from SeqSphere. Small colored circles indicate the origin of isolates. *E. faecium* clades (A1 and B) inferred from the STs and based on Lebreton et al. (2013) are indicated by large dotted circles. Notably, in this analysis clade A2 isolates do not group in a single clade but form at least four distinct clusters that are located in the network and are indicated clade A2_1, A2_2, A2_3 and A2_4.

allowed high-resolution tracing of *E. faecium* outbreaks and performed equally well as an SNP-based phylogenetic approach, but has the advantage that data can be collated in an online database and a standardized nomenclature for clones can be used.

E. faecalis

Enterococcus faecalis appears to be the most widespread and abundant species of *Enterococcus* and can be found in the intestines of humans, farm, companion, and wild animals and in the environment (Mundt, 1963a,b; Devriese et al., 1992a,b; Tannock and Cook, 2002; Ruiz-Garbajosa et al., 2009). Initial molecular epidemiological studies, using MLST, showed the existence of CCs, in which isolates originating from different hosts are contained (Ruiz-Garbajosa et al., 2006; McBride et al., 2007). While many *E. faecalis* clones are shared between hospitalized patients and other reservoirs, some CCs (specifically CC2, CC9, and CC87) and STs (ST6 and ST16) seemed to be enriched among isolates from hospitalized patients (Ruiz-Garbajosa et al., 2006; McBride et al., 2007; Freitas et al., 2009; Kuch et al., 2012). Similarly, in isolates collected from across the globe, antibiotic resistance is more prevalent in strains that belong to CC2, CC8, CC9, CC16, and CC87 (Kawalec et al., 2007; McBride et al., 2007; Freitas et al., 2009; Kuch et al., 2012).

Analysis of the extent of congruence between the topologies of the seven different MLST gene trees revealed that that all 42 pairwise comparisons of the sequences of the MLST loci were incongruent. This observation suggests that *E. faecalis* is a highly recombinogenic species (Ruiz-Garbajosa et al., 2006). For this reason Tedim et al. (2015) used BAPS to investigate the *E. faecalis* population structure based on MLST data of hospitalized ($n = 133$) and non-hospitalized individuals ($n = 173$) isolated from feces. A hierarchical BAPS clustering analysis partitioned the *E. faecalis* population into 5 BAPS groups. Contrary to what has been described in *E. faecium* and to eBURST based analyses of *E. faecalis* MLST data, no significant association was found between isolates from hospitalized patients and particular *E. faecalis* BAPS groups (Tedim et al., 2015).

The first comprehensive *E. faecalis* phylogenomic study involved 18 *E. faecalis* genomes representing clinical ($n = 10$), human commensal ($n = 3$), and animal isolates ($n = 3$), and isolates of unknown origin ($n = 2$) (Palmer et al., 2012). The analysis revealed that the phylogenetic diversity of *E. faecalis* was limited compared to *E. faecium*, with variations in the ANI in a narrow range between 97.8 and 99.5%. There was also no clear structure in the phylogenetic core gene tree, with no distinct clustering of isolates according to source. Similarly, clustering of these 18 isolates supplemented with 3 other published *E. faecalis* genomes from a clinical site, human feces and pig based on an *ad hoc* developed *E. faecalis* cgMLST scheme using tree-like network methods also revealed no distinct clustering of isolates according to their source (**Figure 2**). A recent genome-based study of 515 *E. faecalis* genomes, mainly isolated from clinical settings on the British Isles, revealed three dominant lineages of hospital-associated lineages (L1, L2, and L3). Isolates in L1 and L3 originated from both the UK and USA (Raven et al., 2016). These data clearly indicate independent clonal expansion with subsequent national dissemination. The data may also point toward the existence of specific hospital-associated or hospital-adapted lineages. However, in order to unequivocally infer hospital-adapted lineages it is essential to correct the data for clonal outbreaks, by including outbreak clones only once in the analysis. The fact that only a limited number of STs (3, 2, and 2 STs in lineage 1, 2, and 3, respectively) were represented in these three lineages may suggest that these lineages contain multiple replicates of outbreaks clones. Analyses for recombination frequencies in the core genome of L1, L2 and L3 isolates indicated low levels of recombination, which seems to contrast with previous MLST-based analyses (Ruiz-Garbajosa et al., 2006; Raven et al., 2016). Possibly, recombination drove the initial diversification of *E. faecalis* but contributed less to the relatively recent evolution of the three dominant *E. faecalis* lineages described by Raven and co-workers (Raven et al., 2016).

ADAPTIVE ELEMENTS IN NOSOCOMIAL LINEAGES

Enterococci are ubiquitous in nature, where they act as commensals and opportunistic pathogens. A consequence of these different lifestyles is that enterococci need to adapt to different micro-environments, each one exerting strong selective pressures. A strategy for bacterial species to survive when confronted with a wide range of selective pressures is to specialize in particular fitness peaks. *E. faecium* probably followed such an evolutionary trajectory, resulting in the emergence of specific hospital-adapted lineages. Successful hospital-adapted clones in *E. faecium* have the ability to exchange mobile genetic elements, carrying antimicrobial resistance, and virulence determinants, by horizontal gene transfer (Leavis et al., 2003; Hegstad et al., 2010; Palmer et al., 2010; van Schaik et al., 2010). The cumulative acquisition of adaptive elements has been named "genetic capitalism", in which the acquisition of a particular adaptive element by a particular clone enhances its fitness, thereby increasing the likelihood of acquiring a second adaptive element which finally can lead to the emergence of high-risk multidrug-resistant clones (Baquero, 2004). This cumulative acquisition of adaptive elements by the hospital-associated *E. faecium* clade A1 strains is reflected by their larger genome sizes (2,843 ± 159 genes; 2.98 ± 0.15 Mbp), compared to the genomes of strains in clade A2 (2,597 ± 153 genes; 2.75 ± 0.14 Mbp) or clade B (2,718 ± 120 genes; 2.84 ± 0.1 Mbp) (Lebreton et al., 2013).

The first adaptive element that was described as being specific for hospital-associated *E. faecium* strains was Esp (Willems et al., 2001), a surface protein with a signal sequence for transport and a LPxTG-like motif for cell wall anchoring. The *E. faecium esp* gene is located on an integrative conjugative element, called ICE*Efm1*, and contributes to biofilm formation, and UTIs and endocarditis in animal models (Leavis et al., 2004; Heikens et al., 2007, 2011; Leendertse et al., 2009; Sava et al., 2010; van Schaik et al., 2010; Top et al., 2011).

In addition to Esp, other determinants were found to be specific or significantly enriched among hospital-associated isolates. Characteristics of these genes have been previously discussed (Leavis et al., 2003; Heikens et al., 2008; Hendrickx et al., 2009, 2013; Zhang et al., 2013; Paganelli et al., 2016). Comparative genomic analyses of 73 *E. faecium* strains revealed major differences in gene content between clinical (clade A1), animal (clade A2) and non-clinical (clade B) strains, particularly through gain and loss of gene clusters with predicted roles in carbohydrate metabolism (Kim and Marco, 2013; Lebreton et al., 2013). Many of the clade B-specific genes have a predicted role in the utilization of complex carbohydrates from dietary sources, which were replaced with genes that were associated with the utilization of amino sugars (e.g., galactosamine), which occur on epithelial cell surfaces and in mucin. This metabolic switch may have contributed to the niche specialization of *E. faecium*.

As mentioned above the population structure of *E. faecalis* seems to be significantly different from *E. faecium* as no clearly defined *E. faecalis* ecotypes appear to exist. Correspondingly, previous studies have reported shared antibiotic resistance and virulence genes, such as the *E. faecalis* pathogenicity island, *esp*, capsule polysaccharide genes, and genes encoding for gelatinase, aggregation substance, cytolysin, and Ace, among *E. faecalis* isolates from a wide variety of different niches like the hospital, animals, food products and the environment (Eaton and Gasson, 2001; Creti et al., 2004; McBride et al., 2007; Freitas et al., 2009; Solheim et al., 2009; La Rosa et al., 2015). Note that the absence of clearly defined ecotypes in *E. faecalis* can be interpreted as a high multiplicity of closely related fine-grained ecotypes, none of them reaching predominance.

Comparative genomic analysis of 18 *E. faecalis* genomes showed that the accretion of mobile genetic elements in multiple *E. faecalis* lineages appears to be a major source of genome diversity (Palmer et al., 2012). Although antibiotic resistance and pathogenicity island traits have converged in some *E. faecalis* lineages, substantial differences in gene content exist, indicating that specialization toward specific ecotypes is not apparent in this species, in contrast to *E. faecium*. In addition, a hierarchical

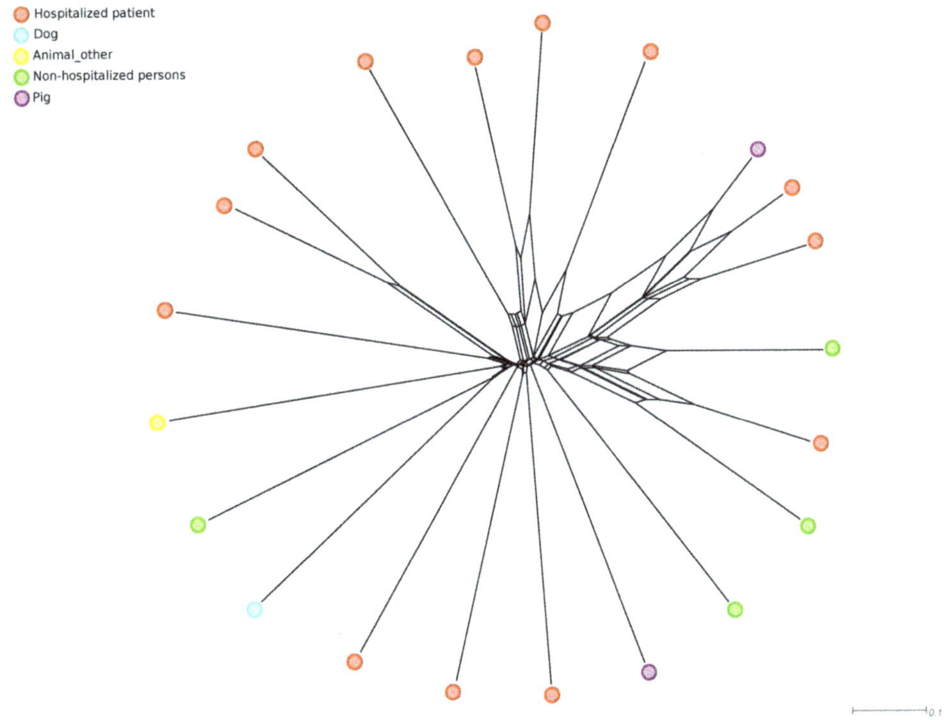

FIGURE 2 | NeighborNet phylogenetic network to visualize the relationships between 21 *E. faecalis* isolates. The distance matrix underlying the network was built from all pairwise allelic profile comparisons of 1298 loci of an *ad hoc E. faecalis* cgMLST scheme using SeqSphere (http://www.ridom.de/seqsphere/). Small colored circles indicate the origin of isolates.

clustering based on genome content of 38 *E. faecalis* genomes from diverse sources including clinical and non-clinical isolates, unearthed no evidence of distinct lineages in *E. faecalis* and no genes were found to be significantly enriched among clinical or non-clinical strains (Kim and Marco, 2013). A pangenome analysis of 168 *E. faecalis* isolates revealed no genes, or even homoplastic non-synonymous single nucleotide polymorphisms, that were ubiquitous in the three dominant lineages of hospital isolates, while being absent from all other sporadic lineages (Raven et al., 2016). These observations are all in accordance with high levels of genetic exchange between ecologically diverse *E. faecalis* clones.

In bacteria, mechanisms that preclude the acquisition of foreign DNA, include the Clustered Regularly Interspaced Short Palindromic Repeats (CRISPR)-Cas system (Horvath and Barrangou, 2010) and restriction and anti-restriction modification (RM-, antiRM-) systems (Clewell et al., 2014). *E. faecalis* strains lacking CRISPR-Cas appear to more readily acquire DNA through horizontal gene transfer and, consequently have a larger genome size than *E. faecalis* strains that carry CRISPR-Cas (Palmer et al., 2010, 2012). This indicates that the CRISPR-Cas status of *E. faecalis* strains may contribute to ecological adaptation. In *E. faecium* CRISPR-Cas is only found in a subset of clade B strains, again highlighting the general lack of barriers to horizontal gene exchange in this species (Lebreton et al., 2013). Restriction and anti-restriction systems located on mobile elements or enterococcal chromosomes can also determine the flow of adaptive traits among populations of the same and/or different species (Clewell et al., 2014). Anti-RM systems include analogs of ArdA (alleviation of restriction of DNA) that are located on conjugative transposons (CTns) that are found widely spread in enterococci. ArdA proteins act against Type I restriction systems (detected in Tn*916* and CTn*6000*) and other genes presumptively involved in methylation (CTn*6000*) (Clewell et al., 2014). Recently, a type IIS restriction-modification (R-M) system SfaNI was described in *E. faecalis* (Furmanek-Blaszk and Sektas, 2015).

Other strategies seem to have exerted positive as well as negative selective pressures on *E. faecalis* clones. *E. faecalis* strain V583 uses phage particles to establish and maintain dominance of its intestinal niche in the presence of closely related competing strains (Duerkop et al., 2012). Furthermore, five of the seven prophages in the same strain (phage01, phage03, phage04, phage05, and phage07) can be excised from the bacterial chromosome and four of them produced infective virions that may promote gene dissemination among isolates and increase pathogenicity (Matos et al., 2013; La Rosa et al., 2015). Conversely, Gaca and Gilmore recently demonstrated how the accretion of mobile genetic elements in *E. faecalis* V583, renders it unable to co-exist with native enterococci in healthy human fecal flora (Gaca and Gilmore, 2016).

CONCLUSION

The ubiquitous nature of enterococci, the flexibility of their genomes and the widespread use of antibiotics in human and veterinary medicine, are important factors that drive the current emergence of *E. faecium* and *E. faecalis* as MDR nosocomial pathogens.

The population structure of *E. faecium* is characterized by a deep phylogenetic split that separates human commensal isolates (clade B) from farm animal and hospital isolates (clade A). It is tempting to speculate that the deep phylogenetic split in *E. faecium* is driven by anthropogenic influences since strains from farm animals (clade A2) and clinical isolates (clade A1), share a common feature as they both originate from environments where mammalian hosts are in close contact with each other and usage of antibiotics is high. The evolutionary trajectory of *E. faecium*, which has led to a clear clade structure, suggests that *E. faecium* is colonizing rugged fitness landscapes, in which characters of particular clones in clade A and B impose strong fitness differences. Under these circumstances, gene exchange between diverging populations is reduced over large genomic regions, as a collateral effect of strong divergent selection on genes involved in local adaptation. This mechanism was named divergence hitchhiking (DH) (Via, 2012). DH facilitates the divergence of genes linked to genomic regions that are most involved in local adaptation, and may explain the evolution of clones in populations with extensive gene flow. In *E. faecium* clade A, DH may have contributed to the fixation of ecologically important genomic traits, like antibiotic resistance or genes linked to pathogenicity. Indeed, using a neutral model incorporating microepidemics and migration, which mimics a situation where ecological factors may limit transmission between subpopulations, showed that *E. faecium* hospital isolates with extensive genotype relatedness markedly deviated from this neutral model compared with other common nosocomial bacteria like *Staphylococcus aureus* and *Staphylococcus epidermidis*, indicating that these hospital isolates represent a subpopulation adapted to the hospital environment (Numminen et al., 2016).

The *E. faecalis* population analyzed in the studies described above included human commensal, farm animal and clinical isolates. It is an open question why this anthropogenic divergence, as observed in *E. faecium*, is not seen in *E. faecalis*. An explanation for this may reside in differences in population size of both species. Indeed, *E. faecalis* seems to be more widespread and abundant in the intestines of cattle, pigs, dogs, horses and poultry (Devriese et al., 1987, 1991, 1992a,b, 1994). *E. faecalis* has also been more frequently isolated from wild mammals, reptiles, birds, insects, and wild plants than *E. faecium* (Mundt, 1963a,b; Martin and Mundt, 1972). Finally, in healthy humans *E. faecalis* is also more frequently found than *E. faecium* (Tannock and Cook, 2002). Moreover, it seems that *E. faecalis* clones are maintained longer in the gut than *E. faecium* clones (Ruiz-Garbajosa et al., 2009). The higher population density of *E. faecalis* relative to *E. faecium*, in combination with its broader host-range, should provide *E. faecalis* with more opportunities for genetic exchange and diversification than *E. faecium*. High population densities also assure frequent transmission events, favoring gene flow among strains. This may act as a unifying force preventing the evolution of distinct clades in *E. faecalis*. Using the same neutral model described above for *E. faecium*, showed that the population structure of *E. faecalis* could be reflective of the evolutionary dynamics of a generalist organism which regularly experiences a high level of drift and gene flow between different host species (Numminen et al., 2016). Different strategies influence interactions of *E. faecalis* with populations of the same or of different species. A mechanism of substantial genetic exchange in *E. faecalis* has been postulated by Manson and co-workers, who described conjugative transfer of chromosomal fragments driven by integration of pheromone responsive conjugative plasmids carrying IS*256* insertions and recombination across IS*256* copies in the genome (Manson et al., 2010). The authors demonstrated that such an event could include transfer and recombination of up to 857 kb ($\sim$ 25% of the genome) and show that essentially every chromosomal location can be transferred by this mechanism. Other strategies may include phages, mobile genetic elements and the generation of polymicrobial biofilms (Weigel et al., 2007; La Rosa et al., 2015, 2016; Gaca and Gilmore, 2016).

High-level genetic connectivity of strains between different niches and hosts might favor dissemination of adaptive elements (e.g., virulence and resistance genes) that are acquired by *E. faecalis* populations in one niche (e.g., animals raised for food production) and finally end up being selected for in another environment (e.g., the hospital). In fact, the panmictic structure of *E. faecalis* favors its behavior as a collective evolutionary individual, which has promoted the generalist lifestyle of *E. faecalis*. The high level of genetic connectivity of *E. faecalis* strains in different niches, in combination with a relative small genome size, is in accordance with the 'Black Queen Hypothesis'. According to this hypothesis collective genes shared in a highly recombinogenic structure act as "common goods", and may favor the loss of genes in a particular niche in individual strains, provided they are not lost entirely from the community (Morris et al., 2012). While there is evidence in favor of this hypothesis, there is still a distinct lack of studies using WGS to describe the evolution, genetic diversity and population structure of *E. faecalis* and these are urgently needed to validate or falsify the current view of *E. faecalis* as a generalist.

In conclusion, *E. faecium* and *E. faecalis* are widely distributed among humans, animals and the environment. The social behavior of enterococci, which includes the accretion of plasmids, phages and CTns (Werner et al., 2013; Clewell et al., 2014), but also the ability to maintain microbial communities (Barnes et al., 2012; Paganelli et al., 2013), has allowed them to rapidly acquire antibiotic resistance genes and genetic elements that increase their ability to colonize and infect patients. *E. faecium* and *E. faecalis* differ in their population structure. *E. faecium* has clearly defined sub-populations, one of which contains the majority of strains that are currently causing nosocomial infections. In contrast, *E. faecalis* is a generalist, with a highly panmictic population where genes involved in patient colonization or virulence are widely spread across niches.

AUTHOR CONTRIBUTIONS

Most of the writing for this manuscript was done by AG, WS, and RW, with MR contributing analyses for both figures and TC, FB, and JC contributing concepts and text in sections 3 and 4. All authors have contributed to and revised the manuscript prior to submission. In our opinion, all authors meet the requirement of authorship as outlined in the Frontiers author guidelines.

ACKNOWLEDGMENTS

WS is funded through a NWO-Vidi grant (917.13.357). TC and FB are supported by grants PI12-01581 and PI15-01307 from the ISCIII-Ministry of Economy and Competitiveness of Spain (to TC), S2010/BMD2414 from the Regional Government of Madrid (to FB) and the CIBERESP Network for Biomedical Research in Epidemiology and Public Health (Instituto Carlos III, Spanish Ministry of Science and Innovation, reference number CB06/02/0053).

REFERENCES

Aarestrup, F. M., Seyfarth, A. M., Emborg, H. D., Pedersen, K., Hendriksen, R. S., and Bager, F. (2001). Effect of abolishment of the use of antimicrobial agents for growth promotion on occurrence of antimicrobial resistance in fecal enterococci from food animals in Denmark. *Antimicrob. Agents Chemother.* 45, 2054–2059. doi: 10.1128/AAC.45.7.2054-2059.2001

Agudelo Higuita, N. I., and Huycke, M. M. (2014). "Enterococcal disease, epidemiology, and implications for treatment," in *Enterococci: From Commensals to Leading Causes of Drug Resistant Infection*, eds M. S. Gilmore, D. B. Clewell, Y. Ike, and N. Shankar (Boston: Massachusetts Eye and Ear Infirmary). Available at: http://www.ncbi.nlm.nih.gov/books/NBK190429/ [Accessed February 21, 2016].

Aksoy, D. Y., and Unal, S. (2008). New antimicrobial agents for the treatment of Gram-positive bacterial infections. *Clin. Microbiol. Infect. Off. Publ. Eur. Soc. Clin. Microbiol. Infect. Dis.* 14, 411–420. doi: 10.1111/j.1469-0691.2007.01933.x

Arias, C. A., and Murray, B. E. (2012). The rise of the *Enterococcus*: beyond vancomycin resistance. *Nat. Rev. Microbiol.* 10, 266–278. doi: 10.1038/nrmicro2761

Arias, C. A., Panesso, D., McGrath, D. M., Qin, X., Mojica, M. F., Miller, C., et al. (2011). Genetic basis for in vivo daptomycin resistance in enterococci. *N. Engl. J. Med.* 365, 892–900. doi: 10.1056/NEJMoa1011138

Arthur, M., Molinas, C., Depardieu, F., and Courvalin, P. (1993). Characterization of Tn1546, a Tn3-related transposon conferring glycopeptide resistance by synthesis of depsipeptide peptidoglycan precursors in *Enterococcus faecium* BM4147. *J. Bacteriol.* 175, 117–127.

Aslangul, E., Massias, L., Meulemans, A., Chau, F., Andremont, A., Courvalin, P., et al. (2006). Acquired gentamicin resistance by permeability impairment in *Enterococcus faecalis*. *Antimicrob. Agents Chemother.* 50, 3615–3621. doi: 10.1128/AAC.00390-06

Bager, F., Madsen, M., Christensen, J., and Aarestrup, F. M. (1997). Avoparcin used as a growth promoter is associated with the occurrence of vancomycin-resistant *Enterococcus faecium* on Danish poultry and pig farms. *Prev. Vet. Med.* 31, 95–112. doi: 10.1016/S0167-5877(96)01119-1

Baquero, F. (2004). From pieces to patterns: evolutionary engineering in bacterial pathogens. *Nat. Rev. Microbiol.* 2, 510–518. doi: 10.1038/nrmicro909

Barber, K. E., Smith, J. R., Raut, A., and Rybak, M. J. (2016). Evaluation of tedizolid against *Staphylococcus aureus* and enterococci with reduced susceptibility to vancomycin, daptomycin or linezolid. *J. Antimicrob. Chemother.* 71, 152–155. doi: 10.1093/jac/dkv302

Barnes, A. M. T., Ballering, K. S., Leibman, R. S., Wells, C. L., and Dunny, G. M. (2012). *Enterococcus faecalis* produces abundant extracellular structures containing DNA in the absence of cell lysis during early biofilm formation. *mbio* 3:e193. doi: 10.1128/mBio.00193-112

Beabout, K., Hammerstrom, T. G., Perez, A. M., Magalhães, B. F., Prater, A. G., Clements, T. P., et al. (2015). The ribosomal S10 protein is a general target for decreased tigecycline susceptibility. *Antimicrob. Agents Chemother.* 59, 5561–5566. doi: 10.1128/AAC.00547-15

Bender, J. K., Kalmbach, A., Fleige, C., Klare, I., Fuchs, S., and Werner, G. (2016). Population structure and acquisition of the vanB resistance determinant in German clinical isolates of Enterococcus *faecium* ST192. *Sci. Rep.* 6:21847. doi: 10.1038/srep21847

Bourdon, N., Fines-Guyon, M., Thiolet, J.-M., Maugat, S., Coignard, B., Leclercq, R., et al. (2011). Changing trends in vancomycin-resistant enterococci in French hospitals, 2001-08. *J. Antimicrob. Chemother.* 66, 713–721. doi: 10.1093/jac/dkq524

Boyd, D. A., Willey, B. M., Fawcett, D., Gillani, N., and Mulvey, M. R. (2008). Molecular characterization of *Enterococcus faecalis* N06-0364 with low-level vancomycin resistance harboring a novel D-Ala-D-Ser gene cluster, vanL. *Antimicrob. Agents Chemother.* 52, 2667–2672. doi: 10.1128/AAC.01516-07

Brandl, K., Plitas, G., Mihu, C. N., Ubeda, C., Jia, T., Fleisher, M., et al. (2008). Vancomycin-resistant enterococci exploit antibiotic-induced innate immune deficits. *Nature* 455, 804–807. doi: 10.1038/nature07250

Brodrick, H. J., Raven, K. E., Harrison, E. M., Blane, B., Reuter, S., Török, M. E., et al. (2016). Whole-genome sequencing reveals transmission of vancomycin-resistant *Enterococcus faecium* in a healthcare network. *Genome Med.* 8:4. doi: 10.1186/s13073-015-0259-7

Brötz-Oesterhelt, H., Beyer, D., Kroll, H.-P., Endermann, R., Ladel, C., Schroeder, W., et al. (2005). Dysregulation of bacterial proteolytic machinery by a new class of antibiotics. *Nat. Med.* 11, 1082–1087. doi: 10.1038/nm1306

Byappanahalli, M. N., Nevers, M. B., Korajkic, A., Staley, Z. R., and Harwood, V. J. (2012). Enterococci in the environment. *Microbiol. Mol. Biol. Rev.* 76, 685–706. doi: 10.1128/MMBR.00023-12

Cai, J., Wang, Y., Schwarz, S., Lv, H., Li, Y., Liao, K., et al. (2015). Enterococcal isolates carrying the novel oxazolidinone resistance gene optrA from hospitals in Zhejiang, Guangdong, and Henan, China, 2010-2014. *Clin. Microbiol. Infect. Off. Publ. Eur. Soc. Clin. Microbiol. Infect. Dis.* 21, e1–e4. doi: 10.1016/j.cmi.2015.08.007

Cattoir, V., and Giard, J.-C. (2014). Antibiotic resistance in *Enterococcus faecium* clinical isolates. *Expert Rev. Anti Infect. Ther.* 12, 239–248. doi: 10.1586/14787210.2014.870886

Cattoir, V., Isnard, C., Cosquer, T., Odhiambo, A., Bucquet, F., Guérin, F., et al. (2015). Genomic analysis of reduced susceptibility to tigecycline in *Enterococcus faecium*. *Antimicrob. Agents Chemother.* 59, 239–244. doi: 10.1128/AAC.04174-14

Chen, C., Sun, J., Guo, Y., Lin, D., Guo, Q., Hu, F., et al. (2015). High Prevalence of vanM in Vancomycin-Resistant *Enterococcus faecium* Isolates from Shanghai, China. *Antimicrob. Agents Chemother.* 59, 7795–7798. doi: 10.1128/AAC.01732-15

Chen, H., Wu, W., Ni, M., Liu, Y., Zhang, J., Xia, F., et al. (2013). Linezolid-resistant clinical isolates of enterococci and *Staphylococcus cohnii* from a multicentre study in China: molecular epidemiology and resistance mechanisms. *Int. J. Antimicrob. Agents* 42, 317–321. doi: 10.1016/j.ijantimicag.2013.06.008

Chow, J. W. (2000). Aminoglycoside resistance in enterococci. *Clin. Infect. Dis. Off. Publ. Infect. Dis. Soc. Am.* 31, 586–589. doi: 10.1086/313949

Clewell, D. B., Weaver, K. E., Dunny, G. M., Coque, T. M., Francia, M. V., and Hayes, F. (2014). "Extrachromosomal and mobile elements in enterococci: transmission, maintenance, and epidemiology," in *Enterococci: From Commensals to Leading Causes of Drug Resistant Infection*, eds M. S. Gilmore, D. B. Clewell, Y. Ike, and N. Shankar (Boston: Massachusetts Eye and Ear Infirmary). Available at: http://www.ncbi.nlm.nih.gov/books/NBK190430/ [Accessed March 19, 2016].

Colca, J. R., McDonald, W. G., Waldon, D. J., Thomasco, L. M., Gadwood, R. C., Lund, E. T., et al. (2003). Cross-linking in the living cell locates the site of action of oxazolidinone antibiotics. *J. Biol. Chem.* 278, 21972–21979. doi: 10.1074/jbc.M302109200

Coombs, G. W., Pearson, J. C., Daley, D. A., Le, T., Robinson, O. J., Gottlieb, T., et al. (2014). Molecular epidemiology of enterococcal bacteremia in Australia. *J. Clin. Microbiol.* 52, 897–905. doi: 10.1128/JCM.03286-13

Corander, J., Connor, T. R., O'Dwyer, C. A., Kroll, J. S., and Hanage, W. P. (2012). Population structure in the Neisseria, and the biological significance of fuzzy species. *J. R. Soc. Interface R. Soc.* 9, 1208–1215. doi: 10.1098/rsif.2011.0601

Cordina, C., Hill, R., Deshpande, A., Hood, J., and Inkster, T. (2012). Tigecycline-resistant *Enterococcus faecalis* associated with omeprazole use in a surgical patient. *J. Antimicrob. Chemother.* 67, 1806–1807. doi: 10.1093/jac/dks122

Costa, Y., Galimand, M., Leclercq, R., Duval, J., and Courvalin, P. (1993). Characterization of the chromosomal aac(6')-Ii gene specific for *Enterococcus faecium*. *Antimicrob. Agents Chemother.* 37, 1896–1903. doi: 10.1128/AAC.37.9.1896

Courvalin, P. (2006). Vancomycin resistance in gram-positive cocci. *Clin. Infect. Dis. Off. Publ. Infect. Dis. Soc. Am.* 42(Suppl. 1), S25–S34. doi: 10.1086/491711

Creti, R., Imperi, M., Bertuccini, L., Fabretti, F., Orefici, G., Di Rosa, R., et al. (2004). Survey for virulence determinants among *Enterococcus faecalis* isolated from different sources. *J. Med. Microbiol.* 53, 13–20. doi: 10.1099/jmm.0.05353-0

Crotty, M. P., Krekel, T., Burnham, C. A., and Ritchie, D. J. (2016). New gram-positive agents: the next generation of oxazolidinones and lipoglycopeptides. *J. Clin. Microbiol.* doi: 10.1128/JCM.03395-15 [Epub ahead of print].

Dahl, K. H., Lundblad, E. W., Rokenes, T. P., Olsvik, O., and Sundsfjord, A. (2000). Genetic linkage of the vanB2 gene cluster to Tn5382 in vancomycin-resistant enterococci and characterization of two novel insertion sequences. *Microbiol. Read. Engl.* 146(Pt 6), 1469–1479. doi: 10.1099/00221287-146-6-1469

Davlieva, M., Zhang, W., Arias, C. A., and Shamoo, Y. (2013). Biochemical characterization of cardiolipin synthase mutations associated with daptomycin resistance in enterococci. *Antimicrob. Agents Chemother.* 57, 289–296. doi: 10.1128/AAC.01743-12

de Been, M., Pinholt, M., Top, J., Bletz, S., Mellmann, A., van Schaik, W., et al. (2015). A core genome MLST scheme for high-resolution typing of *Enterococcus faecium*. *J. Clin. Microbiol.* 53, 3788–3797. doi: 10.1128/JCM.01946-15

de Been, M., van Schaik, W., Cheng, L., Corander, J., and Willems, R. J. (2013). Recent recombination events in the core genome are associated with adaptive evolution in *Enterococcus faecium*. *Genome Biol. Evol.* 5, 1524–1535. doi: 10.1093/gbe/evt111

de Kraker, M. E. A., Jarlier, V., Monen, J. C. M., Heuer, O. E., van de Sande, N., and Grundmann, H. (2013). The changing epidemiology of bacteraemias in Europe: trends from the European Antimicrobial Resistance Surveillance System. *Clin. Microbiol. Infect. Off. Publ. Eur. Soc. Clin. Microbiol. Infect. Dis.* 19, 860–868. doi: 10.1111/1469-0691.12028

De Vecchi, E., Sitia, S., Romanò, C. L., Ricci, C., Mattina, R., and Drago, L. (2013). Aetiology and antibiotic resistance patterns of urinary tract infections in the elderly: a 6-month study. *J. Med. Microbiol.* 62, 859–863. doi: 10.1099/jmm.0.056945-0

Devriese, L. A., Colque, J. I. C., De Herdt, P., and Haesebrouck, F. (1992a). Identification and composition of the tonsillar and anal enterococcal and streptococcal flora of dogs and cats. *J. Appl. Microbiol.* 73, 421–425. doi: 10.1111/j.1365-2672.1992.tb04998.x

Devriese, L. A., Hommez, J., Pot, B., and Haesebrouck, F. (1994). Identification and composition of the streptococcal and enterococcal flora of tonsils, intestines and faeces of pigs. *J. Appl. Bacteriol.* 77, 31–36. doi: 10.1111/j.1365-2672.1994.tb03040.x

Devriese, L. A., Hommez, J., Wijfels, R., and Haesebrouck, F. (1991). Composition of the enterococcal and streptococcal intestinal flora of poultry. *J. Appl. Bacteriol.* 71, 46–50. doi: 10.1111/j.1365-2672.1991.tb04585.x

Devriese, L. A., Ieven, M., Goossens, H., Vandamme, P., Pot, B., Hommez, J., et al. (1996). Presence of vancomycin-resistant enterococci in farm and pet animals. *Antimicrob. Agents Chemother.* 40, 2285–2287.

Devriese, L. A., Kerckhove, A. V. D., Kilpper-Bälz, R., and Schleifer, K. H. (1987). Characterization and Identification of Enterococcus Species Isolated from the Intestines of Animals. *Int. J. Syst. Bacteriol.* 37, 257–259. doi: 10.1099/00207713-37-3-257

Devriese, L. A., Laurier, L., De Herdt, P., and Haesebrouck, F. (1992b). Enterococcal and streptococcal species isolated from faeces of calves, young cattle and dairy cows. *J. Appl. Bacteriol.* 72, 29–31. doi: 10.1111/j.1365-2672.1992.tb05182.x

Diaz, L., Kiratisin, P., Mendes, R. E., Panesso, D., Singh, K. V., and Arias, C. A. (2012). Transferable plasmid-mediated resistance to linezolid due to cfr in a human clinical isolate of *Enterococcus faecalis*. *Antimicrob. Agents Chemother.* 56, 3917–3922. doi: 10.1128/AAC.00419-12

Didelot, X., and Falush, D. (2007). Inference of bacterial microevolution using multilocus sequence data. *Genetics* 175, 1251–1266. doi: 10.1534/genetics.106.063305

Domingo, M.-C., Huletsky, A., Giroux, R., Picard, F. J., and Bergeron, M. G. (2007). vanD and vanG-Like Gene Clusters in a Ruminococcus Species Isolated from Human Bowel Flora. *Antimicrob. Agents Chemother.* 51, 4111–4117. doi: 10.1128/AAC.00584-07

Donskey, C. J., Chowdhry, T. K., Hecker, M. T., Hoyen, C. K., Hanrahan, J. A., Hujer, A. M., et al. (2000). Effect of antibiotic therapy on the density of vancomycin-resistant enterococci in the stool of colonized patients. *N. Engl. J. Med.* 343, 1925–1932. doi: 10.1056/NEJM200012283432604

Duerkop, B. A., Clements, C. V., Rollins, D., Rodrigues, J. L. M., and Hooper, L. V. (2012). A composite bacteriophage alters colonization by an intestinal commensal bacterium. *Proc. Natl. Acad. Sci. U.S.A.* 109, 17621–17626. doi: 10.1073/pnas.1206136109

Eaton, T. J., and Gasson, M. J. (2001). Molecular screening of Enterococcus virulence determinants and potential for genetic exchange between food and medical isolates. *Appl. Environ. Microbiol.* 67, 1628–1635. doi: 10.1128/AEM.67.4.1628-1635.2001

Edelsberg, J., Weycker, D., Barron, R., Li, X., Wu, H., Oster, G., et al. (2014). Prevalence of antibiotic resistance in US hospitals. *Diagn. Microbiol. Infect. Dis.* 78, 255–262. doi: 10.1016/j.diagmicrobio.2013.11.011

Feil, E. J., Li, B. C., Aanensen, D. M., Hanage, W. P., and Spratt, B. G. (2004). eBURST: inferring patterns of evolutionary descent among clusters of related bacterial genotypes from multilocus sequence typing data. *J. Bacteriol.* 186, 1518–1530. doi: 10.1128/JB.186.5.1518-1530.2004

Fernández-Guerrero, M. L., Herrero, L., Bellver, M., Gadea, I., Roblas, R. F., and de Górgolas, M. (2002). Nosocomial enterococcal endocarditis: a serious hazard for hospitalized patients with enterococcal bacteraemia. *J. Intern. Med.* 252, 510–515. doi: 10.1046/j.1365-2796.2002.01061.x

Fiedler, S., Bender, J. K., Klare, I., Halbedel, S., Grohmann, E., Szewzyk, U., et al. (2015). Tigecycline resistance in clinical isolates of *Enterococcus faecium* is mediated by an upregulation of plasmid-encoded tetracycline determinants tet(L) and tet(M). *J. Antimicrob. Chemother.* 71, 871–881. doi: 10.1093/jac/dkv420

Francisco, A. P., Bugalho, M., Ramirez, M., and Carriço, J. A. (2009). Global optimal eBURST analysis of multilocus typing data using a graphic matroid approach. *BMC Bioinformatics* 10:152. doi: 10.1186/1471-2105-10-152

Freitas, A. R., Novais, C., Correia, R., Monteiro, M., Coque, T. M., and Peixe, L. (2011). Non-susceptibility to tigecycline in enterococci from hospitalised patients, food products and community sources. *Int. J. Antimicrob. Agents* 38, 174–176. doi: 10.1016/j.ijantimicag.2011.04.014

Freitas, A. R., Novais, C., Ruiz-Garbajosa, P., Coque, T. M., and Peixe, L. (2009). Clonal expansion within clonal complex 2 and spread of vancomycin-resistant plasmids among different genetic lineages of *Enterococcus faecalis* from Portugal. *J. Antimicrob. Chemother.* 63, 1104–1111. doi: 10.1093/jac/dkp103

Furmanek-Blaszk, B., and Sektas, M. (2015). The SfaNI restriction-modification system from *Enterococcus faecalis* NEB215 is located on a putative mobile genetic element. *FEMS Microbiol. Lett.* 362:fnv028. doi: 10.1093/femsle/fnv028

Gaca, A. O., and Gilmore, M. S. (2016). Killing of VRE *Enterococcus faecalis* by commensal strains: evidence for evolution and accumulation of mobile elements in the absence of competition. *Gut Microbes* 7, 90–96. doi: 10.1080/19490976.2015.1127482

Galimand, M., Schmitt, E., Panvert, M., Desmolaize, B., Douthwaite, S., Mechulam, Y., et al. (2011). Intrinsic resistance to aminoglycosides in *Enterococcus faecium* is conferred by the 16S rRNA m5C1404-specific methyltransferase EfmM. *RNA N. Y. N* 17, 251–262. doi: 10.1261/rna.2233511

Galloway-Peña, J., Roh, J. H., Latorre, M., Qin, X., and Murray, B. E. (2012). Genomic and SNP analyses demonstrate a distant separation of the hospital and community-associated clades of *Enterococcus faecium*. *PLoS ONE* 7:e30187. doi: 10.1371/journal.pone.0030187

Galloway-Peña, J. R., Nallapareddy, S. R., Arias, C. A., Eliopoulos, G. M., and Murray, B. E. (2009). Analysis of clonality and antibiotic resistance among early clinical isolates of *Enterococcus faecium* in the United States. *J. Infect. Dis.* 200, 1566–1573. doi: 10.1086/644790

Galloway-Peña, J. R., Rice, L. B., and Murray, B. E. (2011). Analysis of PBP5 of early U.S. isolates of *Enterococcus faecium*: sequence variation alone does not explain increasing ampicillin resistance over time. *Antimicrob. Agents Chemother.* 55, 3272–3277. doi: 10.1128/AAC.00099-11

Gastmeier, P., Schröder, C., Behnke, M., Meyer, E., and Geffers, C. (2014). Dramatic increase in vancomycin-resistant enterococci in Germany. *J. Antimicrob. Chemother.* 69, 1660–1664. doi: 10.1093/jac/dku035

Gilmore, M. S., Lebreton, F., and van Schaik, W. (2013). Genomic transition of enterococci from gut commensals to leading causes of multidrug-resistant hospital infection in the antibiotic era. *Curr. Opin. Microbiol.* 16, 10–16. doi: 10.1016/j.mib.2013.01.006

Grayson, M. L., Eliopoulos, G. M., Wennersten, C. B., Ruoff, K. L., De Girolami, P. C., Ferraro, M. J., et al. (1991). Increasing resistance to beta-lactam antibiotics among clinical isolates of *Enterococcus faecium*: a 22-year review at one institution. *Antimicrob. Agents Chemother.* 35, 2180–2184. doi: 10.1128/AAC.35.11.2180

Guardabassi, L., Perichon, B., van Heijenoort, J., Blanot, D., and Courvalin, P. (2005). Glycopeptide resistance vana operons in paenibacillus strains isolated from soil. *Antimicrob. Agents Chemother.* 49, 4227–4233. doi: 10.1128/AAC.49.10.4227-4233.2005

Hanrahan, J., Hoyen, C., and Rice, L. B. (2000). Geographic distribution of a large mobile element that transfers ampicillin and vancomycin resistance between *Enterococcus faecium* strains. *Antimicrob. Agents Chemother.* 44, 1349–1351. doi: 10.1128/AAC.44.5.1349-1351.2000

Hegstad, K., Mikalsen, T., Coque, T. M., Werner, G., and Sundsfjord, A. (2010). Mobile genetic elements and their contribution to the emergence of antimicrobial resistant *Enterococcus faecalis* and *Enterococcus faecium*. *Clin. Microbiol. Infect. Off. Publ. Eur. Soc. Clin. Microbiol. Infect. Dis.* 16, 541–554. doi: 10.1111/j.1469-0691.2010.03226.x

Heikens, E., Bonten, M. J., and Willems, R. J. (2007). Enterococcal Surface Protein Esp Is Important for Biofilm Formation of *Enterococcus faecium* E1162. *J. Bacteriol.* 189, 8233–8240. doi: 10.1128/JB.01205-07

Heikens, E., Singh, K. V., Jacques-Palaz, K. D., van Luit-Asbroek, M., Oostdijk, E. A. N., Bonten, M. J. M., et al. (2011). Contribution of the enterococcal surface protein Esp to pathogenesis of *Enterococcus faecium* endocarditis. *Microbes Infect. Inst. Pasteur* 13, 1185–1190. doi: 10.1016/j.micinf.2011.08.006

Heikens, E., van Schaik, W., Leavis, H. L., Bonten, M. J., and Willems, R. J. (2008). Identification of a novel genomic island specific to hospital-acquired clonal complex 17 *Enterococcus faecium* isolates. *Appl. Environ. Microbiol.* 74, 7094–7097. doi: 10.1128/AEM.01378-08

Hendrickx, A. P., Willems, R. J., Bonten, M. J., and van Schaik, W. (2009). LPxTG surface proteins of enterococci. *Trends Microbiol.* 17, 423–430. doi: 10.1016/j.tim.2009.06.004

Hendrickx, A. P. A., van Schaik, W., and Willems, R. J. L. (2013). The cell wall architecture of *Enterococcus faecium*: from resistance to pathogenesis. *Fut. Microbiol.* 8, 993–1010. doi: 10.2217/fmb.13.66

Hoban, D. J., Reinert, R. R., Bouchillon, S. K., and Dowzicky, M. J. (2015). Global in vitro activity of tigecycline and comparator agents: tigecycline Evaluation and Surveillance Trial 2004-2013. *Ann. Clin. Microbiol. Antimicrob.* 14:27. doi: 10.1186/s12941-015-0085-1

Homan, W. L., Tribe, D., Poznanski, S., Li, M., Hogg, G., Spalburg, E., et al. (2002). Multilocus sequence typing scheme for *Enterococcus faecium*. *J. Clin. Microbiol.* 40, 1963–1971. doi: 10.1128/JCM.40.6.1963-1971.2002

Horvath, P., and Barrangou, R. (2010). CRISPR/Cas, the immune system of bacteria and archaea. *Science* 327, 167–170. doi: 10.1126/science.1179555

Howden, B. P., Holt, K. E., Lam, M. M. C., Seemann, T., Ballard, S., Coombs, G. W., et al. (2013). Genomic insights to control the emergence of vancomycin-resistant enterococci. *mBio* 4, e412–e413. doi: 10.1128/mBio.00412-13

Humphries, R. M., Pollett, S., and Sakoulas, G. (2013). A current perspective on daptomycin for the clinical microbiologist. *Clin. Microbiol. Rev.* 26, 759–780. doi: 10.1128/CMR.00030-13

Jensen, L. B., Ahrens, P., Dons, L., Jones, R. N., Hammerum, A. M., and Aarestrup, F. M. (1998). Molecular analysis of Tn1546 in *Enterococcus faecium* isolated from animals and humans. *J. Clin. Microbiol.* 36, 437–442.

Jones, R. N., Ross, J. E., Castanheira, M., and Mendes, R. E. (2008). United States resistance surveillance results for linezolid (LEADER Program for 2007). *Diagn. Microbiol. Infect. Dis.* 62, 416–426. doi: 10.1016/j.diagmicrobio.2008.10.010

Jones, R. N., Sader, H. S., and Flamm, R. K. (2013). Update of dalbavancin spectrum and potency in the USA: report from the SENTRY Antimicrobial Surveillance Program (2011). *Diagn. Microbiol. Infect. Dis.* 75, 304–307. doi: 10.1016/j.diagmicrobio.2012.11.024

Jordan, S., Junker, A., Helmann, J. D., and Mascher, T. (2006). Regulation of LiaRS-dependent gene expression in *Bacillus subtilis*: identification of inhibitor proteins, regulator binding sites, and target genes of a conserved cell envelope stress-sensing two-component system. *J. Bacteriol.* 188, 5153–5166. doi: 10.1128/JB.00310-06

Kamboj, M., Blair, R., Bell, N., Sun, J., Eagan, J., and Sepkowitz, K. (2014). What is the source of bloodstream infection due to vancomycin-resistant enterococci in persons with mucosal barrier injury? *Infect. Control Hosp. Epidemiol.* 35, 99–101. doi: 10.1086/674406

Karlowsky, J. A., Nichol, K., and Zhanel, G. G. (2015). Telavancin: mechanisms of action, in vitro activity, and mechanisms of resistance. *Clin. Infect. Dis. Off. Publ. Infect. Dis. Soc. Am.* 61(Suppl. 2), S58–S68. doi: 10.1093/cid/civ534

Kawalec, M., Pietras, Z., Daniłowicz, E., Jakubczak, A., Gniadkowski, M., Hryniewicz, W., et al. (2007). Clonal structure of *Enterococcus faecalis* isolated from Polish hospitals: characterization of epidemic clones. *J. Clin. Microbiol.* 45, 147–153. doi: 10.1128/JCM.01704-06

Kelesidis, T., Tewhey, R., and Humphries, R. M. (2013). Evolution of high-level daptomycin resistance in *Enterococcus faecium* during daptomycin therapy is associated with limited mutations in the bacterial genome. *J. Antimicrob. Chemother.* 68, 1926–1928. doi: 10.1093/jac/dkt117

Kim, E. B., and Marco, M. L. (2013). Non-clinical and clinical *Enterococcus faecium* but not *Enterococcus faecalis* have distinct structural and functional genomic features. *Appl. Environ. Microbiol.* 80, 154–165.

Klare, I., Badstübner, D., Konstabel, C., Böhme, G., Claus, H., and Witte, W. (1999). Decreased incidence of VanA-type vancomycin-resistant enterococci isolated from poultry meat and from fecal samples of humans in the community after discontinuation of avoparcin usage in animal husbandry. *Microb. Drug Resist. Larchmt. N* 5, 45–52. doi: 10.1089/mdr.1999.5.45

Klare, I., Heier, H., Claus, H., Böhme, G., Marin, S., Seltmann, G., et al. (1995). *Enterococcus faecium* strains with vanA-mediated high-level glycopeptide resistance isolated from animal foodstuffs and fecal samples of humans in the community. *Microb. Drug Resist. Larchmt. N* 1, 265–272. doi: 10.1089/mdr.1995.1.265

Kuch, A., Willems, R. J. L., Werner, G., Coque, T. M., Hammerum, A. M., Sundsfjord, A., et al. (2012). Insight into antimicrobial susceptibility and population structure of contemporary human *Enterococcus faecalis* isolates from Europe. *J. Antimicrob. Chemother.* 67, 551–558. doi: 10.1093/jac/dkr544

La Rosa, S. L., Montealegre, M. C., Singh, K. V., and Murray, B. E. (2016). *Enterococcus faecalis* Ebp Pili are important for cell-cell aggregation and intraspecies gene transfer. *Microbiol. Read. Engl.* 162, 798–802. doi: 10.1099/mic.0.000276

La Rosa, S. L., Snipen, L.-G., Murray, B. E., Willems, R. J. L., Gilmore, M. S., Diep, D. B., et al. (2015). A genomic virulence reference map of *Enterococcus faecalis* reveals an important contribution of phage03-like elements in nosocomial genetic lineages to pathogenicity in a *Caenorhabditis elegans* infection model. *Infect. Immun.* 83, 2156–2167. doi: 10.1128/IAI.02801-14

Leavis, H., Top, J., Shankar, N., Borgen, K., Bonten, M., Van Embden, J., et al. (2004). a novel putative enterococcal pathogenicity island linked to the esp virulence gene of *Enterococcus faecium* and associated with epidemicity. *J. Bacteriol.* 186, 672–682. doi: 10.1128/JB.186.3.672-682.2004

Leavis, H. L., Willems, R. J., Top, J., Spalburg, E., Mascini, E. M., Fluit, A. C., et al. (2003). Epidemic and nonepidemic multidrug-resistant *Enterococcus faecium*. *Emerg. Infect. Dis.* 9, 1108–1115. doi: 10.3201/eid0909.020383

Lebreton, F., Depardieu, F., Bourdon, N., Fines-Guyon, M., Berger, P., Camiade, S., et al. (2011). D-Ala-d-Ser VanN-type transferable vancomycin resistance in *Enterococcus faecium*. *Antimicrob. Agents Chemother.* 55, 4606–4612. doi: 10.1128/AAC.00714-11

Lebreton, F., van Schaik, W., McGuire, A. M., Godfrey, P., Griggs, A., Mazumdar, V., et al. (2013). Emergence of epidemic multidrug-resistant *Enterococcus faecium* from animal and commensal strains. *mBio* 4:e534. doi: 10.1128/mBio.00534-13

Leclercq, R., Derlot, E., Duval, J., and Courvalin, P. (1988). Plasmid-mediated resistance to vancomycin and teicoplanin in *Enterococcus faecium*. *N. Engl. J. Med.* 319, 157–161. doi: 10.1056/NEJM1988072131 90307

Leclercq, R., Oberlé, K., Galopin, S., Cattoir, V., Budzinski, H., and Petit, F. (2013). Changes in enterococcal populations and related antibiotic resistance along a medical center-wastewater treatment plant-river continuum. *Appl. Environ. Microbiol.* 79, 2428–2434. doi: 10.1128/AEM. 03586-12

Leendertse, M., Heikens, E., Wijnands, L. M., van Luit-Asbroek, M., Teske, G. J., Roelofs, J. J., et al. (2009). Enterococcal surface protein transiently aggravates *Enterococcus faecium*-induced urinary tract infection in mice. *J. Infect. Dis.* 200, 1162–1165. doi: 10.1086/605609

Lellek, H., Franke, G. C., Ruckert, C., Wolters, M., Wolschke, C., Christner, M., et al. (2015). Emergence of daptomycin non-susceptibility in colonizing vancomycin-resistant *Enterococcus faecium* isolates during daptomycin therapy. *Int. J. Med. Microbiol. IJMM* 305, 902–909. doi: 10.1016/j.ijmm.2015.09.005

Ling, L. L., Schneider, T., Peoples, A. J., Spoering, A. L., Engels, I., Conlon, B. P., et al. (2015). A new antibiotic kills pathogens without detectable resistance. *Nature* 517, 455–459. doi: 10.1038/nature 14098

Lübbert, C., Rodloff, A. C., and Hamed, K. (2015). Real-world treatment of enterococcal infections with daptomycin: insights from a large european registry (EU-CORE). *Infect. Dis. Ther.* 4, 259–271. doi: 10.1007/s40121-015-0072-z

Manson, J. M., Hancock, L. E., and Gilmore, M. S. (2010). Mechanism of chromosomal transfer of *Enterococcus faecalis* pathogenicity island, capsule, antimicrobial resistance, and other traits. *Proc. Natl. Acad. Sci. U.S.A.* 107, 12269–12274. doi: 10.1073/pnas.1000139107

Martin, J. D., and Mundt, J. O. (1972). Enterococci in insects. *Appl. Microbiol.* 24, 575–580.

Matos, R. C., Lapaque, N., Rigottier-Gois, L., Debarbieux, L., Meylheuc, T., Gonzalez-Zorn, B., et al. (2013). *Enterococcus faecalis* prophage dynamics and contributions to pathogenic traits. *PLoS Genet.* 9:e1003539. doi: 10.1371/journal.pgen.1003539

McBride, S. M., Fischetti, V. A., Leblanc, D. J., Moellering, R. C. Jr., and Gilmore, M. S. (2007). Genetic diversity among *Enterococcus faecalis*. *PLoS ONE* 2:e582. doi: 10.1371/journal.pone.0000582

Miller, W. R., Munita, J. M., and Arias, C. A. (2014). Mechanisms of antibiotic resistance in enterococci. *Expert Rev. Anti Infect. Ther.* 12, 1221–1236. doi: 10.1586/14787210.2014.956092

Montealegre, M. C., Singh, K. V., and Murray, B. E. (2015). Gastrointestinal tract colonization dynamics by different *Enterococcus faecium* clades. *J. Infect. Dis.* doi: 10.1093/infdis/jiv597 [Epub ahead of print].

Montero, C. I., Stock, F., and Murray, P. R. (2008). Mechanisms of resistance to daptomycin in *Enterococcus faecium*. *Antimicrob. Agents Chemother.* 52, 1167–1170. doi: 10.1128/AAC.00774-07

Morris, J. J., Lenski, R. E., and Zinser, E. R. (2012). The Black Queen Hypothesis: evolution of dependencies through adaptive gene loss. *mBio* 3:e00036. doi: 10.1128/mBio.00036-12

Morrison, D., Woodford, N., Barrett, S. P., Sisson, P., and Cookson, B. D. (1999). DNA banding pattern polymorphism in vancomycin-resistant *Enterococcus faecium* and criteria for defining strains. *J. Clin. Microbiol.* 37, 1084–1091.

Mundt, J. O. (1963a). Occurrence of enterococci in animals in a wild environment. *Appl. Microbiol.* 11, 136–140.

Mundt, J. O. (1963b). Occurrence of enterococci on plants in a wild environment. *Appl. Microbiol.* 11, 141–144.

Munita, J. M., Mishra, N. N., Alvarez, D., Tran, T. T., Diaz, L., Panesso, D., et al. (2014). Failure of high-dose daptomycin for bacteremia caused by daptomycin-susceptible *Enterococcus faecium* harboring LiaSR substitutions. *Clin. Infect. Dis. Off. Publ. Infect. Dis. Soc. Am.* 59, 1277–1280. doi: 10.1093/cid/ciu642

Munita, J. M., Panesso, D., Diaz, L., Tran, T. T., Reyes, J., Wanger, A., et al. (2012). Correlation between Mutations in liaFSR of *Enterococcus faecium* and MIC of daptomycin: revisiting daptomycin breakpoints. *Antimicrob. Agents Chemother.* 56, 4354–4359. doi: 10.1128/AAC.00509-12

National Nosocomial Infections Surveillance System (2004). National Nosocomial Infections Surveillance (NNIS) System Report, data summary from January 1992 through June 2004, issued October 2004. *Am. J. Infect. Control* 32, 470–485. doi: 10.1016/S0196655304005425

Neelakanta, A., Sharma, S., Kesani, V. P., Salim, M., Pervaiz, A., Aftab, N., et al. (2015). Impact of changes in the NHSN catheter-associated urinary tract infection (CAUTI) surveillance criteria on the frequency and epidemiology of CAUTI in intensive care units (ICUs). *Infect. Control Hosp. Epidemiol.* 36, 346–349. doi: 10.1017/ice.2014.67

Niebel, M., Quick, J., Prieto, A. M. G., Hill, R. L. R., Pike, R., Huber, D., et al. (2015). Deletions in a ribosomal protein-coding gene are associated with tigecycline resistance in *Enterococcus faecium*. *Int. J. Antimicrob. Agents* 46, 572–575. doi: 10.1016/j.ijantimicag.2015.07.009

Numminen, E., Gutmann, M., Shubin, M., Marttinen, P., Méric, G., van Schaik, W., et al. (2016). The impact of host metapopulation structure on the population genetics of colonizing bacteria. *J. Theor. Biol.* 396, 53–62. doi: 10.1016/j.jtbi.2016.02.019

Paganelli, F. L., Huebner, J., Singh, K. V., Zhang, X., van Schaik, W., Wobser, D., et al. (2016). Genome-wide screening identifies PTS permease BepA to be involved in *Enterococcus faecium* endocarditis and biofilm formation. *J. Infect. Dis.* doi: 10.1093/infdis/jiw108 [Epub ahead of print].

Paganelli, F. L., Willems, R. J. L., Jansen, P., Hendrickx, A., Zhang, X., Bonten, M. J. M., et al. (2013). *Enterococcus faecium* biofilm formation: identification of major autolysin AtlAEfm, associated Acm surface localization, and atlaefm-independent extracellular DNA release. *mBio* 4:e154. doi: 10.1128/mBio.00154-13

Palmer, K. L., Daniel, A., Hardy, C., Silverman, J., and Gilmore, M. S. (2011). Genetic basis for daptomycin resistance in enterococci. *Antimicrob. Agents Chemother.* 55, 3345–3356. doi: 10.1128/AAC.00207-11

Palmer, K. L., Godfrey, P., Griggs, A., Kos, V. N., Zucker, J., Desjardins, C., et al. (2012). Comparative genomics of enterococci: variation in *Enterococcus faecalis*, clade structure in *E. faecium*, and defining characteristics of *E. gallinarum* and *E. casseliflavus*. *mBio* 3:e318. doi: 10.1128/mBio.00318-11

Palmer, K. L., Kos, V. N., and Gilmore, M. S. (2010). Horizontal gene transfer and the genomics of enterococcal antibiotic resistance. *Curr. Opin. Microbiol.* 13, 632–639. doi: 10.1016/j.mib.2010.08.004

Parte, A. C. (2014). LPSN–list of prokaryotic names with standing in nomenclature. *Nucleic Acids Res.* 42, D613–D616. doi: 10.1093/nar/gkt1111

Patterson, J. E., Sweeney, A. H., Simms, M., Carley, N., Mangi, R., Sabetta, J., et al. (1995). An analysis of 110 serious enterococcal infections. Epidemiology, antibiotic susceptibility, and outcome. *Medicine (Baltimore)* 74, 191–200. doi: 10.1097/00005792-199507000-00003

Pinholt, M., Larner-Svensson, H., Littauer, P., Moser, C. E., Pedersen, M., Lemming, L. E., et al. (2015). Multiple hospital outbreaks of vanA *Enterococcus faecium* in Denmark, 2012–13, investigated by WGS, MLST and PFGE. *J. Antimicrob. Chemother.* 70, 2474–2482. doi: 10.1093/jac/dkv142

Polidori, M., Nuccorini, A., Tascini, C., Gemignani, G., Iapoce, R., Leonildi, A., et al. (2011). Vancomycin-resistant *Enterococcus faecium* (VRE) bacteremia in infective endocarditis successfully treated with combination daptomycin and tigecycline. *J. Chemother. Florence Italy* 23, 240–241. doi: 10.1179/joc.2011.23.4.240

Raven, K. E., Reuter, S., Goulouris, T., Reynolds, R., Russell, J. E., Brown, N. M., et al. (2016). Genome-based characterization of hospital-adapted Enterococcus faecalis lineages. *Nat. Microbiol.* 15033. doi: 10.1038/nmicrobiol.2015.33

Reuter, S., Ellington, M. J., Cartwright, E. J. P., Köser, C. U., Török, M. E., Goulouris, T., et al. (2013). Rapid bacterial whole-genome sequencing to enhance diagnostic and public health microbiology. *JAMA Intern. Med.* 173, 1397–1404. doi: 10.1001/jamainternmed.2013.7734

Rice, L. B., Carias, L. L., Hutton-Thomas, R., Sifaoui, F., Gutmann, L., and Rudin, S. D. (2001). Penicillin-binding protein 5 and expression of ampicillin resistance in *Enterococcus faecium*. *Antimicrob. Agents Chemother.* 45, 1480–1486. doi: 10.1128/AAC.45.5.1480-1486.2001

Rice, L. B., Carias, L. L., Rudin, S., Lakticová, V., Wood, A., and Hutton-Thomas, R. (2005). *Enterococcus faecium* low-affinity pbp5 is a transferable determinant. *Antimicrob. Agents Chemother.* 49, 5007–5012. doi: 10.1128/AAC.49.12.5007-5012.2005

Ruiz-Garbajosa, P., Bonten, M. J. M., Robinson, D. A., Top, J., Nallapareddy, S. R., Torres, C., et al. (2006). Multilocus sequence typing scheme for *Enterococcus faecalis* reveals hospital-adapted genetic complexes in a background of high rates of recombination. *J. Clin. Microbiol.* 44, 2220–2228. doi: 10.1128/JCM.02596-05

Ruiz-Garbajosa, P., de Regt, M., Bonten, M., Baquero, F., Coque, T. M., Cantón, R., et al. (2012). High-density fecal *Enterococcus faecium* colonization in hospitalized patients is associated with the presence of the polyclonal subcluster CC17. *Eur. J. Clin. Microbiol. Infect. Dis. Off. Publ. Eur. Soc. Clin. Microbiol.* 31, 519–522. doi: 10.1007/s10096-011-1342-7

Ruiz-Garbajosa, P., del Campo, R., Coque, T. M., Asensio, A., Bonten, M., Willems, R., et al. (2009). Longer intestinal persistence of *Enterococcus faecalis* compared to *Enterococcus faecium* clones in intensive-care-unit patients. *J. Clin. Microbiol.* 47, 345–351. doi: 10.1128/JCM.01597-08

Sader, H. S., Farrell, D. J., Flamm, R. K., and Jones, R. N. (2014). Daptomycin activity tested against 164457 bacterial isolates from hospitalised patients: summary of 8 years of a Worldwide Surveillance Programme (2005-2012). *Int. J. Antimicrob. Agents* 43, 465–469. doi: 10.1016/j.ijantimicag.2014.01.018

Salipante, S. J., SenGupta, D. J., Cummings, L. A., Land, T. A., Hoogestraat, D. R., and Cookson, B. T. (2015). Application of whole-genome sequencing for bacterial strain typing in molecular epidemiology. *J. Clin. Microbiol.* 53, 1072–1079. doi: 10.1128/JCM.03385-14

Sarti, M., Campanile, F., Sabia, C., Santagati, M., Gargiulo, R., and Stefani, S. (2012). Polyclonal diffusion of beta-lactamase-producing *Enterococcus faecium*. *J. Clin. Microbiol.* 50, 169–172. doi: 10.1128/JCM.05640-11

Sava, I. G., Heikens, E., Kropec, A., Theilacker, C., Willems, R., and Huebner, J. (2010). Enterococcal surface protein contributes to persistence in the host but is not a target of opsonic and protective antibodies in *Enterococcus faecium* infection. *J. Med. Microbiol.* 59, 1001–1004. doi: 10.1099/jmm.0.020578-0

Scheetz, M. H., Knechtel, S. A., Malczynski, M., Postelnick, M. J., and Qi, C. (2008). Increasing incidence of linezolid-intermediate or -resistant, vancomycin-resistant *Enterococcus faecium* strains parallels increasing linezolid consumption. *Antimicrob. Agents Chemother.* 52, 2256–2259. doi: 10.1128/AAC.00070-08

Shen, J., Wang, Y., and Schwarz, S. (2013). Presence and dissemination of the multiresistance gene cfr in Gram-positive and Gram-negative bacteria. *J. Antimicrob. Chemother.* 68, 1697–1706. doi: 10.1093/jac/dkt092

Sievert, D. M., Ricks, P., Edwards, J. R., Schneider, A., Patel, J., Srinivasan, A., et al. (2013). Antimicrobial-resistant pathogens associated with healthcare-associated infections: summary of data reported to the National Healthcare Safety Network at the Centers for Disease Control and Prevention, 2009–2010. *Infect. Control Hosp. Epidemiol.* 34, 1–14. doi: 10.1086/668770

Silva-Del Toro, S. L., Greenwood-Quaintance, K. E., and Patel, R. (2016). In vitro activity of tedizolid against linezolid-resistant staphylococci and enterococci. *Diagn. Microbiol. Infect. Dis.* 85, 102–104. doi: 10.1016/j.diagmicrobio.2016.02.008

Simner, P. J., Adam, H., Baxter, M., McCracken, M., Golding, G., Karlowsky, J. A., et al. (2015). Epidemiology of vancomycin-resistant enterococci in Canadian hospitals (CANWARD study, 2007 to 2013). *Antimicrob. Agents Chemother.* 59, 4315–4317. doi: 10.1128/AAC.00384-15

Solheim, M., Aakra, A., Snipen, L. G., Brede, D. A., and Nes, I. F. (2009). Comparative genomics of *Enterococcus faecalis* from healthy Norwegian infants. *BMC Genomics* 10:194. doi: 10.1186/1471-2164-10-194

Stobberingh, E., Van Den Bogaard, A., London, N., Driessen, C., Top, J., and Willems, R. (1999). Enterococci with glycopeptide resistance in turkeys, turkey farmers, turkey slaughterers, and (sub)urban residents in the south of The Netherlands: evidence for transmission of vancomycin resistance from animals to humans? *Antimicrob. Agents Chemother.* 43, 2215–2221.

Sydnor, E. R. M., and Perl, T. M. (2011). Hospital epidemiology and infection control in acute-care settings. *Clin. Microbiol. Rev.* 24, 141–173. doi: 10.1128/CMR.00027-10

Tannock, G. W., and Cook, G. (2002). "Enterococci as members of the Intestinal microflora of humans," in *The Enterococci: Pathogenesis, Molecular Biology, and Antibiotic Resistance*, ed. M. S. Gilmore (Washington, DC: ASM Press), 101–132.

Tedim, A. P., Ruiz-Garbajosa, P., Corander, J., Rodríguez, C. M., Cantón, R., Willems, R. J., et al. (2015). Population biology of intestinal Enterococcus isolates from hospitalized and nonhospitalized individuals in different age groups. *Appl. Environ. Microbiol.* 81, 1820–1831. doi: 10.1128/AEM.03661-14

Thomas, J. C., Zhang, L., and Robinson, D. A. (2014). Differing lifestyles of Staphylococcus epidermidis as revealed through Bayesian clustering of multilocus sequence types. *Infect. Genet. Evol. J. Mol. Epidemiol. Evol. Genet. Infect. Dis.* 22, 257–264. doi: 10.1016/j.meegid.2013.06.020

Tomayko, J. F., Zscheck, K. K., Singh, K. V., and Murray, B. E. (1996). Comparison of the beta-lactamase gene cluster in clonally distinct strains of *Enterococcus faecalis*. *Antimicrob. Agents Chemother.* 40, 1170–1174.

Top, J., Sinnige, J. C., Majoor, E. A., Bonten, M. J., Willems, R. J., and van Schaik, W. (2011). The Recombinase IntA is required for excision of esp-containing ICEEfm1 in *Enterococcus faecium*. *J. Bacteriol.* 193, 1003–1006. doi: 10.1128/JB.00952-10

Top, J., Willems, R., and Bonten, M. (2008). Emergence of CC17 *Enterococcus faecium*: from commensal to hospital-adapted pathogen. *FEMS Immunol. Med. Microbiol.* 52, 297–308. doi: 10.1111/j.1574-695X.2008.00383.x

Tran, T. T., Panesso, D., Gao, H., Roh, J. H., Munita, J. M., Reyes, J., et al. (2013a). Whole-genome analysis of a daptomycin-susceptible *Enterococcus faecium* strain and its daptomycin-resistant variant arising during therapy. *Antimicrob. Agents Chemother.* 57, 261–268. doi: 10.1128/AAC.01454-12

Tran, T. T., Panesso, D., Mishra, N. N., Mileykovskaya, E., Guan, Z., Munita, J. M., et al. (2013b). Daptomycin-resistant *Enterococcus faecalis* diverts the antibiotic molecule from the division septum and remodels cell membrane phospholipids. *mBio* 4:e00281. doi: 10.1128/mBio.00281-13

Turner, K. M. E., Hanage, W. P., Fraser, C., Connor, T. R., and Spratt, B. G. (2007). Assessing the reliability of eBURST using simulated populations with known ancestry. *BMC Microbiol.* 7:30. doi: 10.1186/1471-2180-7-30

Ubeda, C., Taur, Y., Jenq, R. R., Equinda, M. J., Son, T., Samstein, M., et al. (2010). Vancomycin-resistant Enterococcus domination of intestinal microbiota is enabled by antibiotic treatment in mice and precedes bloodstream invasion in humans. *J. Clin. Invest.* 120, 4332–4341. doi: 10.1172/JCI43918

Uttley, A. H., Collins, C. H., Naidoo, J., and George, R. C. (1988). Vancomycin-resistant enterococci. *Lancet Lond. Engl.* 332, 57–58. doi: 10.1016/S0140-6736(88)91037-9

van den Bogaard, A. E., Bruinsma, N., and Stobberingh, E. E. (2000). The effect of banning avoparcin on VRE carriage in The Netherlands. *J. Antimicrob. Chemother.* 46, 146–148. doi: 10.1093/jac/46.1.146

van Hal, S. J., Ip, C. L. C., Ansari, M. A., Wilson, D. J., Espedido, B. A., Jensen, S. O., et al. (2016). Evolutionary dynamics of *Enterococcus faecium* reveals complex genomic relationships between isolates with independent emergence of vancomycin resistance. *Microb. Genomics* 2:48. doi: 10.1099/mgen.0.000048

van Schaik, W., Top, J., Riley, D. R., Boekhorst, J., Vrijenhoek, J. E., Schapendonk, C. M., et al. (2010). Pyrosequencing-based comparative genome analysis of the nosocomial pathogen *Enterococcus faecium* and identification of a large transferable pathogenicity island. *BMC Genomics* 11:239. doi: 10.1186/1471-2164-11-239

Van Tyne, D., and Gilmore, M. S. (2014). Friend turned foe: evolution of enterococcal virulence and antibiotic resistance. *Annu. Rev. Microbiol.* 68, 337–356. doi: 10.1146/annurev-micro-091213-113003

Via, S. (2012). Divergence hitchhiking and the spread of genomic isolation during ecological speciation-with-gene-flow. *Philos. Trans. R. Soc. Lond. B. Biol. Sci.* 367, 451–460. doi: 10.1098/rstb.2011.0260

Wang, Y., Lv, Y., Cai, J., Schwarz, S., Cui, L., Hu, Z., et al. (2015). A novel gene, optrA, that confers transferable resistance to oxazolidinones and phenicols and its presence in *Enterococcus faecalis* and *Enterococcus faecium* of human and animal origin. *J. Antimicrob. Chemother.* 70, 2182–2190. doi: 10.1093/jac/dkv116

Weigel, L. M., Donlan, R. M., Shin, D. H., Jensen, B., Clark, N. C., McDougal, L. K., et al. (2007). High-level vancomycin-resistant *Staphylococcus aureus* isolates associated with a polymicrobial biofilm. *Antimicrob. Agents Chemother.* 51, 231–238. doi: 10.1128/AAC.00576-06

Werner, G., Coque, T. M., Franz, C. M. A. P., Grohmann, E., Hegstad, K., Jensen, L., et al. (2013). Antibiotic resistant enterococci-tales of a drug resistance gene trafficker. *Int. J. Med. Microbiol. IJMM* 303, 360–379. doi: 10.1016/j.ijmm.2013.03.001

Werner, G., Gfrörer, S., Fleige, C., Witte, W., and Klare, I. (2008). Tigecycline-resistant *Enterococcus faecalis* strain isolated from a German intensive care unit patient. *J. Antimicrob. Chemother.* 61, 1182–1183. doi: 10.1093/jac/dkn065

Willems, R. J., Homan, W., Top, J., van Santen-Verheuvel, M., Tribe, D., Manzioros, X., et al. (2001). Variant esp gene as a marker of a distinct genetic lineage of vancomycin-resistant *Enterococcus faecium* spreading in hospitals. *Lancet* 357, 853–855. doi: 10.1016/S0140-6736(00)04205-7

Willems, R. J., Top, J., van Den Braak, N., van Belkum, A., Endtz, H., Mevius, D., et al. (2000). Host specificity of vancomycin-resistant *Enterococcus faecium*. *J. Infect. Dis.* 182, 816–823. doi: 10.1086/315752

Willems, R. J., Top, J., van den Braak, N., van Belkum, A., Mevius, D. J., Hendriks, G., et al. (1999). Molecular diversity and evolutionary relationships of Tn1546-like elements in enterococci from humans and animals. *Antimicrob. Agents Chemother.* 43, 483–491.

Willems, R. J., Top, J., van Santen, M., Robinson, D. A., Coque, T. M., Baquero, F., et al. (2005). Global spread of vancomycin-resistant *Enterococcus faecium* from distinct nosocomial genetic complex. *Emerg. Infect. Dis.* 11, 821–828. doi: 10.3201/1106.041204

Willems, R. J. L., Top, J., van Schaik, W., Leavis, H., Bonten, M., Sirén, J., et al. (2012). Restricted gene flow among hospital subpopulations of *Enterococcus faecium*. *mBio* 3:e151. doi: 10.1128/mBio.00151-12

Woodford, N., Adebiyi, A. M., Palepou, M. F., and Cookson, B. D. (1998). Diversity of VanA glycopeptide resistance elements in enterococci from humans and nonhuman sources. *Antimicrob. Agents Chemother.* 42, 502–508.

Xu, X., Lin, D., Yan, G., Ye, X., Wu, S., Guo, Y., et al. (2010). vanM, a new glycopeptide resistance gene cluster found in *Enterococcus faecium*. *Antimicrob. Agents Chemother.* 54, 4643–4647. doi: 10.1128/AAC.01710-09

Zhang, X., Paganelli, F. L., Bierschenk, D., Kuipers, A., Bonten, M. J. M., Willems, R. J. L., et al. (2012). Genome-wide identification of ampicillin resistance determinants in *Enterococcus faecium*. *PLoS Genet.* 8:e1002804. doi: 10.1371/journal.pgen.1002804

Zhang, X., Top, J., de Been, M., Bierschenk, D., Rogers, M., Leendertse, M., et al. (2013). Identification of a genetic determinant in clinical *Enterococcus faecium* strains that contributes to intestinal colonization during antibiotic treatment. *J. Infect. Dis.* 207, 1780–1786. doi: 10.1093/infdis/jit076

Zorzi, W., Zhou, X. Y., Dardenne, O., Lamotte, J., Raze, D., Pierre, J., et al. (1996). Structure of the low-affinity penicillin-binding protein 5 PBP5fm in wild-type and highly penicillin-resistant strains of *Enterococcus faecium*. *J. Bacteriol.* 178, 4948–4957.

6

Attributes of Carbapenemase Encoding Conjugative Plasmid pNDM-SAL from an Extensively Drug-Resistant *Salmonella Enterica* Serovar Senftenberg

Anirban Sarkar[1], Gururaja P. Pazhani[1,2], Goutam Chowdhury[1], Amit Ghosh[1] and Thandavarayan Ramamurthy[1,3]**

[1] Department of Bacteriology, National Institute of Cholera and Enteric Diseases, Kolkata, India, [2] Centre for Drug Discovery and Development, Sathyabama University, Chennai, India, [3] Center for Human Microbial Ecology, Translational Health Science and Technology Institute, Faridabad, India

***Correspondence:**
Thandavarayan Ramamurthy, Center for Human Microbial Ecology, Translational Health Science and Technology Institute, NCR Biotech Science Cluster, 3rd Milestone, Faridabad-Gurgaon Expressway, Faridabad 121001, Haryana, India
rama1murthy@yahoo.com;
Amit Ghosh, Department of Bacteriology, National Institute of Cholera and Enteric Diseases, P-33, CIT Road, Scheme-XM, Beliaghata, Kolkata 700010, West Bengal, India
amitghosh24@yahoo.com

A carbapenem resistant *Salmonella enterica* serovar Senftenberg isolate BCH 2406 was isolated from a diarrheal child attending an outpatient unit of B.C. Roy Hospital in Kolkata, India. This isolate was positive for the $bla_{\mathrm{NDM-1}}$ in the PCR assay, which was confirmed by amplicon sequencing. Except for tetracycline, this isolate was resistant to all the tested antimicrobials. The $bla_{\mathrm{NDM-1}}$ was found to be located on a 146.13-kb mega plasmid pNDM-SAL, which could be conjugally transferred into *Escherichia coli* and other enteric pathogens such as *Vibrio cholerae* O1 Ogawa and *Shigella flexneri* 2a. However, the expression of β-lactam resistance is not the same in different bacteria. The whole genome sequence of pNDM-SAL was determined and compared with other pNDM plasmids available in public domain. This plasmid is an IncA/C incompatibility type composed of 155 predicted coding sequences and shares homology with plasmids of *E. coli* pNDM-1_Dok01, *Klebsiella* pNDM-KN, and *Citrobacter* pNDM-CIT. In pNDM-SAL, gene cluster containing $bla_{\mathrm{NDM-1}}$ was located between IS*26* and IS*4321* elements. Between the IS*26* element and the $bla_{\mathrm{NDM-1}}$, a truncated IS*Aba125* insertion sequence was identified. Downstream of the $bla_{\mathrm{NDM-1}}$, other genes, such as ble_{MBL}, *trpF*, *tat*, and an IS*CR1* element with class 1 integron containing *aac(6′)-Ib* were detected. Another β-lactacamase gene, $bla_{\mathrm{CMY-4}}$ was found to be inserted in IS*1* element within the type IV conjugative transfer loci of the plasmid. This gene cluster had *blc* and *sugE* downstream of the $bla_{\mathrm{CMY-4}}$. From our findings, it appears that the strain *S.* Senftenberg could have acquired the NDM plasmid from the other members of Enterobacteriaceae. Transfer of NDM plasmids poses a danger in the management of infectious diseases.

Keywords: NDM, carbapenemase, *S.* Senftenberg, enteric pathogens, pNDM-SAL

Introduction

Emergence of carbapenem resistance among Gram-negative bacteria is a major public health problem as they are associated with critical infections. This drug has been considered as one of the vital drugs against pathogens which produce extended spectrum β-lactamases (ESBLs). New Delhi Metallo-β-lactamase-1(NDM-1), which is a recent addition to the carbapenemase has become a major concern worldwide due to its rapid spread across different members of Enterobacteriaceae and other Gram-negative bacteria. The NDM-1 encoding gene (*bla*$_{NDM-1}$) was first detected in *Klebsiella pneumoniae* and *Escherichia coli* recovered from a Swedish patient who had undergone treatment in New Delhi, India (Yong et al., 2009; Ghosh et al., 2014). Thereafter, this gene was identified in different Gram-negative bacteria in several countries including the USA, Canada, France, Sweden, UK, Germany, Japan, Austria, Africa, and Australia (Rolain et al., 2010). Presence of *bla*$_{NDM-1}$ was generally associated with resistance to most of the antimicrobials, including fluoroquinolones, aminoglycosides, and β-lactams. The *bla*$_{NDM}$ was found to be located of different large plasmids, which were readily transferable to other bacterial species (Kumarasamy et al., 2010; Rolain et al., 2010). The spread of the *bla*$_{NDM-1}$ gene acquired by IncA/C MDR plasmids has drastically reduced the therapeutic options available to the physicians (Ghosh et al., 2014). In this study, we report the isolation of a *Salmonella enterica* serovar Senftenberg (S. Senftenberg) isolate carrying the *bla*$_{NDM-1}$ gene on a large plasmid, which also possessed several genes responsible for extensively drug resistance (EDR).

Materials and Methods

Bacterial Strains

A carbapenem resistant *Salmonella enterica* isolate (BCH 2406) was isolated in 2012 from a five years old child who attended the outpatient department of B.C. Roy Memorial Hospital for Children, Kolkata for the treatment of diarrhea. This isolate was serotyped according to White-Kauffmann-Le Minor scheme with commercially available antisera (S&A Reagents Lab Ltd., Bangkok, Thailand). Tetracycline resistant XL1-Blue (TET^R) and sodium azide resistant *E. coli* J53 (Az^R) strains were used for conjugation experiments. In addition, ampicillin sensitive *Vibrio cholerae* O1 Ogawa (IDH 5313) and *Shigella flexneri* 2a (IDH 3077), isolated and identified from diarrheal patients were used as recipients. All the strains were preserved in Luria Bertani (LB) broth (Difco, Sparks, MD, USA) containing 15% glycerol at –80°C. Transconjugants were also maintained in nutrient agar (Difco) stab supplemented with 15 μg/ml ceftriaxone. *E. coli* ATCC 25922 was served as control in antimicrobial susceptibility testing.

Detection of Carbapenem Resistance Encoding Gene

Presence of *bla*$_{NDM}$ was identified by PCR with previously described primers (Chen et al., 2011) using Taq DNA polymerase (Roche, Mannheim, Germany). Amplicons were purified using a PCR product purification kit (Qiagen, Hilden, Germany) and sequenced using the ABI Big Dye terminator cycle sequencing ready reaction kit, version 3.1 (Applied Biosystems, Foster City, CA, USA) in an automated DNA sequencer (ABI 3730, Applied Biosystems). The sequences were assembled and analyzed using DNASTAR software (DNASTAR, Inc., Madison, WI, USA).

Conjugation

To test the mobility and promiscuity of the *bla*$_{NDM-1}$ haboring plasmid, conjugation by broth mating technique was carried out using NDM-positive *Salmonella* isolate as donor with four different recipients, namely *E. coli* XL1-Blue (TET^R), *E. coli* J53 (Az^R), *V. cholerae* O1 Ogawa, and *S. flexneri* 2a. In brief, overnight cultures of the bacteria were diluted in LB broth and allowed to grow as late-exponential phase culture. Cell density was adjusted to 1.5×10^8cells/ml. Donor and recipient cells were mixed at 1:2 donor-to-recipient ratios in 1 ml of LB broth and allowed grow overnight at 37°C. In all cases, the donor and recipient suspensions were also diluted in phosphate buffer saline (PBS) with a dilution of 10^{-3} and 10^{-5} and plated on MacConkey agar (Difco) to confirm the purity and count the colonies.

To recover *bla*$_{NDM-1}$ positive transconjugants, several selective media were used. Transconjugants (conjugally transferred, CT) of XL1-Blue (CT-*E. coli* XL1-Blue) and *S. flexneri* 2a (CT-*S. flexneri*) were selected on xylose lysine desoxycholate (XLD, Difco) agar supplemented with tetracycline (30 μg/ml) and ceftriaxone (5 μg/ml). Similarly, transconjugants of *V. cholerae* (CT-*V. cholerae*) were obtained with selection based on growth on ceftriaxone (5 μg/ml) containing thiosulphate citrate bile salts sucrose (TCBS, Eiken, Tokyo, Japan) agar and for transconjugants of *E. coli* J53 (CT-*E. coli* J53), MacConkey agar containing both ceftriaxone (5 μg/ml) and sodium azide (100 μg/ml) was used. Transconjugants were confirmed as *bla*$_{NDM-1}$ positive by PCR analysis followed by PCR amplicon sequencing. The transfer frequencies were expressed as the number of transconjugants per donor cell.

Antimicrobial Susceptibility Testing

To confirm the transfer of resistance phenotype, antibiotic susceptibility patterns of the donor, recipients, and transconjugants were determined after their growth on Mueller-Hinton (MH, Difco) agar by disk diffusion method in accordance with Clinical and Laboratory Standards Institute [CLSI] (2014) using commercially available disks (Becton Dickinson Company, Sparks, MD) namely, ampicillin (AMP), cefuroxime (CXM), ceftriaxone (CRO), cefotaxime (CTX), cefotaxime/clavulanic acid (CTX-CLA), ceftazidime (CAZ), ceftazidime/clavulanic acid (CAZ-CLA), chloramphenicol (CHL), nalidixic acid (NA), ciprofloxacin (CIP), ofloxacin (OFX), norfloxacin (NOR), imipenem (IPM), streptomycin (STR), azithromycin (AZM) tetracycline (TET), trimethoprim/sulfamethoxazole (SXT). To measure the increase of resistance in the transconjugants, MICs of antibiotics (ceftriaxone, imipenem, tetracycline, and sulfamethoxazole/trimethoprim) were determined by *E* test (AB bioMérieux, Solna, Sweden) in comparison with the wild NDM-positive *Salmonella* isolate.

Plasmid analysis

Plasmid DNA was extracted from donor, recipients, and transconjugants by Kado and Liu's (1981) method and analyzed by gel electrophoresis using 0.8% agarose. Presence of *bla*$_{NDM-1}$

in the purified plasmid of the donor and transconjugants was determined by PCR. For size determination, plasmid analysis by S1 nuclease-pulsed-field gel electrophoresis (PFGE) (Barton et al., 1995) was performed with the donor and CT-*E. coli* J53. Total bacterial DNA prepared in agarose plugs was digested with S1 nuclease (Fermentas, Waltham, MA, USA) and separated using a CHEF-Mapper PFGE system (Bio-Rad, Hercules CA, USA), as reported previously (Kumarasamy et al., 2010). The PFGE conditions were run time 18 h gradient 6 V/cm, temperature 14°C, included angle 120° and initial and final pulses conducted for 2.16 s and 54.17 s, respectively. DNA of *S. enterica* serovar Braenderup H9812 digested with *XbaI* (Roche) was included as control size marker. Incompatibility group of plasmid was determined for the *Salmonella* isolate and *E. coli* J53 transconjugant (as *E. coli* J53 Az^R was devoid of plasmids) using the PCR-based replicon typing (PBRT) method as described by Carattoli et al. (2005).

Sequencing

Plasmid DNA was prepared from the *E. coli* J53 tranconjugant using Qiagen Maxiprep kit (Qiagen). Paired-end libraries (300–500 bp fragments) were constructed by using the Illumina *H* TruSeqTM DNA Sample preparation kit (Illumina). Each library was deposited onto a HiSeq Flow Cell and sequenced using an Illumina HiSeq-2000 next-generation DNA sequencer. The distributions of "base quality" "base composition" and %GC of the plasmid were checked. Based on these distributions, the first 15 bases and last one base were trimmed to avoid specific sequence bias and poor quality bases. Errors were corrected in the sequence data using the HiTEC tool. Contig assembly and predicted gaps were then confirmed and filled by PCR-based gap closure, confirmed by DNA sequencing of the amplicons (Applied Biosystems). To assemble the contigs, ABySS and Edena software were used by varying the parameter 'k' and 'overlap cut-off', respectively. This step produced several contigs for each parameter setting. The contigs were then integrated using contig integrator for sequence assembly (CISA). A blastN search was made against the 'nt' database for each contig and two contigs were retained for further downstream analysis. The open reading frames (ORFs) from the contigs were generated by CISA using Glimmer-MG program. For all these ORFs, the nucleotide sequence and amino acid sequences were obtained and translated in the appropriate frame. The predicted ORFs were annotated using an in-house pipeline (CANoPI-Contig Annotator Pipeline) that also includes blastX search for each ORF sequence against 'nr' database provided by NCBI. ORF search and gene prediction was performed for the complete plasmid sequence with Lasergene software (DNASTAR, Inc., Madison, WI, USA) and pairwise alignment was analyzed by blastN and blastP homology search using the NCBI database (http://www.ncbi.nlm.nih.gov/blast).

Nucleotide Sequence Accession Number

The complete sequence of plasmid pNDM-SAL was submitted to GenBank under accession number KP742988.1

Results

Identification and Characterization of *bla*$_{NDM-1}$-Positive Isolate

The isolated *Salmonella enterica* was found to be positive for *bla*$_{NDM-1}$, which was confirmed by amplicon sequencing. The sequence of the *bla*$_{NDM-1}$ showed 100% homology with those reported previously (Chen et al., 2011; Ho et al., 2011; Sekizuka et al., 2011; Bonnin et al., 2012; Carattoli et al., 2012; Fu et al., 2012; Dolejska et al., 2013). By serology, this isolate was identified as *S.* Senftenberg presenting antigenic formula as 1,3,19 : g,s,t : -. The isolate was found to be resistant to almost all antibiotics, including nalidixic acid, ciprofloxacin, norfloxacin, ofloxacin, chloramphenicol, streptomycin, azithromycin, sulfamethoxazole/trimethoprim, ampicillin, cefuroxime, ceftriaxone, cefotaxime, ceftazidime, and also to β-lactamase inhibitor combinations (cefotaxime/clavulanate and ceftazidime/clavulanate). However, it was susceptible to tetracycline and showed reduced susceptibility towards imipenem.

Conjugation and Transfer of Resistance

Plasmid harboring the *bla*$_{NDM-1}$ was transferable to *E. coli* strains J53, XL1-Blue and to other enteric pathogens like *V. cholerae* O1 Ogawa and *S. flexneri* 2a isolated from the diarrheal patients. Conjugation frequencies were observed at higher rates ($\sim10^{-5}$ tranconjugants per donor cell) while using *E. coli* J53 and *S. flexneri* 2a as recipients than *E. coli* XL1-Blue ($\sim10^{-6}$). The transfer of NDM plasmid into *V. cholerae* O1 Ogawa was even less efficient ($\sim10^{-8}$) compared to the other strains (**Table 1**). In addition, the transconjugants acquired additional resistance against β-lactam antibiotics, namely ampicillin, cefuroxime, ceftriaxone, cefotaxime, ceftazidime, and also toward cephalosporin inhibitor combinations like cefotaxime/clavulanate and ceftazidime/clavulanate (**Table 1**). Interestingly, sulfamethoxazole resistance was noticed only in *S. flexneri* transconjugant. Though the CT-*E. coli* J53 was susceptible to sulfamethoxazole/trimethoprim, its MIC was found to increase by 2.7-fold. In CT-*S. flexneri*, the MIC of SXT rose by 31.6-fold, suggesting that sulfamethoxazole resistance was greatly expressed in CT-*S. flexneri* but not in CT-*E. coli* strains.

The CT-*E. coli* showed reduced susceptibility to carbapenems (**Table 1**). On the contrary, the transconjugants of clinical strains (CT-*V. cholerae* and CT-*S. flexneri*) showed high level resistance to imipenem, far greater than the donor strain *S.* Senftenberg (>5-fold). The MIC values for imipenem were found to be higher in CT-*S. flexneri* (128-fold) than those in CT-*V. cholerae* (>21-fold), CT-*E. coli* XL1-Blue (24-fold), and CT-*E. coli* J53 (12-fold). For ceftriaxone, more than 300-fold rise in MIC was observed for all CTs compared to the respective recipients.

Plasmid Analysis

Plasmid profiles of the donor and transconjugants (**Figure 1**) revealed the transfer of a mega plasmid from *S.* Senftenberg which conferred β-lactam resistance to the recipient strains of *E. coli, V. cholerae*, and *S. flexneri*. NDM-1 PCR assay confirmed

TABLE 1 | Antibiotic susceptibilities of donor, recipients, and transconjugants.

Strain	Bacteria	Resistance profile	MIC value (μg/ml)			
			TET	CRO	IPM	SXT
BCH 2406 (Donor)	*Salmonella* Senftenberg	NA, NOR, CIP, OFX, SXT, CHL, AZM, STR, AMP, CXM, CRO, CTX, CTX-CLA, CAZ, CAZ-CLA, IPM (I)	3	>256	6	>32
XL1-Blue (Recipient)	*Escherichia coli*	NA, TET	192	0.5	0.125	
CT-*E. coli* XL1-Blue (Transconjugant)	*E. coli* Transfer frequency*: 3.6×10^{-6}	NA, TET, AMP, CXM, CRO, CTX, CTX-CLA, CAZ, CAZ-CLA	96	>256	3	
J53 (Recipient)	*E. coli*	AZD		0.5	0.125	0.047
CT-*E. coli* J53 (Transconjugant)	*E. coli* Transfer frequency*: 1.9×10^{-5}	AMP, CXM, CRO, CTX, CTX-CLA, CAZ, CAZ-CLA, AZD		>256	1.5	0.125
IDH 5313 (Recipient)	*Vibrio cholerae* O1 Ogawa	NA, SXT, STR		0.5	1.5	>32
CT-*V. cholerae* (Transconjugant)	*V. cholerae* O1 Ogawa Transfer frequency*: 1.5×10^{-8}	NA, SXT, STR, AMP, CXM, CRO, CTX, CTX-CLA, CAZ, CAZ-CLA, IPM		>256	>32	>32
IDH 3077 (Recipient)	*Shigella flexneri* 2a	NA, STR, TET		0.75	0.25	0.38
CT-*S. flexneri* (Transconjugant)	*S. flexneri* 2a Transfer frequency*: 3.7×10^{-5}	NA, TET, STR, SXT, AMP, CXM, CRO, CTX, CTX-CLA, CAZ, CAZ-CLA, IPM		>256	32	12

NA, nalidixic acid; CIP, ciprofloxacin; NOR, norfloxacin; OFX, ofloxacin; AZM, azithromycin; STR, streptomycin; AMP, ampicillin; CXM, cefuroxime; CRO, ceftriaxone; CTX, cefotaxime; CAZ, ceftazidime; CTX-CLA, cefotaxime/clavulanate; CAZ-CLA, ceftazidime/clavulanate; IPM, Imipenem; SXT, sulfamethoxazole/trimethoprim; CHL, chloramphenicol; TET, tetracycline; AZD, azide; (I) = intermediate resistance.

**Frequency of transfer was calculated as number of transconjugants per donor cell.*

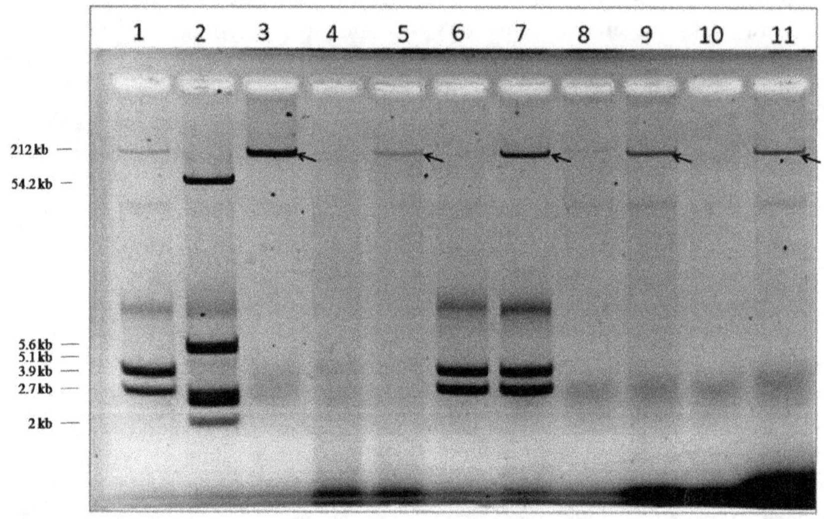

FIGURE 1 | Plasmid profile of donor, recipient, and transconjugants. Lanes 1 and 2, *Shigella flexneri* YSH 6000 and *Escherichia coli* K12 V517 were used as reference standards, respectively; Lane 3, BCH 2406: *Salmonella* Senftenberg (donor); Lane 4, IDH 5313: *Vibrio cholerae* O1 Ogawa (recipient); Lane 5, CT-*V. cholerae* (transconjugant); Lane 6, IDH 3077: *S. flexneri* 2a (recipient); Lane 7, CT-*S. flexneri* (transconjugant); Lane 8, XL1-Blue, *E. coli* (recipient); Lane 9, CT-*E. coli* XL1-Blue (transconjugant); Lane 10, *E. coli* J53 (recipient); Lane 11, CT-*E. coli* J53 (transconjugant).

the presence of bla_{NDM-1} in all the transconjugants. S1-PFGE clearly showed that the resistance phenotype was associated with the transfer of single plasmid of about 146-kb (**Figure 2**) and the NDM-encoding gene was detected by PCR on the plasmid isolated from the transconjugants. Plasmid pNDM-SAL was assigned to the IncA/C incompatibility group using the PCR-based PBRT method.

Sequence Analysis of Plasmid pNDM-SAL

Whole plasmid showed the plasmid pNDM-SAL to consist of 146.13-kb with an average %GC content of 51.7. In addition to its own replication machinery with partitioning system, it has two separate gene clusters for bla_{NDM-1} and bla_{CMY-4} (**Figure 3**). It contained 155 predicted coding sequences (CDSs). This plasmid shared extensive homology (99% identity with 98% query coverage) with an IncA/C plasmid, pNDM-1_Dok01 (195.56-kb, described for *E. coli*, accession no. AP012208) including the complete array of genes for replication, type IV conjugative transfer machinery, partition and stabilization, except for the flanking region around bla_{NDM-1} and bla_{CMY-4}. In addition, it had sequence similarity with *Klebsiella* plasmid pNDM-KN (JN157804) and *Citrobacter* plasmid pNDM-CIT (JX182975)

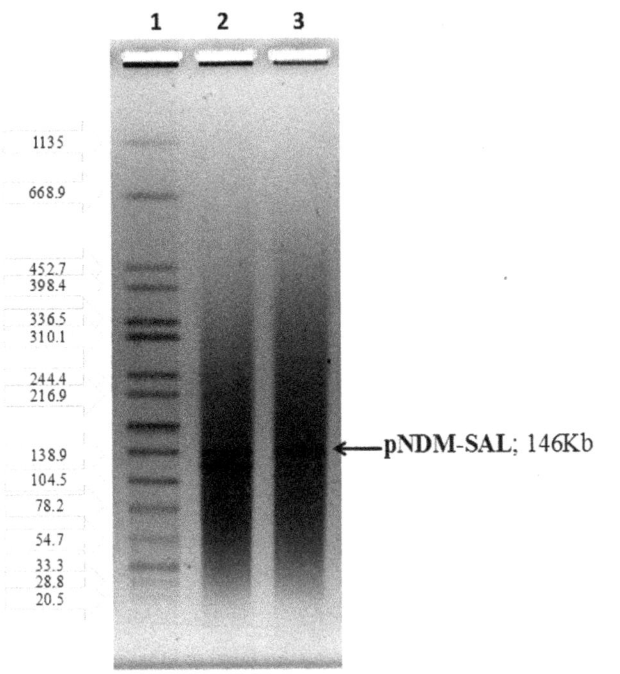

FIGURE 2 | S1-PFGE indicating the transfer of single plasmid harboring bla_{NDM-1}: Lane1, H-9812 (PFGE marker strain, *S.* Braenderup H9812); Lane 2, BCH 2406 *S.* Senftenberg (donor); Lane 3, CT-*E. coli* J53 (transconjugant).

showing 99% identity with 87% query coverage and 99% identity with 87% query coverage, respectively.

Analysis and Comparison of Gene Organization Around bla_{NDM-1} and between Other Plasmids

The bla_{NDM-1} gene was localized in a multidrug resistance region of 15.3-kb (**Figure 4**). This region was bracketed by two different copies of IS elements in inverted orientation, suggesting that the gene was acquired as a composite transposon. Unlike, pNDM-1_Dok01 and pGUE-NDM (87-kb, described in *E. coli*, accession no JQ36496) where bla_{NDM-1} gene was flanked by IS*903* and IS*26*, respectively, pNDM-SAL possessed IS*26* and IS*4321* upstream and downstream of the bla_{NDM-1}. This orientation is much similar to that found in the pNDM-CIT. In between the IS*26* element and the bla_{NDM-1} gene, a remnant of IS*Aba125* insertion sequence was identified which comprised -35 promoter sequences leading to the high expression of NDM.

Downstream of the bla_{NDM-1}, the ble_{MBL} gene encoding resistance to bleomycin followed by a truncated phosphorybosilanthranilate isomerase gene (*trpF*), and a twin-arginine translocation pathway signal protein gene (*tat*) was identified. Such an arrangement has been reported previously in other plasmid scaffolds (Sekizuka et al., 2011). Following this, a IS*CR1* element was identified trailed by a class 1 integron containing *aac(6')-Ib* as its variable resistance gene cassette. Such genetic arrangement has been found in a complex class 1 integron containing IS*CR1* element of ESBL positive *E. coli* (EF450247) and *V. cholerae* (DQ310703). IS*CR1* and IS*26* are known to mediate transposition and/or expression of multiple resistance genes in their close proximity (Bae et al., 2007; Kiiru et al., 2013). Similarly, class 1 integron bearing IS*CR1* was detected in the previously described NDM plasmids, namely pGUE-NDM (JQ364967) and pNDM-CIT (JX182975). The only difference being that the other plasmids harbored *dfrA12-ofrF-aadA2* as their integron resistance gene cassettes (**Figure 4**). However, in all the cases *qacE*Δ1 and *sul*1 gene are present in the integron structure. The *intI*1 gene was followed by the *tnpR* and *tnpA*, the genes of transposon Tn*21* and then by a copy of insertion element IS*4321*, as previously identified in pNDM-CIT (JX182975). Like pNDM-SAL, linkage between Tn*21* transposon bearing class 1 integron and IS*26* element has been seen in NDM negative *E. coli* of human and animal origin (Dawes et al., 2010) and in *E. coli* D22 (EU914098). From the *in silico* analysis, it was not possible to determine the genetic events that led to the formation of this heterogeneous genetic structure. However, it is likely that multiple genetic events contributed to the acquisition of the bla_{NDM-1} containing locus by the plasmid.

The bla_{CMY} Module Region

In addition to the bla_{NDM-1}, pNDM-SAL carried an additional β-lactam resistance gene, bla_{CMY-4}, distantly located from the NDM harboring composite transposon (**Figure 5**). The bla_{CMY} gene was preceded by the IS*1* instead of IS*Ecp1*, as reported previously for other NDM plasmids, pNDM-1_Dok01 and pKP1-NDM-1 (KF992018). Remarkably, bla_{CMY} gene associated with IS*1* element was detected before in NDM-negative *E. coli* (DQ173300) with inverted repeats (IRs) of IS*Ecp1* (Hopkins et al., 2006). This makes it tempting to speculate that an intact copy of IS*EcP1* was responsible for the early transposition and mobility of bla_{CMY-4} followed by insertion of the IS*1* element; a process similar to what has been reported previously for *E. coli* (DQ173300). Downstream of the bla_{CMY-4} gene, *blc* gene encoding outer membrane lipoprotein was detected followed by the *sugE* encoding quaternary ammonium compound resistance protein (**Figure 5**), a feature common in many of the bla_{CMY} regions in other NDM-positive plasmids (Kang et al., 2006; Sekizuka et al., 2011). This IS*1*-bla_{CMY} module was located within the *tra* locus. Similar insertion with IS*Ecp1* has been demonstrated before (Poole et al., 2009).

Discussion

The present study revealed EDR features of the pNDM-SAL isolated from *S.* Senftenberg. This finding highlights the ease with which the resistance determinants can move to other enteric pathogens. Firstly, the bla_{NDM-1} is located on a broad host range IncA/C plasmid, providing a possible explanation why the pNDM-SAL could be transferred by conjugation to *E. coli* (*E. coli* J53 and XL1-Blue) as well as to wild-type strains of *V. cholerae* O1 Ogawa and *S. flexneri* 2a easily. In repeated experiments, CT-*V. cholerae* and CT-*S. flexneri* showed a higher level of

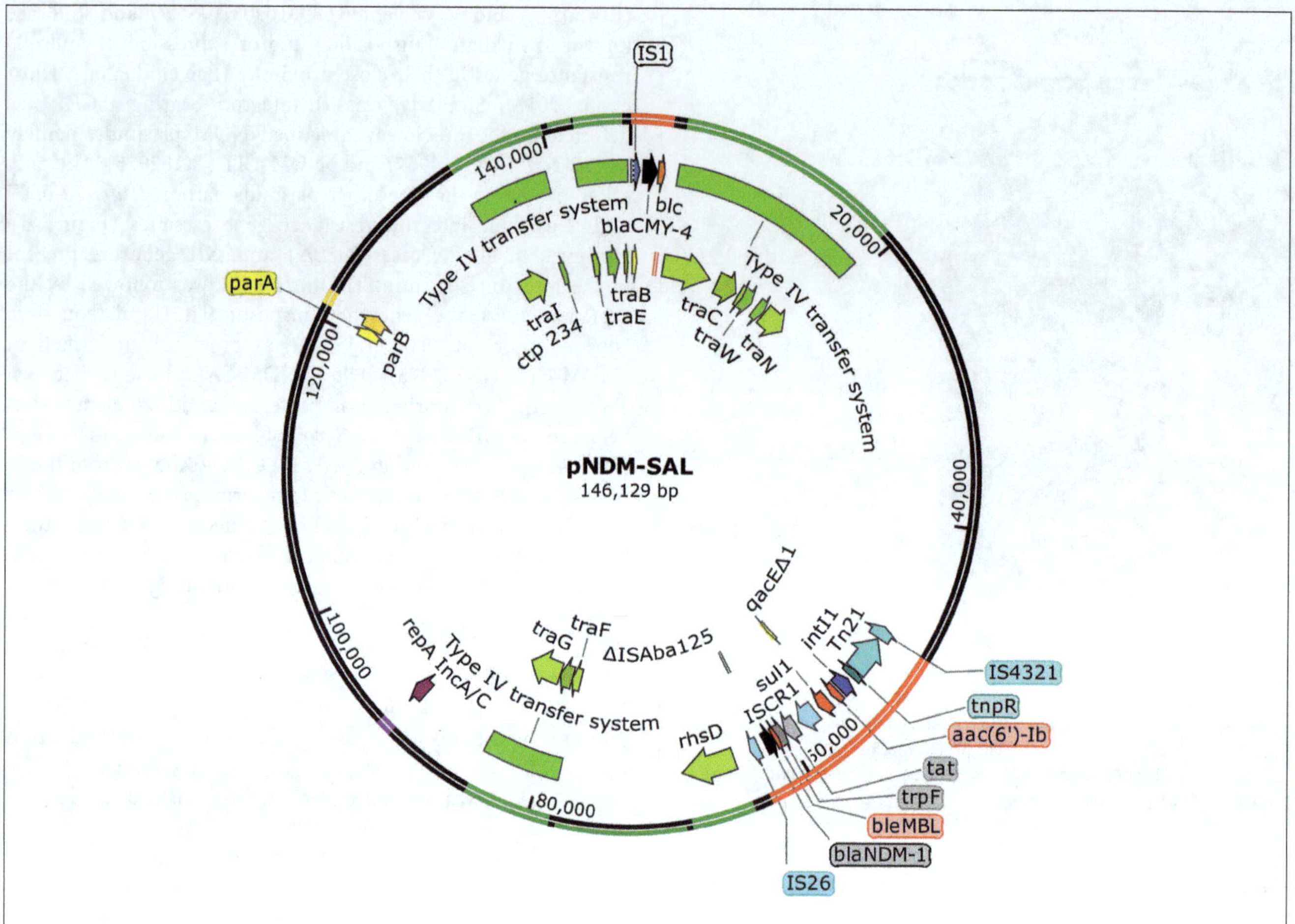

FIGURE 3 | Salient features of $bla_{\text{NDM}-1}$ and $bla_{\text{CMY}-4}$ encoding plasmid pNDM-SAL with transposons and type IV conjugative transfer system. Starting from the outside, the first circle indicates the coordinate of the complete plasmid circle. Transfer machinery and variable region are showed in green and red, respectively. *rep* and *par* loci are showed in purple and yellow, respectively. The open reading frames (ORFs) were annotated in the inner circle with arrows representing the direction of transcription. β-lactam resistance genes are denoted by black color and other resistance genes as red color. All transposons and IS elements are indicated in shades of blue. ORFs responsible for conjugation are denoted by shades of green.

imipenem resistance compared to the donor strains. This could be hypothesized to be due to "possibly" a higher copy number of the plasmid carrying the gene in the recipient than in the donor strains. In other words due to "gene dosage" effect. There could be other possible explanations as well. Examples of such phenomenon, transconjugants showing higher drug resistance are documented in the literature (Ouellette et al., 1988; Petroni et al., 2002).

Since IncA/C plasmids possess highly mobile nature, it has commonly been found in NDM harboring bacteria, especially among the members of Enterobacteriaceae (Ghosh et al., 2014). The size of plasmid varies considerably, ranging from 35.9 to 400-kb (Ghosh et al., 2014) with certain shared genetic background. A recent study has shown that the NDM carrying plasmids from Enterobacteriaceae at New Delhi, shared the replicon type IncA/C with140-kb in size (Walsh et al., 2011). Large plasmids belonging to the IncA/C group have received increased attention, primarily due to their ability to confer resistance to a diverse group of antimicrobial agents (Fernandez-Alarcon et al., 2011). Transmission of pNDM-SAL was also found to be closely associated with transfer of multi drug resistance. Since clinical strains of *Vibrio* or *Shigella* are typically resistant to some antimicrobials, transfer of pNDM-SAL carrying $bla_{\text{NDM}-1}$ and other genes could make them multi resistance. This situation is of major concern in the clinical management of infections. Since the pNDM-SAL harbored antibiotic resistance genes are clustered in a few integration hotspots within complex genetic structures, they may be able to acquire novel antibiotic resistance genes through homologous recombination (Doublet et al., 2012). In the pNDM-SAL, *rhs* locus remains near $bla_{\text{NDM}-1}$ with a phage-integrase and hence they are likely to be the hotspots for integration of accessory genes within the IncA/C plasmids (Carattoli et al., 2012).

In pNDM-SAL, $bla_{\text{NDM}-1}$ was located between IS*26* and IS*4321*. Similar genetic arrangements have previously been observed in other $bla_{\text{NDM}-1}$ bearing plasmids (Sekizuka et al., 2011; Bonnin et al., 2012; Dolejska et al., 2013). Comparison of the flanking regions of $bla_{\text{NDM}-1}$ present in different plasmids suggests that different genetic events may have supported acquisition of this gene in different plasmids (Ghosh et al.,

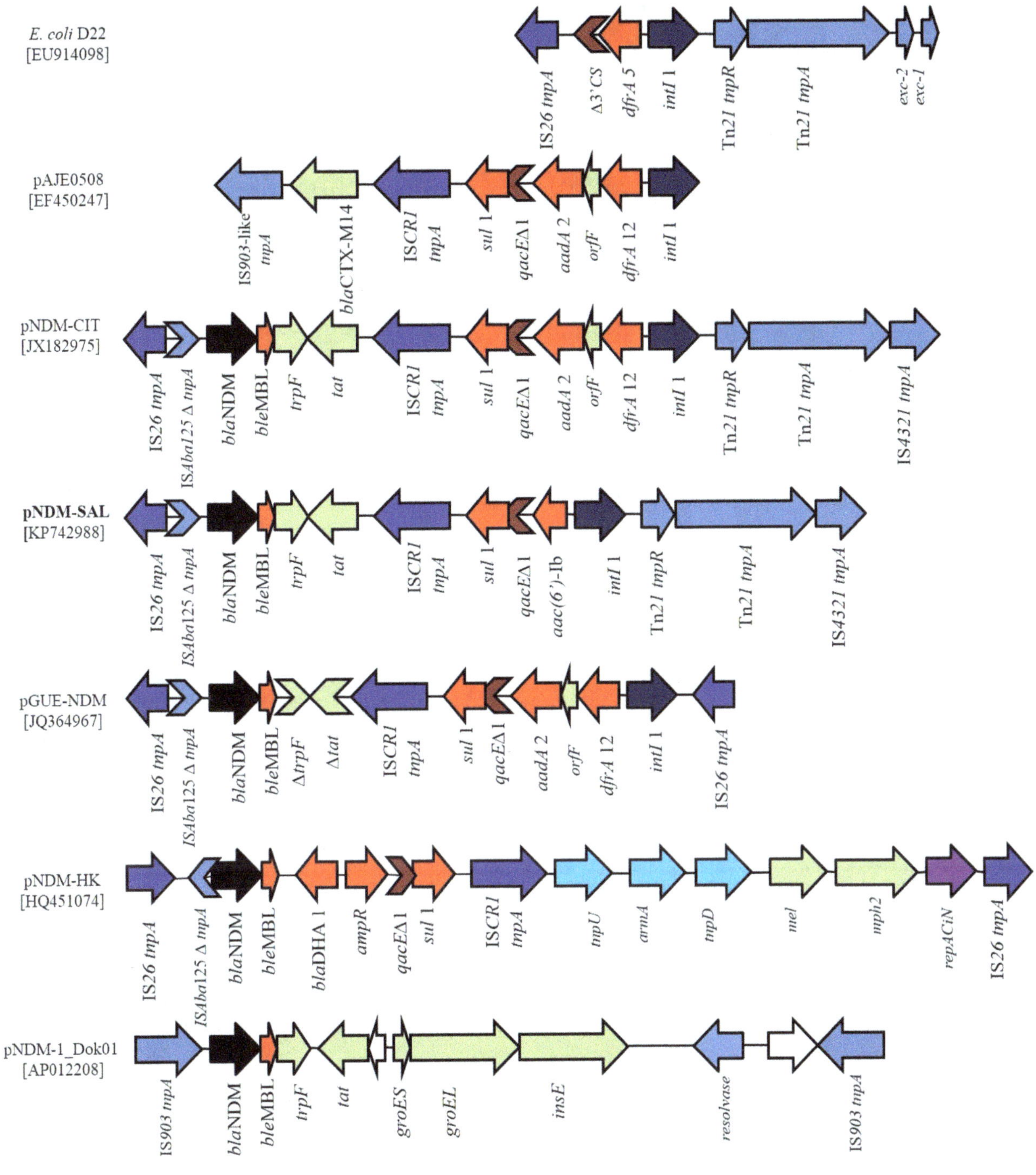

FIGURE 4 | Schematic representation of the DNA sequences surrounding the bla_{NDM-1} genes in pNDM-SAL compared to the sequences in *E. coli* D22 (EU914098), pAJE0508 (EF450247), pNDM-CIT (JX182975), pGUE-NDM (JQ36496), pNDM-HK (HQ451074), and pNDM-1_Dok01 (AP012208). Arrows showed the direction of transcription. Gaps between the ORFs are indicated by straight line. The same color and label are used to represent homologous genes. bla_{NDM-1} are denoted by black color and other resistance genes as red color. All transposons and IS elements are indicated in shades of blue. Other ORFs are denoted by light gray and white. Violet arrow indicates the *repAciN* gene.

2014). In pNDM-SAL, the linkage of bla_{NDM-1} with IS*26* creates a condition favorable for the mobilization of bla_{NDM-1}. Interestingly, in many NDM negative *E. coli* strains, arrangement of class 1 integron between Tn*21* transposon and IS*26* was a characteristic feature (Dawes et al., 2010) as noticed in the pNDM-SAL. The IS*26* element, a member of the IS*6* family, is widespread among Enterobacteriaceae and exists adjacent to the β-lactamases region, which is a part of transposon-like structure

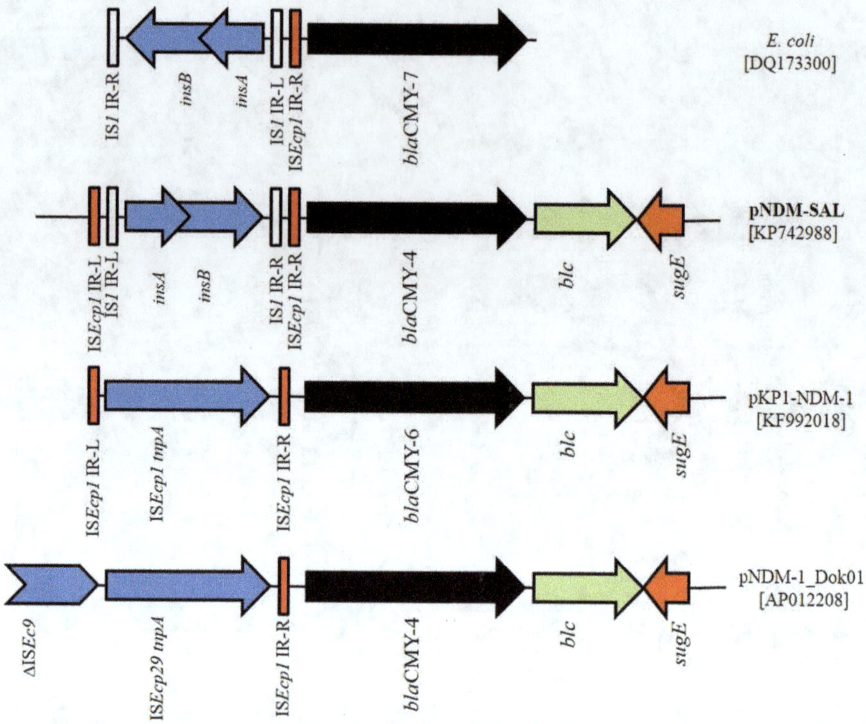

FIGURE 5 | Schematic representation of the DNA sequences surrounding the *bla*$_{CMY-4}$ genes in pNDM-SAL and comparison with the sequences in *E. coli* (DQ173300), pKP1-NDM-1 (KF992018), and pNDM-1_Dok01 (AP012208). Arrows indicating *bla*$_{CMY-4}$ (black), *blc* (gray) and *sugE* (orange) are showed according the direction of transcription. Gaps between the ORFs are indicated by straight line. The same color and label are used to represent homologous genes. IS elements are indicated in shades of blue. Red and white color indicates the inverted repeat (IR) of IS*Ecp1* and IS*1*, respectively.

in many plasmids (Yu et al., 2006; Ho et al., 2011). Similar to the other plasmids such as *E. coli* DVR22 (JF922606), pNDM-HK (HQ451074), pKpANDM-1 (FN396877) and p271A (JF785549), a truncated IS*Aba125* has been identified upstream of *bla*$_{NDM-1}$ in pNDM-SAL. Interestingly, in *Acinetobacter baumanii* isolates of African origin, the NDM encoding gene was found within the Tn*125*- like or IS*Aba125* (Bonnin et al., 2013). This suggests that the IS*Aba125* insertion sequence has possibly played an important role in the mobilization of *bla*$_{NDM-1}$ from a common progenitor and that event was followed by subsequent transfer events mediated by the other insertion elements like IS*26* as seen in *Acinetobacter* spp. and Enterobacteriaceae (Ghosh et al., 2014). In pGUE-NDM (*E. coli*, JQ364967) and pNDM-MAR (*Klebseilla*, JN420336) NDM regions are fully bracketed by IS*26*. But pNDM-SAL possesses IS*4321* and Tn*21* tranposases in one end similar to what is found in the pNDM-CIT (*Citrobacter*, JX182975). Instead of *bla*$_{DHA-1}$ gene commonly found downstream of many NDM plasmids, eg. pNDM-HK (*E. coli*, HQ451074) the pNDM-SAL IS*CR1* bearing class 1 integron has been inversely located (**Figure 4**). The association between *ble*$_{MBL}$ and *bla*$_{NDM}$ appears to be strong as they are expressed by a common promoter. This may be the reason why both the genes transfer *en bloc* (Fu et al., 2012; Ghosh et al., 2014).

Based on the genetic environment of *bla*$_{NDM-1}$ in pNDM-SAL, we hypothesize that (i) a common transposon structure with IS*Aba125* could be responsible for the early acquisition of *bla*$_{NDM-1}$, (ii) by extensive genetic rearrangement; it was captured in a Tn*21* linked complex class 1 integron bearing IS*CR1*, (iii) later on it was bracketed by two IS elements, namely IS*4321* preceding the Tn*21* transposase and IS*26* truncating the IS*Aba125*.

The plasmid pNDM-SAL has similarity to other NDM-plasmids, which harbor *bla*$_{CMY-4}$ and the complex class 1 integron carrying several antibiotic resistance-conferring genes (Toleman et al., 2006a,b). AmpC-like cephalosporinase (*bla*$_{CMY}$) genes that have been frequently mobilized by IncA/C-type plasmids, could be identified in *E. coli* and *Salmonella* isolates not only of human but also of animal origins in the United States, Canada, and Europe (Winokur et al., 2001; Carattoli et al., 2012). The evolutionary relationship between NDM-1-bearing plasmids and the IncA/C plasmids carrying *bla*$_{CMY}$ suggests that IncA/C *bla*$_{CMY}$-carrying plasmid could have acquired the *bla*$_{NDM-1}$ within its scaffold due to a secondary event (Carattoli et al., 2012).

From the data presented in this paper it can be seen that plasmid pNDM-SAL possesses many interesting features as it contain gene encoding the metallo-β-lactamase NDM-1, which is positively associated with multidrug resistance (**Figure 3**). Besides, pNDM-SAL harbors a large arsenal of genetic elements (integrons, transposons, and IS*CRs*), giving it the ability to acquire and disseminate antibiotic resistance genes. However, the extent of carbapenem resistance due to the presence of NDM plasmid varied in different hosts. The trend of antimicrobial susceptibility is shifting toward old generation antibiotics, which are less commonly used in recent years (Meziane-Cherif and Courvalin, 2014). In many studies, it was shown that NDM producers were susceptible to old generation antibiotics such as chloramphenicol, tetracycline etc. (Abdul Rahim et al., 2015; Uppu et al., 2015). In our finding, excerpt for tetracycline, the NDM-positive *S.* Senftenberg were resistant to most of the old generation antibiotics such as ampicillin, trimethoprim, sulfamethoxazole, streptomycin, nalidixic acid, and chloramphenicol. It appears that the maintenance of resistance to any given antibiotic may vary from species to species.

From our data it can be inferred that the strain *S.* Senftenberg probably was not the natural host for this NDM plasmid but once acquired, the plasmid it had the ability to transfer it to a broad range of pathogenic and non-pathogenic Gram-negative bacteria, an observation which merits serious and important attention to our findings.

Author Contributions

AS, GP, and GC isolated and identified the pathogens, performed phenotypic and genetic analysis. AG and TR analyzed the data, conceived the idea, and wrote the manuscript. All authors were involved in the compilation of the report and approved the final version.

Funding

Our work was funded in part by the Indian Council of Medical Research and Indian National Science Academy, New Delhi.

Acknowledgments

We thank George A. Jacoby, Lehey Hospital and Medical Center, Burlington, MA, USA for kindly providing the *Escherichia coli* J53 (Az^R) strain.

References

Abdul Rahim, N., Cheah, S. E., Johnson, M. D., Yu, H., Sidjabat, H. E., Boyce, J., et al. (2015). Synergistic killing of NDM-producing MDR *Klebsiella pneumoniae* by two 'old' antibiotics-polymyxin B and chloramphenicol. *J. Antimicrob. Chemother.* 70, 2589–2597. doi: 10.1093/jac/dkv135

Bae, I. K., Lee, Y. N., Lee, W. G., Lee, S. H., and Jeong, S. H. (2007). Novel complex class 1 integron bearing an ISCR1 element in an *Escherichia coli* isolate carrying the blaCTX-M-14 gene. *Antimicrob. Agents Chemother.* 51, 3017–3019. doi: 10.1128/AAC.00279-07

Barton, B. M., Harding, G. P., and Zuccarelli, A. J. (1995). A general method for detecting and sizing large plasmids. *Anal. Biochem.* 226, 235–240. doi: 10.1006/abio.1995.1220

Bonnin, R. A., Cuzon, G., Poirel, L., and Nordmann, P. (2013). Multidrug-resistant *Acinetobacter baumannii* clone, France. *Emerg. Infect. Dis.* 19, 822–823. doi: 10.3201/eid1905.121618

Bonnin, R. A., Poirel, L., Carattoli, A., and Nordmann, P. (2012). Characterization of an IncFII plasmid encoding NDM-1 from *Escherichia coli* ST131. *PLoS ONE* 7:e34752. doi: 10.1371/journal.pone.0034752

Carattoli, A., Bertini, A., Villa, L., Falbo, V., Hopkins, K. L., and Threlfall, E. J. (2005). Identification of plasmids by PCR-based replicon typing. *J. Microbiol. Methods* 63, 219–228. doi: 10.1016/j.mimet.2005.03.018

Carattoli, A., Villa, L., Poirel, L., Bonnin, R. A., and Nordmann, P. (2012). Evolution of IncA/C blaCMY-2 carrying plasmids by acquisition of the blaNDM-1 carbapenemase gene. *Antimicrob. Agents Chemother.* 56, 783–786. doi: 10.1128/AAC.05116-11

Chen, Y., Zhou, Z., Jiang, Y., and Yu, Y. (2011). Emergence of NDM-1 producing *Acinetobacter baumannii* in China. *J. Antimicrob. Chemother.* 66, 1255–1259. doi: 10.1093/jac/dkr082

CLSI [Clinical and Laboratory Standards Institute]. (2014). *Performance Standards for Antimicrobial Susceptibility Testing: Twenty-Fourth Informational Supplement. CLSI Document M*100-S24. Wayne, PA: Clinical and Laboratory Standards Institute.

Dawes, F. E., Kuzevski, A., Bettelheim, K. A., Hornitzky, M. A., Djordjevic, S. P., and Walker, M. J. (2010). Distribution of class 1 integrons with IS26-mediated deletions in their 3'-conserved segments in *Escherichia coli* of human and animal origin. *PLoS ONE* 5:e12754. doi: 10.1371/journal.pone.0012754

Dolejska, M., Villa, L., Poirel, L., Nordmann, P., and Carattoli, A. (2013). Complete sequencing of an IncHI1 plasmid encoding the carbapenemase NDM-1, the ArmA 16S RNA methylase and a resistance-nodulation-cell division/multidrug efflux pump. *J. Antimicrob. Chemother.* 68, 34–39. doi: 10.1093/jac/dks357

Doublet, B., Boyd, D., Douard, G., Praud, K., Cloeckaert, A., and Mulvey, M. R. (2012). Complete nucleotide sequence of the multidrug resistance IncA/C plasmid pR55 from *Klebsiella pneumoniae* isolated in 1969. *J. Antimicrob. Chemother.* 67, 2354–2360. doi: 10.1093/jac/dks251

Fernandez-Alarcon, C., Singer, R. S., and Johnson, T. J. (2011). Comparative genomics of multidrug resistance-encoding IncA/C plasmids from commensal and pathogenic *Escherichia coli* from multiple animal sources. *PLoS ONE* 6:e23415. doi: 10.1371/journal.pone.0023415

Fu, Y. L., Du, X., Ji, J., Chen, Y., Jiang, Y., and Yu, Y. (2012). Epidemiological characteristics and genetic structure of blaNDM-1 in non-baumannii *Acinetobacter* spp. *in China. J. Antimicrob. Chemother.* 67, 2114–2122. doi: 10.1093/jac/dks192

Ghosh, A., Sarkar, A., Chowdhury, G., Pazhani, G. P., and Ramamurthy, T. (2014). An overview on the New Delhi Mettalo β-lactamase (NDM)-producers. *Proc. Indian Natl. Sci. Acad.* 80, 547–563. doi: 10.16943/ptinsa/2014/v80i3/55132

Ho, P. L., Lo, W. U., Yeung, M. K., Lin, C. H., Chow, K. H., Ang, I., et al. (2011). Complete sequencing of pNDM-HK encoding NDM-1 carbapenemase from a multidrug-resistant *Escherichia coli* strain isolated in Hong Kong. *PLoS ONE* 6:e17989. doi: 10.1371/journal.pone.0017989

Hopkins, K. L., Batchelor, M. J., Liebana, E., Deheer-Graham, A. P., and Threlfall, E. J. (2006). Characterisation of CTX-M and AmpC genes in human isolates of *Escherichia coli* identified between 1995 and 2003 in England and Wales. *Int. J. Antimicrob. Agents* 28, 180–192. doi: 10.1016/j.ijantimicag.2006.03.027

Kado, C. I., and Liu, S. T. (1981). Rapid procedure for detection and isolation of large and small plasmids. *J. Bacteriol.* 145, 1365–1373.

Kang, M. S., Besser, T. E., and Call, D. R. (2006). Variability in the region downstream of the blaCMY-2-lactamase gene in *Escherichia coli* and *Salmonella enterica* plasmids. *Antimicrob. Agents Chemother.* 50, 1590–1593. doi: 10.1128/AAC.50.4.1590-1593.2006

Kiiru, J., Butaye, P., Goddeeris, B. M., and Kariuki, S. (2013). Analysis for prevalence and physical linkages amongst integrons, ISEcp1, ISCR1, Tn21 and Tn7 encountered in *Escherichia coli* strains from hospitalized and non-hospitalized patients in Kenya during a 19-year period (1992-2011). *BMC Microbiol.* 13:109. doi: 10.1186/1471-2180-13-109

Kumarasamy, K. K., Toleman, M. A., Walsh, T. R., Bagaria, J., Butt, F., Balakrishnan, R., et al. (2010). Emergence of a new antibiotic resistance mechanism in India, Pakistan, and the UK: a molecular, biological, and epidemiological study. *Lancet Infect. Dis.* 10, 597–602. doi: 10.1016/S1473-3099(10)70143-2

Meziane-Cherif, D., and Courvalin, P. (2014). Antibiotic resistance: to the rescue of old drugs. *Nature* 510, 477–478. doi: 10.1038/510477a

Ouellette, M., Gerbaud, G., and Courvalin, P. (1988). Genetic, biochemical and molecular characterization of strains of *Vibrio cholerae* multiresistant to antibiotics. *Ann. Inst. Pasteur Microbiol.* 139, 105–113.

Petroni, A., Corso, A., Melano, R., Cacace, M. L., Bru, A. M., Rossi, A., et al. (2002). Plasmidic extended-spectrum beta-lactamases in *Vibrio cholerae* O1 El Tor isolates in Argentina. *Antimicrob. Agents Chemother.* 46, 1462–1468. doi: 10.1128/AAC.46.5.1462-1468.2002

Poole, T. L., Edrington, T. S., Brichta-Harhay, D. M., Carattoli, A., Anderson, R. C., and Nisbet, D. J. (2009). Conjugative transferability of the A/C plasmids from *Salmonella enterica* isolates that possess or lack blaCMY in the A/C plasmid backbone. *Foodborne Pathog. Dis.* 6, 1185–1194. doi: 10.1089/fpd.2009.0316

Rolain, J. M., Parola, P., and Cornaglia, G. (2010). New Delhi metallo-β-lactamase (NDM-1): towards a new pandemia? *Clin. Microbiol. Infect.* 16, 1699–1701. doi: 10.1111/j.1469-0691.2010.03385.x

Sekizuka, T., Matsui, M., Yamane, K., Takeuchi, F., Ohnishi, M., Hishinuma, A., et al. (2011). Complete sequencing of the blaNDM-1-positive IncA/C plasmid from *Escherichia coli* ST38 isolate suggests a possible origin from plant pathogens. *PLoS ONE* 6:e25334. doi: 10.1371/journal.pone.0025334

Toleman, M. A., Bennett, P. M., and Walsh, T. R. (2006a). Common regions e.g. *orf*513 and antibiotic resistance: IS91-like elements evolving complex class 1 integrons. *J. Antimicrob. Chemother.* 58, 1–6. doi: 10.1093/jac/dkl204

Toleman, M. A., Bennett, P. M., and Walsh, T. R. (2006b). ISCR elements: novel gene-capturing systems of the 21st century? *Microbiol. Mol. Biol. Rev* 70, 296–316. doi: 10.1128/MMBR.00048-05

Uppu, D. S., Manjunath, G. B., Yarlagadda, V., Kaviyil, J. E., Ravikumar, R., Paramanandham, K., et al. (2015). Membrane-active macromolecules resensitize NDM-1 gram-negative clinical isolates to chloramphenicol antibiotics. *PLoS ONE* 10:e0119422. doi: 10.1371/journal.pone.0119422

Walsh, T. R., Weeks, J., Livermore, D. M., and Toleman, M. A. (2011). Dissemination of NDM-1 positive bacteria in the New Delhi environment and its implications for human health: an environmental point prevalence study. *Lancet Infect. Dis.* 11, 355–362. doi: 10.1016/S1473-3099(11)70059-7

Winokur, P. L., Vonstein, D. L., Hoffman, L. J., Uhlenhopp, E. K., and Doern, G. V. (2001). Evidence for transfer of CMY-2 AmpC β-lactamase plasmids between *Escherichia coli* and *Salmonella* isolates from food animals and humans. *Antimicrob. Agents Chemother.* 45, 2716–2722. doi: 10.1128/AAC.45.10.2716-2722.2

Yong, D., Toleman, M. A., Giske, C. G., Cho, H. S., Sundman, K., Lee, K., et al. (2009). Characterization of a new metallo-β-lactamase gene, blaNDM-1, and a novel erythromycin esterase gene carried on a unique genetic structure in *Klebsiella pneumoniae* sequence type 14 from India. *Antimicrob. Agents Chemother.* 53, 5046–5054. doi: 10.1128/AAC.00774-09

Yu, W. L., Chen, S. C., Hung, S. W., Chuang, Y. C., Chung, J. G., Chen, I. C., et al. (2006). Genetic association of blaSHV-5 with transposable elements IS26 and IS5 in *Klebsiella pneumoniae* from Taiwan. *Clin. Microbiol. Infect.* 12, 806–809. doi: 10.1111/j.1469-0691.2006.01488.x

7

Functional Categorization of Carbapenemase-Mediated Resistance by a Combined Genotyping and Two-Tiered Modified Hodge Test Approach

*Marcus H. Wong[1,2], Yi Li[3], Edward W. Chan[1,2] and Sheng Chen[1,2]**

[1] Shenzhen Key Laboratory for Food Biological Safety Control, Food Safety and Technology Research Centre, Shenzhen Research Institute, The Hong Kong Polytechnic University, Shenzhen, China, [2] State Key Laboratory of Chirosciences, Department of Applied Biology and Chemical Technology, The Hong Kong Polytechnic University, Hong Kong, China, [3] Laboratory of Clinical Inspection, Henan Provincial People's Hospital, Zhengzhou, China

Correspondence:
Sheng Chen,
State Key Laboratory of Chirosciences, Department of Applied Biology and Chemical Technology, The Hong Kong Polytechnic University, Hong Kong, China
sheng.chen@polyu.edu.hk

The functional relationship between the detection of carbapenemase activity and phenotypic resistance in Gram-negative bacterial pathogens is often ill-defined. To address this issue, we developed a two-tiered Modified Hodge Test approach for carbapenemase detection and typing, in which the use of *Pseudomonas aeruginosa* strain PAO1 and *Escherichia coli* as indicator strains conferred two levels of sensitivities to carbapenemases. When applied alongside PCR genotyping tests for existence of known carbapenemase genes in 92 carbapenem resistant clinical isolates, this method is extremely useful in elucidating the relative role by which different enzymes contributed to the prevalent carbapenem-resistance phenotypes. With this study approach, we showed that the proportion of *P. aeruginosa* and *Acinetobacter baumannii* strains whose carbapenem resistance phenotypes could at least be partially attributed to carbapenemase were 34 and 89%, respectively. Our data also facilitates detailed functional categorization of carbapenem resistance phenotypes on the basis of the types and activities of detectable carbapenemase produced by the test organism. For example, six *A. baumannii* isolates harboring the $bla_{OXA-51/23}$-like gene without detectable enzymatic activities were identified, suggesting that other resistance mechanisms may be involved. On the other hand, there were seven *P. aeruginosa* strains which produced carbapenemase phenotype without harboring known carbapenemase genes, inferring the existence of some hitherto unknown resistance determinants. Findings in this work therefore provide a comprehensive view on the cellular basis of carbapenem resistance phenotypes in major Gram-negative bacterial species, paving the way for development of novel strategies to reverse the effects of the major resistance mechanisms concerned.

Keywords: *Pseudomonas aeruginosa, Acinetobacter baumannii*, carbapenem resistance, two-tiered Modified Hodge Test, genotyping

Introduction

Dissemination of carbapenem-resistant organisms continues to cause an increasing number of severe and often untreatable bacterial infections in nosocomial settings (Woodford et al., 2014). Production of active carbapenemase is a major carbapenem resistance mechanism among clinical Gram-negative isolates. There are several major classes of carbapenemase, including Class A serine carbapenemases such as *Klebsiella pneumoniae* carbapenemase (KPC), Class B metallo-β-lactamase such as New Delhi metallo-β-lactamse (NDM), Verona integron-encoded metallo-β-lactamase (VIM), IMP, and Carbapenem-hydrolyzing Class D β-lactamases (CHDL) such as OXA-48, OXA-23, OXA-24, etc. Nordmann et al. (2011). In addition, OXA-51 has been reported to be intrinsic to *Acinetobacter baumannii* and it normally does not lead to carbapenem resistance unless an Insertion Sequence, IS*Aba1*, is introduced into its upstream region, enhancing the strength of the promoter and causing over-expression of the *oxa-51* gene. Even if this happens, it has been reported that the CHDL OXA-51 produced exhibits only weak carbapenemase activity (Turton et al., 2006). Determination of the presence of carbapenemase activity is pivotal for clinicians to devise a suitable treatment regimen. To date, several enzymatic- and inhibitor-based phenotypic assays have been developed for detection of bacterial carbapenemase activities. Enzymatic assays, including Modified Hodge Test (MHT) and Carba NP test, involve the direct observation of carbapenmase hydrolysing phenotype in the presence of an indicator organism or a pH indicator (Nordmann et al., 2012b). Inhibitor assays which include commercially available *E* test MBL Strip (Nordmann et al., 2012a) and the Imipenem/EDTA inhibition test, deploy chemicals such as EDTA, a metal chelator, to inhibit enzyme activity, and are mainly used for detecting Metallo-β-lactamase (MBL) activity. Above all, MHT is a recommended method for identifying carbapenemase producers (CLSI, 2013). However, this method is not ideal for detection of CHDL, including OXA-51, OXA-23, and OXA-58 in *Acinetobacter* (Dortet et al., 2014). OXA-51 is intrinsic to *A. baumannii* and normally does not lead to carbapenem resistance, thus it is regarded as a less active carbapenemase, unless it is over-expressed by introduction of a strong promoter in its upstream region as a result of insertion of the insertion sequence IS*Aba1*(Turton et al., 2006). Detection of other CHDLs such as OXA-23 and OXA-58 in *A. baumannii* has always been a challenge due to the low membrane permeability of the organisms (Dortet et al., 2014). Failure to detect carbapenemases produced by clinical strains is a serious concern since false categorization of the corresponding resistant organisms may result in wrong diagnosis and ineffective treatment. Although performance of *Pseudomonas aeruginosa* as an indicator strain in MHT to detect MBLs was found to be unsatisfactory previously due to the low specificity of this method and its high tendency to produce intermediate results (Pasteran et al., 2011), the suitability of this test approach to detect CHDLs has not been evaluated. In this work, replacement of *Escherichia coli* ATCC25922 by *P. aeruginosa* PAO1 as the indicator strain was found to significantly improve sensitivity in detection of less active carbapenemase producers. We showed that combined analysis of results of carbapenemase detection tests with MHT and PAO1-MHT, which cover detection sensitivity toward CHDLs and other carbapenemases, as well as those of PCR genotyping of carbapenemase genes, allows highly accurate assessment of the type and relative role of the carbapenemase involved in conferring the observable phenotype. This analytical approach shall therefore help better define the molecular basis of clinical carbapenem resistance and establish a profile of carbapenemase-encoding elements commonly harbored by each of the key Gram-negative pathogens. Such data are essential for future development of novel antimicrobial strategies, especially those involved in carbapenemase inhibition.

Materials and Methods

Bacterial Isolates

A total of 102 clinical isolates were collected from patients in The Prince of Wales Hospital, Hong Kong ($n = 27$), and Henan Provincial People's Hospital, Henan, China ($n = 75$). The collection comprised *A. baumannii* ($n = 68$), *P. aeruginosa* ($n = 32$), *Enterobacter cloacae* ($n = 1$) and *E. coli* ($n = 1$). Four *P. otitidis* isolates recovered from food products were also included as control (**Table 1**). The isolates were identified by the Vitek Bacterial Identification System, followed by 16SrRNA sequencing.

Antimicrobial Susceptibility Testing

Antimicrobial susceptibility assay was performed and interpreted according to the CLSI guidelines (CLSI, 2013). Antimicrobial susceptibility testing was carried out by the agar dilution method with meropenem and imipenem as the test agents. Results were interpreted following the CLSI guidelines (CLSI, 2013). *E. coli* strains ATCC 25922, 35218, and *P. aeruginosa* strain ATCC 27853 were used as quality control.

PCR Detection of Carbapenemase Genes

Possession of carbapenemase genes by isolates was investigated by PCR as described previously (Woodford et al., 2006; Poirel et al., 2011; Thaller et al., 2011; Lopes et al., 2012). The PCR assays included detection of bla_{OXA-23}, bla_{OXA-24}, bla_{OXA-51}, bla_{OXA-58}, bla_{IMP}, bla_{VIM}, bla_{OXA}, bla_{SIM}, bla_{KPC}, bla_{SPM}, bla_{DIM}, bla_{GIM}, bla_{NDM}, and bla_{POM}. Primers used in this study are shown in **Table 1**. Genomic DNA was obtained by using the Purelink Genomic Mini Kit (Life Technologies). 1 μl of DNA template was added into PCR reaction mix (1X PCR Buffer, 3 mM $MgCl_2$, 0.2 mM dNTP, 0.25 μM primer, 1U Takara rTaq). The assays were validated by including a pET15B vector carrying target constructs as positive control, and amplicons were subjected to sequencing.

Two-Tiered Modified Hodge Test

Modified Hodge Test for meropenem was performed on all isolates as described using the *E. coli* strain ATCC25922 and *P. aeruginosa* strain PAO1 as indicator organisms (designated as MHT and MHT-PAO1, respectively), and a bla_{IMP}-producing *E. coli* strain as positive control. Meropenem susceptibility testing disks were obtained from Oxoid, and Mueller-Hinton Agar

TABLE 1 | Primers used in this study.

β-lactamase	Forward sequence (5′–3′)	Reverse sequence (5′–3′)	Amplicon size (bp)	Reference
bla_{OXA-51}	TAATGCTTTGATCGGCCTTG	TGGATTGCACTTCATCTTGG	353	Woodford et al. (2006)
bla_{OXA-23}	GATCGGATTGGAGAACCAGA	ATTTCTGACCGCATTTCCAT	501	
bla_{OXA-24}	GGTTAGTTGGCCCCCTTAAA	AGTTGAGCGAAAAGGGGATT	246	
bla_{OXA-58}	AAGTATTGGGGCTTGTGCTG	CCCCTCTGCGCTCTACATAC	599	
bla_{IMP}	GGAATAGAGTGGCTTAAYTCTC	GGTTTAAYAAAACAACCACC	232	Poirel et al. (2011)
bla_{VIM}	GATGGTGTTTGGTCGCATA	CGAATGCGCAGCACCAG	390	
bla_{OXA}	GCGTGGTTAAGGATGAACAC	CATCAAGTTCAACCCAACCG	438	
bla_{SIM}	TACAAGGGATTCGGCATCG	TAATGGCCTGTTCCCATGTG	570	
bla_{KPC}	CGTCTAGTTCTGCTGTCTTG	CTTGTCATCCTTGTTAGGCG	798	
bla_{SPM}	AAAATCTGGGTACGCAAACG	ACATTATCCGCTGGAACAGG	271	
bla_{DIM}	GCTTGTCTTCGCTTGCTAACG	CGTTCGGCTGGATTGATTTG	699	
bla_{GIM}	TCGACACACCTTGGTCTGAA	AACTTCCAACTTTGCCATGC	477	
bla_{NDM}	GGTTTGGCGATCTGGTTTTC	CGGAATGGCTCATCACGATC	621	
bla_{POM}	ACGTCGCTGATGCTCAG	CCTGCGTCATCAGAGACCTC	1143	Thaller et al. (2011)

(MHA) was obtained from BD. Indicator organisms and test isolates were grown overnight on MHA. Bacterial suspension of indicated organism at 0.5 McFarland Standard was prepared by saline and spread on MHA plates on which meropenem disks (10 μg, Oxoid) were placed at center. Inoculated plates were allowed to dry for 10 min followed by streaking a line of test strains with a 10 μl inoculation loop from the drug disk to the edge of agar plate. Results were recorded after 18 h incubation at 37°C. Production of active carbapenemases was regarded as positive upon observation of "clover-leaf" appearance due to enhanced growth of indicator strain toward the meropenem disk alongside the test strain.

Results and Discussion

A total of 106 strains (102 clinical isolates and four food isolates) were first subjected to antibiotic susceptibility tests. Among the 68 *A. baumannii* isolates tested, 52 were resistant to meropenem and imipenem. The rest of the test strains, including 32 *P. aeruginosa* isolates, four *P. Otitis* isolates, one *Enterobacter cloacae* strain, and one *E. coli* strain, were all carbapenem resistant. PCR genotyping experiments showed that known carbapenemase genes were not detectable in as many as 28 carbapenem resistant *P. aeruginosa* strains. The other four strains were found to contain the bla_{IMP} (three strains) and bla_{KPC} genes (one strain), respectively. MHT tests were then performed to assess the degree of correlation between existence of carbapenemase genes and the detection of carbapenemase activity, with results indicating that the carbapenemase bla_{IMP} was detectable by both MHT tests, yet the bla_{KPC} enzyme was detectable by MHT-PAO1 only (**Figure 1**). Despite such discrepancy, the four host strains concerned all exhibited high level meropenem resistance (≥32 μg/ml), suggesting that active carbapenemase production might not be the sole basis of carbapenem resistance in these organisms. Likewise, 5 of the 28 *P. aeruginosa* strains which did not harbor any known carbapenemase gene were found to produce a carbapenemase phenotype detectable by MHT–PAO1 only; in addition, another two strains were found to produce an enzyme detectable by both MHT tests (**Figure 1**). Although false positives have been described in MHT to detect carbapenemases in *P. aeruginosa* previously, it can not rule out the possibility that other unknown carbapenem resistance mechanisms are present in these seven strains (Vasoo et al., 2013). **Table 2** summarizes the results of genotyping, MHT tests and antibiotic susceptibility levels of the 106 strains studied in this work. For carbapenem-resistant *P. aeruginosa,* MHT–POA1 can detect 100% (4/4) of know carbapenemase producers and 25% (7/28) of isolates without known carbapenemase genes, while MHT can detect 75% (3/4) of know carbapenemase producers and 7% (2/28) of isolates without known carbapenemase genes.

Of the 68 *A. baumannii* isolates tested, 17 were found to contain the intrinsic bla_{OXA-51}-like CHDL-producing gene only, among which were the 14 carbapenem-sensitive strains. Interestingly, the IS*Aba1* element was detectable in the upstream region of a bla_{OXA-51}-like gene in the other three of these 17 strains, which were also found to produce a carbapenemase detectable only by MHT–PAO1, corroborating with the fact that the IS*Aba1* element is essential for over-expression of the bla_{OXA-51}-like gene. The remaining 51 isolates were found to harbor other carbapenemase-producing genetic elements, with the vast majority (49 strains) containing the $bla_{OXA-51/23}$-like gene. Among these 49 strains, whose meropenem MIC ranged from 8 to ≥32 μg/ml, 16 were positive in both MHT tests, 27 were positive in the MHT-PAO1 only, and six strains did not produce any detectable carbapenemases (**Figure 1**). These findings appear to suggest that $bla_{OXA-51/23}$-like gene product may not be the sole factor responsible for the development of the carbapenem resistance phenotypes of the organisms concerned.

The other two of the 51 isolates which harbored carbapenemase genes included one highly resistant strain (meropenem MIC ≥ 32 μg/ml) which contained a $bla_{OXA-51/24}$-like element. However, this strain was tested positive only in MHT-PAO1, again suggesting the involvement of non-carbapenemase

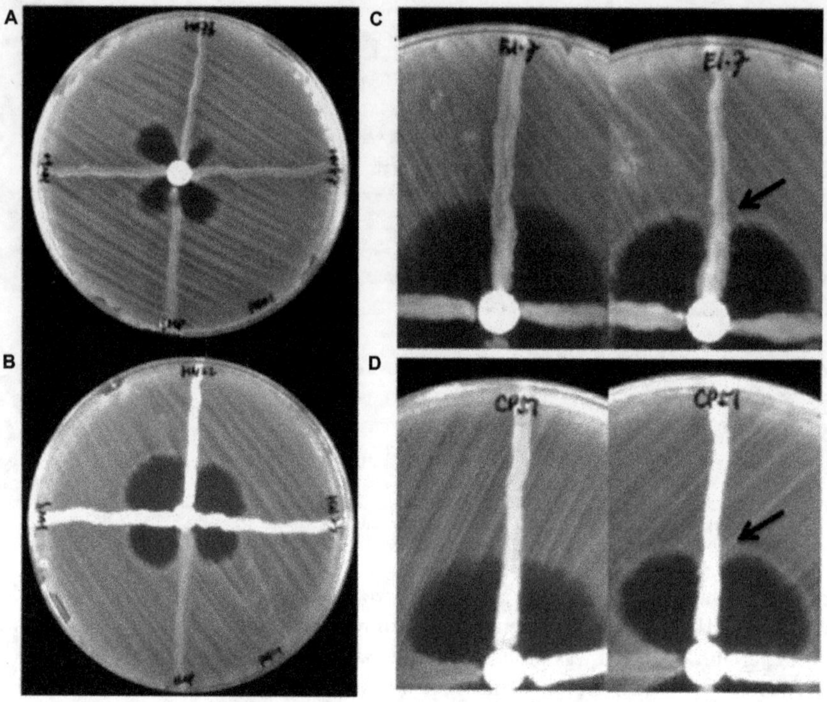

FIGURE 1 | Result of Modified Hodge Test for representative isolates. (A) Modified Hodge Test (MHT–PAO1 of three bla_{IMP}–borne *Pseudomonas aeruginosa* strains (left, upper, right) with an *Escherichia coli* strain harboring the bla_{IMP} gene as positive control (bottom). **(B)** MHT–PAO1 of three bla_{OXA-23}-borne *Acinetobacter baumannii* strains (left, upper, right) with an *E. coli* strain harboring the bla_{IMP} gene as positive control (bottom). **(C)** MHT (Left) and MHT-PAO1 (right) of a bla_{POM}-borne *P. otitidis* strain (E1-7). **(D)** MHT (Left) and MHT-PAO1 (right) of a bla_{OXA-23}-borne *A. baumannii* strain (CP51). Arrows indicate enhanced growth of indicator organism.

mechanisms. Likewise, another strain which contained both the bla_{OXA-51}-like and bla_{NDM} genes, exhibited a meropenem MIC of ≥32 μg/ml, which is significantly higher than that observed in the three strains containing the bla_{OXA-51}-like gene alone, plus an IS*Aba1* element (8 μg/ml). Such discrepancy highlighted the additive effects of different enzymes on the strength of phenotypic carbapenem resistance. It should be noted that the phenomenon of high level carbapenem resistance and low level enzyme production was also observable in a meropenem-resistant *E. coli* strain harboring the bla_{NDM} gene in this study, as well as several *E. coli* and *K. pneumoniae* isolates described previously (Girlich et al., 2012). This phenomenon suggests the possibility that the level of carbapenem resistance exhibited by a given clinical isolate is more closely related to the type rather than level of carbapenemase produced by the organism. Alternatively, these findings could also indicate that the bla_{NDM} enzyme might only confer relatively low level carbapenem resistance, and other factors including altered membrane permeability and expression of efflux may be responsible for the high level carbapenem resistance phenotypes exhibited by the host strains concerned. This possibility is also raised by the finding that the meropenem MIC of an *Enterobacter cloacae* strain, which harbored the bla_{NDM} and bla_{KPC} genes and exhibited a positive result in both MHT tests, was only16 μg/ml. On the other hand, bla_{POM}, encoding an intrinsic Metallo-β-Lactamase, was detected in all the four *P. otitidis* isolates (**Table 2**). This enzyme, which was consistently detectable by the MHT–PAO1 test only, was associated with a meropenem MIC of 16–≥32 μg/ml. These findings suggest a need to investigate the role of each specific carbapenemase in conferring carbapenem resistance in the host strain. In summary, within 60 isolates of carbapenem-resistant *P. otitidis, h coli,* and *K. pneumonia,* 90% (54/60) were positive for MHT–POA1 test, while only 26% (16/60) were positive for MHT testing.

In this work, MHT–PAO1 yielded 63 carbapenemase-positive cases whereas MHT yielded 22 positive cases, with MHT–PAO1 being much more sensitive in detecting CHDL in *A. baumannii*, as well as the bla_{NDM} enzyme. Detection of CHDLs produced by *Acinetobacter* by MHT has always been a challenge due to the relative weak activity of enzymes and low membrane permeability of the organism. Unsatisfactory CHDL detection by MHT, including those produced by bla_{OXA-23}, bla_{OXA-24}, and bla_{OXA-58}, has been documented (Bonnin et al., 2012). Previous studies have also shown that false negative results of MHT in detection of bla_{NDM} in *Acinetobacter* and *Enterobacteriaceae* were not uncommon, which may be due to the variation in zinc content in MHA produced by different suppliers (Pasteran et al., 2011; Girlich et al., 2012). Although performance of *P. aeruginosa* ATCC27853 as indicator strain in MHT to detect certain *P. aeruginosa* MBLs has been attempted by another group which reported a low sensitivity and specificity (Pasteran et al., 2011), our data confirmed that the use of MHT–PAO1 can successfully identify the enzymes encoded by bla_{IMP}, bla_{NDM}, and

TABLE 2 | Summary of genotypic and phenotypic characteristics of 106 Gram-negative bacterial isolates tested in this study.

Bacterial species tested (Carbapenemases)	Meropenem MIC(μg/ml)	No. of isolates	No. of positive Modified Hodge Test (MHT)	
			***Pseudomonas aeruginosa* PAO1**	***Escherichia coli* ATCC25922**
P. aeruginosa				
bla$_{IMP}$	≥32	3	3	3
bla$_{KPC}$	≥32	1	1	0
Not Detected	16 – ≥32	28	7	2†
P. otitidis				
bla$_{POM}$	16 – ≥32	4	4	0
Acinetobacter baumannii				
bla$_{OXA-51like}$	<0.5	14*	0	0
IS*Aba1* + *bla*$_{OXA-51like}$	8	3	3	0
bla$_{OXA-51/23like}$	8 – ≥32	49	43	16†
bla$_{OXA-51/24like}$	≥32	1	1	0
bla$_{OXA-51like}$ + *bla*$_{NDM}$	≥32	1	1	0
Enterobacter cloacae				
bla$_{NDM}$ + *bla*$_{KPC}$	16	1	1	1
E. coli				
bla$_{NDM}$	16	1	1	0
Total		106	63	22

Carbapenem-susceptible. †Also positive in MHT–PAO1 test.

TABLE 3 | Categorization of carbapenem susceptibility phenotypes.

Category	Carbapenem susceptibility phenotypes	MHT (*E. coli*)	MHT-PAO1	PCR-genotyping	Implication	Examples
I	S	–	–	+	The carbapenemase gene detectable by genotyping does not produce carbapenemase	14 carbapenem sensitive *A. baumannii* strains harboring the *bla*$_{OXA-51}$ like gene
II	R	–	–	–	Carbapenem resistance not due to carbapenemase	21 carbapenem resistant *P. aeruginosa* strains
III	R	+/–	+	–	Carbapenem resistance due to a combination of resistance mechanisms including an unknown carbapenemase or overexpression of AmpC	Seven *P. aeruginosa* strains
IV	R	+	+	+	Carbapenem resistance due to carbapenemase encoded by the gene detectable by genotyping	Various *P. aeruginosa* and *A. baumannii* strains
V	R	–	+	+	Carbapenem resistance due to a combination of resistance mechanisms including a weakly detected carbapenemase	*P. otitidis* and *A. baumannii* strains with a meropenem MIC ≥32 μg/ml
VI	I	–	+	+	Carbapenem resistance due to a weakly detected carbapenemase encoded by the gene detectable by genotyping	Three *A. baumannii* strains harboring the IS*Aba1*+ *bla*$_{OXA-51like}$ elements

Categorization according to results of the two-tiered MHT tests and genotyping of carbapenemase genes. S, carbapenem sensitive; R, carbapenem resistant; I, intermediate or low level carbapenem resistance.

bla$_{POM}$ in *P. aeruginosa*, *E. coli*, and *P. otitidis*, respectively. Currently, testing carbapenemase activity by MHT or other assays has been widely used to determine the carbapenem resistance potential of clinical Gram-negative bacterial pathogens to guide the treatment decision. False negative detection of CHDLs by traditional MHT may result in wrong diagnosis and ineffective treatment. The improved two-tiered MHT can achieve much less false negative, around 9% (6/64) in isolates carrying known carbapenemase genes compared to 71% (45/64) by MHT test, suggesting the high reliability of the two-tiered MHT in the clinical screening of carbapenem-resistant Gram-negative pathogens in particular CHDL producers. If the discrepancy in results of conventional MHT and MHT–PAO1 observable in this study is indeed due to the low enzyme levels as discussed above, future research efforts should be focused on investigating the underlying basis of high level carbapenem resistance in organisms producing less active carbapenemase.

Findings of this study allow us to systematically categorize the test strains according to their carbapenem susceptibility phenotypes and results of the MHT and genotyping tests. As shown in **Table 3** which lists the scenarios which give rise to at least six possible combinations of the phenotypic and genotypic test results, we can not only better define the relative role of each carbapenemase in the production of a resistance phenotype on the basis of the test results, but also identify strains whose resistance phenotypes were also due to other resistance mechanisms. For example, if a highly resistant strain with positive genotyping result is found to produce a weakly detected carbapenamase (positive in MHT–PAO1 test), it is likely that the resistance phenotype is due to multiple mechanisms including production of active carbapenemase. Here we showed that 34% of the clinical carbapenem resistant *P. aeruginosa* strains tested were at least partially due to carbapenemase, yet the corresponding rate for *A. baumannii* was as high as 89%. Detailed delineation of the molecular basis of carbapenem resistance in these Gram-negative pathogens, which were previously not possible to perform on the basis of a single phenotypic or molecular test, shall greatly facilitate development of novel strategies to treat bacterial infections, especially those aimed at devising a carbapenemase inhibition approach.

Conclusion

A two-tiered MHT approach which conferred two levels of sensitivities for carbapenemase detection can be applied alongside PCR genotyping of known carbapenemase genes to determine the degree of contribution of specific carbapenemase in expression of phenotypic carbapenem resistance in Gram-negative bacterial pathogens. This analytical approach therefore facilitates categorization of carbapenem resistance phenotypes and identification of the major mechanisms concerned in bacterial species in which clinical carbapenem resistance is common.

Acknowledgments

This work was supported by the Chinese National Key Basic Research and Development (973) Program (2013CB127200) and the Health and Medical Research Fund of the Food and Health Bureau, The Government of Hong Kong (12111612 to SC).

References

Bonnin, R. A., Naas, T., Poirel, L., and Nordmann, P. (2012). Phenotypic, biochemical, and molecular techniques for detection of metallo-beta-lactamase NDM in *Acinetobacter baumannii*. *J. Clin. Microbiol.* 50, 1419–1421. doi: 10.1128/JCM.06276-11

CLSI. (2013). *Performance Standards for Antimicrobial Susceptibility Testing: 23th Informational Supplement M100-S23*. Wayne, PA: Clinical and Laboratory Standards Institute.

Dortet, L., Poirel, L., Errera, C., and Nordmann, P. (2014). CarbAcineto NP Test for rapid detection of carbapenemase-producing *Acinetobacter* spp. *J. Clin. Microbiol.* 52, 2359–2364. doi: 10.1128/JCM.00594-14

Girlich, D., Poirel, L., and Nordmann, P. (2012). Value of the modified Hodge test for detection of emerging carbapenemases in Enterobacteriaceae. *J. Clin. Microbiol.* 50, 477–479. doi: 10.1128/JCM.05247-11

Lopes, B. S., Evans, B. A., and Amyes, S. G. (2012). Disruption of the blaOXA-51-like gene by ISAba16 and activation of the blaOXA-58 gene leading to carbapenem resistance in *Acinetobacter baumannii* Ab244. *J. Antimicrob. Chemother.* 67, 59–63. doi: 10.1093/jac/dkr415

Nordmann, P., Gniadkowski, M., Giske, C. G., Poirel, L., Woodford, N., Miriagou, V., et al. (2012a). Identification and screening of carbapenemase-producing Enterobacteriaceae. *Clin. Microbiol. Infect.* 18, 432–438. doi: 10.1111/j.1469-0691.2012.03815.x

Nordmann, P., Poirel, L., and Dortet, L. (2012b). Rapid detection of carbapenemase-producing Enterobacteriaceae. *Emerg. Infect. Dis.* 18, 1503–1507. doi: 10.3201/eid1809.120355

Nordmann, P., Naas, T., and Poirel, L. (2011). Global spread of Carbapenemase-producing Enterobacteriaceae. *Emerg. Infect. Dis.* 17, 1791–1798. doi: 10.3201/eid1710.110655

Pasteran, F., Veliz, O., Rapoport, M., Guerriero, L., and Corso, A. (2011). Sensitive and specific modified Hodge test for KPC and metallo-beta- lactamase detection in *Pseudomonas aeruginosa* by use of a novel indicator strain, *Klebsiella pneumoniae* ATCC 700603. *J. Clin. Microbiol.* 49, 4301–4303. doi: 10.1128/JCM.05602-11

Poirel, L., Walsh, T. R., Cuvillier, V., and Nordmann, P. (2011). Multiplex PCR for detection of acquired carbapenemase genes. *Diagn. Microbiol. Infect. Dis.* 70, 119–123. doi: 10.1016/j.diagmicrobio.2010.12.002

Thaller, M. C., Borgianni, L., Di Lallo, G., Chong, Y., Lee, K., Dajcs, J., et al. (2011). Metallo-beta-lactamase production by *Pseudomonas otitidis*: a species-related trait. *Antimicrob. Agents Chemother.* 55, 118–123. doi: 10.1128/AAC.01062-10

Turton, J. F., Ward, M. E., Woodford, N., Kaufmann, M. E., Pike, R., Livermore, D. M., et al. (2006). The role of ISAba1 in expression of OXA carbapenemase genes in *Acinetobacter baumannii*. *FEMS Microbiol. Lett.* 258, 72–77. doi: 10.1111/j.1574-6968.2006.00195.x

Vasoo, S., Cunningham, S. A., Kohner, P. C., Simner, P. J., Mandrekar, J. N., Lolans, K., et al. (2013). Comparison of a novel, rapid chromogenic biochemical assay, the Carba NP test, with the modified Hodge test for detection of carbapenemase-producing gram-negative bacilli. *J. Clin. Microbiol.* 51, 3097–3101. doi: 10.1128/JCM.00965-13

Woodford, N., Ellington, M. J., Coelho, J. M., Turton, J. F., Ward, M. E., Brown, S., et al. (2006). Multiplex PCR for genes encoding prevalent OXA carbapenemases in *Acinetobacter spp. Int. J. Antimicrob. Agents* 27, 351–353. doi: 10.1016/j.ijantimicag.2006.01.004

Woodford, N., Wareham, D. W., Guerra, B., and Teale, C. (2014). Carbapenemase-producing Enterobacteriaceae and non-Enterobacteriaceae from animals and the environment: an emerging public health risk of our own making? *J. Antimicrob. Chemother.* 69, 287–291. doi: 10.1093/jac/dkt392

8

Multilocus Sequence Typing and Antimicrobial Resistance of *Campylobacter jejuni* Isolated from Dairy Calves in Austria

Daniela Klein-Jöbstl[1], Dmitri Sofka[2], Michael Iwersen[1], Marc Drillich[1] and Friederike Hilbert[2]*

[1] Clinical Unit for Herd Health Management, University Clinic for Ruminants, Department for Farm Animals and Veterinary Public Health, University of Veterinary Medicine Vienna, Vienna, Austria, [2] Institute of Meat Hygiene, Meat Technology and Food Science, Department for Farm Animals and Veterinary Public Health, University of Veterinary Medicine Vienna, Vienna, Austria

***Correspondence:**
Daniela Klein-Jöbstl
daniela.klein@vetmeduni.ac.at

Human campylobacteriosis is primarily associated with poultry but also cattle. In this study, 55 *Campylobacter jejuni* strains isolated from 382 dairy calves' feces were differentiated by multilocus sequence typing and tested for antimicrobial resistance. The most prevalent sequence type (ST) was ST883 (20.0%), followed by ST48 (14.5%), and ST50 (9.1%). In contrast to ST48 and ST50, ST883 has rarely been described in cattle previously. Furthermore, risk factor analysis was performed for the presence of the most prevalent STs in these calves. Multiple regression analysis revealed that the type of farm (organic vs. conventional) and calf housing (place, and individual vs. group) were identified as significantly ($p < 0.05$) associated with the presence of ST883 in calves, and ST50 was associated with calf diarrhea. Antimicrobial resistance was detected in 58.2% of the isolates. Most of the resistant isolates (81.3%) were resistant to more than one antimicrobial. Most frequently, resistance to ciprofloxacin (49.1%), followed by nalidixic acid (42.8%), and tetracycline (14.5%) was observed. The results of the present study support the hypothesis that dairy calves may serve as a potential reservoir for *C. jejuni* and pose a risk for transmission, including antimicrobial resistant isolates to the environment and to humans.

Keywords: *Campylobacter jejuni*, dairy calf, MLST, antimicrobial resistance

INTRODUCTION

Campylobacter jejuni is one of the most common causes of bacterial gastro-enteritis in humans and is of significant public health concern worldwide. Human campylobacteriosis is primarily associated with poultry, followed by cattle (French et al., 2009; de Haan et al., 2010b; Mughini Gras et al., 2012). Main risk factors are consumption of contaminated food, particularly poultry meat, raw milk, and water, as well as close contact to animals (Schildt et al., 2006; Heuvelink et al., 2009; Mughini Gras et al., 2012).

To distinguish between different *Campylobacter* strains, various methods have been applied, whereby multilocus sequence typing (MLST) has been identified as one of the best methods for application in epidemiological studies (Dingle et al., 2001; Korczak et al., 2009). The most commonly identified *C. jejuni* clonal complexes (CC) in bovines are CC21 and CC61. Sequence types (ST) of CC21 are not only typical for bovines, but also for other ruminants (sheep), poultry, and humans. In contrast, CC61 STs have been described as cattle associated (Kärenlampi et al., 2007; French et al., 2009; Grove-White et al., 2010; Bianchini et al., 2014), but are also frequently identified in humans, but not in poultry, suggesting that cattle may be an important source of human infection by contamination of food and water (French et al., 2009; Grove-White et al., 2010; Bianchini et al., 2014). Furthermore, CC42, CC45, CC48, and CC403 are frequently detected in cattle. Most of these CCs (CC42, CC45, and CC48) are also frequently identified in humans (French et al., 2009). These findings underline the importance of cattle in the epidemiology of human campylobacteriosis.

Campylobacter has been classified by the European Union as a zoonotic pathogen to be screened for antimicrobial resistance (Council directive, 2003/99/EC). However, this screening is limited to chicken and turkey isolates and does not include isolates from cattle. Key reasons for this are missing extensive European wide information on the risk of cattle isolates for human disease and low rates in antimicrobial resistance reported for cattle isolates (Aarestrup et al., 1997).

The aim of the present study was to evaluate different genotypes of *C. jejuni* in feces of preweaned calves in Austrian dairy herds by use of MLST and to investigate their antimicrobial resistance.

MATERIALS AND METHODS

Samples

Fecal samples were collected from preweaned calves on 100 dairy farms in two Austrian regions (Lower Austria and Styria) during 2009 and 2010. This study was part of a study designed to examine differences between calves and farms with and without diarrhea (Klein et al., 2013; Klein-Jobstl et al., 2014). For selection of farms, local veterinarians were asked to provide lists with dairy farms with a documented problem of calf diarrhea during the last year. A farm with diarrhea problems was defined as a farm with multiple treatments for calf diarrhea by the veterinarian. Out of these lists, farms were randomly chosen. Additionally, farms from the same geographical region and of similar structure, with no history of calf diarrhea problems and no diarrheic calf at the time of sampling, were examined. In herds with five or less preweaned calves (which was the case on 62 of all farms), all calves were tested. In herds with more than five preweaned calves, five animals were randomly chosen. On the assumption of an inter-herd prevalence of over 40% (Ellis-Iversen et al., 2009) five samples were required from each herd to detect one positive calf with 95% confidence (calculation by use of Win Episcope 2.0[1]). Samples were taken directly from the rectum. Feces were placed in sterile plastic tubes and transported to the laboratory in coolers. Farm management characteristics were evaluated by a face to face interview by use of a questionnaire during the farm visit (**Table 1**; Klein et al., 2013). All sampled calves were examined clinically according to the clinical examination of ruminants (Radostits et al., 2007). Feces was evaluated as described by Larson et al. (1977), where score 3 and 4 were categorized as diarrheic. Furthermore, the calf rearing areas were inspected and hygiene was estimated by evaluation of calf housings (bedding and pen walls) and the calves themselves according to Lundborg et al. (2005).

Results regarding risk factors for the presence of *C. jejuni* in calves were published elsewhere (Klein et al., 2013).

This study was discussed and approved by the institutional ethics committee of the University of Veterinary Medicine Vienna in accordance with Good Scientific Practice and national legislation.

Laboratory Procedures

All fecal samples were processed within one day, held chilled until processing. Samples were prepared for detection and isolation of thermophilic *Campylobacter* according to standards described by ISO-10272-(2002). Briefly, after enrichment in Bolton Broth (Oxoid, Basingstoke, England) at 42°C for 48 h under microaerophilic conditions (10% CO_2, 5% O_2, and 85% N_2), the samples were plated on two selective agars, modified CCDA (charcoal cefoperazon deoxycholate; Oxoid, Basingstoke, England) and CampyFoodAgar (Bio Merieux, Marcy l'Etoile, France) and incubated at 42°C for 48 h under microaerophilic conditions. Additionally, all fecal samples were directly streaked onto the two selective agars without prior enrichment. One morphological typical colony per sample was differentiated by aerobic incubation, PCR (Linton et al., 1997) and 16S-rRNA-gene sequencing on selected strains.

MLST

The MLST analysis was carried out as described by Dingle et al. (2001). Genomic DNA was extracted using a QIAamp DNA mini kit (Qiagen, Venlo, The Netherlands). The seven housekeeping

TABLE 1 | Variables surveyed on farm.

Area of interest	Variable
Farm characteristics	Region; production (organic vs. conventional); number of cattle and cows on farm; other farm animals than cattle on farm; if yes, which other farm animals, contact to other farm animals; workers on farm; other animals (companion animals) with access to the cows' and calves' stable; water source
Housing	Housing of cows; pasture; calving area; calf housing (location, type, bedding)
Calf feeding	Colostrum management; milk feeding; feeding of hay, and concentrates; water
Hygiene	Cleaning and disinfection of different areas and barns; feed hygiene; cleaning of feeding equipment; milking hygiene
Miscellaneous	Dry off regime; dry period length
Individually sampled calves	Age (days); housing; feeding; diseases; treatments; treatment with antimicrobials; feeding of non-saleable milk

[1]http://www.winepi.net/uk/index.html

loci defined by Dingle et al. (2001) as are *aspA*, *glnA*, *gltA*, *glyA*, *pgm*, *tkt*, and *uncA* were amplified using primers and protocols as described (Dingle et al., 2001). Sequencing was carried out by BigDye Terminator v3.1 cycle sequencing kit and an Applied Biosystems 310 ABI Prism genetic analyser. Sequence data were analyzed for MLST Types using the Campylobacter Multi Locus Sequence Typing website[2] developed by Jolley and Maiden (2010) and funded by the Wellcome Trust.

Antimicrobial Resistance Testing

Antimicrobial resistance was determined using CLSI M45-A[3] for antimicrobial dilution and disk susceptibility testing of infrequently isolated or fastidious bacteria. Isolates were tested using disk susceptibility and the minimal inhibitory concentration was determined by antimicrobial dilution against ampicillin, amoxicillin/clavulanate, chloramphenicol, ciprofloxacin, colistin, erythromycin, gentamicin, nalidixic acid, neomycin, streptomycin, and tetracycline. As clinical breakpoints for *C. jejuni* are documented only for ciprofloxacin, erythromycin, and tetracycline (EUCAST) epidemiological cut-off-values, which have been determined by the European Committee on Antimicrobial Susceptibility Testing[4], were applied on all *C. jejuni* isolates. For colistin and neomycin, no epidemiological cut-off-values have been determined. For these two antimicrobial substances cut-off were evaluated by comparing against values given in the literature and according to the distribution of our isolates were 16 mg/l for colistin and 4 mg/l for neomycin (El-Adawy et al., 2012; Ghimire et al., 2014).

Detection of *gyrA* Mutations

In all quinolone resistant isolates the quinolone resistance-determining region (QRDR) of the *gyrA* gene a 220Bp PCR product was amplified with primers GyrA-for 5′-gctatgcaaaatgatgaggc-3′ and GyrA-rev 5′-cagtataacgcatcg cagcgg-3′ to detect the responsible point mutation at codon 86. Genomic DNA used for amplification was extracted as described above (MLST).

Statistical Analysis

Data were statistically analyzed using PASW, version 20.0 (IBM Cooperation, New York, NY, USA).

The presence of each MLST ST was summarized in a binary variable. The presence of each ST was given as an individual variable, where presence of the given ST was categorized as one and not present as zero. Similarly, resistances against antimicrobials were categorized as either resistant (=1) or not resistant (=0). Different STs, resistances as well as farm characteristics or management factors were compared with the most prevalent STs and resistances against antimicrobials. Depending on the independent variable either Fisher's exact test, Chi square test, binary logistic regression, or *t*-test was calculated. Correlation between *C. jejuni* MLST-types and resistance against antimicrobials were tested by Spearman correlation coefficient. The level of significance was set at a p-value of <0.05.

The presence of the most prevalent STs was associated with farm characteristics and management in a two-step process. First, the presence of the ST was compared to the different independent variables as described above by either Fisher's exact test, Chi square test, binary logistic regression or *t*-test. All variables were tested for correlation among each other by Spearman correlation coefficient before entering the model. If a correlation between two variables >0.60 was given, one of the covariates was discarded taking biological plausibility into account. In a second step, variables with a p-value ≤ 0.20 were entered in a multiple logistic regression model. Confounding was monitored by the change in the coefficient of a variable after removing another variable (Dohoo et al., 2009). If the change of the estimates was $\geq 25\%$ the removed variable was considered to have a potential confounding effect and was consequently forced into the model. Model fit was evaluated with the Hosmer–Lemeshow test for 10 groups.

RESULTS

In total, 382 calves were sampled on 100 farms. Mean herd size was 40 ± 29 dairy cows (varying between 5 and 223 cows). Mean herd size did not differ between *C. jejuni* positive and negative farms ($p = 0.67$).

The median age of the sampled animals was 17 days [25 and 75% interquartile range (IR) 10–28]. *C. jejuni* positive calves were as young as 3 days and up to 67 days (median 18, IR 11-36). From 382 fecal samples, 55 (14.4%) were positive for *C. jejuni*. Another four samples (1.0%) were positive for *C. coli*. On farm level, on 30 of the 100 farms at least one animal shed *C. jejuni*, whereas only on five of these 30 farms all sampled animals were positive.

MLST of *C. jejuni* Isolates from Calves

The 55 *C. jejuni* isolates yielded 19 STs of which two were previously unreported. The isolates were assigned to eight clonal complexes (CC), dominated by three CCs (CC21, CC48, and CC206) that accounted for 74.5%. Half of the isolates (50.9%) belonged to CC21. The most prevalent STs were ST883 (20.0%), followed by ST48 (14.5%), and ST50 (9.1%; **Table 2**).

On 24 of the 30 *C. jejuni* positive farms (80%) one ST was present. On five farms two and on one farm three different types were isolated.

MLST Types and Risk Factors

The presence of *C. jejuni* ST883 in calves compared to the presence of other STs in calves was significantly associated with season, the presence of calf diarrhea on farm, the type of farm (organic vs. conventional), workers on farm, the feeding of waste milk, the separation of the calf from its dam after birth, calf feeding, and calf housing (individual versus group, and within cows' barn versus outside the barn). These variables were entered in the multiple logistic regression model. As farm had a confounding effect, this variable was forced into the model. In the final model, the type of farm and calf housing pointed out to be significantly associated with the presence of ST883 in calves

[2]http://pubmlst.org/campylobacter/
[3]www.clsi.org
[4]www.eucast.org

TABLE 2 | *Campylobacter jejuni* MLST types among 55 isolates from preweand dairy calves.

CC	ST	N positive samples	N positive farms
21	21	4	4
	47	1	1
	50	5	4
	864	4	2
	883	11	6
	1943	3	1
22	22	1	1
	2497	3	2
42	42	2	1
	2580	1	1
45	45	1	1
48	48	8	6
206	122	2	1
	572	2	1
	6021	1	1
353	356	1	1
354	4899	3	2
	Unknown	2	2

(**Table 3**). On conventional farms, the risk for calves to shed ST883 was lower compared with organic farms. Housing calves in groups inside the cows' barn was identified as a risk for shedding ST883.

No significant associations were found with regard to the presence of ST48.

The presence of ST50 strains compared to other STs were associated with farm, type of farm (organic vs. conventional), farm size (number of cows on farm), the presence of poultry on farm, diarrhea, calf feeding, and antibiotic treatment in the calf. Farm was left in the multiple logistic model as it had a confounding effect. Finally, only one variable stayed significant in the final model. Calves suffering from diarrhea at the time of sampling had a higher risk to be ST50 positive than calves not suffering from diarrhea (OR 23.21, 95%CI 23.21-248.87, $p = 0.01$).

Antimicrobial Resistance in *C. jejuni*

Of the 55 *C. jejuni* strains, 32 (58.2%) were resistant to at least one of the tested antimicrobials. Strains were resistant to ampicillin, ciprofloxacin, nalidixic acid, neomycin, streptomycin, and tetracycline (**Figure 1**). Twenty-six of the isolates (47.3%) were resistant to at least two of the tested antimicrobials. Seven of these isolates were resistant against three to five antimicrobials. Most frequently, resistance to ciprofloxacin was observed (49.1%), followed by nalidixic acid with 42.8%, and tetracycline (14.5%). No resistance was observed against amoxicillin/clavulanate, chloramphenicol, colistin, erythromycin, and gentamicin.

Associations could be observed in resistance against ciprofloxacin and nalidixic acid ($p = 0.01$), as well as against ciprofloxacin and ampicillin ($p = 0.02$). All isolates that were resistant against nalidixic acid or ampicillin were also resistant against ciprofloxacin.

Genetic Identification of Quinolone Resistant Isolates

As almost half of the isolates were identified to be quinolone resistant in all of these isolates the QRDR of the *gyrA* gene was sequenced to detect the responsible point mutation at codon 86 the most important mechanism in *C. jejuni* for quinolone resistance. In all but one of the 27 isolates the point mutation at codon 86 (ACA to ATA) resulting in a substitution of isoleucine for threonine in gyrase A has been detected. One sensitive isolate sequenced confirmed the original sequence (ACA) as seen in sensitive isolates. Another mutation detected in three of the isolates at location codon 110 did not change the amino acid composition of the enzyme. One isolate resistant to ciprofloxacin and nalidixic acid did not have any mutation in the sequenced region.

Antimicrobial Resistance and MLST-Types

Of the 11 STs that were detected repeatedly, in four STs (ST122, ST572, ST883, and ST4899) all isolates were ciprofloxacin resistant, whereas in ST883 this association was significant ($p = 0.01$). In ST42, ST864, ST1943, and ST2497, no resistance against ciprofloxacin were observed. Similar findings were obtained for nalidixic acid resistances, see **Table 4**.

Testing for correlation between *C. jejuni* MLST-types and antimicrobial resistance revealed significant associations between ST48 and ciprofloxacin resistance (coefficient = −0.33; $p = 0.01$), between ST572 and resistance against ampicillin (coefficient = −0.61; $p < 0.01$), between ST864 and ciprofloxacin (coefficient = −0.28; $p = 0.04$) and tetracycline (coefficient = 0.28; $p = 0.04$) resistance, and between ST883 and ciprofloxacin and nalidixic acid, respectively (coefficient = 0.51; $p < 0.01$).

Only two of the 55 *C. jejuni* positive calves were previously treated with antimicrobials. Both of these calves shed antimicrobial resistant *C. jejuni* (ST50). Both strains were

TABLE 3 | Variables significantly associated with the presence of *C. jejuni* sequence type (ST) 883 in the final multiple logistic regression with farm forced into the model as a confounder.

Variable		ST				
		ST883	Others	OR[1]	95% CI[2]	p
Farm						0.45
Type of farm	Organic	6	6	1		
	Conventional	38	5	0.62	0.01-0.64	0.02
Calf housing	Individual	40	4	1		
	Group	4	7	23.16	2.08-257.43	0.01
	Within cows' barn	15	9	1		
	Outside cows' barn	29	2	0.78	0.01-0.93	0.04

[1]*OR = odds ratio*
[2]*CI = confidence interval*
Hosmer–Lemeshow for the model p = 0.77.

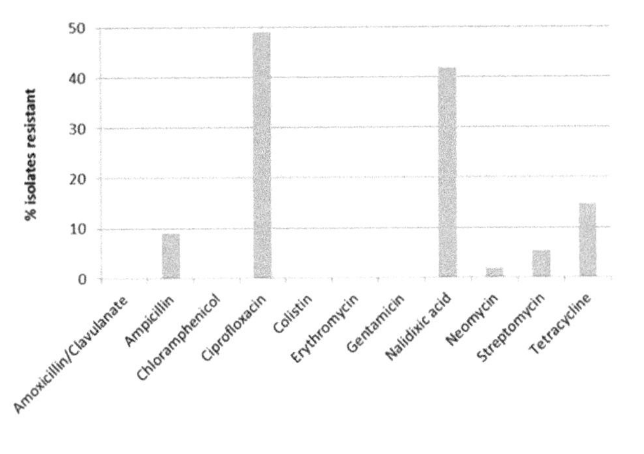

FIGURE 1 | Antimicrobial resistance of *Campylobacter jejuni* (*n* = 55) isolated from preweaned dairy calves.

resistant to ciprofloxacin and nalidixic acid, one additionally against tetracycline.

There was no association between the evaluated use of antimicrobials on farm and resistances.

DISCUSSION

To the knowledge of the authors studies examining *C. jejuni* MLST types in cattle have been limited to quinolone-resistant *C. jejuni* in Austria (Kovac et al., 2015) and studies in dairy calves are sparse.

Campylobacter jejuni isolates detected in the calves of our study were dominated by CC21, a CC regularly associated with cattle (Manning et al., 2003; Kwan et al., 2008; Ragimbeau et al., 2008; de Haan et al., 2010b; Bianchini et al., 2014; Jonas et al., 2015). Ten of the 19 *C. jejuni* STs detected were previously described in cattle. ST 883, the most common ST in our study, has only been described in cattle sporadically (de Haan et al., 2010b; Bianchini et al., 2014), and was furthermore sporadically associated with poultry and human campylobacteriosis (Wirz et al., 2010; Kittl et al., 2013b). In the study by Kovac et al. (2015) where ciprofloxacin-resistant *C. jejuni* of 17 cattle from Austria were examined, only one was ST883. In contrast, ST48 and ST50, the second and third most prevalent STs in the present study, were commonly associated with bovines, including Austria (Kwan et al., 2008; Ragimbeau et al., 2008; Rapp et al., 2014; Kovac et al., 2015). This ST was also associated with humans and other species, e.g., poultry (Ragimbeau et al., 2008; de Haan et al., 2010a). In the PubMLST database so far 32 human isolates and 33 chicken isolates from Austria have been downloaded between 2008 and 2014. Interestingly CC21 the CC most prevalent in claves in our study was also dominant in human isolates in Austria (21% of all isolates) whereas the clonal complex CC464, CC353, and CC354 dominated in chicken isolates.

Different studies emphasized the role of cattle in human campylobacteriosis. Following poultry, bovines were frequently associated with human infections (Wilson et al., 2009; de Haan et al., 2010a; Mughini Gras et al., 2012). Risk factors that have been described are direct contact to cattle and to cattle feces, as well as consumption of raw milk (Eberhart-Phillips et al., 1997; Smith et al., 2004; Schildt et al., 2006). Direct contact and consumption of raw milk is given not only for farmers but also, e.g., during farm vacation which is popular in Austria with nearly 10,000 farms offering this service (Grüner Bericht, n.d.). Furthermore, contaminated food or water can play a role in human infection (Clark et al., 2005; Levesque et al., 2008; Mughini Gras et al., 2012). The three dominating CCs (CC21, CC48, and CC206) and STs (ST883, ST48, and ST50) of our study were also recovered from infected humans (Ragimbeau et al., 2008; Mullner et al., 2009; Sheppard et al., 2009; de Haan et al., 2010a). This result indicates that calves may be a potential source of human infection, but this cannot be proven by this study. Gilpin et al. (2008) found indistinguishable *C. jejuni* genotypes in dairy calves and humans, using Penner serotyping and pulsed field gel electrophoreses, and came to the same conclusion, that calves may be a source of human campylobacteriosis.

TABLE 4 | Antimicrobial resistance of the 55 *C. jejuni* MLST types obtained from dairy preweaned calves.

CC	ST	N isolates	N resistances	Antimicrobial resistance
21	21	2	2	CIP, NAL
	21	1	3	AMP, CIP, NAL
	21	1	1	STREP
	47	1	1	TET
	50	2	3	CIP, NAL, TET
	50	2	2	CIP, NAL
	50	1	0	
	864	2	1	TET
	864	2	0	
	883	9	2	CIP, NAL
	883	1	3	CIP, NAL, STREP
	883	1	3	AMP, CIP, NAL
	1943	3	0	
22	22	1	0	
	2497	3	0	
42	42	2	0	
	2580	1	0	
45	45	1	0	
48	48	7	0	
	48	1	2	CIP, NAL
206	122	2	2	CIP, NAL
	572	1	2	AMP, CIP
	572	1	3	AMP, CIP, NAL
	6021	1	5	AMP, CIP, NEO, STREP, TET
353	356	1	0	
354	4899	2	2	CIP, NAL
	4899	1	1	CIP
	Unknown	1	0	
	Unknown	1	1	TET

CIP, ciprofloxacin; NAL, nalidixic acid; TET, tetracycline; AMP, ampicillin; STREP, streptomycin; NEO, neomycin.

Early campylobacter infection in calves might be due to a high level of environmental contamination, as well as direct contact with feces and ingestion of milk (Bianchini et al., 2014). Contamination of calf housings represents a risk for early infection with *C. jejuni* ST883. Calves housed within the cows' barn were at higher risk to shed ST883 than calves housed in a special barn for calves and young animals or outside the barn. Additionally, grouping of animals was a risk factor for shedding ST883. These findings suggest that close contact to adult as well as to other young cattle lead to higher infection pressure and mutual infection between animals. Furthermore, grouping can be a stress factor and consequently lead to a higher rate of campylobacter shedding and infection (Rapp et al., 2014).

Furthermore, ST883 was associated with the type of farm. On conventional farms, the risk for preweaned calves to be ST883 positive was lower than on organic farms, a finding that can hardly be explained. Possibly a certain clone is circulating on these farms and has been established.

The presence of ST50 was associated with diarrhea at the time of sampling in calves. Although some authors (Al-Mashat and Taylor, 1983; Diker et al., 1990; Schulze, 1992) suggested a possible role of *Campylobacter* in calf diarrhea, in other studies no association between *C jejuni* and disease was given (De Rycke et al., 1986; Acha et al., 2004), as was also true for the calves of the present study (Klein et al., 2013) and more likely other pathogens generally associated with calf diarrhea may be the cause of disease.

More than half of the calves (55%) originated from farms where also poultry was kept. Most of the STs detected in the calves of our study were also described in poultry. In the present study, specific types like ST50, frequently detected in poultry (de Haan et al., 2010b; Griekspoor et al., 2010; Kovanen et al., 2014), were associated with the presence of poultry on farm, suggesting cross-contamination between the two species. In contrast, other STs (e.g., ST21, ST48, and ST883) also regularly detected in poultry (Wirz et al., 2010; Kittl et al., 2013b) appeared in calves independently of the presence of poultry.

Antimicrobial resistance, particularly multidrug resistance is of public health concern. In the present study, 58.2% of the *C. jejuni* isolates were resistant to at least one and 47.3% against at least two of the tested antimicrobials. Because only two of the 55 *C. jejuni* positive calves were previously treated with antibiotics, no valid conclusion can be drawn if resistance to STs in this study was associated with previous antibiotic treatment. Nevertheless, these two treated animals shed *C. jejuni* resistant to two and three of the tested antimicrobials, respectively. In the present study, antimicrobial resistance to quinolones was detected most often. This has also been described for isolates originating from other sources (Oporto et al., 2009; Kittl et al., 2013a), explained by the fact that a single mutation is sufficient to cause resistance (Wang et al., 1993). This was confirmed in our study as all but one resistant isolates harbored a point mutation in codon 86 of the *gyrA* gene. Quinolone resistance has been described to be associated with specific ST types as detected in our study (Kittl et al., 2013a; Kovac et al., 2015). A survey performed in Austria revealed that quinolones are frequently used by Austrian veterinarians for treatment of cattle (Mayrhofer et al., 2006). This might explain a high level of quinolone resistance in *C. jejuni* isolates from calves.

CONCLUSION

The results of the present study support the hypothesis that cattle including dairy calves may be a reservoir for *C. jejuni* and represent a risk for transmission of these bacteria to the environment and to humans. Cattle have not been recognized as an important source for antimicrobial resistant *Campylobacter* sp. or other bacteria, yet. Nevertheless, high resistance rates found in this and other recent studies point out that screening for antimicrobial resistance in cattle is necessary to better understand the epidemiology of resistance and its spread.

AUTHOR CONTRIBUTIONS

DK-J designed the study, took all samples, performed statistical analysis, and drafted the manuscript. DS performed laboratory work and analysis. MI advised statistical analysis and interpretation, and reviewed the manuscript. MD supported the statistical analysis and reviewed the manuscript. FH designed the study together with DK-J and supervised the study, supervised, and performed laboratory work and analysis, and provided valuable references and suggestions during the preparation of the manuscript.

ACKNOWLEDGMENT

This study was financed by the University of Veterinary Medicine Vienna by a start-up project (Profillinie 2).

REFERENCES

Aarestrup, F. M., Nielsen, E. M., Madsen, M., and Engberg, J. (1997). Antimicrobial susceptibility patterns of thermophilic *Campylobacter* spp. from humans, pigs, cattle, and broilers in Denmark. *Antimicrob. Agents Chemother.* 41, 2244–2250.

Acha, S. J., Kuhn, I., Jonsson, P., Mbazima, G., Katouli, M., and Mollby, R. (2004). Studies on calf diarrhoea in Mozambique: prevalence of bacterial pathogens. *Acta Vet. Scand.* 45, 27–36. doi: 10.1186/1751-014 7-45-27

Al-Mashat, R. R., and Taylor, D. J. (1983). Production of enteritis in calves by the oral inoculation of pure cultures of *Campylobacter* fetus subspecies intestinalis. *Vet. Rec.* 112, 54–58. doi: 10.1136/vr.112.7.141

Bianchini, V., Borella, L., Benedetti, V., Parisi, A., Miccolupo, A., Santoro, E., et al. (2014). Prevalence in bulk tank milk and epidemiology of *Campylobacter jejuni* in dairy herds in Northern Italy. *Appl. Environ. Microbiol.* 80, 1832–1837. doi: 10.1128/AEM.03784-13

Clark, C. G., Bryden, L., Cuff, W. R., Johnson, P. L., Jamieson, F., Ciebin, B., et al. (2005). Use of the oxford multilocus sequence typing protocol and sequencing of the flagellin short variable region to characterize isolates from a large outbreak of waterborne *Campylobacter* sp. strains in Walkerton, Ontario, Canada. *J. Clin. Microbiol.* 43, 2080–2091. doi: 10.1128/JCM.43.5.2080-2091.2005

Council directive (2003/99/EC). *DIRECTIVE 2003/99/EC of the European Parliament and of the Council on the Monitoring of Zoonoses and Zoonotic Agents, Amending Council Decision 90/424/EEC and Repealing Council Directive 92/117/EEC.* Available at: http://eur-lex.europa.eu/legal-content/EN/TXT/?uri=celex:32003L0099.

de Haan, C. P., Kivisto, R., Hakkinen, M., Rautelin, H., and Hanninen, M. L. (2010a). Decreasing trend of overlapping multilocus sequence types between human and chicken *Campylobacter jejuni* isolates over a decade in Finland. *Appl. Environ. Microbiol.* 76, 5228–5236. doi: 10.1128/AEM.00581-10

de Haan, C. P., Kivisto, R. I., Hakkinen, M., Corander, J., and Hanninen, M. L. (2010b). Multilocus sequence types of Finnish bovine *Campylobacter jejuni* isolates and their attribution to human infections. *BMC Microbiol.* 10:200. doi: 10.1186/1471-2180-10-200

De Rycke, J., Bernard, S., Laporte, J., Naciri, M., Popoff, M. R., and Rodolakis, A. (1986). Prevalence of various enteropathogens in the feces of diarrheic and healthy calves. *Ann. Rech. Vet.* 17, 159–168.

Diker, K. S., Diker, S., and Ozlem, M. B. (1990). Bovine diarrhea associated with *Campylobacter hyointestinalis. Zentralbl. Veterinarmed. B.* 37, 158–160.

Dingle, K. E., Colles, F. M., Wareing, D. R., Ure, R., Fox, A. J., Bolton, F. E., et al. (2001). Multilocus sequence typing system for *Campylobacter jejuni. J. Clin. Microbiol.* 39, 14–23. doi: 10.1128/JCM.39.1.14-23.2001

Dohoo, I., Martin, W., and Stryhn, H. (2009). *Veterinary Epidemiologic Research.* Charlottettown, PE: Atlantic Veterinary College.

Eberhart-Phillips, J., Walker, N., Garrett, N., Bell, D., Sinclair, D., Rainger, W., et al. (1997). *Campylobacteriosis* in New Zealand: results of a case-control study. *J. Epidemiol. Commun. Health* 51, 686–691. doi: 10.1136/jech.51.6.686

El-Adawy, H., Hotzel, H., Dupre, S., Tomaso, H., Neubauer, H., and Hafez, H. M. (2012). Determination of antimicrobial sensitivities of *Campylobacter jejuni* isolated from commercial turkey farms in Germany. *Avian Dis.* 56, 685–692. doi: 10.1637/10135-031912-Reg.1

Ellis-Iversen, J., Cook, A. J., Smith, R. P., Pritchard, G. C., and Nielen, M. (2009). Temporal patterns and risk factors for *Escherichia coli* O157 and *Campylobacter* spp, in young cattle. *J. Food Prot.* 72, 490–496.

French, N. P., Midwinter, A., Holland, B., Collins-Emerson, J., Pattison, R., Colles, F., et al. (2009). Molecular epidemiology of *Campylobacter jejuni* isolates from wild-bird fecal material in children's playgrounds. *Appl. Environ. Microbiol.* 75, 779–783. doi: 10.1128/AEM.01979-08

Ghimire, L., Singh, D. K., Basnet, H. B., Bhattarai, R. K., Dhakal, S., and Sharma, B. (2014). Prevalence, antibiogram and risk factors of thermophilic *Campylobacter* spp. in dressed porcine carcass of Chitwan, Nepal. *BMC Microbiol.* 14:85. doi: 10.1186/1471-2180-14-85

Gilpin, B. J., Thorrold, B., Scholes, P., Longhurst, R. D., Devane, M., Nicol, C., et al. (2008). Comparison of *Campylobacter jejuni* genotypes from dairy cattle and human sources from the Matamata-Piako District of New Zealand. *J. Appl. Microbiol.* 105, 1354–1360. doi: 10.1111/j.1365-2672.2008.03863.x

Griekspoor, P., Engvall, E. O., Olsen, B., and Waldenstrom, J. (2010). Multilocus sequence typing of *Campylobacter jejuni* from broilers. *Vet. Microbiol.* 140, 180–185. doi: 10.1016/j.vetmic.2009.07.022

Grove-White, D. H., Leatherbarrow, A. J., Cripps, P. J., Diggle, P. J., and French, N. P. (2010). Temporal and farm-management-associated variation in the faecal-pat prevalence of *Campylobacter jejuni* in ruminants. *Epidemiol. Infect.* 138, 549–558. doi: 10.1017/S0950268809991051

Grüner Bericht (n.d.). Available at: http://www.gruenerbericht.at/cm3/download/summary/82-gruener-bericht-oesterreich/1392-gruener-bericht-2014.html

Heuvelink, A. E., van Heerwaarden, C., Zwartkruis-Nahuis, A., Tilburg, J. J., Bos, M. H., Heilmann, F. G., et al. (2009). Two outbreaks of *campylobacteriosis* associated with the consumption of raw cows' milk. *Int. J. Food Microbiol.* 134, 70–74. doi: 10.1016/j.ijfoodmicro.2008.12.026

Jolley, K. A., and Maiden, M. C. (2010). BIGSdb: scalable analysis of bacterial genome variation at the population level. *BMC Bioinformatics* 11:595. doi: 10.1186/1471-2105-11-595

Jonas, R., Kittl, S., Overesch, G., and Kuhnert, P. (2015). Genotypes and antibiotic resistance of bovine *Campylobacter* and their contribution to human campylobacteriosis. *Epidemiol. Infect.* 143, 2373–2380. doi: 10.1017/S0950268814003410

Kärenlampi, R., Rautelin, H., Schonberg-Norio, D., Paulin, L., and Hanninen, M. L. (2007). Longitudinal study of Finnish *Campylobacter jejuni* and *C. coli* isolates from humans, using multilocus sequence typing, including comparison with epidemiological data and isolates from poultry and cattle. *Appl. Environ. Microbiol.* 73, 148–155. doi: 10.1128/AEM.01488-06

Kittl, S., Heckel, G., Korczak, B. M., and Kuhnert, P. (2013a). Source attribution of human *Campylobacter* isolates by MLST and fla-typing and association of genotypes with quinolone resistance. *PLoS ONE* 8:e81796. doi: 10.1371/journal.pone.0081796

Kittl, S., Korczak, B. M., Niederer, L., Baumgartner, A., Buettner, S., Overesch, G., et al. (2013b). Comparison of genotypes and antibiotic resistances of *Campylobacter jejuni* and *Campylobacter coli* on chicken retail meat and at slaughter. *Appl. Environ. Microbiol.* 79, 3875–3878. doi: 10.1128/AEM.00493-13

Klein, D., Alispahic, M., Sofka, D., Iwersen, M., Drillich, M., and Hilbert, F. (2013). Prevalence and risk factors for shedding of thermophilic *Campylobacter* in calves with and without diarrhea in Austrian dairy herds. *J. Dairy Sci.* 96, 1203–1210. doi: 10.3168/jds.2012-5987

Klein-Jobstl, D., Iwersen, M., and Drillich, M. (2014). Farm characteristics and calf management practices on dairy farms with and without diarrhea: a case-control study to investigate risk factors for calf diarrhea. *J. Dairy Sci.* 97, 5110–5119. doi: 10.3168/jds.2013-7695

Korczak, B. M., Zurfluh, M., Emler, S., Kuhn-Oertli, J., and Kuhnert, P. (2009). Multiplex strategy for multilocus sequence typing, fla typing, and genetic determination of antimicrobial resistance of *Campylobacter jejuni* and *Campylobacter* coli isolates collected in Switzerland. *J. Clin. Microbiol.* 47, 1996–2007. doi: 10.1128/JCM.00237-09

Kovac, J., Cadez, N., Stessl, B., Stingl, K., Gruntar, I., Ocepek, M., et al. (2015). High genetic similarity of ciprofloxacin-resistant *Campylobacter jejuni* in central Europe. *Front. Microbiol.* 6:1169. doi: 10.3389/fmicb.2015.01169

Kovanen, S. M., Kivisto, R. I., Rossi, M., and Hanninen, M. L. (2014). A combination of MLST and CRISPR typing reveals dominant *Campylobacter jejuni* types in organically farmed laying hens. *J. Appl. Microbiol.* 117, 249–257. doi: 10.1111/jam.12503

Kwan, P. S., Birtles, A., Bolton, F. J., French, N. P., Robinson, S. E., Newbold, L. S., et al. (2008). Longitudinal study of the molecular epidemiology of *Campylobacter jejuni* in cattle on dairy farms. *Appl. Environ. Microbiol.* 74, 3626–3633. doi: 10.1128/AEM.01669-07

Larson, L. L., Owen, F. G., Albright, R. D., Appleman, R. D., Lamb, R. C., and Miller, L. D. (1977). Guidelines toward more uniformity in measuring and reporting calf experimental data. *J. Dairy Sci.* 60, 989–991. doi: 10.3168/jds.S0022-0302(77)83975-1

Levesque, S., Frost, E., Arbeit, R. D., and Michaud, S. (2008). Multilocus sequence typing of *Campylobacter jejuni* isolates from humans, chickens, raw milk, and environmental water in Quebec, Canada. *J. Clin. Microbiol.* 46, 3404–3411. doi: 10.1128/JCM.00042-08

Linton, D., Lawson, A. J., Owen, R. J., and Stanley, J. (1997). PCR detection, identification to species level, and fingerprinting of *Campylobacter jejuni* and *Campylobacter coli* direct from diarrheic samples. *J. Clin. Microbiol.* 35, 2568–2572.

Lundborg, G. K., Svensson, E. C., and Oltenacu, P. A. (2005). Herd-level risk factors for infectious diseases in Swedish dairy calves aged 0-90 days. *Prev. Vet. Med.* 68, 123–143. doi: 10.1016/j.prevetmed.2004.11.014

Manning, G., Dowson, C. G., Bagnall, M. C., Ahmed, I. H., West, M., and Newell, D. G. (2003). Multilocus sequence typing for comparison of veterinary and human isolates of *Campylobacter jejuni. Appl. Environ. Microbiol.* 69, 6370–6379. doi: 10.1128/AEM.69.11.6370-6379.2003

Mayrhofer, S., Paulsen, P., Smulders, F. J., and Hilbert, F. (2006). Antimicrobial resistance in commensal *Escherichia coli* isolated from muscle foods as related to the veterinary use of antimicrobial agents in food-producing animals in Austria. *Microb. Drug Resist.* 12, 278–283. doi: 10.1089/mdr.2006.12.278

Mughini Gras, L., Smid, J. H., Wagenaar, J. A., de Boer, A. G., Havelaar, A. H., Friesema, I. H., et al. (2012). Risk factors for campylobacteriosis of chicken, ruminant, and environmental origin: a combined case-control and source attribution analysis. *PLoS ONE* 7:e42599. doi: 10.1371/journal.pone.0042599

Mullner, P., Spencer, S. E., Wilson, D. J., Jones, G., Noble, A. D., Midwinter, A. C., et al. (2009). Assigning the source of human *campylobacteriosis* in New Zealand: a comparative genetic and epidemiological approach. *Infect. Genet. Evol.* 9, 1311–1319. doi: 10.1016/j.meegid.2009.09.003

Oporto, B., Juste, R. A., and Hurtado, A. (2009). Phenotypic and genotypic antimicrobial resistance profiles of *Campylobacter jejuni* isolated from cattle, sheep, and free-range poultry faeces. *Int. J. Microbiol.* 2009:456573. doi: 10.1155/2009/456573

Radostits, O. M., Gay, C. C., Hinchcliff, K. W., and Constable, P. D. (eds). (2007). "Examination of the patient," in *Veterinary Medicine. A Textbook of the Diseases of Cattle, Horses, Sheep, Pigs, and Goats*, 10th Edn (Philadelphia, PA: Saunders/Elsevier Limited), 8–21.

Ragimbeau, C., Schneider, F., Losch, S., Even, J., and Mossong, J. (2008). Multilocus sequence typing, pulsed-field gel electrophoresis, and fla short variable region typing of clonal complexes of *Campylobacter jejuni* strains of human, bovine, and poultry origins in Luxembourg. *Appl. Environ. Microbiol.* 74, 7715–7722. doi: 10.1128/AEM.00865-08

Rapp, D., Ross, C. M., Cave, V., and Muirhead, R. W. (2014). Prevalence, concentration and genotypes of *Campylobacter jejuni* in faeces from dairy herds managed in farm systems with or without housing. *J. Appl. Microbiol.* 116, 1035–1043. doi: 10.1111/jam.12425

Schildt, M., Savolainen, S., and Hanninen, M. L. (2006). Long-lasting *Campylobacter jejuni* contamination of milk associated with gastrointestinal illness in a farming family. *Epidemiol. Infect.* 134, 401–405. doi: 10.1017/S0950268805005029

Schulze, F. (1992). [*Campylobacter* as the cause of diarrhea in calves]. *Dtsch. Tierarztl. Wochenschr.* 99, 458–461.

Sheppard, S. K., Dallas, J. F., MacRae, M., McCarthy, N. D., Sproston, E. L., Gormley, F. J., et al. (2009). *Campylobacter* genotypes from food animals, environmental sources and clinical disease in Scotland 2005/6. *Int. J. Food Microbiol.* 134, 96–103. doi: 10.1016/j.ijfoodmicro.2009.02.010

Smith, K. E., Stenzel, S. A., Bender, J. B., Wagstrom, E., Soderlund, D., Leano, F. T., et al. (2004). Outbreaks of enteric infections caused by multiple pathogens associated with calves at a farm day camp. *Pediatr. Infect. Dis. J.* 23, 1098–1104.

Wang, Y., Huang, W. M., and Taylor, D. E. (1993). Cloning and nucleotide sequence of the *Campylobacter jejuni* gyrA gene and characterization of quinolone resistance mutations. *Antimicrob. Agents Chemother.* 37, 457–463. doi: 10.1128/AAC.37.3.457

Wilson, D. J., Gabriel, E., Leatherbarrow, A. J., Cheesbrough, J., Gee, S., Bolton, E., et al. (2009). Rapid evolution and the importance of recombination to the gastroenteric pathogen *Campylobacter jejuni*. *Mol. Biol. Evol.* 26, 385–397. doi: 10.1093/molbev/msn264

Wirz, S. E., Overesch, G., Kuhnert, P., and Korczak, B. M. (2010). Genotype and antibiotic resistance analyses of *Campylobacter* isolates from ceca and carcasses of slaughtered broiler flocks. *Appl. Environ. Microbiol.* 76, 6377–6386. doi: 10.1128/AEM.00813-10

9

Characterization of Carbapenem-Resistant *Acinetobacter baumannii* Isolates in a Chinese Teaching Hospital

Yaowen Chang[1†], Guangxin Luan[2†], Ying Xu[3], Yanhong Wang[2], Min Shen[2], Chi Zhang[1], Wei Zheng[1], Jinwei Huang[4], Jingni Yang[2], Xu Jia[2] and Baodong Ling[1*]*

[1] *Small Molecule Drugs Sichuan Key Laboratory, Institute of Materia Medica, Chengdu Medical College, Chengdu, China,* [2] *Non-coding RNA and Drug Discovery Laboratory, Chengdu Medical College, Chengdu, China,* [3] *Clinical Laboratory, The First Affiliated Hospital, Chengdu Medical College, Chengdu, China,* [4] *Institute of Antibiotics, The Fifth Affiliated Hospital, Wenzhou Medical University, Lishui, China*

***Correspondence:**
*Baodong Ling,
Small Molecule Drugs Sichuan Key Laboratory, Institute of Materia Medica, Chengdu Medical College, Chengdu 610500, China
lingbaodong@cmc.edu.cn;
Xu Jia,
Non-coding RNA and Drug Discovery Laboratory, Chengdu Medical College, Chengdu 610500, China
fdjiaxu@gmail.com*

[†] *These authors have contributed equally to this work.*

Carbapenem-resistant *Acinetobacter baumannii* (CRAB) presents a serious therapeutic and infection control challenge. In this study, we investigated the epidemiological and molecular differences of CRAB and the threatening factors for contributing to increased CRAB infections at a hospital in western China. A total of 110 clinical isolates of *A. baumannii*, collected in a recent 2-year period, were tested for carbapenem antibiotic susceptibility, followed by a molecular analysis of carbapenemase genes. Genetic relatedness of the isolates was characterized by multilocus sequence typing. Sixty-seven of the 110 isolates (60.9%) were resistant to carbapenems, 80.60% (54/67) of which carried the bla_{OXA-23} gene. Most of these CRAB isolates (77.62%) were classified as clone complex 92 (CC92), and sequence type (ST) 92 was the most prevalent STs, followed by ST195, ST136, ST843, and ST75. One CRAB isolate of ST195 harbored plasmid pAB52 from a Chinese patient without travel history. This plasmid contains toxin–antitoxin elements related to adaptation for growth, which might have emerged as a common vehicle indirectly mediating the spread of OXA-23 in CRAB. Thus, CC92 *A. baumannii* carrying OXA-23 is a major drug-resistant strain spreading in China. Our findings indicate that rational application of antibiotics is indispensable for minimizing widespread of drug resistance.

Keywords: *Acinetobacter baumannii*, carbapenem resistance, OXA-23, plasmid, CC92

Introduction

Acintobacter baumannii is an opportunistic Gram-negative pathogen, which has recently been successfully spreading worldwide nosocomial infections and causing outbreaks of hospital-acquired infections, primarily due to its prominent ability to acquire antibiotic resistance (Peleg et al., 2008; Kempf and Rolain, 2012). *A. baumannii* is developing into multidrug resistant (MDR), extensively drug resistant (XDR), and pandrug resistant (PDR) bacteria, and its adaptation to the environment with drug resistance has previously been reported (Durante-Mangoni and Zarrilli, 2011). Carbapenems were regarded as the most powerful antibiotics because of its extremely effective antibacterial activity and low toxicity, but the emergence of carbapenem resistance in *A. baumannii* has become a global concern recently (Peleg et al., 2008; Tiwari et al., 2012a,b).

Resistance of *A. baumannii* to expanded-spectrum cephalosporins and carbapenems is rapidly growing over the years (Bradford, 2001; Peleg et al., 2008). For example, surveillance by the CHINET project in China revealed that the rate of resistance of *A. baumannii* to the carbapenem, such as imipenem, doubled from 30.1% in 2006 to 62.8% in 2013 (Wang, 2008; Fupin et al., 2014).

Various types of β-lactam antibiotics, for example carbapenems, contain β-lactam rings in their structures and can be inactivated by β-lactamase enzymes. The β-lactamases are classified into four different molecular groups, Ambler classes A through D, according to amino acid sequence identities (Bush et al., 1995). Class A, C, and D (OXA enzymes) of β-lactamases contain a catalytically active serine residue that cleaves the lactam ring of antibiotics (Bush et al., 1995; Tiwari and Moganty, 2014). Class B of β-lactamases is a metallo-enzyme that requires zinc for their catalytic activity, and therefore, has a completely different mechanism for enzyme activity (Tiwari and Moganty, 2013). Carbapenems, such as imipenem and meropenem, have an exceedingly broad spectrum of activity and are able to resist hydrolysis by most of the β-lactamases, including extended-spectrum and derepressed class C chromosomal AmpC β-lactamases (Bush et al., 1995). However, most of the metallo-β-lactamases and a number of class A and D β-lactamases are able to hydrolyse broad spectrum carbapenems, such as imipenem and meropenem (Walther-Rasmussen and Hoiby, 2006; Evans and Amyes, 2014). Thus, outbreaks of OXA-23-producing *A. baumannii* have been reported from various regions of the world (Carvalho et al., 2009). It is generally believed that OXA-23 is responsible for carbapenem antibiotic resistance. Previously, Liu et al. (2015) reported the dissemination of MDR OXA-23-producing *A. baumannii* clones throughout multiple cities in China, but little is known about the molecular mechanisms of resistance to carbapenems in western China.

The multilocus sequence typing (MLST) has been widely applied in genotyping of bacteria, including *A. baumannii* (Bartual et al., 2005). Molecular epidemiological research of *A. baumannii* indicates that CC92 has played an important role in nosocomial infection outbreak and spread nationwide (Runnegar et al., 2010).

This study aimed to report the dissemination of *A. baumannii* harboring carbapenemase genes throughout a university hospital in western China, then identify the risk factors for carbapenem-resistant *Acinetobacter baumannii* (CRAB) infections, and finally perform a comprehensive evaluation and comparison of their genetic diversity.

Materials and Methods

Bacteria Isolates

A total of 110 consecutive and non-duplicated *A. baumannii* clinical isolates were collected from different departments [intensive care unit (ICU), gastroenterology, respiratory, neurosurgery and other wards] at the First Affiliated Hospital of Chengdu Medical College, Chengdu, Sichuan, China from 2012 to 2013. Isolates were identified by standard laboratory methods and ATB New (bioMérieux, France). *A. baumannii* was further verified when two PCR products were yielded as reported: a 425-bp internal control amplicon corresponding to the *recA* gene of *Acinetobacter* spp. and the 208-bp fragment of the 16S rRNA (Chiang et al., 2011) intergenic spacer region of *A. baumannii* (**Table 1**). All strains were stored at −80°C, and bacteria were grown on tryptose agar or Mueller–Hinton broth or agar (Oxoid, England).

Minimal Inhibitory Concentration (MIC)

The minimal inhibitory concentration (MIC) of carbapenems including imipenem and meropenem for *A. baumannii* were determined by the agar dilution method as previously described in the guidelines from the Clinical and Laboratory Standards Institute (CLSI, 2014). *Escherichia coli* ATCC25922 and *A. baumannii* ATCC19606 were used as quality control strain. The results were interpreted according to the CLSI guidelines (CLSI, 2014), i.e., CRAB was defined as an *A. baumannii* isolate that was resistant to both imipenem and meropenem (i.e., ≥8 μg/ml as resistant), whereas carbapenem-susceptible *A. baumannii* (CSAB) possessed carbapenem MIC of ≤2 μg/ml and carbapenem-intermediate *A. baumannii* (CIAB) has MIC of 4 μg/ml).

PCR Experiments

The genes encoding carbapenemases class A [e.g., *Klebsiella pneumoniae* carbapenemase gene, bla_{KPC} (Li et al., 2014)], class B [e.g., the metallo-β-lactamases, bla_{IMP} (Valenzuela et al., 2007), bla_{VIM-1} (Tsakris et al., 2000), bla_{SIM} (Ellington et al., 2007), and bla_{NDM-1} (Yong et al., 2009)], class C [e.g., bla_{AmpC} (Bou and Martinez-Beltran, 2000)], and class D [e.g., bla_{OXA-23} (Woodford et al., 2006), bla_{OXA-24} (Woodford et al., 2006), bla_{OXA-51} (Fu et al., 2010), bla_{OXA-58} (Netsvyetayeva et al., 2011), and $bla_{OXA-235}$ (Higgins et al., 2013)] were investigated by polymerase chain reaction (PCR).

To prepare DNA templates, each isolate was grown on tryptose agar (Oxoid, England) plate overnight. Two to three colonies were suspended in 100 μl of sterile distilled water in a 1.5-ml Eppendorf tube, and the suspension was heated at 100°C for 15 min, followed by centrifugation at 12,000 *g* for 10 min to pellet the debris. The resultant supernatant was used as the DNA template in the PCRs, which were carried out in a 50-μl volume containing 0.2 mM each deoxynucleotide, 0.5 μM each primer, 1.25 U of Taq polymerase, and 5 μl of 10× buffer (Thermo, China). Primers (**Table 1**) were synthesized by Sangon Company (Sangon, China). Reaction conditions of PCR were 94°C for 5 min and 30 cycles of 94°C for 30 s, 55°C for 30 s (57°C for bla_{AmpC} amplification), and 72°C for 30 s, followed by a final extension at 72°C for 5 min.

Multilocus Sequence Typing

MLST was used to describe the genetic backgrounds of CRAB and CSAB in all clinical isolates (Diancourt et al., 2010; Hamouda et al., 2010; Pournaras et al., 2014). eBURST was performed to cluster sequence types (STs) into clonal complexes (CCs) and infer evolutionary descent. MLST was carried out as described

TABLE 1 | Primers used in this study.

Locus	Primer	Oligonucleotides (5′ → 3′)	Expected size (bp)	Source
16S	F	CAT TAT CAC GGT AAT TAG TG	208	Chiang et al. (2011)
	R	AGA GCA CTG TGC ACT TAA G		
RecA	F	CCT GAA TCT TCT GGT AAA AC	425	Chiang et al. (2011)
	R	GTT TCT GGG CTG CCA AAC ATT AC		
KPC	F	GCT CAG GCG CAA CTG TAA GT	823	Li et al. (2014)
	R	GTC CAG ACG GAA CGT GGT AT		
IMP	F	CTA CCG CAG AGT CTT TG	587	Valenzuela et al. (2007)
	R	AAC CAG TTT TGC CTT ACC AT		
VIM-1	F	AGT GGT GAG TAT CCG ACA G	261	Tsakris et al. (2000)
	R	ATG AAA GTG CGT GGA GAC		
17SIM	F	TAC AAG GGA TTC GGC ATC G	570	Ellington et al. (2007)
	R	TAA TGG CCT GTT CCC ATG TG		
NDM-1	F	TCT CGA CAT GCC GGG TTT CGG	475	Yong et al. (2009)
	R	ACC GAG ATT GCC GAG CGA CTT		
AmpC	F	ACT TAC TTC AAC TCG CGA CG	663	Bou and Martinez-Beltran (2000)
	R	TAA ACA CCA CAT ATG TTC CG		
OXA-23	F	GAT CGG ATT GGA GAA CCA GA	501	Woodford et al. (2006)
	R	ATT TCT GAC CGC ATT TCC AT		
OXA-24	F	CAA GAG CTT GCA AGA CGG ACT	420	Woodford et al. (2006)
	R	TCC AAG ATT TTC TAG CRA CTT ATA		
OXA-51	F	TAA TGC TTT GAT CGG CCT TG	353	Fu et al. (2010)
	R	TGG ATT GCA CTT CAT CTT GG		
OXA-58	F	TCG ATC AGA ATG TTC AAG CGC	530	Netsvyetayeva et al. (2011)
	R	ACG ATT CTC CCC TCT GCG C		
OXA-235	F	TTG TTG CCT TTA CTT AGT TGC	831	Higgins et al. (2013)
	R	CAA AAT TTT AAG ACG GAT CG		

by Bartual et al. (2005) for all isolates. In brief, internal fragments of seven housekeeping genes, *gltA*, *gyrB*, *gdhB*, *recA*, *cpn60*, *gpi*, and *rpoD*, were amplified, purified, and sequenced (Fu et al., 2010; Ruan et al., 2013). And eBURST (version 3, http://eburst.mlst.net/) was used to assign STs to CCs and define the genetic relatedness of STs with the most stringent definition of the groups by sharing the same alleles at ≥6 of 7 loci (Feil et al., 2004).

Plasmid Conjugation, Extraction, Sequencing, and Analysis

Plasmid conjugations were performed by using meropenem-resistant *A. baumannii* as donors and an azide-resistant *E. coli* J53 as the recipient (MIC > 200 μg/ml; Bogaerts et al., 2006). Matings were performed as described previously (Wang et al., 2003), and transconjugants were recovered on L-agar plates containing meropenem (8 μg/ml) and azide (150 μg/ml).

Plasmid DNA was extracted using a Plasmid Midi Kit (Omega, USA) according to the manufacturer's instructions. The plasmid was cut by EcoRI, then the fragments were ligated to the vector pET21a. The complete sequencing work was performed by the ABI-3730 XL system (Biosune Biotechnology Company, Shanghai, China) and sequences were assembled by Life Technologies (BigDye V3.1, four fluorescence reagent kit). The sequenced plasmid was annotated by the RAST server (Aziz et al., 2008), then all of the predicted proteins were further compared against the NCBI non-redundant protein database using the BLASTP program. In addition, the DNAMAN software (Lynnon Corporation, USA) was used to generate a circular map of plasmid pAB52.

Nucleotide Sequence Accession Numbers

The complete nucleotide sequences (Supplementary Table S1) of plasmid pAB52 in this study were submitted to GenBank with assigned accession number KR030046.

Results

Characterization of Clinical Isolates

According to our previous study, 67 of the total 110 *A. baumannii* isolates exhibited resistance to carbapenems, imipenem and meropenem, with MIC values of ≥8 μg/ml. These CRAB isolates were from different types of specimens including sputum ($n = 49$), throat swabs ($n = 5$), blood ($n = 1$), douche ($n = 4$), and others ($n = 8$). Epidemiological analysis of the 67 patients with CRAB revealed that 27 were ≥70 years old, 15 were between 60 and 70 years old, 24 were between 21 and 60 years old, and one was ≤20 years old. Among 40 males (59.7%) and 27 females (40.3%), 24 isolates were collected from the ICU, and 17 isolates were collected from the respiratory department. Clinical characteristics of the 67 patients with CRAB are summarized in **Table 2**.

TABLE 2 | Clinical characteristics of 67 patients with Carbapenem-resistant *Acinetobacter baumannii* (CRAB).

Gender	Total number	Source	Total number	Department	Total number
Male	40 (59.70%)	Blood	1 (1.49%)	Intensive care unit	24 (35.82%)
Female	27 (40.30%)	Sputum	49 (73.13%)	Respiratory	17 (25.37%)
Age (years)	**Total number**	Swab	5 (7.46%)	Neurosurgery	9 (13.43%)
0–20	1	Secretion	5 (7.46%)	Infectious diseases department	2 (2.99%)
21–50	11	Douche	4 (5.98%)	Other wards	15 (22.39%)
51–60	13	Cerebrospinal fluid	2 (2.99%)		
70–more	27	Other	2 (2.99%)		

Expression of Carbapenemase Genes in *A. baumannii* Isolates

All isolates were screened for the presence of β-lactamases genes (**Table 3**). In the 67 CRAB isolates, all of them were positive to *recA*, 16S rRNA conservative region and bla_{OXA-51}. The major carbapenemase gene bla_{OXA-23} was detected in 80.60% of all isolates. In addition, 65 isolates (97.10%) were found to carry bla_{AmpC}. Other β-lactamase genes including bla_{KPC}, bla_{IMP}, bla_{VIM-1}, bla_{SIM}, bla_{NDM-1}, bla_{OXA-24}, bla_{OXA-58}, and $bla_{OXA-235}$ were undetectable in all isolates. Thus, the expression of bla_{OXA-23} is still the dominant carbapenem resistance mechanism in *A. baumannii* among the isolates investigated aside from the inherent bla_{OXA-51} gene. One susceptible isolate carried bla_{OXA-72}.

Genetic Analysis of STs

To investigate whether the isolates were genetically related, MLST was performed to characterize the CRAB, CIAB and CSAB (**Figure 1**). Each ST was represented by a dot that was proportionally sized to the number of *A. baumannii* containing that ST. The MLST analysis revealed a total of 32 different STs in 67 CRAB, including 6 existing STs and 26 novel STs. 77.62% of the 67 CRAB isolates were classified to CC92, ST92 and its 4 novel single-locus variants (SLVs), ST195, ST136, ST843, and ST75, were the predominant STs, found in 34.33, 7.46, 5.97, 4.48, and 2.99% of isolates, respectively. Among these five predominant STs, 82.35% of carbapenem-resistant isolates carried the bla_{OXA-23} gene, in which ST92 is one of the most widespread STs across the world and belongs to the CC92 in *A. baumannii* MLST databases (Wang et al., 2013). Seventeen of the 32 CRAB STs belong to CC92 (**Figure 1A**). In this study, the number of polymorphic sites in *gpi* ($n = 8$), *rpoD* ($n = 4$), *cpn60* ($n = 3$), *recA* ($n = 2$), and the other three loci (*gdhB*, *gyrB*, and *gltA*) indicated that all were the same in CC92. Seven of the eight CIAB were CC92 (**Figure 1B**), but only 3 of the 35 CSAB were CC92 (**Figure 1C**). All these suggested that CC92 represents the most widely distributed carbapenem-resistant CC identified in the hospital.

Plasmid Analysis

To detect the nosocomial infection mediated by *A. baumannii* plasmid DNA, a plasmid conjugation assay was performed. However, none of the carbapenem-resistant isolates transferred plasmids from *A. baumannii* isolates to *E. coli* J53 successfully. This could be due to multiple reasons. Carbapenem resistance genes were located in the genome, and plasmids were not conjugative or unable to replicate in the *E. coli* host. Of note, with no. 52 CRAB isolate (which harbored pAB52 plasmid) as the donor strain, a few colonies with azide resistance and displaying a similar morphology as the *E. coli* J53 recipient were obtained from the conjugation. Yet, these colonies were found to be arisen from *A. baumannii* based on the amplification and sequence analysis of 16S rRNA and seven housekeeping genes *gltA*, *gyrB*, *gdhB*, *recA*, *cpn60*, *gpi*, and *rpoD*. Likely, the generation of azide-resistant *A. baumannii* was a result of spontaneous mutation(s). In addition, plasmid extraction assay confirmed no plasmid-containing *E. coli* J53 colonies, indicating that pAB52 could not be transferred from *A. baumannii* to *E. coli* J53.

RAST analysis showed that the plasmid pAB52 was 8,893 bp in size and contained 10 open reading frames (ORFs; **Figure 2A**), with an average G+C content of 34.5%. Three genes with a putative role in virulence were detected in pAB52: a septicolysin-like gene coding for a pore-forming toxin (Rosado et al., 2008; Gan et al., 2012), a TonB-dependent receptor gene coding for an outer membrane protein involved in iron uptake and virulence (Acosta et al., 2011; Gan et al., 2012), and a toxin–antitoxin (TA) gene coding for is a functional TA system and that its toxin, SplT (Gan et al., 2012; Jurenaite et al., 2013), inhibits translation (**Figure 2A**). One ORF (orf 9) codes for proteins that share similarity to hypothetical proteins encoded by plasmid genes found in other bacteria (Gan et al., 2012), while the predicted products of three others (orf 1, orf 2, and orf 3) do not match any known sequences. orf 9, which is adjacent to the replication initiation protein, is similar to replication protein of other genomes (**Figure 2B**). Therefore, it might contribute to a replication ability. Further analysis indicated that this plasmid

TABLE 3 | Positive rates of β-lactamase genes in CRAB.

No. of strains	Rate of gene (%)										
	KPC	IMP	VIM-1	SIM	NDM-1	AmpC	OXA-23	OXA-24	OXA-51	OXA-58	OXA-235
67	0	0	0	0	0	97.10	80.60	0	100	0	0

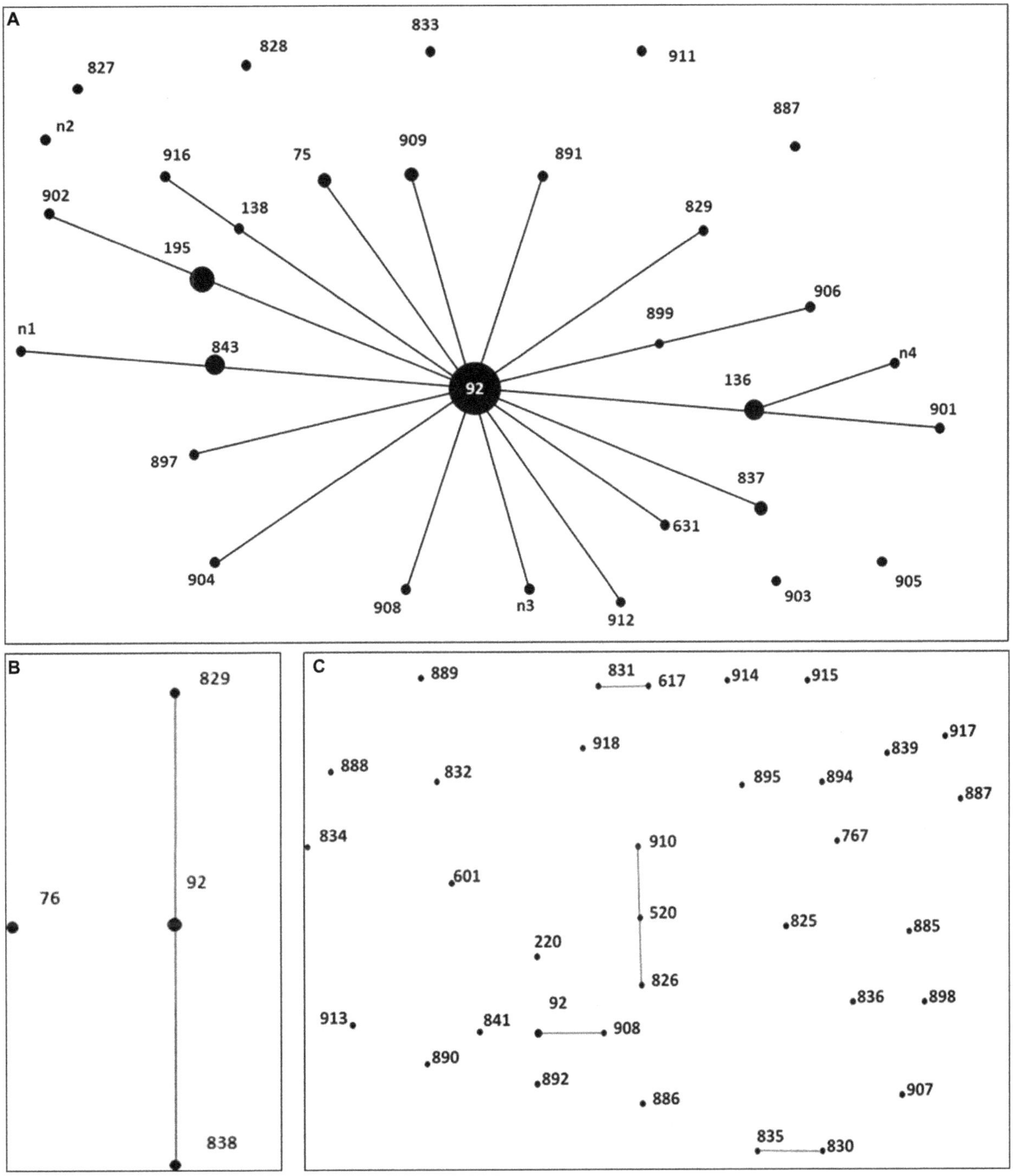

FIGURE 1 | Distribution of genes in 67 Carbapenem-resistant *Acinetobacter baumannii* (CRAB), 8 carbapenem-intermediate *A. baumannii* (CIAB), and 35 carbapenem-susceptible *A. baumannii* (CSAB). Minimum spanning tree analysis by eBURST algorithm of CRAB **(A)**, CIAB **(B)** and CSAB **(C)** isolates based on MLST data. Each circle represents a specific sequence type (ST). The size of each circle homologizes to a different number of isolates, with larger sizes representing higher frequency of occurrence. The solid lines connecting the circles indicate the relationship between different STs.

shared most nucleotide identity with previously described isolates of *A. baumannii*, WM99C, TCDC-AB0715, AB210, AB0057, 1656-2, AYE, UMB001, and ATCC17978, *Acinetobacter sp.* 6013150 and 6013113 (**Figure 2B**).

Discussion

Clinical characterization of 67 patients with CRAB indicated that compromised immunity (e.g., elderly patients suffering

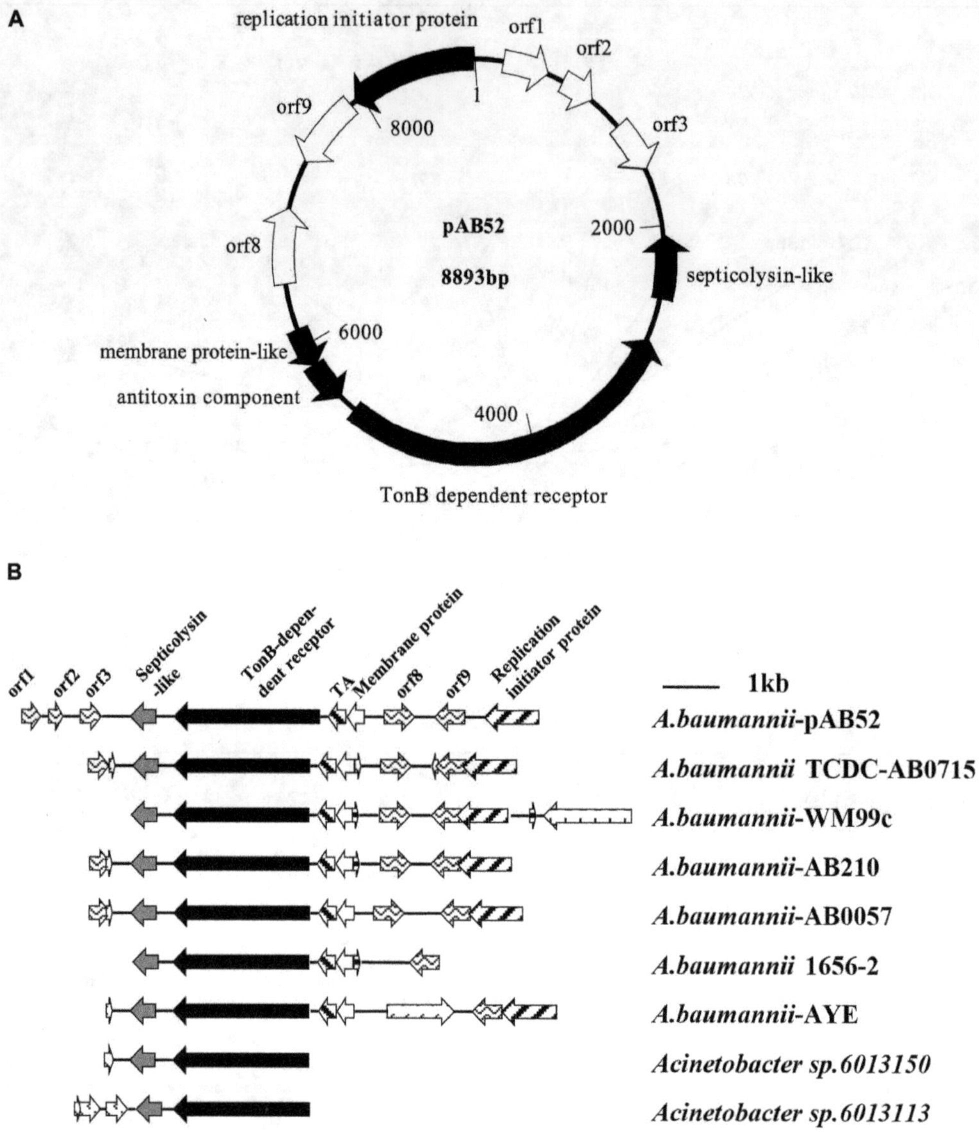

FIGURE 2 | (A) Diagram of plasmid pAB52 (GenBank accession number KR03004). The open reading frames (ORFs), represented by arrows pointing in the direction of transcription, are white, except for that for the reported genes, which is black. **(B)** Comparative analysis of the genetic environment of the TonB-dependent receptor in plasmid DNA by RAST software. Regions with 100% nucleotide sequence identity or similar biological function are filled with the same pattern.

from respiratory failure, encephalorrhagia, extensive burns, and chronic obstructive pulmonary disease) increases the risk of *A. baumannii* infection. In addition, males appear to have a greater risk of infection than females.

Surveillance by the CHINET project demonstrated that the *A. baumannii* rate of imipenem resistance increased over twofold between 2006 and 2013 to 62.8% in China (Fupin et al., 2014). In contrast, the resistance rate of *Acinetobacter* spp. to cefoperazone/sulbactam was only 36.4% (Fupin et al., 2014), far below the rates of imipenem resistance. Mirroring the national trend, the *A. baumannii* resistance rate of imipenem and cefoperazone/sulbactamat in our hospital was 60.9 and 14.75%, respectively (Ying et al., 2013).

We observed that OXA-23 was also the major carbapenemase mechanism responsible for the resistance phenotype (**Table 3**), as the gene was expressed in most CRAB isolates. This result is consistent with previous reports in which the CRAB is often caused due to the expression of OXA enzymes, particularly OXA-23 (Fu et al., 2010; Merino et al., 2014; Zowawi et al., 2015).

In our experiments, we unexpectedly found one particular carbapenem-susceptible isolate, which belonged to ST220 and carried the bla_{OXA-72} gene, a member of the bla_{OXA-24} subgroup, but this isolate did not carry any other resistance genes. This is the first time that bla_{OXA-24} has been detected in western China. Of note, this isolate was not only susceptible to meropenem and imipenem, but also susceptible to other β-lactams, aminoglycosides, tetracyclines, fluoroquinolone, rifampicin, and trimethoprim–sulfamethoxazole (data not shown). Our results are in contrast with previous reports that nosocomial infections were triggered by CRAB-carrying plasmids encoding OXA-72 carbapenemase via plasmid DNA from Lithuanian (Povilonis et al., 2013) and Taiwan (Lu et al., 2009). There are several possible reasons for the different results. For instance, the resistance genes may not get expressed because of different or changed genetic environment. Further investigation is needed why the presence of bla_{OXA-72} does not produce β-lactam resistance.

MLST is an unambiguous and utility technique that identifies accurate and portable nucleotide sequences of internal fragments from multiple loci housekeeping genes and was used to assess the genetic background of populations in an ocean of bacteria (Maiden et al., 1998). Based on our existing epidemiological data and the eBURST arithmetic, it was shown that CC92 was widespread and this major clone was resistant to carbapenem antibiotics in our hospital. Ruan et al. (2013) explained that CC92 is a widespread variant that has advantages with respect to acquiring resistance determinants and surviving in the nosocomial environment, which renders it preferentially selected under antibiotic pressure. This may be the primary reason for the current epidemiological situation of CC92 in our study. In addition, 82.35% of CRAB from the five predominant STs in CC92 carried the bla_{OXA-23} gene. We propose that the bla_{OXA-23} gene is tied to the genetic background of CC92, representing an important setting in which to study the emergence of carbapenem resistance. A novel SLVs of CC92, ST843 was identified in three isolates, and sharing six same alleles with ST92 and n1 (**Figure 1A**). We could infer that it may be one of the evolutions of *A. baumannii* to adapt nosocomial environment.

We used eBURST to analyze the distribution of the STs of CSAB, and found it was more dispersive than the distribution of CRAB (**Figure 1**). This may confirm that CSAB has a more diverse genetic background compared to CRAB. The high genetic diversity in CSAB might be attributed to the decreased survival rate of antibiotic treatment. None of them emerge more than once among 35 carbapenem-susceptible isolates from May 2012 to October 2013 except for 3 isolates belonging to CC92. These data also imply that CSAB may have a more diverse genetic source compared to the CRAB, and suggest CC92 is an evolution for *A. baumannii* to survive while antibiotics existed.

We also observed that a few *A. baumannii* isolates could acquire azide resistance after the bacteria were mixed and incubated with *E. coli* J53 together overnight. This can be totally expected since spontaneous mutation(s) conferring azide resistance could arise and be selected in the presence of azide, similar to those *in vitro* selections of drug-resistant mutants. This can be considered as a process of adaptation to a selection pressure. Indeed, mutations conferring resistance to azide have been reported in both *E. coli* (Fortin et al., 1990) and *Acinetobacter* spp. (Elkeles et al., 1994).

The cryptic plasmid pAB52 from CRAB isolate no. 52 does not carry any carbapenem resistance genes. It matches best with pAC12 which was reported in 2012 in Malaysia (Gan et al., 2012), and the ST of this pAB52 CRAB is ST195 (1-3-3-2-2-96-3) in parallel with *A. baumannii* AC12. According to the results of pAB52 sequence blast in NCBI (date not shown), all plasmids carry the same genes with over 99.9% identity over the full length, and belong to the *A. baumannii* plasmid group (GenBank accession number CP008850, CP007535, CP007550, CP001183, CP007578, KJ586856, KJ477078, and CP006964). Thus, plasmid pAB52 is larger with an additional segment of TonB gene **Figure 2B**), but these plasmids may have the same function. For example, the septicolysin-like protein encoded in these plasmids suggests that it may be involved in survival of bacteria in the lungs and blood (Acosta et al., 2011). The TA systems have been suggested to mediate bacterial persistence by generating slowly growing cells tolerant to antibiotics and environmental changes Jurenaite et al., 2013). In particular, the pAB52-containing strain was isolated from the sputum of a 74-year-old male who lived in a rural district in western China and had not traveled to any areas before admission to the hospital's ICU. This individual was found to have acquired an XDR *A. baumannii* infection during his hospital stay and eventually, he died from respiratory failure. The observation suggests that probably plasmid-borne *A. baumannii* TA systems contribute for the evolution of antibiotic resistance in this opportunistic pathogen.

In summary, this investigation confirms that the CC92 producing bla_{OXA-23} was the leading reason for the dramatic increase in the carbapenem resistance rates in China. Awareness of carbapenem-resistant organisms and their development in hospitals has crucial implications in optimizing infection control practices, establishing antimicrobial stewardship programs within the hospital, and finally establishing active regional surveillance systems (Zowawi et al., 2015).

Acknowledgments

This work was supported by the National Natural Science Foundation of China (grants 81373454, 31300659, 31470246, and 31401099), Applied Basic Research Programs of Sichuan Province (grant 2013jy0065), Scientific Research and Innovation Team of Sichuan Province (grant 15TD0025), and Preeminent Youth Fund of Sichuan Province (grant 2015JQO019). We thank Drs. Gang Liu (Xiamen University) and Xian-Zhi Li (Health Canada, Ottawa) for insightful revision of this manuscript.

References

Acosta, J., Merino, M., Viedma, E., Poza, M., Sanz, F., Otero, J. R., et al. (2011). Multidrug-resistant *Acinetobacter baumannii* Harboring OXA-24 carbapenemase, Spain. *Emerg. Infect. Dis.* 17, 1064–1067. doi: 10.3201/eid/1706.091866.

Aziz, R. K., Bartels, D., Best, A. A., Dejongh, M., Disz, T., Edwards, R. A., et al. (2008). The RAST Server: rapid annotations using subsystems technology. *BMC Genomics* 9:75. doi: 10.1186/1471-2164-9-75

Bartual, S. G., Seifert, H., Hippler, C., Luzon, M. A., Wisplinghoff, H., and Rodriguez-Valera, F. (2005). Development of a multilocus sequence typing scheme for characterization of clinical isolates of *Acinetobacter baumannii. J. Clin. Microbiol.* 43, 4382–4390. doi: 10.1128/jcm.43.9.4382-4390.2005

Bogaerts, P., Naas, T., Wybo, I., Bauraing, C., Soetens, O., Pierard, D., et al. (2006). Outbreak of infection by carbapenem-resistant *Acinetobacter baumannii* producing the carbapenemase OXA-58 in Belgium. *J. Clin. Microbiol.* 44, 4189–4192. doi: 10.1128/jcm.00796-06

Bou, G., and Martinez-Beltran, J. (2000). Cloning, nucleotide sequencing, and analysis of the gene encoding an AmpC beta-lactamase in *Acinetobacter baumannii. Antimicrob. Agents Chemother.* 44, 428–432. doi: 10.1128/AAC.44.2.428-432.2000

Bradford, P. A. (2001). Extended-spectrum beta-lactamases in the 21st century: characterization, epidemiology, and detection of this important resistance threat. *Clin. Microbiol. Rev.* 14, 933–951. doi: 10.1128/cmr.14.4.933-951.2001

Bush, K., Jacoby, G. A., and Medeiros, A. A. (1995). A functional classification scheme for beta-lactamases and its correlation with molecular structure. *Antimicrob. Agents Chemother.* 39, 1211–1233. doi: 10.1128/AAC.39.6.1211

Carvalho, K. R., Carvalho-Assef, A. P., Peirano, G., Santos, L. C., Pereira, M. J., and Asensi, M. D. (2009). Dissemination of multidrug-resistant *Acinetobacter baumannii* genotypes carrying bla(OXA-23) collected from hospitals in Rio de Janeiro, Brazil. *Int. J. Antimicrob. Agents* 34, 25–28. doi: 10.1016/j.ijantimicag.2008.12.009

Chiang, M. C., Kuo, S. C., Chen, Y. C., Lee, Y. T., Chen, T. L., and Fung, C. P. (2011). Polymerase chain reaction assay for the detection of *Acinetobacter baumannii* in endotracheal aspirates from patients in the intensive care unit. *J. Microbiol. Immunol. Infect.* 44, 106–110. doi: 10.1016/j.jmii.2010.04.003

CLSI. (2014). *Performance Standards for Antimicrobial Susceptibility Testing; Twenty-Fourth Informational Supplement. CLSI Document M100-S24.* Wayne, PA: Clinical and Laboratory Standards Institute.

Diancourt, L., Passet, V., Nemec, A., Dijkshoorn, L., and Brisse, S. (2010). The population structure of *Acinetobacter baumannii*: expanding multiresistant clones from an ancestral susceptible genetic pool. *PLoS ONE* 5:e10034. doi: 10.1371/journal.pone.0010034

Durante-Mangoni, E., and Zarrilli, R. (2011). Global spread of drug-resistant *Acinetobacter baumannii*: molecular epidemiology and management of antimicrobial resistance. *Future Microbiol.* 6, 407–422. doi: 10.2217/fmb.11.23

Elkeles, A., Rosenberg, E., and Ron, E. Z. (1994). Azide-resistant mutants in Acinetobacter calcoaceticus A2 are defective in protein secretion. *FEMS Microbiol. Lett.* 116, 221–224. doi: 10.1111/j.1574-6968.1994.tb06704.x

Ellington, M. J., Kistler, J., Livermore, D. M., and Woodford, N. (2007). Multiplex PCR for rapid detection of genes encoding acquired metallo-beta-lactamases. *J. Antimicrob Chemother.* 59, 321–322. doi: 10.1093/jac/dkl481

Evans, B. A., and Amyes, S. G. (2014). OXA beta-lactamases. *Clin. Microbiol. Rev.* 27, 241–263. doi: 10.1128/cmr.00117-113

Feil, E. J., Li, B. C., Aanensen, D. M., Hanage, W. P., and Spratt, B. G. (2004). eBURST: inferring patterns of evolutionary descent among clusters of related bacterial genotypes from multilocus sequence typing data. *J. Bacteriol.* 186, 1518–1530. doi: 10.1128/JB.186.5.1518-1530.2004

Fortin, Y., Phoenix, P., and Drapeau, G. R. (1990). Mutations conferring resistance to azide in *Escherichia coli* occur primarily in the secA gene. *J. Bacteriol.* 172, 6607–6610.

Fu, Y., Zhou, J., Zhou, H., Yang, Q., Wei, Z., Yu, Y., et al. (2010). Wide dissemination of OXA-23-producing carbapenem-resistant *Acinetobacter baumannii* clonal complex 22 in multiple cities of China. *J. Antimicrob. Chemother.* 65, 644–650. doi: 10.1093/jac/dkq027

Fupin, H., Demei, Z., Fu, W., Xiaofei, J., Ziyong, S., Zhongju, C., et al. (2014). CHINET 2013 surveillance of bacterial resistance in China. *Chin. J. Infect. Chemother.* 14, 365–374.

Gan, H. M., Lean, S. S., Suhaili, Z., Thong, K. L., and Yeo, C. C. (2012). Genome sequence of *Acinetobacter baumannii* AC12, a polymyxin-resistant strain isolated from Terengganu, Malaysia. *J. Bacteriol.* 194, 5979–5980. doi: 10.1128/jb.01466-1412

Hamouda, A., Evans, B. A., Towner, K. J., and Amyes, S. G. (2010). Characterization of epidemiologically unrelated *Acinetobacter baumannii* isolates from four continents by use of multilocus sequence typing, pulsed-field gel electrophoresis, and sequence-based typing of bla(OXA-51-like) genes. *J. Clin. Microbiol.* 48, 2476–2483. doi: 10.1128/jcm.0243 1-2439

Higgins, P. G., Perez-Llarena, F. J., Zander, E., Fernandez, A., Bou, G., and Seifert, H. (2013). OXA-235, a novel class D beta-lactamase involved in resistance to carbapenems in *Acinetobacter baumannii. Antimicrob. Agents Chemother.* 57, 2121–2126. doi: 10.1128/aac.02413-2412

Jurenaite, M., Markuckas, A., and Suziedeliene, E. (2013). Identification and characterization of type II toxin-antitoxin systems in the opportunistic pathogen *Acinetobacter baumannii. J. Bacteriol.* 195, 3165–3172. doi: 10.1128/JB.00237-213

Kempf, M., and Rolain, J. M. (2012). Emergence of resistance to carbapenems in *Acinetobacter baumannii* in Europe: clinical impact and therapeutic options. *Int. J. Antimicrob. Agents* 39, 105–114. doi: 10.1016/j.ijantimicag.2011.10.004

Li, Y., Guo, Q., Wang, P., Zhu, D., Ye, X., Wu, S., et al. (2014). Clonal dissemination of extensively drug-resistant *Acinetobacter baumannii* producing an OXA-23 beta-lactamase at a teaching hospital in Shanghai, China. *J. Microbiol. Immunol. Infect.* 48, 101–108. doi: 10.1016/j.jmii.2014.04.005

Liu, L. L., Ji, S. J., Ruan, Z., Fu, Y., Fu, Y. Q., Wang, Y. F., et al. (2015). Dissemination of blaOXA-23 in *Acinetobacter* spp. in China: main roles of conjugative plasmid pAZJ221 and Transposon Tn2009. *Antimicrob. Agents Chemother.* 59, 1998–2005. doi: 10.1128/AAC.04574-4514

Lu, P. L., Doumith, M., Livermore, D. M., Chen, T. P., and Woodford, N. (2009). Diversity of carbapenem resistance mechanisms in *Acinetobacter baumannii* from a Taiwan hospital: spread of plasmid-borne OXA-72 carbapenemase. *J. Antimicrob. Chemother.* 63, 641–647. doi: 10.1093/jac/dkn553

Maiden, M. C., Bygraves, J. A., Feil, E., Morelli, G., Russell, J. E., Urwin, R., et al. (1998). Multilocus sequence typing: a portable approach to the identification of clones within populations of pathogenic microorganisms. *Proc. Natl. Acad. Sci. U.S.A.* 95, 3140–3145. doi: 10.1073/pnas.95.6.3140

Merino, M., Poza, M., Roca, I., Barba, M. J., Sousa, M. D., Vila, J., et al. (2014). Nosocomial outbreak of a multiresistant *Acinetobacter baumannii* expressing OXA-23 carbapenemase in Spain. *Microb. Drug Resist.* 20, 259–263. doi: 10.1089/mdr.2013.0127

Netsvyetayeva, I., Sikora, M., Golas, M., Swoboda-Kopec, E., De Walthoffen, S.W., Dembicka, O., et al. (2011). *Acinetobacter baumannii* multidrug-resistant strain occurrence in liver recipients with reference to other high-risk groups. *Transplant Proc.* 43, 3116–3120. doi: 10.1016/j.transproceed.2011.07.008

Peleg, A. Y., Seifert, H., and Paterson, D. L. (2008). *Acinetobacter baumannii*: emergence of a successful pathogen. *Clin. Microbiol. Rev.* 21, 538–582. doi: 10.1128/cmr.00058-57

Pournaras, S., Gogou, V., Giannouli, M., Dimitroulia, E., Dafopoulou, K., Tsakris, A., et al. (2014). Single-locus-sequence-based typing of blaOXA-51-like genes for rapid assignment of *Acinetobacter baumannii* clinical isolates to international clonal lineages. *J. Clin. Microbiol.* 52, 1653–1657. doi: 10.1128/jcm.03565-3513

Povilonis, J., Seputiene, V., Krasauskas, R., Juskaite, R., Miskinyte, M., Suziedelis, K., et al. (2013). Spread of carbapenem-resistant *Acinetobacter baumannii* carrying a plasmid with two genes encoding OXA-72 carbapenemase in *Lithuanian hospitals. J. Antimicrob. Chemother.* 68, 1000–1006. doi: 10.1093/jac/dks499

Rosado, C. J., Kondos, S., Bull, T. E., Kuiper, M. J., Law, R. H., Buckle, A. M., et al. (2008). The MACPF/CDC family of pore-forming toxins. *Cell Microbiol.* 10, 1765–1774. doi: 10.1111/j.1462-5822.2008.01191.x

Ruan, Z., Chen, Y., Jiang, Y., Zhou, H., Zhou, Z., Fu, Y., et al. (2013). Wide distribution of CC92 carbapenem-resistant and OXA-23-producing *Acinetobacter baumannii* in multiple provinces of China. *Int. J. Antimicrob. Agents* 42, 322–328. doi: 10.1016/j.ijantimicag.2013.06.019

Runnegar, N., Sidjabat, H., Goh, H. M., Nimmo, G. R., Schembri, M. A., and Paterson, D. L. (2010). Molecular epidemiology of multidrug-resistant

Acinetobacter baumannii in a single institution over a 10-year period. *J. Clin. Microbiol.* 48, 4051–4056. doi: 10.1128/jcm.01208-1210

Tiwari, V., Kapil, A., and Moganty, R. R. (2012a). Carbapenem-hydrolyzing oxacillinase in high resistant strains of *Acinetobacter baumannii* isolated from India. *Microb. Pathog.* 53, 81–86. doi: 10.1016/j.micpath.2012.05.004

Tiwari, V., Vashistt, J., Kapil, A., and Moganty, R. R. (2012b). Comparative proteomics of inner membrane fraction from carbapenem-resistant *Acinetobacter baumannii* with a reference strain. *PLoS ONE* 7:e39451. doi: 10.1371/journal.pone.0039451

Tiwari, V., and Moganty, R. R. (2013). Structural studies on New Delhi Metallo-beta-lactamase (NDM-2) suggest old beta-lactam, penicillin to be better antibiotic for NDM-2-harbouring *Acinetobacter baumanni*. *J. Biomol. Struct. Dyn.* 31, 591–601. doi: 10.1080/07391102.2012.706075

Tiwari, V., and Moganty, R. R. (2014). Conformational stability of OXA-51 beta-lactamase explains its role in carbapenem resistance of *Acinetobacter baumannii*. *J. Biomol. Struct. Dyn.* 32, 1406–1420. doi: 10.1080/07391102.2013.819789

Tsakris, A., Pournaras, S., Woodford, N., Palepou, M.-F. I., Babini, G. S., Douboyas, J., et al. (2000). Outbreak of infections caused by *Pseudomonas aeruginosa* producing VIM-1 carbapenemase in Greece. *J. Clin. Microbiol.* 38, 1290–1292.

Valenzuela, J. K., Thomas, L., Partridge, S. R., Van Der Reijden, T., Dijkshoorn, L., and Iredell, J. (2007). Horizontal gene transfer in a polyclonal outbreak of carbapenem-resistant *Acinetobacter baumannii*. *J. Clin. Microbiol.* 45, 453–460. doi: 10.1128/jcm.01971-1976

Walther-Rasmussen, J., and Hoiby, N. (2006). OXA-type carbapenemases. *J. Antimicrob. Chemother.* 57, 373–383. doi: 10.1093/jac/dki482

Wang, F. (2008). CHINET 2006 surveillance of bacterial resistance in China. *Chin. J. Infect. Chemother.* 8, 1–9.

Wang, M., Tran, J. H., Jacoby, G. A., Zhang, Y., Wang, F., and Hooper, D. C. (2003). Plasmid-mediated quinolone resistance in clinical isolates of *Escherichia coli* from Shanghai, China. *Antimicrob. Agents Chemother.* 47, 2242–2248. doi: 10.1128/AAC.47.7.2242-2248.2003

Wang, X., Qiao, F., Yu, R., Gao, Y., and Zong, Z. (2013). Clonal diversity of *Acinetobacter baumannii* clinical isolates revealed by a snapshot study. *BMC Microbiol.* 13:234. doi: 10.1186/1471-2180-13-234

Woodford, N., Ellington, M. J., Coelho, J. M., Turton, J. F., Ward, M. E., Brown, S., et al. (2006). Multiplex PCR for genes encoding prevalent OXA carbapenemases in *Acinetobacter* spp. *Int. J. Antimicrob. Agents* 27, 351–353. doi: 10.1016/j.ijantimicag.2006.01.004

Ying, X. U., Wang, D., Wang, J. Q., and Ling, B. D. (2013). Drug resistance surveillance of common bacteria in a hospital between 2010 and 2013 years. *J. Chengdu Med. College* 8, 506–511.

Yong, D., Toleman, M. A., Giske, C. G., Cho, H. S., Sundman, K., Lee, K., et al. (2009). Characterization of a new metallo-beta-lactamase gene, bla(NDM-1), and a novel erythromycin esterase gene carried on a unique genetic structure in *Klebsiella pneumoniae* sequence type 14 from India. *Antimicrob. Agents Chemother.* 53, 5046–5054. doi: 10.1128/AAC.00774-779

Zowawi, H. M., Sartor, A. L., Sidjabat, H. E., Balkhy, H. H., Walsh, T. R., Al Johani, S. M., et al. (2015). Molecular epidemiology of carbapenem-resistant *Acinetobacter baumannii* isolates in the gulf cooperation council states: dominance of oxa-23-type producers. *J. Clin. Microbiol.* 53, 896–903. doi: 10.1128/jcm.02784-2714

10

Vancomycin Modifies the Expression of the *agr* System in Multidrug-Resistant *Staphylococcus aureus* Clinical Isolates

Vicenta Cázares-Domínguez[1], Sara A. Ochoa[1], Ariadnna Cruz-Córdova[1], Gerardo E. Rodea[1], Gerardo Escalona[1], Alma L. Olivares[2], José de Jesús Olivares-Trejo[3], Norma Velázquez-Guadarrama[2]* and Juan Xicohtencatl-Cortes[1]*

[1] Laboratorio de Investigación en Bacteriología Intestinal, Hospital Infantil de México Federico Gómez, México DF, Mexico, [2] Laboratorio de Infectología, Hospital Infantil de México Federico Gómez, México DF, Mexico, [3] Posgrado en Ciencias Genómicas, Universidad Autónoma de la Ciudad de México, México DF, Mexico

***Correspondence:**
Juan Xicohtencatl-Cortes and Norma Velázquez-Guadarrama, Laboratorio de Investigación en Bacteriología Intestinal and Laboratorio de Infectología, Hospital Infantil de México Federico Gómez, Doctor Márquez 162, Colonia Doctores, Delegación Cuauhtémoc, 06720 México DF, Mexico
juanxico@yahoo.com;
normave@yahoo.com

Staphylococcus aureus is an opportunistic pathogen that colonizes human hosts and causes a wide variety of diseases. Two interacting regulatory systems called *agr* (accessory gene regulator) and *sar* (staphylococcal accessory regulator) are involved in the regulation of virulence factors. The aim of this study was to evaluate the effect of vancomycin on *hld* and *spa* gene expression during the exponential and post-exponential growth phases in multidrug-resistant (MDR) *S. aureus*.

Methods: Antibiotic susceptibility was evaluated by the standard microdilution method. The phylogenetic profile was obtained by pulsed-field gel electrophoresis (PFGE). Polymorphisms of *agr* and SCC*mec* (staphylococcal cassette chromosome *mec*) were analyzed by multiplex polymerase chain reaction (PCR). The expression levels of *hld* and *spa* were analyzed by reverse transcription-PCR. An enzyme-linked immunosorbent assay (ELISA) was performed to detect protein A, and biofilm formation was analyzed via crystal violet staining.

Results: In total, 60.60% (20/33) of *S. aureus* clinical isolates were MDR. Half (10/20) of the MDR *S. aureus* isolates were distributed in subcluster 10, with >90% similarity among them. In the isolates of this subcluster, a high prevalence (100%) for the *agr*II and the cassette SCC*mec* II polymorphisms was found. Our data showed significant increases in *hld* expression during the post-exponential phase in the presence and absence of vancomycin. Significant increases in *spa* expression, protein A production and biofilm formation were observed during the post-exponential phase when the MDR *S. aureus* isolates were challenged with vancomycin.

Conclusion: The polymorphism *agr*II, which is associated with nosocomial isolates, was the most prevalent polymorphism in MDR *S. aureus*. Additionally, under our study conditions, vancomycin modified *hld* and *spa* expression in these clinical isolates. Therefore, vancomycin may regulate alternative systems that jointly participate in the regulation of these virulence factors.

Keywords: *Staphylococcus aureus*, multidrug-resistant, vancomycin, *agr* system, antibiotics

Introduction

Staphylococcus aureus is an opportunistic pathogen capable of causing a wide variety of diseases in humans, ranging from localized infections of the skin and soft tissues to life-threatening systemic infections (Archer, 1998; Shopsin and Kreiswirth, 2001; David and Daum, 2010; Sowash and Uhlemann, 2014). An infection is initiated when *S. aureus* has access to subcutaneous tissues and is disseminated by the circulatory system, and it infects nearly every organ, leading to severe osteomyelitis, sepsis, abscesses, endocarditis, pneumonia, and toxic shock syndrome (Archer, 1998; Shopsin and Kreiswirth, 2001; David and Daum, 2010; Sowash and Uhlemann, 2014). *S. aureus* can produce a remarkable array of wall surface and secreted virulence factors that contribute to the establishment and maintenance of infection (Novick et al., 1993). These cell surface virulence factors include microbial surface components that recognize extracellular matrix proteins, such as fibrinogen, laminin, plasminogen, vitronectin, fibronectin, thrombospondin, and bone sialoprotein (Falord et al., 2011; Yamamoto et al., 2013). The secreted virulence factors are generally produced during the post-exponential or stationary phase, and they include several extracellular toxins (i.e., alpha-, beta-, gamma-, and delta-hemolysin, enterotoxins, exfoliative toxins A and B, and toxic shock syndrome toxin-1) and exoenzymes (i.e., lipase, nucleases, proteases, hyaluronate lyase, and metalloproteases; Chan and Foster, 1998; Kolar et al., 2013).

The expression of virulence genes in *S. aureus* is regulated under the partial control of the two-component quorum-sensing system encoded by genes at the *agr* locus (Bronner et al., 2004). The P2 transcript (RNAII) encodes the quorum-sensing system, which consists of the following four proteins: AgrB (the secreted protein responsible for the export and processing of AgrD to its active form), AgrD (a signaling peptide), and AgrC-ArgA (a two-component system in which AgrC is the transmembrane receptor histidine kinase and AgrA is the DNA-binding response regulator; Novick et al., 1995; Gilot et al., 2002; Novick, 2003; Gilot and van Leeuwen, 2004). A high cell population density causes the activation of AgrA, which induces the transcription of the P3 promoter. Next, P3 drives the transcription of RNAIII, a regulatory RNA that is both a positive and negative regulator of virulence factor production. The activation of RNAIII transcription in response to an increase in cell population density induces a transition in gene expression correlated with metabolic changes and stress adaptations. Toxin- and extracellular enzyme-encoding genes are positively regulated by the *agr-hld* (δ-lysin gene) system, and the genes coding for protein A (*spa*) and coagulase are negatively regulated (Morfeldt et al., 1988). Protein A has a molecular weight of 42 kDa and is covalently anchored to the peptidoglycan of *S. aureus* (Palmqvist et al., 2002). Ninety percent of the molecule is localized in the cell wall, and 10% is in the bacterial cytoplasm. Protein A is an important virulence factor of *S. aureus* based on its ability to bind to a variety of ligands, including the Fc region of IgG, the von Willebrand factor, tumor necrosis factor receptor-1 (TNFR-1), the Fab-heavy chains of the Vh3 subclass, and the epidermal growth factor receptor (EGFR; Cedergren et al., 1993; Viau and Zouali, 2005; Gómez et al., 2006; O'Seaghdha et al., 2006). An increase in protein A during the post-exponential phase has been observed in *agr*-deleted *S. aureus* strains (Novick, 2003).

In addition, AgrA activation leads to increased transcription of the δ-lysin gene (*hld*), which is located immediately upstream of the *agr* operon (Janzon and Arvidson, 1990). δ-lysin is a small polypeptide of only 26 amino acids. It is secreted without a signal peptide, and it makes cation-selective channels in the phospholipid bilayers (Lee and Birkbeck, 1984). δ-lysin is a virulence factor with lytic activity in a wide range of cells, such as neutrophils, macrophages, mammalian erythrocytes, and bacterial protoplasts, as well as in cellular organelles (Julander et al., 1983).

The aim of this study was to assess *agr* system expression by quantifying *hld* and *spa* expression in multidrug-resistant (MDR) *S. aureus* clinical isolates cultured from the exponential to post-exponential growth phases in the presence of vancomycin. In addition, the *agr* group I–IV polymorphisms were evaluated as a factor that predisposes the permanence and survival of MDR *S. aureus* clinical isolates during nosocomial or hospital-acquired infections in the Hospital Infantil de México Federico Gómez (HIMFG).

Materials and Methods

Bacterial Isolates

Thirty-three *S. aureus* from different infections were obtained from a clinical isolates collection at the Central Laboratory of the HIMFG. The *S. aureus* isolates were collected from January 2006 to June 2007. They were cultured on 5% sheep blood agar plates (Becton Dickinson, East Rutherford, NJ, USA) at 37°C under 5% CO_2 for 24 h and kept at −70°C in skim milk (Becton Dickinson, East Rutherford, NJ, USA).

Diagnostic Tests for Identifying *S. aureus*

The bacteria were grown on blood agar, and identification was performed using conventional bacteriological techniques, such as colony morphology examination, catalase assays, coagulase assays, Gram staining (Sigma-Aldrich, St. Louis, MO, USA), mannitol fermentation, and Brain Heart Infusion (BHI) broth growth assays (Becton Dickinson, East Rutherford, NJ, USA) with 15% NaCl (MacFadin, 1996).

Antimicrobial Susceptibility

The antibiotic susceptibility profiles of the *S. aureus* isolates were determined by the Minimum Inhibitory Concentration (MIC) technique with the microdilution method in Mueller-Hinton broth (MH; Becton Dickinson, East Rutherford, NJ, USA), as recommended by the Clinical and Laboratory Standards Institute (2014). The MIC tests were conducted with vancomycin, ciprofloxacin, erythromycin (MP Biomedicals, Solon, OH, USA), clarithromycin (Grünenthal Gmbh, Aachen, Germany), oxacillin, clindamycin, linezolid (Sigma-Aldrich, St. Louis, MO, USA), meropenem (AstraZeneca Pharmaceuticals LP, Wilmington, DE, USA), trimethoprim, sulfamethoxazole (Roche, Basel, Switzerland), and gentamicin (Schering-Plough Pharmaceuticals,

Kenilworth, NJ, USA). To identify methicillin-resistant *S. aureus* clinical isolates, the bacteria were tested for oxacillin resistance by the oxacillin-salt screening method. Oxacillin is a more stable antibiotic than methicillin, although they are chemically identical. *S. aureus* strain ATCC 29213 (American Type Culture Collection, Manassas, VA, USA) was used as a positive control.

Molecular Genotyping Assays

Pulsed-field gel electrophoresis (PFGE) was performed using a previously described protocol (Pereira et al., 2014). The chromosomal DNA from MDR and sensitive *S. aureus* isolates was digested with the Sma I restriction enzyme (Thermo Fisher Scientific Inc., Life Technologies, Grand Island, NY, USA) and subjected to electrophoresis on 1% agarose gels (Promega; Madison, WI, USA) using the following parameters: 200 V (6 v/cm) at 14°C for 21.5 h, with an initial change of 5 s and a final change of 40 s. The gels were stained with 0.5 μg/mL ethidium bromide solution (Sigma-Aldrich, St. Louis, MO, USA) and visualized using a gel imaging system (ChemiDoc^{TM} MP System, Biorad, Hercules, CA, USA). The DNA fragment patterns generated by PFGE were analyzed with NTSY-pc software (version 2.0, Applied Biostatistics, Inc., Port Jefferson, NY, USA; Ramazanzadeh et al., 2013) using the Sørensen–Dice similarity coefficient and the unweighted pair group method with arithmetic mean (UPGMA) clustering approach (Dice, 1945).

Multiplex Polymerase Chain Reaction (PCR) Conditions

The *S. aureus* clinical isolates were recovered from frozen stock onto BHI agar plates and incubated at 37°C for 18–24 h. Genomic DNA extraction was performed with a Wizard Genomic DNA Purification Kit (Promega, Madison, WI, USA) from a bacterial culture grown in BHI broth. Briefly, the bacterial culture pellet was mixed with TE buffer (10 mM Tris HCl and 1 mM EDTA, pH 8.0), lysozyme (0.25 mg/mL; Sigma-Aldrich, St. Louis, MO, USA), proteinase K (0.0125 mg/mL; Sigma-Aldrich, St. Louis, MO, USA), and lysostaphin (0.062 mg/mL; Sigma-Aldrich, St. Louis, MO, USA). Multiplex polymerase chain reaction (PCR) assays for detecting *agr* polymorphisms (*agr*I, *agr*II, *agr*III, and *agr*IV; **Table 1**) were prepared according to the protocol for Go Taq Green Master Mix (Promega, Madison, WI, USA). Multiplex PCR reactions were prepared in a final volume of 25 μL as follows: 12.5 μL of Go Taq Green Master Mix 2x (Promega, Madison, WI, USA), 2 μL of bacterial DNA (100 ng/μL), 5 μL of *agr* primers (*agr*I, *agr*II, *agr*III, and *agr*IV) at 10 pg/μL, and 5.5 μL of nuclease-free water. The DNA amplification was performed in a Veriti 96-Well Thermal Cycler-Life Technologies (Applied Biosystems, Foster City, CA, USA) with the following parameters: an initial denaturation at 94°C for 5 min followed by 26 amplification cycles (denaturation at 94°C for 30 s, annealing at 55°C for 30 s, and extension at 72°C for 60 s), ending with a final extension at 72°C for 7 min. An external positive control [DNA extracted from *S. aureus* strains USA300 (*agr*I), 1749 (*agr*II), and ATCC 25923 (*agr*III)] and an external negative control (DNase/RNase-free distilled water) were included with each run. PCR amplicons (10 μL) were loaded into a 1.5% (wt/v) agarose gel (Promega, Madison, WI, USA) using a 100 bp DNA ladder (Promega, Madison, WI, USA), and electrophoresis was performed in 1x TAE buffer at 100 V for 1 h. The bands were visualized using a gel imaging system (ChemiDoc^{TM} MP System, Biorad, Hercules, CA, USA).

The SCC*mec* genes (I, II, III, and IVa) were characterized by multiplex PCR according to Cázares-Domínguez et al. (2015).

TABLE 1 | Primers used in *agr* and SCC*mec* typing by multiplex PCR.

Gene	Primer	Sequence (5′–3′)	Amplicon size (bp)	Reference
*agr*I	*agr* I-F *agr* I-R	ATGCACATGGTGCACATGC GTCACAAGTACTATAAGCTGCGAT	441	Gilot et al. (2002)
*agr*II	*agr* II-F *agr* II-R	ATGCACATGGTGCACATGC TATTACTAATTGAAAAGTGGCCATAGC	575	Gilot et al. (2002)
*agr*III	*agr* III-F *agr* III-R	ATGCACATGGTGCACATGC GTAATGTAATAGCTTGTAAAAAGTGGCCATAGC	323	Gilot et al. (2002)
*agr*IV	*agr* IV-F *agr* IV-R	ATGCACATGGTGCACATGC CGATAATGCCGTAATACCCG	659	Gilot et al. (2002)
SCC*mec* I	*mec* I-F *mec* I-R	GCTTTAAAGAGTGTCGTTACAGG GTTCTCTCATAGTATGACGTCC	613	Zhang et al. (2005)
SCC*mec* II	*mec* II-F *mec* II-R	CGTTGAAGATGATGAAGCG CGAAATCAATGGTTAATGGACC	398	Zhang et al. (2005)
SCC*mec* III	*mec* III-F *mec* III-R	CCATATTGTGTACGATGCG CCTTAGTTGTCGTAACAGATCG	280	Zhang et al. (2005)
SCC*mec* IVa	*mec* IVa-F *mec* IVa-R	GCCTTATTCGAAGAAACCG CTACTCTTCTGAAAAGCGTCG	776	Zhang et al. (2005)
SCC*mec* V	*mec* V-F *mec* V-R	GAACATTGTTACTTAAATGAGCG TGAAAGTTGTACCCTTGACACC	325	Zhang et al. (2005)
mecA	*mec* 147-F *mec* 147-R	GTGAAGATATACCAAGTGATT ATGCGCTATAGATTGAAAGGAT	147	Zhang et al. (2005)

F, forward; R, reverse.

RNA Extraction from *S. aureus* Clinical Isolates

Staphylococcus aureus isolates that were grown overnight were adjusted to an optical density of 0.05 at 600 nm and incubated in BHI with 1 μg/mL vancomycin. The bacterial cultures were grown for 4 h until reaching the exponential phase (OD_{600} of 0.6–0.8) and for 11 h just in the post-exponential phase (OD_{600} of 1.2–1.4). Aliquots of bacterial cultures were harvested by centrifugation at 10,000 × *g* for 3 min at 4°C. Each pellet was washed in an equal volume of TE buffer (10 mM Tris HCl and 1 mM EDTA, pH 8.0) three times and lysed with the same buffer supplemented with 0.25 mg/mL lysozyme, 0.0125 mg/mL proteinase K, and 0.062 mg/mL lysostaphin. Total bacterial RNA was isolated (TRIzol, Life Technologies, Carlsbad, CA, USA) according to the manufacturer's directions. After purification, contaminating DNA was removed with RNase-free DNase I (2 U/10 μg of total bacterial RNA at 37°C for 30 min). The RNA was then re-purified with RNeasy Minicolumns (Qiagen Incorporated, Ln Valencia, CA, USA). The amount of recovered RNA was determined spectrophotometrically, and the samples were then stored at −80°C.

Transcriptional Expression Analyses of the *spa* and *hld* Genes

The relative expression levels of the *spa* and *hld* genes were determined by cDNA-PCR. The purified RNA of all *S. aureus* clinical isolates was employed for reverse transcription (RT)-PCR assays with the GeneAmp RNA PCR Kit (Applied Biosystems, Foster City, CA, USA), using specific primers for the *spa* (encoding protein A) and *hld* (encoding a delta toxin) genes (**Table 2**). A GeneAmp RNA PCR Kit was used with 0.2 μg of total RNA per reaction as a template for PCR amplification. Reactions containing *S. aureus* cells alone, only RNA, or no reverse transcriptase were used as negative controls. Specific primers were used for the amplification of 16S RNA, which was used as an internal control (**Table 2**).

The expression levels of *spa* and *hld* transcripts from *S. aureus* clinical isolates were quantified by densitometric analysis with Bio-Rad image software (Bio-Rad chemi-doc, Quantity one 4.4.1). The data are expressed as the mean ± standard error of the means. A *p*-value of less than 0.05 was considered significant. All experiments were repeated at least three times, and a representative result is shown for each experiment.

Quantitative Measurements of Protein A

A qualitative screening test for the production of protein A was conducted by an enzyme linked immunosorbent assay (ELISA). Briefly, 96-well plates containing 200 μL of BHI were inoculated with 10 μL (1.5 × 10^8 bacteria/mL) of bacterial suspensions and incubated at 37°C in the presence or absence of vancomycin to the exponential phase (4 h) and post-exponential phase (11 h). Cell wall-associated protein A was identified using anti-protein A, followed by mouse anti-IgG antibodies, *o*-phenylenediamine (OPD) compounds, and ELISA as previously described (Ohkaru et al., 1995).

Quantitative Determination by Biofilm Assays

Biofilm formation was quantitatively analyzed according to the protocol described by Erdem et al. (2008). MDR and sensitive *S. aureus* clinical isolates were grown in BHI broth overnight at 37°C. Then, 96-well plates containing 200 μL of BHI were inoculated with 10 μL (1.5 × 10^8 bacteria/mL) of bacterial suspensions and grown at 37°C in the presence or absence of vancomycin in the exponential phase (4 h) and post-exponential phase (11 h). The biofilms that developed on the surfaces of the wells were gently washed three times with 1x phosphate-buffered saline (PBS; pH 7.4) and fixed with 2% formaldehyde at 4°C overnight. Wells with fixed biofilms were decanted, washed three times with PBS and stained with 200 μL of 1% crystal violet for 30 min. The excess crystal violet was removed, and the plates were washed twice with water. Crystal violet was subsequently solubilized in 70% methanol, and the absorbance was determined at 620 nm. Assays were performed in triplicate and repeated three consecutive times.

Results

The Antimicrobial Susceptibility Testing of *S. aureus* Clinical Isolates

Thirty-three *S. aureus* clinical isolates were tested for antimicrobial susceptibility. In total, 60.60% (20/33) of *S. aureus* clinical isolates were MDR, 100% (33/33) were sensitive to four antibiotics (vancomycin, trimethoprim/sulfamethoxazole, gentamicin, and linezolid), and 39.39% (13/33) were sensitive to all antimicrobials (**Table 3**).

Molecular Typing Analysis of the MDR and Sensitive *S. aureus* Isolates by PFGE

A total of 19 DNA pulsotypes grouped in four clusters (I–IV) were identified, revealing patterns that consisted of 11–18 DNA fragments ranging in size from 48.5 to 339.5 Kb (**Figure 1**). In total, 3.03% (1/33) of the *S. aureus* clinical isolates were identified as pulsotype F, and cluster I showed 38% genetic similarity when compared with other pulsotypes. Four pulsotypes (E, I, K, and M) belonging to cluster II included 21.21% (7/33) of the *S. aureus* clinical isolates with 64% genetic similarity. In addition, the nine pulsotypes (H, J, L, N, O, P, Q, R, and S) organized in cluster III included 54.55% (18/33) of the *S. aureus* clinical isolates with

TABLE 2 | Primers used in *hld* and *spa* expression analysis by RT-PCR.

Gene	Primer sequence (5′–3′)	Product size (bp)	Reference
spa	TATCTGGTGGCGTAACACCTG GATGAAGCCGTTACGTTGTTC	322	Goerke et al. (2000)
hld	GAAGGAGTGTTTCAATGG TAAGAAAATACATAGCACTGAG	260	Goerke et al. (2000)
16S	TCCGGAATTATTGGGCGTAA CCACTTTCCTCTTCTGCACTCA	121	Goerke et al. (2000)

TABLE 3 | MIC analysis for the MDR and sensitive *S. aureus* clinical isolates.

Clinical isolates	Antibiotics (µg/mL)										
	CEC	**CEP**	**VAN**	**CLA**	**CIP**	**STX**	**OXA**	**GEN**	**ERI**	**MEM**	**LZN**
488H, 428H, 10H, 714H	≥128	≥128	1	≥128	64	0.5	16	1	≥128	64	2
242H, 143H	≥128	≥128	1	≥128	≥128	2	≥128	0.5	≥128	32	4
246H, 175H	≥128	≥128	1	≥128	≥128	1	≥128	0.5	≥128	16	4
882HR	≥128	≥128	1	≥128	64	1	≥128	1	≥128	32	2
425LCR	≥128	≥128	1	≥128	64	0.5	16	0.5	≥128	64	2
330H	≥128	≥128	1	≥128	≥128	2	≥128	1	≥128	32	4
783H	≥128	≥128	1	≥128	≥128	2	≥128	1	≥128	32	4
902H	≥128	≥128	1	≥128	64	1	≥128	2	≥128	32	4
54H	≥128	≥128	0.5	≥128	64	0.5	16	0.5	≥128	32	2
A-32	≥128	≥128	0.5	≥128	≥128	0.5	≥128	0.5	≥128	16	2
828H	8	≥128	1	≥128	64	0.5	16	0.5	≥128	64	2
931H	8	≥128	1	≥128	64	0.5	16	1	≥128	64	2
260H	0.125	64	0.5	32	0.25	0.5	64	0.5	≥128	8	0.5
622H	0.125	≥128	1	0.06	0.25	1	32	0.5	0.25	4	4
299H	0.060	≥128	1	≥128	32	0,5	8	0.5	≥128	0.12	0.5
679H, 318LCR, 573H, 770H, 633H, 779H, 291H, 336H, 18H, 108H, 780H, 493H, 198H	≤0.25	≤8	≤1	≤0.25	≤2	≤1	≤1	≤1	≤1	≤0.125	≤4
CVR*	**≥32**	**≥32**	**≥16**	**≥8**	**≥4**	**≥4**	**≥4**	**≥16**	**≥8**	**≥16**	**≥8**

*CEC, cefaclor; CEP, cephalothin; VAN, vancomycin; CLA, clarithromycin; CIP, ciprofloxacin; STX, trimethoprim/sulfametoxazole; OXA, oxacillin; GEN, gentamicin; ERI, erythromycin; MEM, meropenem; LZN, linezolid. The MDR S. aureus clinical isolates are marked in gray. *Cut-off values for resistance to MIC (µg/mL; CVR).*

56% genetic similarity. In this cluster, the *S. aureus* clinical isolates with >90% genetic similarity were grouped in subcluster 10 and distributed in the following order: two isolates in pulsotype Q, one isolate in pulsotype R, and eight isolates in pulsotype J (**Figure 1**). In addition, 21.21% (7/33) of the *S. aureus* clinical isolates were classified as cluster IV with 62% genetic similarity and distributed as pulsotypes A, B, C, D, and G (**Figure 1**).

Only one *S. aureus* clinical isolate sensitive to all antibiotics was identified (pulsotype F of cluster I; **Figure 1**). Four MDR and three sensitive *S. aureus* clinical isolates were distributed in the four pulsotypes (E, I, K, and M) of cluster II (**Figure 1**). Fourteen MDR and four sensitive *S. aureus* clinical isolates were distributed into the nine pulsotypes (H, J, L, N, O, P, Q, R, and S) of cluster III. Interestingly, seven MDR and one sensitive *S. aureus* clinical isolates that were distributed in pulsotype J were organized in subcluster 10. In addition, two MDR and five sensitive *S. aureus* clinical isolates were distributed over the five pulsotypes (A, B, C, D, and G) of cluster IV (**Figure 1**).

Distributing *agr*, SCC*mec*, *hld*, and *spa* Genes by Multiplex PCR Endpoint Analysis

Multiplex PCR amplification assays were performed for *agr*-specific group identification in the selected MDR and sensitive *S. aureus* clinical isolates. Our results showed three different *agr* specificity groups (**Figures 2A,B**), which were identified according to the expected product sizes (**Table 1**). Briefly, the sensitive *S. aureus* clinical isolates were 30.76% (4/13) *agr*I, 30.76% (4/13) *agr*II, and 38.46% (5/13) *agr*III (**Figures 2B,C**). The MDR *S. aureus* clinical isolates were 70% (14/20) *agr*II, 10% (2/20) *agr*I, and 10% (2/20) *agr*III (**Figures 2A,C**). Furthermore, both the *agr*I/II and *agr*II/III polymorphism groups were identified in a single (1/20; 5%) clinical isolate of MDR *S. aureus* (**Figures 2A,C**). The expression of the *agr*IV was not identified in either the MDR or sensitive *S. aureus* clinical isolates.

A 398 bp product corresponding to the SCC*mec* group II polymorphism was present in 60.60% (20/33) of the MDR *S. aureus* clinical isolates that were distributed in the fourth cluster as determined by PFGE. A SCC*mec* II polymorphism was observed in 12.12% (4/33) of the isolates in cluster II, 42.42% (14/33) of the isolates in cluster III, and 6.06% (2/33) of the isolates in cluster IV. However, the SCC*mec* II polymorphism was not identified in cluster I (**Figure 1**). Moreover, SCC*mec* polymorphism types I, III, and IV were not identified in the MDR *S. aureus* clinical isolates (**Figure 1**). It is important to emphasize that the presence of the SCC*mec* gene explains methicillin (oxacillin) resistance; therefore, it was not identified in sensitive *S. aureus* clinical isolates (**Figure 1**).

The frequencies of the *hld* and *spa* genes were determined using specific primers to be 100% (33/33) in MDR and sensitive *S. aureus* clinical isolates. Our results showed two bands, which corresponded to *hld* at 260 bp and *spa* at 322 bp (data not shown).

hld and *spa* Gene Expression as Determined by RT-PCR

The specific transcripts of the *spa* and *hld* genes produced by the ten MDR *S. aureus* isolates that were distributed in subcluster 10 of cluster III were quantified by RT-PCR-densitometry. These MDR *S. aureus* isolates were treated with and without vancomycin during two different stages of the growth phase

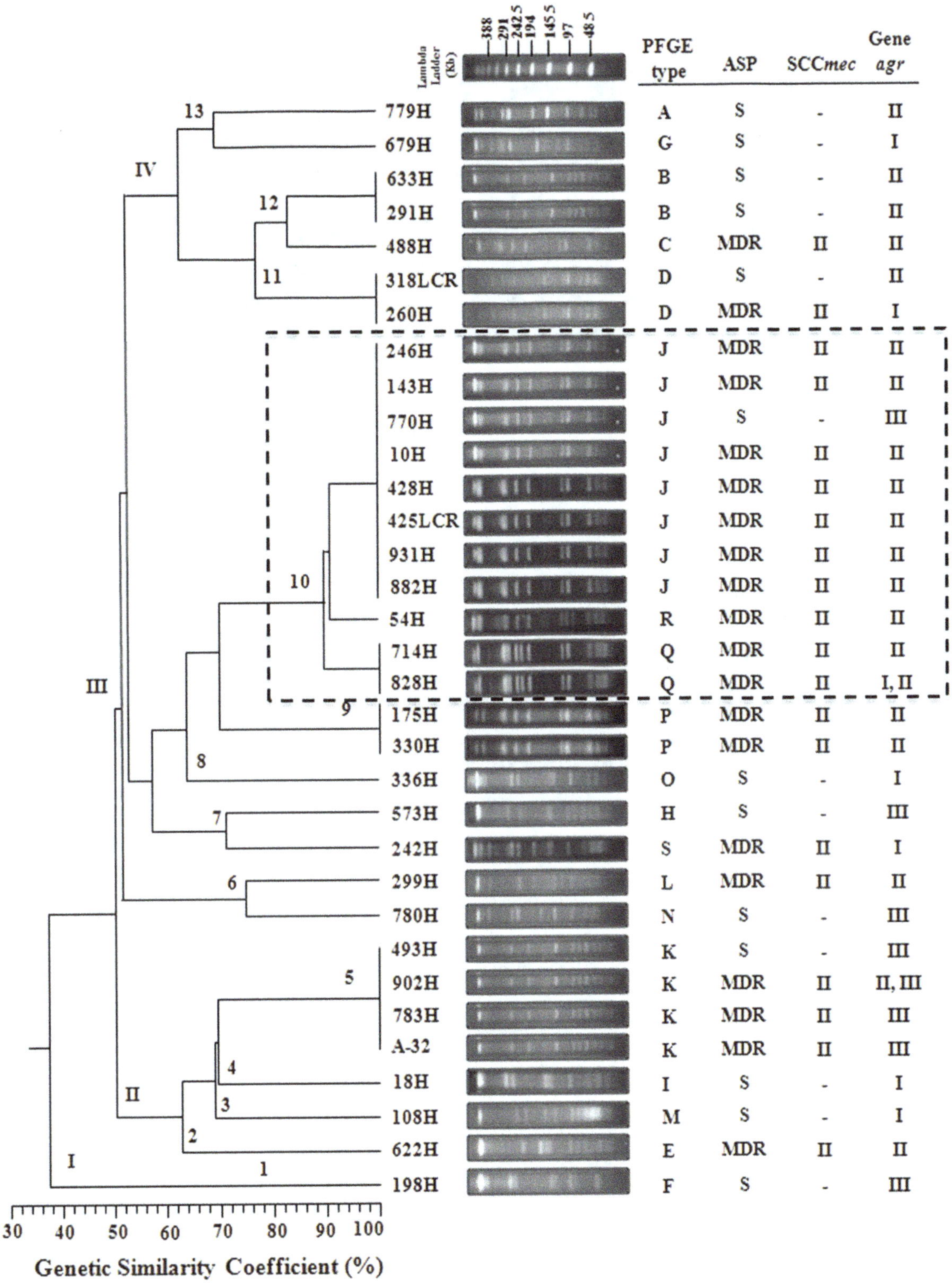

FIGURE 1 | Dendrogram analysis of PFGE results showing the genetic relationships among the PFGE profiles and the presence of *agr* and SCC*mec* type among the 33 *S. aureus* clinical isolates. A phylogenetic analysis was conducted using the Sørensen–Dice similarity coefficient in association with the UPGMA algorithm as the grouping method. The dendrogram was evaluated by obtaining the cophenetic correlation coefficient with the Mantel test, which yielded an *r*-value of 0.9077. Antimicrobial susceptibility profile (ASP), multidrug-resistant (MDR), sensitive (S). The dotted line indicates that the clinical isolates that belong to subcluster 10 with >90% genetic similarity.

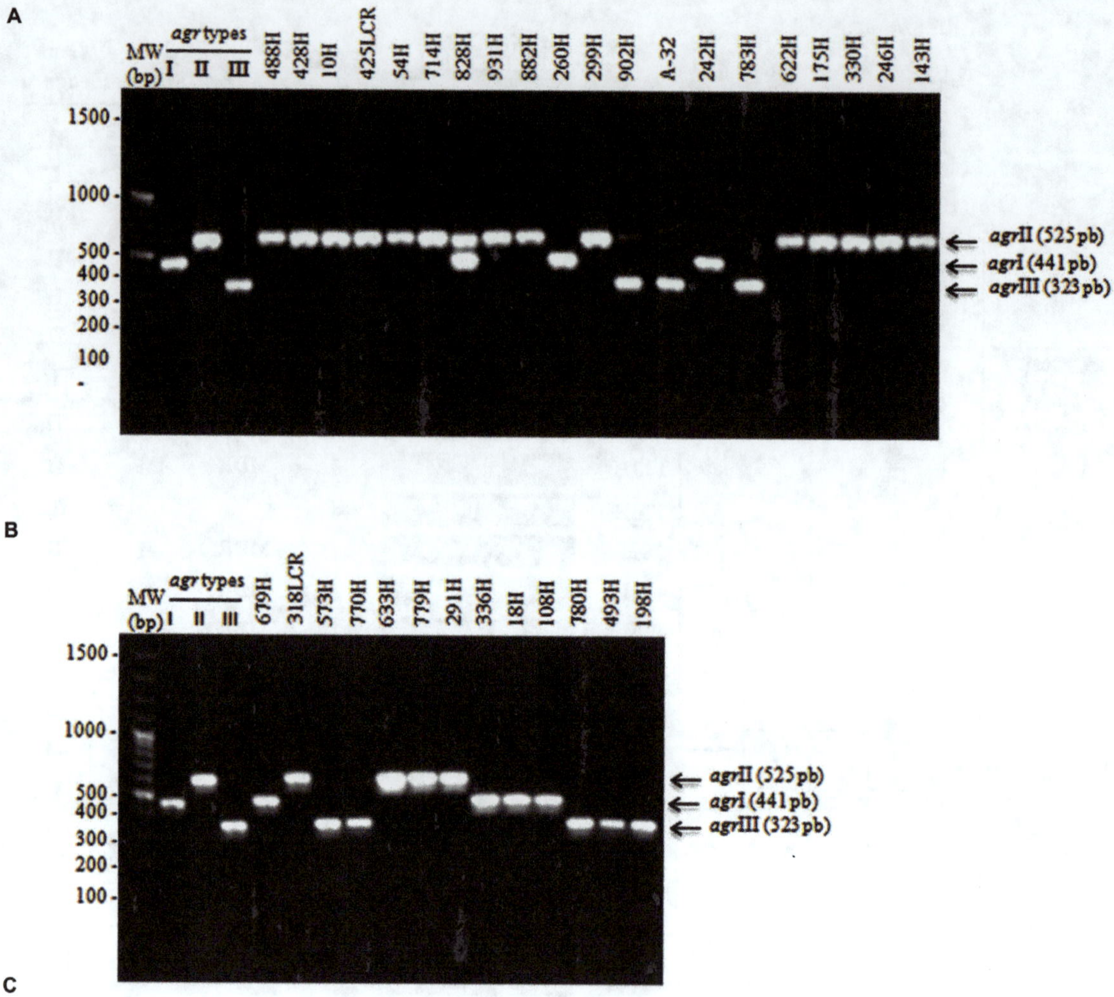

C

Isolates	I	II	III	I, II	II,III
MDR	2/20 (10%)	14/20 (70%)	2/20 (10%)	1/20 (5%)	1/20 (5%)
Sensitive	4/13 (30.76%)	4/13 (30.76%)	5/13 (38.46%)		

FIGURE 2 | Multiplex PCR assays for identifying *agr* types. (A) MDR *S. aureus* clinical isolates showing the *agrI* (441 bp), *agrII* (575 bp), and *agrIII* (323 bp) types. **(B)** Sensitive *S. aureus* clinical isolates. **(C)** The percentages of *agr* types that were identified in MDR and sensible *S. aureus* clinical isolates. *S. aureus* strain USA300 (*agrI*), 1749 (*agrII*), and ATCC 25923 (*agrIII*) were used as reference strains. MW, molecular weight (bp).

(exponential and post-exponential). The *hld* expression in MDR *S. aureus* clinical isolates showed a significant increase (1.68-fold; $p = 0.0001$) when cultured from the exponential to the post-exponential growth phase in the absence of vancomycin as well as a significant increase (2.04-fold; $p = 0.0001$) when grown in the presence of vancomycin (**Figure 3A**). Interestingly, under the same test conditions, a significant increase (2.07-fold; $p = 0.0001$) in *hld* expression was observed in MDR *S. aureus* clinical isolates in the exponential growth phase treated with vancomycin compared to those treated without vancomycin (**Figure 3A**). In addition, a significant increase (2.53-fold; $p = 0.0001$) in *hld* expression was observed in the MDR *S. aureus* clinical isolates cultured to the post-exponential growth phase when challenged with 1 μg/mL vancomycin compared to without vancomycin challenge (**Figure 3A**). The *hld* expression in sensitive *S. aureus* isolates showed increases from the exponential to the post-exponential growth phase in both the presence and absence of vancomycin (data not shown).

The *spa* expression level showed a significant reduction (4.01-fold; $p = 0.0001$) when the MDR *S. aureus* clinical isolates were cultured from the exponential to post-exponential growth phases in the absence of vancomycin (**Figure 3B**). Interestingly, a remarkable increase (3.67-fold; $p = 0.0001$) in *spa* expression was observed from the exponential to the post-exponential

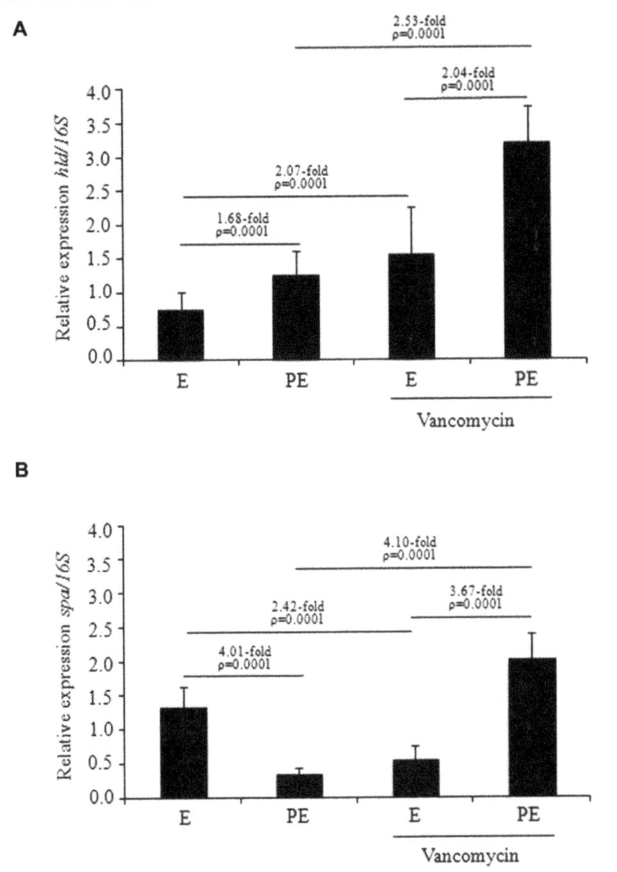

FIGURE 3 | Expression levels of the *hld* and *spa* genes determined by RT-PCR in MDR *S. aureus* clinical isolates cultured from the exponential to post-exponential growth phases in the presence of vancomycin. (A) *hld* expression; **(B)** *spa* expression. The densitometric results of the transcripts were quantified using the Quantity One program and analyzed by paired Student's *t*-tests. E, exponential; PE, post-exponential.

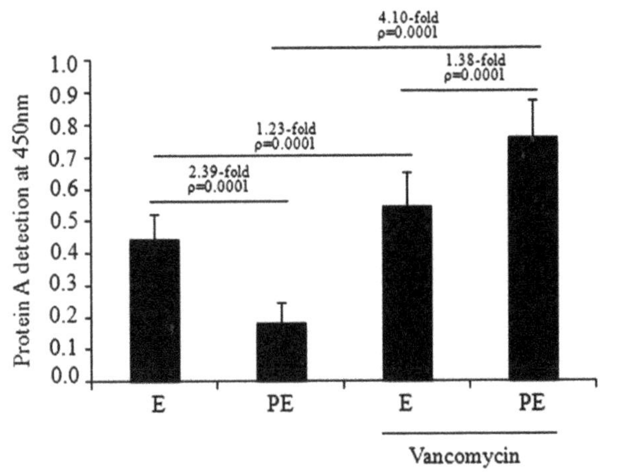

FIGURE 4 | Protein A production in the MDR *S. aureus* clinical isolates cultured from the exponential to post-exponential growth phases in the presence of vancomycin. A significant reduction of 2.39-fold ($p = 0.0001$) was observed in the MDR *S. aureus* clinical isolates when they were cultured from the exponential to post-exponential growth phases in the absence of vancomycin and there was a significant increase of 1.38-fold ($p = 0.0001$) when cultured from the exponential to post-exponential growth phase when challenged with 1 μg/mL vancomycin.

phases in MDR *S. aureus* clinical isolates cultured in the presence of vancomycin (**Figure 3B**). During the exponential phase, a significant reduction in *spa* expression (2.42-fold; $p = 0.0001$) was observed when the MDR *S. aureus* clinical isolates were challenged with vancomycin compared to those grown in the absence of vancomycin; the MDR *S. aureus* clinical isolates in the post-exponential growth phase showed a significant increase (4.10-fold; $p = 0.0001$) in *spa* expression when challenged with vancomycin compared to those grown in the absence of vancomycin (**Figure 3B**). The *spa* expression level did not show a significant changes (reduction and/or increase) when the sensitive *S. aureus* clinical isolates were cultured in the exponential and post-exponential growth phases in the presence or absence of vancomycin (data not shown).

Protein A Immunodetection by ELISA

Protein A immunodetection assays were performed for the MDR *S. aureus* clinical isolates distributed in subcluster 10 of cluster III. Quantitative analysis showed a significant reduction (2.39-fold; $p = 0.0001$) in protein A expression in the post-exponential phase compared to the exponential phase in the absence of vancomycin (**Figure 4**). Furthermore, a significant increase (1.38-fold; $p = 0.0001$) in protein A expression was observed from the exponential to the post-exponential growth phase in the MDR *S. aureus* clinical isolates cultured the presence of 1 μg/mL vancomycin. Interestingly, a significant increase (4.10-fold; $p = 0.0001$) in protein A expression was observed in these clinical isolates cultured to the post-exponential growth phase in the presence of vancomycin compared to in the absence of vancomycin (**Figure 4**). Likewise, a significant increase (1.23-fold; $p = 0.0001$) in protein A expression was observed in the MDR *S. aureus* clinical isolates that were cultured to the exponential growth phase in the presence of vancomycin compared to in the absence of vancomycin.

Clinical Strains of MDR and Sensitive *S. aureus* Isolates Produce Biofilms

A crystal-violet biofilm assay was performed for the MDR *S. aureus* clinical isolates that were distributed in subcluster 10 of cluster III. Quantitative analysis showed no difference in biofilm formation when the MDR *S. aureus* clinical isolates were cultured to the exponential and post-exponential growth phases in the absence of vancomycin. In the presence of vancomycin, a significant increase (1.42-fold; $p = 0.0001$) was observed in the biofilm formation of the MDR *S. aureus* clinical isolates cultured to the exponential growth phase and a significant increase (1.85-fold; $p = 0.0001$) was observed in the isolates cultured to the post-exponential growth phase. Furthermore, a significant increase (1.35-fold; $p = 0.0001$) in the biofilm formation of these isolates was observed from the exponential to

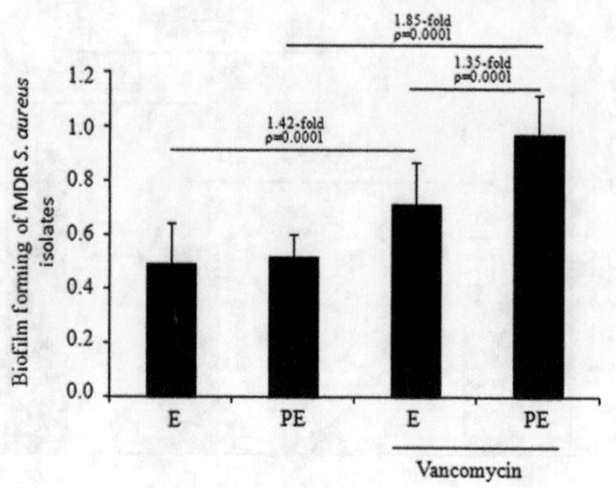

FIGURE 5 | Biofilm assays of the MDR *S. aureus* clinical isolates when cultured from the exponential to post-exponential growth phases in the presence of vancomycin. Interestingly, a significant increase (1.35-fold; $p = 0.0001$) in biofilm formation was observed in the MDR *S. aureus* clinical isolates cultured from the exponential to post-exponential growth phases in the presence of the vancomycin.

post-exponential growth phases in the presence of vancomycin (**Figure 5**).

Discussion

In this study, 100% of *S. aureus* clinical isolates showed sensitive to vancomycin, trimethoprim/sulfamethoxazole, gentamicin, and linezolid. Low rates of resistance to tetracycline, gentamicin, and trimethoprim/sulfamethoxazole in clinical methicillin-resistant staphylococcus aureus (MRSA) isolates have been described in other studies (Bordon et al., 2010; Davis et al., 2013). However, the impressive ability of *S. aureus* to develop resistance against various antibiotics through point mutations, transposons, plasmids, and resistance cassettes is the most likely reason for the rapid increase in antimicrobial resistance worldwide in recent years (Sakoulas and Moellering, 2008). High levels of erythromycin resistance and increasing ciprofloxacin resistance in MRSA clinical isolates have been observed (Davis et al., 2013). The clinical condition, the administration route, and the resistance pattern of the organism are also risk factors that are considered in treating MRSA infections when drugs such as vancomycin, linezolid, daptomycin, clindamycin, and mupirocin are administered (Liu et al., 2011). The increase of vancomycin-resistant *S. aureus* (VRSA) isolates in the United States could be explained by the selective pressure originating from the excessive use of vancomycin to treat MRSA infections (David and Daum, 2010). The high frequency of resistance to the antibiotics tested in our study may reflect the fact that the public health service in Mexico prescribes all of these antibiotics for treating pediatric patients; these drugs are considered to be essential in this health sector.

According to the PFGE analysis, 33.33% (11/33) of the *S. aureus* clinical isolates distributed in subcluster 10 shared 90% similarity. In particular, 50% of the MDR *S. aureus* clinical isolates were distributed in subcluster 10 and showed closely related pulsotypes with three clones that were assigned as J, Q, and R according to the criteria of Tenover et al. (1995). These results suggest that MDR *S. aureus* isolates are associated with patients with infections acquired during their current hospital stay. MRSA isolates with a highly related PFGE type have been associated with an MDR profile to β-lactams, gentamicin, ciprofloxacin, clindamycin, and erythromycin (Velazquez-Meza et al., 2004). In total, 90.90% (10/11) of the *S. aureus* clinical isolates belonging to subcluster 10 were resistant to six antibiotics, namely cephalothin, clarithromycin, ciprofloxacin, oxacillin, erythromycin, and meropenem. Furthermore, 72.72% (8/11) of these clinical isolates were resistant to cefaclor.

Our data showed a high prevalence of the *agr* group II polymorphism, with a PCR-amplified product of 575 kb, in the MDR *S. aureus* clinical isolates belonging to subcluster 10 by PFGE analysis. Several studies performed in Japan and the USA described the *agr* group II polymorphism as the *agr* type that was most frequent in MDR *S. aureus* clinical isolates, and it has been associated with nosocomial infections from pediatric patients (Sakoulas et al., 2003). Similar findings indicated that all *S. aureus* clinical isolates from diverse geographic origins and those recovered from patients undergoing intubation showed the *agr* group II polymorphism (Sakoulas et al., 2002; Goerke et al., 2003; Moise et al., 2004). Furthermore, the *agr* group II polymorphism in MRSA predicts the failure of vancomycin therapy (Moise et al., 2007).

We further evaluated *agr* expression through indirect mechanisms by quantification of *hld* and *spa* expression in MDR *S. aureus* clinical isolates during the exponential and post-exponential growth phases and upon vancomycin challenge. Our RT-PCR analysis showed that *hld* activation occurs during the post-exponential growth phase, resulting in an increase of 1.68-fold without vancomycin and 2.04-fold with vancomycin compared with the exponential phase. The increased *hld* expression in the MDR *S. aureus* clinical isolates of subcluster 10 suggested that the activation of this gene is influenced by the growth phase and vancomycin challenge. Other studies have shown significant increases in *hld* expression at the end of the exponential growth phase by an *agr*-dependent mechanism that is involved in the regulation of virulence genes. *agr*A codes for a protein that can activate *hld* transcription as a response to the growth phase (Janzon and Arvidson, 1990). Our data indicated that the MDR *S. aureus* isolates were stimulated by vancomycin at subinhibitory concentrations, as indicated by increases in *hld* expression. These results indicate that the antibiotic stimulates the *agr* system. In a previous study, the over-expression of the *hld* transcript in vancomycin-resistant *S. aureus* strains associated with activation of the sigma factor was observed when these isolates were exposed to subinhibitory antibiotic concentrations (Chen et al., 2011). Recently, community-associated (CA) MRSA was challenged with subinhibitory concentrations of tetracycline and clindamycin, which had a strong stimulatory effect on the activity of the *agr* operon (Joo et al., 2010). These results, together with the data obtained in our studies, suggest that vancomycin also exerts a strong stimulatory effect on the activity of the *agr* operon. It is important to note that *agr* controls many virulence factors of

S. aureus and that vancomycin is frequently used in Mexico for treating MRSA infections in pediatric patients.

Interestingly, a reduction in *spa* expression was observed during the exponential to post-exponential growth phases when MDR *S. aureus* clinical isolates were cultured without vancomycin. In contrast, a significant increase (3.67-fold) in *spa* expression was observed when MDR *S. aureus* clinical isolates were only cultured with vancomycin during the post-exponential growth phase. These data suggest that *spa* activation or repression is regulated by the growth phase and that expression of *spa* mRNA is modified by drugs such as vancomycin. In previous studies, CA-MRSA isolates treated with daptomycin and vancomycin showed no change in *spa* mRNA and SpA protein levels (Subrt et al., 2011; Otto et al., 2013). However, in a different study, the expression levels of different virulence factors in the CA-MRSA isolates were suppressed in the presence of clindamycin and linezolid (Otto et al., 2013).

The expression levels of the *spa* gene are directly correlated with protein A production because most of the gene transcript is translated as protein A. The localization of protein A in the cell wall could contribute to biofilm formation in MRSA clinical isolates during the post-exponential phase and when challenged with vancomycin. The XdrA regulator has almost as strong an activating effect on *spa* as SarS, and it acts on the same *spa* operator regions (identified potential *cis*-acting regulatory regions) as SarS or on closely overlapping regions (McCallum et al., 2010). The current evidence suggests that XdrA directly regulates *spa* transcripts independently of other well-characterized regulators. The over-expression of *spa* transcripts and protein A in the MDR *S. aureus* clinical isolates challenged with vancomycin could also be related to the action of the regulator XdrA, which should be examined in future studies. In addition, the variation in the levels of *spa* transcripts during different growth phases is likely due to the influence of other regulators and/or to the characteristics of clinical isolates.

Conclusion

Staphylococcus aureus clinical strains are influenced by intercellular signaling through the *agr* system, which modulates the activation and/or repression of many outer membrane proteins associated with biofilms (Cafiso et al., 2012). Vancomycin is the antibiotic of choice for treating nosocomial infections of MDR *S. aureus* clinical isolates in pediatric patients at the HIMFG. These data provide a direct evidence for evaluating the role of virulence genes (*hld* and *spa*) associated with *agr* regulation and suggest the participation of other regulatory elements; which are include an operon that activates many virulence factors that could be important during infection by this nosocomial pathogen. In our study, the polymorphism *agr*II was associated with nosocomial isolates and was the most prevalent polymorphism in MDR *S. aureus*. Our finding showed that vancomycin modified the *hld* and *spa* expression in the MDR *S. aureus* clinical isolates; suggesting that, vancomycin may regulate alternative systems that jointly participate in the regulation of virulence factors involved in bacteria pathogenesis, which allows spread and adaptation into a hospital environmental.

Author Contributions

Designed and conceived the experiments: VCD, SAO, ACC, NVG, and JXC. Performed the experiments: VCD, GE, and SAO. Analyzed the data: VCD, SAO, ALO, and GER. Contributed reagents/materials/analysis tools: VCD, SAO, ACC, JJOT, and JXC. Wrote and reviewed the manuscript: VCD, SAO, NVG, and JXC.

Acknowledgments

This work was supported by grant number CONACYT 133451 and public Federal Funds HIM/2007/004. The study was reviewed and approved by the Research Committee (Dr. Onofre Muñoz Hernández), Ethics Committee (Dr. Amparo Faure Fontenla), and Biosecurity Committee (Dr. Herlinda Vera Hermosillo) of HIMFG, under permit number HIM/2007/004. We thank Karina Espinosa-Mazariego and Juana Rodriguez for their technical assistance with this work. The Central Laboratory of HIMFG provided the *S. aureus* clinical isolates for this study. The anonymous clinical information presented in this manuscript prior to analysis was obtained from patient medical records, and the records were chosen by considering the diagnosis and sample type. According to the institutional ethical, biosecurity and investigation evaluation, informed consent was not required.

References

Archer, G. L. (1998). *Staphylococcus aureus*: a well-armed pathogen. *Clin. Infect. Dis.* 26, 1179–1181. doi: 10.1086/520289

Bordon, J., Master, R. N., Clark, R. B., Duvvuri, P., Karlowsky, J. A., Ayesu, K., et al. (2010). Methicillin-resistant *Staphylococcus aureus* resistance to non-beta-lactam antimicrobials in the United States from 1996 to 2008. *Diagn. Microbiol. Infect. Dis.* 67, 395–398. doi: 10.1016/j.diagmicrobio.2010.03.006

Bronner, S., Monteil, H., and Prévost, G. (2004). Regulation of virulence determinants in *Staphylococcus aureus*: complexity and applications. *FEMS Microbiol. Rev.* 28, 183–200. doi: 10.1016/j.femsre.2003.09.003

Cafiso, V., Bertuccio, T., Spina, D., Purrello, S., Campanile, F., Di Pietro, C., et al. (2012). Modulating activity of vancomycin and daptomycin on the expression of autolysis cell-wall turnover and membrane charge genes in hVISA and VISA strains. *PLoS ONE* 7:e29573. doi: 10.1371/journal.pone.0029573

Cázares-Domínguez, V., Cruz-Córdova, A., Ochoa, S. A., Escalona, G., Espinosa, M. S., Jiménez, R., et al. (2015). Thickening of cell wall as an effect to vancomycin tolerantce of methicillin-resistant *Staphylococcus aureus* strains from pediatric patients. *PLoS ONE* 10:e0118791. doi: 10.1371/journal.pone.0118791

Cedergren, L., Andersson, R., Jansson, B., Uhlén, M., and Nilsson, B. (1993). Mutational analysis of the interaction between staphylococcal protein A and human IgG1. *Protein Eng.* 6, 441–448. doi: 10.1093/protein/6.4.441

Chan, P. F., and Foster, S. J. (1998). The role of environmental factors in the regulation of virulence-determinant expression in *Staphylococcus aureus* 8325-4. *Microbiol. Read. Engl.* 144, 2469–2479. doi: 10.1099/00221287-144-9-2469

Chen, H.-Y., Chen, C.-C., Fang, C.-S., Hsieh, Y.-T., Lin, M.-H., and Shu, J.-C. (2011). Vancomycin activates σ(B) in vancomycin-resistant *Staphylococcus aureus* resulting in the enhancement of cytotoxicity. *PLoS ONE* 6:e24472. doi: 10.1371/journal.pone.0024472

Clinical and Laboratory Standards Institute. (2014). *Performance Standards for Antimicrobial Susceptibility Testing: Twenty Fourth Informational Supplement*, M100-S24, Vol. 34. Wayne, PA: Clinical and Laboratory Standards Institute.

David, M. Z., and Daum, R. S. (2010). Community-associated methicillin-resistant *Staphylococcus aureus*: epidemiology and clinical consequences of an emerging epidemic. *Clin. Microbiol. Rev.* 23, 616–687. doi: 10.1128/CMR.00081-09

Davis, M. F., Peterson, A. E., Julian, K. G., Greene, W. H., Price, L. B., Nelson, K., et al. (2013). Household risk factors for colonization with multidrug-resistant *Staphylococcus aureus* isolates. *PLoS ONE* 8:e54733. doi: 10.1371/journal.pone.0054733

Dice, L. R. (1945). Measures of the amount of ecologic association between species. *Ecology* 26, 297–302. doi: 10.2307/1932409

Erdem, A. K., Sanli-Yürüdü, N. O., Arslan-Aydoğdu, E. Ö., Dogruoz, N., Zeybek, Z., Türetgen, I., et al. (2008). Quantitative microbiological analysis of biofilm communities from the surfaces of different cooling tower materials. *IUFS J. Biol.* 67, 9–16.

Falord, M., Mäder, U., Hiron, A., Débarbouillé, M., and Msadek, T. (2011). Investigation of the *Staphylococcus aureus* GraSR regulon reveals novel links to virulence, stress response and cell wall signal transduction pathways. *PLoS ONE* 6:e21323. doi: 10.1371/journal.pone.0021323

Gilot, P., Lina, G., Cochard, T., and Poutrel, B. (2002). Analysis of the genetic variability of genes encoding the RNA III-activating components Agr and TRAP in a population of *Staphylococcus aureus* strains isolated from cows with mastitis. *J. Clin. Microbiol.* 40, 4060–4067. doi: 10.1128/JCM.40.11.4060-4067.2002

Gilot, P., and van Leeuwen, W. (2004). Comparative analysis of *agr* locus diversification and overall genetic variability among bovine and human *Staphylococcus aureus* isolates. *J. Clin. Microbiol.* 42, 1265–1269. doi: 10.1128/JCM.42.3.1265-1269.2004

Goerke, C., Campana, S., Bayer, M. G., Döring, G., Botzenhart, K., and Wolz, C. (2000). Direct quantitative transcript analysis of the *agr* regulon of *Staphylococcus aureus* during human infection in comparison to the expression profile in vitro. *Infect. Immun.* 68, 1304–1311. doi: 10.1128/IAI.68.3.1304-1311.2000

Goerke, C., Kümmel, M., Dietz, K., and Wolz, C. (2003). Evaluation of intraspecies interference due to *agr* polymorphism in *Staphylococcus aureus* during infection and colonization. *J. Infect. Dis.* 188, 250–256. doi: 10.1086/376450

Gómez, M. I., O'Seaghdha, M., Magargee, M., Foster, T. J., and Prince, A. S. (2006). *Staphylococcus aureus* protein A activates TNFR1 signaling through conserved IgG binding domains. *J. Biol. Chem.* 281, 20190–20196. doi: 10.1074/jbc.M601956200

Janzon, L., and Arvidson, S. (1990). The role of the delta-lysin gene (*hld*) in the regulation of virulence genes by the accessory gene regulator (*agr*) in *Staphylococcus aureus*. *EMBO J.* 9, 1391–1399.

Joo, H.-S., Chan, J. L., Cheung, G. Y. C., and Otto, M. (2010). Subinhibitory concentrations of protein synthesis-inhibiting antibiotics promote increased expression of the *agr* virulence regulator and production of phenol-soluble modulin cytolysins in community-associated methicillin-resistant *Staphylococcus aureus*. *Antimicrob. Agents Chemother.* 54, 4942–4944. doi: 10.1128/AAC.00064-10

Julander, I. G., Granström, M., Hedström, S. A., and Möllby, R. (1983). The role of antibodies against alpha-toxin and teichoic acid in the diagnosis of staphylococcal infections. *Infection* 11, 77–83. doi: 10.1007/BF01641071

Kolar, S. L., Ibarra, J. A., Rivera, F. E., Mootz, J. M., Davenport, J. E., Stevens, S. M., et al. (2013). Extracellular proteases are key mediators of *Staphylococcus aureus* virulence via the global modulation of virulence-determinant stability. *Microbiologyopen* 2, 18–34. doi: 10.1002/mbo3.55

Lee, K. Y., and Birkbeck, T. H. (1984). In vitro synthesis of the delta-lysin of *Staphylococcus aureus*. *Infect. Immun.* 44, 434–438.

Liu, C., Bayer, A., Cosgrove, S. E., Daum, R. S., Fridkin, S. K., Gorwitz, R. J., et al. (2011). Clinical practice guidelines by the infectious diseases society of America for the treatment of methicillin-resistant *Staphylococcus aureus* infections in adults and children: executive summary. *Clin. Infect. Dis.* 52, 285–292. doi: 10.1093/cid/cir034

MacFadin, J. F. (1996). *Biochemical Tests for Identification of Medical Bacteria*, 3rd Edn. Baltimore: Williams and Wilkins.

McCallum, N., Hinds, J., Ender, M., Berger-Bächi, B., and Meier, P. S. (2010). Transcriptional profiling of XdrA, a new regulator of *spa* transcription in *Staphylococcus aureus*. *J. Bacteriol.* 192, 5151–5164. doi: 10.1128/JB.00491-10

Moise, P. A., Sakoulas, G., Eliopoulos, G. M., Schentag, J. J., Forrest, A., and Moellering, R. C. (2004). Accessory gene regulator group II polymorphism in methicillin-resistant *Staphylococcus aureus* is predictive of failure of vancomycin therapy. *Clin. Infect. Dis. Off. Publ. Infect. Dis. Soc. Am.* 38, 1700–1705. doi: 10.1086/421092

Moise, P. A., Sakoulas, G., Forrest, A., and Schentag, J. J. (2007). Vancomycin in vitro bactericidal activity and its relationship to efficacy in clearance of methicillin-resistant *Staphylococcus aureus* bacteremia. *Antimicrob. Agents Chemother.* 51, 2582–2586. doi: 10.1128/AAC.00939-06

Morfeldt, E., Janzon, L., Arvidson, S., and Löfdahl, S. (1988). Cloning of a chromosomal locus (exp) which regulates the expression of several exoprotein genes in *Staphylococcus aureus*. *Mol. Gen. Genet.* 211, 435–440. doi: 10.1007/BF00425697

Novick, R. P. (2003). Autoinduction and signal transduction in the regulation of *Staphylococcal* virulence. *Mol. Microbiol.* 48, 1429–1449. doi: 10.1046/j.1365-2958.2003.03526.x

Novick, R. P., Projan, S. J., Kornblum, J., Ross, H. F., Ji, G., Kreiswirth, B., et al. (1995). The *agr* P2 operon: an autocatalytic sensory transduction system in *Staphylococcus aureus*. *Mol. Gen. Genet.* 248, 446–458. doi: 10.1007/BF02191645

Novick, R. P., Ross, H. F., Projan, S. J., Kornblum, J., Kreiswirth, B., and Moghazeh, S. (1993). Synthesis of staphylococcal virulence factors is controlled by a regulatory RNA molecule. *EMBO J.* 12, 3967–3975.

Ohkaru, Y., Asayama, K., Ishii, H., Nishimura, S., Sunahara, N., Tanaka, T., et al. (1995). Development of a sandwich enzyme-linked immunosorbent assay for the determination of human heart type fatty acid-binding protein in plasma and urine by using two different monoclonal antibodies specific for human heart fatty acid-binding protein. *J. Immunol. Methods* 178, 99–111. doi: 10.1016/0022-1759(94)00248-U

O'Seaghdha, M., van Schooten, C. J., Kerrigan, S. W., Emsley, J., Silverman, G. J., Cox, D., et al. (2006). *Staphylococcus aureus* protein A binding to von Willebrand factor A1 domain is mediated by conserved IgG binding regions. *FEBS J.* 273, 4831–4841. doi: 10.1111/j.1742-4658.2006.05482.x

Otto, M. P., Martin, E., Badiou, C., Lebrun, S., Bes, M., Vandenesch, F., et al. (2013). Effects of subinhibitory concentrations of antibiotics on virulence factor expression by community-acquired methicillin-resistant *Staphylococcus aureus*. *J. Antimicrob. Chemother.* 68, 1524–1532. doi: 10.1093/jac/dkt073

Palmqvist, N., Foster, T., Tarkowski, A., and Josefsson, E. (2002). Protein A is a virulence factor in *Staphylococcus aureus* arthritis and septic death. *Microb. Pathog.* 33, 239–249. doi: 10.1006/mpat.2002.0533

Pereira, V. C., Riboli, D., and da Cunha, M. (2014). Characterization of the clonal profile of MRSA isolated in neonatal and pediatric intensive care units of a University Hospital. *Ann. Clin. Microbiol. Antimicrob.* 13:50. doi: 10.1186/s12941-014-0050-4

Ramazanzadeh, R., Zamani, S., and Zamani, S. (2013). Genetic diversity in clinical isolates of *Escherichia coli* by enterobacterial repetitive intergenic consensus (ERIC)-PCR technique in Sanandaj hospitals. *Iran J. Microbiol.* 5, 126–131.

Sakoulas, G., Eliopoulos, G. M., Moellering, R. C., Novick, R. P., Venkataraman, L., Wennersten, C., et al. (2003). *Staphylococcus aureus* accessory gene regulator (*agr*) group II: is there a relationship to the development of intermediate-level glycopeptide resistance? *J. Infect. Dis.* 187, 929–938. doi: 10.1086/368128

Sakoulas, G., Eliopoulos, G. M., Moellering, R. C., Wennersten, C., Venkataraman, L., Novick, R. P., et al. (2002). Accessory gene regulator (*agr*) locus in geographically diverse *Staphylococcus aureus* isolates with reduced susceptibility to vancomycin. *Antimicrob. Agents Chemother.* 46, 1492–1502. doi: 10.1128/AAC.46.5.1492-1502.2002

Sakoulas, G., and Moellering, R. C. (2008). Increasing antibiotic resistance among methicillin-resistant *Staphylococcus aureus* strains. *Clin. Infect. Dis. Off. Publ. Infect. Dis. Soc. Am.* 46(Suppl. 5), S360–S367. doi: 10.1086/533592

Shopsin, B., and Kreiswirth, B. N. (2001). Molecular epidemiology of methicillin-resistant *Staphylococcus aureus*. *Emerg. Infect. Dis.* 7, 323–326. doi: 10.3201/eid0702.700323

Sowash, M. G., and Uhlemann, A.-C. (2014). Community-associated methicillin-resistant *Staphylococcus aureus* case studies. *Methods Mol. Biol.* 1085, 25–69. doi: 10.1007/978-1-62703-664-1_2

Subrt, N., Mesak, L. R., and Davies, J. (2011). Modulation of virulence gene expression by cell wall active antibiotics in *Staphylococcus aureus*. *J. Antimicrob. Chemother.* 66, 979–984. doi: 10.1093/jac/dkr043

Tenover, F. C., Arbeit, R. D., Goering, R. V., Mickelsen, P. A., Murray, B. E., Persing, D. H., et al. (1995). Interpreting chromosomal DNA restriction patterns produced by pulsed-field gel electrophoresis: criteria for bacterial strain typing. *J. Clin. Microbiol.* 33, 2233–2239.

Velazquez-Meza, M. E., Aires de Sousa, M., Echaniz-Aviles, G., Solorzano-Santos, F., Miranda-Novales, G., Silva-Sanchez, J., et al. (2004). Surveillance of methicillin-resistant *Staphylococcus aureus* in a pediatric hospital in Mexico city during a 7-year period (1997 to 2003): clonal evolution and impact of infection control. *J. Clin. Microbiol.* 42, 3877–3880. doi: 10.1128/JCM.42.8.3877-3880.2004

Viau, M., and Zouali, M. (2005). Effect of the B cell superantigen protein A from *S. aureus* on the early lupus disease of (NZBxNZW) F1 mice. *Mol. Immunol.* 42, 849–855. doi: 10.1016/j.molimm.2004.07.047

Yamamoto, W, C. H., Takano T. A. N., and Takano, W. (2013). Genetic nature and virulence of community associated methicillin-resistant *Staphylococcus aureus*. *BioMedicine* 3, 2–18. doi: 10.1016/j.biomed.2012.12.001

Zhang, K., McClure, J.-A., Elsayed, S., Louie, T., and Conly, J. M. (2005). Novel multiplex PCR assay for characterization and concomitant subtyping of staphylococcal cassette chromosome mec types I to V in methicillin-resistant *Staphylococcus aureus*. *J. Clin. Microbiol.* 43, 5026–5033. doi: 10.1128/JCM.43.10.5026-5033.2005

11

Global Dissemination of Carbapenemase-Producing *Klebsiella pneumoniae*: Epidemiology, Genetic Context, Treatment Options and Detection Methods

Chang-Ro Lee[1†], Jung Hun Lee[1†], Kwang Seung Park[1], Young Bae Kim[2], Byeong Chul Jeong[1] and Sang Hee Lee[1*]

[1] National Leading Research Laboratory of Drug Resistance Proteomics, Department of Biological Sciences, Myongji University, Yongin, South Korea, [2] Division of STEM, North Shore Community College, Danvers, MA, USA

***Correspondence:**
Sang Hee Lee
sangheelee@mju.ac.kr

[†]These authors have contributed equally to this work.

The emergence of carbapenem-resistant Gram-negative pathogens poses a serious threat to public health worldwide. In particular, the increasing prevalence of carbapenem-resistant *Klebsiella pneumoniae* is a major source of concern. *K. pneumoniae* carbapenemases (KPCs) and carbapenemases of the oxacillinase-48 (OXA-48) type have been reported worldwide. New Delhi metallo-β-lactamase (NDM) carbapenemases were originally identified in Sweden in 2008 and have spread worldwide rapidly. In this review, we summarize the epidemiology of *K. pneumoniae* producing three carbapenemases (KPCs, NDMs, and OXA-48-like). Although the prevalence of each resistant strain varies geographically, *K. pneumoniae* producing KPCs, NDMs, and OXA-48-like carbapenemases have become rapidly disseminated. In addition, we used recently published molecular and genetic studies to analyze the mechanisms by which these three carbapenemases, and major *K. pneumoniae* clones, such as ST258 and ST11, have become globally prevalent. Because carbapenemase-producing *K. pneumoniae* are often resistant to most β-lactam antibiotics and many other non-β-lactam molecules, the therapeutic options available to treat infection with these strains are limited to colistin, polymyxin B, fosfomycin, tigecycline, and selected aminoglycosides. Although, combination therapy has been recommended for the treatment of severe carbapenemase-producing *K. pneumoniae* infections, the clinical evidence for this strategy is currently limited, and more accurate randomized controlled trials will be required to establish the most effective treatment regimen. Moreover, because rapid and accurate identification of the carbapenemase type found in *K. pneumoniae* may be difficult to achieve through phenotypic antibiotic susceptibility tests, novel molecular detection techniques are currently being developed.

Keywords: carbapenemase, *Klebsiella pneumoniae*, epidemiology, KPC, NDM, OXA-48-like

INTRODUCTION

The increasing prevalence of antibiotic resistance and the lack of new antibiotic drug development has gradually reduced the treatment options for bacterial infections (Lee et al., 2013a; Nathan and Cars, 2014). In 2013, the Centers for Disease Control and Prevention (CDC) named three microorganisms that pose an urgent threat to public health: carbapenem-resistant (CR) *Enterobacteriaceae* (CRE), drug-resistant *Neisseria gonorrhoeae*, and *Clostridium difficile* (Zowawi et al., 2015). Carbapenems (imipenem, meropenem, biapenem, ertapenem, and doripenem) are antibiotics used for the treatment of severe infections caused by multi-resistant *Enterobacteriaceae*, such as *Klebsiella pneumoniae* and *Escherichia coli* (Nordmann et al., 2009). However, over the past 10 years, CRE have increasingly been reported worldwide (Nordmann et al., 2011a). In particular, *K. pneumoniae* have acquired carbapenemases, which are enzymes capable of breaking down most β-lactams including carbapenems, and thus conferring resistance to these drugs (Jeon et al., 2015). High mortality rates have been reported in patients with bloodstream infections caused by CR *K. pneumoniae* (Munoz-Price et al., 2013). Carbapenemases can be divided according to their dependency on divalent cations for enzyme activation into metallo-carbapenemases (zinc-dependent class B) and non-metallo-carbapenemases (zinc-independent classes A, C, and D; Jeon et al., 2015). The class A carbapenemases, such as the *K. pneumoniae* carbapenemase (KPC) enzymes, have been identified worldwide in *K. pneumoniae* (Tangden and Giske, 2015). Various class B and D carbapenemases have also been detected in hospital-acquired multi-resistant *K. pneumoniae* (Nordmann et al., 2011a), whereas class C carbapenemases have rarely been reported. In this review, we summarize the epidemiology of the major four classes of carbapenemases and discuss their molecular genetics, methods used for their detection, and the therapeutic options available for their treatment.

THE EPIDEMIOLOGY, GENETIC CONTEXT, TREATMENT OPTIONS, AND DETECTION METHODS OF CARBAPENEM-RESISTANT *K. pneumoniae*

Class A Carbapenemases

Epidemiology

Various class A carbapenemases forming six distantly related branches have been identified (Jeon et al., 2015). While some carbapenemases are chromosome-encoded (IMI-1, NMC-A, SME enzymes, SHV-38, and SFC-1), others are plasmid-encoded (KPC enzymes, GES enzymes, and IMI-2). KPCs have been the most frequently observed class A carbapenemases since their first description in the eastern the USA in 1996 (Yigit et al., 2001). Of the many different KPC family variants (KPC-1 to KPC-22), the most well-characterized variants are KPC-2 and KPC-3. KPCs are mostly plasmid-encoded enzymes and bacteria producing these enzymes are susceptible to only a few antibiotics such as colistin, aminoglycosides, and tigecycline. Therefore, the mortality of the patient's bloodstream infections caused by these bacteria is very high (Munoz-Price et al., 2013).

The epidemiology of *K. pneumoniae* producing KPCs varies geographically. The endemic spread of these bacteria has been reported in the USA, China, Italy, Poland, Greece, Israel, Brazil, Argentina, Colombia, and Taiwan (Munoz-Price et al., 2013; **Figure 1**). Sporadic spread of KPC-producing *K. pneumoniae* has also been observed in many European countries including Spain, France, Germany, the Netherlands, the UK, Ireland, Belgium, Sweden, and Finland, and in several countries in the Asia-Pacific region, including India, South Korea, and Australia (Munoz-Price et al., 2013; Nordmann and Poirel, 2014). In the USA, the transmission of CR *K. pneumoniae* is primarily driven by the spread of organisms carrying KPC enzymes (Kaiser et al., 2013), but other carbapenemase enzymes, such as the New-Delhi metallo-β-lactamase (NDM), have also emerged (Lascols et al., 2013). Within the USA, the prevalence of KPC-positive isolates was relatively stable between 2007 and 2009 (5.9% in 2007, 4.9% in 2008, and 5.7% in 2009; Kaiser et al., 2013), and KPC-2 and KPC-3 were the most frequently identified carbapenemases in *K. pneumoniae* (Deshpande et al., 2006).

The outbreaks caused by KPC-producing *K. pneumoniae* have been reported in the USA (Woodford et al., 2004) and Israel (Leavitt et al., 2007), but recently, similar outbreaks associated with patients traveling to endemic areas have also been reported in many European counties. Since KPC-producing *K. pneumoniae* was identified in France, Italy, and Greece (Naas et al., 2005; Tsakris et al., 2008; Giani et al., 2009), the sporadic spread of KPC-producing *K. pneumoniae* has been observed in many European countries including Spain (Robustillo Rodela et al., 2012), France (Carbonne et al., 2010), Germany (Wendt et al., 2010; Steinmann et al., 2011), the Netherlands (Weterings et al., 2015), the UK (Woodford et al., 2008; Virgincar et al., 2011), Ireland (Roche et al., 2009; Morris et al., 2012), Belgium (Bogaerts et al., 2010), Sweden (Samuelsen et al., 2009), and Finland (Osterblad et al., 2012; Kanerva et al., 2015). KPC-producing *K. pneumoniae* were also recently detected in eastern European countries including the Czech Republic (Hrabak et al., 2013b), Hungary (Toth et al., 2010), and Croatia (Bedenic et al., 2012).

In Greece, KPC-producing *K. pneumoniae* was first isolated in August 2007 (Tsakris et al., 2008), and the prevalence of KPC-producers among carbapenemase-producing *K. pneumoniae* isolates collected at a tertiary Greek hospital increased from 0% in 2003 to 38.3% in 2010 (Zagorianou et al., 2012). Most of the genotyped KPC-producing *K. pneumoniae* in Greece harbored KPC-2 (Zagorianou et al., 2012). While many carbapenemase-producing *K. pneumoniae* in the USA and Greece had KPC enzymes (Nordmann et al., 2009; Zagorianou et al., 2012), several studies in Spain showed that most carbapenemase-producing *K. pneumoniae* harbored OXA-48-like or class B carbapenemases, and the rate of KPC- producing *K. pneumoniae* was very low (2–3%; Oteo et al., 2013b; Palacios-Baena et al., 2016). These results indicate that the prevalent genotype of carbapenemase-producing *K. pneumoniae* varies geographically. For example, in

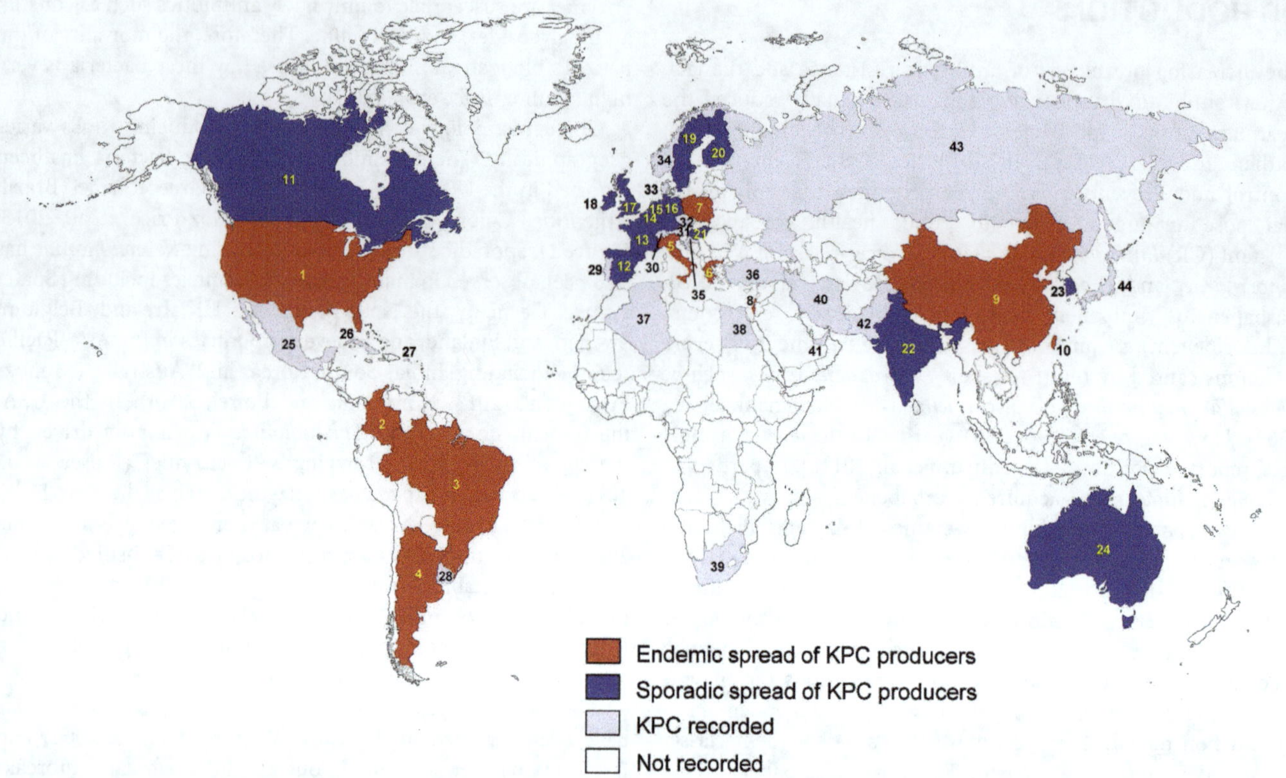

FIGURE 1 | Epidemiological features of KPC-producing *Klebsiella pneumoniae*. (1) USA; (2) Colombia; (3) Brazil; (4) Argentina; (5) Italy; (6) Greece; (7) Poland; (8) Israel; (9) China; (10) Taiwan; (11) Canada; (12) Spain; (13) France; (14) Belgium; (15) Netherlands; (16) Germany; (17) UK; (18) Ireland; (19) Sweden; (20) Finland; (21) Hungary; (22) India; (23) South Korea; (24) Australia; (25) Mexico; (26) Cuba; (27) Puerto Rico; (28) Uruguay; (29) Portugal; (30) Switzerland; (31) Austria; (32) Czech Republic; (33) Denmark; (34) Norway; (35) Croatia; (36) Turkey; (37) Algeria; (38) Egypt; (39) South Africa; (40) Iran; (41) United Arab Emirates; (42) Pakistan; (43) Russia; (44) Japan.

Italy which is a representative southern European country where KPC is becoming endemic, 89.5% of carbapenemase producers have been reported to have KPC-type enzymes, followed by VIM-1 (9.2%) and OXA-48 (1.3%; Giani et al., 2013).

In America, the endemic spread of KPCs has been reported in Colombia (Villegas et al., 2006; Rojas et al., 2013), Brazil (Peirano et al., 2009; Fehlberg et al., 2012), and Argentina (Pasteran et al., 2008; Gomez et al., 2011). In Canada, KPC-producing *K. pneumoniae* has sporadically been reported (Goldfarb et al., 2009; Lefebvre et al., 2015), and since plasmid-mediated KPC-producing *K. pneumoniae* was first detected in Ottawa in Goldfarb et al. (2009), a laboratory surveillance program found a high frequency (89.3%) of KPC-type enzymes among carbapenemase producers between 2010 and 2012 (Lefebvre et al., 2015). The emergence of KPCs in Argentina was characterized by two patterns of dispersion: the first was the irregular occurrence of diverse *Enterobacteriaceae* harboring bla_{KPC-2} in the IncL/M transferable plasmid in distant regions and the second was the sudden clonal spread of *K. pneumoniae* ST258 harboring bla_{KPC-2} in Tn*4401a* (Gomez et al., 2011). KPC-producing *K. pneumoniae* was recently also detected in Cuba (Quinones et al., 2014), Mexico (Garza-Ramos et al., 2014), Uruguay (Marquez et al., 2014), and Puerto Rico (Gregory et al., 2010).

In the Asia-Pacific region, the endemic dissemination of KPC-producing *K. pneumoniae* has been reported in China (Li et al., 2014) and Taiwan (Tseng et al., 2015), and the sporadic spread has been reported in India (Shanmugam et al., 2013), South Korea (Yoo et al., 2013), and Australia (Partridge et al., 2015). A novel KPC-15 variant which is closely homologous with KPC-4 was discovered in China (Wang et al., 2014b) and its enzymatic activity and phenotype was characterized (Wang et al., 2014a). In China, the frequency of KPC-type enzymes among carbapenemase producers was high (63%; Li et al., 2014). While ST258 is the predominant clone observed in European countries and the USA (Giani et al., 2013; Chen et al., 2014e; Bowers et al., 2015), ST11, which is closely related to ST258, is the prevalent clone associated with the spread of KPC-producing *K. pneumoniae* in Asia (particularly in China and Taiwan; Qi et al., 2011; Yang et al., 2013; Tseng et al., 2015). KPC-producing ST11 strain has also been reported in Latin America (Munoz-Price et al., 2013). Although it is unknown why ST11 is prevalent, a recent report showed that the KPC-producing *K. pneumoniae* ST11 clone was resistant to serum killing (Chiang et al., 2016). In a Chinese hospital, another nosocomial outbreak of KPC-2-producing *K. pneumoniae* was caused by multiple *K. pneumoniae* strains including ST37, ST392, ST395, and ST11, implying the horizontal transfer of bla_{KPC-2} gene between different *K. pneumoniae* clones in China (Yang et al., 2013). In Taiwan, two novel KPC variants were identified; KPC-16 and KPC-17 differed from KPC-2 by two (P202S and F207L) and a single (F207L) amino acid substitutions, respectively (Yu et al., 2015). A nationwide survey in Taiwan between 2011 and 2013 reported the national spread of KPC-2 and KPC-17 (Tseng et al., 2015). KPC-producing *K. pneumoniae* was recently also detected in Japan (Saito et al., 2014), Pakistan (Pesesky et al., 2015), Iran (Nobari et al., 2014), and United Arab Emirates (Sonnevend et al., 2015a). In the Arabian Peninsula, the prevalence of KPC-producing *K. pneumoniae* was very low in comparison to NDM-1 and OXA-48-like carbapenemases (Sonnevend et al., 2015b). Sonnevend et al. (2015a), two *K. pneumoniae* ST14 strains producing KPC-2 were first identified in the United Arab Emirates of the Arabian Peninsula. In Africa, several countries such as South Africa (Brink et al., 2012), Algeria (Bakour et al., 2015b), and Egypt (Metwally et al., 2013), have also isolated KPC-producing *K. pneumoniae*.

The coexistence of KPCs and other carbapenemases in *K. pneumoniae* was frequently reported worldwide, including in Italy (KPC-3/VIM-2 and KPC-2/VIM-1; Richter et al., 2012; Perilli et al., 2013), Colombia (KPC-2/VIM-24; Rojas et al., 2013), Brazil (KPC-2/NDM-1; Pereira et al., 2015), China (KPC-2/NDM-1, KPC-2/CMY-2, and KPC-2/IMP-4; Hu et al., 2014; Dong et al., 2015; Liu et al., 2015), Canada (KPC-3/CMY-2; Leung et al., 2012), and Greece (KPC-2/VIM-1; Giakkoupi et al., 2009), indicating the worldwide prevalence of *K. pneumoniae* co-harboring two carbapenemases.

Aside from KPC-type carbapenemases, other class A carbapenemases, such as GES-2, GES-4, GES-5, GES-6, GES-11, GES-14, GES-18, SFC-1, SHV-38, NMC-A, SME-1, and IMI-type enzymes, were rarely found in *K. pneumoniae* (**Table 1**).

Molecular and Genetic Context

The bla_{KPC} in *K. pneumoniae* has been reported on numerous plasmid types, such as IncF, IncI2, IncX, IncA/C, IncR, and ColE1 (Garcia-Fernandez et al., 2012; Chen et al., 2014e; Pitout et al., 2015), but the predominant plasmid type is IncF with FII_K replicons (Pitout et al., 2015). IncF often contains several additional genes responsible for resistance to other antibiotics, including aminoglycosides, tetracyclines, quinolones, trimethoprim, and sulfonamides (Pitout et al., 2015). Many bla_{KPC} genes are associated with a promiscuous transposon-related structure Tn*4401*, which is approximately 10 kb in size and consists of a transposase gene, a resolvase gene, the bla_{KPC} gene, and two insertion sequences, IS*Kpn6* and IS*Kpn7* (**Figure 2A**; Naas et al., 2008). This transposon has jumped to numerous plasmids that are commonly conjugative (Chen et al., 2014e). In China, a novel genetic environment was detected (Shen et al., 2009). It contains an integration structure consisting of a Tn*3*-based transposon and partial Tn*4401* segment, with the gene order Tn*3*-transposase, Tn*3*-resolvase, IS*Kpn8*, the bla_{KPC-2}

TABLE 1 | The epidemiology of various carbapenemases in *Klebsiella pneumoniae*.

Molecular class	Carbapenemase	Geographical distribution
A	SME types	Not found
	IMI types	Not found
	GES types	Greece (Vourli et al., 2004), Finland (Osterblad et al., 2012), Brazil (Picao et al., 2010), and South Korea (Jeong et al., 2005; Bae et al., 2007)
	SFC-1, SHV-38, and NMC-A	France (Poirel et al., 2003) and Brazil (Tollentino et al., 2011)
B	OXA-23, OXA-24/40, OXA-51, OXA-58, OXA-134, OXA-143, OXA-211, OXA-213, OXA-214, OXA-229, and OXA-235	Not found
C	DHA-1	Taiwan (Lee et al., 2012b), South Korea (Park et al., 2013), and China (Hu et al., 2014)
	CMY-2 and CMY-10	China (Hu et al., 2014), Canada (Leung et al., 2012), and Greece (Pournaras et al., 2010b)
	ADC-68	Not found
D	IMP types	Malaysia (Hamzan et al., 2015), Taiwan (Tseng et al., 2015), China (Chen et al., 2015), Thailand (Rimrang et al., 2012), Ireland (Morris et al., 2016), Greece (Lascols et al., 2013), Spain (Lascols et al., 2013), Italy (Lascols et al., 2013), Turkey (Lascols et al., 2013), Austria (Zarfel et al., 2011), the USA (Limbago et al., 2011; Rojas et al., 2013), and Mexico (Gales et al., 2012)
	VIM types	Greece (Pournaras et al., 2010b), Ireland (Morris et al., 2016), Spain (Pena et al., 2014), Australia (Lascols et al., 2013), Croatia (Zujic Atalic et al., 2014), the Czech Republic (Hrabak et al., 2013b), Hungary (Melegh et al., 2014), Italy (Giani et al., 2013), Norway (Naseer et al., 2012), Austria (Zarfel et al., 2011), Finland (Osterblad et al., 2012), Germany (Steinmann et al., 2011), France (Birgy et al., 2011), China (Liu et al., 2015), India (Castanheira et al., 2011), Philippines (Lascols et al., 2013), Iran (Rajabnia et al., 2015), Taiwan (Tseng et al., 2015), Colombia (Rojas et al., 2013), Mexico (Gales et al., 2012), and Algeria (Rodriguez-Martinez et al., 2010)
	GIM-1, KHM-1, and SPM-1	Not found

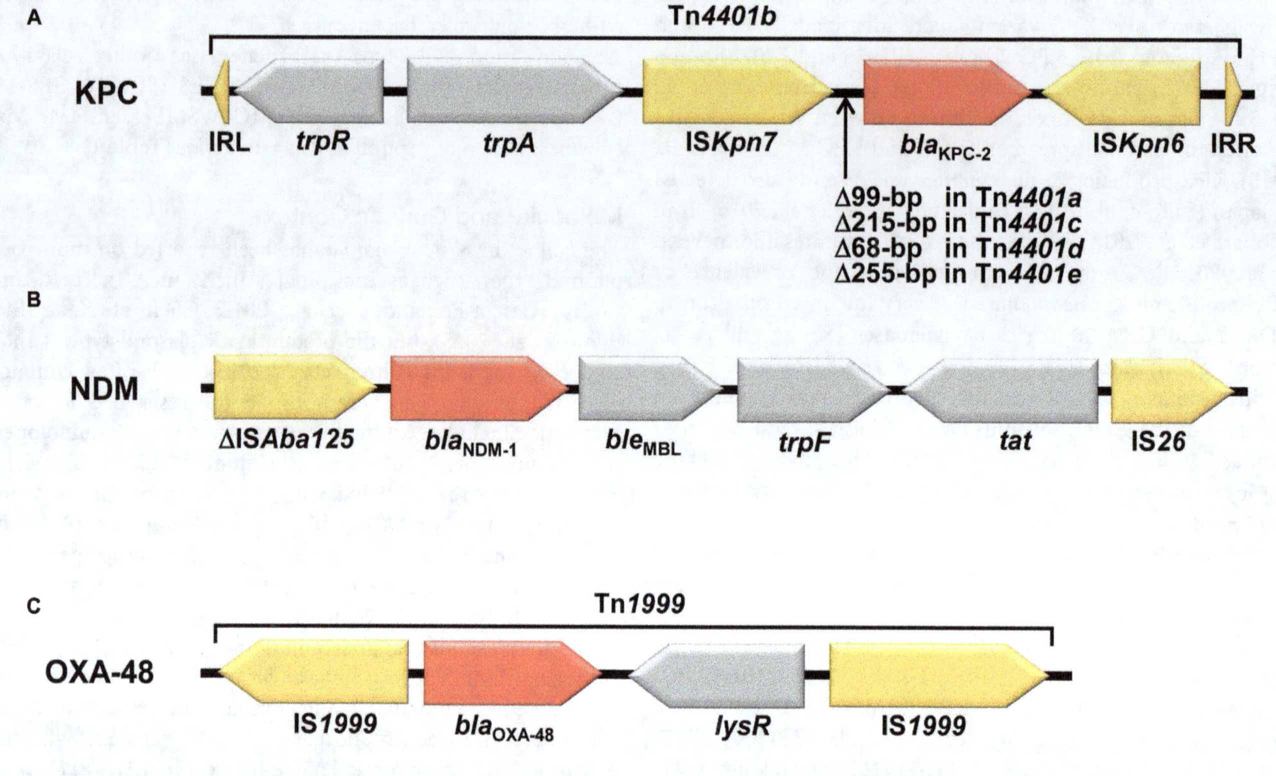

FIGURE 2 | Structural features of representative genetic environments of *bla*$_{KPC-2}$, *bla*$_{NDM-1}$, and *bla*$_{OXA-48}$ genes. (A) The *bla*$_{KPC-2}$-containing Tn*4401* transposon from the plasmid pNYC (GenBank accession no. EU176011) is shown in horizontal arrows. Two inverted repeat sequences (IRL and IRR) of Tn*4401* are depicted in triangles at either end. Tn*4401* has five isoforms which differ by deletions (68–255 bp) just upstream of the *bla*$_{KPC}$ gene [(a) deletion of 99 bp; (b) no deletion; (c) deletion of 215 bp; (d) deletion of 68 bp; (e) deletion of 255 bp]. **(B)** The *bla*$_{NDM-1}$ genetic context of pNDM_MGR194 (GenBank accession no. KF220657) is shown in horizontal arrows. **(C)** The *bla*$_{OXA-48}$-containing Tn*1999* transposon from the plasmid pOXA-48 (GenBank accession no. JN626286) is shown in horizontal arrows.

gene, and the IS*Kpn6*-like element (Shen et al., 2009). This genetic structure is the chimera form of several transposon-associated elements. This transposon was also identified in many other countries (Chen et al., 2014e), and several variants with various fragment insertions between the IS*Kpn8* and *bla*$_{KPC}$ gene have been found among *Enterobacteriaceae* in China (Shen et al., 2009; Li et al., 2011; Qi et al., 2011). Tn*4401* has five isoforms which differ by deletions (68–255 bp) just upstream of the *bla*$_{KPC}$ gene [(a) deletion of 99 bp; (b) no deletion; (c) deletion of 215 bp; (d) deletion of 68 bp; (e) deletion of 255 bp; Chen et al., 2014e]. Notably, in many cases, different Tn*4401* isoforms was associated with different *bla*$_{KPC}$-harboring plasmids. Tn*4401a* was frequently found in the *bla*$_{KPC-3}$-harboring IncFII$_{K2}$ plasmids (Leavitt et al., 2010; Garcia-Fernandez et al., 2012; Chen et al., 2014d), and Tn*4401b* and Tn*4401d* were often associated with the IncN and IncFIA plasmids, respectively (Chen et al., 2013a, 2014c,e). Up to now, more than 30 *bla*$_{KPC}$-harboring plasmids obtained from *K. pneumoniae* have been sequenced (Gootz et al., 2009; Shen et al., 2009; Jiang et al., 2010; Leavitt et al., 2010; Almeida et al., 2012; Chen et al., 2013a,b, 2014b,c,e). One of common features shared by these sequenced plasmids is the presence of the *tra* operon, which encodes the plasmid conjugation machinery proteins that induce the spread of plasmids (Chen et al., 2014e). These genes may be important for the successful dissemination of *bla*$_{KPC}$-harboring plasmids. The IncFII plasmids are one of predominant *bla*$_{KPC}$-harboring plasmids. pKpQIL, which is an IncFII$_{K2}$ plasmid harboring Tn*4401a*, was initially identified in Israel in 2006 (Leavitt et al., 2010), and then this plasmid and its variants are believed to have spread to Italy, Poland, the UK, Colombia, the Czech Republic, the USA, and other countries (Baraniak et al., 2011; Garcia-Fernandez et al., 2012; Hidalgo-Grass et al., 2012; Warburg et al., 2012; Hrabak et al., 2013b; Chen et al., 2014d,e), suggesting the wide dissemination of this plasmid. The *bla*$_{KPC}$ gene has also been identified in other non-Tn*4401* mobile elements that mostly have partial IS*Kpn6* genes (Shen et al., 2009; Gomez et al., 2011). Based on the insertion sequence upstream of the *bla*$_{KPC}$ gene, they can be divided into three groups: group I, no insertion (Shen et al., 2009; Liu et al., 2012; Chen et al., 2014g); group II, insertion of truncated *bla*$_{TEM}$ (Gomez et al., 2011); group III, insertion of Tn*5563*/IS*6100* (Wolter et al., 2009). These non-Tn*4401* genetic elements harboring *bla*$_{KPC}$ sometimes have an IS*26* transposon (Liu et al., 2012; Chen et al., 2014g).

Carbapenemase genes often spread worldwide through clonal expansion in several successful pathogenic strains (Chen et al., 2014e). For example, the dissemination of KPC-producing *K. pneumoniae* in most countries including the USA and

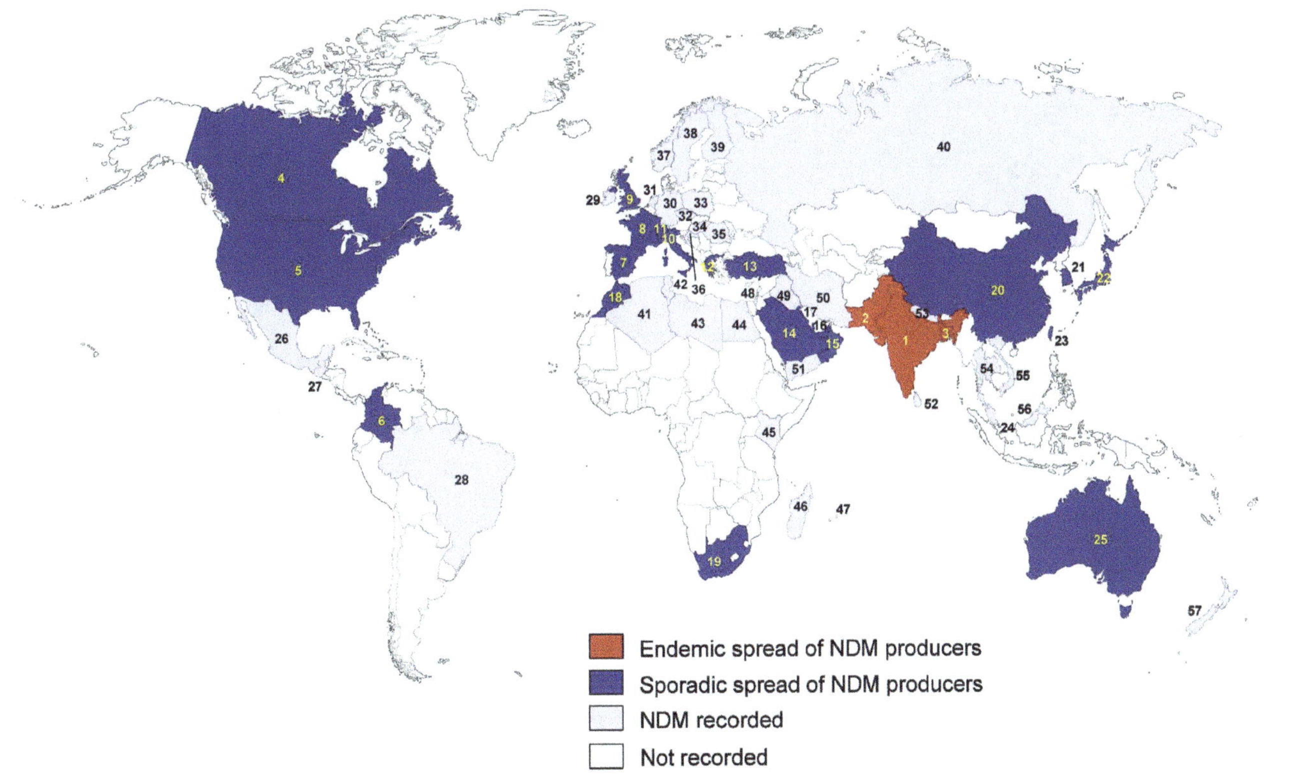

FIGURE 3 | Epidemiological features of NDM-producing *K. pneumoniae*. (1) India; (2) Pakistan; (3) Bangladesh; (4) Canada; (5) USA; (6) Colombia; (7) Spain; (8) France; (9) UK; (10) Italy; (11) Switzerland; (12) Greece; (13) Turkey; (14) Saudi Arabia; (15) Oman; (16) United Arab Emirates; (17) Kuwait; (18) Morocco; (19) South Africa; (20) China; (21) South Korea; (22) Japan; (23) Taiwan; (24) Singapore; (25) Australia; (26) Mexico; (27) Guatemala; (28) Brazil; (29) Ireland; (30) Germany; (31) Netherlands; (32) Czech Republic; (33) Poland; (34) Hungary; (35) Romania; (36) Croatia; (37) Norway; (38) Sweden; (39) Finland; (40) Russia; (41) Algeria; (42) Tunisia; (43) Libya; (44) Egypt; (45) Kenya; (46) Madagascar; (47) Mauritius; (48) Israel; (49) Iraq; (50) Iran; (51) Yemen; (52) Sri Lanka; (53) Nepal; (54) Thailand; (55) Vietnam; (56) Malaysia, (57) New Zealand.

European countries is largely caused by expansion of a single dominant strain, ST258 (Giani et al., 2013; Chen et al., 2014e; Bowers et al., 2015). This strain is a prototype of a high-risk clone of *K. pneumoniae*. Recent data from Israel showed that the KPC-producing *K. pneumoniae* ST258 clone remains the predominant clone, representing 80% of the KPC-producing *K. pneumoniae* population (Adler et al., 2015). ST258 may be a hybrid clone that was created by a large recombination event between ST11 and ST442 (Pitout et al., 2015). It is unknown why the ST258 lineage is the most prevalent clone of KPC-producing *Klebsiella* species. The ST258 clone is highly susceptible to serum killing in animal models and lacks well-known *K. pneumoniae* virulence factors, such as aerobactin genes, K1, K2, and K5 capsular antigen genes, and the regulator of the mucoid phenotype gene *rmpA* (Tzouvelekis et al., 2013; Pitout et al., 2015). Two recent reports revealed that the ST258 strains consist of two distinct genetic clades and genetic differentiation between the two clades (*-1* and *cps-2*) results from an approximately 215-kb region of divergence that includes *cps* genes involved in capsule polysaccharide synthesis (Chen et al., 2014e; Deleo et al., 2014). Multiplex PCR for the identification of two capsular types in *K. pneumoniae* ST258 strains revealed a significant association between the *cps* type and KPC variant: the *cps-1* clade is largely associated with KPC-2, while the *cps-2* clade is primarily associated with KPC-3 (Chen et al., 2014a). Because the capsule polysaccharide can help *K. pneumoniae* to evade phagocytosis, the global success of this strain may involve the capsule polysaccharide biosynthesis regions *cps-1* and *cps-2*. A recent report revealed a relationship between the integrative conjugative element ICEKp258.2 and the global success of the ST258 clone (Adler et al., 2014). ICEKp258.2 contains two specific gene clusters, a type IV pilus gene cluster (i.e., *pilV*) associated with the uptake and exchange of plasmids and adherence to living and non-living surfaces, and a gene cluster of a type III restriction-modification system determining host specificity in the exchange of certain compatible plasmids or mobile elements (Adler et al., 2014). Because these genes associated with the restriction of plasmids and specific mobile elements were present only in ST258 and genetically related sequence types, this difference may explain the divergence of ST258 predominantly harboring KPC and ST11, another high-risk clone that lacks ICEKp258.2, harboring a broad range of plasmids and carbapenemases, including KPC, NDM, OXA-48, VIM, and IMP (Chen et al., 2014f; Pitout et al., 2015). Although the ICEKp258.2 of ST258 strains may contribute to global success, the precise reason for the predominance of the ST258 strain in KPC-producing *K. pneumoniae* is still not entirely

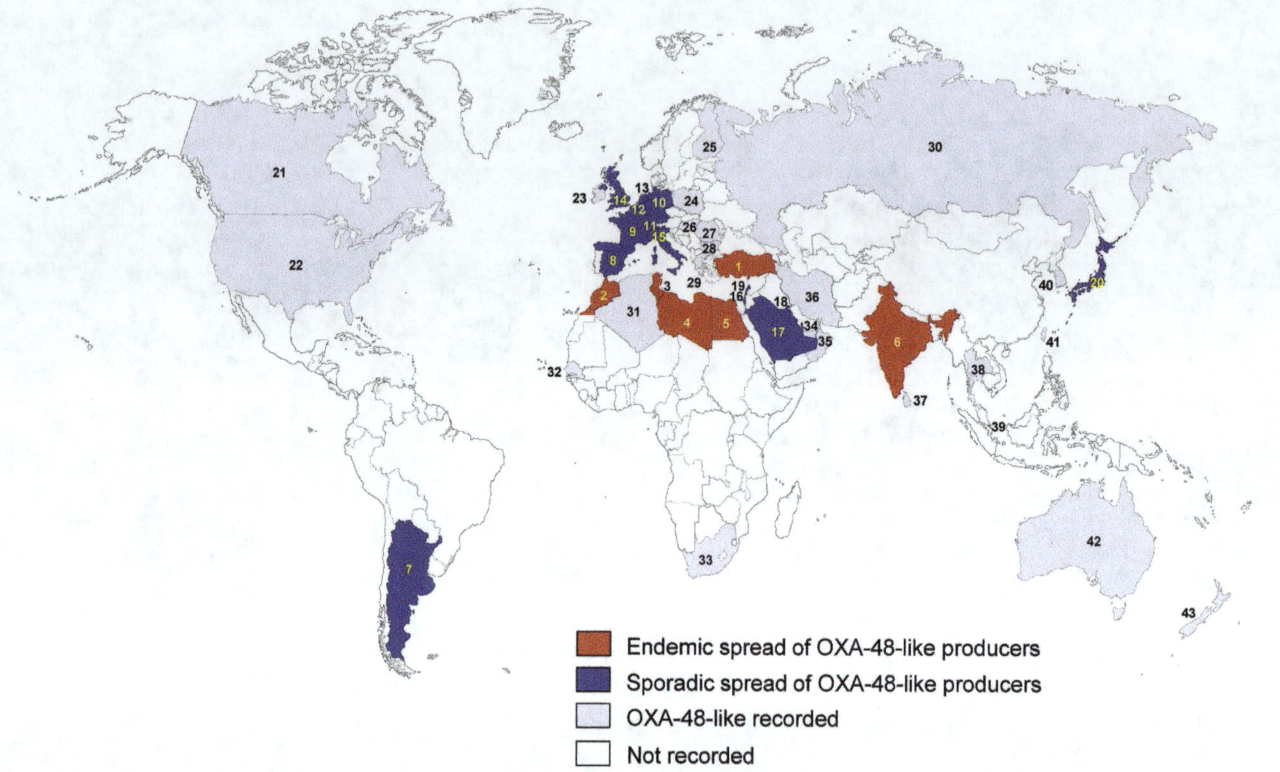

FIGURE 4 | Epidemiological features of OXA-48-like-producing *K. pneumoniae*. (1) Turkey; (2) Morocco; (3) Tunisia; (4) Libya; (5) Egypt; (6) India; (7) Argentina; (8) Spain; (9) France; (10) Germany; (11) Switzerland; (12) Belgium; (13) Netherlands; (14) UK; (15) Italy; (16) Israel; (17) Saudi Arabia; (18) Kuwait; (19) Lebanon; (20) Japan; (21) Canada; (22) USA; (23) Ireland; (24) Poland; (25) Finland; (26) Hungary; (27) Romania; (28) Bulgaria; (29) Greece; (30) Russia; (31) Algeria; (32) Senegal; (33) South Africa; (34) United Arab Emirates; (35) Oman; (36) Iran; (37) Sri Lanka; (38) Thailand; (39) Singapore; (40) South Korea; (41) Taiwan; (42) Australia; (43) New Zealand.

understood. Recently, an outbreak of non-ST258 KPC-producing *K. pneumoniae* clones has been reported in the USA and Europe (Ruiz-Garbajosa et al., 2013; Bonura et al., 2015; Garbari et al., 2015).

The habitat of *K. pneumoniae* is not limited to humans but extends to the ecological environment, such as soil, water, and sewage, and *K. pneumoniae* can survive in extreme environments for long periods of time (Pitout et al., 2015). Therefore, *K. pneumoniae* producing KPCs were detected in various nosocomial environments, such as gowns and gloves (Rock et al., 2014) and wastewater (Chagas et al., 2011; Galler et al., 2014). The frequency of KPC-producing *K. pneumoniae* contamination of gowns and gloves of healthcare workers is similar to that of contamination with methicillin-resistant *Staphylococcus aureus* and vancomycin-resistant *Enterococcus* (Rock et al., 2014), indicating fast transmission of KPC-producing *Klebsiella* species in a nosocomial environment. A long-term observation in a hospital with low-frequency outbreaks of KPC-producing *K. pneumoniae* suggested the possible role of a persisting environmental reservoir of resistant strains in the maintenance of this long-term outbreak (Tofteland et al., 2013). After discharge from the hospital, long-term (>3 years) carriage of KPC-producing *K. pneumoniae* is also possible (Lübbert et al., 2014), and lateral gene transfer of KPC among *Enterobacteriaceae* colonizing the human intestine appears frequent, for example from *K. pneumoniae* to *E. coli* (Richter et al., 2011; Gona et al., 2014). Therefore, reservoirs in healthcare workers, patients, or the hospital environment may be a principle mode of spread in nosocomial outbreaks.

Treatment Options

Carbapenemase-producing *K. pneumoniae* strains are currently one of the most important nosocomial pathogens. Hospital outbreaks of KPC-producing *K. pneumoniae* mainly affect severely ill patients and are associated with an increased risk of death (Ducomble et al., 2015; Tumbarello et al., 2015). KPC-producing *K. pneumoniae* bloodstream infections in intensive care unit (ICU) have also been associated with increased mortality (Chang et al., 2015).

Because carbapenemase-producing *K. pneumoniae* are mostly resistant to several important antibiotic classes (β-lactam drugs, fluoroquinolones, and aminoglycosides), antibiotics, such as polymyxin B, colistin (polymyxin E), fosfomycin, tigecycline, and sometimes selected aminoglycosides, are the last-resort agents. KPC-producing *K. pneumoniae* are usually resistant to all β-lactam antibiotics, but temocillin can be active against some KPC-producing *K. pneumoniae*, particularly in the case of lower urinary tract infections (Adams-Haduch et al., 2009).

To maximize bacterial killing and minimize bacterial resistance, combined therapy is sometimes recommended. Combination therapy including a carbapenem, such as a combination of tigecycline, colistin, and meropenem, was strongly effective in the treatment of KPC-producing *K. pneumoniae*, including colistin-resistant isolates (Tumbarello et al., 2012, 2015; Giamarellou et al., 2013; Hong et al., 2013; Daikos et al., 2014). The synergistic combination of colistin and rifampin was also effective in the treatment of colistin-resistant KPC-producing *K. pneumoniae* by slowing the selection of heteroresistant subpopulations during colistin therapy (Tascini et al., 2013). However, several reports have shown that combination therapy was not superior to monotherapy (de Oliveira et al., 2015; Toledo et al., 2015). Thus, extensive studies will be required to assess the effectiveness of combination therapy. A triple combination of colistin-doripenem-ertapenem was effective only in isolates with high levels of OmpK35 and OmpK36 porin expression (Hong et al., 2013). The expression level of OmpK36 was also involved in the rapid induction of high-level carbapenem resistance in heteroresistant KPC-producing *K. pneumoniae* populations (Adams-Sapper et al., 2015). Therefore, molecular characterization of the KPC-producing *K. pneumoniae* strain, such as the determination of the expression level of OmpK35 and OmpK36, can be used to identify effective combination regimens. However, as a minor effect of OmpK35 and OmpK36 on carbapenem resistance of *K. pneumoniae* was also reported (Zhang et al., 2014), more extensive studies on the role of these proteins on *K. pneumoniae* carbapenem resistance are also required.

Colistin (polymyxin E), an agent discovered more than 60 years ago, is a key component of the combination of antimicrobial regimens used for the treatment of severe KPC-producing *K. pneumoniae* infections (Cannatelli et al., 2014b). Since the global spread of KPC-producing *K. pneumoniae*, the emergence of colistin resistance in KPC-producing *K. pneumoniae* have been reported in many countries, including Italy (Cannatelli et al., 2014b; Giani et al., 2015), the USA (Bogdanovich et al., 2011), Greece (Kontopoulou et al., 2010), Hungary (Toth et al., 2010), and Turkey (Labarca et al., 2014). The increasing prevalence of colistin-resistant *K. pneumoniae* producing KPC poses a threat to public health because colistin resistance increases the mortality due to KPC-producing *K. pneumoniae* bloodstream infections and reduces therapeutic options. A multicenter case-control-control study in Italy showed that the rate of colistin resistance among KPC-producing *K. pneumoniae* blood isolates increased more than threefold during the 4.5-years study period, and the 30-days mortality due to colistin-resistant KPC-producing *K. pneumoniae* bloodstream infections was approximately 51% (Giacobbe et al., 2015). Data collected from 21 hospital laboratories in Italy between 2013 and 2014 also showed that 43% of carbapenemase-producing *K. pneumoniae* were resistant to colistin, 6% were resistant to tigecycline, 16% were resistant to gentamicin, 82% were resistant to trimethoprim-sulfamethoxazole, and 1% were resistant to all four antibiotics, and colistin-resistant isolates were detected in all participating hospital laboratories (Monaco et al., 2014). The progressive increase of colistin resistance was also reported elsewhere (Pena et al., 2014; Bonura et al., 2015; Parisi et al., 2015). These results indicate that the strict rules for colistin use are required to diminish the spread of colistin resistance in the endemic regions of KPC-producing *K. pneumoniae*.

Molecular and biochemical studies have shown that insertional inactivation of the *mgrB* gene, encoding a negative-feedback regulator of the PhoQ–PhoP signaling system, can be responsible for colistin resistance in KPC-producing *K. pneumoniae*, due to the resulting up-regulation of the Pmr lipopolysaccharide modification system (Cannatelli et al., 2013, 2014b). A recent study analyzing a series of colistin-resistant *K. pneumoniae* isolates of worldwide origin identified a single amino acid change (T157P) in the PmrB protein as being responsible for the overexpression of *pmrCAB* and *pmrHFIJKLM* operons involved in lipopolysaccharide modification, leading to colistin resistance (Jayol et al., 2014). The relationship between colistin resistance and inactivation of the *mgrB* gene was further supported by analysis of clinical colistin-resistant *K. pneumoniae* isolates producing KPC (Bonura et al., 2015; Giani et al., 2015). The emergence of colistin resistance was also associated with low-dosage colistin treatment (Cannatelli et al., 2014a). A recent report showed that the plasmid carrying the *mcr-1* gene, which encodes a phosphoethanolamine transferase enzyme catalyzing the addition of phosphoethanolamine to lipid A, is a major contributor to colistin resistance in Gram-negative bacteria and is spread through horizontal gene transfer (Liu et al., 2016). This *mcr-1*-harboring plasmid was also detected in *E. coli* isolates collected from 78 (15%) of 523 samples of raw meat, 166 (21%) of 804 animals, and 16 (1%) of 1322 samples from inpatients with infection, indicating the emergence of this plasmid-mediated colistin resistance mechanism (Liu et al., 2016).

Fosfomycin is a broad-spectrum antibiotic that inhibits bacterial cell wall biogenesis by inactivating the enzyme UDP-*N*-acetylglucosamine-3-enolpyruvyltransferase, also known as MurA (Brown et al., 1995). Fosfomycin has been used to treat KPC-producing *K. pneumoniae*, but recently, a high fosfomycin resistance rate was observed is in countries with higher usage (Giske, 2015). Only 43.4% of KPC-producing *K. pneumoniae* strains retained susceptibility to fosfomycin in a Chinese university hospital (Li et al., 2012) and a similar fosfomycin susceptibility rate (39.2%) was observed in KPC-producing *K. pneumoniae* collected from 12 hospitals in China (Jiang et al., 2015). Like colistin, fosfomycin resistance seems to be associated with the plasmid containing the *fosA3* gene which encodes glutathione S-transferase to modify fosfomycin and was characterized first in CTX-M-producing *E. coli* in Japan (Wachino et al., 2010). In China, the *fosA3*-harboring plasmid was attributed to 55.6% of fosfomycin-resistant KPC-producing *K. pneumoniae* strains (Jiang et al., 2015). Although the *fosA3* gene is mainly associated with the bla_{CTX-M} gene, the *fosA3* gene has also been characterized in atypical bla_{KPC}-harboring plasmids (Jiang et al., 2015; Li et al., 2015). In pFOS18 (Jiang et al., 2015) and pKP1034 (Xiang et al., 2015), the *fosA3* and bla_{KPC-2} genes were located on different transposon systems, whereas in pHS102707 belonging to the IncP1 group (Li et al.,

2015), two genes were co-located in the same Tn*1721*-Tn*3*-like transposon.

Tigecycline, a derivative of minocycline, is the first member of the glycylcycline class that acts as a protein synthesis inhibitor by blocking the interaction of aminoacyl-tRNA with the A site of the ribosome (Rose and Rybak, 2006). Due to the increased clinical use of tigecycline for treatment of KPC-producing *K. pneumoniae*, increased tigecycline resistance was reported (Zagorianou et al., 2012; Papadimitriou-Olivgeris et al., 2014; Weterings et al., 2015). In the Netherlands, all KPC-producing *K. pneumoniae* isolates exhibited reduced susceptibility to tigecycline (Weterings et al., 2015). Another report showed that during ICU stay, 17.9% (39/257) of patients became colonized by tigecycline-resistant KPC-producing *K. pneumoniae* during their stay (Papadimitriou-Olivgeris et al., 2014). In a Greek tertiary hospital during 2004 to 2010, 11.3% (34/301) of KPC-producing isolates were resistant to tigecycline (Zagorianou et al., 2012). Overproduction of efflux pumps such as AcrAB and overexpression of RamA, a positive regulator of the AcrAB efflux system, seem to be major factors for decreased sensitivity of *K. pneumoniae* strains to tigecycline (Ruzin et al., 2005; Rosenblum et al., 2011; Sun et al., 2013). A recent report in China showed that the OqxAB efflux pump was also contributed to tigecycline resistance in *K. pneumoniae* isolates (Zhong et al., 2014).

Because KPC-producing *K. pneumoniae* sometimes remains susceptible to several aminoglycosides such as gentamicin (Tzouvelekis et al., 2014), aminoglycosides can be used alone or in combination therapy to treat KPC-producing *K. pneumoniae* infections. Actually, gentamicin monotherapy or in combination with tigecycline was recently reported to reduce the mortality from sepsis caused by *K. pneumoniae* ST512 clone producing KPC-3, SHV-11, or TEM-1 (Gonzalez-Padilla et al., 2015). New weapons are always indispensable for combating KPC-producing *K. pneumoniae* infections (Lee et al., 2007, 2015a). The effectiveness of some antibiotics in development was also estimated against KPC-producing *K. pneumoniae*. Potent inhibitors of serine β-lactamases, such as avibactam and MK7655, were effective against KPC-producing *K. pneumoniae* infections (Temkin et al., 2014). Combination therapy with avibactam and ceftazidime exhibited significant synergetic effects against organisms with combinations of extended-spectrum β-lactamases (ESBLs), AmpCs, and KPC-2 (Wang et al., 2014e). Plazomicin (a novel aminoglycoside) also exhibited significant activity against KPC-producing *K. pneumoniae* (Temkin et al., 2014). The novel polymyxin derivatives with lower nephrotoxicity are under development (Vaara, 2010). A recent report suggested that synthetic peptides with antimicrobial and antibiofilm activities are a promising strategy in the treatment of infections caused by KPC-producing *K. pneumoniae* (Ribeiro et al., 2015). The *in vitro* activity of the next-generation aminoglycoside plazomicin alone and in combination with colistin, meropenem, fosfomycin or tigecycline was tested against carbapenemase-producing *Enterobacteriaceae* (CPE) strains. When plazomicin was combined with meropenem, colistin or fosfomycin, synergy was observed against CPE isolates (Rodriguez-Avial et al., 2015).

Detection Methods

Because a delay in the appropriate antibiotic therapy for severe infections of KPC-producing *K. pneumoniae* is strongly associated with unfavorable prognosis and increased mortality rates (Karaiskos and Giamarellou, 2014), rapid detection of CR strains is essential for the effective management of these infections (Lee et al., 2005b, 2006b, 2016). Various methods for the identification of KPCs have been developed, including multiplex PCR assay (Spanu et al., 2012; Adler et al., 2014), real-time PCR assay (Wang et al., 2012; Lee et al., 2015c), DNA microarray (Peter et al., 2012; Braun et al., 2014), Raman spectroscopic analysis (Willemse-Erix et al., 2012), single-colony whole-genome sequencing (Koser et al., 2014), matrix-assisted laser desorption ionization-time-of-flight mass spectrometry (MALDI-TOF MS; Carvalhaes et al., 2014), loop-mediated isothermal amplification (LAMP) method (Nakano et al., 2015), chromogenic medium (Vrioni et al., 2012), and a new phenotypic test, called the carbapenem inactivation method (CIM; van der Zwaluw et al., 2015). False positive results can occur when the modified Hodge test is used to detect carbapenemases in carbapenemase-negative *K. pneumoniae* clinical isolates (Wang et al., 2011). Therefore, to improve the efficiency in the phenotypic detection of KPC-producing *K. pneumoniae* isolates, the modified Hodge test can be combined with an EDTA disk test (Yan et al., 2014) or a disk test using boronic acid compounds (Pournaras et al., 2010a). These methods enhanced the sensitivity and specificity of KPC detection in *K. pneumoniae* isolates. In a new developed phenotypic test, called the CIM, a susceptibility-testing disk containing carbapenem was immersed in the suspension made by suspending an inoculation loop of bacterial culture (van der Zwaluw et al., 2015). After incubation, the disk was placed on an agar plate inoculated with a susceptible *E. coli* indicator strain. If the bacterial isolate produces carbapenemase, the susceptibility-testing disk will allow the growth of the susceptible indicator strain. This method showed high concordance with results obtained by PCR (van der Zwaluw et al., 2015). Nevertheless, these culture-based phenotypic tests are time-consuming and cannot easily detect ESBLs and carbapenemases produced by *Enterobacteriaceae*, owing to varied levels of enzyme expression and the poor specificity of some antibiotic combinations (Okeke et al., 2011; Swayne et al., 2013). To overcome these limitations of phenotypic methods, various molecular-based diagnostic methods have been developed. Especially, the direct detection of the carbapenemase gene using multiplex PCR, real-time PCR, and DNA microarray method can increase the speed and accuracy of detecting CR strains (Okeke et al., 2011; Solanki et al., 2014). Because of the high genetic diversity of genes coding for carbapenemase, the precise design of primers or probes is necessary for correctly amplifying or detecting only expected carbapenemase genes. Therefore, already developed methods for *bla* gene detection are restricted to the detection of only limited types of *bla* genes (Lee et al., 2015b). However, the $_{\text{large-scale}}$*bla*Finder was recently developed on the basis of multiplex PCR, and this large-scale detection method can detect almost all *bla* genes, including KPCs, NDMs, OXA-48-likes, present in bacterial pathogens (Lee et al., 2015b).

Recently, mass spectrometry-based methods, such as MALDI-TOF MS and ultra-performance liquid chromatography–tandem mass spectrometry (UPLC-MS/MS), have been shown to be capable of characterizing carbapenemase-producing bacteria (Carricajo et al., 2014; Carvalhaes et al., 2014). These methods are fast and accurate to routinely identify bacterial isolates with great specificity and sensitivity (Hrabak et al., 2011; Patel, 2013; Carvalhaes et al., 2014; Lasserre et al., 2015), but these systems do not accurately provide the carbapenem minimum inhibitory concentrations (MICs) for carbapenemase-producing *K. pneumoniae* (Patel, 2013). Several experiments showed that this method is more rapid and accurate for detection of carbapenemase activity in Gram-negative bacteria than some methods including the modified Hodge test (Lee et al., 2013b; Chong et al., 2015). The LAMP method has emerged as a powerful gene amplification assay for the rapid identification of microbial infections (Notomi et al., 2000). This method employs a DNA polymerase and a set of four specially designed primers that recognize a total of six distinct sequences on the target DNA. The assay amplifies the DNA under isothermal conditions (63–65°C) with high degrees of specificity, efficiency, and speed (Reddy et al., 2010). The cycling reaction continues with accumulation of 10^9 copies of target in less than an hour. The assay can be conducted in a water bath or heating block instead of the thermal cycling using a PCR machine (Notomi et al., 2000). The LAMP assay can be applied for detection of KPC producers in the clinical laboratory (Nakano et al., 2015) and has greater sensitivity, specificity, and rapidity compared to the phenotypic methods and PCR for the detection of KPC-producing *K. pneumoniae* (Solanki et al., 2013).

Colistin has often been used as a therapeutic option for the treatment of CR *K. pneumoniae* infections. However, the imprudent use of colistin has caused rapid spread of colistin resistance in *K. pneumoniae* producing carbapenemases, particularly the KPC-type carbapenemases (Monaco et al., 2014; Giacobbe et al., 2015). This situation demonstrates the need for the development of accurate and reliable methods for detecting colistin resistance. Recently, several methods for the identification of colistin resistance were reported, including various routine colistin MIC testing methods, such as BMD, BMD-P80, AD, Etest, MTS, and Vitek2 (Dafopoulou et al., 2015; Humphries, 2015); capillary electrophoresis method according to characteristic surface properties of bacteria (Sautrey et al., 2015); and the micromax assay based on evaluation of the efficacy of antibiotics that affect cell wall integrity (Tamayo et al., 2013). Because a recent report showed that the *mcr-1* gene, involved in the modification of lipid A, is a major contributor to colistin resistance in Gram-negative bacteria (Liu et al., 2016), detection of this gene may be important in the detection of colistin resistance.

Class B Carbapenemases

Epidemiology

Class B β-lactamases are metallo-β-lactamases that require zinc or another heavy metal for catalysis. Class B β-lactamases have a broad substrate spectrum and can catalyze the hydrolysis of virtually all β-lactam antibiotics including carbapenems except for monobactams (Jeon et al., 2015). Class B carbapenemases were mostly identified in *Enterobacteriaceae* and include VIMs, IMPs, and the emerging NDM group (Jeon et al., 2015). Among them, NDM (New Delhi metallo-β-lactamase) is one of the most clinically significant carbapenemases. NDM-1 was first detected in 2008 in *K. pneumoniae* and *E. coli* in a patient returning to Sweden from India and has since spread worldwide (Yong et al., 2009; Jeon et al., 2015). Thus far, 15 NDM variants have been assigned (Jeon et al., 2015), and most of them originated from Asia (Nordmann and Poirel, 2014). NDMs shares very little identity with other metallo-β-lactamases (Nordmann and Poirel, 2014).

Since 2008, *K. pneumoniae* producing NDMs rapidly spread in many countries (Berrazeg et al., 2014; **Figure 3**). NDM-producing *K. pneumoniae* are considered to be endemic in the Indian subcontinent, including India, Pakistan, and Bangladesh (Nordmann et al., 2011b; Nordmann and Poirel, 2014). The sporadic spread has been reported in the USA (Rasheed et al., 2013; Centers for Disease Control and Prevention, 2014; Doi et al., 2014), Canada (Mulvey et al., 2011; Borgia et al., 2012; Lowe et al., 2013), Colombia (Escobar Perez et al., 2013; Ocampo et al., 2015), Spain (Oteo et al., 2013b; Seara et al., 2015), France (Arpin et al., 2012; Robert et al., 2014), Switzerland (Poirel et al., 2011g; Spyropoulou et al., 2016), Italy (Gaibani et al., 2011), the UK (Kumarasamy et al., 2010; Giske et al., 2012), Greece (Voulgari et al., 2014; Spyropoulou et al., 2016), Turkey (Poirel et al., 2014; Kilic and Baysallar, 2015), Morocco (Poirel et al., 2011b; Barguigua et al., 2013), South Africa (Brink et al., 2012; de Jager et al., 2015), Singapore (Chen et al., 2012; Balm et al., 2013; Ling et al., 2015), Saudi Arabia (Shibl et al., 2013; Zowawi et al., 2014), Oman (Poirel et al., 2011a; Zowawi et al., 2014), United Arab Emirates (Sonnevend et al., 2013; Dash et al., 2014), Kuwait (Jamal et al., 2012, 2013; Sonnevend et al., 2015b), China (Qin et al., 2014; Jin et al., 2015; Liu et al., 2015), Japan (Yamamoto et al., 2013; Nakano et al., 2014), Taiwan (Chiu et al., 2013; Tseng et al., 2015), South Korea (Kim et al., 2012; Cho et al., 2015), and Australia (Shoma et al., 2014; Wailan et al., 2015). In India, NDM-1 was the most common carbapenemase type detected and accounted for 75.22% of the carbapenemase-producing isolates (Kazi et al., 2015). In Singapore and the United Arab Emirates, NDM-1 also was the most common carbapenemase type observed (44.4 and 100%, respectively; Dash et al., 2014; Ling et al., 2015). The endemic spread of NDM-producing *K. pneumoniae* has also been reported in the UK, which has close relationships with India and Pakistan (Nordmann and Poirel, 2014). In China, NDM-1 has been found mostly in *Acinetobacter* spp., but data obtained from patients between June 2011 and July 2012 showed that 33.3% of the CRE isolates, including *K. pneumoniae*, had NDM-1, suggesting the possible transmission of bla_{NDM-1}-containing sequences from *Acinetobacter* spp. to *Enterobacteriaceae* (Qin et al., 2014). These findings reveal the emergence and active transmission of NDM-1-producing *K. pneumoniae* in China. Comparative analyses of the conserved NDM-1-encoding region among different plasmids from *K. pneumoniae* and *E. coli* suggested that the transposable elements and two unknown inverted repeat-associated elements flanking the NDM-1-encoding region aided the spreading of this resistance determinant (Chen et al., 2012). Recently, in China, eight *K. pneumoniae* isolates producing NDM-1 were identified in the neonatal ward of a teaching hospital (Zhang et al., 2015), and four diverse types (NDM-1, KPC-2, VIM-2, and IMP-4) of carbapenemase of *K. pneumoniae* clones were identified in a single hospital in China (Liu et al., 2015).

The Balkan states (Livermore et al., 2011; Voulgari et al., 2014), the Arabian Peninsula (Nordmann and Poirel, 2014), and North African countries (Dortet et al., 2014c), have also been recently considered as an additional reservoir of NDM producers. In the Arabian Peninsula, NDM-1 was the most frequently encountered carbapenemase (46.5%) followed by OXA-48-like carbapenemases (32.5%; Sonnevend et al., 2015b). In Greece, among 132 non-repetitive CRE isolates between 2010 and 2013, 78 *K. pneumoniae* isolates with the bla_{NDM-1} gene were identified (Voulgari et al., 2014). In the USA, KPC-producing *K. pneumoniae* have been responsible for much of the increase in carbapenemase-producing bacteria detection, but recent increases in NDM-producing *K. pneumoniae* have the potential to add to this burden (Rasheed et al., 2013; Centers for Disease Control and Prevention, 2014).

The movement of patients between countries may be a trigger for the international spread of carbapenemase-producing *K. pneumoniae* (Berrazeg et al., 2014). Carbapenemase-producing Gram-negative bacteria including *K. pneumoniae*, obtained from patients that had recently traveled outside Canada between 2010 and 2013, were found to be NDM-producing *K. pneumoniae*, belonging to various sequence types (ST15, ST16, ST147, ST258, ST340, ST512, and ST972) with different plasmids (IncF, IncA/C, and IncL/M), and were imported from India to Canada (Peirano et al., 2014). Therefore, more careful attention is required when treating patients with a recent history of foreign hospitalization in countries where carbapenemase-producing bacteria are prevalent. The international transportations of patients between countries recently resulted in the detection of NDM-producing *K. pneumoniae* in Mexico (Barrios et al., 2014), Guatemala (Pasteran et al., 2012), Brazil (Quiles et al., 2015), the Netherlands (Bathoorn et al., 2015), Ireland (McDermott et al., 2012), Poland (Baraniak et al., 2016), Czech Republic (Studentova et al., 2015), Croatia (Kocsis et al., 2016), Russia (Ageevets et al., 2014), Tunisia (Ben Nasr et al., 2013), Romania (Lixandru et al., 2015), Egypt (Bathoorn et al., 2013), Kenya (Poirel et al., 2011f), Madagascar (Chereau et al., 2015), Iraq (Poirel et al., 2011e), Yemen (Gharout-Sait et al., 2014), Iran (Shahcheraghi et al., 2013), Mauritius (Poirel et al., 2012b), Sri Lanka (Dortet et al., 2013), Thailand (Rimrang et al., 2012), Nepal (Tada et al., 2013), Vietnam (Hoang et al., 2013), Malaysia (Al-Marzooq et al., 2015), and New Zealand (Williamson et al., 2012).

The global dissemination of NDM-producing *K. pneumoniae* also has a serious impact on neonatal mortality rates, particularly in low-income countries where the burden of neonatal sepsis is high (Zaidi et al., 2005). Colonization of NDM-producing *K. pneumoniae* isolates in pregnant women in the community in Madagascar highlighted the potential for mother-to-child transmission (Chereau et al., 2015). In India, analysis of *Enterobacteriaceae*, including *K. pneumoniae*, isolated from the blood of septicaemic neonates, indicted that 14% of the isolates possessed bla_{NDM-1}, and there was a significantly higher incidence of sepsis caused by NDM-1-harboring isolates (Datta et al., 2014). In China, an outbreak of bla_{NDM-1}-producing *K. pneumoniae* ST20 and ST17 isolates was identified in a neonatal unit (Jin et al., 2015). In Turkey, the spread of NDM-1-producing *K. pneumoniae* in a neonatal ICU was reported (Poirel et al., 2014). In Colombia, NDM-1-producing *K. pneumoniae* strains were identified from an outbreak that affected six neonatal patients (Escobar Perez et al., 2013).

As with KPC, coexistence of NDMs and other carbapenemases in *K. pneumoniae* has also been reported worldwide, in Brazil (NDM-1/KPC-2; Pereira et al., 2015), Malaysia (NDM-1/OXA-232; Al-Marzooq et al., 2015), South Korea (NDM-5/OXA-181; Cho et al., 2015), China (NDM-1/IMP-4; Chen et al., 2015), India (NDM-1/OXA-232; Al-Marzooq et al., 2015), Turkey (NDM-1/OXA-48; Kilic and Baysallar, 2015), Pakistan (NDM-1/KPC-2; Sattar et al., 2014), Switzerland (NDM-1/OXA-48; Seiffert et al., 2014), United Arab Emirates (NDM-1/OXA-48-like; Dash et al., 2014), Australia (NDM-1/OXA-48; Sidjabat et al., 2015), Morocco (NDM-1/OXA-48; Barguigua et al., 2013), Singapore (NDM-1/OXA-181 and NDM-5/OXA-181; Balm et al., 2013), and the USA (NDM-1/OXA-232; Doi et al., 2014).

Besides NDM-type carbapenemases, the IMP and VIM groups have also been detected worldwide in *K. pneumoniae*, but other carbapenemases, such as GIM-1, KHM-1, and SPM-1, have been not found in *K. pneumoniae* (**Table 1**). Since IMP and VIP were first detected in *Serratia marcescens* in 1991 and in *Pseudomonas aeruginosa* in 1996, respectively (Osano et al., 1994; Lauretti et al., 1999), IMP- and VIM-producing *K. pneumoniae* have spread in Europe and Asia, but were rarely reported in other regions, such as America and Africa (**Table 1**). Although in this review we have focused only on carbapenemase-producing *K. pneumoniae*, the VIM group is one of the most commonly reported carbapenemases worldwide if we consider all bacteria species (the VIM groups have been mainly identified in *P. aeruginosa*; Poirel et al., 2007).

Molecular and Genetic Context

The bla_{NDM} gene is frequently observed in the transposon Tn*125* (with two flanking IS*Aba125* elements) within NDM-producing species of the genus *Acinetobacter* (Partridge and Iredell, 2012; Wailan et al., 2015). The bla_{NDM} gene was hypothesized to originate in the genus *Acinetobacter* (Toleman et al., 2012). In *Enterobacteriaceae*, the IS*Aba125* elements of the Tn*125* structure carrying bla_{NDM} are frequently truncated (Tn*125*) at various lengths and the Tn*125* structure frequently has different IS elements (**Figure 2B**; Partridge and Iredell, 2012; Wailan et al., 2015). The bla_{NDM} genes in *K. pneumoniae* have been reported on numerous broad-host-range plasmid types, including IncA/C (Hudson et al., 2014), IncF (Hishinuma et al., 2013), IncR (Studentova et al., 2015), IncH (Villa et al., 2012), IncN (Chen et al., 2014a), IncL/M (Peirano et al., 2014), and IncX types (Wang et al., 2014d). The predominant plasmid type responsible for spreading bla_{NDM-1} is the IncA/C type plasmids (Poirel et al., 2011d; Pitout et al., 2015). Many IncA/C plasmids with bla_{NDM-1} also carry various antibiotic resistance genes including 16S rRNA methylases (RmtA and RmtC), associated with aminoglycoside resistance; CMY-type β-lactamases, associated with broad-spectrum cephalosporin resistance; and QnrA, associated with quinolone resistance (Pitout et al., 2015). Consequently, many NDM-producing *K. pneumoniae* were susceptible only to colistin, fosfomycin, and tigecycline (Nordmann and Poirel, 2014). A novel NDM-1 variant (NDM-9) located on a novel IncH variant plasmid was recently identified in a clinical *K. pneumoniae* isolate in China (Wang et al., 2014c). Because the bla_{NDM} genes are located in numerous broad-host-range plasmids, the spread of NDM-1 is facilitated by horizontal gene transfer between bacteria. In China, CRE 21 strains harboring the bla_{NDM-1} gene were found to consist of multiple *Enterobacteriaceae* species including nine *Enterobacter cloacae*, three *E. coli*, three *Citrobacter freundii*, two *K. pneumoniae*, two *K. oxytoca*, and two *Morganella morganii* strains (Wang et al., 2015).

Unlike KPC, the bla_{NDM} genes were detected in various *K. pneumoniae* clones. ST11, a major high-risk sequence type of KPC-producing *K. pneumoniae* in Asia, has also been associated with bla_{NDM-1} in *K. pneumoniae* identified in many countries, such as the USA (Rasheed et al., 2013), Greece (Voulgari et al., 2014), Australia (Shoma et al., 2014), Switzerland (Seiffert et al., 2014), the Czech Republic (Studentova et al., 2015), Spain (Oteo et al., 2013b), and Thailand (Netikul et al., 2014). The recent outbreak of NDM-producing *K. pneumoniae* ST11 in Poland was caused by a clone similar to an isolate identified in the Czech Republic in 2013 (Baraniak et al., 2016), indicating the local spread of this clone. ST11 has also been associated with OXA-48-like enzymes from isolates found in Argentina, Turkey, and Spain (Lascols et al., 2013; Oteo et al., 2013b). ST14, ST147, and ST340 have sometimes been associated with bla_{NDM} in *K. pneumoniae* in many countries (Peirano et al., 2011, 2014; Poirel et al., 2011d; Giske et al., 2012; Osterblad et al., 2012; Lascols et al., 2013; Gharout-Sait et al., 2014; Lee et al., 2014; Shoma et al., 2014; Wang et al., 2014d; Izdebski et al., 2015; Sonnevend et al., 2015b). Over 50% of NDM-producing *K. pneumoniae* isolates from India belonged to either ST11 or ST147 (Lascols et al., 2013). The analysis of clinical isolates of NDM-1-producing *K. pneumoniae* from India, the UK, and Sweden, showed that the most frequently detected sequence types were ST14, ST11, ST149, ST231, and ST147 (Giske et al., 2012). Although ST258 is a high-risk KPC-producing *K. pneumoniae* clone, ST258 harboring bla_{NDM} has never been reported, to the best of our knowledge. This phenomenon may result from the integrative conjugative element ICEKp258.2, present only in ST258 and genetically related sequence types (Adler et al., 2014). This genetic locus contains a type IV pilus gene cluster and a type III restriction-modification system. Because these genes are associated with the restriction of plasmids and specific mobile elements, most ST258 may predominantly harbor plasmids with the bla_{KPC} gene. Therefore, PCR for this unique region (ICEKp258.2) can provide a reliable tool for the rapid detection of the ST258 clone.

High-resolution genomic analysis of multidrug-resistant hospital outbreaks of *K. pneumoniae* through whole-genome sequencing revealed the emergence of a capsule switching NDM-1 bearing *K. pneumoniae* ST15 strain, suggesting that further studies should concentrate on the diversity and spread of this specific clone (Chung The et al., 2015). ST15 harboring bla_{NDM-1} has often been reported in many countries, including Spain (Ruiz-Garbajosa et al., 2013), Croatia (Kocsis et al., 2016), Thailand (Netikul et al., 2014), Canada (Peirano et al., 2014), China (Hu et al., 2013), France (Arpin et al., 2012), and Morocco (Poirel et al., 2011b). In Bulgaria, this clone was responsible for the clonal dissemination of KPC-2-producing *K. pneumoniae* (Markovska et al., 2015). Another whole-genome sequencing analysis of CR *K. pneumoniae* strains, which were isolated from 26 individuals involved in infections in a Nepali neonatal unit, showed that three temporally separated cases were caused by a single NDM-producing *K. pneumoniae* strain with four conserved plasmids including a plasmid carrying bla_{NDM-1} (Stoesser et al., 2014). The plasmids contained a large number of antimicrobial resistance and plasmid maintenance genes, which may explain their persistence (Stoesser et al., 2014). These reports suggest that whole-genome sequencing analysis play an important role in the elucidation of the factors that allow emergence and persistence of resistance.

Treatment Options

New-Delhi metallo-β-lactamase-producing *K. pneumoniae* are usually resistant to most β-lactam antibiotics but remain susceptible to aztreonam (Nordmann et al., 2012a). As with the case of KPC-producing *K. pneumoniae*, the effect of combination therapy was tested in the treatment of NDM-producing *K. pneumoniae* infections. When double- and triple-antibiotic combinations of aztreonam, ciprofloxacin, colistin, daptomycin, fosfomycin, meropenem, rifampin, telavancin, tigecycline, and vancomycin were used in patients infected with two NDM-producing *K. pneumoniae* strains susceptible to colistin, the combination of rifampin-meropenem-colistin was the most effective regimen against these strains (Tangden et al., 2014). The *in vitro* synergetic effect of the combination therapy of colistin and fosfomycin against NDM-producing *K. pneumoniae* has also been reported (Bercot et al., 2011). The combination of polymyxin B and chloramphenicol used against NDM-producing *K. pneumoniae* substantially enhanced bacterial killing and suppressed the emergence of polymyxin resistance (Abdul Rahim et al., 2015). Combination therapy including aztreonam and avibactam (a novel inhibitor of serine β-lactamases under development) was effective in the treatment of metallo-β-lactamase-producing bacterial infections (Wang et al., 2014e).

In the case of NDM-1-producing *Enterobacteriaceae* infections, carbapenems have been suggested to still represent a viable treatment option (Wiskirchen et al., 2014b). Despite unfavorable *in vitro* MICs of NDM-producing *K. pneumoniae*, recent *in vivo* studies have demonstrated the efficacy of carbapenems against NDM-1-producing isolates in immunocompetent-mouse and neutropenic-mouse thigh infection models (Wiskirchen et al., 2013, 2014b; MacVane et al., 2014). Although high-dose, prolonged infusions of ertapenem or doripenem induced reduction in bacterial density, bacterial density was also reduced in standard infusions of ertapenem at 1 g every 24 h or of doripenem at 500 mg every 8 h (Wiskirchen et al., 2013, 2014b). Notably, these efficacies were observed only against NDM-1-producing *K. pneumoniae* (Wiskirchen et al., 2013, 2014b). In addition to carbapenems, this discordance between *in vitro* and *in vivo* activities against NDM-1-producing *K. pneumoniae* was also observed in human simulated regimens of ceftazidime at 2 g every 8 h or ceftazidime/avibactam at 2,000/500 mg every 8 h (MacVane et al., 2014). Despite experiments in immunocompetent-mouse and neutropenic-mouse thigh infection models, these results show that standard infusions of ertapenem and doripenem could reduce bacterial density. Therefore, further experiments in human are required to determine whether carbapenems are sometimes a viable treatment option for NDM-1-producing *K. pneumoniae* infections.

The copy number of bla_{NDM-1} was assessed using Southern blotting and quantitative PCR under different conditions. The bla_{NDM-1} sequence was maintained under antibiotic selection; however, removal of the antibiotic selection led to the emergence of susceptible bacterial populations with a reduced copy number or even the complete loss of the bla_{NDM-1} gene (Huang et al., 2013). The dynamic nature of the copy number of bla_{NDM-1} provides a strong argument for the prudent use of clinically important antibiotics to reduce the development and dissemination of antibiotic resistance among pathogenic bacteria (Huang et al., 2013).

Detection Methods

Because NDM-producing *K. pneumoniae* infections are also associated with significant in-hospital mortality (de Jager et al., 2015), the rapid and accurate detection of NDM-producing *K. pneumoniae* is becoming a major issue in limiting the spread of CR bacteria. Several methods recently developed to detect NDM-producing *K. pneumoniae* include Xpert® Carba-R based on real time PCR (Anandan et al., 2015), the Carba NP test based on the color change of a pH indicator (Nordmann et al., 2012b), and its derivatives (Pires et al., 2013; Dortet et al., 2014b), and a method based on MALDI-TOF (Hrabak et al., 2013a). Xpert® Carba-R based on real time PCR effectively identified various carbapenemases of KPC, NDM, IMP, and VIM, with 100% sensitivity and 77% specificity (Anandan et al., 2015). However, this method failed to detect OXA-48-like carbapenemases, in contrast to multiplex PCR (Anandan et al., 2015).

The Carba NP test using chromogenic medium is based on the color change of a pH indicator (Nordmann et al., 2012b). The enzymatic hydrolysis of the β-lactam ring of a carbapenem (usually imipenem) causes the acidification of an indicator solution (phenol red for the Carba NP test) that changes its color due to pH modification (Garcia-Fernandez et al., 2016). This method could rapidly detect KPC, IMP, VIM, NDM, and OXA-48-like producers with sensitivity and specificity of 97.9 and 100%, respectively, directly from spiked blood cultures (Dortet et al., 2014a). Comparative evaluation of the Carba NP test with other detection methods was tested in many reports (Vasoo et al., 2013; Huang et al., 2014; Yusuf et al., 2014; Gallagher et al., 2015; Lifshitz et al., 2016). When the Carba NP test and the modified Hodge test were compared, the Carba NP test was more specific (100% versus 80%) and faster (Vasoo et al., 2013; Huang et al., 2014; Yusuf et al., 2014). When it was compared the performance of the Carba NP test and the commercially available imipenem hydrolysis-based rapid test (the Rosco Rapid CARB screen kit) for detecting CPE and *P. aeruginosa*, the Carba NP test showed superior specificity and sensitivity (Huang et al., 2014; Yusuf et al., 2014; Gallagher et al., 2015). A novel simplified protocol of the Carba NP test designed for carbapenemase detection direct from bacterial cultures (instead of bacterial extracts) showed enhanced detection of carbapenemase producers (Pasteran et al., 2015).

However, several reports showed that false-negative results in the Carba NP test were associated with mucoid strains or linked to enzymes with low carbapenemase activity, particularly OXA-48-like (Tijet et al., 2013; Osterblad et al., 2014). To overcome these problems, several derivatives of the Carba NP test were developed, such as the Rapidec Carba NP test (bioMérieux; Poirel and Nordmann, 2015), the CarbAcineto NP test for rapid detection of carbapenemase-producing *Acinetobacter* spp. (Dortet et al., 2014b), the Rapid CARB Screen (Rosco Diagnostica; Dortet et al., 2015), the Blue-Carba test using bromothymol blue as a pH indicator solution (Garcia-Fernandez et al., 2016), a modified Carba NP test (Bakour et al., 2015a), and the BYG Carba test based on an electro-active polymer biosensing technology (Bogaerts et al., 2016). Many studies evaluated the performance of the Rapidec Carba NP test (bioMérieux), which was introduced into the market for the detection of carbapenemase production (Dortet et al., 2015; Hombach et al., 2015; Kabir et al., 2016; Lifshitz et al., 2016). These report showed that this method was user-friendly and had a high overall performance, making it an attractive option for clinical laboratories (Kabir et al., 2016). Recently, performance evaluation of two biochemical rapid tests commercialized (the Rapidec Carba NP test and the Rapid CARB Screen) was reported and compared with the home-made Carba NP test (Dortet et al., 2015). The Rapidec CARBA NP test possesses the best performance for rapid and efficient detection of CPE (Dortet et al., 2015). The BYG Carba test based on a new and original electrochemical method detects the variations of conductivity of a polyaniline (an electro-sensing polymer)-coated electrode which is highly sensitive to the modifications of pH and of redox activity occurring during the imipenem enzymatic hydrolysis reaction (Bogaerts et al., 2016). In comparison with PCR results, the BYG Carba test displayed sensitivity of 95% and specificity of 100% versus 89% and 100%, respectively, for the Carba NP test (Bogaerts et al., 2016). The development of these detection methods based on inexpensive and affordable techniques can limit the spread of CR bacteria.

Class D Carbapenemases

Epidemiology

Class D β-lactamases were referred to as oxacillinases (OXAs) because they commonly hydrolyze isoxazolylpenicillins (oxacillin, cloxacillin, and dicloxacillin) much faster than benzylpenicillin (Jeon et al., 2015). Of over 400 Class D β-lactamases, only some variants actually possess carbapenemase activity. Based on their amino acid sequence, class D carbapenemases were recently reclassified into 12 subgroups: OXA-23, OXA-24/40, OXA-48, OXA-51, OXA-58, OXA-134a, OXA-143, OXA-211, OXA-213, OXA-214, OXA-229, and OXA-235 (Jeon et al., 2015). Among them, only several subgroups such as OXA-23, OXA-48, OXA-51, and OXA-58 are reported in *K. pneumoniae* (Evans and Amyes, 2014). OXA-48 is the most efficient class D carbapenemase for imipenem and is one of the most prevalent class D carbapenemases (Jeon et al., 2015). The OXA-48 was first identified in *K. pneumoniae* in Turkey in 2003 (Poirel et al., 2004), and thus far, 10 variants of the bla_{OXA-48} gene have been identified (Jeon et al., 2015). Turkey may be one of the main reservoirs of OXA-48-producing *K. pneumoniae* (Nordmann and Poirel, 2014). Since 2003, the endemic spread of these bacteria has been reported in countries such as Turkey, Morocco, Libya, Egypt, Tunisia, and India (Nordmann and Poirel, 2014; **Figure 4**). The sporadic spread has been reported in France (Liapis et al., 2014; Semin-Pelletier et al., 2015), Spain (Oteo et al., 2013b; Pena et al., 2014), Italy (Giani et al., 2013, 2014), Belgium (Cuzon et al., 2008; Huang et al., 2013), the Netherlands (Kalpoe et al., 2011; Dautzenberg et al., 2014), the UK (Dimou et al., 2012; Thomas et al., 2013), Germany (Pfeifer et al., 2012; Kola et al., 2015), Switzerland (Potron et al., 2012; Seiffert et al., 2014), Argentina (Poirel et al., 2011c; Lascols et al., 2013), Lebanon (Beyrouthy et al., 2014), Israel (Adler et al., 2013, 2015), Kuwait (Poirel et al., 2012a; Zowawi et al., 2014), Saudi Arabia (Shibl et al., 2013; Liu et al., 2015), and Japan (Nagano et al., 2013; Hashimoto et al., 2014). The prevalence of OXA-48 carbapenemases among carbapenemase-producing *K. pneumoniae* in Spain and France was particularly high (74 and 78%, respectively; Robert et al., 2014; Palacios-Baena et al., 2016). In Africa, OXA-48-producing *K. pneumoniae* have been mainly reported in the northern countries, such as Morocco, Libya, Egypt, Tunisia, and Algeria (**Figure 4**). In the Arabian Peninsula, the prevalence of OXA-48-like carbapenemases among carbapenemase-producing *K. pneumoniae* was also high (32.5–56%; Zowawi et al., 2014; Sonnevend et al., 2015b). Among CRE isolates in Lebanon, 88% produced OXA-48 carbapenemase (Beyrouthy et al., 2014). In Saudi Arabia, 78% of carbapenemase-producing *K. pneumoniae* isolates harbored bla_{OXA-48}, and three strains of 47 bla_{OXA-48}-positive *K. pneumoniae* isolates were resistant to colistin, suggesting that colistin resistance is emerging in Saudi Arabia (Shibl et al., 2013).

In France, examination of the epidemiologic features of an outbreak of OXA-48-producing *K. pneumoniae* in an ICU revealed that the outbreak was caused by environmental persistence of OXA-48-producing *K. pneumoniae* over 20 months (Pantel et al., 2016). This report emphasizes the importance of early environmental screening to interrupt the transmission of carbapenemase-producing *K. pneumoniae* (Pantel et al., 2016). Similarly, a large outbreak of OXA-48 carbapenemase-producing *K. pneumoniae* in a French university hospital was recently attributed to the late implementation of successive cohort units and a high level of staff movement between the infectious diseases and internal medicine ward (Semin-Pelletier et al., 2015). These results suggest that practical guidelines are required to help hospitals confronting uncontrolled outbreaks. Because the gut of colonized patients is the main source of CPE, accurate and stringent hygiene of endoscopic instruments is also important. A recent report partially attributed an outbreak of OXA-48-producing *K. pneumoniae* in a German University hospital to complex instruments such as the duodenoscope (Kola et al., 2015). Strict hygiene regulations for various nosocomial environments, including endoscopic instruments as well as gowns and gloves, are required.

OXA-48-producing *K. pneumoniae* was recently also detected in Canada (Ellis et al., 2013), the USA (Mathers et al., 2013), Ireland (Wrenn et al., 2014), Poland (Izdebski et al., 2015), Hungary (Janvari et al., 2014), Greece (Voulgari et al., 2013), Romania (Lixandru et al., 2015), Bulgaria (Markovska et al., 2015), Finland (Osterblad et al., 2012), Russia (Fursova et al., 2015), Algeria (Cuzon et al., 2015), United Arab Emirates (Ahn et al., 2015), Iran (Azimi et al., 2014), South Africa (Brink et al., 2013), Senegal (Moquet et al., 2011), Taiwan (Ma et al., 2015), Singapore (Ling et al., 2015), South Korea (Jeong et al., 2015), and Australia (Espedido et al., 2013; **Figure 4**). In North America, the frequency of OXA-48-like enzymes among carbapenemase-producing *K. pneumoniae* isolates was very low (11%; Lascols et al., 2013). A recent report in Romania showed that among 65 carbapenemase-producing *K. pneumoniae*, the most frequently identified gene was the bla_{OXA-48} gene (78%), 12% were positive for bla_{NDM-1} gene, 6% had the bla_{KPC-2} gene (Lixandru et al., 2015). The recent spread of OXA-48 and OXA-244 carbapenemase genes in Russia was reported among *Proteus mirabilis*, *E. aerogenes*, and *E. cloacae* as well as *K. pneumoniae* (Fursova et al., 2015). In South Africa, the emergence of a colistin-resistant OXA-181-producing *K. pneumoniae* isolate was also reported (Brink et al., 2013), and in the Netherlands, an OXA-48-producing *K. pneumoniae* was reported to infect two patients (Kalpoe et al., 2011), and a hospital-wide outbreak was successfully controlled (Dautzenberg et al., 2014).

OXA-181, a derivative of OXA48 with the substitution of a single amino acid, was first identified in India in Potron et al. (2011), and then has been spread to many different countries, such as the UK (Dimou et al., 2012), Romania (Székely et al., 2013), Canada (Peirano et al., 2014), Oman (Potron et al., 2011), Singapore (Balm et al., 2013), Sri Lanka (Hall et al., 2014), South Korea (Cho et al., 2015), Australia (Sidjabat et al., 2015), Japan (Kayama et al., 2015), and New Zealand (Williamson et al., 2011). However, in many cases, the infections have been associated with India (Potron et al., 2011; Williamson et al., 2011; Dimou et al., 2012; Hall et al., 2014; Peirano et al., 2014). Other OXA-48-derivatives, such as OXA-204 (Potron et al., 2013), OXA-232 (Potron et al., 2013), and OXA-163 (Poirel et al., 2011b), were recently identified in Tunisia, France, and Argentina, respectively, and OXA-244 and

OXA-245 were first reported in Spain (Oteo et al., 2013a). All of these countries are regions with a high prevalence of OXA-48-producing *K. pneumoniae* (**Figure 4**). In addition, OXA-232-producing *K. pneumoniae* has been reported in various countries, such as the USA (Doi et al., 2014), Singapore (Teo et al., 2013), India (Al-Marzooq et al., 2015), and South Korea (Jeong et al., 2015). These results indicate that besides OXA-48, its many derivatives also spread worldwide. OXA-163, which differs from OXA-48 by a four amino acid deletion and a single amino acid substitution, has lower carbapenemase activity than OXA-48, but this enzyme exhibits extended activity against expanded-spectrum cephalosporins, and its activity is partially inhibited by clavulanic acid, a β-lactamase inhibitor (Poirel et al., 2011c). OXA-247, a derivative of OXA-163 with a single amino acid substitution, was recently reported in Argentina (Gomez et al., 2013), where OXA-163-producing *K. pneumoniae* were often reported (Lascols et al., 2013).

The coexistence of OXA-48-like and other carbapenemases in *K. pneumoniae* was also frequently reported worldwide, such as in Turkey (OXA-48/NDM-1; Kilic and Baysallar, 2015), Switzerland (OXA-48/NDM-1; Seiffert et al., 2014), United Arab Emirates (OXA-48-like/NDM-1; Dash et al., 2014), Australia (OXA-48/ NDM-1; Sidjabat et al., 2015), Morocco (OXA-48/NDM-1; Barguigua et al., 2013), India (OXA-181/VIM-5; Castanheira et al., 2011), Singapore (OXA-181/NDM-1 and OXA-181/NDM-5; Balm et al., 2013), the USA (OXA-232/NDM-1; Doi et al., 2014), and India (OXA-232/NDM-1; Al-Marzooq et al., 2015).

Other class D carbapenemases, such as OXA-23, OXA-24/40, OXA-51, OXA-58, OXA-134a, OXA-143, OXA-211, OXA-213, OXA-214, OXA-229, and OXA-235, were mainly identified in *Acinetobacter* species such as *A. baumannii* but not in *K. pneumoniae* (**Table 1**; Evans and Amyes, 2014).

Molecular and Genetic Context

Unlike KPCs and NDMs, one highly transferable IncL group plasmid (pOXA-48a) was mainly responsible for the spread of the bla_{OXA-48} gene in *K. pneumoniae* (**Figure 2C**; Pitout et al., 2015). The molecular epidemiology of OXA-48 in European and North African countries showed that in 92.5% of the isolates, the bla_{OXA-48} gene was located on this self-conjugative IncL/M type plasmid (Potron et al., 2013). The bla_{OXA-48} gene was found only in the IncL group of IncL/M type plasmids (Carattoli et al., 2015). In contrast to the IncA/C plasmids of NDM-1, the pOXA-48a plasmid contains bla_{OXA-48}, a unique antibiotic resistance gene (Pitout et al., 2015). The conjugation rate of the pOXA-48a plasmid was very high (1×10^{-1}); therefore, this self-conjugative plasmid can conjugate at a very high frequency to *Enterobacteriaceae* (Potron et al., 2014). Inactivation of the *tir* gene, which is known to encode a transfer inhibition protein, was recently reported to be responsible for a 50- to 100-fold increase in the efficiency of transfer of the pOXA-48a plasmid (Potron et al., 2014), explaining the spread of the pOXA-48a plasmid with bla_{OXA-48}. Recently, the $bla_{OXA-48-like}$ gene has also been reported on other plasmids and genetic elements, such as IncA/C types (Ma et al., 2015), IncH types (Wang et al., 2014c), and Tn*1999* (Poirel et al., 2012a). In comparison to the pOXA-48a plasmid, the transmission frequency of the Tn*1999* composite transposon was very low ($<1.0 \times 10^{-7}$; Aubert et al., 2006).

OXA-48 has most often been found in *K. pneumoniae*, but OXA-48 was also found in various *Enterobacteriaceae*, because of the high conjugation rate of the pOXA-48a (Potron et al., 2014). Molecular and epidemiological analyses in a German hospital showed the horizontal gene transfer of the OXA-48-containing plasmid from *K. pneumoniae* to *E. coli* (Gottig et al., 2015). Besides *E. coli*, OXA-48 has also been identified in *K. oxytoca*, *Enterobacter* spp., *Providencia rettgeri*, *C. freundii*, and *S. marcescens* (Poirel et al., 2012c; Berger et al., 2013). The bla_{OXA-48} gene was identified in all *Enterobacteriaceae* isolates from the index case in Spain, indicating the capacity of OXA-48 carbapenemase to spread among *Enterobacteriaceae* by the horizontal gene transfer (Arana et al., 2015).

Similar to NDM, the $bla_{OXA-48-like}$ genes were detected in various *K. pneumoniae* clones. ST11 has often been associated with $bla_{OXA-48-like}$ in *K. pneumoniae* isolated in many countries, such as Spain (Oteo et al., 2013a,b, 2015; Ruiz-Garbajosa et al., 2013; Branas et al., 2015), Taiwan (Ma et al., 2015), Libya (Lafeuille et al., 2013), Turkey (Lascols et al., 2013), Argentina (Lascols et al., 2013), and Greece (Voulgari et al., 2013). In Spain, a large outbreak was initiated in 2013 by a OXA-48-producing *K. pneumoniae* ST11 clone, and this strain was detected in 44 patients (Branas et al., 2015). The ST11 isolates carried various carbapenemases, including NDM-1, VIM-1, OXA-48, KPC-2, and OXA-245 (Lascols et al., 2013; Oteo et al., 2013b, 2015). In addition, ST14, ST15, ST101, ST147, and ST405 harboring $bla_{OXA-48-like}$ have often been reported in many countries, such as the USA (Lascols et al., 2013), Spain (Oteo et al., 2013b, 2015; Ruiz-Garbajosa et al., 2013; Arana et al., 2015; Cubero et al., 2015), the Czech Republic (Hrabak et al., 2015), Libya (Lafeuille et al., 2013), India (Lascols et al., 2013), Germany (Gottig et al., 2015), Finland (Osterblad et al., 2012), France (Liapis et al., 2014), and Japan (Hashimoto et al., 2014). Recent results from 83 hospitals in Spain showed that OXA-48 (71.5%) and VIM-1 (25.3%) were the most frequently detected carbapenemases, and the most prevalent sequence types were ST11 and ST405 for *K. pneumoniae* (Oteo et al., 2015). However, the molecular epidemiology of OXA-48-producing enterobacterial isolates collected from European and north-African countries between 2001 and 2011 indicated that ST101 was the most commonly observed sequence type in *K. pneumoniae* isolates, accounting for 17 out of 67 isolates (25.4%), followed by ST395 and ST15 (each seven isolates, 10.5%; Potron et al., 2013). Two outbreaks of OXA-48-producing *K. pneumoniae* ST101 clones were reported in Spain (Pitart et al., 2011; Cubero et al., 2015). OXA-48-producing *K. pneumoniae* appear to vary geographically. As with NDM, to the best of our knowledge, ST258 harboring $bla_{OXA-48-like}$ has never been reported.

Treatment Options

OXA-48-like-producing *K. pneumoniae* are usually resistant to most β-lactam antibiotics, but OXA-48-producing *K. pneumoniae* without ESBLs remain susceptible to the expanded-spectrum cephalosporins (Nordmann et al., 2012a; Munoz-Price et al., 2013). In addition, OXA-48-like-producing

K. pneumoniae sometimes remains susceptible to several aminoglycosides such as gentamicin (Tzouvelekis et al., 2014). Unlike for NDMs (Wiskirchen et al., 2013, 2014b; MacVane et al., 2014), carbapenems may not be a reliable treatment option for OXA-48 producer infection (Wiskirchen et al., 2014a). Combination therapy with sulbactam, meropenem, and colistin was more effective in isolates producing NDM carbapenemase than those producing OXA-48-like carbapenemases, suggesting that the identification of the carbapenemase type helps determine the combination most likely to clear the infection (Laishram et al., 2015). The combination of fosfomycin with imipenem, meropenem, and tigecycline was also synergistic against OXA 48-positive *K. pneumoniae* strains *in vitro* with the ratios of 42, 33, and 33%, respectively (Evren et al., 2013). Similarly, the *in vitro* assays indicate that imipenem-containing combinations were effective against serine-β-lactamase producers (KPC, OXA-48), while no synergy was observed for all NDM-1 producers (Poirel et al., 2016). However, because carbapenems were effective against NDM-1-producing isolates *in vivo* (Wiskirchen et al., 2013, 2014a; MacVane et al., 2014), the effect of carbapenem combination therapy on carbapenemase-producing isolates should be determined *in vivo*, particularly in the case of NDM-1-producing bacteria.

Despite the effectiveness of combination therapies, the prognosis for bloodstream infections caused by OXA-48-producing *Enterobacteriaceae* remains poor, and the 30-days mortality reached 50% (Navarro-San Francisco et al., 2013). A similar result was reported in OXA-48-producing *K. pneumoniae* infections in a tertiary hospital in Spain (Pano-Pardo et al., 2013). Although OXA-48-producing *K. pneumoniae* were susceptible to several antibiotics, including amikacin (97.2% susceptible), colistin (90.1%), tigecycline (73%), and fosfomycin (66.2%), in-hospital mortality among patients with OXA-48-producing *K. pneumoniae* infections was 43.5% (Pano-Pardo et al., 2013). Therefore, to prevent delay in diagnosis and initiation of optimal antimicrobial therapy, rapid identification of OXA-48-producing isolates is required.

Detection Methods

As with KPCs and NDMs, various detection methods were developed to identify OXA-48-like carbapenemases (Tsakris et al., 2010; Giske et al., 2011; Girlich et al., 2013; Naas et al., 2013; Lee et al., 2015c). Accurate differentiation of the various carbapenemase types, such as KPC-type, NDM-type, and OXA-48-type enzymes, is crucial for controlling the spread of carbapenem resistance among *Enterobacteriaceae* (Nordmann et al., 2012a). Many phenotypic detection methods to allow differentiation between class A and class B carbapenemases were developed using boronic acid derivatives and EDTA or dipicolinic acid (Tsakris et al., 2010; Giske et al., 2011). Recently, a specific phenotypic method to differentiate a single OXA-48 producer from those producing other carbapenemase types (e.g., KPC-types, NDM-types) was also developed (Tsakris et al., 2015). This method was based on an imipenem disk and two blank disks adjacent to the imipenem disk, loaded with the tested strain and impregnated with EDTA and EDTA plus phenyl boronic acid, respectively (Tsakris et al., 2015). This novel method exhibited 96.3% sensitivity and 97.7% specificity (Tsakris et al., 2015). ChromID OXA-48, based on chromogenic media, have been commercialized for the direct isolation of CPE from clinical samples (Girlich et al., 2013), with an estimated 91% sensitivity and 100% specificity (Girlich et al., 2013).

Emerging Class C Carbapenemases

Class C β-lactamases confer resistance to penicillins, cephalosporins, and cephamycins (cefoxitin and cefotetan) and are not significantly inhibited by clinically applied β-lactamase inhibitors such as clavulanic acid (Jeon et al., 2015). Although four class C carbapenemases (ACT-1, CMY-2, CMY-10, and ADC-68) have been reported, ACT-1 and CMY-2 exhibit reduced susceptibility to carbapenems, particularly ertapenem, only when combined with permeability defects, due to their low catalytic efficiencies (K_{cat}/K_m) for imipenem (0.007 and 0.04 $M^{-1}{\cdot}S^{-1}$, respectively; Mammeri et al., 2010). CMY-10 with the catalytic efficiency of 0.14 $M^{-1}{\cdot}S^{-1}$ for imipenem was the first reported carbapenemase among plasmidic class C β-lactamases (Kim et al., 2006), and this enzyme was also a class C extended-spectrum β-lactamase with extended substrate specificity for extended-spectrum cephalosporins (Lee et al., 2009, 2012a; Jeon et al., 2015). Among the chromosomal class C β-lactamases, ADC-68 identified in *A. baumannii* was the first reported enzyme possessing both class C extended-spectrum β-lactamase and carbapenemase activities (Jeon et al., 2014, 2015), and its catalytic efficiency for imipenem is 0.17 M^{-1} S^{-1}. CMY-10-producing *K. pneumoniae* was identified only in South Korea (Lee et al., 2005a, 2006a) and ADC-68-producing *K. pneumoniae* was never reported.

Many reports showed that carbapenem resistance can be triggered by the loss of two major porins, OmpK35 and OmpK36, in combination with ESBLs or Ambler class C AmpC cephalosporinases, and the production of carbapenemase (Wang et al., 2009; Shin et al., 2012; Tsai et al., 2013). A recently genetically engineered mutant of *K. pneumoniae* showed that several carbapenems (imipenem, meropenem, and doripenem) remain effective against these carbapenemase-independent CR strains (Tsai et al., 2013). Therefore, laboratory testing for susceptibility to imipenem, meropenem, and doripenem can improve the accuracy of identification of these isolates (Tsai et al., 2013).

CONCLUSION

We analyzed the epidemiology of *K. pneumoniae* producing true carbapenemases (Ambler molecular class A, B, D, and several carbapenemases of class C) responsible for non-susceptibility to carbapenems without additional permeability defects. Many types of CR *K. pneumoniae* have been identified worldwide (**Figures 1, 3,** and **4**). During the past 3 years, many countries have reported the arrival of carbapenemases previously unreported in those countries. For example, although *K. pneumoniae* producing KPC- and NDM-type carbapenemases have been extensively reported in the USA, a *K. pneumoniae* producing OXA-48-type carbapenemases was recently detected in the USA (Mathers et al., 2013). Since a first report of NDM-1 in 2008, this carbapenemase has rapidly spread worldwide, and NDM-producing *K. pneumoniae* still continues to be found in new countries, implying that NDM-producing *K. pneumoniae* is still spreading quickly. Despite the global dissemination of KPC, NDM, and OXA-48, the prevalence of carbapenemases varies geographically. The frequency of KPC- and NDM-producing strains was significantly higher in the USA, Canada, Greece, Taiwan, Colombia, and China, whereas OXA-48-producing strains were rarely found in those countries (**Figures 1, 3,** and **4**). In Argentina, despite the extensive spread of KPC- and OXA-48-producing strains, no NDM-producing *K. pneumoniae* has been reported. Although NDM- and OXA-48-producing *K. pneumoniae* significantly spread in Turkey, KPC-producing strains have rarely been reported there. In Brazil, KPC-producing *K. pneumoniae* has been mainly reported. In the Arabian Peninsula, OXA-48 and NDM producers are common, whereas KPC-type, VIM-type, or IMP-type producers are rare. In India, Spain, France, Italy, and the UK, all three types of carbapenemases have been frequently reported (**Figures 1, 3,** and **4**).

ST258 is an important strain responsible for the extensive global spread of KPC-producing *K. pneumoniae*. Although the precise reason for the predominance of the ST258 strain in KPC-producing *K. pneumoniae* is not fully understood, recent molecular studies unveiled the genetic characteristics of this strain. The ST258 strains consists of two distinct genetic clades (*cps*-1 and *cps*-2) derived from genetic differentiation in genes involved in capsule polysaccharide biosynthesis. In addition, the integrative conjugative element ICEKp258.2, present only in ST258 and genetically related sequence types, may be linked to the global success of the ST258 clone. ICEKp258.2 contains a type IV pilus gene cluster and a type III restriction-modification system. The type IV pilus gene cluster ICEKp258.2, particularly *pilV*, may contribute to the global success of ST258 clone. The type III restriction-modification system associated with the restriction of plasmids and specific mobile elements may explain the differences observed between ST258 predominantly harboring KPC and ST11, another high-risk clone that lacks ICEKp258.2, harboring various carbapenemases, such as NDM-1, OXA-48, KPC-2, VIM-1, and OXA-245. IncF with FII_K replicons, a plasmid most commonly identified in ST258 with bla_{KPC}, often contains several genes associated with resistance to other antibiotics, such as aminoglycosides, tetracyclines, quinolones, trimethoprim, and sulfonamides. The features of this plasmid may also play an important role in the current global success of ST258.

The rapid global spread of NDM-type carbapenemases may be partly attributed to the dissemination of various epidemic broad-host-range plasmids bearing the bla_{NDM} genes. NDM-type carbapenemases were found in various plasmids such as IncA/C, IncF, IncR, IncH, IncN, IncL/M, and IncX types. The IncA/C type plasmids, most common plasmids associated with spread of the bla_{NDM} genes, often have various antibiotic resistance genes, such as 16S rRNA methylases associated with aminoglycoside resistance, CMY-type β-lactamases associated with broad-spectrum cephalosporin resistance, and QnrA associated with quinolone resistance. These features may be linked to the current global success of NDM-producing *K. pneumoniae*.

The current spread of OXA-48-producing bacteria is attributed to the pOX-48a plasmid, which belongs to the IncL group of IncL/M type plasmids. Although the pOXA-48a plasmid contains bla_{OXA-48}, a unique antibiotic resistance gene, the conjugation rate of the pOXA-48a plasmid was very high, which may be responsible for its global spread in *K. pneumoniae*. The high pOXA-48a conjugation rate was recently attributed to mutations in the *tir* gene known to encode a transfer inhibition protein, which may lead to a 50- to 100-fold increase in the efficiency of transfer of the pOXA-48a plasmid. Furthermore, the pOXA-48a plasmid is self-conjugative. Therefore, these specific features of the pOXA-48a plasmid may explain the global dissemination of OXA-48-type carbapenemases.

ST11, ST14, ST101, ST147, and ST258 are major carbapenemase-producing *K. pneumoniae* clones. ST258 was mainly found in KPC-producing *K. pneumoniae*, whereas other clones were found in various carbapenemase-producing *K. pneumoniae* regardless of carbapenemase types. Well-designed epidemiological and molecular studies will be required to understand the dynamics of transmission, risk factors, and reservoirs of these *K. pneumoniae* clones. This will provide information essential for preventing infections and the spread of these risky sequence types.

Most currently available antibiotics may be not sufficiently effective for the treatment of all types of carbapenemase producers in monotherapy. Combination therapy of carbapenems with polymyxin B, colistin, rifampin, fosfomycin, or tigecycline has been reported to effectively treat carbapenemase-producing *K. pneumoniae*. Despite these data supporting the use of combination therapy for treatment of severe carbapenemase-producing *K. pneumoniae* infections, current clinical evidence for treatment guidelines are limited and more accurate randomized controlled *in vivo* studies are required. Moreover, considerable caution is required when applying these therapies. For example, temocillin can actively treat against some KPC-producing *K. pneumoniae*, particularly lower urinary tract infections, and NDM-producing *K. pneumoniae* is often susceptible to aztreonam. OXA-48-producing *K. pneumoniae* remain susceptible to the expanded-spectrum cephalosporins in approximately 20% of cases without ESBLs. In the case of NDM-1-producing *K. pneumoniae*, carbapenems were recently reported to represent a viable treatment option for infections caused by these bacteria, despite unfavorable *in vitro* MICs. Because these results imply that carbapenems can sometimes be a viable treatment option for infection with carbapenemase producers, more extensive studies on the effect of carbapenem monotherapy will be required in the case of NDM-producing *K. pneumoniae* infections.

The accurate and rapid detection of the genotype of carbapenemases can minimize the delay to appropriate prescription of antibiotics. Many detection kits based on various phenotypic or molecular techniques, such as multiplex PCR assay, real-time PCR assay, DNA microarray, Raman spectroscopic analysis, single-colony whole-genome sequencing,

MALDI-TOF MS, loop-mediated isothermal amplification method, chromogenic medium, and new phenotypic test methods, have been developed. Through the imprudent use of colistin which is a key component used for the treatment of severe carbapenemase-producing *K. pneumoniae* infections, the rapid spread of colistin resistance was recently reported in *K. pneumoniae* producing carbapenemases, particularly KPC-type carbapenemases. This situation strongly demonstrates the need for the development of novel accurate and reliable methods for detecting resistance to clinically important antimicrobial agents, such as colistin. Hospital interventions can effectively reduce the spread of carbapenemase-producing *K. pneumoniae*. Standard infection control guidelines should be implemented upon the detection of carbapenemase-producing *K. pneumoniae*, and carbapenemase-producing *K. pneumoniae* positive patients should be individually isolated and treated according to strict standard guidelines.

AUTHOR CONTRIBUTIONS

C-RL, JL, and SL contributed to the conception and the design of the review and C-RL, JL, KP, YK, BJ, and SL researched and wrote the review.

FUNDING

This review was supported by the National Research Lab Program through the National Research Foundation of Korea (NRF) funded by the Ministry of Science, ICT & Future Planning (No. 2011-0027928); the Cooperative Research Program for Agriculture Science & Technology Development (No. PJ01103103) of Rural Development Administration in South Korea; and the National Research Foundation of the Ministry of Education, South Korea (NRF-2015R1C1A1A02037470).

REFERENCES

Abdul Rahim, N., Cheah, S. E., Johnson, M. D., Yu, H., Sidjabat, H. E., Boyce, J., et al. (2015). Synergistic killing of NDM-producing MDR *Klebsiella pneumoniae* by two 'old' antibiotics-polymyxin B and chloramphenicol. *J. Antimicrob. Chemother.* 70, 2589–2597. doi: 10.1093/jac/dkv135

Adams-Haduch, J. M., Potoski, B. A., Sidjabat, H. E., Paterson, D. L., and Doi, Y. (2009). Activity of temocillin against KPC-producing *Klebsiella pneumoniae* and *Escherichia coli*. *Antimicrob. Agents Chemother.* 53, 2700–2701. doi: 10.1128/AAC.00290-09

Adams-Sapper, S., Nolen, S., Donzelli, G. F., Lal, M., Chen, K., Justo da Silva, L. H., et al. (2015). Rapid induction of high-level carbapenem resistance in heteroresistant KPC-producing *Klebsiella pneumoniae*. *Antimicrob. Agents Chemother.* 59, 3281–3289. doi: 10.1128/AAC.05100-14

Adler, A., Hussein, O., Ben-David, D., Masarwa, S., Navon-Venezia, S., Schwaber, M. J., et al. (2015). Persistence of *Klebsiella pneumoniae* ST258 as the predominant clone of carbapenemase-producing Enterobacteriaceae in post-acute-care hospitals in Israel, 2008-13. *J. Antimicrob. Chemother.* 70, 89–92. doi: 10.1093/jac/dku333

Adler, A., Khabra, E., Chmelnitsky, I., Giakkoupi, P., Vatopoulos, A., Mathers, A. J., et al. (2014). Development and validation of a multiplex PCR assay for identification of the epidemic ST-258/512 KPC-producing *Klebsiella pneumoniae* clone. *Diagn. Microbiol. Infect. Dis.* 78, 12–15. doi: 10.1016/j.diagmicrobio.2013.10.003

Adler, A., Solter, E., Masarwa, S., Miller-Roll, T., Abu-Libdeh, B., Khammash, H., et al. (2013). Epidemiological and microbiological characteristics of an outbreak caused by OXA-48-producing Enterobacteriaceae in a neonatal intensive care unit in Jerusalem. *Israel J. Clin. Microbiol.* 51, 2926–2930. doi: 10.1128/JCM.01049-13

Ageevets, V. A., Partina, I. V., Lisitsyna, E. S., Ilina, E. N., Lobzin, Y. V., Shlyapnikov, S. A., et al. (2014). Emergence of carbapenemase-producing Gram-negative bacteria in Saint Petersburg. *Russ. Int. J. Antimicrob. Agents* 44, 152–155. doi: 10.1016/j.ijantimicag.2014.05.004

Ahn, C., Butt, A. A., Rivera, J. I., Yaqoob, M., Hag, S., Khalil, A., et al. (2015). OXA-48-producing Enterobacteriaceae causing bacteremia, United Arab Emirates. *Int. J. Infect. Dis.* 30, 36–37. doi: 10.1016/j.ijid.2014.11.008

Al-Marzooq, F., Ngeow, Y. F., and Tay, S. T. (2015). Emergence of *Klebsiella pneumoniae* producing dual carbapenemases (NDM-1 and OXA-232) and 16S rRNA methylase (armA) isolated from a Malaysian patient returning from India. *Int. J. Antimicrob. Agents* 45, 445–446. doi: 10.1016/j.ijantimicag.2014.12.013

Almeida, A. C., de Sa Cavalcanti, F. L., Vilela, M. A., Gales, A. C., and de Morais, M. A. Jr. (2012). *Escherichia coli* ST502 and *Klebsiella pneumoniae* ST11 sharing an IncW plasmid harbouring the blaKPC-2 gene in an intensive care unit patient. *Int. J. Antimicrob. Agents* 40, 374–376. doi: 10.1016/j.ijantimicag.2012.05.022

Anandan, S., Damodaran, S., Gopi, R., Bakthavatchalam, Y. D., and Veeraraghavan, B. (2015). Rapid screening for carbapenem resistant organisms: current results and future approaches. *J. Clin. Diagn. Res.* 9, DM01–DM03. doi: 10.7860/JCDR/2015/14246.6530

Arana, D. M., Saez, D., Garcia-Hierro, P., Bautista, V., Fernandez-Romero, S., Angel de la Cal, M., et al. (2015). Concurrent interspecies and clonal dissemination of OXA-48 carbapenemase. *Clin. Microbiol. Infect.* 21, 148.e141–148.e144. doi: 10.1016/j.cmi.2014.07.008

Arpin, C., Noury, P., Boraud, D., Coulange, L., Manetti, A., Andre, C., et al. (2012). NDM-1-producing *Klebsiella pneumoniae* resistant to colistin in a French community patient without history of foreign travel. *Antimicrob. Agents Chemother.* 56, 3432–3434. doi: 10.1128/AAC.00230-12

Aubert, D., Naas, T., Heritier, C., Poirel, L., and Nordmann, P. (2006). Functional characterization of IS1999, an IS4 family element involved in mobilization and expression of b-lactam resistance genes. *J. Bacteriol.* 188, 6506–6514. doi: 10.1128/JB.00375-06

Azimi, L., Nordmann, P., Lari, A. R., and Bonnin, R. A. (2014). First report of OXA-48-producing *Klebsiella pneumoniae* strains in Iran. *GMS Hyg. Infect. Control.* 9:Doc07. doi: 10.3205/dgkh000227

Bae, I. K., Lee, Y. N., Jeong, S. H., Hong, S. G., Lee, J. H., Lee, S. H., et al. (2007). Genetic and biochemical characterization of GES-5, an extended-spectrum class A b-lactamase from *Klebsiella pneumoniae*. *Diagn. Microbiol. Infect. Dis.* 58, 465–468. doi: 10.1016/j.diagmicrobio.2007.02.013

Bakour, S., Garcia, V., Loucif, L., Brunel, J. M., Gharout-Sait, A., Touati, A., et al. (2015a). Rapid identification of carbapenemase-producing Enterobacteriaceae, *Pseudomonas aeruginosa* and *Acinetobacter baumannii* using a modified Carba NP test. *New Microbes New Infect.* 7, 89–93. doi: 10.1016/j.nmni.2015.07.001

Bakour, S., Sahli, F., Touati, A., and Rolain, J. M. (2015b). Emergence of KPC-producing *Klebsiella pneumoniae* ST512 isolated from cerebrospinal fluid of a child in Algeria. *New Microbes New Infect.* 3, 34–36. doi: 10.1016/j.nmni.2014.09.001

Balm, M. N., La, M. V., Krishnan, P., Jureen, R., Lin, R. T., and Teo, J. W. (2013). Emergence of *Klebsiella pneumoniae* co-producing NDM-type and OXA-181 carbapenemases. *Clin. Microbiol. Infect.* 19, E421–E423. doi: 10.1111/1469-0691.12247

Baraniak, A., Grabowska, A., Izdebski, R., Fiett, J., Herda, M., Bojarska, K., et al. (2011). Molecular characteristics of KPC-producing Enterobacteriaceae at the early stage of their dissemination in Poland, 2008-2009. *Antimicrob. Agents Chemother.* 55, 5493–5499. doi: 10.1128/AAC.05118-11

Baraniak, A., Izdebski, R., Fiett, J., Gawryszewska, I., Bojarska, K., Herda, M., et al. (2016). NDM-producing Enterobacteriaceae in Poland, 2012-14: inter-regional outbreak of *Klebsiella pneumoniae* ST11 and sporadic cases. *J. Antimicrob. Chemother.* 7, 85–91. doi: 10.1093/jac/dkv282

Barguigua, A., El Otmani, F., El Yaagoubi, F. L., Talmi, M., Zerouali, K., and Timinouni, M. (2013). First report of a *Klebsiella pneumoniae* strain coproducing NDM-1, VIM-1 and OXA-48 carbapenemases isolated in Morocco. *APMIS* 121, 675–677. doi: 10.1111/apm.12034

Barrios, H., Silva-Sanchez, J., Reyna-Flores, F., Sanchez-Perez, A., Sanchez-Francia, D., Aguirre-Torres, J. A., et al. (2014). Detection of a NDM-1-producing *Klebsiella pneumoniae* (ST22) clinical isolate at a pediatric hospital in Mexico. *Pediatr. Infect. Dis. J.* 33:335. doi: 10.1097/INF.0000000000000173

Bathoorn, E., Friedrich, A. W., Zhou, K., Arends, J. P., Borst, D. M., Grundmann, H., et al. (2013). Latent introduction to the Netherlands of multiple antibiotic resistance including NDM-1 after hospitalisation in Egypt, August 2013. *Euro. Surveill.* 18:20610. doi: 10.2807/1560-7917.ES2013.18.42.20610

Bathoorn, E., Rossen, J. W., Lokate, M., Friedrich, A. W., and Hammerum, A. M. (2015). Isolation of an NDM-5-producing ST16 *Klebsiella pneumoniae* from a Dutch patient without travel history abroad, August 2015. *Euro Surveill.* 20:30040. doi: 10.2807/1560-7917.ES.2015.20.41.30040

Bedenic, B., Mazzariol, A., Plecko, V., Bosnjak, Z., Barl, P., Vranes, J., et al. (2012). First report of KPC-producing *Klebsiella pneumoniae* in Croatia. *J. Chemother.* 24, 237–239. doi: 10.1179/1973947812Y.0000000017

Ben Nasr, A., Decre, D., Compain, F., Genel, N., Barguellil, F., and Arlet, G. (2013). Emergence of NDM-1 in association with OXA-48 in *Klebsiella pneumoniae* from Tunisia. *Antimicrob. Agents Chemother.* 57, 4089–4090. doi: 10.1128/AAC.00536-13

Bercot, B., Poirel, L., Dortet, L., and Nordmann, P. (2011). In vitro evaluation of antibiotic synergy for NDM-1-producing Enterobacteriaceae. *J. Antimicrob. Chemother.* 66, 2295–2297. doi: 10.1093/jac/dkr296

Berger, S., Alauzet, C., Aissa, N., Henard, S., Rabaud, C., Bonnet, R., et al. (2013). Characterization of a new blaOXA-48-carrying plasmid in Enterobacteriaceae. *Antimicrob. Agents Chemother.* 57, 4064–4067. doi: 10.1128/AAC.02550-12

Berrazeg, M., Diene, S., Medjahed, L., Parola, P., Drissi, M., Raoult, D., et al. (2014). New Delhi Metallo-b-lactamase around the world: an eReview using Google Maps. *Euro. Surveill.* 19:20809. doi: 10.2807/1560-7917.ES2014.19.20.20809

Beyrouthy, R., Robin, F., Dabboussi, F., Mallat, H., Hamze, M., and Bonnet, R. (2014). Carbapenemase and virulence factors of Enterobacteriaceae in North Lebanon between 2008 and 2012: evolution via endemic spread of OXA-48. *J. Antimicrob. Chemother.* 69, 2699–2705. doi: 10.1093/jac/dku181

Birgy, A., Doit, C., Mariani-Kurkdjian, P., Genel, N., Faye, A., Arlet, G., et al. (2011). Early detection of colonization by VIM-1-producing *Klebsiella pneumoniae* and NDM-1-producing *Escherichia coli* in two children returning to France. *J. Clin. Microbiol.* 49, 3085–3087. doi: 10.1128/JCM.00540-11

Bogaerts, P., Montesinos, I., Rodriguez-Villalobos, H., Blairon, L., Deplano, A., and Glupczynski, Y. (2010). Emergence of clonally related *Klebsiella pneumoniae* isolates of sequence type 258 producing KPC-2 carbapenemase in Belgium. *J. Antimicrob. Chemother.* 65, 361–362. doi: 10.1093/jac/dkp453

Bogaerts, P., Yunus, S., Massart, M., Huang, T. D., and Glupczynski, Y. (2016). Evaluation of the BYG Carba test, a new electrochemical assay for rapid laboratory detection of carbapenemase-producing enterobacteriaceae. *J. Clin. Microbiol.* 54, 349–358. doi: 10.1128/JCM.02404-15

Bogdanovich, T., Adams-Haduch, J. M., Tian, G. B., Nguyen, M. H., Kwak, E. J., Muto, C. A., et al. (2011). Colistin-resistant, *Klebsiella pneumoniae* carbapenemase (KPC)-producing *Klebsiella pneumoniae* belonging to the international epidemic clone ST258. *Clin. Infect. Dis.* 53, 373–376. doi: 10.1093/cid/cir401

Bonura, C., Giuffre, M., Aleo, A., Fasciana, T., Di Bernardo, F., Stampone, T., et al. (2015). An Update of the evolving epidemic of blaKPC carrying *Klebsiella pneumoniae* in Sicily, Italy, 2014: emergence of multiple Non-ST258 Clones. *PLoS ONE* 10:e0132936. doi: 10.1371/journal.pone.0132936

Borgia, S., Lastovetska, O., Richardson, D., Eshaghi, A., Xiong, J., Chung, C., et al. (2012). Outbreak of carbapenem-resistant enterobacteriaceae containing blaNDM-1, Ontario, Canada. *Clin. Infect. Dis.* 55, e109-17. doi: 10.1093/cid/cis737

Bowers, J. R., Kitchel, B., Driebe, E. M., MacCannell, D. R., Roe, C., Lemmer, D., et al. (2015). Genomic analysis of the emergence and rapid global dissemination of the clonal group 258 *Klebsiella pneumoniae* Pandemic. *PLoS ONE* 10:e0133727. doi: 10.1371/journal.pone.0133727

Branas, P., Villa, J., Viedma, E., Mingorance, J., Orellana, M. A., and Chaves, F. (2015). Molecular epidemiology of carbapenemase-producing *Klebsiella pneumoniae* in a hospital in madrid: successful establishment of an OXA-48 ST11 clone. *Int. J. Antimicrob. Agents* 46, 111–116. doi: 10.1016/j.ijantimicag.2015.02.019

Braun, S. D., Monecke, S., Thurmer, A., Ruppelt, A., Makarewicz, O., Pletz, M., et al. (2014). Rapid identification of carbapenemase genes in gram-negative bacteria with an oligonucleotide microarray-based assay. *PLoS ONE* 9:e102232. doi: 10.1371/journal.pone.0102232

Brink, A. J., Coetzee, J., Clay, C. G., Sithole, S., Richards, G. A., Poirel, L., et al. (2012). Emergence of New Delhi metallo-b-lactamase (NDM-1) and *Klebsiella pneumoniae* carbapenemase (KPC-2) in South Africa. *J. Clin. Microbiol.* 50, 525–527. doi: 10.1128/JCM.05956-11

Brink, A. J., Coetzee, J., Corcoran, C., Clay, C. G., Hari-Makkan, D., Jacobson, R. K., et al. (2013). Emergence of OXA-48 and OXA-181 carbapenemases among Enterobacteriaceae in South Africa and evidence of in vivo selection of colistin resistance as a consequence of selective decontamination of the gastrointestinal tract. *J. Clin. Microbiol.* 51, 369–372. doi: 10.1128/JCM.02234-12

Brown, E. D., Vivas, E. I., Walsh, C. T., and Kolter, R. (1995). MurA (MurZ), the enzyme that catalyzes the first committed step in peptidoglycan biosynthesis, is essential in *Escherichia coli*. *J. Bacteriol.* 177, 4194–4197.

Cannatelli, A., D'Andrea, M. M., Giani, T., Di Pilato, V., Arena, F., Ambretti, S., et al. (2013). In vivo emergence of colistin resistance in *Klebsiella pneumoniae* producing KPC-type carbapenemases mediated by insertional inactivation of the PhoQ/PhoP mgrB regulator. *Antimicrob. Agents Chemother.* 57, 5521–5526. doi: 10.1128/AAC.01480-13

Cannatelli, A., Di Pilato, V., Giani, T., Arena, F., Ambretti, S., Gaibani, P., et al. (2014a). In vivo evolution to colistin resistance by PmrB sensor kinase mutation in KPC-producing *Klebsiella pneumoniae* is associated with low-dosage colistin treatment. *Antimicrob. Agents Chemother.* 58, 4399–4403. doi: 10.1128/AAC.02555-14

Cannatelli, A., Giani, T., D'Andrea, M. M., Di Pilato, V., Arena, F., Conte, V., et al. (2014b). MgrB inactivation is a common mechanism of colistin resistance in KPC-producing *Klebsiella pneumoniae* of clinical origin. *Antimicrob. Agents Chemother.* 58, 5696–5703. doi: 10.1128/AAC.03110-14

Carattoli, A., Seiffert, S. N., Schwendener, S., Perreten, V., and Endimiani, A. (2015). Differentiation of IncL and IncM plasmids associated with the spread of clinically relevant antimicrobial resistance. *PLoS ONE* 10:e0123063. doi: 10.1371/journal.pone.0123063

Carbonne, A., Thiolet, J. M., Fournier, S., Fortineau, N., Kassis-Chikhani, N., Boytchev, I., et al. (2010). Control of a multi-hospital outbreak of KPC-producing *Klebsiella pneumoniae* type 2 in France, September to October 2009. *Euro. Surveill* 15, 19734.

Carricajo, A., Verhoeven, P. O., Guezzou, S., Fonsale, N., and Aubert, G. (2014). Detection of carbapenemase-producing bacteria by using an ultra-performance liquid chromatography-tandem mass spectrometry method. *Antimicrob. Agents Chemother.* 58, 1231–1234. doi: 10.1128/AAC.01540-13

Carvalhaes, C. G., Cayo, R., Visconde, M. F., Barone, T., Frigatto, E. A., Okamoto, D., et al. (2014). Detection of carbapenemase activity directly from blood culture vials using MALDI-TOF MS: a quick answer for the right decision. *J. Antimicrob. Chemother.* 69, 2132–2136. doi: 10.1093/jac/dku094

Castanheira, M., Deshpande, L. M., Mathai, D., Bell, J. M., Jones, R. N., and Mendes, R. E. (2011). Early dissemination of NDM-1- and OXA-181-producing Enterobacteriaceae in Indian hospitals: report from the SENTRY Antimicrobial Surveillance Program, 2006-2007. *Antimicrob. Agents Chemother.* 55, 1274–1278. doi: 10.1128/AAC.01497-10

Centers for Disease Control and Prevention (2014). Notes from the Field: New Delhi metallo-β-lactamase-producing *Escherichia coli* associated with endoscopic retrograde cholangiopancreatography - Illinois, 2013. *MMWR Morb. Mortal. Wkly. Rep.* 62:1051.

Chagas, T. P., Seki, L. M., da Silva, D. M., and Asensi, M. D. (2011). Occurrence of KPC-2-producing *Klebsiella pneumoniae* strains in hospital wastewater. *J. Hosp. Infect.* 77:281. doi: 10.1016/j.jhin.2010.10.008

Chang, Y. Y., Chuang, Y. C., Siu, L. K., Wu, T. L., Lin, J. C., Lu, P. L., et al. (2015). Clinical features of patients with carbapenem nonsusceptible *Klebsiella pneumoniae* and *Escherichia coli* in intensive care units: a nationwide multicenter study in Taiwan. *J. Microbiol. Immunol. Infect.* 48, 219–225. doi: 10.1016/j.jmii.2014.05.010

Chen, C. J., Wu, T. L., Lu, P. L., Chen, Y. T., Fung, C. P., Chuang, Y. C., et al. (2014a). Closely related NDM-1-encoding plasmids from *Escherichia coli* and *Klebsiella pneumoniae* in Taiwan. *PLoS ONE* 9:e104899. doi: 10.1371/journal.pone.0104899

Chen, L., Chavda, K. D., Al Laham, N., Melano, R. G., Jacobs, M. R., Bonomo, R. A., et al. (2013a). Complete nucleotide sequence of a blaKPC-harboring IncI2 plasmid and its dissemination in New Jersey and New York hospitals. *Antimicrob. Agents Chemother.* 57, 5019–5025. doi: 10.1128/AAC.01397-13

Chen, L., Chavda, K. D., Findlay, J., Peirano, G., Hopkins, K., Pitout, J. D., et al. (2014b). Multiplex PCR for identification of two capsular types in epidemic KPC-producing *Klebsiella pneumoniae* sequence type 258 strains. *Antimicrob. Agents Chemother.* 58, 4196–4199. doi: 10.1128/AAC.02673-14

Chen, L., Chavda, K. D., Fraimow, H. S., Mediavilla, J. R., Melano, R. G., Jacobs, M. R., et al. (2013b). Complete nucleotide sequences of blaKPC-4- and blaKPC-5-harboring IncN and IncX plasmids from *Klebsiella pneumoniae* strains isolated in New Jersey. *Antimicrob. Agents Chemother.* 57, 269–276. doi: 10.1128/AAC.01648-12

Chen, L., Chavda, K. D., Melano, R. G., Hong, T., Rojtman, A. D., Jacobs, M. R., et al. (2014c). Molecular survey of the dissemination of two blaKPC-harboring IncFIA plasmids in New Jersey and New York hospitals. *Antimicrob. Agents Chemother.* 58, 2289–2294. doi: 10.1128/AAC.02749-13

Chen, L., Chavda, K. D., Melano, R. G., Jacobs, M. R., Koll, B., Hong, T., et al. (2014d). Comparative genomic analysis of KPC-encoding pKpQIL-like plasmids and their distribution in New Jersey and New York Hospitals. *Antimicrob. Agents Chemother.* 58, 2871–2877. doi: 10.1128/AAC.00120-14

Chen, L., Mathema, B., Chavda, K. D., DeLeo, F. R., Bonomo, R. A., and Kreiswirth, B. N. (2014e). Carbapenemase-producing *Klebsiella pneumoniae*: molecular and genetic decoding. *Trends Microbiol.* 22, 686–696. doi: 10.1016/j.tim.2014.09.003

Chen, L., Mathema, B., Pitout, J. D., DeLeo, F. R., and Kreiswirth, B. N. (2014f). Epidemic *Klebsiella pneumoniae* ST258 is a hybrid strain. *MBio.* 5, e1355-14. doi: 10.1128/mBio.01355-14

Chen, Y. T., Lin, A. C., Siu, L. K., and Koh, T. H. (2012). Sequence of closely related plasmids encoding blaNDM-1 in two unrelated *Klebsiella pneumoniae* isolates in Singapore. *PLoS ONE* 7:e48737. doi: 10.1371/journal.pone.0048737

Chen, Y. T., Lin, J. C., Fung, C. P., Lu, P. L., Chuang, Y. C., Wu, T. L., et al. (2014g). KPC-2-encoding plasmids from *Escherichia coli* and *Klebsiella pneumoniae* in Taiwan. *J. Antimicrob. Chemother.* 69, 628–631. doi: 10.1093/jac/dkt409

Chen, Z., Wang, Y., Tian, L., Zhu, X., Li, L., Zhang, B., et al. (2015). First report in China of Enterobacteriaceae clinical isolates coharboring blaNDM-1 and blaIMP-4 drug resistance genes. *Microb. Drug Resist.* 21, 167–170. doi: 10.1089/mdr.2014.0087

Chereau, F., Herindrainy, P., Garin, B., Huynh, B. T., Randrianirina, F., Padget, M., et al. (2015). Colonization of extended-spectrum-β-lactamase- and NDM-1-producing Enterobacteriaceae among pregnant women in the community in a low-income country: a potential reservoir for transmission of multiresistant Enterobacteriaceae to neonates. *Antimicrob. Agents Chemother.* 59, 3652–3655. doi: 10.1128/AAC.00029-15

Chiang, T. T., Yang, Y. S., Yeh, K. M., Chiu, S. K., Wang, N. C., Lin, T. Y., et al. (2016). Quantification and comparison of virulence and characteristics of different variants of carbapenemase-producing *Klebsiella pneumoniae* clinical isolates from Taiwan and the United States. *J. Microbiol. Immunol. Infect.* 49, 83–90. doi: 10.1016/j.jmii.2015.08.011

Chiu, S. K., Wu, T. L., Chuang, Y. C., Lin, J. C., Fung, C. P., Lu, P. L., et al. (2013). National surveillance study on carbapenem non-susceptible *Klebsiella pneumoniae* in Taiwan: the emergence and rapid dissemination of KPC-2 carbapenemase. *PLoS ONE* 8:e69428. doi: 10.1371/journal.pone.0069428

Cho, S. Y., Huh, H. J., Baek, J. Y., Chung, N. Y., Ryu, J. G., Ki, C. S., et al. (2015). *Klebsiella pneumoniae* co-producing NDM-5 and OXA-181 carbapenemases, South Korea. *Emerg. Infect. Dis.* 21, 1088–1089. doi: 10.3201/eid2106.150048

Chong, P. M., McCorrister, S. J., Unger, M. S., Boyd, D. A., Mulvey, M. R., and Westmacott, G. R. (2015). MALDI-TOF MS detection of carbapenemase activity in clinical isolates of Enterobacteriaceae spp., *Pseudomonas aeruginosa*, and *Acinetobacter baumannii* compared against the Carba-NP assay. *J. Microbiol. Methods* 111, 21–23. doi: 10.1016/j.mimet.2015.01.024

Chung The, H., Karkey, A., Pham Thanh, D., Boinett, C. J., Cain, A. K., Ellington, M., et al. (2015). A high-resolution genomic analysis of multidrug-resistant hospital outbreaks of *Klebsiella pneumoniae*. *EMBO Mol. Med.* 7, 227–239. doi: 10.15252/emmm.201404767

Cubero, M., Cuervo, G., Dominguez, M. A., Tubau, F., Marti, S., Sevillano, E., et al. (2015). Carbapenem-resistant and carbapenem-susceptible isogenic isolates of *Klebsiella pneumoniae* ST101 causing infection in a tertiary hospital. *BMC Microbiol.* 15:177. doi: 10.1186/s12866-015-0510-9

Cuzon, G., Bentchouala, C., Vogel, A., Hery, M., Lezzar, A., Smati, F., et al. (2015). First outbreak of OXA-48-positive carbapenem-resistant *Klebsiella pneumoniae* isolates in Constantine, Algeria. *Int. J. Antimicrob. Agents* 46, 725–727. doi: 10.1016/j.ijantimicag.2015.08.005

Cuzon, G., Naas, T., Bogaerts, P., Glupczynski, Y., Huang, T. D., and Nordmann, P. (2008). Plasmid-encoded carbapenem-hydrolyzing β-lactamase OXA-48 in an imipenem-susceptible *Klebsiella pneumoniae* strain from Belgium. *Antimicrob. Agents Chemother.* 52, 3463–3464. doi: 10.1128/AAC.00543-08

Dafopoulou, K., Zarkotou, O., Dimitroulia, E., Hadjichristodoulou, C., Gennimata, V., Pournaras, S., et al. (2015). Comparative evaluation of colistin susceptibility testing methods among carbapenem-nonsusceptible *Klebsiella pneumoniae* and *Acinetobacter baumannii* clinical isolates. *Antimicrob. Agents Chemother.* 59, 4625–4630. doi: 10.1128/AAC.00868-15

Daikos, G. L., Tsaousi, S., Tzouvelekis, L. S., Anyfantis, I., Psichogiou, M., Argyropoulou, A., et al. (2014). Carbapenemase-producing *Klebsiella pneumoniae* bloodstream infections: lowering mortality by antibiotic combination schemes and the role of carbapenems. *Antimicrob. Agents Chemother.* 58, 2322–2328. doi: 10.1128/AAC.02166-13

Dash, N., Panigrahi, D., Zarouni, M. A., Darwish, D., Ghazawi, A., Sonnevend, A., et al. (2014). High incidence of New Delhi metallo-β-lactamase-producing *Klebsiella pneumoniae* isolates in Sharjah, United Arab Emirates. *Microb. Drug Resist.* 20, 52–56. doi: 10.1089/mdr.2013.0040

Datta, S., Roy, S., Chatterjee, S., Saha, A., Sen, B., Pal, T., et al. (2014). A five-year experience of carbapenem resistance in Enterobacteriaceae causing neonatal septicaemia: predominance of NDM-1. *PLoS ONE* 9:e112101. doi: 10.1371/journal.pone.0112101

Dautzenberg, M. J., Ossewaarde, J. M., de Kraker, M. E., van der Zee, A., van Burgh, S., de Greeff, S. C., et al. (2014). Successful control of a hospital-wide outbreak of OXA-48 producing Enterobacteriaceae in the Netherlands, 2009 to 2011. *Euro. Surveill.* 19:20723. doi: 10.2807/1560-7917.ES2014.19.9.20723

de Jager, P., Chirwa, T., Naidoo, S., Perovic, O., and Thomas, J. (2015). Nosocomial outbreak of New Delhi Metallo-β-Lactamase-1-Producing Gram-Negative BACTERIA in South Africa: a case-control study. *PLoS ONE* 10:e0123337. doi: 10.1371/journal.pone.0123337

Deleo, F. R., Chen, L., Porcella, S. F., Martens, C. A., Kobayashi, S. D., Porter, A. R., et al. (2014). Molecular dissection of the evolution of carbapenem-resistant multilocus sequence type 258 *Klebsiella pneumoniae*. *Proc. Natl. Acad. Sci. U.S.A.* 111, 4988–4993. doi: 10.1073/pnas.1321364111

de Oliveira, M. S., de Assis, D. B., Freire, M. P., Boas do Prado, G. V., Machado, A. S., Abdala, E., et al. (2015). Treatment of KPC-producing Enterobacteriaceae: suboptimal efficacy of polymyxins. *Clin. Microbiol. Infect.* 21, 179.e1–179.e7. doi: 10.1016/j.cmi.2014.07.010

Deshpande, L. M., Rhomberg, P. R., Sader, H. S., and Jones, R. N. (2006). Emergence of serine carbapenemases (KPC and SME) among clinical strains of Enterobacteriaceae isolated in the United States medical centers: report from the MYSTIC Program (1999-2005). *Diagn. Microbiol. Infect. Dis.* 56, 367–372. doi: 10.1016/j.diagmicrobio.2006.07.004

Dimou, V., Dhanji, H., Pike, R., Livermore, D. M., and Woodford, N. (2012). Characterization of Enterobacteriaceae producing OXA-48-like carbapenemases in the UK. *J. Antimicrob. Chemother.* 67, 1660–1665. doi: 10.1093/jac/dks124

Doi, Y., O'Hara, J. A., Lando, J. F., Querry, A. M., Townsend, B. M., Pasculle, A. W., et al. (2014). Co-production of NDM-1 and OXA-232 by *Klebsiella pneumoniae*. *Emerg. Infect. Dis.* 20, 163–165. doi: 10.3201/eid2001.130904

Dong, D., Liu, W., Li, H., Wang, Y., Li, X., Zou, D., et al. (2015). Survey and rapid detection of *Klebsiella pneumoniae* in clinical samples targeting the rcsA gene in Beijing, China. *Front. Microbiol.* 6:519. doi: 10.3389/fmicb.2015.00519

Dortet, L., Agathine, A., Naas, T., Cuzon, G., Poirel, L., and Nordmann, P. (2015). Evaluation of the RAPIDECR CARBA NP, the Rapid CARB ScreenR and the Carba NP test for biochemical detection of carbapenemase-producing Enterobacteriaceae. *J. Antimicrob. Chemother.* 70, 3014–3022. doi: 10.1093/jac/dkv213

Dortet, L., Brechard, L., Grenet, K., Nguessan, M. S., and Nordmann, P. (2013). Sri Lanka, another country from the Indian subcontinent with NDM-1-producing Enterobacteriaceae. *J. Antimicrob. Chemother.* 68, 2172–2173. doi: 10.1093/jac/dkt145

Dortet, L., Brechard, L., Poirel, L., and Nordmann, P. (2014a). Rapid detection of carbapenemase-producing Enterobacteriaceae from blood cultures. *Clin. Microbiol. Infect.* 20, 340–344. doi: 10.1111/1469-0691.12318

Dortet, L., Poirel, L., Errera, C., and Nordmann, P. (2014b). CarbAcineto NP test for rapid detection of carbapenemase-producing *Acinetobacter* spp. *J. Clin. Microbiol.* 52, 2359–2364. doi: 10.1128/JCM.00594-14

Dortet, L., Poirel, L., and Nordmann, P. (2014c). Worldwide dissemination of the NDM-type carbapenemases in Gram-negative bacteria. *Biomed Res. Int.* 2014, 249856. doi: 10.1155/2014/249856

Ducomble, T., Faucheux, S., Helbig, U., Kaisers, U. X., Konig, B., Knaust, A., et al. (2015). Large hospital outbreak of KPC-2-producing *Klebsiella pneumoniae*: investigating mortality and the impact of screening for KPC-2 with polymerase chain reaction. *J. Hosp. Infect.* 89, 179–185. doi: 10.1016/j.jhin.2014.11.012

Ellis, C., Chung, C., Tijet, N., Patel, S. N., Desjardins, M., Melano, R. G., et al. (2013). OXA-48-like carbapenemase-producing Enterobacteriaceae in Ottawa. Canada. *Diagn. Microbiol. Infect. Dis.* 76, 399–400. doi: 10.1016/j.diagmicrobio.2013.04.017

Escobar Perez, J. A., Olarte Escobar, N. M., Castro-Cardozo, B., Valderrama Marquez, I. A., Garzon Aguilar, M. I., Martinez de la Barrera, L., et al. (2013). Outbreak of NDM-1-producing *Klebsiella pneumoniae* in a neonatal unit in Colombia. *Antimicrob. Agents Chemother.* 57, 1957–1960. doi: 10.1128/AAC.01447-12

Espedido, B. A., Steen, J. A., Ziochos, H., Grimmond, S. M., Cooper, M. A., Gosbell, I. B., et al. (2013). Whole genome sequence analysis of the first Australian OXA-48-producing outbreak-associated *Klebsiella pneumoniae* isolates: the resistome and in vivo evolution. *PLoS ONE* 8:e59920. doi: 10.1371/journal.pone.0059920

Evans, B. A., and Amyes, S. G. (2014). OXA β-lactamases. *Clin. Microbiol. Rev.* 27, 241–263. doi: 10.1128/CMR.00117-13

Evren, E., Azap, O. K., Colakoğlu, S., and Arslan, H. (2013). *In vitro* activity of fosfomycin in combination with imipenem, meropenem, colistin and tigecycline against OXA 48-positive *Klebsiella pneumoniae* strains. *Diagn. Microbiol. Infect. Dis.* 76, 335–338. doi: 10.1016/j.diagmicrobio.2013.04.004

Fehlberg, L. C., Carvalho, A. M., Campana, E. H., Gontijo-Filho, P. P., and Gales, A. C. (2012). Emergence of *Klebsiella pneumoniae*-producing KPC-2 carbapenemase in Paraiba, Northeastern Brazil. *Braz. J. Infect. Dis.* 16, 577–580. doi: 10.1016/j.bjid.2012.07.001

Fursova, N. K., Astashkin, E. I., Knyazeva, A. I., Kartsev, N. N., Leonova, E. S., Ershova, O. N., et al. (2015). The spread of blaOXA-48 and blaOXA-244 carbapenemase genes among *Klebsiella pneumoniae*, Proteus mirabilis and Enterobacter spp. isolated in Moscow, Russia. *Ann. Clin. Microbiol. Antimicrob.* 14:46. doi: 10.1186/s12941-015-0108-y

Gaibani, P., Ambretti, S., Berlingeri, A., Cordovana, M., Farruggia, P., Panico, M., et al. (2011). Outbreak of NDM-1-producing Enterobacteriaceae in northern Italy, July to August 2011. *Euro. Surveill.* 16:20027.

Gales, A. C., Castanheira, M., Jones, R. N., and Sader, H. S. (2012). Antimicrobial resistance among Gram-negative bacilli isolated from Latin America: results from SENTRY Antimicrobial Surveillance Program (Latin America, 2008-2010). *Diagn. Microbiol. Infect. Dis.* 73, 354–360. doi: 10.1016/j.diagmicrobio.2012.04.007

Gallagher, L. C., Roundtree, S. S., Lancaster, D. P., Rudin, S. D., Bard, J. D., Roberts, A. L., et al. (2015). Performance of the CLSI Carba NP and the rosco carb screen assays Using North American Carbapenemase-Producing Enterobacteriaceae and *Pseudomonas aeruginosa* isolates. *J. Clin. Microbiol.* 53, 3370–3373. doi: 10.1128/JCM.01547-15

Galler, H., Feierl, G., Petternel, C., Reinthaler, F. F., Haas, D., Grisold, A. J., et al. (2014). KPC-2 and OXA-48 carbapenemase-harbouring Enterobacteriaceae detected in an Austrian wastewater treatment plant. *Clin. Microbiol. Infect.* 20, O132–O134. doi: 10.1111/1469-0691.12336

Garbari, L., Busetti, M., Dolzani, L., Petix, V., Knezevich, A., Bressan, R., et al. (2015). pKBuS13, a KPC-2-encoding plasmid from *Klebsiella pneumoniae* sequence type 833, carrying Tn4401b inserted into an Xer site-specific recombination locus. *Antimicrob. Agents Chemother.* 59, 5226–5231. doi: 10.1128/AAC.04543-14

Garcia-Fernandez, A., Villa, L., Carta, C., Venditti, C., Giordano, A., Venditti, M., et al. (2012). *Klebsiella pneumoniae* ST258 producing KPC-3 identified in italy carries novel plasmids and OmpK36/OmpK35 porin variants. *Antimicrob. Agents Chemother.* 56, 2143–2145. doi: 10.1128/AAC.05308-11

Garcia-Fernandez, S., Morosini, M. I., Gijon, D., Beatobe, L., Ruiz-Garbajosa, P., Dominguez, L., et al. (2016). Detection of carbapenemase production in a collection of Enterobacteriaceae with characterized resistance mechanisms from clinical and environmental origins by use of both Carba NP and Blue-Carba Tests. *J. Clin. Microbiol.* 54, 464–466. doi: 10.1128/JCM.02580-15

Garza-Ramos, U., Barrios, H., Reyna-Flores, F., Sanchez-Perez, A., Tamayo-Legorreta, E., Ibarra-Pacheco, A., et al. (2014). Characteristics of KPC-2-producing *Klebsiella pneumoniae* (ST258) clinical isolates from outbreaks in 2 Mexican medical centers. *Diagn. Microbiol. Infect. Dis.* 79, 483–485. doi: 10.1016/j.diagmicrobio.2014.05.010

Gharout-Sait, A., Alsharapy, S. A., Brasme, L., Touati, A., Kermas, R., Bakour, S., et al. (2014). Enterobacteriaceae isolates carrying the New Delhi metallo-β-lactamase gene in Yemen. *J. Med. Microbiol.* 63 (Pt 8) , 1316–1323. doi: 10.1099/jmm.0.073767-0

Giacobbe, D. R., Del Bono, V., Trecarichi, E. M., De Rosa, F. G., Giannella, M., Bassetti, M., et al. (2015). Risk factors for bloodstream infections due to colistin-resistant KPC-producing *Klebsiella pneumoniae*: results from a multicenter case-control-control study. *Clin. Microbiol. Infect.* 21, 1106.e1–1106.e8. doi: 10.1016/j.cmi.2015.08.001

Giakkoupi, P., Pappa, O., Polemis, M., Vatopoulos, A. C., Miriagou, V., Zioga, A., et al. (2009). Emerging *Klebsiella pneumoniae* isolates coproducing KPC-2 and VIM-1 carbapenemases. *Antimicrob. Agents Chemother.* 53, 4048–4050. doi: 10.1128/AAC.00690-09

Giamarellou, H., Galani, L., Baziaka, F., and Karaiskos, I. (2013). Effectiveness of a double-carbapenem regimen for infections in humans due to carbapenemase-producing pandrug-resistant *Klebsiella pneumoniae*. *Antimicrob. Agents Chemother.* 57, 2388–2390. doi: 10.1128/AAC.02399-12

Giani, T., Arena, F., Vaggelli, G., Conte, V., Chiarelli, A., Henrici De Angelis, L., et al. (2015). Large nosocomial outbreak of colistin-resistant, carbapenemase-producing *Klebsiella pneumoniae* traced to clonal expansion of an mgrB deletion mutant. *J. Clin. Microbiol.* 53, 3341–3344. doi: 10.1128/JCM.01017-15

Giani, T., Conte, V., Mandala, S., D'Andrea, M. M., Luzzaro, F., Conaldi, P. G., et al. (2014). Cross-infection of solid organ transplant recipients by a multidrug-resistant *Klebsiella pneumoniae* isolate producing the OXA-48 carbapenemase, likely derived from a multiorgan donor. *J. Clin. Microbiol.* 52, 2702–2705. doi: 10.1128/JCM.00511-14

Giani, T., D'Andrea, M. M., Pecile, P., Borgianni, L., Nicoletti, P., Tonelli, F., et al. (2009). Emergence in Italy of *Klebsiella pneumoniae* sequence type 258 producing KPC-3 carbapenemase. *J. Clin. Microbiol.* 47, 3793–3794. doi: 10.1128/JCM.01773-09

Giani, T., Pini, B., Arena, F., Conte, V., Bracco, S., Migliavacca, R., et al. (2013). Epidemic diffusion of KPC carbapenemase-producing *Klebsiella pneumoniae* in Italy: results of the first countrywide survey, 15 May to 30 June 2011. *Euro. Surveill.* 18:20489.

Girlich, D., Anglade, C., Zambardi, G., and Nordmann, P. (2013). Comparative evaluation of a novel chromogenic medium (chromID OXA-48) for detection of OXA-48 producing Enterobacteriaceae. *Diagn. Microbiol. Infect. Dis.* 77, 296–300. doi: 10.1016/j.diagmicrobio.2013.08.015

Giske, C. G. (2015). Contemporary resistance trends and mechanisms for the old antibiotics colistin, temocillin, fosfomycin, mecillinam and nitrofurantoin. *Clin. Microbiol. Infect.* 21, 899–905. doi: 10.1016/j.cmi.2015.05.022

Giske, C. G., Froding, I., Hasan, C. M., Turlej-Rogacka, A., Toleman, M., Livermore, D., et al. (2012). Diverse sequence types of *Klebsiella pneumoniae* contribute to the dissemination of blaNDM-1 in India, Sweden, and the United Kingdom. *Antimicrob. Agents Chemother.* 56, 2735–2738. doi: 10.1128/AAC.06142-11

Giske, C. G., Gezelius, L., Samuelsen, O., Warner, M., Sundsfjord, A., and Woodford, N. (2011). A sensitive and specific phenotypic assay for detection of metallo-β-lactamases and KPC in *Klebsiella pneumoniae* with the use of meropenem disks supplemented with aminophenylboronic acid, dipicolinic acid and cloxacillin. *Clin. Microbiol. Infect.* 17, 552–556. doi: 10.1111/j.1469-0691.2010.03294.x

Goldfarb, D., Harvey, S. B., Jessamine, K., Jessamine, P., Toye, B., and Desjardins, M. (2009). Detection of plasmid-mediated KPC-producing *Klebsiella pneumoniae* in Ottawa, Canada: evidence of intrahospital transmission. *J. Clin. Microbiol.* 47, 1920–1922. doi: 10.1128/JCM.00098-09

Gomez, S., Pasteran, F., Faccone, D., Bettiol, M., Veliz, O., De Belder, D., et al. (2013). Intrapatient emergence of OXA-247: a novel carbapenemase found in a patient previously infected with OXA-163-producing *Klebsiella pneumoniae*. *Clin. Microbiol. Infect.* 19, E233–E235. doi: 10.1111/1469-0691.12142

Gomez, S. A., Pasteran, F. G., Faccone, D., Tijet, N., Rapoport, M., Lucero, C., et al. (2011). Clonal dissemination of *Klebsiella pneumoniae* ST258 harbouring KPC-2 in Argentina. *Clin. Microbiol. Infect.* 17, 1520–1524. doi: 10.1111/j.1469-0691.2011.03600.x

Gona, F., Barbera, F., Pasquariello, A. C., Grossi, P., Gridelli, B., Mezzatesta, M. L., et al. (2014). *In vivo* multiclonal transfer of bla_{KPC-3} from *Klebsiella pneumoniae* to *Escherichia coli* in surgery patients. *Clin. Microbiol. Infect.* 20, O633–O635. doi: 10.1111/1469-0691.12577

Gonzalez-Padilla, M., Torre-Cisneros, J., Rivera-Espinar, F., Pontes-Moreno, A., Lopez-Cerero, L., Pascual, A., et al. (2015). Gentamicin therapy for sepsis due to carbapenem-resistant and colistin-resistant *Klebsiella pneumoniae*. *J. Antimicrob. Chemother.* 70, 905–913. doi: 10.1093/jac/dku432

Gootz, T. D., Lescoe, M. K., Dib-Hajj, F., Dougherty, B. A., He, W., Della-Latta, P., et al. (2009). Genetic organization of transposase regions surrounding blaKPC carbapenemase genes on plasmids from *Klebsiella* strains isolated in a New York City hospital. *Antimicrob. Agents Chemother.* 53, 1998–2004. doi: 10.1128/AAC.01355-08

Gottig, S., Gruber, T. M., Stecher, B., Wichelhaus, T. A., and Kempf, V. A. (2015). In vivo horizontal gene transfer of the carbapenemase OXA-48 during a nosocomial outbreak. *Clin. Infect. Dis.* 60, 1808–1815. doi: 10.1093/cid/civ191

Gregory, C. J., Llata, E., Stine, N., Gould, C., Santiago, L. M., Vazquez, G. J., et al. (2010). Outbreak of carbapenem-resistant *Klebsiella pneumoniae* in Puerto Rico associated with a novel carbapenemase variant. *Infect. Control Hosp. Epidemiol.* 31, 476–484. doi: 10.1086/651670

Hall, J. M., Corea, E., Sanjeewani, H. D., and Inglis, T. J. (2014). Molecular mechanisms of β-lactam resistance in carbapenemase-producing *Klebsiella pneumoniae* from Sri Lanka. *J. Med. Microbiol.* 63(Pt 8), 1087–1092. doi: 10.1099/jmm.0.076760-0

Hamzan, N. I., Yean, C. Y., Rahman, R. A., Hasan, H., and Rahman, Z. A. (2015). Detection of blaIMP4 and blaNDM1 harboring *Klebsiella pneumoniae* isolates in a university hospital in Malaysia. *Emerg. Health Threats J.* 8, 26011. doi: 10.3402/ehtj.v8.26011

Hashimoto, A., Nagamatsu, M., Ohmagari, N., Hayakawa, K., Kato, Y., and Kirikae, T. (2014). Isolation of OXA-48 carbapenemase-producing *Klebsiella pneumoniae* ST101 from an overseas traveler returning to Japan. *Jpn. J. Infect. Dis.* 67, 120–121. doi: 10.7883/yoken.67.120

Hidalgo-Grass, C., Warburg, G., Temper, V., Benenson, S., Moses, A. E., Block, C., et al. (2012). KPC-9, a novel carbapenemase from clinical specimens in Israel. *Antimicrob. Agents Chemother.* 56, 6057–6059. doi: 10.1128/AAC.01156-12

Hishinuma, A., Yoshida, A., Suzuki, H., Okuzumi, K., and Ishida, T. (2013). Complete sequencing of an IncFII NDM-1 plasmid in *Klebsiella pneumoniae* shows structural features shared with other multidrug resistance plasmids. *J. Antimicrob. Chemother.* 68, 2415–2417. doi: 10.1093/jac/dkt190

Hoang, T. H., Wertheim, H., Minh, N. B., Duong, T. N., Anh, D. D., Phuong, T. T., et al. (2013). Carbapenem-resistant *Escherichia coli* and *Klebsiella pneumoniae* strains containing New Delhi metallo-β-lactamase isolated from two patients in Vietnam. *J. Clin. Microbiol.* 51, 373–374. doi: 10.1128/JCM.02322-12

Hombach, M., von Gunten, B., Castelberg, C., and Bloemberg, G. V. (2015). Evaluation of the rapidec carba NP Test for detection of carbapenemases in Enterobacteriaceae. *J. Clin. Microbiol.* 53, 3828–3833. doi: 10.1128/JCM.02327-15

Hong, J. H., Clancy, C. J., Cheng, S., Shields, R. K., Chen, L., Doi, Y., et al. (2013). Characterization of porin expression in *Klebsiella pneumoniae* Carbapenemase (KPC)-producing *K. pneumoniae* identifies isolates most susceptible to the combination of colistin and carbapenems. *Antimicrob. Agents Chemother.* 57, 2147–2153. doi: 10.1128/AAC.02411-12

Hrabak, J., Chudackova, E., and Walkova, R. (2013a). Matrix-assisted laser desorption ionization-time of flight (maldi-tof) mass spectrometry for detection of antibiotic resistance mechanisms: from research to routine diagnosis. *Clin. Microbiol. Rev.* 26, 103–114. doi: 10.1128/CMR.00058-12

Hrabak, J., Papagiannitsis, C. C., Studentova, V., Jakubu, V., Fridrichova, M., Zemlickova, H., et al. (2013b). Carbapenemase-producing *Klebsiella pneumoniae* in the Czech Republic in 2011. *Euro. Surveill.* 18:20626. doi: 10.2807/1560-7917.ES2013.18.45.20626

Hrabak, J., Studentova, V., Jakubu, V., Adamkova, V., Dvorakova, L., Balejova, M., et al. (2015). Prevalence study on carbapenemase-producing *Escherichia coli* and *Klebsiella pneumoniae* isolates in Czech hospitals–results from Czech Part of European Survey on Carbapenemase–Producing Enterobacteriaceae (EuSCAPE). *Epidemiol. Mikrobiol. Imunol.* 64, 87–91.

Hrabak, J., Walkova, R., Studentova, V., Chudackova, E., and Bergerova, T. (2011). Carbapenemase activity detection by matrix-assisted laser desorption ionization-time of flight mass spectrometry. *J. Clin. Microbiol.* 49, 3222–3227. doi: 10.1128/JCM.00984-11

Hu, L., Zhong, Q., Shang, Y., Wang, H., Ning, C., Li, Y., et al. (2014). The prevalence of carbapenemase genes and plasmid-mediated quinolone resistance determinants in carbapenem-resistant Enterobacteriaceae from five teaching hospitals in central China. *Epidemiol. Infect.* 142, 1972–1977. doi: 10.1017/S0950268813002975

Hu, L., Zhong, Q., Tu, J., Xu, Y., Qin, Z., Parsons, C., et al. (2013). Emergence of blaNDM-1 among *Klebsiella pneumoniae* ST15 and novel ST1031 clinical isolates in China. *Diagn. Microbiol. Infect. Dis.* 75, 373–376. doi: 10.1016/j.diagmicrobio.2013.01.006

Huang, T. D., Berhin, C., Bogaerts, P., and Glupczynski, Y. (2014). Comparative evaluation of two chromogenic tests for rapid detection of carbapenemase in Enterobacteriaceae and in *Pseudomonas aeruginosa* isolates. *J. Clin. Microbiol.* 52, 3060–3063. doi: 10.1128/JCM.00643-14

Huang, T. W., Chen, T. L., Chen, Y. T., Lauderdale, T. L., Liao, T. L., Lee, Y. T., et al. (2013). Copy number change of the NDM-1 sequence in a multidrug-resistant *Klebsiella pneumoniae* clinical isolate. *PLoS ONE* 8:e62774. doi: 10.1371/journal.pone.0062774

Hudson, C. M., Bent, Z. W., Meagher, R. J., and Williams, K. P. (2014). Resistance determinants and mobile genetic elements of an NDM-1-encoding *Klebsiella pneumoniae* strain. *PLoS ONE* 9:e99209. doi: 10.1371/journal.pone.0099209

Humphries, R. M. (2015). Susceptibility testing of the polymyxins: where are we now? *Pharmacotherapy* 35, 22–27. doi: 10.1002/phar.1505

Izdebski, R., Bojarska, K., Baraniak, A., Literacka, E., Herda, M., Zabicka, D., et al. (2015). NDM-1- or OXA-48-producing Enterobacteriaceae colonising Polish tourists following a terrorist attack in Tunis, March 2015. *Euro. Surveill.* 20:20489. doi: 10.2807/1560-7917.ES2015.20.23.21150

Jamal, W., Rotimi, V. O., Albert, M. J., Khodakhast, F., Nordmann, P., and Poirel, L. (2013). High prevalence of VIM-4 and NDM-1 metallo-β-lactamase among carbapenem-resistant Enterobacteriaceae. *J. Med. Microbiol.* 62, 1239–1244. doi: 10.1099/jmm.0.059915-0

Jamal, W., Rotimi, V. O., Albert, M. J., Khodakhast, F., Udo, E. E., and Poirel, L. (2012). Emergence of nosocomial New Delhi metallo-β-lactamase-1 (NDM-1)-producing *Klebsiella pneumoniae* in patients admitted to a tertiary care hospital in Kuwait. *Int. J. Antimicrob. Agents* 39, 183–184. doi: 10.1016/j.ijantimicag.2011.10.002

Janvari, L., Damjanova, I., Lazar, A., Racz, K., Kocsis, B., Urban, E., et al. (2014). Emergence of OXA-162-producing *Klebsiella pneumoniae* in Hungary. *Scand. J. Infect. Dis.* 46, 320–324. doi: 10.3109/00365548.2013.879993

Jayol, A., Poirel, L., Brink, A., Villegas, M. V., Yilmaz, M., and Nordmann, P. (2014). Resistance to colistin associated with a single amino acid change in protein PmrB among *Klebsiella pneumoniae* isolates of worldwide origin. *Antimicrob. Agents Chemother.* 58, 4762–4766. doi: 10.1128/AAC.00084-14

Jeon, J. H., Hong, M. K., Lee, J. H., Lee, J. J., Park, K. S., Karim, A. M., et al. (2014). Structure of ADC-68, a novel carbapenem-hydrolyzing class C extended-spectrum b-lactamase isolated from *Acinetobacter baumannii*. *Acta Crystallogr. D Biol. Crystallogr.* 70, 2924–2936. doi: 10.1107/S13990047140 19543

Jeon, J. H., Lee, J. H., Lee, J. J., Park, K. S., Karim, A. M., Lee, C. R., et al. (2015). Structural basis for carbapenem-hydrolyzing mechanisms of carbapenemases conferring antibiotic resistance. *Int. J. Mol. Sci.* 16, 9654–9692. doi: 10.3390/ijms16059654

Jeong, S. H., Bae, I. K., Kim, D., Hong, S. G., Song, J. S., Lee, J. H., et al. (2005). First outbreak of *Klebsiella pneumoniae* clinical isolates producing GES-5 and SHV-12 extended-spectrum β-lactamases in Korea. *Antimicrob. Agents Chemother.* 49, 4809–4810. doi: 10.1128/AAC.49.11.4809-4810.2005

Jeong, S. H., Lee, K. M., Lee, J., Bae, I. K., Kim, J. S., Kim, H. S., et al. (2015). Clonal and horizontal spread of the blaOXA-232 gene among Enterobacteriaceae in a Korean hospital. *Diagn. Microbiol. Infect. Dis.* 82, 70–72. doi: 10.1016/j.diagmicrobio.2015.02.001

Jiang, Y., Shen, P., Wei, Z., Liu, L., He, F., Shi, K., et al. (2015). Dissemination of a clone carrying a fosA3-harbouring plasmid mediates high fosfomycin resistance rate of KPC-producing *Klebsiella pneumoniae* in China. *Int. J. Antimicrob. Agents* 45, 66–70. doi: 10.1016/j.ijantimicag.2014.08.010

Jiang, Y., Yu, D., Wei, Z., Shen, P., Zhou, Z., and Yu, Y. (2010). Complete nucleotide sequence of *Klebsiella pneumoniae* multidrug resistance plasmid pKP048, carrying blaKPC-2, blaDHA-1, qnrB4, and armA. *Antimicrob. Agents Chemother.* 54, 3967–3969. doi: 10.1128/AAC.00137-10

Jin, Y., Shao, C., Li, J., Fan, H., Bai, Y., and Wang, Y. (2015). Outbreak of multidrug resistant NDM-1-producing *Klebsiella pneumoniae* from a neonatal unit in Shandong Province, China. *PLoS ONE* 10:e0119571. doi: 10.1371/journal.pone.0119571

Kabir, M. H., Meunier, D., Hopkins, K. L., Giske, C. G., and Woodford, N. (2016). A two-centre evaluation of RAPIDECR CARBA NP for carbapenemase detection in Enterobacteriaceae, *Pseudomonas aeruginosa* and *Acinetobacter* spp. *J. Antimicrob. Chemother.* 71, 1213–1216. doi: 10.1093/jac/dkv468

Kaiser, R. M., Castanheira, M., Jones, R. N., Tenover, F., and Lynfield, R. (2013). Trends in *Klebsiella pneumoniae* carbapenemase-positive K. *pneumoniae* in US hospitals: report from the 2007-2009 SENTRY antimicrobial surveillance program. *Diagn. Microbiol. Infect. Dis.* 76, 356–360. doi: 10.1016/j.diagmicrobio.2013.03.032

Kalpoe, J. S., Al Naiemi, N., Poirel, L., and Nordmann, P. (2011). Detection of an ambler class D OXA-48-type β-lactamase in a *Klebsiella pneumoniae* strain in the netherlands. *J. Med. Microbiol.* 60, 677–678. doi: 10.1099/jmm.0.028308-0

Kanerva, M., Skogberg, K., Ryynanen, K., Pahkamaki, A., Jalava, J., Ollgren, J., et al. (2015). Coincidental detection of the first outbreak of carbapenemase-producing *Klebsiella pneumoniae* colonisation in a primary care hospital, Finland, 2013. *Euro. Surveill.* 20:21172. doi: 10.2807/1560-7917.ES2015.20.26.21172

Karaiskos, I., and Giamarellou, H. (2014). Multidrug-resistant and extensively drug-resistant Gram-negative pathogens: current and emerging therapeutic approaches. *Exp. Opin. Pharmacother.* 15, 1351–1370. doi: 10.1517/14656566.2014.914172

Kayama, S., Koba, Y., Shigemoto, N., Kuwahara, R., Kakuhama, T., Kimura, K., et al. (2015). Imipenem-susceptible, meropenem-resistant *Klebsiella pneumoniae* producing OXA-181 in Japan. *Antimicrob. Agents Chemother.* 59, 1379–1380. doi: 10.1128/AAC.04330-14

Kazi, M., Drego, L., Nikam, C., Ajbani, K., Soman, R., Shetty, A., et al. (2015). Molecular characterization of carbapenem-resistant Enterobacteriaceae at a tertiary care laboratory in Mumbai. *Eur. J. Clin. Microbiol. Infect. Dis.* 34, 467–472. doi: 10.1007/s10096-014-2249-x

Kilic, A., and Baysallar, M. (2015). The First *Klebsiella pneumoniae* isolate Co-Producing OXA-48 and NDM-1 in Turkey. *Ann. Lab. Med.* 35, 382–383. doi: 10.3343/alm.2015.35.3.382

Kim, J. Y., Jung, H. I., An, Y. J., Lee, J. H., Kim, S. J., Jeong, S. H., et al. (2006). Structural basis for the extended substrate spectrum of CMY-10, a plasmid-encoded class C β-lactamase. *Mol. Microbiol.* 60, 907–916. doi: 10.1111/j.1365-2958.2006.05146.x

Kim, M. N., Yong, D., An, D., Chung, H. S., Woo, J. H., Lee, K., et al. (2012). Nosocomial clustering of NDM-1-producing *Klebsiella pneumoniae* sequence type 340 strains in four patients at a South Korean tertiary care hospital. *J. Clin. Microbiol.* 50, 1433–1436. doi: 10.1128/JCM.06855-11

Kocsis, E., Guzvinec, M., Butic, I., Kresic, S., Crnek, S. S., Tambic, A., et al. (2016). bla Carriage on IncR Plasmid in Enterobacteriaceae Strains. *Microb. Drug Resist.* 22, 123–128. doi: 10.1089/mdr.2015.0083

Kola, A., Piening, B., Pape, U. F., Veltzke-Schlieker, W., Kaase, M., Geffers, C., et al. (2015). An outbreak of carbapenem-resistant OXA-48 - producing *Klebsiella pneumoniae* associated to duodenoscopy. *Antimicrob. Resist. Infect. Control.* 4:8. doi: 10.1186/s13756-015-0049-4

Kontopoulou, K., Protonotariou, E., Vasilakos, K., Kriti, M., Koteli, A., Antoniadou, E., et al. (2010). Hospital outbreak caused by *Klebsiella pneumoniae* producing KPC-2 β-lactamase resistant to colistin. *J. Hosp. Infect.* 76, 70–73. doi: 10.1016/j.jhin.2010.03.021

Koser, C. U., Fraser, L. J., Ioannou, A., Becq, J., Ellington, M. J., Holden, M. T., et al. (2014). Rapid single-colony whole-genome sequencing of bacterial pathogens. *J. Antimicrob. Chemother.* 69, 1275–1281. doi: 10.1093/jac/dkt494

Kumarasamy, K. K., Toleman, M. A., Walsh, T. R., Bagaria, J., Butt, F., Balakrishnan, R., et al. (2010). Emergence of a new antibiotic resistance mechanism in India, Pakistan, and the UK: a molecular, biological, and epidemiological study. *Lancet Infect. Dis.* 10, 597–602. doi: 10.1016/S1473-3099(10)70143-2

Labarca, J., Poirel, L., Ozdamar, M., Turkoglu, S., Hakko, E., and Nordmann, P. (2014). KPC-producing *Klebsiella pneumoniae*, finally targeting Turkey. *New Microbes New Infect.* 2, 50–51. doi: 10.1002/nmi2.42

Lafeuille, E., Decre, D., Mahjoub-Messai, F., Bidet, P., Arlet, G., and Bingen, E. (2013). OXA-48 carbapenemase-producing *Klebsiella pneumoniae* isolated from Libyan patients. *Microb. Drug Resist.* 19, 491–497. doi: 10.1089/mdr.2012.0219

Laishram, S., Anandan, S., Devi, B. Y., Elakkiya, M., Priyanka, B., Bhuvaneshwari, T., et al. (2015). Determination of synergy between sulbactam, meropenem and colistin in carbapenem-resistant *Klebsiella pneumoniae* and *Acinetobacter baumannii* isolates and correlation with the molecular mechanism of resistance. *J. Chemother* doi: 10.1179/1973947815Y.0000000079 [Epub ahead of print].

Lascols, C., Peirano, G., Hackel, M., Laupland, K. B., and Pitout, J. D. (2013). Surveillance and molecular epidemiology of *Klebsiella pneumoniae* isolates that produce carbapenemases: first report of OXA-48-like enzymes in North America. *Antimicrob. Agents Chemother.* 57, 130–136. doi: 10.1128/AAC.01686-12

Lasserre, C., De Saint Martin, L., Cuzon, G., Bogaerts, P., Lamar, E., Glupczynski, Y., et al. (2015). Efficient detection of carbapenemase activity in enterobacteriaceae by matrix-assisted laser desorption ionization-time of flight mass spectrometry in less than 30 minutes. *J. Clin. Microbiol.* 53, 2163–2171. doi: 10.1128/JCM.03467-14

Lauretti, L., Riccio, M. L., Mazzariol, A., Cornaglia, G., Amicosante, G., Fontana, R., et al. (1999). Cloning and characterization of blaVIM, a new integron-borne metallo-β-lactamase gene from a *Pseudomonas aeruginosa* clinical isolate. *Antimicrob. Agents Chemother.* 43, 1584–1590.

Leavitt, A., Chmelnitsky, I., Carmeli, Y., and Navon-Venezia, S. (2010). Complete nucleotide sequence of KPC-3-encoding plasmid pKpQIL in the epidemic *Klebsiella pneumoniae* sequence type 258. *Antimicrob. Agents Chemother.* 54, 4493–4496. doi: 10.1128/AAC.00175-10

Leavitt, A., Navon-Venezia, S., Chmelnitsky, I., Schwaber, M. J., and Carmeli, Y. (2007). Emergence of KPC-2 and KPC-3 in carbapenem-resistant *Klebsiella pneumoniae* strains in an Israeli hospital. *Antimicrob. Agents Chemother.* 51, 3026–3029. doi: 10.1128/AAC.00299-07

Lee, C. R., Cho, I. H., Jeong, B. C., and Lee, S. H. (2013a). Strategies to minimize antibiotic resistance. *Int. J. Environ. Res. Public Health* 10, 4274–4305. doi: 10.3390/ijerph10094274

Lee, C. S., Vasoo, S., Hu, F., Patel, R., and Doi, Y. (2014). *Klebsiella pneumoniae* ST147 coproducing NDM-7 carbapenemase and RmtF 16S rRNA methyltransferase in Minnesota. *J. Clin. Microbiol.* 52, 4109–4110. doi: 10.1128/JCM.01404-14

Lee, J. H., Bae, I. K., and Lee, S. H. (2012a). New definitions of extended-spectrum β-lactamase conferring worldwide emerging antibiotic resistance. *Med. Res. Rev.* 32, 216–232. doi: 10.1002/med.20210

Lee, J. H., Jeong, S. H., Cha, S. S., and Lee, S. H. (2007). A lack of drugs for antibiotic-resistant Gram-negative bacteria. *Nat. Rev. Drug Discov.* 6, 938–939. doi: 10.1038/nrd2201-c1

Lee, J. H., Jeong, S. H., Cha, S. S., and Lee, S. H. (2009). New disturbing trend in antimicrobial resistance of gram-negative pathogens. *PLoS Pathog.* 5:e1000221. doi: 10.1371/journal.ppat.1000221

Lee, J. H., Lee, J. J., Park, K. S., and Lee, S. H. (2015a). Urgent need for b-lactam-β-lactamase inhibitors. *Lancet Infect. Dis.* 15, 876–877. doi: 10.1016/S1473-3099(15)00143-7

Lee, J. H., Park, A. M., Karim, C. R., Lee, C. R., and Lee, S. H. (2016). How to minimise antibiotic resistance. *Lancet Infect. Dis* 16, 17–18. doi: 10.1016/S1473-3099(15)00467-3

Lee, J. J., Lee, J. H., Kwon, D. B., Jeon, J. H., Park, K. S., Lee, C. R., et al. (2015b). Fast and accurate large-scale detection of β-lactamase genes conferring antibiotic resistance. *Antimicrob. Agents Chemother.* 59, 5967–5975. doi: 10.1128/AAC.04634-14

Lee, K., Hong, S. G., Park, Y. J., Lee, H. S., Song, W., Jeong, J., et al. (2005a). Evaluation of phenotypic screening methods for detecting plasmid-mediated AmpC β-lactamases-producing isolates of *Escherichia coli* and *Klebsiella pneumoniae*. *Diagn. Microbiol. Infect. Dis.* 53, 319–323. doi: 10.1016/j.diagmicrobio.2005.07.004

Lee, K., Lee, M., Shin, J. H., Lee, M. H., Kang, S. H., Park, A. J., et al. (2006a). Prevalence of plasmid-mediated AmpC β-lactamases in *Escherichia coli* and *Klebsiella pneumoniae* in Korea. *Microb. Drug Resist.* 12, 44–49. doi: 10.1089/mdr.2006.12.44

Lee, N. Y., Wu, J. J., Lin, S. H., Ko, W. C., Tsai, L. H., and Yan, J. J. (2012b). Characterization of carbapenem-nonsusceptible *Klebsiella pneumoniae* bloodstream isolates at a Taiwanese hospital: clinical impacts of lowered breakpoints for carbapenems. *Eur. J. Clin. Microbiol. Infect. Dis.* 31, 1941–1950. doi: 10.1007/s10096-011-1525-2

Lee, S. H., Jeong, S. H., and Cha, S. S. (2005b). Minimising antibiotic resistance. *Lancet Infect. Dis.* 5, 668–670. doi: 10.1016/S1473-3099(05)70247-4

Lee, S. H., Jeong, S. H., and Cha, S. S. (2006b). Screening for carbapenem-resistant Gram-negative bacteria. *Lancet Infect. Dis.* 6, 682–684. doi: 10.1016/S1473-3099(06)70607-7

Lee, T. D., Adie, K., McNabb, A., Purych, D., Mannan, K., Azana, R., et al. (2015c). Rapid Detection of KPC, NDM, and OXA-48-Like carbapenemases by Real-Time PCR from rectal swab surveillance samples. *J. Clin. Microbiol.* 53, 2731–2733. doi: 10.1128/JCM.01237-15

Lee, W., Chung, H. S., Lee, Y., Yong, D., Jeong, S. H., Lee, K., et al. (2013b). Comparison of matrix-assisted laser desorption ionization-time-of-flight mass spectrometry assay with conventional methods for detection of IMP-6, VIM-2, NDM-1, SIM-1, KPC-1, OXA-23, and OXA-51 carbapenemase-producing *Acinetobacter* spp., *Pseudomonas aeruginosa*, and *Klebsiella pneumoniae*. *Diagn. Microbiol. Infect. Dis.* 77, 227–230. doi: 10.1016/j.diagmicrobio.2013.07.005

Lefebvre, B., Levesque, S., Bourgault, A. M., Mulvey, M. R., Mataseje, L., Boyd, D., et al. (2015). Carbapenem non-susceptible enterobacteriaceae in Quebec, Canada: results of a laboratory surveillance program (2010-2012). *PLoS ONE* 10:e0125076. doi: 10.1371/journal.pone.0125076

Leung, V., Loo, V. G., Frenette, C., Domingo, M. C., Bourgault, A. M., Mulvey, M. R., et al. (2012). First Canadian outbreak of Enterobacteriaceae-expressing *Klebsiella pneumoniae* carbapenemase type 3. *Can. J. Infect. Dis. Med. Microbiol.* 23, 117–120.

Li, G., Wei, Q., Wang, Y., Du, X., Zhao, Y., and Jiang, X. (2011). Novel genetic environment of the plasmid-mediated KPC-3 gene detected in *Escherichia coli* and *Citrobacter freundii* isolates from China. *Eur. J. Clin. Microbiol. Infect. Dis.* 30, 575–580. doi: 10.1007/s10096-010-1124-7

Li, G., Zhang, Y., Bi, D., Shen, P., Ai, F., Liu, H., et al. (2015). First report of a clinical, multidrug-resistant Enterobacteriaceae isolate coharboring fosfomycin resistance gene fosA3 and carbapenemase gene blaKPC-2 on the same transposon, Tn1721. *Antimicrob. Agents Chemother.* 59, 338–343. doi: 10.1128/AAC.03061-14

Li, H., Zhang, J., Liu, Y., Zheng, R., Chen, H., Wang, X., et al. (2014). Molecular characteristics of carbapenemase-producing Enterobacteriaceae in China from 2008 to 2011: predominance of KPC-2 enzyme. *Diagn. Microbiol. Infect. Dis.* 78, 63–65. doi: 10.1016/j.diagmicrobio.2013.10.002

Li, J. J., Sheng, Z. K., Deng, M., Bi, S., Hu, F. S., Miao, H. F., et al. (2012). Epidemic of *Klebsiella pneumoniae* ST11 clone coproducing KPC-2 and 16S rRNA methylase RmtB in a Chinese University Hospital. *BMC Infect. Dis.* 12:373. doi: 10.1186/1471-2334-12-373

Liapis, E., Pantel, A., Robert, J., Nicolas-Chanoine, M. H., Cavalie, L., van der Mee-Marquet, N., et al. (2014). Molecular epidemiology of OXA-48-producing *Klebsiella pneumoniae* in France. *Clin. Microbiol. Infect.* 20, O1121–O1123. doi: 10.1111/1469-0691.12727

Lifshitz, Z., Adler, A., and Carmeli, Y. (2016). Comparative study of a novel biochemical assay, the rapidec carba NP Test, for detecting carbapenemase-producing Enterobacteriaceae. *J. Clin. Microbiol.* 54, 453–456. doi: 10.1128/JCM.02626-15

Limbago, B. M., Rasheed, J. K., Anderson, K. F., Zhu, W., Kitchel, B., Watz, N., et al. (2011). IMP-producing carbapenem-resistant *Klebsiella pneumoniae* in the United States. *J. Clin. Microbiol.* 49, 4239–4245. doi: 10.1128/JCM.05297-11

Ling, M. L., Tee, Y. M., Tan, S. G., Amin, I. M., How, K. B., Tan, K. Y., et al. (2015). Risk factors for acquisition of carbapenem resistant Enterobacteriaceae in an acute tertiary care hospital in Singapore. *Antimicrob. Resist. Infect. Control.* 4:26. doi: 10.1186/s13756-015-0066-3

Liu, P., Li, P., Jiang, X., Bi, D., Xie, Y., Tai, C., et al. (2012). Complete genome sequence of *Klebsiella pneumoniae* subsp. *pneumoniae HS*11286, a multidrug-resistant strain isolated from human sputum. *J. Bacteriol.* 194, 1841–1842. doi: 10.1128/JB.00043-12

Liu, Y., Wan, L. G., Deng, Q., Cao, X. W., Yu, Y., and Xu, Q. F. (2015). First description of NDM-1-, KPC-2-, VIM-2- and IMP-4-producing *Klebsiella pneumoniae* strains in a single Chinese teaching hospital. *Epidemiol. Infect.* 143, 376–384. doi: 10.1017/S0950268814000995

Liu, Y. Y., Wang, Y., Walsh, T. R., Yi, L. X., Zhang, R., Spencer, J., et al. (2016). Emergence of plasmid-mediated colistin resistance mechanism MCR-1 in animals and human beings in China: a microbiological and molecular biological study. *Lancet Infect. Dis.* 16, 161–168. doi: 10.1016/S1473-3099(15)00424-7

Livermore, D. M., Walsh, T. R., Toleman, M., and Woodford, N. (2011). Balkan NDM-1: escape or transplant? *Lancet Infect. Dis.* 11:164. doi: 10.1016/S1473-3099(11)70048-2

Lixandru, B. E., Cotar, A. I., Straut, M., Usein, C. R., Cristea, D., Ciontea, S., et al. (2015). Carbapenemase-Producing *Klebsiella pneumoniae* in Romania: a Six-Month survey. *PLoS ONE* 10:e0143214. doi: 10.1371/journal.pone.0143214

Lowe, C. F., Kus, J. V., Salt, N., Callery, S., Louie, L., Khan, M. A., et al. (2013). Nosocomial transmission of New Delhi metallo-b-lactamase-1-producing *Klebsiella pneumoniae* in Toronto, Canada. *Infect. Control Hosp. Epidemiol.* 34, 49–55. doi: 10.1086/668778

Lübbert, C., Lippmann, N., Busch, T., Kaisers, U. X., Ducomble, T., Eckmanns, T., et al. (2014). Long-term carriage of *Klebsiella pneumoniae* carbapenemase-2-producing *K. pneumoniae* after a large single-center outbreak in Germany. *Am. J. Infect. Control* 42, 376–380. doi: 10.1016/j.ajic.2013.12.001

Ma, L., Wang, J. T., Wu, T. L., Siu, L. K., Chuang, Y. C., Lin, J. C., et al. (2015). Emergence of OXA-48-producing *Klebsiella pneumoniae* in Taiwan. *PLoS ONE* 10:e0139152. doi: 10.1371/journal.pone.0139152

MacVane, S. H., Crandon, J. L., Nichols, W. W., and Nicolau, D. P. (2014). Unexpected in vivo activity of ceftazidime alone and in combination with avibactam against New Delhi metallo-β-lactamase-producing Enterobacteriaceae in a murine thigh infection model. *Antimicrob. Agents Chemother.* 58, 7007–7009. doi: 10.1128/AAC.02662-14

Mammeri, H., Guillon, H., Eb, F., and Nordmann, P. (2010). Phenotypic and biochemical comparison of the carbapenem-hydrolyzing activities of five plasmid-borne AmpC b-lactamases. *Antimicrob. Agents Chemother.* 54, 4556–4560. doi: 10.1128/AAC.01762-09

Markovska, R., Stoeva, T., Schneider, I., Boyanova, L., Popova, V., Dacheva, D., et al. (2015). Clonal dissemination of multilocus sequence type ST15 KPC-2-producing *Klebsiella pneumoniae* in Bulgaria. *APMIS* 123, 887–894. doi: 10.1111/apm.12433

Marquez, C., Ingold, A., Echeverria, N., Acevedo, A., Vignoli, R., Garcia-Fulgueiras, V., et al. (2014). Emergence of KPC-producing *Klebsiella pneumoniae* in Uruguay: infection control and molecular characterization. *New Microbes New Infect.* 2, 58–63. doi: 10.1002/nmi2.40

Mathers, A. J., Hazen, K. C., Carroll, J., Yeh, A. J., Cox, H. L., Bonomo, R. A., et al. (2013). First clinical cases of OXA-48-producing carbapenem-resistant *Klebsiella pneumoniae* in the United States: the "menace" arrives in the new world. *J. Clin. Microbiol.* 51, 680–683. doi: 10.1128/JCM.02580-12

McDermott, H., Morris, D., McArdle, E., O'Mahony, G., Kelly, S., Cormican, M., et al. (2012). Isolation of NDM-1-producing *Klebsiella* pnemoniae in Ireland, July 2011. *Euro. Surveill* 17, 20087.

Melegh, S., Kovacs, K., Gam, T., Nyul, A., Patko, B., Toth, A., et al. (2014). Emergence of VIM-4 metallo-b-lactamase-producing *Klebsiella pneumoniae* ST15 clone in the Clinical Centre University of Pecs, Hungary. *Clin. Microbiol. Infect.* 20, O27–O29. doi: 10.1111/1469-0691.12293

Metwally, L., Gomaa, N., Attallah, M., and Kamel, N. (2013). High prevalence of *Klebsiella pneumoniae* carbapenemase-mediated resistance in *K. pneumoniae* isolates from Egypt. *East Mediterr. Health J.* 19, 947–952.

Monaco, M., Giani, T., Raffone, M., Arena, F., Garcia-Fernandez, A., Pollini, S., et al. (2014). Colistin resistance superimposed to endemic carbapenem-resistant *Klebsiella pneumoniae*: a rapidly evolving problem in Italy,

November 2013 to April 2014. *Euro. Surveill.* 19:20939. doi: 10.2807/1560-7917.ES2014.19.42.20939

Moquet, O., Bouchiat, C., Kinana, A., Seck, A., Arouna, O., Bercion, R., et al. (2011). Class D OXA-48 carbapenemase in multidrug-resistant enterobacteria, Senegal. *Emerg. Infect. Dis.* 17, 143–144. doi: 10.3201/eid1701.100244

Morris, D., Boyle, F., Morris, C., Condon, I., Delannoy-Vieillard, A. S., Power, L., et al. (2012). Inter-hospital outbreak of *Klebsiella pneumoniae* producing KPC-2 carbapenemase in Ireland. *J. Antimicrob. Chemother.* 67, 2367–2372. doi: 10.1093/jac/dks239

Morris, D., O'Connor, M., Izdebski, R., Corcoran, M., Ludden, C. E., Mc, G. E., et al. (2016). Dissemination of clonally related multidrug-resistant *Klebsiella pneumoniae* in Ireland. *Epidemiol. Infect.* 144, 443–448. doi: 10.1017/S0950268815001041

Mulvey, M. R., Grant, J. M., Plewes, K., Roscoe, D., and Boyd, D. A. (2011). New Delhi metallo-β-lactamase in *Klebsiella pneumoniae* and *Escherichia coli*, Canada. *Emerg. Infect. Dis.* 17, 103–106. doi: 10.3201/eid1701.101358

Munoz-Price, L. S., Poirel, L., Bonomo, R. A., Schwaber, M. J., Daikos, G. L., Cormican, M., et al. (2013). Clinical epidemiology of the global expansion of *Klebsiella pneumoniae* carbapenemases. *Lancet Infect. Dis.* 13, 785–796. doi: 10.1016/S1473-3099(13)70190-7

Naas, T., Cotellon, G., Ergani, A., and Nordmann, P. (2013). Real-time PCR for detection of blaOXA-48 genes from stools. *J. Antimicrob. Chemother.* 68, 101–104. doi: 10.1093/jac/dks340

Naas, T., Cuzon, G., Villegas, M. V., Lartigue, M. F., Quinn, J. P., and Nordmann, P. (2008). Genetic structures at the origin of acquisition of the β-lactamase bla KPC gene. *Antimicrob. Agents Chemother.* 52, 1257–1263. doi: 10.1128/AAC.01451-07

Naas, T., Nordmann, P., Vedel, G., and Poyart, C. (2005). Plasmid-mediated carbapenem-hydrolyzing β-lactamase KPC in a *Klebsiella pneumoniae* isolate from France. *Antimicrob. Agents Chemother.* 49, 4423–4424. doi: 10.1128/AAC.49.10.4423-4424.2005

Nagano, N., Endoh, Y., Nagano, Y., Toyama, M., Matsui, M., Shibayama, K., et al. (2013). First report of OXA-48 carbapenemase-producing *Klebsiella pneumoniae* and *Escherichia coli* in Japan from a patient returned from Southeast Asia. *Jpn. J. Infect. Dis.* 66, 79–81. doi: 10.7883/yoken.66.79

Nakano, R., Nakano, A., Hikosaka, K., Kawakami, S., Matsunaga, N., Asahara, M., et al. (2014). First report of metallo-β-lactamase NDM-5-producing *Escherichia coli* in Japan. *Antimicrob. Agents Chemother.* 58, 7611–7612. doi: 10.1128/AAC.04265-14

Nakano, R., Nakano, A., Ishii, Y., Ubagai, T., Kikuchi-Ueda, T., Kikuchi, H., et al. (2015). Rapid detection of the *Klebsiella pneumoniae* carbapenemase (KPC) gene by loop-mediated isothermal amplification (LAMP). *J. Infect. Chemother.* 21, 202–206. doi: 10.1016/j.jiac.2014.11.010

Naseer, U., Eriksen, B. O., Sundsfjord, A., and Samuelsen, O. (2012). Fecal colonization of VIM-1-producing *Klebsiella pneumoniae* and in vivo transfer of multidrug-resistant IncN plasmid in a renal transplant patient. *Diagn. Microbiol. Infect. Dis.* 72, 363–366. doi: 10.1016/j.diagmicrobio.2011.12.010

Nathan, C., and Cars, O. (2014). Antibiotic resistance–problems, progress, and prospects. *N. Engl. J. Med.* 371, 1761–1763. doi: 10.1056/NEJMp1408040

Navarro-San Francisco, C., Mora-Rillo, M., Romero-Gómez, M. P., Moreno-Ramos, F., Rico-Nieto, A., Ruiz-Carrascoso, G., et al. (2013). Bacteraemia due to OXA-48-carbapenemase-producing *Enterobacteriaceae*: a major clinical challenge. *Clin. Microbiol. Infect.* 19, E72–E79. doi: 10.1111/1469-0691.12091

Netikul, T., Sidjabat, H. E., Paterson, D. L., Kamolvit, W., Tantisiriwat, W., Steen, J. A., et al. (2014). Characterization of an IncN2-type blaNDM-1-carrying plasmid in *Escherichia coli* ST131 and *Klebsiella pneumoniae* ST11 and ST15 isolates in Thailand. *J. Antimicrob. Chemother.* 69, 3161–3163. doi: 10.1093/jac/dku275

Nobari, S., Shahcheraghi, F., Rahmati Ghezelgeh, F., and Valizadeh, B. (2014). Molecular characterization of carbapenem-resistant strains of *Klebsiella pneumoniae* isolated from Iranian patients: first identification of blaKPC gene in Iran. *Microb. Drug Resist.* 20, 285–293. doi: 10.1089/mdr.2013.0074

Nordmann, P., Cuzon, G., and Naas, T. (2009). The real threat of *Klebsiella pneumoniae* carbapenemase-producing bacteria. *Lancet Infect. Dis.* 9, 228–236. doi: 10.1016/S1473-3099(09)70054-4

Nordmann, P., Dortet, L., and Poirel, L. (2012a). Carbapenem resistance in Enterobacteriaceae: here is the storm! *Trends Mol. Med.* 18, 263–272. doi: 10.1016/j.molmed.2012.03.003

Nordmann, P., Naas, T., and Poirel, L. (2011a). Global spread of Carbapenemase-producing Enterobacteriaceae. *Emerg. Infect. Dis.* 17, 1791–1798. doi: 10.3201/eid1710.110655

Nordmann, P., and Poirel, L. (2014). The difficult-to-control spread of carbapenemase producers among Enterobacteriaceae worldwide. *Clin. Microbiol. Infect.* 20, 821–830. doi: 10.1111/1469-0691.12719

Nordmann, P., Poirel, L., and Dortet, L. (2012b). Rapid detection of carbapenemase-producing Enterobacteriaceae. *Emerg. Infect. Dis.* 18, 1503–1507. doi: 10.3201/eid1809.120355

Nordmann, P., Poirel, L., Walsh, T. R., and Livermore, D. M. (2011b). The emerging NDM carbapenemases. *Trends Microbiol.* 19, 588–595. doi: 10.1016/j.tim.2011.09.005

Notomi, T., Okayama, H., Masubuchi, H., Yonekawa, T., Watanabe, K., Amino, N., et al. (2000). Loop-mediated isothermal amplification of DNA. *Nucleic Acids Res.* 28:E63. doi: 10.1093/nar/28.12.e63

Ocampo, A. M., Chen, L., Cienfuegos, A. V., Roncancio, G., Chavda, K. D., Kreiswirth, B. N., et al. (2015). High frequency of non-CG258 clones of carbapenem-resistant *Klebsiella pneumoniae* with distinct clinical characteristics: a two-year surveillance in five Colombian tertiary care hospitals. *Antimicrob. Agents Chemother.* 60, 332–342. doi: 10.1128/AAC.01775-15

Okeke, I. N., Peeling, R. W., Goossens, H., Auckenthaler, R., Olmsted, S. S., de Lavison, J. F., et al. (2011). Diagnostics as essential tools for containing antibacterial resistance. *Drug Resist. Updat.* 14, 95–106. doi: 10.1016/j.drup.2011.02.002

Osano, E., Arakawa, Y., Wacharotayankun, R., Ohta, M., Horii, T., Ito, H., et al. (1994). Molecular characterization of an enterobacterial metallo β-lactamase found in a clinical isolate of *Serratia marcescens* that shows imipenem resistance. *Antimicrob. Agents Chemother.* 38, 71–78. doi: 10.1128/AAC.38.1.71

Osterblad, M., Hakanen, A. J., and Jalava, J. (2014). Evaluation of the Carba NP test for carbapenemase detection. *Antimicrob. Agents Chemother.* 58, 7553–7556. doi: 10.1128/AAC.02761-13

Osterblad, M., Kirveskari, J., Hakanen, A. J., Tissari, P., Vaara, M., and Jalava, J. (2012). Carbapenemase-producing Enterobacteriaceae in Finland: the first years (2008–2011). *J. Antimicrob. Chemother.* 67, 2860–2864. doi: 10.1093/jac/dks299

Oteo, J., Hernandez, J. M., Espasa, M., Fleites, A., Saez, D., Bautista, V., et al. (2013a). Emergence of OXA-48-producing *Klebsiella pneumoniae* and the novel carbapenemases OXA-244 and OXA-245 in Spain. *J. Antimicrob. Chemother.* 68, 317–321. doi: 10.1093/jac/dks383

Oteo, J., Saez, D., Bautista, V., Fernandez-Romero, S., Hernandez-Molina, J. M., Perez-Vazquez, M., et al. (2013b). Carbapenemase-producing enterobacteriaceae in Spain in 2012. *Antimicrob. Agents Chemother.* 57, 6344–6347. doi: 10.1128/AAC.01513-13

Oteo, J., Ortega, A., Bartolome, R., Bou, G., Conejo, C., Fernandez-Martinez, M., et al. (2015). Prospective multicenter study of carbapenemase-producing Enterobacteriaceae from 83 hospitals in Spain reveals high in vitro susceptibility to colistin and meropenem. *Antimicrob. Agents Chemother.* 59, 3406–3412. doi: 10.1128/AAC.00086-15

Palacios-Baena, Z. R., Oteo, J., Conejo, C., Larrosa, M. N., Bou, G., Fernandez-Martinez, M., et al. (2016). Comprehensive clinical and epidemiological assessment of colonisation and infection due to carbapenemase-producing Enterobacteriaceae in Spain. *J. Infect.* 72, 152–160. doi: 10.1016/j.jinf.2015.10.008

Pano-Pardo, J. R., Ruiz-Carrascoso, G., Navarro-San Francisco, C., Gomez-Gil, R., Mora-Rillo, M., Romero-Gomez, M. P., et al. (2013). Infections caused by OXA-48-producing *Klebsiella pneumoniae* in a tertiary hospital in Spain in the setting of a prolonged, hospital-wide outbreak. *J. Antimicrob. Chemother.* 68, 89–96. doi: 10.1093/jac/dks364

Pantel, A., Richaud-Morel, B., Cazaban, M., Bouziges, N., Sotto, A., and Lavigne, J. P. (2016). Environmental persistence of OXA-48-producing *Klebsiella pneumoniae* in a French intensive care unit. *Am. J. Infect. Control* 44, 366–368. doi: 10.1016/j.ajic.2015.09.021

Papadimitriou-Olivgeris, M., Christofidou, M., Fligou, F., Bartzavali, C., Vrettos, T., Filos, K. S., et al. (2014). The role of colonization pressure in the dissemination of colistin or tigecycline resistant KPC-producing *Klebsiella pneumoniae* in critically ill patients. *Infection* 42, 883–890. doi: 10.1007/s15010-014-0653-x

Parisi, S. G., Bartolini, A., Santacatterina, E., Castellani, E., Ghirardo, R., Berto, A., et al. (2015). Prevalence of *Klebsiella pneumoniae* strains producing carbapenemases and increase of resistance to colistin in an Italian teaching hospital from January 2012 To December 2014. *BMC Infect. Dis.* 15:244. doi: 10.1186/s12879-015-0996-7

Park, M. J., Kim, T. K., Song, W., Kim, J. S., Kim, H. S., and Lee, J. (2013). An Increase in the clinical isolation of acquired AmpC β-lactamase-producing *Klebsiella pneumoniae* in Korea from 2007 to 2010. *Ann. Lab. Med.* 33, 353–355. doi: 10.3343/alm.2013.33.5.353

Partridge, S. R., Ginn, A. N., Wiklendt, A. M., Ellem, J., Wong, J. S., Ingram, P., et al. (2015). Emergence of blaKPC carbapenemase genes in Australia. *Int. J. Antimicrob. Agents* 45, 130–136. doi: 10.1016/j.ijantimicag.2014.10.006

Partridge, S. R., and Iredell, J. R. (2012). Genetic contexts of blaNDM-1. *Antimicrob. Agents Chemother.* 56, 6065–6067. doi: 10.1128/AAC.00117-12 author reply 6071,

Pasteran, F., Albornoz, E., Faccone, D., Gomez, S., Valenzuela, C., Morales, M., et al. (2012). Emergence of NDM-1-producing *Klebsiella pneumoniae* in Guatemala. *J. Antimicrob. Chemother.* 67, 1795–1797. doi: 10.1093/jac/dks101

Pasteran, F., Tijet, N., Melano, R. G., and Corso, A. (2015). Simplified protocol for carba NP Test for enhanced detection of carbapenemase producers directly from bacterial cultures. *J. Clin. Microbiol.* 53, 3908–3911. doi: 10.1128/JCM.02032-15

Pasteran, F. G., Otaegui, L., Guerriero, L., Radice, G., Maggiora, R., Rapoport, M., et al. (2008). *Klebsiella pneumoniae* carbapenemase-2, buenos aires, argentina. *Emerg. Infect. Dis.* 14, 1178–1180. doi: 10.3201/eid1407.070826

Patel, R. (2013). Matrix-assisted laser desorption ionization-time of flight mass spectrometry in clinical microbiology. *Clin. Infect. Dis.* 57, 564–572. doi: 10.1093/cid/cit247

Peirano, G., Ahmed-Bentley, J., Fuller, J., Rubin, J. E., and Pitout, J. D. (2014). Travel-related carbapenemase-producing Gram-negative bacteria in Alberta, Canada: the first 3 years. *J. Clin. Microbiol.* 52, 1575–1581. doi: 10.1128/JCM.00162-14

Peirano, G., Pillai, D. R., Pitondo-Silva, A., Richardson, D., and Pitout, J. D. (2011). The characteristics of NDM-producing *Klebsiella pneumoniae* from Canada. *Diagn. Microbiol. Infect. Dis.* 71, 106–109. doi: 10.1016/j.diagmicrobio.2011.06.013

Peirano, G., Seki, L. M., Val Passos, V. L., Pinto, M. C., Guerra, L. R., and Asensi, M. D. (2009). Carbapenem-hydrolysing β-lactamase KPC-2 in *Klebsiella pneumoniae* isolated in Rio de Janeiro. *Braz. J. Antimicrob. Chemother.* 63, 265–268. doi: 10.1093/jac/dkn484

Pena, I., Picazo, J. J., Rodriguez-Avial, C., and Rodriguez-Avial, I. (2014). Carbapenemase-producing Enterobacteriaceae in a tertiary hospital in Madrid, Spain: high percentage of colistin resistance among VIM-1-producing *Klebsiella pneumoniae* ST11 isolates. *Int. J. Antimicrob. Agents* 43, 460–464. doi: 10.1016/j.ijantimicag.2014.01.021

Pereira, P. S., Borghi, M., Albano, R. M., Lopes, J. C., Silveira, M. C., Marques, E. A., et al. (2015). Coproduction of NDM-1 and KPC-2 in Enterobacter hormaechei from Brazil. *Microb. Drug Resist.* 21, 234–236. doi: 10.1089/mdr.2014.0171

Perilli, M., Bottoni, C., Grimaldi, A., Segatore, B., Celenza, G., Mariani, M., et al. (2013). Carbapenem-resistant *Klebsiella pneumoniae* harbouring blaKPC-3 and blaVIM-2 from central Italy. *Diagn. Microbiol. Infect. Dis.* 75, 218–221. doi: 10.1016/j.diagmicrobio.2012.10.008

Pesesky, M. W., Hussain, T., Wallace, M., Wang, B., Andleeb, S., Burnham, C. A., et al. (2015). KPC and NDM-1 genes in related Enterobacteriaceae strains and plasmids from Pakistan and the United States. *Emerg. Infect. Dis.* 21, 1034–1037. doi: 10.3201/eid2106.141504

Peter, H., Berggrav, K., Thomas, P., Pfeifer, Y., Witte, W., Templeton, K., et al. (2012). Direct detection and genotyping of *Klebsiella pneumoniae* carbapenemases from urine by use of a new DNA microarray test. *J. Clin. Microbiol.* 50, 3990–3998. doi: 10.1128/JCM.00990-12

Pfeifer, Y., Schlatterer, K., Engelmann, E., Schiller, R. A., Frangenberg, H. R., Stiewe, D., et al. (2012). Emergence of OXA-48-type carbapenemase-producing Enterobacteriaceae in German hospitals. *Antimicrob. Agents Chemother.* 56, 2125–2128. doi: 10.1128/AAC.05315-11

Picao, R. C., Santos, A. F., Nicoletti, A. G., Furtado, G. H., and Gales, A. C. (2010). Detection of GES-5-producing *Klebsiella pneumoniae* in Brazil. *J. Antimicrob. Chemother.* 65, 796–797. doi: 10.1093/jac/dkq024

Pires, J., Novais, A., and Peixe, L. (2013). Blue-carba, an easy biochemical test for detection of diverse carbapenemase producers directly from bacterial cultures. *J. Clin. Microbiol.* 51, 4281–4283. doi: 10.1128/JCM.01634-13

Pitart, C., Sole, M., Roca, I., Fabrega, A., Vila, J., and Marco, F. (2011). First outbreak of a plasmid-mediated carbapenem-hydrolyzing OXA-48 β-lactamase in *Klebsiella pneumoniae* in Spain. *Antimicrob. Agents Chemother.* 55, 4398–4401. doi: 10.1128/AAC.00329-11

Pitout, J. D., Nordmann, P., and Poirel, L. (2015). Carbapenemase-Producing *Klebsiella pneumoniae*, a Key pathogen set for global nosocomial dominance. *Antimicrob. Agents Chemother.* 59, 5873–5884. doi: 10.1128/AAC.01019-15

Poirel, L., Al Maskari, Z., Al Rashdi, F., Bernabeu, S., and Nordmann, P. (2011a). NDM-1-producing *Klebsiella pneumoniae* isolated in the Sultanate of Oman. *J. Antimicrob. Chemother.* 66, 304–306. doi: 10.1093/jac/dkq428

Poirel, L., Benouda, A., Hays, C., and Nordmann, P. (2011b). Emergence of NDM-1-producing *Klebsiella pneumoniae* in Morocco. *J. Antimicrob. Chemother.* 66, 2781–2783. doi: 10.1093/jac/dkr384

Poirel, L., Castanheira, M., Carrer, A., Rodriguez, C. P., Jones, R. N., Smayevsky, J., et al. (2011c). OXA-163, an OXA-48-related class D β-lactamase with extended activity toward expanded-spectrum cephalosporins. *Antimicrob. Agents Chemother.* 55, 2546–2551. doi: 10.1128/AAC.00022-11

Poirel, L., Dortet, L., Bernabeu, S., and Nordmann, P. (2011d). Genetic features of blaNDM-1-positive Enterobacteriaceae. *Antimicrob. Agents Chemother.* 55, 5403–5407. doi: 10.1128/AAC.00585-11

Poirel, L., Fortineau, N., and Nordmann, P. (2011e). International transfer of NDM-1-producing *Klebsiella pneumoniae* from Iraq to France. *Antimicrob. Agents Chemother.* 55, 1821–1822. doi: 10.1128/AAC.01761-10

Poirel, L., Revathi, G., Bernabeu, S., and Nordmann, P. (2011f). Detection of NDM-1-producing *Klebsiella pneumoniae* in Kenya. *Antimicrob. Agents Chemother.* 55, 934–936. doi: 10.1128/AAC.01247-10

Poirel, L., Schrenzel, J., Cherkaoui, A., Bernabeu, S., Renzi, G., and Nordmann, P. (2011g). Molecular analysis of NDM-1-producing enterobacterial isolates from Geneva, Switzerland. *J. Antimicrob. Chemother.* 66, 1730–1733. doi: 10.1093/jac/dkr174

Poirel, L., Carbonnelle, E., Bernabeu, S., Gutmann, L., Rotimi, V., and Nordmann, P. (2012a). Importation of OXA-48-producing *Klebsiella pneumoniae* from Kuwait. *J. Antimicrob. Chemother.* 67, 2051–2052. doi: 10.1093/jac/dks167

Poirel, L., Lascols, C., Bernabeu, S., and Nordmann, P. (2012b). NDM-1-producing *Klebsiella pneumoniae* in mauritius. *Antimicrob. Agents Chemother.* 56, 598–599. doi: 10.1128/AAC.05639-11

Poirel, L., Potron, A., and Nordmann, P. (2012c). OXA-48-like carbapenemases: the phantom menace. *J. Antimicrob. Chemother.* 67, 1597–1606. doi: 10.1093/jac/dks121

Poirel, L., Heritier, C., Podglajen, I., Sougakoff, W., Gutmann, L., and Nordmann, P. (2003). Emergence in *Klebsiella pneumoniae* of a chromosome-encoded SHV β-lactamase that compromises the efficacy of imipenem. *Antimicrob. Agents Chemother.* 47, 755–758. doi: 10.1128/AAC.47.2.755-758.2003

Poirel, L., Heritier, C., Tolun, V., and Nordmann, P. (2004). Emergence of oxacillinase-mediated resistance to imipenem in *Klebsiella pneumoniae*. *Antimicrob. Agents Chemother.* 48, 15–22. doi: 10.1128/AAC.48.1.15-22.2004

Poirel, L., Kieffer, N., and Nordmann, P. (2016). In vitro evaluation of dual carbapenem combinations against carbapenemase-producing Enterobacteriaceae. *J. Antimicrob. Chemother.* 71, 156–161. doi: 10.1093/jac/dkv294

Poirel, L., and Nordmann, P. (2015). Rapidec carba NP Test for rapid detection of carbapenemase producers. *J. Clin. Microbiol.* 53, 3003–3008. doi: 10.1128/JCM.00977-15

Poirel, L., Pitout, J. D., and Nordmann, P. (2007). Carbapenemases: molecular diversity and clinical consequences. *Future Microbiol.* 2, 501–512. doi: 10.2217/17460913.2.5.501

Poirel, L., Yilmaz, M., Istanbullu, A., Arslan, F., Mert, A., Bernabeu, S., et al. (2014). Spread of NDM-1-producing Enterobacteriaceae in a neonatal intensive care unit in Istanbul, Turkey. *Antimicrob. Agents Chemother.* 58, 2929–2933. doi: 10.1128/AAC.02047-13

Potron, A., Nordmann, P., Lafeuille, E., Al Maskari, Z., Al Rashdi, F., and Poirel, L. (2011). Characterization of OXA-181, a carbapenem-hydrolyzing class D β-lactamase from *Klebsiella pneumoniae. Antimicrob. Agents Chemother.* 55, 4896–4899. doi: 10.1128/AAC.00481-11

Potron, A., Poirel, L., and Nordmann, P. (2014). Derepressed transfer properties leading to the efficient spread of the plasmid encoding carbapenemase OXA-48. *Antimicrob. Agents Chemother.* 58, 467–471. doi: 10.1128/AAC.01344-13

Potron, A., Poirel, L., Rondinaud, E., and Nordmann, P. (2013). Intercontinental spread of OXA-48 β-lactamase-producing Enterobacteriaceae over a 11-year period, 2001 to 2011. *Euro. Surveill.* 18:20549. doi: 10.2807/1560-7917.ES2013.18.31.20549

Potron, A., Schrenzel, J., Poirel, L., Renzi, G., Cherkaoui, A., and Nordmann, P. (2012). Emergence of OXA-48-producing Enterobacteriaceae in Switzerland. *Int. J. Antimicrob. Agents* 40, 563–564. doi: 10.1016/j.ijantimicag.2012.07.003

Pournaras, S., Poulou, A., and Tsakris, A. (2010a). Inhibitor-based methods for the detection of KPC carbapenemase-producing Enterobacteriaceae in clinical practice by using boronic acid compounds. *J. Antimicrob. Chemother.* 65, 1319–1321. doi: 10.1093/jac/dkq124

Pournaras, S., Poulou, A., Voulgari, E., Vrioni, G., Kristo, I., and Tsakris, A. (2010b). Detection of the new metallo-β-lactamase VIM-19 along with KPC-2, CMY-2 and CTX-M-15 in *Klebsiella pneumoniae. J. Antimicrob. Chemother.* 65, 1604–1607. doi: 10.1093/jac/dkq190

Qi, Y., Wei, Z., Ji, S., Du, X., Shen, P., and Yu, Y. (2011). ST11, the dominant clone of KPC-producing *Klebsiella pneumoniae* in China. *J. Antimicrob. Chemother.* 66, 307–312. doi: 10.1093/jac/dkq431

Qin, S., Fu, Y., Zhang, Q., Qi, H., Wen, J. G., Xu, H., et al. (2014). High incidence and endemic spread of NDM-1-positive Enterobacteriaceae in Henan Province, China. *Antimicrob. Agents Chemother.* 58, 4275–4282. doi: 10.1128/AAC.02813-13

Quiles, M. G., Rocchetti, T. T., Fehlberg, L. C., Kusano, E. J., Chebabo, A., Pereira, R. M., et al. (2015). Unusual association of NDM-1 with KPC-2 and armA among Brazilian Enterobacteriaceae isolates. *Braz. J. Med. Biol. Res.* 48, 174–177. doi: 10.1590/1414-431X20144154

Quinones, D., Hart, M., Espinosa, F., Garcia, S., Carmona, Y., Ghosh, S., et al. (2014). Emergence of *Klebsiella pneumoniae* clinical isolates producing KPC-2 carbapenemase in Cuba. *New Microbes New Infect.* 2, 123–126. doi: 10.1002/nmi2.54

Rajabnia, R., Asgharpour, F., Ferdosi Shahandashti, E., and Moulana, Z. (2015). Nosocomial emerging of (VIM1) carbapenemase-producing isolates of in North of Iran. *Iran J. Microbiol.* 7, 88–93.

Rasheed, J. K., Kitchel, B., Zhu, W., Anderson, K. F., Clark, N. C., Ferraro, M. J., et al. (2013). New Delhi metallo-β-lactamase-producing Enterobacteriaceae, United States. *Emerg. Infect. Dis.* 19, 870–878. doi: 10.3201/eid1906.121515

Reddy, A. K., Balne, P. K., Reddy, R. K., Mathai, A., and Kaur, I. (2010). Development and evaluation of loop-mediated isothermal amplification assay for rapid and inexpensive detection of cytomegalovirus DNA in vitreous specimens from suspected cases of viral retinitis. *J. Clin. Microbiol.* 48, 2050–2052. doi: 10.1128/JCM.02248-09

Ribeiro, S. M., de la Fuente-Nunez, C., Baquir, B., Faria-Junior, C., Franco, O. L., and Hancock, R. E. (2015). Antibiofilm peptides increase the susceptibility of carbapenemase-producing *Klebsiella pneumoniae* clinical isolates to b-lactam antibiotics. *Antimicrob. Agents Chemother.* 59, 3906–3912. doi: 10.1128/AAC.00092-15

Richter, S. N., Frasson, I., Bergo, C., Parisi, S., Cavallaro, A., and Palù, G. (2011). Transfer of KPC-2 Carbapenemase from *Klebsiella pneumoniae* to *Escherichia coli* in a patient: first case in Europe. *J. Clin. Microbiol.* 49, 2040–2042. doi: 10.1128/JCM.00133-11

Richter, S. N., Frasson, I., Franchin, E., Bergo, C., Lavezzo, E., Barzon, L., et al. (2012). KPC-mediated resistance in *Klebsiella pneumoniae* in two hospitals in Padua, Italy, June 2009-December 2011: massive spreading of a KPC-3-encoding plasmid and involvement of non-intensive care units. *Gut Pathog.* 4:7. doi: 10.1186/1757-4749-4-7

Rimrang, B., Chanawong, A., Lulitanond, A., Wilailuckana, C., Charoensri, N., Sribenjalux, P., et al. (2012). Emergence of NDM-1- and IMP-14a-producing Enterobacteriaceae in Thailand. *J. Antimicrob. Chemother.* 67, 2626–2630. doi: 10.1093/jac/dks267

Robert, J., Pantel, A., Merens, A., Lavigne, J. P., Nicolas-Chanoine, M. H., and Group, O. N. S. C. R. S. (2014). Incidence rates of carbapenemase-producing Enterobacteriaceae clinical isolates in France: a prospective nationwide study in 2011-12. *J. Antimicrob. Chemother.* 69, 2706–2712. doi: 10.1093/jac/dku208

Robustillo Rodela, A., Diaz-Agero Perez, C., Sanchez Sagrado, T., Ruiz-Garbajosa, P., Pita Lopez, M. J., and Monge, V. (2012). Emergence and outbreak of carbapenemase-producing KPC-3 *Klebsiella pneumoniae* in Spain, September 2009 to February 2010: control measures. *Euro. Surveill.* 17:20086.

Roche, C., Cotter, M., O Connell, N., and Crowley, B. (2009). First identification of class A carbapenemase-producing *Klebsiella pneumoniae* in the Republic of Ireland. *Euro. Surveill.* 14:19163.

Rock, C., Thom, K. A., Masnick, M., Johnson, J. K., Harris, A. D., and Morgan, D. J. (2014). Frequency of *Klebsiella pneumoniae* carbapenemase (KPC)-producing and non-KPC-producing *Klebsiella* species contamination of healthcare workers and the environment. *Infect. Control Hosp. Epidemiol.* 35, 426–429. doi: 10.1086/675598

Rodriguez-Avial, I., Pena, I., Picazo, J. J., Rodriguez-Avial, C., and Culebras, E. (2015). In vitro activity of the next-generation aminoglycoside plazomicin alone and in combination with colistin, meropenem, fosfomycin or tigecycline against carbapenemase-producing Enterobacteriaceae strains. *Int. J. Antimicrob. Agents* 46, 616–621. doi: 10.1016/j.ijantimicag.2015.07.021

Rodriguez-Martinez, J. M., Nordmann, P., Fortineau, N., and Poirel, L. (2010). VIM-19, a metallo-β-lactamase with increased carbapenemase activity from *Escherichia coli* and *Klebsiella pneumoniae. Antimicrob. Agents Chemother.* 54, 471–476. doi: 10.1128/AAC.00458-09

Rojas, L. J., Mojica, M. F., Blanco, V. M., Correa, A., Montealegre, M. C., De La Cadena, E., et al. (2013). Emergence of *Klebsiella pneumoniae* Coharboring KPC and VIM Carbapenemases in Colombia. *Antimicrob. Agents Chemother.* 57, 1101–1102. doi: 10.1128/AAC.01666-12

Rose, W. E., and Rybak, M. J. (2006). Tigecycline: first of a new class of antimicrobial agents. *Pharmacotherapy* 26, 1099–1110. doi: 10.1592/phco.26.8.1099

Rosenblum, R., Khan, E., Gonzalez, G., Hasan, R., and Schneiders, T. (2011). Genetic regulation of the ramA locus and its expression in clinical isolates of *Klebsiella pneumoniae. Int. J. Antimicrob. Agents* 38, 39–45. doi: 10.1016/j.ijantimicag.2011.02.012

Ruiz-Garbajosa, P., Curiao, T., Tato, M., Gijon, D., Pintado, V., Valverde, A., et al. (2013). Multiclonal dispersal of KPC genes following the emergence of non-ST258 KPC-producing *Klebsiella pneumoniae* clones in Madrid, Spain. *J. Antimicrob. Chemother.* 68, 2487–2492. doi: 10.1093/jac/dkt237

Ruzin, A., Visalli, M. A., Keeney, D., and Bradford, P. A. (2005). Influence of transcriptional activator RamA on expression of multidrug efflux pump AcrAB and tigecycline susceptibility in *Klebsiella pneumoniae. Antimicrob. Agents Chemother.* 49, 1017–1022. doi: 10.1128/AAC.49.3.1017-1022.2005

Saito, R., Takahashi, R., Sawabe, E., Koyano, S., Takahashi, Y., Shima, M., et al. (2014). First report of KPC-2 Carbapenemase-producing *Klebsiella pneumoniae* in Japan. *Antimicrob. Agents Chemother.* 58, 2961–2963. doi: 10.1128/AAC.02072-13

Samuelsen, O., Naseer, U., Tofteland, S., Skutlaberg, D. H., Onken, A., Hjetland, R., et al. (2009). Emergence of clonally related *Klebsiella pneumoniae* isolates of sequence type 258 producing plasmid-mediated KPC carbapenemase in Norway and Sweden. *J. Antimicrob. Chemother.* 63, 654–658. doi: 10.1093/jac/dkp018

Sattar, H., Toleman, M., Nahid, F., and Zahra, R. (2014). Co-existence of blaNDM-1 and blaKPC-2 in clinical isolates of *Klebsiella pneumoniae* from Pakistan. *J. Chemother.* doi: 10.1179/1973947814Y.0000000223 [Epub ahead of print]

Sautrey, G., Duval, R. E., Chevalley, A., Fontanay, S., and Clarot, I. (2015). Capillary electrophoresis for fast detection of heterogeneous population in colistin-resistant Gram-negative bacteria. *Electrophoresis* 36, 2630–2633. doi: 10.1002/elps.201500064

Seara, N., Oteo, J., Carrillo, R., Perez-Blanco, V., Mingorance, J., Gomez-Gil, R., et al. (2015). Interhospital spread of NDM-7-producing *Klebsiella pneumoniae* belonging to ST437 in Spain. *Int. J. Antimicrob. Agents* 46, 169–173. doi: 10.1016/j.ijantimicag.2015.04.001

Seiffert, S. N., Marschall, J., Perreten, V., Carattoli, A., Furrer, H., and Endimiani, A. (2014). Emergence of *Klebsiella pneumoniae* co-producing NDM-1, OXA-48, CTX-M-15, CMY-16, QnrA and ArmA in Switzerland. *Int. J. Antimicrob. Agents* 44, 260–262. doi: 10.1016/j.ijantimicag.2014.05.008

Semin-Pelletier, B., Cazet, L., Bourigault, C., Juvin, M. E., Boutoille, D., Raffi, F., et al. (2015). Challenges of controlling a large outbreak of OXA-48 carbapenemase-producing *Klebsiella pneumoniae* in a French university hospital. *J. Hosp. Infect.* 89, 248–253. doi: 10.1016/j.jhin.2014.11.018

Shahcheraghi, F., Nobari, S., Rahmati Ghezelgeh, F., Nasiri, S., Owlia, P., Nikbin, V. S., et al. (2013). First report of New Delhi metallo-β-lactamase-1-producing *Klebsiella pneumoniae* in Iran. *Microb. Drug Resist.* 19, 30–36. doi: 10.1089/mdr.2012.0078

Shanmugam, P., Meenakshisundaram, J., and Jayaraman, P. (2013). blaKPC gene detection in clinical isolates of carbapenem resistant enterobacteriaceae in a tertiary care hospital. *J. Clin. Diagn. Res.* 7, 2736–2738. doi: 10.7860/JCDR/2013/7759.3747

Shen, P., Wei, Z., Jiang, Y., Du, X., Ji, S., Yu, Y., et al. (2009). Novel genetic environment of the carbapenem-hydrolyzing β-lactamase KPC-2 among Enterobacteriaceae in China. *Antimicrob. Agents Chemother.* 53, 4333–4338. doi: 10.1128/AAC.00260-09

Shibl, A., Al-Agamy, M., Memish, Z., Senok, A., Khader, S. A., and Assiri, A. (2013). The emergence of OXA-48- and NDM-1-positive *Klebsiella pneumoniae* in Riyadh, Saudi Arabia. *Int. J. Infect. Dis.* 17:e1130-3. doi: 10.1016/j.ijid.2013.06.016

Shin, S. Y., Bae, I. K., Kim, J., Jeong, S. H., Yong, D., Kim, J. M., et al. (2012). Resistance to carbapenems in sequence type 11 *Klebsiella pneumoniae* is related to DHA-1 and loss of OmpK35 and/or OmpK36. *J. Med. Microbiol.* 61, 239–245. doi: 10.1099/jmm.0.037036-0

Shoma, S., Kamruzzaman, M., Ginn, A. N., Iredell, J. R., and Partridge, S. R. (2014). Characterization of multidrug-resistant *Klebsiella pneumoniae* from Australia carrying blaNDM-1. *Diagn. Microbiol. Infect. Dis.* 78, 93–97. doi: 10.1016/j.diagmicrobio.2013.08.001

Sidjabat, H. E., Townell, N., Nimmo, G. R., George, N. M., Robson, J., Vohra, R., et al. (2015). Dominance of IMP-4-producing enterobacter cloacae among carbapenemase-producing Enterobacteriaceae in Australia. *Antimicrob. Agents Chemother.* 59, 4059–4066. doi: 10.1128/AAC.04378-14

Solanki, R., Vanjari, L., Ede, N., Gungi, A., Soory, A., and Vemu, L. (2013). Evaluation of LAMP assay using phenotypic tests and conventional PCR for detection of blaNDM-1 and blaKPC genes among carbapenem-resistant clinical Gram-negative isolates. *J. Med. Microbiol.* 62, 1540–1544. doi: 10.1099/jmm.0.059907-0

Solanki, R., Vanjari, L., Subramanian, S., B, A., E, N., and Lakshmi, V. (2014). Comparative evaluation of multiplex PCR and routine laboratory phenotypic methods for detection of carbapenemases among gram negative bacilli. *J. Clin. Diagn. Res* 8, DC23–DC26. doi: 10.7860/JCDR/2014/10794.5322

Sonnevend, A., Al Baloushi, A., Ghazawi, A., Hashmey, R., Girgis, S., Hamadeh, M. B., et al. (2013). Emergence and spread of NDM-1 producer Enterobacteriaceae with contribution of IncX3 plasmids in the United Arab Emirates. *J. Med. Microbiol.* 62, 1044–1050. doi: 10.1099/jmm.0.059014-0

Sonnevend, A., Ghazawi, A., Darwish, D., AlDeesi, Z., Kadhum, A. F., and Pal, T. (2015a). Characterization of KPC-type carbapenemase-producing *Klebsiella pneumoniae* strains isolated in the Arabian Peninsula. *J. Antimicrob. Chemother.* 70, 1592–1593. doi: 10.1093/jac/dku576

Sonnevend, A., Ghazawi, A. A., Hashmey, R., Jamal, W., Rotimi, V. O., Shibl, A. M., et al. (2015b). Characterization of carbapenem-resistant enterobacteriaceae with high rate of autochthonous transmission in the Arabian Peninsula. *PLoS ONE* 10:e0131372. doi: 10.1371/journal.pone.0131372

Spanu, T., Fiori, B., D'Inzeo, T., Canu, G., Campoli, S., Giani, T., et al. (2012). Evaluation of the new NucliSENS EasyQ KPC test for rapid detection of *Klebsiella pneumoniae* carbapenemase genes (blaKPC). *J. Clin. Microbiol.* 50, 2783–2785. doi: 10.1128/JCM.00284-12

Spyropoulou, A., Bartzavali, C., Vamvakopoulou, S., Marangos, M., Anastassiou, E. D., Spiliopoulou, I., et al. (2016). The first NDM Metallo-β-lactamase producing *Klebsiella pneumoniae* isolate in a University Hospital of Southwestern Greece. *J. Chemother.* doi: 10.1179/1973947815Y.0000000003 [Epub ahead of print].

Steinmann, J., Kaase, M., Gatermann, S., Popp, W., Steinmann, E., Damman, M., et al. (2011). Outbreak due to a *Klebsiella pneumoniae* strain harbouring KPC-2 and VIM-1 in a German university hospital, July 2010 to January 2011. *Euro. Surveill.* 16:19944.

Stoesser, N., Giess, A., Batty, E. M., Sheppard, A. E., Walker, A. S., Wilson, D. J., et al. (2014). Genome sequencing of an extended series of NDM-producing *Klebsiella pneumoniae* isolates from neonatal infections in a Nepali hospital characterizes the extent of community- versus hospital-associated transmission in an endemic setting. *Antimicrob. Agents Chemother.* 58, 7347–7357. doi: 10.1128/AAC.03900-14

Studentova, V., Dobiasova, H., Hedlova, D., Dolejska, M., Papagiannitsis, C. C., and Hrabak, J. (2015). Complete nucleotide sequences of two NDM-1-encoding plasmids from the same sequence type 11 *Klebsiella pneumoniae* strain. *Antimicrob. Agents Chemother.* 59, 1325–1328. doi: 10.1128/AAC.04095-14

Sun, Y., Cai, Y., Liu, X., Bai, N., Liang, B., and Wang, R. (2013). The emergence of clinical resistance to tigecycline. *Int. J. Antimicrob. Agents* 41, 110–116. doi: 10.1016/j.ijantimicag.2012.09.005

Swayne, R., Ellington, M. J., Curran, M. D., Woodford, N., and Aliyu, S. H. (2013). Utility of a novel multiplex TaqMan PCR assay for metallo-β-lactamase genes plus other TaqMan assays in detecting genes encoding serine carbapenemases and clinically significant extended-spectrum β-lactamases. *Int. J. Antimicrob. Agents* 42, 352–356. doi: 10.1016/j.ijantimicag.2013.06.018

Székely, E., Damjanova, I., Jánvári, L., Vas, K. E., Molnár, S., Bilca, D. V., et al. (2013). First description of bla_{NDM-1}, bla_{OXA-48}, $bla_{OXA-181}$ producing *Enterobacteriaceae* strains in Romania. *Int. J. Med. Microbiol.* 303, 697–700. doi: 10.1016/j.ijmm.2013.10.001

Tada, T., Miyoshi-Akiyama, T., Dahal, R. K., Mishra, S. K., Ohara, H., Shimada, K., et al. (2013). Dissemination of multidrug-resistant *Klebsiella pneumoniae* clinical isolates with various combinations of carbapenemases (NDM-1 and OXA-72) and 16S rRNA methylases (ArmA, RmtC and RmtF) in Nepal. *Int. J. Antimicrob. Agents* 42, 372–374. doi: 10.1016/j.ijantimicag.2013.06.014

Tamayo, M., Santiso, R., Otero, F., Bou, G., Lepe, J. A., McConnell, M. J., et al. (2013). Rapid determination of colistin resistance in clinical strains of *Acinetobacter baumannii* by use of the micromax assay. *J. Clin. Microbiol.* 51, 3675–3682. doi: 10.1128/JCM.01787-13

Tangden, T., and Giske, C. G. (2015). Global dissemination of extensively drug-resistant carbapenemase-producing Enterobacteriaceae: clinical perspectives on detection, treatment and infection control. *J. Intern. Med.* 277, 501–512. doi: 10.1111/joim.12342

Tangden, T., Hickman, R. A., Forsberg, P., Lagerback, P., Giske, C. G., and Cars, O. (2014). Evaluation of double- and triple-antibiotic combinations for VIM- and NDM-producing *Klebsiella pneumoniae* by in vitro time-kill experiments. *Antimicrob. Agents Chemother.* 58, 1757–1762. doi: 10.1128/AAC.00741-13

Tascini, C., Tagliaferri, E., Giani, T., Leonildi, A., Flammini, S., Casini, B., et al. (2013). Synergistic activity of colistin plus rifampin against colistin-resistant KPC-producing *Klebsiella pneumoniae*. *Antimicrob. Agents Chemother.* 57, 3990–3993. doi: 10.1128/AAC.00179-13

Temkin, E., Adler, A., Lerner, A., and Carmeli, Y. (2014). Carbapenem-resistant Enterobacteriaceae: biology, epidemiology, and management. *Ann. N. Y. Acad. Sci.* 1323, 22–42. doi: 10.1111/nyas.12537

Teo, J. W. P., Kurup, A., Lin, R. T. P., and Hsien, K. T. (2013). Emergence of clinical *Klebsiella pneumoniae* producing OXA-232 carbapenemase in Singapore. *New Microbes New Infect.* 1, 13–15. doi: 10.1002/2052-2975.4

Thomas, C. P., Moore, L. S., Elamin, N., Doumith, M., Zhang, J., Maharjan, S., et al. (2013). Early (2008–2010) hospital outbreak of *Klebsiella pneumoniae* producing OXA-48 carbapenemase in the UK. *Int. J. Antimicrob. Agents* 42, 531–536. doi: 10.1016/j.ijantimicag.2013.08.020

Tijet, N., Boyd, D., Patel, S. N., Mulvey, M. R., and Melano, R. G. (2013). Evaluation of the Carba NP test for rapid detection of carbapenemase-producing Enterobacteriaceae and *Pseudomonas aeruginosa*. *Antimicrob. Agents Chemother.* 57, 4578–4580. doi: 10.1128/AAC.00878-13

Tofteland, S., Naseer, U., Lislevand, J. H., Sundsfjord, A., and Samuelsen, O. (2013). A long-term low-frequency hospital outbreak of KPC-producing *Klebsiella pneumoniae* involving Intergenus plasmid diffusion and a persisting environmental reservoir. *PLoS ONE* 8:e59015. doi: 10.1371/journal.pone.0059015

Toledo, P. V., Aranha Junior, A. A., Arend, L. N., Ribeiro, V., Zavascki, A. P., and Tuon, F. F. (2015). Activity of antimicrobial combinations against KPC-2-producing *Klebsiella pneumoniae* in a rat model and time-kill assay. *Antimicrob. Agents Chemother.* 59, 4301–4304. doi: 10.1128/AAC.00323-15

Toleman, M. A., Spencer, J., Jones, L., and Walsh, T. R. (2012). blaNDM-1 is a chimera likely constructed in *Acinetobacter baumannii*. *Antimicrob. Agents Chemother.* 56, 2773–2776. doi: 10.1128/AAC.06297-11

Tollentino, F. M., Polotto, M., Nogueira, M. L., Lincopan, N., Neves, P., Mamizuka, E. M., et al. (2011). High prevalence of blaCTX-M extended spectrum β-lactamase genes in *Klebsiella pneumoniae* isolates from a tertiary care hospital: first report of blaSHV-12, blaSHV-31, blaSHV-38, and blaCTX-M-15 in Brazil. *Microb. Drug Resist.* 17, 7–16. doi: 10.1089/mdr.2010.0055

Toth, A., Damjanova, I., Puskas, E., Janvari, L., Farkas, M., Dobak, A., et al. (2010). Emergence of a colistin-resistant KPC-2-producing *Klebsiella pneumoniae* ST258 clone in Hungary. *Eur. J. Clin. Microbiol. Infect. Dis.* 29, 765–769. doi: 10.1007/s10096-010-0921-3

Tsai, Y. K., Liou, C. H., Fung, C. P., Lin, J. C., and Siu, L. K. (2013). Single or in combination antimicrobial resistance mechanisms of *Klebsiella pneumoniae* contribute to varied susceptibility to different carbapenems. *PLoS ONE* 8:e79640. doi: 10.1371/journal.pone.0079640

Tsakris, A., Kristo, I., Poulou, A., Markou, F., Ikonomidis, A., and Pournaras, S. (2008). First occurrence of KPC-2-possessing *Klebsiella pneumoniae* in a Greek hospital and recommendation for detection with boronic acid disc tests. *J. Antimicrob. Chemother.* 62, 1257–1260. doi: 10.1093/jac/dkn364

Tsakris, A., Poulou, A., Bogaerts, P., Dimitroulia, E., Pournaras, S., and Glupczynski, Y. (2015). Evaluation of a new phenotypic OXA-48 disk test for differentiation of OXA-48 carbapenemase-producing Enterobacteriaceae clinical isolates. *J. Clin. Microbiol.* 53, 1245–1251. doi: 10.1128/JCM.03318-14

Tsakris, A., Poulou, A., Pournaras, S., Voulgari, E., Vrioni, G., Themeli-Digalaki, K., et al. (2010). A simple phenotypic method for the differentiation of metallo-β-lactamases and class A KPC carbapenemases in Enterobacteriaceae clinical isolates. *J. Antimicrob. Chemother.* 65, 1664–1671. doi: 10.1093/jac/dkq210

Tseng, I. L., Liu, Y. M., Wang, S. J., Yeh, H. Y., Hsieh, C. L., Lu, H. L., et al. (2015). Emergence of carbapenemase producing *Klebsiella pneumoniae* and spread of KPC-2 and KPC-17 in Taiwan: a nationwide study from 2011 to 2013. *PLoS ONE* 10:e0138471. doi: 10.1371/journal.pone.0138471

Tumbarello, M., Trecarichi, E. M., De Rosa, F. G., Giannella, M., Giacobbe, D. R., Bassetti, M., et al. (2015). Infections caused by KPC-producing *Klebsiella pneumoniae*: differences in therapy and mortality in a multicentre study. *J. Antimicrob. Chemother.* 70, 2133–2143. doi: 10.1093/jac/dkv086

Tumbarello, M., Viale, P., Viscoli, C., Trecarichi, E. M., Tumietto, F., Marchese, A., et al. (2012). Predictors of mortality in bloodstream infections caused by *Klebsiella pneumoniae* carbapenemase-producing *K. pneumoniae*: importance of combination therapy. *Clin. Infect. Dis.* 55, 943–950. doi: 10.1093/cid/cis588

Tzouvelekis, L. S., Markogiannakis, A., Piperaki, E., Souli, M., and Daikos, G. L. (2014). Treating infections caused by carbapenemase-producing Enterobacteriaceae. *Clin. Microbiol. Infect.* 20, 862–872. doi: 10.1111/1469-0691.12697

Tzouvelekis, L. S., Miriagou, V., Kotsakis, S. D., Spyridopoulou, K., Athanasiou, E., Karagouni, E., et al. (2013). KPC-producing, multidrug-resistant *Klebsiella pneumoniae* sequence type 258 as a typical opportunistic pathogen. *Antimicrob. Agents Chemother.* 57, 5144–5146. doi: 10.1128/AAC.01052-13

Vaara, M. (2010). Polymyxins and their novel derivatives. *Curr. Opin. Microbiol.* 13, 574–581. doi: 10.1016/j.mib.2010.09.002

van der Zwaluw, K., de Haan, A., Pluister, G. N., Bootsma, H. J., de Neeling, A. J., and Schouls, L. M. (2015). The carbapenem inactivation method (CIM), a simple and low-cost alternative for the Carba NP test to assess phenotypic carbapenemase activity in gram-negative rods. *PLoS ONE* 10:e0123690. doi: 10.1371/journal.pone.0123690

Vasoo, S., Cunningham, S. A., Kohner, P. C., Simner, P. J., Mandrekar, J. N., Lolans, K., et al. (2013). Comparison of a novel, rapid chromogenic biochemical assay, the Carba NP test, with the modified Hodge test for detection of carbapenemase-producing Gram-negative bacilli. *J. Clin. Microbiol.* 51, 3097–3101. doi: 10.1128/JCM.00965-13

Villa, L., Poirel, L., Nordmann, P., Carta, C., and Carattoli, A. (2012). Complete sequencing of an IncH plasmid carrying the blaNDM-1, blaCTX-M-15 and qnrB1 genes. *J. Antimicrob. Chemother.* 67, 1645–1650. doi: 10.1093/jac/dks114

Villegas, M. V., Lolans, K., Correa, A., Suarez, C. J., Lopez, J. A., Vallejo, M., et al. (2006). First detection of the plasmid-mediated class A carbapenemase KPC-2 in clinical isolates of *Klebsiella pneumoniae* from South America. *Antimicrob. Agents Chemother.* 50, 2880–2882. doi: 10.1128/AAC.00186-06

Virgincar, N., Iyer, S., Stacey, A., Maharjan, S., Pike, R., Perry, C., et al. (2011). *Klebsiella pneumoniae* producing KPC carbapenemase in a district general hospital in the UK. *J. Hosp. Infect.* 78, 293–296. doi: 10.1016/j.jhin.2011.03.016

Voulgari, E., Gartzonika, C., Vrioni, G., Politi, L., Priavali, E., Levidiotou-Stefanou, S., et al. (2014). The balkan region: NDM-1-producing *Klebsiella pneumoniae* ST11 clonal strain causing outbreaks in Greece. *J. Antimicrob. Chemother.* 69, 2091–2097. doi: 10.1093/jac/dku105

Voulgari, E., Zarkotou, O., Ranellou, K., Karageorgopoulos, D. E., Vrioni, G., Mamali, V., et al. (2013). Outbreak of OXA-48 carbapenemase-producing *Klebsiella pneumoniae* in Greece involving an ST11 clone. *J. Antimicrob. Chemother.* 68, 84–88. doi: 10.1093/jac/dks356

Vourli, S., Giakkoupi, P., Miriagou, V., Tzelepi, E., Vatopoulos, A. C., and Tzouvelekis, L. S. (2004). Novel GES/IBC extended-spectrum β-lactamase variants with carbapenemase activity in clinical enterobacteria. *FEMS Microbiol. Lett.* 234, 209–213. doi: 10.1016/j.femsle.2004.03.028

Vrioni, G., Daniil, I., Voulgari, E., Ranellou, K., Koumaki, V., Ghirardi, S., et al. (2012). Comparative evaluation of a prototype chromogenic medium (ChromID CARBA) for detecting carbapenemase-producing Enterobacteriaceae in surveillance rectal swabs. *J. Clin. Microbiol.* 50, 1841–1846. doi: 10.1128/JCM.06848-11

Wachino, J., Yamane, K., Suzuki, S., Kimura, K., and Arakawa, Y. (2010). Prevalence of fosfomycin resistance among CTX-M-producing *Escherichia coli* clinical isolates in Japan and identification of novel plasmid-mediated fosfomycin-modifying enzymes. *Antimicrob. Agents Chemother.* 54, 3061–3064. doi: 10.1128/AAC.01834-09

Wailan, A. M., Paterson, D. L., Kennedy, K., Ingram, P. R., Bursle, E., and Sidjabat, H. E. (2015). Genomic characteristics of NDM-producing Enterobacteriaceae in Australia and their blaNDM genetic contexts. *Antimicrob. Agents Chemother.* 60, 136–141. doi: 10.1128/AAC.01243-15

Wang, D., Chen, J., Yang, L., Mou, Y., and Yang, Y. (2014a). Phenotypic and enzymatic comparative analysis of the KPC variants, KPC-2 and its recently discovered variant KPC-15. *PLoS ONE* 9:e111491. doi: 10.1371/journal.pone.0111491

Wang, D., Hou, W., Chen, J., Mou, Y., Yang, L., Yang, L., et al. (2014b). Characterization of the blaKPC-2 and blaKPC-3 genes and the novel blaKPC-15 gene in *Klebsiella pneumoniae. J. Med. Microbiol.* 63, 981–987. doi: 10.1099/jmm.0.073841-0

Wang, L., Gu, H., and Lu, X. (2012). A rapid low-cost real-time PCR for the detection of *Klebsiella pneumoniae* carbapenemase genes. *Ann. Clin. Microbiol. Antimicrob.* 11:9. doi: 10.1186/1476-0711-11-9

Wang, P., Chen, S., Guo, Y., Xiong, Z., Hu, F., Zhu, D., et al. (2011). Occurrence of false positive results for the detection of carbapenemases in carbapenemase-negative *Escherichia coli* and *Klebsiella pneumoniae* isolates. *PLoS ONE* 6:e26356. doi: 10.1371/journal.pone.0026356

Wang, X. D., Cai, J. C., Zhou, H. W., Zhang, R., and Chen, G. X. (2009). Reduced susceptibility to carbapenems in *Klebsiella pneumoniae* clinical isolates associated with plasmid-mediated β-lactamase production and OmpK36 porin deficiency. *J. Med. Microbiol.* 58(Pt 9), 1196–1202. doi: 10.1099/jmm.0.008094-0

Wang, X., Chen, G., Wu, X., Wang, L., Cai, J., Chan, E. W., et al. (2015). Increased prevalence of carbapenem resistant Enterobacteriaceae in hospital setting due to cross-species transmission of the blaNDM-1 element and clonal spread of progenitor resistant strains. *Front. Microbiol.* 6:595. doi: 10.3389/fmicb.2015.00595

Wang, X., Li, H., Zhao, C., Chen, H., Liu, J., Wang, Z., et al. (2014c). Novel NDM-9 metallo-β-lactamase identified from a ST107 *Klebsiella pneumoniae* strain isolated in China. *Int. J. Antimicrob. Agents* 44, 90–91. doi: 10.1016/j.ijantimicag.2014.04.010

Wang, X., Xu, X., Li, Z., Chen, H., Wang, Q., Yang, P., et al. (2014d). An outbreak of a nosocomial NDM-1-producing *Klebsiella pneumoniae* ST147 at a teaching hospital in mainland China. *Microb. Drug Resist.* 20, 144–149. doi: 10.1089/mdr.2013.0100

Wang, X., Zhang, F., Zhao, C., Wang, Z., Nichols, W. W., Testa, R., et al. (2014e). In vitro activities of ceftazidime-avibactam and aztreonam-avibactam against 372 Gram-negative bacilli collected in 2011 and 2012 from 11 teaching hospitals in China. *Antimicrob. Agents Chemother.* 58, 1774–1778. doi: 10.1128/AAC.02123-13

Warburg, G., Hidalgo-Grass, C., Partridge, S. R., Tolmasky, M. E., Temper, V., Moses, A. E., et al. (2012). A carbapenem-resistant *Klebsiella pneumoniae* epidemic clone in Jerusalem: sequence type 512 carrying a plasmid encoding aac(6')-Ib. *J. Antimicrob. Chemother.* 67, 898–901. doi: 10.1093/jac/dkr552

Wendt, C., Schutt, S., Dalpke, A. H., Konrad, M., Mieth, M., Trierweiler-Hauke, B., et al. (2010). First outbreak of *Klebsiella pneumoniae* carbapenemase (KPC)-producing *K. pneumoniae* in Germany. *Eur. J. Clin. Microbiol. Infect. Dis.* 29, 563–570. doi: 10.1007/s10096-010-0896-0

Weterings, V., Zhou, K., Rossen, J. W., van Stenis, D., Thewessen, E., Kluytmans, J., et al. (2015). An outbreak of colistin-resistant *Klebsiella pneumoniae* carbapenemase-producing *Klebsiella pneumoniae* in the Netherlands (July to December 2013), with inter-institutional spread. *Eur. J. Clin. Microbiol. Infect. Dis.* 34, 1647–1655. doi: 10.1007/s10096-015-2401-2

Willemse-Erix, D., Bakker-Schut, T., Slagboom-Bax, F., Jachtenberg, J. W., Lemmens-den Toom, N., Papagiannitsis, C. C., et al. (2012). Rapid typing of extended-spectrum β-lactamase- and carbapenemase-producing *Escherichia coli* and *Klebsiella pneumoniae* isolates by use of SpectraCell RA. *J. Clin. Microbiol.* 50, 1370–1375. doi: 10.1128/JCM.05423-11

Williamson, D. A., Heffernan, H., Sidjabat, H., Roberts, S. A., Paterson, D. L., Smith, M., et al. (2011). Intercontinental transfer of OXA-181-producing *Klebsiella pneumoniae* into New Zealand. *J. Antimicrob. Chemother.* 66, 2888–2890. doi: 10.1093/jac/dkr396

Williamson, D. A., Sidjabat, H. E., Freeman, J. T., Roberts, S. A., Silvey, A., Woodhouse, R., et al. (2012). Identification and molecular characterisation of New Delhi metallo-β-lactamase-1 (NDM-1)- and NDM-6-producing Enterobacteriaceae from New Zealand hospitals. *Int. J. Antimicrob. Agents* 39, 529–533. doi: 10.1016/j.ijantimicag.2012.02.017

Wiskirchen, D. E., Nordmann, P., Crandon, J. L., and Nicolau, D. P. (2013). Efficacy of humanized carbapenem exposures against New Delhi metallo-β-lactamase (NDM-1)-producing enterobacteriaceae in a murine infection model. *Antimicrob. Agents Chemother.* 57, 3936–3940. doi: 10.1128/AAC.00708-13

Wiskirchen, D. E., Nordmann, P., Crandon, J. L., and Nicolau, D. P. (2014a). Efficacy of humanized carbapenem and ceftazidime regimens against Enterobacteriaceae producing OXA-48 carbapenemase in a murine infection model. *Antimicrob. Agents Chemother.* 58, 1678–1683. doi: 10.1128/AAC.01947-13

Wiskirchen, D. E., Nordmann, P., Crandon, J. L., and Nicolau, D. P. (2014b). In vivo efficacy of human simulated regimens of carbapenems and comparator agents against NDM-1-producing Enterobacteriaceae. *Antimicrob. Agents Chemother.* 58, 1671–1677. doi: 10.1128/AAC.01946-13

Wolter, D. J., Kurpiel, P. M., Woodford, N., Palepou, M. F., Goering, R. V., and Hanson, N. D. (2009). Phenotypic and enzymatic comparative analysis of the novel KPC variant KPC-5 and its evolutionary variants, KPC-2 and KPC-4. *Antimicrob. Agents Chemother.* 53, 557–562. doi: 10.1128/AAC.00734-08

Woodford, N., Tierno, P. M. Jr., Young, K., Tysall, L., Palepou, M. F. I., Ward, E., et al. (2004). Outbreak of *Klebsiella pneumoniae* producing a new carbapenem-hydrolyzing class a β-Lactamase, KPC-3, in a New York medical center. *Antimicrob. Agents Chemother.* 48, 4793–4799. doi: 10.1128/AAC.48.12.4793-4799.2004

Woodford, N., Zhang, J., Warner, M., Kaufmann, M. E., Matos, J., Macdonald, A., et al. (2008). Arrival of *Klebsiella pneumoniae* producing KPC carbapenemase in the United Kingdom. *J. Antimicrob. Chemother.* 62, 1261–1264. doi: 10.1093/jac/dkn396

Wrenn, C., O'Brien, D., Keating, D., Roche, C., Rose, L., Ronayne, A., et al. (2014). Investigation of the first outbreak of OXA-48-producing *Klebsiella pneumoniae* in Ireland. *J. Hosp. Infect.* 87, 41–46. doi: 10.1016/j.jhin.2014.03.001

Xiang, D. R., Li, J. J., Sheng, Z. K., Yu, H. Y., Deng, M., Bi, S., et al. (2015). Complete sequence of a novel IncR-F33:A-:B- Plasmid, pKP1034, Harboring fosA3, blaKPC-2, blaCTX-M-65, blaSHV-12, and rmtB from an epidemic *Klebsiella pneumoniae* Sequence Type 11 Strain in China. *Antimicrob. Agents Chemother.* 60, 1343–1348. doi: 10.1128/AAC.01488-15

Yamamoto, T., Takano, T., Fusegawa, T., Shibuya, T., Hung, W. C., Higuchi, W., et al. (2013). Electron microscopic structures, serum resistance, and plasmid restructuring of New Delhi metallo-β-lactamase-1 (NDM-1)-producing ST42 *Klebsiella pneumoniae* emerging in Japan. *J. Infect. Chemother.* 19, 118–127. doi: 10.1007/s10156-012-0470-z

Yan, Y., Yang, H., Pan, L., Sun, K., Fan, H., Lu, Y., et al. (2014). Improving the efficiency of the modified Hodge test in KPC-producing *Klebsiella pneumoniae* isolates by incorporating an EDTA disk. *Curr. Microbiol.* 69, 47–52. doi: 10.1007/s00284-014-0552-5

Yang, J., Ye, L., Guo, L., Zhao, Q., Chen, R., Luo, Y., et al. (2013). A nosocomial outbreak of KPC-2-producing *Klebsiella pneumoniae* in a Chinese hospital: dissemination of ST11 and emergence of ST37, ST392 and ST395. *Clin. Microbiol. Infect.* 19, E509–E515. doi: 10.1111/1469-0691.12275

Yigit, H., Queenan, A. M., Anderson, G. J., Domenech-Sanchez, A., Biddle, J. W., Steward, C. D., et al. (2001). Novel carbapenem-hydrolyzing β-lactamase, KPC-1, from a carbapenem-resistant strain of *Klebsiella pneumoniae. Antimicrob. Agents Chemother.* 45, 1151–1161. doi: 10.1128/AAC.45.4.1151-1161.2001

Yong, D., Toleman, M. A., Giske, C. G., Cho, H. S., Sundman, K., Lee, K., et al. (2009). Characterization of a new metallo-β-lactamase gene, blaNDM-1, and a novel erythromycin esterase gene carried on a unique genetic structure in *Klebsiella pneumoniae* sequence type 14 from India. *Antimicrob. Agents Chemother.* 53, 5046–5054. doi: 10.1128/AAC.00774-09

Yoo, J. S., Kim, H. M., Yoo, J. I., Yang, J. W., Kim, H. S., Chung, G. T., et al. (2013). Detection of clonal KPC-2-producing *Klebsiella pneumoniae* ST258 in Korea during nationwide surveillance in 2011. *J. Med. Microbiol.* 62, 1338–1342. doi: 10.1099/jmm.0.059428-0

Yu, W. L., Lee, M. F., Tang, H. J., Chang, M. C., Walther-Rasmussen, J., and Chuang, Y. C. (2015). Emergence of KPC new variants (KPC-16 and KPC-17) and ongoing outbreak in southern Taiwan. *Clin. Microbiol. Infect.* 21, 375.e5–375.e8. doi: 10.1016/j.cmi.2014.11.030

Yusuf, E., Van Der Meeren, S., Schallier, A., and Pierard, D. (2014). Comparison of the Carba NP test with the Rapid CARB screen Kit for the detection of carbapenemase-producing Enterobacteriaceae and *Pseudomonas aeruginosa. Eur. J. Clin. Microbiol. Infect. Dis.* 33, 2237–2240. doi: 10.1007/s10096-014-2199-3

Zagorianou, A., Sianou, E., Iosifidis, E., Dimou, V., Protonotariou, E., Miyakis, S., et al. (2012). Microbiological and molecular characteristics of carbapenemase-producing *Klebsiella pneumoniae* endemic in a tertiary Greek hospital during 2004-2010. *Euro. Surveill.* 17:20088.

Zaidi, A. K., Huskins, W. C., Thaver, D., Bhutta, Z. A., Abbas, Z., and Goldmann, D. A. (2005). Hospital-acquired neonatal infections in developing countries. *Lancet* 365, 1175–1188. doi: 10.1016/S0140-6736(05)71881-X

Zarfel, G., Hoenigl, M., Wurstl, B., Leitner, E., Salzer, H. J., Valentin, T., et al. (2011). Emergence of carbapenem-resistant Enterobacteriaceae in Austria, 2001-2010. *Clin. Microbiol. Infect.* 17, E5–E8. doi: 10.1111/j.1469-0691.2011.03659.x

Zhang, X., Li, X., Wang, M., Yue, H., Li, P., Liu, Y., et al. (2015). Outbreak of NDM-1-producing *Klebsiella pneumoniae* causing neonatal infection in a teaching hospital in mainland China. *Antimicrob. Agents Chemother.* 59, 4349–4351. doi: 10.1128/AAC.03868-14

Zhang, Y., Jiang, X., Wang, Y., Li, G., Tian, Y., Liu, H., et al. (2014). Contribution of β-lactamases and porin proteins OmpK35 and OmpK36 to carbapenem resistance in clinical isolates of KPC-2-producing *Klebsiella pneumoniae. Antimicrob. Agents Chemother.* 58, 1214–1217. doi: 10.1128/AAC.02045-12

Zhong, X., Xu, H., Chen, D., Zhou, H., Hu, X., and Cheng, G. (2014). First emergence of acrAB and oqxAB mediated tigecycline resistance in clinical isolates of *Klebsiella pneumoniae* pre-dating the use of tigecycline in a Chinese hospital. *PLoS ONE* 9:e115185. doi: 10.1371/journal.pone.0115185

Zowawi, H. M., Forde, B. M., Alfaresi, M., Alzarouni, A., Farahat, Y., Chong, T. M., et al. (2015). Stepwise evolution of pandrug-resistance in *Klebsiella pneumoniae. Sci. Rep.* 5:15082. doi: 10.1038/srep15082

Zowawi, H. M., Sartor, A. L., Balkhy, H. H., Walsh, T. R., Al Johani, S. M., AlJindan, R. Y., et al. (2014). Molecular characterization of carbapenemase-producing *Escherichia coli* and *Klebsiella pneumoniae* in the countries of the Gulf cooperation council: dominance of OXA-48 and NDM producers. *Antimicrob. Agents Chemother.* 58, 3085–3090. doi: 10.1128/AAC.02050-13

Zujic Atalic, V., Bedenic, B., Kocsis, E., Mazzariol, A., Sardelic, S., Barisic, M., et al. (2014). Diversity of carbapenemases in clinical isolates of Enterobacteriaceae in Croatia–the results of a multicentre study. *Clin. Microbiol. Infect.* 20, O894–O903. doi: 10.1111/1469-0691.12635

12

An Assessment on DNA Microarray and Sequence-based Methods for the Characterization of Methicillin-Susceptible *Staphylococcus aureus* from Nigeria

Adebayo O. Shittu [1,2*], *Omotayo Oyedara* [3], *Kenneth Okon* [4], *Adeola Raji* [5], *Georg Peters* [6], *Lutz von Müller* [2†], *Frieder Schaumburg* [6], *Mathias Herrmann* [2] and *Ulla Ruffing* [2]

[1] Department of Microbiology, Obafemi Awolowo University, Ile-Ife, Nigeria, [2] Institute of Medical Microbiology and Hygiene, Saarland University, Homburg, Germany, [3] Department of Biological Sciences, College of Science, Engineering and Technology, Osun State University, Osogbo, Nigeria, [4] Department of Medical Laboratory Services, Federal Medical Centre, Makurdi, Nigeria, [5] Department of Microbiology and Immunology, Alfaisal University, Riyadh, Saudi Arabia, [6] Institute of Medical Microbiology, University Hospital Münster, Münster, Germany

***Correspondence:**
Adebayo O. Shittu
bayo_shittu@yahoo.com

†Present Address:
Lutz von Müller,
Institute of Laboratory Medicine,
Microbiology and Hygiene (LMH),
Christophorus-Kliniken GmbH,
Coesfeld, Germany

Staphylococcus aureus is an important human pathogen causing nosocomial and community-acquired infections worldwide. In the characterization of this opportunistic pathogen, DNA microarray hybridization technique is used as an alternative to sequence based genotyping to obtain a comprehensive assessment on the virulence, resistance determinants, and population structure. The objective of this study was to characterize a defined collection of *S. aureus* isolates from Nigeria using the microarray technique, and to assess the extent that it correlates with sequence-based genotyping methods. The clonal diversity and genomic content of 52 methicillin-susceptible *Staphylococcus aureus* (MSSA) were investigated by *spa* typing, MLST and DNA microarray hybridization. More than half (55.8%) of these isolates were associated with clonal complexes (CCs) typically associated with methicillin-resistant *S. aureus* (MRSA) clones i.e., CC1, CC5, CC8, CC30, and CC45. Certain genes linked with virulence (*hlgA* and *clfA*) and adherence (*ebpS, fnbA*, *sspA*, *sspB,* and *sspP*) were detected in all isolates. A number of genes or gene clusters were associated with distinct clonal types. The enterotoxin gene cluster (*egc*) was linked with CC5, CC25, CC30, CC45, and CC121, enterotoxin H gene (*seh*) with CC1, exfoliative toxin D gene (*etd*) with CC25 and CC80, and the epidermal cell differentiation inhibitor B gene (*edinB*) with CC25, CC80, and CC152. The excellent agreement between data from DNA microarray and MLST in the delineation of Nigerian MSSA isolates indicates that the microarray technique is a useful tool to provide information on antibiotic resistance, clonal diversity and virulence factors associated with infection and disease.

Keywords: *Staphylococcus aureus*, microarray, MLST, genotyping, Nigeria

---

Abbreviations: Agr, accessory gene regulator; CC, Clonal complex; CLSI, Clinical Laboratory Standards Institute; MSSA, Methicillin susceptible *Staphylococcus aureus*; MLST, Multilocus sequence typing; PVL, Panton-Valentine Leukocidin; S.aureus, *Staphylococcus aureus*; SCC*mec*, Staphylococcal chromosome cassette *mec*; *spa*, *Staphylococcus aureus* protein A; ST, Sequence Type.

INTRODUCTION

Staphylococcus aureus is implicated in a variety of human infections with high rates of morbidity and mortality (Lowy, 1998; Corey, 2009). In infection, *S. aureus* exhibits a coordinated and regulated expression for a wide variety of cell and surface-associated virulence factors (Foster and Höök, 1998; Novick, 2006). These factors mediate adherence to host cells and damaged tissue, facilitate tissue destruction and spreading, promote iron uptake and evasion of host immune system, as well as tissue damage (Skaar and Schneewind, 2004; Grumann et al., 2014). Recent studies in Cameroon (Kihla et al., 2014), Egypt (Ahmed et al., 2014), Gabon (Alabi et al., 2013), Nigeria (Jido and Garba, 2012; Oladeinde et al., 2013), South Africa (Groome et al., 2012; Naidoo et al., 2013), and Tanzania (Kayange et al., 2010; Mhada et al., 2012) have identified *S. aureus* as the main etiological agent for various infections in Africa. Moreover, this species has been recognized as one main cause of community-acquired neonatal sepsis in Africa (Waters et al., 2011). These studies clearly establish the important role of this major human pathogen in tropical Africa.

In many health care institutions in sub-Saharan Africa, the lack of skilled laboratory manpower and resources is a major constraint in the identification of bacterial pathogens from clinical samples. If such analysis can be provided at all, identification of *S. aureus* typically relies on phenotypic methods precluding in-depth strain characterization. Molecular analysis of clonal attribution and presence of single genes contained in *S. aureus* isolates have emerged in pilot studies from select African centers, areas and populations (Ateba Ngoa et al., 2012; Shittu et al., 2012; Seni et al., 2013; Aiken et al., 2014; Egyir et al., 2014; Oosthuysen et al., 2014; Conceição et al., 2015; De Boeck et al., 2015; Kraef et al., 2015; Schaumburg et al., 2015). Nevertheless, in view of the impact of *S. aureus* disease in sub-Saharan Africa, the clonal characterization in concert with a comprehensive analysis of the hitherto ill-described virulence factor armamentarium of *S. aureus* isolates from this region is urgently warranted. Such analyses should target a broad spectrum of variable staphylococcal factors such as genes or gene clusters conferring antibiotic resistance, toxins, virulence, adhesion or immune evasion factors. These analyses have not been performed on a collection of *S. aureus* isolates in Nigeria, and reports from African countries are limited and only addressed a limited and select analytical spectrum (Raji et al., 2013; Aiken et al., 2014; Rovira et al., 2015).

The DNA microarray used for this analysis is a unique and comprehensive genotyping technique based on the analysis of 334 target sequences corresponding to approximately 170 distinct genes and their allelic variants. It enables the simultaneous identification of various gene classes including species markers, genes encoding resistance and virulence properties, exotoxin and adhesion factors, accessory gene regulator (*agr*), capsule, and SCC*mec* types (Monecke et al., 2011). Based on the observation of a high level of genetic diversity from previous investigations on methicillin-susceptible *S. aureus* (MSSA) in Nigeria (Shittu et al., 2011, 2012; Kolawole et al., 2013), we studied MSSA isolates obtained from various clinical sources in Nigeria using this comprehensive, array-based approach to provide an insight on the major factors associated with infection and disease.

MATERIALS AND METHODS

Identification and Antibiotic Susceptibility Testing of *S. aureus* Isolates

The isolates ($n = 52$) were obtained from samples processed as part of surveillance activities in the microbiology laboratories of six health care institutions located in Ado-Ekiti, Ile-Ife, Osogbo, Lagos, and Ibadan in South-West Nigeria, and Maiduguri in North-East Nigeria. The duration of collection of isolates was from March 2009 to April, 2010. Only the isolates were analyzed in this study. Preliminary verification as *S. aureus* was based on colony characteristics on blood agar, positive results for catalase, coagulase and DNase tests. Twelve isolates from a previous study (Shittu et al., 2011) were also included in this investigation. Identification was confirmed by Matrix-Assisted Laser Desorption/Ionization-Time Of Flight analysis (MALDI-TOF). Susceptibility testing to penicillin (10 units), cefoxitin (30 μg), doxycycline (30 μg), erythromycin (15 μg), clindamycin (2 μg), gentamicin (10 μg), chloramphenicol (30 μg), and trimethoprim-sulfamethoxazole (1.25/23.75 μg) were determined using the disk diffusion method according to the Clinical Laboratory Standards Institute guidelines (Clinical and Laboratory Standards Institute (CLSI), 2009).

DNA Extraction

S. aureus genomic DNA was extracted from an 18–24 h old culture on sheep blood agar using lysis buffer and lysis enhancer (StaphyType Kit, Alere Technologies GmbH, Jena, Germany) and processed using a DNeasy tissue kit (Qiagen, Hilden, Germany).

Molecular Typing of the Isolates

Typing of *S. aureus* was based on sequencing of the hypervariable region of the protein A gene (*spa*). The *spa* types were determined using the Ridom StaphType software (Ridom GmbH, Würzburg, Germany, version 2.1.1) (Harmsen et al., 2003). Multilocus sequence typing (MLST) was performed for one isolate of each *spa* type (Enright et al., 2000), as a *spa* type usually belongs to one sequence type (ST) with few exceptions due to homoplasies (Basset et al., 2009, 2012). The allelic profiles and STs were assigned using the MLST *S. aureus* database (www.mlst.net), and the sequence types of the remaining isolates were inferred from the derived MLST data.

DNA Microarray Hybridization

The DNA microarray of the StaphyType™ kit (Alere Technologies GmbH, Jena, Germany) was used in this study according to previously established protocols (Monecke et al., 2008). The isolates were grouped with various clonal complexes (CCs) by the imaging software Iconoclust based on comparison of hybridization profiles to a collection of reference strains previously characterized by MLST.

Splits Graph Construction

The SplitsTree algorithm (Huson and Bryant, 2006) and software was used to analyze the similarities between hybridization patterns, and network tree construction was performed using SplitsTree 4.10 on default settings (characters transformation, uncorrected P; distance transformation, Neighbor-Net; and variance, ordinary least squares).

RESULTS

Identification of *S. aureus* Isolates

A total of 52 MSSA (3 and 49 isolates from nasal and clinical sources, respectively) were analyzed (**Table 1**). The clinical isolates were obtained from wounds and associated infections ($n = 29$; 59.2%), urinary tract infections ($n = 6$; 12.2%), semen/infertility diagnosis ($n = 4$; 8.2%), ocular infections ($n = 3$; 6.1%), and pneumonia ($n = 2$; 4.1%). One isolate each was from otitis media, and blood related infections, while information on three isolates was not available. The clinical isolates were obtained from health care institutions located in Ile-Ife ($n = 26$; 53.1%), Osogbo ($n = 11$; 22.4%), Maiduguri ($n = 5$; 10.2%), Lagos ($n = 4$; 8.2%), Ibadan, and Ado-Ekiti ($n = 2$ isolates each: 4.1%).

Antibiotic Susceptibility Testing

All the isolates were susceptible to cefoxitin and 98.1% ($n = 51$) were resistant to penicillin. Only two isolates each exhibited resistance to chloramphenicol and gentamicin, and four to doxycycline. Intermediate susceptibility to clindamycin and erythromycin were identified in six and 21 isolates, respectively. The predominant antibiotype was resistance only to penicillin ($n = 23$; 44.2%), and resistance to penicillin with intermediate susceptibility to erythromycin ($n = 10$; 19.2%) (**Table 1**).

Sequence based Typing (*spa* and MLST)

A total of 26 *spa* types were identified among the 52 MSSA isolates and the most common were t318 ($n = 7$), t311 ($n = 5$), t084, t127, and t2304 ($n = 4$ each). Based on MLST, the MSSA were classified into 13 sequence types (STs) (**Table 1**).

DNA Microarray Analysis

The assay confirmed the identity of the isolates (*S. aureus*) by positive results for specific markers including rndD1 (domain 1 of 23S rRNA), protein A (*spa*), glyceraldehyde 3-phosphate dehydrogenase (*gapA*), catalase A (*katA*), thermostable nuclease (*nuc*), and staphylococcal accessory regulator A (*sarA*) (Supplementary Materials 1, 2). The hybridization profiles revealed that the 52 MSSA isolates clustered in 12 different CCs. More than half (55.8%) of the CCs were associated with the genetic background common to the major methicillin-resistant *S. aureus* (MRSA) clones i.e., CC1 ($n = 6$ isolates), CC5 ($n = 9$), CC8 ($n = 4$), CC30 ($n = 8$), and CC45 ($n = 2$). The rest were assigned with CC7 ($n = 1$), CC15 ($n = 7$), CC25 ($n = 2$), CC80 ($n = 1$), CC97 ($n = 1$), CC121 ($n = 8$), and CC152 ($n = 3$).

Antibiotic Resistance Genes

A total of 69.2% ($n = 36$) of the isolates yielded a hybridization signal for the beta-lactamase gene (*blaZ*) and only 10 and three isolates were positive for the tetracycline resistance genes (*tetK* and *tetM*), respectively. The two MSSA in CC8 which exhibited phenotypic resistance to chloramphenicol and gentamicin possessed the corresponding resistance genes (*cat* and *aacA-aphD*). In addition, the single CC80 isolate was positive for the lincosamide resistance gene (*lnuA*).

Accessory Gene Regulator and Capsular Typing

The distribution of *agr*/CCs/capsule types for the MSSA is indicated in **Figure 1**. Overall, 13 (25%) isolates assigned to different clonal lineages (CC7, CC8, CC25, CC45, CC97, and CC152) were associated with *agr* group I, 16 (30.2%; CC5 and CC15) with group II, and 15 (28.8%; CC1, CC30 and CC80) with group III. CC121 was the only representative for *agr* group IV ($n = 8$; 15.4%) (**Table 1**). The capsule type 8 was the most frequent and detected in 33 (63.5%) isolates affiliated with CC1, CC7, CC15, CC30, CC45, CC80, and CC121. The remaining isolates (20; 38.5%) belonged to capsule type 5 (assigned with CC5, CC8, CC25, CC97, and CC152).

Enterotoxin Genes

PVL-positive isolates ($n = 27$) belonged to CC1, CC5, CC15, CC30, CC80, CC121, and CC152 (Supplementary Material 1). Moreover, the *lukF* gene (haemolysin gamma; component B) was universally detected in all the CCs and the *lukE* genes was a common feature except with MSSA isolates in CC30, CC45, and CC152 (Supplementary Material 3). With respect to the carriage of superantigen genes, only three MSSA (one isolate in CC1 and two in CC45) tested positive for the toxic shock syndrome toxin gene (*tst-1*) (Supplementary Material 1). All the isolates in this study lacked a hybridization signal for the enterotoxin E gene (see Supplementary Material 2) and the enterotoxin genes were not detected in MSSA assigned with CC80, CC97, and CC152. In the haemolysin gene family, almost all (98.1%) the isolates in the various CCs possessed the haemolysin alpha and delta genes (*hla*, *hld*), while the haemolysin beta gene (*hlb*) was identified in the various CCs except in CC15, CC45, and CC152.

Microbial Surface Components Recognizing Adhesive Matrix Molecule (MSCRAMM) Genes

All the isolates were negative for the surface protein involved in biofilm production (*bap*), but possessed the genes for the intercellular adhesion protein (*icaA/C/D*) (CC152 isolates were *icaC* negative). The genes for clumping factor A (*clfA*), cell surface elastin binding protein (*ebpS*), fibronectin-binding protein A (*fnbA*) and proteases (*sspA*, *sspB*, and *sspP*) were detected in all the isolates (Supplementary Materials 1, 3).

Splits Tree Analysis

The analysis identified four main clusters (CC5/CC25; CC8/CC97; CC1/CC7/CC80, and CC30/CC45) indicating the phylogenetic relationship between the isolates (**Figure 2**).

TABLE 1 | Characterization of the methicillin-susceptible *S. aureus* (MSSA) from Nigeria based on antibiotyping, microarray analysis, *spa* typing, and MLST.

Isolate Number	Location	Sample/Clinical diagnosis	Antibiogram	Score (%) (Alere)	*agr*/Clonal complex (Alere)	*spa* type	MLST
11486_24	Ile-Ife	Wound Infection	PEN	93.8	agr_III/CC1	t127¶	ST1
AB5_28	Osogbo	UTI	PEN, ERY(i)	92.8	agr_III/CC1	t127	ST1
Aro_29	Osogbo	Semen	PEN	94.3	agr_III/CC1	t127	ST1
MD16_4	Not available	Not available	PEN	94.3	agr_III/CC1	t127	ST1
MD20_8*	Maiduguri	Wound infection	PEN, ERY(i), CC(i)	93.5	agr_III/CC1	t321¶	ST1
6056_34	Osogbo	Urine	PEN	93.9	agr_III/CC1	t10433¶	ST1
5675_6	Ile-Ife	Abscess	PEN	91.8	agr_II/CC5	t311	ST5
5221_7	Ile-Ife	Urine	PEN, ERY(i), SXT(i)	93.8	agr_II/CC5	t311	ST5
D23_15	Ile-Ife	Pneumonia	PEN	92.8	agr_II/CC5	t311	ST5
D42_17	Ile-Ife	Adenocarcinoma	PEN, ERY(i)	92.4	agr_II/CC5	t311	ST5
D46_18	Ile-Ife	Wound Infection	PEN, ERY(i)	92.2	agr_II/CC5	t311¶	ST5
1423_36	Osogbo	Urine	PEN, ERY(i)	93.8	agr_II/CC5	t442¶	ST5
D19_14	Ile-Ife	Not available	PEN	93.5	agr_II/CC5	t688¶	ST5
Asu29_27	Osogbo	Otitis media	PEN, DO, ERY(i)	91.9	agr_II/CC5	t1277¶	ST5
3211_30	Osogbo	Wound Infection	PEN	92.9	agr_II/CC5	t3235¶	ST5
6773_11	Ile-Ife	Wound Infection	PEN	93.6	agr_I/CC7	t091¶	ST789
N37_19	Ile-Ife	Erythematous lesion	PEN, SXT	90	agr_I/CC8	t064¶	ST2427
UC45_37	Ibadan	Eye swab	PEN, GM, CHL, SXT	91.3	agr_I/CC8	t2658¶	ST2427
55_40	Ado-Ekiti	Wound Infection	PEN, DO(i), GM, CHL, SXT	90.3	agr_I/CC8	t2658	ST2427
OS39_13*	Lagos	Semen/Infertility	PEN, DO(i), SXT	91.7	agr_I/CC8	t951¶	ST8
11450_23	Ile-Ife	Sputum	PEN	92.9	agr_II/CC15	t084	ST15
5189_1	Ile-Ife	Advanced Cancer	PEN	94	agr_II/CC15	t084	ST15
189_2	Ile-Ife	Blood	PEN, DO(i), ERY(i)	93.9	agr_II/CC15	t084¶	ST15
4013_14*	Ile-Ife	Wound infection	PEN	94.9	agr_II/CC15	t084	ST15
5828_5	Ile-Ife	Abscess	susceptible to all antibiotics tested	94.4	agr_II/CC15	t2216¶	ST15
MD7_3*	Maiduguri	Semen/Infertility	PEN, ERY(i)	94.6	agr_II/CC15	t2216	ST15
MD19_11*	Maiduguri	Wound infection	PEN	94.4	agr_II/CC15	t2216	ST15
S13_6*	Lagos	Urinary Tract Infection	PEN, ERY(i), SXT	93.1	agr_I/CC25	t3772¶	ST25
3925_32	Osogbo	Wound Infection	PEN, ERY(i), SXT	91.4	agr_I/CC25	t10183¶	ST25
6073_3	Not available	Not available	PEN, DO	91.7	agr_III/CC30	t017¶	ST30
D30_16	Ile-Ife	Cholecystitis	PEN	94.7	agr_III/CC30	t318¶	ST30
6506_2	Osogbo	Wound Infection	PEN, ERY(i), CC(i)	91.4	agr_III/CC30	t318	ST30
NS7708_22	Ile-Ife	Nasal swab/screening	PEN, ERY(i)	94.7	agr_III/CC30	t318	ST30
54_39	Ado-Ekiti	Wound Infection	PEN	94.4	agr_III/CC30	t318	ST30
S12_7*	Lagos	Wound infection	PEN, ERY(i)	93.8	agr_III/CC30	t318	ST30
OS41_10*	Lagos	Wound infection	PEN	93.1	agr_III/CC30	t318	ST30
6330_4	Ile-Ife	Osteomyelitis	PEN	94.3	agr_III/CC30	t318	ST30
NS2907_21	Ile-Ife	Nasal swab/screening	PEN, ERY(i), CC(i)	91.8	agr_I/CC45	t095¶	ST508
3950_33	Osogbo	Urine	PEN	91.5	agr_I/CC45	t10434¶	ST508
GDC_35	Osogbo	Semen	PEN	94.9	agr_III /CC80	t934¶	ST80
MD14_2*	Maiduguri	Wound infection	PEN, DO(i)	92.9	agr_I/CC97	t458¶	ST97
ZU_26	Ile-Ife	Unavailable	PEN, ERY(i)	89.3	agr_IV/CC121	t159¶	ST121
UC47_38	Ibadan	Eye swab	PEN, DO, ERY(i), CC(i)	92.1	agr_IV/CC121	t159	ST121
W10_5*	Ile-Ife	Wound infection	PEN, ERY(i)	91.8	agr_IV/CC121	t314¶	ST121
MD_9*	Maiduguri	Wound infection	PEN, ERY(i), CC(i)	92.1	agr_IV/CC121	t314	ST121
6376_3	Ile-Ife	Abscess	PEN, DO(i)	93.1	agr_IV/CC121	t2304¶	ST121
6540_10	Ile-Ife	Bone Marrow Infection	PEN	93.5	agr_IV/CC121	t2304	ST121
NS2986_20	Ile-Ife	Nasal swab/screening	PEN, DO, ERY(i), CC(i), SXT(i)	92.8	agr_IV/CC121	t2304	ST121
3920_31	Osogbo	Aspirate	PEN	92.8	agr_IV/CC121	t2304	ST121
D3_12	Ile-Ife	Cervical cancer	PEN, ERY(i)	94.6	agr_I/ST152	t355¶	ST152

(Continued)

TABLE 1 | Continued

Isolate Number	Location	Sample/Clinical diagnosis	Antibiogram	Score (%) (Alere)	*agr*/Clonal complex (Alere)	*spa* type	MLST
D12_13	Ile-Ife	Ocular infection	PEN	94.6	agr_I/ST152	t355	ST152
W7.2_4*	Ile-Ife	Wound infection	PEN	96.3	agr_I/ST152	t355	ST152

**S. aureus analyzed in a previous study; (i): intermediate susceptibility; agr, accessory gene regulator; PEN, Penicillin; DO, Doxycycline; ERY, Erythromycin; CLI, Clindamycin; GEN, Gentamicin; CHL, Chloramphenicol; SXT, trimethoprim/sulfamethoxazole; CC, Clonal Complex; ST, Sequence type.*

[¶]spa types selected for Multilocus sequence typing (MLST); Sequence types (STs) of the remaining isolates were inferred from the derived MLST data.

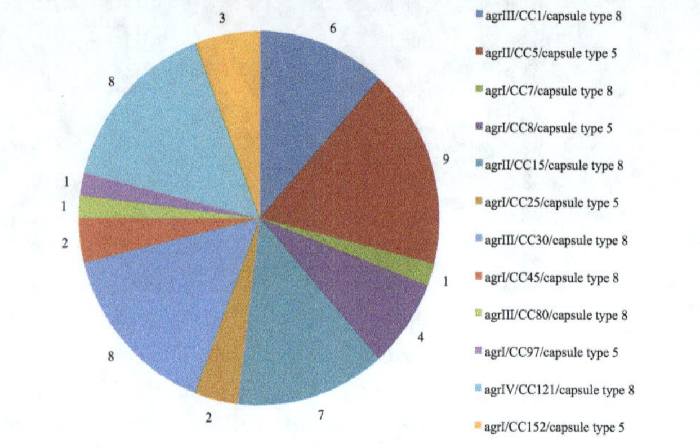

FIGURE 1 | Distribution of *agr*/CCs/capsule type of MSSA in Nigeria.

DISCUSSION

We observed a complete agreement between DNA microarray analysis and MLST in the delineation of the isolates (**Table 1**), showing that the hybridization profile could be used to predict the lineages. Furthermore, the heterogeneous and divergent nature of the isolates observed in this study provided evidence on the overall higher diversity of MSSA compared with MRSA (Deurenberg and Stobberingh, 2008; Goering et al., 2008; Ghasemzadeh-Moghaddam et al., 2011; Ruffing et al., 2012; Blomfeldt et al., 2013; Rasmussen et al., 2013, 2014). In Nigeria, many diagnostic microbiology laboratories rely on the disc diffusion technique for antibiotic susceptibility testing, but this protocol does not provide information on the nature of resistance genes. The antibiotic susceptibility results observed in this study were in accordance with the corresponding resistance gene profiles by DNA microarray. MSSA isolates that exhibited full resistance to trimethoprim-sulfamethoxazole clustered with CC8 and CC25, but were *dfrS1* negative indicating that a different mechanism could be attributed to resistance. A recent study (Nurjadi et al., 2014) has provided strong evidence that the *dfrG* gene is the predominant trimethoprim resistance determinant on *S. aureus* in Africa. Overall, resistant determinants for antibiotics, heavy metal and quaternary ammonium compounds were observed more often in CC8 than other CCs (Supplementary Materials 1, 3).

The accessory gene regulator (*agr*) and capsule typing methods are useful front-line tools for the characterization of *S. aureus* (Goerke et al., 2005). Hybridization signals for *agr* type I and IV were observed for one, three, and four isolates grouped with CC25, CC152, and CC121, respectively (Supplementary Materials 1, 2). This could be attributed to possible cross-hybridization as the alleles for the two *agr* types are closely related (Monecke et al., 2010). Our observations on CCs and *agr* groups were similar to previous reports on MSSA in five major African towns (Breurec et al., 2010), Gabon (Ateba Ngoa et al., 2012), and Nigeria (Ghebremedhin et al., 2009; Kolawole et al., 2013). In addition, our study also support the view (Wright et al., 2005; Holtfreter et al., 2007; Rasmussen et al., 2014) that an *agr* type may be detected in isolates which are assigned to genetically diverse CCs, whereas, it is also associated with specific CCs. The dominance of capsule type 8 in MSSA is consistent with data from Gabon (Schaumburg et al., 2011), Norway (Blomfeldt et al., 2013), and Sweden (Rasmussen et al., 2013, 2014).

Staphylococcal enterotoxins are typically encoded by genes located on mobile genetic elements (Baba et al., 2002). The *egc* cluster (*seg*+*sei*+*sem*+*sen*+*seo*+*seu*) is located on the genomic island vSAβ and reported to be associated with specific clonal types regardless of the geographical strain distribution (Lindsay and Holden, 2006). In this investigation, the *egc*-enterotoxin gene cluster was a unique feature for CC5, CC25, CC30, CC45, and CC121. Previous studies have indicated that the cluster is predominantly present in MSSA assigned with CC5, CC25, CC30, and CC45 (Van Trijp et al., 2010; Rasmussen et al., 2013). The *seh* gene is linked to the staphylococcal cassette chromosome *mec* (SCC*mec* elements) and reported to be restricted to the CC1 genomic background (Baba et al., 2002). Moreover, the *seh* gene has also been reported mainly in MSSA-CC30 (Blomfeldt et al., 2013). Nevertheless, our observation on *seh*-positive MSSA-CC1 is in agreement with previous reports (Chen et al., 2013; Rasmussen et al., 2013).

The genes associated with staphylococcal complement inhibitor (*scn*) and staphylokinase (*sak*) were also widely distributed across the CCs but CC15 isolates were *sak* gene negative. Virulence associated with the exfoliative toxins has been identified to cause epidermal cleavage in staphylococcal scalded skin syndrome (SSSS) and bullous impetigo (Ladhani et al., 1999). The exfoliative toxin D (ETD) is a 27-kDa protein which causes epidermal blisters in newborn mice (Yamasaki et al., 2006). The epidermal cell differentiation factors (EDIN) target and inhibit the small host protein RhoA, a master regulator of the host cell actin cytoskeleton (Inoue et al., 1991; Jaffe and Hall, 2005; Aktories, 2011). Furthermore, the edin-isoform (*edinB*) and *etd* genes are located in tandem in a *S. aureus etd* pathogenicity island in a chromosome of *etd*-positive *S. aureus* strains (Yamaguchi et al., 2002). A strong association of the *etd* gene with invasive CC25 *S. aureus* isolates has also been

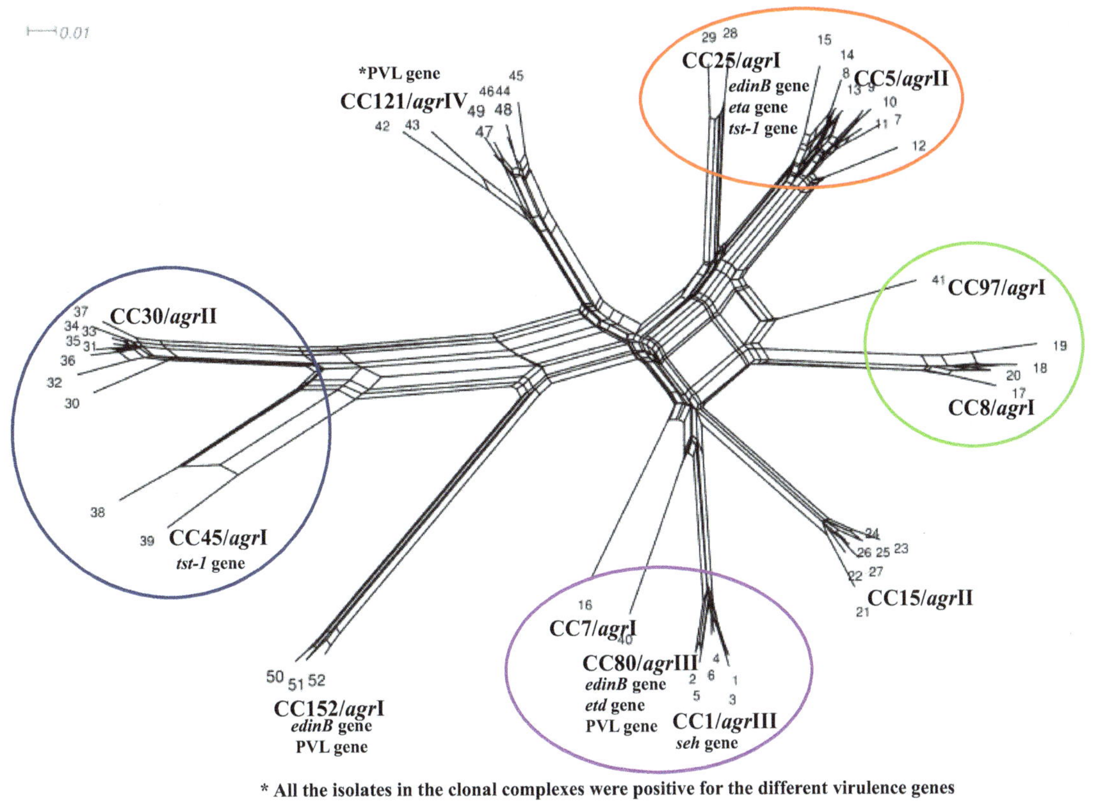

FIGURE 2 | Splits tree graph based on hybridization profile of the MSSA isolates. The results of all array hybridization experiments were arranged in a matrix. Columns represent the target genes and the rows represent the number of experiments; 1, positive; 0, negative; –, ambiguous. Converted to "sequences": 1, c; 0, g; –, c.

reported. In this study, all the isolates assigned with CC25 and CC80 were *etd*-positive, which is in agreement with a previous study in Nigeria (Shittu et al., 2011). Moreover, MSSA grouped with CC25, CC80, and CC152 were *edinB* positive but CC152 isolates were *etd* negative. Our observations were similar to a study on the distribution of the *edin* gene in *S. aureus* from diabetic foot ulcers (Messad et al., 2013). A study in MSSA bacteremia isolates in Sweden showed that the collagen binding protein (Cna) was detected in CC1, CC30, and CC45. Our report identified the gene in isolates assigned with CC1, CC30, CC45, CC121, and CC152.

Our study has a number of limitations. Although all isolates were of human origin, and the large majority was obtained from clinical samples, a clear distinction between commensal and clinical strains could not be made based on the available information. An association of isolates within the context of endemicity i.e., nosocomial vs. community associated infections, is also not clear. Furthermore, whereas the microarray analytical database is exhaustive, well-characterized, and validated with isolates from all continents, the attribution of CCs is based on the hybridization reactions and resulting microarray profile rather than gene sequencing, and a positive signal does not necessarily imply the presence of gene product (e.g., protein). In addition, the microarray method was unable to separate ST8 from ST2427. This might be due to the close phylogenetic relation of both STs as they are single locus variants (ST8: 3–3–1–1–4–4–3 and ST2427: 3–3–297–1–4–4–3). Finally, with a collection of 52 isolates studied, and a large number of genes and genetic profile ascertained by microarray, the potential for individual statistical comparisons is limited. Yet, with this comprehensive genetic-analytical approach performed on a clinical isolate collection obtained from patients of various medical institutions in a sub-Saharan African country, Nigeria, a number of important observations could be made which clearly characterize and demarcate the clonal distribution as well as the virulence gene equipment.

More than one half (55.8%; $n = 29$) of these MSSA isolates were associated with a genetic background which is attributable to classic methicillin-resistant *S. aureus* (MRSA) clones. PVL-positive isolates were identified in seven of the 12 CCs. Moreover, toxin genes were observed to be distributed mainly with certain clonal types, and in agreement with previous investigations (Holtfreter et al., 2007; Monecke et al., 2008). Antibiotic resistance gene profiles of the isolates by the DNA microarray demonstrated concordant results with data on antibiotic susceptibility testing. The array-based, comprehensive approach has been shown to yield such diverse CC and gene specific results on an isolate collection from sub-Saharan Africa.

Overall, microarray analysis proved to be a useful tool to provide useful information on antibiotic resistance, population structure and various virulence factor profiles associated with infection and disease. It is assumed that these findings might be useful for a better understanding of clinical staphylococcal disease presentation, patient care and for assistance in outbreak investigation in health care institutions in a country such as Nigeria. Moreover, our study also underlines the need for further trials employing well-controlled, prospectively collected clinical isolates to delineate the genetic pathogen profile in conjunction with the clinical disease presentation in sub-Saharan Africa.

AUTHOR CONTRIBUTIONS

AS, UR, GP, FS, LM, and MH conceived the study, OO, KO, AR conducted the sample collection and preliminary identification of the isolates. AS performed the microarray technique, AS and UR analyzed the microarray data, and AS wrote the manuscript (with input from all authors). All authors read and approved the final version of the manuscript.

FUNDING

The stay of AS at the Institute of Medical Microbiology and Hygiene, Institute of Medical Microbiology, Saarland University Medical Centre, Homburg/Saar and the University Hospital Münster, Münster, Germany was supported by the Third World Academy of Science and Deutsche Forschungsgemeinschaft (TWAS-DFG award) and the Deutsche Forschungsgemeinschaft (PAK296, Ei 247/8, He 1850/9−1, and He 1850/11−1).

ACKNOWLEDGMENTS

We would like to thank the management and laboratory staff of the hospitals for their support in isolate collection, and Eniola Dolapo for her assistance in the preparation of this manuscript. We gratefully acknowledge the technical assistance of the staff at the Institute of Medical Microbiology, Saarland University Medical Centre, Homburg/Saar, the Institute of Medical Microbiology, University Hospital Münster and the staff at the Sequencing Unit at the Institute of Hygiene, University Hospital Münster, Germany. We thank Professor Iruka Okeke for the critical review of the manuscript.

REFERENCES

Ahmed, E. F., Gad, G. F., Abdalla, A. M., Hasaneen, A. M., and Abdelwahab, S. F. (2014). Prevalence of methicillin resistant *Staphylococcus aureus* among Egyptian patients after surgical interventions. *Surg. Infect. (Larchmt).* 15, 404–411. doi: 10.1089/sur.2013.212

Aiken, A. M., Mutuku, I. M., Sabat, A. J., Akkerboom, V., Mwangi, J., Scott, J. A., et al. (2014). Carriage of *Staphylococcus aureus* in Thika Level 5 Hospital, Kenya: a cross-sectional study. *Antimicrob. Resist. Infect. Control* 3:22. doi: 10.1186/2047-2994-3-22

Aktories, K. (2011). Bacterial protein toxins that modify host regulatory GTPases. *Nat. Rev. Microbiol.* 9, 487–498. doi: 10.1038/nrmicro2592

Alabi, A. S., Frielinghaus, L., Kaba, H., Kösters, K., Huson, M. A., Kahl, B. C., et al. (2013). Retrospective analysis of antimicrobial resistance and bacterial spectrum of infection in Gabon, Central Africa. *BMC Infect. Dis.* 13:455. doi: 10.1186/1471-2334-13-455

Ateba Ngoa, U., Schaumburg, F., Adegnika, A. A., Kosters, K., Moller, T., Fernandes, J. F., et al. (2012). Epidemiology and population structure of *Staphylococcus aureus* in various population groups from a rural and semi urban area in Gabon, Central Africa. *Acta Trop.* 124, 42–47. doi: 10.1016/j.actatropica.2012.06.005

Baba, T., Takeuchi, F., Kuroda, M., Yuzawa, H., Aoki, K., Oguchi, A., et al. (2002). Genome and virulence determinants of high virulence community-acquired MRSA. *Lancet* 359, 1819–1827. doi: 10.1016/S0140-6736(02)08713-5

Basset, P., Hammer, N. B., Kuhn, G., Vogel, V., Sakwinska, O., and Blanc, D. S. (2009). *Staphylococcus aureus clfB* and *spa* alleles of the repeat regions are segregated into major phylogenetic lineages. *Infect. Genet. Evol.* 9, 941–947. doi: 10.1016/j.meegid.2009.06.015

Basset, P., Nübel, U., Witte, W., and Blanc, D. S. (2012). Evaluation of adding a second marker to overcome *Staphylococcus aureus spa* typing homoplasies. *J. Clin. Microbiol.* 50, 1475–1477. doi: 10.1128/JCM.00664-11

Blomfeldt, A., Aamot, H. V., Eskesen, A. N., Müller, F., and Monecke, S. (2013). Molecular characterization of methicillin-sensitive *Staphylococcus aureus* isolates from bacteremic patients in a Norwegian University Hospital. *J. Clin. Microbiol.* 51, 345–347. doi: 10.1128/JCM.02571-12

Breurec, S., Fall, C., Pouillot, R., Boisier, P., Brisse, S., Diene-Sarr, F., et al. (2010). Epidemiology of methicillin-susceptible *Staphylococcus aureus* lineages in five major African towns: high prevalence of Panton-Valentine leukocidin genes. *Clin. Microbiol. Infect.* 17, 633–639. doi: 10.1111/j.1469-0691.2010.03320.x

Chen, X., Wang, W.-K., Han, L.-Z., Liu, Y., Zhang, H., Tang, J., et al. (2013). Epidemiological and genetic diversity of *Staphylococcus aureus* causing bloodstream infection in Shanghai, 2009-2011. *PLoS ONE* 8:e72811. doi: 10.1371/journal.pone.0072811

Clinical and Laboratory Standards Institute (CLSI). (2009). *Performance Standards for Antimicrobial Disk Susceptibility Tests. Approved Standard, 10th Edn.* Wayne, Penn.

Conceição, T., Coelho, C., Santos Silva, I., de Lencastre, H., and Aires-de-Sousa, M. (2015). *Staphylococcus aureus* in former Portuguese colonies from Africa and the far east: missing data to help fill the world map. *Clin. Microbiol. Infect.* 21, 842.e1–842.e10. doi: 10.1016/j.cmi.2015.05.010

Corey, G. R. (2009). *Staphylococcus aureus* bloodstream infections: definitions and treatment. *Clin. Infect. Dis.* 48, S254–S259. doi: 10.1086/598186

De Boeck, H., Vandendriessche, S., Hallin, M., Batoko, B., Alworonga, J. P., Mapendo, B., et al. (2015). *Staphylococcus aureus* nasal carriage among health care workers in Kisangani, the democratic republic of Congo. *Eur. J. Clin. Microbiol. Infect. Dis.* 34, 1567–1572. doi: 10.1007/s10096-015-2387-9

Deurenberg, R. H., and Stobberingh, E. E. (2008). The evolution of *Staphylococcus aureus*. *Infect. Genet. Evol.* 8, 747–763. doi: 10.1016/j.meegid.2008.07.007

Egyir, B., Guardabassi, L., Sørum, M., Nielsen, S. S., Kolekang, A., Frimpong, E., et al. (2014). Molecular epidemiology and antimicrobial susceptibility of clinical *Staphylococcus aureus* from healthcare institutions in Ghana. *PLoS ONE* 9:e89716. doi: 10.1371/journal.pone.0089716

Enright, M. C., Day, N. P., Davies, C. E., Peacock, S. J., and Spratt, B. G. (2000). Multilocus sequence typing for characterization of methicillin-resistant and methicillin-susceptible clones of *Staphylococcus aureus*. *J. Clin. Microbiol.* 38, 1008–1015.

Foster, T. J., and Höök, M. (1998). Surface protein adhesins of *Staphylococcus aureus*. *Trends Microbiol.* 6, 484–488. doi: 10.1016/S0966-842X(98)01400-0

Ghasemzadeh-Moghaddam, H., Ghaznavi-Rad, E., Sekawi, Z., Yun-Khoon, L., Aziz, M. N., Hamat, R. A., et al. (2011). Methicillin susceptible *Staphylococcus aureus* from clinical and community sources are genetically diverse. *Int. J. Med. Microbiol.* 301, 347–353. doi: 10.1016/j.ijmm.2010.10.004

Ghebremedhin, B., Olugbosi, M. O., Raji, A. M., Layer, F., Bakare, R. A., König, B., et al. (2009). Emergence of a community-associated methicillin-resistant *Staphylococcus aureus* strain with a unique resistance profile in Southwest Nigeria. *J. Clin. Microbiol.* 47, 2975–2980. doi: 10.1128/JCM.00648-09

Goering, R. V., Shawar, R. M., Scangarella, N. E., O'Hara, F. P., Amrine-Madsen, H., West, J. M., et al. (2008). Molecular epidemiology of methicillin-resistant and methicillin-susceptible *Staphylococcus aureus* isolates from global clinical trials. *J. Clin. Microbiol.* 46, 2842–2847. doi: 10.1128/JCM.00521-08

Goerke, C., Esser, S., Kümmel, M., and Wolz, C. (2005). *Staphylococcus aureus* strain designation by *agr* and *cap* polymorphism typing and delineation of *agr* diversification by sequence analysis. *Int. J. Med. Microbiol.* 295, 67–75. doi: 10.1016/j.ijmm.2005.01.004

Groome, M. J., Albrich, W. C., Wadula, J., Khoosal, M., and Madhi, S. A. (2012). Community-onset *Staphylococcus aureus* bacteraemia in hospitalized African children: high incidence in HIV-infected children and high prevalence of multidrug resistance. *Paediatr. Int. Child Health.* 32, 140–146. doi: 10.1179/1465328111Y.0000000044

Grumann, D., Nübel, U., and Bröker, B. M. (2014). *Staphylococcus aureus* toxins - their functions and genetics. *Infect. Genet. Evol.* 21, 583–592. doi: 10.1016/j.meegid.2013.03.013

Harmsen, D., Claus, H., Witte, W., Rothgänger, J., Claus, H., Turnwald, D., et al. (2003). Typing of methicillin-resistant *Staphylococcus aureus* in a university setting by using novel software for *spa* repeat determination and database management. *J. Clin. Microbiol.* 41, 5442–5448. doi: 10.1128/JCM.41.12.5442-5448.2003

Holtfreter, S., Grumann, D., Schmudde, M., Nguyen, H. T. T., Eichler, P., Strommenger, B., et al. (2007). Clonal distribution of superantigen genes in clinical *Staphylococcus aureus* isolates. *J. Clin. Microbiol.* 45, 2669–2680. doi: 10.1128/JCM.00204-07

Huson, D. H., and Bryant, D. (2006). Application of phylogenetic networks in evolutionary studies. *Mol. Biol. Evol.* 23, 254–267. doi: 10.1093/molbev/msj030

Inoue, S., Sugai, M., Murooka, Y., Paik, S. Y., Hong, Y. M., Ohgai, H., et al. (1991). Molecular cloning and sequencing of the epidermal cell differentiation inhibitor gene from *Staphylococcus aureus*. *Biochem. Biophys. Res. Commun.* 174, 459–464. doi: 10.1016/0006-291X(91)91438-I

Jaffe, A. B., and Hall, A. (2005). RHO GTPases: biochemistry and biology. *Annu. Rev. Cell Dev. Biol.* 21, 247–269. doi: 10.1146/annurev.cellbio.21.020604.150721

Jido, T., and Garba, I. (2012). Surgical-site infection following cesarean section in Kano, Nigeria. *Ann. Med. Health Sci. Res.* 2, 33–36. doi: 10.4103/2141-9248.96934

Kayange, N., Kamugisha, E., Mwizamholya, D. L., Jeremiah, S., and Mshana, S. E. (2010). Predictors of positive blood culture and deaths among neonates with suspected neonatal sepsis in a tertiary hospital, Mwanza-Tanzania. *BMC Pediatr.* 10:39. doi: 10.1186/1471-2431-10-39

Kihla, A. J., Ngunde, P. J., Evelyn, M. S., Gerard, N., and Ndip, R. N. (2014). Risk factors for wound infection in health care facilities in Buea, Cameroon: aerobic bacterial pathogens and antibiogram of isolates. *Pan. Afr. Med. J.* 18, 6. doi: 10.11604/pamj.2014.18.6.2304

Kolawole, D. O., Adeyanju, A., Schaumburg, F., Akinyoola, A. L., Lawal, O. O., Amusa, Y. B., et al. (2013). Characterization of colonizing *Staphylococcus aureus* isolated from surgical wards' patients in a Nigerian university hospital. *PLoS ONE* 8:e68721. doi: 10.1371/journal.pone.0068721

Kraef, C., Alabi, A. S., Peters, G., Becker, K., Kremsner, P. G., Rossatanga, E. G., et al. (2015). Co-detection of Panton-Valentine leukocidin encoding genes and cotrimoxazole resistance in *Staphylococcus aureus* in Gabon: implications for HIV-patients' care. *Front. Microbiol.* 6:60. doi: 10.3389/fmicb.2015.00060

Ladhani, S., Joannou, C. L., Lochrie, D. P., Evans, R. W., and Poston, S. M. (1999). Clinical, microbial and biochemical aspects of the exfoliative toxins causing staphylococcal scalded-skin syndrome. *Clin. Microbiol. Rev.* 12, 224–242.

Lindsay, J. A., and Holden, M. T. (2006). Understanding the rise of the superbug: investigation of the evolution and genomic variation of *Staphylococcus aureus*. *Funct. Integr. Genomics* 6, 186–201. doi: 10.1007/s10142-005-0019-7

Lowy, F. D. (1998). *Staphylococcus aureus* infections. *N. Engl. J. Med.* 339, 520–532. doi: 10.1056/NEJM199808203390806

Messad, N., Landraud, L., Canivet, B., Lina, G., Richard, J.-L., and Sotto, A. (2013). Distribution of *edin* in *Staphylococcus aureus* isolated from diabetic foot ulcers. *Clin. Microbiol. Infect.* 19, 875–880. doi: 10.1111/1469-0691.12084

Mhada, T. V., Fredrick, F., Matee, M. I., and Massawe, A. (2012). Neonatal sepsis at Muhimbili National Hospital, Dar es Salaam, Tanzania; aetiology, antimicrobial sensitivity pattern and clinical outcome. *BMC Public Health* 12:904. doi: 10.1186/1471-2458-12-904

Monecke, S., Coombs, G., Shore, A. C., Coleman, D. C., Akpaka, P., Borg, M., et al. (2011). A field guide to pandemic, epidemic and sporadic clones of methicillin-resistant *Staphylococcus aureus*. *PLoS ONE* 6:e17936. doi: 10.1371/journal.pone.0017936

Monecke, S., Kanig, H., Rudolph, W., Müller, E., Coombs, G., Hotzel, H., et al. (2010). Characterisation of Australian MRSA strains ST75- and ST883-MRSA-IV and analysis of their accessory gene regulator locus. *PLoS ONE* 5:e14025. doi: 10.1371/journal.pone.0014025

Monecke, S., Slickers, P., and Ehricht, R. (2008). Assignment of *Staphylococcus aureus* isolates to clonal complexes based on microarray analysis and pattern recognition. *FEMS Immunol. Med. Microbiol.* 53, 237–251. doi: 10.1111/j.1574-695X.2008.00426.x

Naidoo, R., Nuttali, J., Whitelaw, A., and Eley, B. (2013). Epidemiology of *Staphylococcus aureus* Bacteraemia at a Tertiary Children's Hospital in Cape Town, South Africa. *PLoS ONE* 8:e78396. doi: 10.1371/journal.pone.0078396

Novick, R. P. (2006). "Staphylococcal pathogenesis and pathogenicity factors: genetics and regulation," in *Gram-Positive Pathogens*, eds V. Fischetti, R. Novick, J. Ferretti, D. Portnoy, and J. Rood (Washington, DC: ASM Press), 496–516.

Nurjadi, D., Olalekan, A. O., Layer, F., Shittu, A. O., Alabi, A., Ghebremedhin, B., et al. (2014). Emergence of trimethoprim resistance gene dfrG in *Staphylococcus aureus* causing human infection and colonization in sub-Saharan Africa and its import to Europe. *J. Antimicrob. Chemother.* 69, 2361–2368. doi: 10.1093/jac/dku174

Oladeinde, B. H., Omoregie, R., Olley, M., Anunibe, J. A., and Onifade, A. A. (2013). A 5-year surveillance of wound infections at a rural tertiary hospital in Nigeria. *Afr. Health Sci.* 13, 351–356. doi: 10.4314/ahs.v13i2.22

Oosthuysen, W. F., Orth, H., Lombard, C. J., Sinha, B., and Wasserman, E. (2014). Population structure analyses of *Staphylococcus aureus* at Tygerberg Hospital, South Africa, reveals a diverse population, high prevalence of Panton-Valentine leukocidin genes and unique local MRSA clones. *Clin. Microbiol. Infect.* 20, 652–659. doi: 10.1111/1469-0691.12452

Raji, A., Ojemhen, O., Umejiburu, U., Ogunleye, A., Blanc, D. S., and Basset, P. (2013). High genetic diversity of *Staphylococcus aureus* in a tertiary care hospital in South west Nigeria. *Diagn. Microbiol. Infect. Dis.* 77, 367–369. doi: 10.1016/j.diagmicrobio.2013.08.030

Rasmussen, G., Monecke, S., Brus, O., Ehricht, R., and Soderquist, B. (2014). Long term molecular epidemiology of methicillin-susceptible *Staphylococcus aureus* bacteremia isolates in Sweden. *PLoS ONE* 9:e114276. doi: 10.1371/journal.pone.0114276

Rasmussen, G., Monecke, S., Ehricht, R., and Soderquist, B. (2013). Prevalence of clonal complexes and virulence genes among commensal and invasive *Staphylococcus aureus* isolates in Sweden. *PLoS ONE* 8:e77477. doi: 10.1371/journal.pone.0077477

Rovira, G., Cooke, E. F., Mucavele, H., Sitoe, A., Madrid, L., Nhampossa, T., et al. (2015). The challenge of diagnosing and treating *Staphylococcus aureus* invasive infections in a resource-limited sub-Saharan Africa setting: a case report. *J. Trop. Pediatr.* 61, 397–402. doi: 10.1093/tropej/fmv045

Ruffing, U., Akulenko, R., Bischoff, M., Helms, V., Herrmann, M., and von Müller, L. (2012). Matched-Cohort DNA microarray diversity analysis of methicillin sensitive and methicillin-resistant *Staphylococcus aureus* isolates from hospital admission patients. *PLoS ONE* 7:e52487. doi: 10.1371/journal.pone.0052487

Schaumburg, F., Ngoa, U. A., Kösters, K., Köck, R., Adegnika, A. A., Kremsner, P. G., et al. (2011). Virulence factors and genotypes of *Staphylococcus aureus* from infection and carriage in Gabon. *Clin. Microbiol. Infect.* 17, 1507–1513. doi: 10.1111/j.1469-0691.2011.03534.x

Schaumburg, F., Pauly, M., Anoh, E., Mossoun, A., Wiersma, L., Schubert, G., et al. (2015). *Staphylococcus aureus* complex from animals and humans in three remote African regions. *Clin. Microbiol. Infect.* 21, e1–e8. doi: 10.1016/j.cmi.2014.12.001

Seni, J., Bwanga, F., Najjuka, C. F., Makobore, P., Okee, M., and Mshana, S. E. (2013). Molecular characterization of *Staphylococcus aureus* from patients with surgical site infections at Mulago Hospital in Kampala, Uganda. *PLoS ONE* 8:e66153. doi: 10.1371/journal.pone.0066153

Shittu, A. O., Okon, K., Adesida, S., Oyedara, O., Witte, W., Strommenger, B., et al. (2011). Antibiotic resistance and molecular epidemiology of *Staphylococcus aureus* in Nigeria. *BMC Microbiol.* 11:92. doi: 10.1186/1471-2180-11-92

Shittu, A., Oyedara, O., Abegunrin, F., Okon, K., Raji, A., Taiwo, S., et al. (2012). Characterization of methicillin-susceptible and -resistant staphylococci in the clinical setting: a multicentre study in Nigeria. *BMC Infect. Dis.* 12:286. doi: 10.1186/1471-2334-12-286

Skaar, E. P., and Schneewind, O. (2004). Iron-regulated surface determinants (Isd) of *Staphylococcus aureus*: stealing iron from heme. *Microbes Infect.* 6, 390–397. doi: 10.1016/j.micinf.2003.12.008

Van Trijp, M. J. C. A., Melles, D. C., Snijders, S. V., Wertheim, H. F. L., Verbrugh, H. A., van Belkum, A., et al. (2010). Genotypes, superantigen gene profiles, and presence of exfoliative toxin genes in clinical methicillin-susceptible *Staphylococcus aureus* isolates. *Diagn. Microbiol. Infect. Dis.* 66, 222–224. doi: 10.1016/j.diagmicrobio.2009.08.021

Waters, D., Jawad, I., Ahmad, A., Luksic, I., Nair, H., Zgaga, L., et al. (2011). Aetiology of community-acquired neonatal sepsis in low- and middle-income countries. *J. Glob. Health* 1, 154–170.

Wright, J. S., Traber, K. E., Corrigan, R., Benson, S. A., Musser, J. M., and Novick, R. P. (2005). The agr radiation: an early event in the evolution of staphylococci. *J. Bacteriol.* 187, 5585–5594. doi: 10.1128/JB.187.16.5585-5594.2005

Yamaguchi, T., Nishifuji, K., Sasaki, M., Fudaba, Y., Aepfelbacher, M., and Takata, T. (2002). Identification of the *Staphylococcus aureus etd* pathogenicity island which encodes a novel exfoliative toxin, ETD, and EDIN-B. *Infect. Immun.* 70, 5835–5845. doi: 10.1128/IAI.70.10.5835-5845.2002

Yamasaki, O., Tristan, A., Yamaguchi, T., Sugai, M., Lina, G., Bes, M., et al. (2006). Distribution of the exfoliative toxin D gene in clinical *Staphylococcus aureus* isolates in France. *Clin. Microbiol. Infect.* 12, 585–588. doi: 10.1111/j.1469-0691.2006.01410.x

13

Comparative Genomics of Two ST 195 Carbapenem-Resistant *Acinetobacter baumannii* with Different Susceptibility to Polymyxin Revealed Underlying Resistance Mechanism

Soo-Sum Lean[1], *Chew Chieng Yeo*[2], *Zarizal Suhaili*[3] *and Kwai-Lin Thong*[1]*

[1] *Faulty of Science, Institute of Biological Sciences, Universiti Malaya, Kuala Lumpur, Malaysia,* [2] *Faculty of Medicine, Biomedical Research Centre, Universiti Sultan Zainal Abidin, Kuala Terengganu, Malaysia,* [3] *Faculty of Bioresources and Food Industry, Universiti Sultan Zainal Abidin, Kuala Terengganu, Malaysia*

***Correspondence:**
Kwai-Lin Thong
thongkl@um.edu.my

Acinetobacter baumannii is a Gram-negative nosocomial pathogen of importance due to its uncanny ability to acquire resistance to most antimicrobials. These include carbapenems, which are the drugs of choice for treating *A. baumannii* infections, and polymyxins, the drugs of last resort. Whole genome sequencing was performed on two clinical carbapenem-resistant *A. baumannii* AC29 and AC30 strains which had an indistinguishable *Apa*I pulsotype but different susceptibilities to polymyxin. Both genomes consisted of an approximately 3.8 Mbp circular chromosome each and several plasmids. AC29 (susceptible to polymyxin) and AC30 (resistant to polymyxin) belonged to the ST195 lineage and are phylogenetically clustered under the International Clone II (IC-II) group. An AbaR4-type resistance island (RI) interrupted the *comM* gene in the chromosomes of both strains and contained the $bla_{\mathrm{OXA-23}}$ carbapenemase gene and determinants for tetracycline and streptomycin resistance. AC29 harbored another copy of $bla_{\mathrm{OXA-23}}$ in a large (~74 kb) conjugative plasmid, pAC29b, but this gene was absent in a similar plasmid (pAC30c) found in AC30. A 7 kb Tn*1548*::*armA* RI which encodes determinants for aminoglycoside and macrolide resistance, is chromosomally-located in AC29 but found in a 16 kb plasmid in AC30, pAC30b. Analysis of known determinants for polymyxin resistance in AC30 showed mutations in the *pmrA* gene encoding the response regulator of the two-component *pmrAB* signal transduction system as well as in the *lpxD*, *lpxC,* and *lpsB* genes that encode enzymes involved in the biosynthesis of lipopolysaccharide (LPS). Experimental evidence indicated that impairment of LPS along with overexpression of *pmrAB* may have contributed to the development of polymyxin resistance in AC30. Cloning of a novel variant of the bla_{AmpC} gene from AC29 and AC30, and its subsequent expression in *E. coli* also indicated its likely function as an extended-spectrum cephalosporinase.

Keywords: ***Acinetobacter baumannii*, carbapenem-resistant, polymyxin, whole genome sequencing, resistance island, resistance mechanisms**

INTRODUCTION

Acinetobacter baumannii is a significant nosocomial pathogen that infects immunocompromised patients, including patients with underlying diseases, especially those who are warded in intensive care units (ICU; Bergogne-Bérézin and Towner, 1996; Nordmann and Poirel, 2008; Camp and Tatum, 2010). These nosocomial infections include ventilator-associated pneumonia, secondary meningitis, endocarditis, urinary tract infections, surgical site infections, and blood stream infections (Camp and Tatum, 2010; Huang et al., 2012). *A. baumannii* is intrinsically resistant to commonly used antibiotics such as aminopenicillins, first- and second-generation cephalosporins and chloramphenicol (Dijkshoorn et al., 2007). The notoriety of this pathogen stems from its ability to develop and acquire resistance to almost all antimicrobial drugs as well as its tolerance to desiccation and ability to survive on inanimate surfaces for prolonged periods of time (Camp and Tatum, 2010; Roca et al., 2012; Doi et al., 2015). In depth analyses of the resistance mechanisms in *A. baumannii* revealed that its multidrug resistance phenotype is mediated by all the major resistance mechanisms that are known to occur in bacteria, including modification of target sites, enzymatic inactivation, active efflux, and decreased influx of drugs (Dijkshoorn et al., 2007).

Carbapenem resistance in *A. baumannii* is usually acquired with the most significant mechanism being the production of carbapenemases (Poirel and Norman, 2006). *A. baumannii* naturally produces the OXA-51-group carbapenemase at a low level. The transposition of an insertion sequence (usually IS*Aba1* or IS*Aba9*) upstream of the $bla_{\mathrm{OXA-51}}$ gene provides a strong promoter for expression of the carbapenemase leading to elevation of carbapenem MICs and thus, resistance (Higgins et al., 2010). *A. baumannii* also readily acquires several OXA-group β-lactamases usually through transposons and plasmids with OXA-23 being the most prevalent (Poirel and Norman, 2006; Roca et al., 2012).

A. baumannii naturally produces the AmpC-type β-lactamase and overexpression of the bla_{AmpC} gene has been associated with the presence of IS*Aba1* providing a strong promoter leading to cephalosporin resistance (Segal et al., 2005; Tian et al., 2011). These *Acinetobacter*-derived cephalosporinases (ADCs) typically hydrolyze penicillins, narrow- and extended-spectrum cephalosporins but not cefepime or carbapenems (Rodríguez-Martínez et al., 2010). Recently, several ADC variants have been identified that are capable of hydrolyzing cefepime and these are termed as extended-spectrum AmpCs (ESACs) (Rodríguez-Martínez et al., 2010; Tian et al., 2011).

It is of great concern when carbapenems, which are the drugs of choice for treatment, are increasingly compromised. As the efficacy of these drugs decreases, polymyxins were re-introduced as the drug of "last resort" (Landman et al., 2008). The renewed interest in polymyxins (polymyxin B and colistin, or polymyxin E) as therapeutic agents was due to the pathogen's outer membrane, polyanionic lipopolysaccharide (LPS) having high affinity toward the cationic polymyxin molecules (Landman et al., 2008). Binding of the LPS and polymyxin molecules resulted in a detergent-like effect that disrupts the outer membrane thereby causing cytoplasm leakage of the pathogen (Landman et al., 2008; Arroyo et al., 2011). Although polymyxins were once discarded due to their neurotoxicity and nephrotoxicity, polymyxins have a more effective antibacterial activity against most Gram negative pathogens, including *A. baumannii* (Landman et al., 2008). Previous studies from Garnacho-Montero et al. (2003) and Moffatt et al. (2010) showed that intravenous polymyxins were safe to use as an effective treatment to *Acinetobacter* infections. However, uncontrolled use or overuse of polymyxins in the hospital environment may lead to the development of polymyxin-resistance in *A. baumannii* (Arroyo et al., 2011).

Polymyxin resistance in *A. baumannii* appeared to develop intrinsically as a result of exposure to this class of drugs. Two major mechanisms of polymyxin resistance have been described for *A. baumannii*. The first is the modification of the lipid A moiety of LPS with phosphoethanolamine as a result of mutations in the *pmrA/pmrB* two-component signal transduction system which leads to the up-regulated expression of the *pmrCAB* operon. Overexpression of *pmrC* which encodes the enzyme responsible for phosphoethanolamine addition to lipid A, impairs the binding of polymyxin to the outer membrane thereby leading to resistance (Adams et al., 2009; Arroyo et al., 2011; Beceiro et al., 2011; Park et al., 2011). The second mechanism is the complete loss of the LPS caused either by mutations or the insertional inactivation of the lipid A biosynthesis genes, namely *lpxA*, *lpxC*, and *lpxD* (Moffatt et al., 2010, 2011). Mutations in the *lpsB* gene that encodes a glycosyltransferase involved in the biosynthesis of the LPS core have also been implicated in polymyxin resistance (Hood et al., 2013).

We have previously characterized 54 *A. baumannii* strains obtained from a tertiary hospital in Terengganu, Malaysia (Lean et al., 2014). Out of these, 39 were carbapenem- and multidrug-resistant (MDR). Among the 39 carbapenem resistant strains, 14 were also resistant to polymyxin B and categorized as extensive-drug resistant (XDR). Two strains, *A. baumannii* AC29 and AC30 were isolated from the wounds of different patients, and shared an identical *Apa*I pulsotype. However, AC29 is susceptible to polymyxin B whereas AC30 was resistant with an MIC value of 128 μg/ml. Here, we report the comparative genomic analyses of these two *A. baumannii* strains, AC29 and AC30, to show that despite sharing an identical pulsotype, there are significant changes in the genome structure particularly in the resistance islands and the plasmid content of the two isolates. We also present experimental evidence to elucidate possible mechanisms for the development of polymyxin resistance in *A. baumannii* AC30 and the likely implication of a novel bla_{AmpC} cephalosporinase gene in resistance to extended-spectrum cephalosporins as well as imipenem in both AC29 and AC30.

MATERIALS AND METHODS

Whole Genome Sequencing

Strains Selection, Antibiotic Resistance Profiles, and DNA Extraction

Two strains of *A. baumannii* AC29 and AC30 from a tertiary hospital in Terengganu, Malaysia were selected for this study (Lean et al., 2014). Both strains were obtained from the wounds of different patients using standard microbiology procedures. *A. baumannii* AC29 and AC30 shared an identical *Apa*I pulsotype and showed resistance to gentamycin, tobramycin, amikacin, ciprofloxacin, levofloxacin, piperacillin-tazobactam, cefotaxime, ceftazidime, cefepime, ampicillin-sulbactam, tetracycline, and doxycycline. Both isolates were also resistant to carbapenem with MIC values of >32 µg/ml for both imipenem and meropenem (Clinical and Laboratory Standards Institute, 2013). However, AC29 was susceptible to polymyxin B whereas AC30 was resistant with an MIC value of 128 µg/ml (Magiorakos et al., 2011; Lean et al., 2014).

Genomic DNA of AC29 and AC30 were extracted using Wizard Genomic DNA Purification Kit (Promega, USA) according to the manufacturer's instructions. Extracted DNA was quantified using the spectrophotometer at OD_{260} and the purity was determined by OD_{260}/OD_{280} ratio (Ausubel et al., 2002).

Genomic Sequencing, Assembly, and Annotations

Whole genome sequencing of *A. baumannii* AC29 and AC30 was carried out by a commercial vendor using the Illumina Genome Analyzer IIx platform. CLC Bio software package was used to assemble the genome sequence data. Open reading frame (ORF) prediction and gene functional assignments were done using Prodigal 2.60 (Hyatt et al., 2010), RNAmmer 1.2 (Lagesen et al., 2007), and tRNAscan-SE (Lowe and Eddy, 1997). Functional annotation of the genome was performed using Blast2Go and the Rapid Annotation using Subsystem Technology (RAST) server (Aziz et al., 2008).

MLST and Phylogenetic Analysis

To determine the sequence types, multilocus sequence typing (MLST) of AC29 and AC30 was performed according to the Bartual or Oxford scheme using the seven housekeeping genes *cpn60*, *gdhB*, *gltA*, *gpi*, *gyrB*, *recA*, and *rpoD* (Bartual et al., 2005) and the Pasteur scheme using the genes *cpn60*, *fusA*, *gltA*, *pyrG*, *recA*, *rplB*, and *rpoB* (Diancourt et al., 2010). The gene sequences were compared to the PubMLST database for *A. baumannii* (http://pubmlst.org/abaumannii/) and assigned to the appropriate sequence types.

The phylogenetic relationship of AC29 and AC30 to sequenced *A. baumannii* strains was inferred using composition vector tree (CVTree) version 2 (Xu and Hao, 2009) based on the concatenated nucleotide sequences of the seven reference genes used in the Bartual MLST scheme. The CVTree web-based server (http://tlife.fudan.edu.cn/cvtree/) generates phylogenetic trees based on *k*-tuple values (Xu and Hao, 2009). The completed *A. baumannii* genomes used in the analysis were downloaded via the NCBI FTP site and the reference gene sequences extracted for phylogenetic analysis.

Comparative Genomics Analyses

Whole genome sequences of these two *A. baumannii* AC29 and AC30 strains were compared to our previously reported *A. baumannii* AC12 genome (Lean et al., 2015) and the completed *A. baumannii* genomes available from NCBI FTP site using Mauve (Darling et al., 2004). Circular chromosome and plasmid map of these strains were constructed using CGView Server developed by the Stothart Research Group (Grant and Stothard, 2008).

Accession Numbers

Genome sequences of *A. baumannii* strains AC29 and AC30 were deposited under the accession number CP007535 and CP007577, respectively. Plasmid sequences of pAC29a, pAC29b, pAC30a, pAC30b, and pAC30c were deposited under the accession number CP008850, CP008851, CP007578, CP007579, and CP007580, respectively. Accession numbers for our previously reported *A. baumannii* strain AC12 (CP007549) and pAC12 (CP007550) were as in (Lean et al., 2015).

Quantitative Real-Time PCR Of *pmrAB*

One of the gene loci that have been implicated in polymyxin resistance is the *pmrCAB* operon with *pmrAB* encoding the two-component signal transduction system and *pmrC* encoding the enzyme that catalyzes the addition of phosphoethanolamine to the lipid A moiety of lipopolysaccharide (LPS; Arroyo et al., 2011; Park et al., 2011; Hood et al., 2013). To determine the transcript levels of *pmrAB* from the polymyxin-susceptible AC29, and the polymyxin-resistant isolates AC30 as well as AC12 (Lean et al., 2015), quantitative real-time PCR (qRT-PCR) was employed. *A. baumannii* ATCC19606 was used as the polymyxin-susceptible control. Total RNA was extracted from *A. baumannii* strains grown in LB broth using Qiagen RNeasy Mini Kit (Qiagen).

The extracted total RNA was subjected to reverse transcription using the Quantitect Reverse Transcription Kit (Qiagen) in a T-Gradient PCR machine from Biometra/Applied Biosystems. The resulting cDNA produced was used as template for quantitative real-time PCR using the SYBR Green PCR Kit (Qiagen) performed in a Rotor-Gene 6000 Real-Time PCR Machine (Corbett Life Science/Qiagen). Prior to the actual amplification cycling conditions, the mixture was subjected to initial heat activation at 95°C for 5 min followed by 35 cycles of the two-step cycling condition of denaturation at 95°C for 10 s and annealing at 60°C for 30 s. The melting curve profile for each amplification reaction and the relevant C_T value were automatically generated using the software provided (Corbett Life Science/Qiagen). The *rpoB* gene was used as the housekeeping gene for normalization (Park et al., 2011). Relative quantification using the $\Delta\Delta C_T$ method (Pfaffl, 2001) was then applied to quantify the expression of *pmrAB*.

Lipopolysaccharide Analysis

Lipopolysaccharide (LPS) analysis was carried out for the polymyxin resistant strains AC30 and AC12 (Lean et al.,

2015), the polymyxin susceptible strain AC29 along with *A. baumannii* ATCC19606 as the polymyxin-susceptible control. LPS extraction was carried out according to Hood et al. (2013). Briefly, bacteria were swabbed from LB agar plates into 154 mM NaCl, and the cell density was adjusted to an OD_{600} of 1.5, pelleted and resuspended in lysis buffer (2% SDS, 4% 2-mercaptoethanol, 10% glycerol, 0.1 M Tris-HCl, pH 6.8). Samples were then boiled for 10 min, cooled to 60°C and treated with proteinase K for 1 h. These samples were electrophoresed through a 15% polyacrylamide gel using standard methods and stained with Pro-Q Emerald 300 LPS stain according to the manufacturer's recommendations (Invitrogen). Molecular weight marker used in the experiment was the Precision Plus Protein Dual Xtra Standards (BioRad). The stained minigels were then viewed and photographed under UV light using a UV Gel Imager (Alpha-Innotech).

Cloning and Expression of the Novel Variant *bla*$_{\text{AmpC}}$ Gene

The novel variant *bla*$_{\text{AmpC}}$ gene was PCR-amplified from the genomes of *A. baumannii* AC29 and AC30 using specific primers ampC_BamHI: 5′-GGATCCATGGCTGTGGGTGTTATTCAA-3′ and ampC_HindIII: 5′-AAGCTTTTATTTCTTTATTGCATTCAGCAC-3′ with the following conditions—initial hold at 95°C for 3 min, followed by 40 cycles of 95°C for 50 s, 55°C for 50 s, 72°C for 1 min, and final extension at 72°C for 90 s. Purified PCR products were then cloned into pGEM-T Easy (Promega, USA) according to the manufacturer's instructions and transformed into *E. coli* JM109. Transformants were selected on LB agar supplemented with 100 mM ampicillin with 100 mM IPTG and 50 mg/ml X-gal. Plasmids were prepared from white colonies and screened by digestion with *Bam*HI and *Hin*dIII which cut at the restriction sites incorporated into the primers used for amplification. Plasmids with the expected restriction fragments were validated by conventional Sanger dideoxy sequencing prior to subcloning into the pET30a expression vector. The resulting pET30a-*ampC* recombinant plasmids were then transformed into *E. coli* BL21 (DE3/pLysS) and expression of the cloned *bla*$_{\text{AmpC}}$ gene carried out by inducing the cells with 0.1 mM IPTG. After that, the *E. coli* BL21 recombinant clones were tested for their MIC values for extended-spectrum cephalosporins (i.e., ceftazidime and cefepime) as well as aztreonam and imipenem at concentrations of 2, 4, 8, 16, and 32 μg/ml using the agar dilution method.

RESULTS

Strain Characteristics and Genome Analyses

Basic Genome Features and Sequence Types

Analyses of the whole genome sequencing indicated that the *A. baumannii* AC29 and AC30 genomes were 3,935,134 and 3,925,274 bp, with GC content of 38.84 and 38.98%, respectively. Predicted ORFs from the genomes of *A. baumannii* AC29 and AC30 were 3728 and 3646, respectively. Both strains contained only one chromosome with varying number of plasmids, as featured in **Table 1**. The Bartual or Oxford scheme for MLST utilizes seven housekeeping genes *cpn60, gdhB, gltA, gpi, gyrB, recA*, and *rpoD* to determine the sequence type (ST) of *A. baumannii* (Bartual et al., 2005). Both AC29 and AC30 were designated ST195 (derived from *cpn60*-2, *gdhB*-3, *gltA*-1, *gpi*-96, *gyrB*-3, *recA*-2, and *rpoD*-3) whereas using the Pasteur MLST scheme (Diancourt et al., 2010), both *A. baumannii* strains were designated ST2 (*cpn60*-2, *fusA*-2, *gltA*-2, *pyrG*-2, *recA*-2, *rplB*-2, and *rpoB*-2). Thus, AC29 and AC30 were categorized as belonging to the International Clone-II (or IC-II, also referred to as Global Clonal 2, or GC2) group of strains.

TABLE 1 | General genomic features of the whole genome sequences of *A. baumannii* strains AC29 and AC30.

Feature	Strains	
	AC29	AC30
Accumulated length	3,935,134 bp	3,925,274 bp
Average GC content	38.84%	38.98%
Number of contigs	102	91
Number of ORF	3728	3646
Number of tRNA	66	58
Number of rRNA	4	3
Number of plasmid	2	3
Plasmid size	pAC29a (8737 bp) pAC29b (74,749 bp)	pAC30a (8729 bp) pAC30b (16,236 bp) pAC30c (71,433 bp)

Phylogenetic analysis of the *A. baumannii* AC29 and AC30 genomes were performed with other available *A. baumannii* genomes from the NCBI's FTP site and using *A. baumannii* SDF as an outgroup to root the phylogenetic tree based on the concatenated nucleotide sequences of the seven reference genes utilized in the Bartual MLST scheme (Farrugia et al., 2013). Based on the phylogenetic tree (**Figure 1**), *A. baumannii* AC29 and AC30 were closely related to *A. baumannii* AC12 which was also isolated from the same hospital in Terengganu and was also typed as ST195 (Lean et al., 2015) along with *A. baumannii* M1 which was also isolated from Malaysia (accession no. LAIL0100000). These strains were clustered together with other *A. baumannii* isolates of the IC-II group and this was also reflected in their genome grouping based on genomic BLAST (as indicated at http://www.ncbi.nlm.nih.gov/genome/genomegroups/403? accessed on 19 August 2015).

The AbaR4-Type Resistance Island

The genomes of *A. baumannii* AC29 and AC30 contained an AbaR4-type resistance island (RI) which interrupts the *comM* gene (Nigro and Hall, 2012). The island found in AC29 is 23 kb and designated AC29-RI1 whereas in AC30, it is 26 kb and designated AC30-RI1. Tn*6022* and/or ΔTn*6022* is part of the important features for this type of RI; in AC29-RI1, only the entire *tniC* gene is deleted in ΔTn*6022* whereas AC30-RI1 carries a complete copy of Tn*6022* (**Figure 2**). In AC29-RI1, the ΔTn*6022* is missing the *tniC* gene on its far left end, but full-length *tniD* and *tniE* genes are present in the transposon.

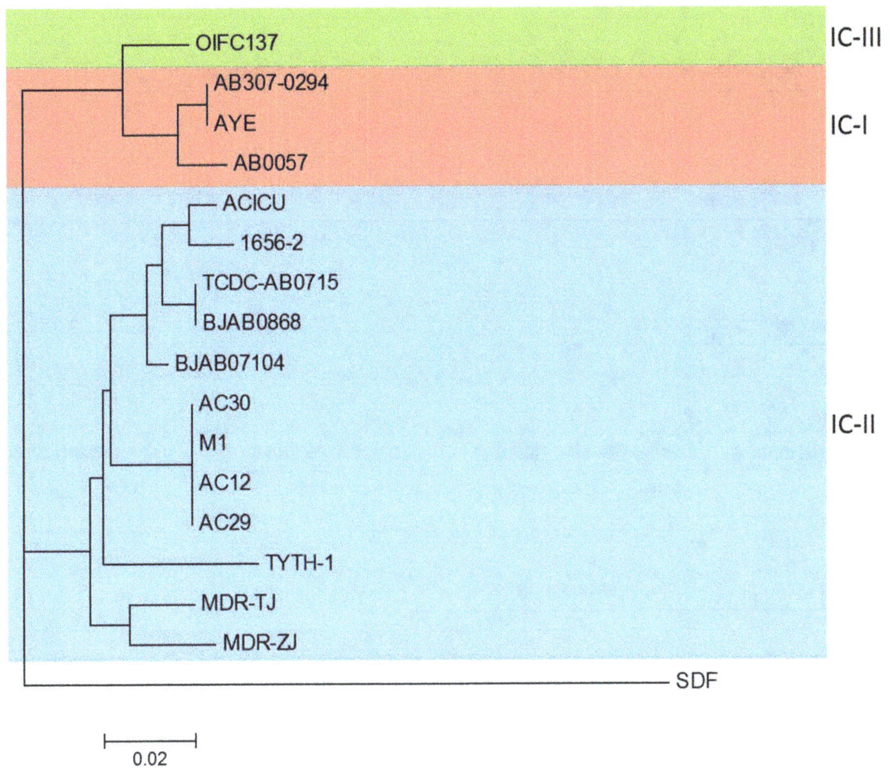

FIGURE 1 | Phylogenetic analysis of *A. baumannii* AC12, AC29, AC30, and completed *A. baumannii* genomes obtained from NCBI. The phylogenetic tree was constructed by composition vector tree version 2 (CVTree 2) using concatenated nucleotide sequences of the seven reference genes based on the Bartual/Oxford MLST scheme (*cpn60*, *gdhB*, *gltA*, *gpi*, *gyrB*, *recA,* and *rpoD*). Strains belonging to the international clonal groups IC-I, IC-II, and IC-III are indicated. The accession nos. of the strains used in the analysis are as follows: AB307-0294 (CP001172.1), AYE (CU459141.1), AB0057 (CP001182.1), ACICU (CP000863.1), 1656-2 (CP001921.1), TCDC-AB0715 (CP002522.2), BJAB0868 (CP003849.1), BJAB07104 (CP003846.1), M1 (LAIL01000001.1), TYTH-1 (CP003856.1), MDR-TJ (CP003500.1), MDR-ZJ (CP001937.1), OIFC137 (NZ_AFDK01000002.1), SDF (NC_010400.1).

AC30-RI1 contains the full length Tn*6022*. In both AC29-RI1 and AC30-RI1 islands, only one copy of either ΔTn*6022* or Tn*6022* is found.

Drug resistance genes found in the AC29-RI1 and ACRI30-1 islands are *sulI* (conferring sulphonamide resistance), bla_{OXA-23} (conferring carbapenem resistance), *tetA* and *tetB* (conferring tetracycline resistance), and *strA* with *strB* (conferring streptomycin resistance). The bla_{OXA-23} gene is flanked by two copies of the IS*Aba1* insertion element in a composite transposon structure similar to Tn*2006* (**Figure 2**). Tn*2006* comprises of the bla_{OXA-23} carbapenemase-encoding gene, a DEAD/DEAH box helicase-like gene and an ATPase gene (*yeeA*) flanked by two copies of IS*Aba1*.

Plasmids of *A. baumannii* AC29 and AC30

A Small 8.7 kb Plasmid Found in Both AC29 and AC30

Sequence analysis revealed the presence of a small 8.7 kb plasmid in both the genomes of *A. baumannii* AC29 and AC30, designated pAC29a and pAC30a, respectively (**Table 1**). These plasmids contained a total of nine coding sequences (CDS) (**Figure 3**). No antibiotic resistance gene was found on these plasmids. Two plasmid replication genes, designated *repA* and *repB*, were found on these plasmids along with an iteron sequence made up of four direct repeats of 5′-ATA TGT CCA CGT TTA CCT TGC A-3′ located 53 nucleotides upstream of the *repB* gene. Other features on this cryptic plasmid are a Sel1 repeat protein-encoding gene (sel1) which is flanked by XerC/XerD-like recombination sites in an inverted repeat formation, an outer membrane TonB-dependent receptor gene, a gene encoding for putative septicolysin and two hypothetical protein-encoding genes. A toxin-antitoxin (TA) system designated AbkB/AbkA (Mosqueda et al., 2014) was also encoded on these 8.7 kb plasmids.

pAC30b, a 16.2 kb Resistance Plasmid Found in *A. baumannii* AC30

Plasmid pAC30b is 16,236 bp in size and is found only in *A. baumannii* AC30. pAC30b contained 11 CDS and is a resistance plasmid, as indicated by the presence of the 16S rRNA methylase gene (*arm*A) and aminoglycoside 3′-phosphotransferase gene (*aph*A1) which confer resistance to aminoglycosides, along with macrolide 2′-phosphotransferase (*mph2*) and macrolide efflux protein-coding (*mel*) genes which confer resistance to macrolides (**Figure 4**; Zhou et al., 2011). The *armA*, *mph2,* and *mel* genes along with adjacent putative transposase genes *tnpD* and *tnpU* make up a small 7 kb RI designated Tn*1548*::*armA* which have been reported in several other plasmids from *Enterobacteriaceae* (Dolejska et al., 2013).

FIGURE 2 | Structures of resistance islands present in *A. baumannii* AC12, AC30, and AC29 as compared to similar islands in *A. baumannii* MDR-ZJ and MDR-TJ. (A) Structure of resistance island ACRI12-1 present in *A. baumannii* AC12. ΔTn*6022*, Tn*6022*, and Tn*2006* are indicated in the red dotted-line box. Directions of genes and ORFs are indicated by arrows above the central thick line and the names are as given above. Green boxes represent truncated Δ*comM*, turquoise blue boxes represent insertion element, IS*Aba1*, and orange boxes represent genes conferring antibiotic resistance. Numbered boxes in gray represent genes encoding for hypothetical proteins. Comparison between structures of **(A)** AC12-RI1, **(B)** AC30-RI1, **(C)** AC29-RI1, **(D)** AbaR22, and **(E)** RI_{MDR-TJ}, revealed similarities in the islands, as indicated by the same colors and red dotted-line box.

The *tnpU* putative transposase gene located at the 3′-end of Tn*1548::armA* is truncated in pAC30b whereby only the last 480 bp of the 836-bp gene could be found and was thus designated Δ*tnpU*. The pAC30b plasmid encodes a number of putative transposases: Δ*tnpU* and *tnpD* from Tn*1548::armA*, *tnpA*, and *tnpB* transposases from IS66, a 2888-bp *tnpA* from the Tn3 family of transposases (or Tn*1000*-like) and immediately adjacent to that, another smaller 704-bp *tnpA* transposase belonging to IS*26*. The aminoglycoside resistance gene *aphA1* is found downstream from IS*26*.

A Large ca. 70 kb Conjugative Plasmid in the Genomes of *A. baumannii AC29* and AC30

Both *A. baumannii* AC29 and AC30 were found to harbor similar conjugative plasmids of ca. 70 kb in size designated pAC29b and pAC30c, respectively. There are a total of 101 CDS in the 74,749 bp pAC29b whereas pAC30b is 71,433 bp and contains a total of 96 CDS. Both pAC29b and pAC30c encode a complete *tra* locus (**Figure 5**) similar to the *tra* locus found in pACICU2 (Iacono et al., 2008), p2ABTCDC0715 (Chen et al., 2011), pAb-G7-2 (Hamidian and Hall, 2014), and pA85-3 (Hamidian et al., 2014). Genes for the T4SS in pAC30c and pAC29b are clustered in two separate regions: the main *tra* region responsible for mating pair formation spans approximately 20 kb from *traL* to *traG*, and a second smaller region containing *traD* and *traI* (or *trwB* and *trwC*) responsible for plasmid mobilization. In pAC30c, the space between these two regions contains several hypothetical ORFs and a solitary *relE* toxin gene without the corresponding *relB* antitoxin gene (**Figure 5**).

Plasmid pAC29b is larger by 3316 bp when compared to pAC30c. However, there is a 14,374 bp fragment in pAC30c which has no homology with pAC29b. The additional fragment found in pAC29b included the β-lactamase gene bla_{OXA-23} and two transcriptional regulators, *tetR* and *asnC*. Other identifiable genes within this pAC29b-unique fragment include *yeeA* (encoding DNA methyltransferase), *yeeB* (encoding ATP-dependent helicase), *yhbS* (encoding N-acetyltransfease),

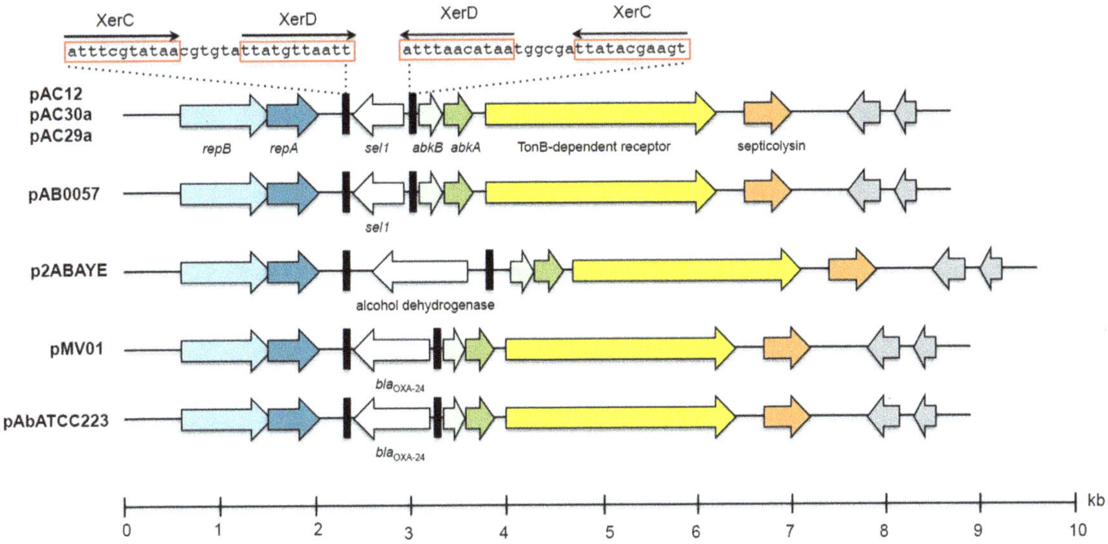

FIGURE 3 | Linear map showing the genetic organization of plasmids pAC29a, pAC30a, and pAC12 in comparison with other similar *A. baumanii* plasmids. The XerC/XerD-like recombination sites are indicated in black rectangular boxes with the nucleotide sequences as indicated above the boxes. The two plasmid replication genes, *repB* and *repA*, are indicated in blue whereas the putative *abkA/abkB* toxin-antitoxin genes are indicated in green arrows. Hypothetical ORFs are shaded gray. The TonB-dependent receptor gene is indicated in yellow whereas the putative septicolysin gene is in orange.

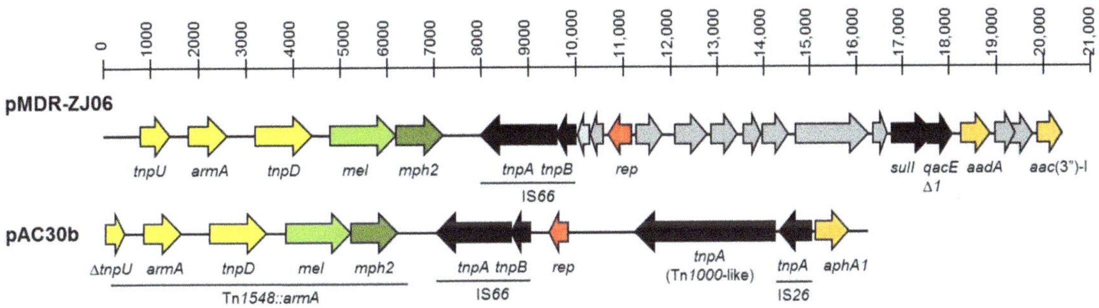

FIGURE 4 | Linear map comparison of the Tn*1548::armA* region in plasmid pAC30b from *A. baumannii* AC30 and pMDR-ZJ06 from *A. baumannii* MDR-ZJ06 (NC_017172). The Tn*1548::armA* region and insertion elements (IS) were underlined and indicated in the diagram.

mscM (encoding mini-conductance mechanosensitive channel), *iutA* (encoding ferric aerobactin receptor), *lrp* (encoding leucine-responsive regulatory protein), and *aroQ* (encoding monofunctional chorismate mutase). Thus, AC29 harbors two copies of bla_{OXA-23}—one within the AbaR4-like AC29RI-1 island in the chromosome, and another encoded on the pAC29b plasmid. However, plasmid pAC30c does not encode for any known resistance determinant.

Investigations into the Possible Contributors of Polymyxin Resistance in *A. baumannii* AC30

To investigate if differential expression of *pmrAB* occurred in the polymyxin-resistant AC30 as well as AC12 strains, relative quantification of the *pmrAB* transcript levels were determined by qRT-PCR with *A. baumannii* ATCC19606 as the polymyxin-susceptible control. When compared to ATCC19606, the relative expression levels of *pmrA* in AC30 was surprisingly lower at about 0.2-folds but levels of *pmrB* were higher at 4.8-folds (**Figure 6**). In contrast, the other polymyxin-resistant strain, AC12 displayed about 8.5-folds higher *pmrA* expression levels and about two-folds higher *pmrB* levels in comparison with ATCC19606. The polymyxin-susceptible AC29 showed dramatically lower *pmrAB* expression, at about 0.05- and 0.03-folds, respectively.

LPS from the two polymyxin-resistant strains, AC12 and AC30, along with two polymxyin-susceptible strains AC29 and ATCC19606 were extracted and analyzed on 15% SDS-polyacrylamide gels (**Figure 7**). SDS-PAGE of the extracted LPS yielded a band of ~10 kD, which was within the expected molecular weight (between 6 and 10 kDa) for LPS. Results indicated that the LPS in the polymyxin resistant strains AC12 and AC30 were not totally absent as had been previously reported

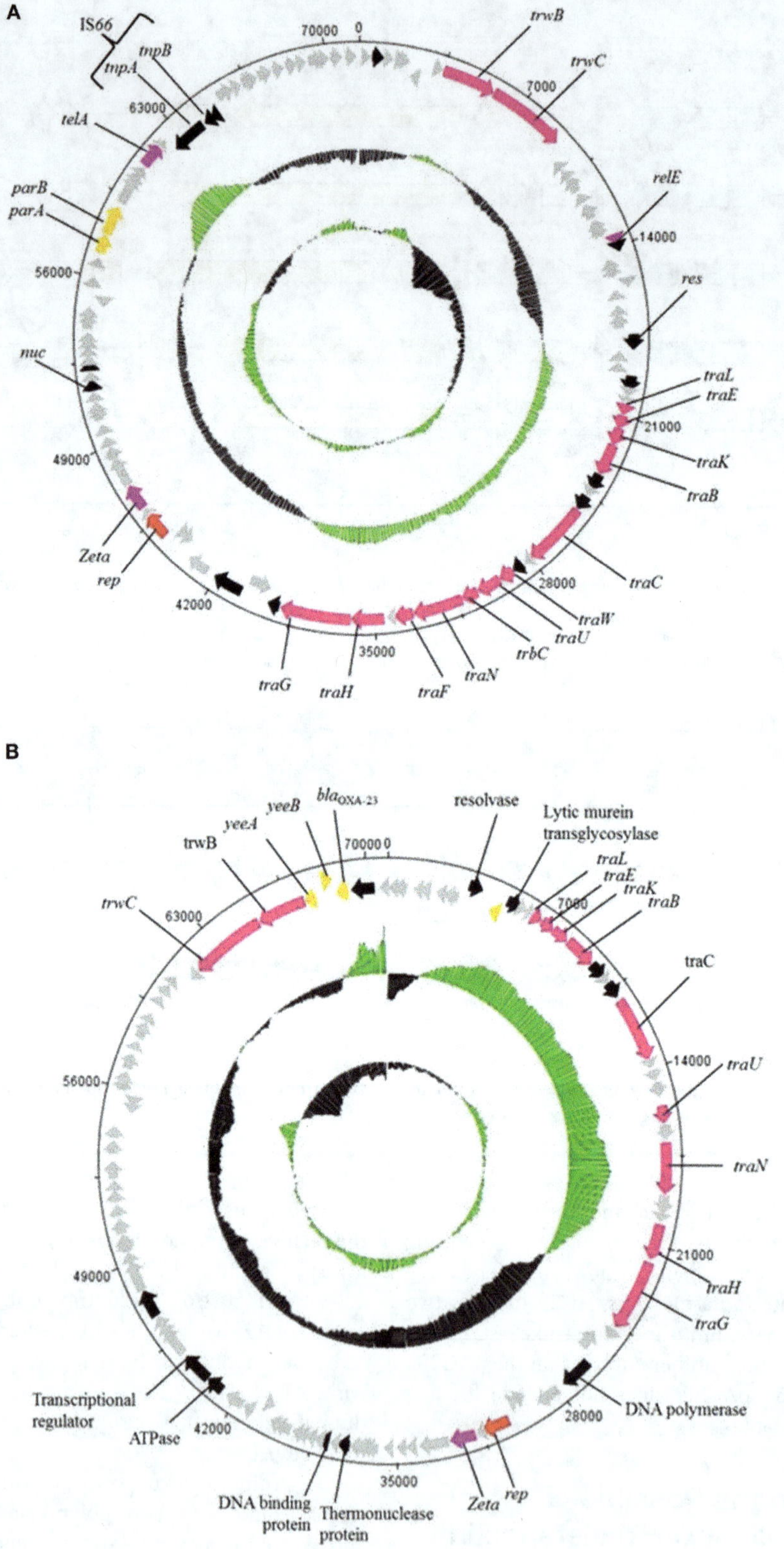

FIGURE 5 | Circular map of the ~70 kb plasmids pAC30c (A) and pAC29b (B). Outer circle of the map represents ORFs found in the plasmid, gray ORFs are genes encoding for hypothetical proteins. The two inner circles represent GC plot and GC skew, whereby the green circle stands for above average and black circle stands for below average G+C content. Red colored arrow represents plasmid replication gene, *rep*; orange colored arrow represents resistant determinants; pink colored arrow represents *tra* genes; black colored arrow represents genes encoding proteins with known functional homologs; purple colored arrow represents putative toxin gene homologs of toxin-antitoxin systems; and gray colored arrow represents genes encoding hypothetical proteins.

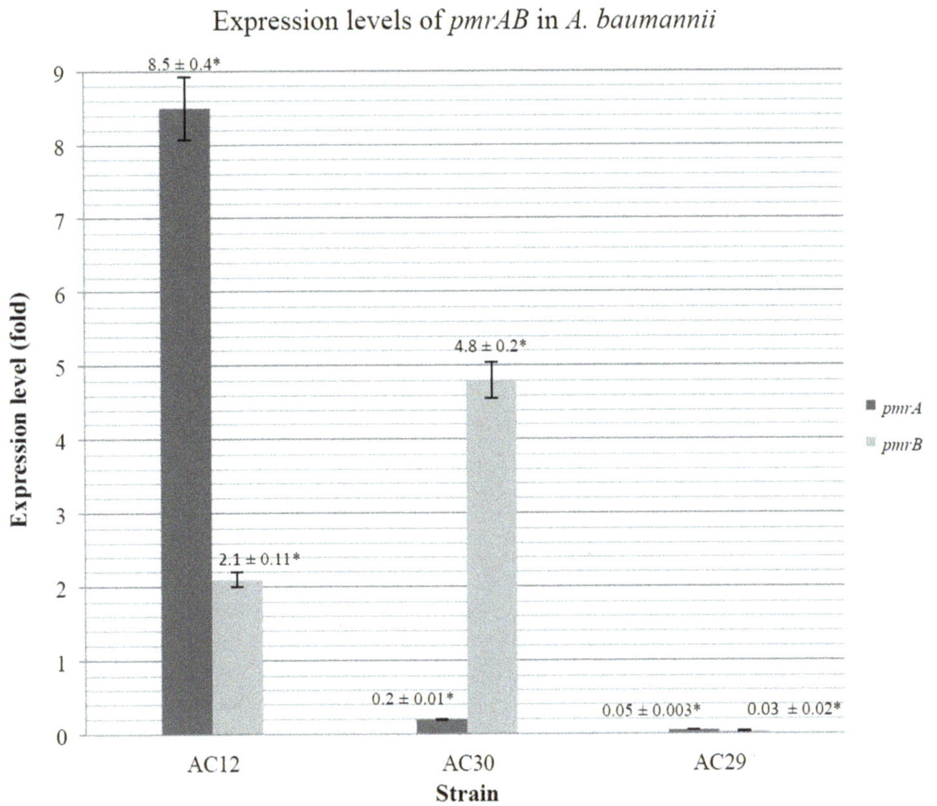

FIGURE 6 | Relative expression levels of *pmrAB* of the polymyxin-resistant *A. baumannii* AC12 and AC30 as compared to the polymyxin-susceptible *A. baumannii* AC29 as determined by Quantitative Real-Time reverse transcriptase-PCR (qRT-PCR). The data represents the mean fold change ± standard deviation (SD; indicated as error bars in the graph) taken from three replicates performed for each qRT-PCR reaction. Asterisk (*) indicate statistical significance, as determined by using two-tailed, unpaired Student's *t*-test with $p < 0.05$.

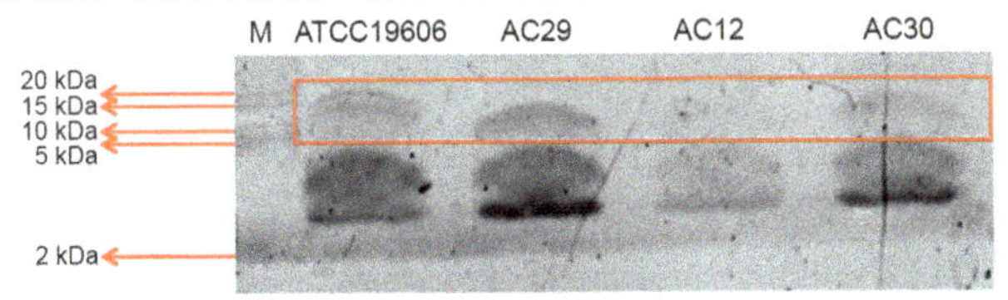

FIGURE 7 | SDS-PAGE analysis of the extracted lipopolysaccharide (LPS) layer of the *A. baumannii* ATCC19606 (control; polymyxin susceptible strain), AC12 (polymyxin resistant strain), AC29 (polymyxin susceptible strain), and AC30 (polymyxin resistant strain). M stands for the protein marker Precision Plus Protein Dual Xtra Standards (BioRad) with the sizes as indicated in kDa.

in other polymyxin-resistant strains (Moffatt et al., 2010, 2011; Hood et al., 2013) but the intensity of the LPS band was considerably less when compared with the LPS band of the polymyxin-susceptible strains.

AmpC-Mediated Resistance to Extended-Spectrum Cephalosporins

The AmpC cephalosporinase encoded by *A. baumannii* AC29 and AC30 displayed identical non-synonymous mutations, R80S and G246S (with reference to ADC-7, the reference *A. baumannii*-encoded AmpC; Hujer et al., 2005; Rodríguez-Martínez et al., 2010) to that found in *A. baumannii* AC12 (Lean et al., 2015) but the effects of these mutations on cephalosporin resistance have not been reported. To investigate if the mutations in the bla_{AmpC} gene harbored by *A. baumannii* AC29 and AC30 has any effect on resistance against β-lactams especially cephalosporins, these genes were cloned and expressed in *E. coli* BL21 through the IPTG-inducible T7 promoter of the pET30a vector. Recombinant *E. coli* BL21 carrying the bla_{AmpC} from AC29 and AC30 displayed resistance to ceftazidime, cefepime, aztreonam, and even imipenem (**Table 2**). Thus, the bla_{AmpC}

TABLE 2 | MIC values of the *ampC* clones in IPTG-induced *E. coli* BL21 toward selected antibiotics.

Clones	MIC value (μg/mL)			
	Aztreonam	Ceftazidime	Imipenem	Cefepime
BL21_ampC-AC12	16	16	32	16
BL21_ampC-AC30	16	16	32	16
BL21_ampC-AC29	16	16	32	16
Control (BL21_pET30a)	2	2	2	2

encoded by AC29 and AC30 is likely an extended-spectrum *Acinetobacter*-derived AmpC (ESAC).

Other Resistance Determinants and Efflux Pumps

Besides the bla_{AmpC}-encoded cephalosporinase (belonging to Ambler Class C β-lactamases) and the Class D carbapenemases encoded by $bla_{\text{OXA-23}}$ and $bla_{\text{OXA-51}}$ that were described in the previous sections, two other β-lactamase genes were found in the genomes of AC29 and AC30 (**Table 3**). These are the bla_{TEM} gene which encodes class A extended-spectrum β-lactamase and a gene identified as belonging to the metalo-β-lactamase family (Ambler Class B). The contributions of these genes to the β-lactam resistance phenotype of AC12 and AC30 are currently unknown.

One of the main mechanisms of fluoroquinolone resistance is mutations that alter the drug targets. The targets of fluoroquinolone action are the bacterial enzymes DNA gyrase (encoded by *gyrA* and *gyrB*) and DNA topoisomerase IV (encoded by *parC* and *parE*), both of which work together in replication, transcription, recombination and repair of DNA (Jacoby, 2009).The AC29 and AC30-encoded *gyrA* and *parC* showed the Ser → Leu amino acid substitutions at positions 83 and 80, respectively, which have been implicated in fluoroquinolone resistance (Wisplinghoff et al., 2003; Fournier

TABLE 3 | Genes conferring antibiotic resistance found in *A. baumannii* AC29 and AC30.

Antibiotics	Genes	Contig location		Products
		AC29	AC30	
Aminoglycosides	*aad*A	3_40	10_40	Adenyltransferase
	*aad*A1a	3_40	10_40	Aminoglycoside adenylyltransferase
	*aph*A1b	88_1	91_1	Aminoglycoside 3′-phosphotransferase
	*str*A	54_7	38_7	Phosphotransferase
	*str*B	54_6	38_6	Phosphotransferase
Beta-lactams	*amp*C	14_1	37_1	Beta-lactamase Class C
	bla_{TEM}	85_1	87_1	Beta-lactamase TEM
	Class A beta-lactamase	34_58	6_58	Beta-lactamase Class A
	MBL	16_33	3_184	Metallo-beta-lactamase family protein
	$bla_{\text{OXA-23}}$	60_16	40_16	Beta-lactamase OXA-23
	$bla_{\text{OXA-51}}$	53_6	60_1	Beta-lactamase OXA-51
Carbapenems	*car*O	16_123	6_123	Putative porin protein associated with imipenem resistance
Chloramphenicol	*cml*A	35_12	47_5	Major facilitator superfamily permease
	*cml*A	35_12	47_5	cmlA transporter
Fluoroquinolones	*par*C	45_19	49_19	DNA topoisomerase IV subunit A
	*par*E	17_100	1_100	DNA topoisomerase IV subunit B
	*gyr*A	16_68	3_149	DNA gyrase subunit A
	*gyr*B	40_38	23_38	DNA gyrase subunit B
Tetracyclines	*tet*A, Class A	54_1	38_1	Tetracycline resistance protein
	*tet*R	54_2	38_2	Tetracycline repressor protein
Trimethoprim	*dhfr*I	33_27	5_27	Dihydrofolate reductase
Sulfonamides	*sul*I	16_16	57_14	Dihydropteroate synthase
Polymyxin B	*pmr*A	15_53	34_55	Polymyxin resistance component, PmrA
	*pmr*B	15_54	34_54	Polymyxin resistance component, PmrB
	*pmr*C	15_55	43_1/39_1	Polymyxin resistance component, PmrC
Colistin	*lpx*A	29_1	13_136	UDP-N-acetylglucosamine acyltransferase
	*lpx*B	38_1	8_113	Lipid-A-disaccharide synthase
	*lpx*C	12_51	1_129	UDP-3-O-[3-hydroxymyristoyl] N-acetylglucosamine deacetylase
	*lpx*D	19_75	13_138	UDP-3-O-[3-hydroxymyristoyl] glucosamine N-acyltransferase
	*lpx*H	17_129	35_87	UDP-2,3-diacylglucosamine hydrolase
	*lpx*K	12_53	8_19	Tetraacyldisaccharide 4′-kinase

et al., 2006; Maragakis and Perl, 2008). However, four additional novel point mutations (G145D, S118G, L644P, and T872A) were observed in the AC29 and AC30-encoded *gyrA*. Whether these mutations contribute to fluoroquinolone resistance in *A. baumannii* AC29 and AC30 would require further investigations.

Multidrug efflux pumps and porins also play important roles in *A. baumannii* antimicrobial resistance (Vila et al., 2007). Among the five major families of bacterial efflux pumps (i.e., RND, MFS, APC, ABC, and MATE; **Table 4**), the RND family is widely disseminated in Gram-negative bacteria (Poole, 2004; Bonomo and Szabo, 2006; Nordmann and Poirel, 2008; Wieczorek et al., 2008). The genome analysis of *A. baumannii* AC29 and AC30 showed the presence of the complete *adeABC* genes along with its two-component regulatory genes, *adeRS*. Another member of the RND family encoded by *adeIJK* was also present in AC29 and AC30. However, a third RND efflux pump encoded by the *adeFGH* operon was absent in both isolates. Overexpression of *adeABC* and to a certain extent, *adeIJK*, has also been associated with the multidrug resistance phenotype in *A. baumannii* (Vila et al., 2007) and it would be of interest to investigate if this is likewise in *A. baumannii* AC29 and AC30.

TABLE 4 | Drug transporter and efflux pumps found in *A. baumannii* AC29 and AC30.

Drug transporters	Genes	Contig location		Gene products
		AC29	AC30	
APC family transporter	*cycA*	16_60	3_157	D-serine/D-alanine/glycine transporter
	cycA2	3_70	10_70	amino acid APC transporter
		3_71	10_71	amino acid APC transporter
	mmuP	67_1	22_32	S-methylmethionine APC transporter
	proY	17_59	1_59	proline-specific permease ProY
RND family transporter	*adeA*	31_11	16_59	membrane-fusion protein
	adeB	31_10	16_60	cation/multidrug efflux pump
	adeC	31_9	16_61	multidrug efflux protein AdeC
	adeI	15_38	34_39	multidrug efflux protein AdeI
	adeJ	15_39	34_40	multidrug efflux protein AdeJ
	adeK	15_40	34_41	multidrug efflux protein AdeK
	adeR	31_21	16_58	AdeR
	adeS	31_13	16_57	AdeS
	fusE	9_83	19_57	Putative FusE-MFP/HlyD membrane fusion protein
	mdtA	20_59	14_55	multidrug ABC transporter
	nolF	35_29	4_13	NolF secretion protein
	RND efflux pumps	35_29	4_13	NolF secretion protein
MFS family transporter	*emrB*	26_33	27_212	major facilitator superfamily multidrug resistance protein
	fsr	25_59	27_36	major facilitator superfamily permease
	bcr	35_12	47_5	MFS superfamily bicyclomycin/multidrug transport protein
	MFS superfamily protein	12_101	13_186	transporter, major facilitator family protein
	MFS transporter	19_98	8_90	major facilitator superfamily permease
	ygaY	8_35	18_33	transporter, major facilitator family protein
	norM	9_13	19_5	multidrug ABC transporter
	cmr	22_155	22_42	major facilitator superfamily multidrug/chloramphenicol efflux transporter
ABC family transporter	ABC efflux pump	9_13	19_5	multidrug ABC transporter
Na+ driven transporters	Na+ driven efflux pump	17_88	1_88	Na+-driven multidrug efflux pump
Efflux pumps	*abeM*	9_13	19_5	multidrug efflux pump AbeM
	dmt	16_81	3_136	EamA-like transporter family protein
	ywfM	56_27	12_128	DMT family permease
	MATE efflux pump	17_88	1_88	MATE efflux family protein
SMR	*smr*	22_155	14_62	multidrug resistance protein, SMR family
		20_66	14_1	Smr protein/MutS2

DISCUSSIONS

Whole Genome Sequencing and Sequence Types

In this study, we applied comparative genomics on two *A. baumannii* strains that are indistinguishable by PFGE but have different antibiograms to better understand their mechanism of resistance. Although these two strains only differed by their susceptibility toward polymyxin B (AC29 was susceptible whereas AC30 was resistant), detailed genomic analysis indicated a number of differences in the RI structures and the plasmid contents. Both *A. baumannii* AC29 and AC30 were typed as ST195 and, not surprisingly, their closest neighbor was *A. baumannii* AC12 which was also another ST195 strain as previously reported from (Lean et al., 2015). Another close neighbor was *A. baumannii* M1, an isolate from Malaysia that was recently deposited and from its genome sequence, was deduced to be also ST195. All these strains were clearly clustered in the IC-II group of global *A. baumannii* clones (**Figure 1**). Strains that were typed as ST195 under the Bartual MLST scheme are grouped into the clonal complex 92 (CC92) which are characterized as under IC-II (Kim et al., 2013; Zarrilli et al., 2013).

The AbaR4-Type Resistance Island

One of the known hotspots for the insertion of resistance islands in *A. baumannii* is the *comM* gene (Post et al., 2010; Zhou et al., 2011; Huang et al., 2012; Kim et al., 2013). The genomes of *A. baumannii* AC29 and AC30 also contained a similar island within *comM* but with slight differences. Previous analysis of the *A. baumannii* AC12 genome showed the presence of a 23 kb AbaR4-type island designated AC12-RI1 within *comM* (Lean et al., 2015) which consisted of a backbone of Tn*6167* and a truncated version of Tn*6022* designated ΔTn*6022*. In all three islands (i.e., AC12RI-1, AC29RI-1, and AC30RI-1), only one copy of either ΔTn*6022* or Tn*6022* is found whereas other similar resistance islands usually contained two copies of ΔTn*6022* and/or Tn*6022*. For instance, in RI_{MDR-TJ} and AbaR22, another copy of the complete Tn*6022* was located next to the *tetA*(B) gene (**Figure 2**). Besides that, the bla_{OXA-23} gene in AC12-RI1, AC29-RI1, and AC30-RI1 was found flanked by two copies of the IS*Aba1* insertion element in a composite transposon structure similar to Tn*2006* (**Figure 2**). The presence of IS*Aba1* upstream of bla_{OXA-23} may contribute to carbapenem resistance by increasing the expression level of the gene by virtue of its outward-directed promoter (Zhu et al., 2013). The Tn*2006*-like structure was inserted immediately downstream from ΔTn*6022* or Tn*6022*, and interestingly in AC12-RI1 and AC29-RI1, the *orf4* gene found at the right-most end of ΔTn*6022* was duplicated, resulting in the ΔTn*2006*-like structure being flanked by *orf4* on either end (**Figure 2**). Such an arrangement has not been reported in similar resistance islands before. Besides that, comparison between the Tn*2006* present in the three RIs and Tn*6167* showed some differences. Tn*6167* harbors Tn*2006*, but at a different location whereas both RI_{MDR-TJ} and AbaR22 do not contain Tn*2006*. However, the right end of the islands were identical with Tn*6167*, RI_{MDR-TJ} and AbaR22 and comprises of *tetA*(B)-*tetR*(B), the small mobile element CR2, *strA-strB* and *orf4b*, a hypothetical ORF related to *orf4* of Tn*6022* and Tn*6022*Δ1 (**Figure 2**).

The structures of AC29-RI1 and AC30-RI1 indicated that they are novel variants of the AbaR4-type RIs that were recently reported in a survey of RIs found in *A. baumannii* strains throughout Asia (Kim et al., 2013). Kim et al. (2013) also reported that AbaR4-type RIs were commonly found among carbapenem-resistant CC92 strains which included the ST195 lineage.

Plasmids of *A. baumannii* AC29 and AC30

A Small 8.7 kb Plasmid Found in Both AC29 and AC30

Plasmid pAC29a and pAC30a were almost identical to pAC12 (Lean et al., 2015), pAB0057 (Adams et al., 2008), p1ABTCDC0715 (Chen et al., 2011), and pPKAB07 (Saranathan et al., 2014). A similar *A. baumannii* plasmid, pABVA01 (8963 bp), was found to harbor the bla_{OXA-24} carbapenese gene flanked by XerC/XerD-like recombination sites (D'Andrea et al., 2009) and this arrangement was also subsequently reported in the 8771 bp plasmid pMMCU1 (Merino et al., 2010). The XerC/XerD-like recombination sites were present in pAC29a/pAC30a but with the *selI* gene instead of bla_{OXA-24} within the potential recombination region. The pAC12 plasmid from *A. baumannii* AC12 (Lean et al., 2015) and pAB0057 from *A. baumannii* AB0057 (Adams et al., 2008) also contained the *sel1* gene in between the XerC/XerD sites. The function of the *sel1* gene is currently unknown. Other similar *Acinetobacter* plasmids harbored different sized fragments between the XerC/XerD recombination sites with p2ABAYE from *A. baumannii* AYE (Fournier et al., 2006) for example, harboring a putative alcohol dehydrogenase gene (**Figure 3**). The bla_{OXA-24} gene is so far reported only in strains from Italy and Spain (D'Andrea et al., 2009; Merino et al., 2010). This was corroborated in a recent multicenter study which showed that the $bla_{OXA-24}/bla_{OXA40-like}$ genes in *A. baumannii* strains isolated from Spain was predominantly carried by small 8–12 kb plasmids with two of the sequenced plasmids, pAbATCC223 and pAbATCC329, harboring the bla_{OXA-24} gene in between the XerC/XerD sites (Mosqueda et al., 2013). Xer recombination is a site-specific recombination mechanism involved in events such as the integration of phage CTX-Φ at the *dif1* site in the *Vibrio cholerae* chromosome (D'Andrea et al., 2009). Moreover, the proteins required for Xer recombination, such as the XerC and XerD recombinases and PepA are reportedly encoded in the chromosome of several *A. baumannii* strains (Merino et al., 2010). Thus, the XerC/XerD-like sites on these 8.7 kb plasmids could act as site-specific recombination targets responsible for mobilization of discrete gene modules such as bla_{OXA-24} and *sel1* within *Acinetobacter* plasmids (D'Andrea et al., 2009; Merino et al., 2010).

Other highlights of this small plasmid are the presence of a toxin-antitoxin (TA) system designated AbkB/AbkA (Mosqueda et al., 2014) and a gene encoding a possible TonB-dependent receptor protein. Most TA systems are usually characterized by two co-transcribed genes with the antitoxin gene preceding the toxin gene (Chan et al., 2012; Hayes and Kêdzierska, 2014). AbkB/AbkA seems to differ from canonical TA systems as the *abkB* toxin gene precedes the *abkA* antitoxin gene. To date, only three characterized TA loci have been reported to display this unusual genetic arrangement: the *mqsRA* (Brown et al., 2009), the *higBA* (Tian et al., 1996), and *hicAB* (Jørgensen et al., 2009) modules. AbkB/AbkA was previously identified as one of the four functional TA systems in *A. baumannii* (it was designated SplT/SplA or DUF497/COG3914) whereby overexpression of the toxin was shown to inhibit growth in *E. coli* and this was overcame by co-expression of the cognate antitoxin (Jurenaite et al., 2013). The AbkB (or SplT) toxin was shown to inhibit translation when overexpressed in *E. coli* with cleavage of *lpp* mRNA and transfer-messenger RNA (tmRNA) demonstrated, thus indicating that the AbkB toxin likely functions as an endoribonuclease or RNA interferase. The AbkB/AbkA locus was found to be highly prevalent in small plasmids of *A. baumannii* clinical strains (88.6% prevalence among 476 clinical isolates from Lithuania; Jurenaite et al., 2013). The presence of a TA system on these plasmids would explain their stability in the absence of any apparent selection pressure, particularly for the small plasmids without the $bla_{OXA-24}/bla_{OXA40-like}$ gene such as pAC30a and pAC29a.

Some TonB-dependent receptors, in particular BauA, play important roles in the acquisition of iron in *A. baumannii* (Dorsey et al., 2004; Mihara et al., 2004) with recent transcriptomic and proteomic analyses indicating approximately 20 TonB-dependent receptors in *A. baumannii*, some of which are regulated by iron (Antunes et al., 2011; Eijkelkamp et al., 2011; Nwugo et al., 2011). TonB-dependent receptors are usually found in the outer membrane where they interact with TonB and associated inner membrane proteins (ExbB and ExbD) that provide energy needed to transport host iron-carrier and iron-siderophore complexes into the periplasm once these complexes are bound to the TonB-dependent receptors (Zimbler et al., 2013). Thus, along with TonB, the TonB-depended receptors play an important role in the virulence of *A. baumannii*. Whether the TonB-dependent receptor protein encoded by these small plasmids plays a similar role in iron acquisition and hence, virulence, awaits further experimentation. Another possible virulence-associated gene encoded in these small plasmids is found downstream of the gene encoding the TonB-dependent receptor. This gene is predicted to encode a 152-amino acids-residue protein homologous to septicolysin, a putative virulence factor (Mosqueda et al., 2014). Septicolysin is a member of thiol-activated cytolysins which have been implicated in the pathogenesis of infections by several Gram-positive pathogens such as *Clostridium perfringens*, *Listeria monocytogenes*, and *Streptococcus pneumoniae* and are characterized by their cytolytic activity for eukaryotic cells (Billington et al., 2000). With two putative virulence factors encoded on these small plasmids, it would therefore be of interest to investigate if these plasmids play a role in the virulence and pathogenesis of *A. baumannii* infections, especially in view of their prevalence among clinical *A. baumannii* isolates.

pAC30b, a 16.2 kb Resistance Plasmid Found in *A. baumannii* AC30

Both pAC30b and pMDR-ZJ06 are resistance plasmids, as indicated by the presence of genes which confer resistance to aminoglycosides and macrolides within the 7 kb Tn*1548*::*armA* resistance island (Zhou et al., 2011). Interestingly, Tn*1548*::*armA* is located in the chromosome of *A. baumannii* AC12 (Lean et al., 2015) as well as in *A. baumannii* AC29. In both cases, Tn*1548*::*armA* was found downstream of a cluster of five genes encoding proteins annotated as paraquat-inducible protein A and protein B. The pMDR-ZJ06 plasmid harbors a class 1 integron encoding the aminoglycoside acetyltransferase (*aacC1*) and adenyltransferase (*aadA1*) along with *sulI* that confers sulphonamide resistance (Zhou et al., 2011) but this integron was absent in pAC30b.

Besides pMDR-ZJ06, pAC30b also shared sequence identity with p3BJAB0868 from *A. baumannii* BJAB0868 and p2BJAB07104 from *A. baumannii* BJAB07104 (Zhu et al., 2013) and this is mainly an approximately 10 kb fragment that spanned Tn*1548*::*armA*, IS*66* and the *rep* gene. Two other parts of pAC30b that were identical with pMDR-ZJ06, p3BJAB0868, and p2BJAB07104 are IS*26* and the *aphA1* gene, both of which were in different locations in pAC30b when compared with pMDR-ZJ06 (**Figure 4**), p3BJAB0868, and p2BJAB07104. The *aphA1* gene is flanked by two copies of IS*26* in a composite transposon-like structure designated Tn*6210* in p3BJAB0868 and p2BJAB07104 (Zhu et al., 2013) as well as pMDR-ZJ06 but in pAC30b, only one copy of IS*26* is found adjacent to *aphA1*. It is possible that a deletion had occurred in pAC30b that took out the other copy of IS*26* as well as part of the *tnpU* gene of Tn*1548*::*armA* which is located adjacent to the IS*26*-*aphA1* structure. All three plasmids pMDR-ZJ06, p3BJAB0868, and p2BJAB07104 which were isolated from *A. baumannii* strains in China, harbored a class 1 integron but this was absent in pAC30b.

It is noteworthy that in contrast to pAC30a and pAC29a, pAC30b has very few similar plasmids in other *A. baumannii* isolates besides pMDR-ZJ06, p3BJAB0868, and p2BJAB07104, suggesting that these plasmids are not so prevalent. What the four *A. baumannii* plasmids (namely, pAC30b, pMDR-ZJ06, p3BJAB0868, and p2BJAB07104) have in common are (1) a number of transposases encoded by IS elements and transposons, and (2) a lack of any plasmid stability genes such as toxin-antitoxin systems. Thus, there is a likelihood that these plasmids are much less stable as compared to the smaller pAC29a/pAC30a plasmid which is endowed with the AbkB/AbkA toxin-antitoxin system. Hence, in *A. baumannii* AC29 (and AC12), portions of pAC30b are chromosomally-located and there is no widespread occurrence of pAC30b-like plasmids in other *A. baumannii* strains.

A Large ca. 70 kb Conjugative Plasmid in the Genomes of *A. baumannii* AC29 and AC30

The *tra* locus in pAC29b and pAC30c encodes for a type IV secretion system (T4SS) which is related to the bacterial conjugation machinery and mediates horizontal gene transfer (Juhas et al., 2008). Plasmids pAb-G7-2, pA85-3, and even pACICU2 have been demonstrated to be transmissible by conjugation to other *A. baumannii* strains (Hamidian and Hall, 2014; Hamidian et al., 2014). It is thus very likely that pAC29b and pCA30c are also conjugative as the *tra* locus in these plasmids is almost identical. In pAb-G7-2, the space between these two regions contained the aminoglycoside resistance transposon, Tn*aphA6* (Hamidian and Hall, 2014) whereas in pA85-3, the AbaR4 resistance island is located in the same area (Hamidian et al., 2014). In pAC30c, this region contained several hypothetical ORFs and a solitary *relE* toxin gene without the corresponding *relB* antitoxin gene. However, the two genes flanking *relE* could perhaps function as the antitoxin as they encode for hypothetical proteins of about the same size as the putative RelE toxin. Toxin-antitoxin pairs are usually about the same size with a few exceptions such as the Zeta toxin which is much larger (~270 amino acids) as compared to their cognate Epsilon antitoxin (~90 amino acids; Chan et al., 2012; Jurenaite et al., 2013). Although toxins usually interact with their cognate antitoxin pair, sometimes mixing and matching between different toxin and antitoxin families do occur (Chan et al., 2012; Hayes and Kêdzierska, 2014). Nevertheless, the functionality of the plasmid pAC30c and pAC29b-encoded *relE* toxin gene needs to be ascertained.

Both pAC30c and pAC29b carry the *rep* gene designated *rep*Aci6, similar to pACICU2, pAb-G7-2, and pA85-3. Another solitary toxin, this time from the Zeta family, is located downstream from the *rep*Aci6 gene in both plasmids. Experimental evidence had suggested that its overexpression is non-toxic to *E. coli* and thus, may not function as a typical toxin (Jurenaite et al., 2013). Its putative antitoxin partner, located upstream of its reading frame, does not bear any homology to the Epsilon antitoxin. These solitary Zeta-like toxins have been observed in several other plasmids in diverse bacterial species and their function is currently unknown (Chan et al., 2012; Jurenaite et al., 2013). In the absence of typical canonical toxin-antitoxin systems, the only other identifiable genes that could contribute to plasmid stability in pAC30c and pAC29b are the *parAB* genes which likely encode proteins that are involved in plasmid partitioning.

Plasmid pAC29b harbors a bla_{OXA-23} gene but unlike the *A. baumannii* A85 plasmid pA85-3 (Hamidian et al., 2014), the gene is not located within an AbaR4 island. Both pAb-G7-2 and pACICU2 harbor the aminoglycoside resistance gene *aphA6* within a composite transposon designated Tn*aphA6* (Hamidian and Hall, 2014; Hamidian et al., 2014). Plasmid pAC30c however, does not encode any known resistance determinant.

Investigations into the Possible Contributors of Polymyxin Resistance in *A. baumannii* AC30

The mechanism for polymyxin resistance in *A. baumannii* has only recently been elucidated and the main mechanism appeared to be either covalent modification of the lipid A portion of LPS (Arroyo et al., 2011; Beceiro et al., 2011) or disruption of LPS biosynthesis (Moffatt et al., 2010; Park et al., 2011). By modification and/or mutations in the amino acid sequences, negative charges on the outer membrane can be reduced, leading to reduction in the affinity of the positively-charged polymyxin component, hence giving rise to polymyxin resistance (Adams et al., 2009; Arroyo et al., 2011; Beceiro et al., 2011). Overexpression of the two-component signal transduction system *pmrAB* and mutations within these genes, especially *pmrB*, were reported to contribute to polymyxin resistance (Adams et al., 2009; Arroyo et al., 2011; Park et al., 2011). These two genes are part of an operon along with *pmrC*, which encodes the enzyme responsible for the covalent addition of phosphoethanolamine to lipid A. Sequence analysis of *pmrCAB* from the polymyxin-resistant AC30 in comparison with the polymyxin-susceptible AC29 and other susceptible strains in the database (including ATCC19606 and ATCC17978) showed a P102H mutation within *pmrA*. Identical mutations were found within *pmrA* of AC12 (Lean et al., 2015) and other polymyxin-resistant isolates from Terengganu (Lean et al., 2014). Polymyxin-resistant isolate AC12 displayed higher *pmrA* expression levels (8.5-folds) and also higher *pmrB* levels (about two-folds) in comparison with ATCC19606. In contrast, the polymyxin-susceptible AC29 isolate showed dramatically lower *pmrAB* expression (0.05- and 0.03 folds). It should be noted that in this case, no isogenic polymyxin-susceptible strains for the polymyxin-resistant strains were available for comparison. Thus, the expression levels for AC29, AC30, and AC12 were compared with the non-isogenic reference strain ATCC19606 and therefore may not yield an accurate picture for the *pmrAB* levels. Nevertheless, the results do indicate upregulation particularly for *pmrB* in both the polymyxin-resistant isolates AC30 and AC12.

The *lpxA*, *lpxC*, and *lpxD* genes encode the first three enzymes in the lipid A biosynthesis pathway (Moffatt et al., 2010, 2011). No mutation was found in *lpxA*. In contrast, *lpxD* showed three amino acid mutations (S102T, V141I, R173G) in AC30 while *lpxC* had a K141R substitution. The *lpsB* gene encodes a glycosyltransferase is involved in the biosynthesis of the LPS core and was recently implicated in *A. baumannii* colistin resistance (Hood et al., 2013). Comparison of *lpsB* sequences indicated a H181Y substitution in AC30. Identical mutations in *lpxD*, *lpxC*, and *lpsB* were reported in the polymyxin-resistant *A. baumannii* AC12 (Lean et al., 2015); likewise, these and other mutations had been previously reported in the polymyxin-resistant *A. baumannii* strains from Terengganu (Lean et al., 2014). Results from the LPS analysis indicated that the LPS in the polymyxin resistant strains AC12 and AC30 were considerably less when compared with the LPS band of the polymyxin-susceptible strains, which has been previously reported in other polymyxin-resistant strains (Moffatt et al., 2010, 2011; Hood et al., 2013). This suggested that the mutations found in the *lpxD*, *lpxC*, and *lpsB* genes in the two polymyxin-resistant strains may have led to impairment but not a total loss of the LPS. It is likely that in the case of AC12 and AC30, polymyxin resistance could be the result of a combination of increased *pmrAB* expression leading to covalent modification of the lipid A moiety of LPS and possibly impaired LPS synthesis as well. Nevertheless, it should be noted that in a recent study (Hood et al., 2013), screening of transposon mutant libraries led to the identification of more than 20 genes that may be involved in inducible colistin resistance in *A. baumannii*. Most of these genes converged on pathways involved in osmotolerance, cell envelope biosynthesis along with protein folding (Hood et al., 2013). The role that these factors may play in the development of polymyxin resistance in the Terengganu *A. baumannii* strains would also need to be investigated.

AmpC-Mediated Resistance to Extended-Spectrum Cephalosporins

Resistance to β-lactam antibiotics via synthesis of β-lactamase encoded by the chromosome and/or plasmids is the most common resistance mechanism observed in *A. baumannii* (Bou and Martínez-Beltrán, 2000). Resistance to broad-spectrum cephalosporins in *A. baumannii* are usually related to the over production of extended spectrum β-lactamases (ESBL), especially AmpC-type β-lactamases designated ADCs (*Acinetobacter*-derived cephalosporinases) (Rodríguez-Martínez et al., 2010). ADCs typically hydrolyze penicillins, narrow- and extended-spectrum cephalosporins but not zwitterionic cephalosporins such as cefepime or carbapenems (Rodríguez-Martínez et al., 2010). However, extended-spectrum AmpCs (ESACs) have been reported in *A. baumannii* that confer reduced susceptibility to all cephalosporins and this includes ADC-33 (Rodríguez-Martínez et al., 2010) and ADC-56 (Tian et al., 2011).

Since the recombinant *E. coli* BL21 carrying the bla_{AmpC} from AC29 and AC30 displayed resistance to ceftazidime, cefepime, aztreonam, and even imipenem, this strongly suggests that the AmpC from AC29 and AC30 is an ESAC cephalosphorinase. ADC-33 possessed a P210R substitution together with a duplication of the Ala residue at position 215 within the Ω loop, both of which are required for extended spectrum activity (Rodríguez-Martínez et al., 2010). ADC-56 possesed an R148Q mutation also within the Ω loop, that enabled the enzyme to hydrolyze cefepime (Tian et al., 2011). Thus, the G246S mutation within the Ω loop of the AmpC from AC29 and AC30 could be responsible for the extended spectrum activity. This could be examined and verified by site-directed mutagenesis of the pET30a recombinant clones. Likewise, the contribution of the R80S mutation toward extended spectrum activity should be investigated even though it is located in a non-active site.

CONCLUSIONS

In this study, we presented the comparative genome analyses of two Malaysian *A. baumannii* strains AC29 and AC30 that belonged to the ST195 lineage and had identical *Apa*I pulsotype but different susceptibilities to polymyxin. Their MLST profiles and phylogenetic clustering based on their concatenated MLST sequences clearly showed these strains belonging to the International Clone II (IC-II) group. Novel resistance island (RI) variants and plasmids were discovered from the genome sequence of these strains. Both strains shared a similar AbaR4-type RI of approximately 22 kb interrupting the *comM* gene insertional hotspot and which contains the carbapenem resistance gene, $bla_{\mathrm{OXA-23}}$ within a composite transposon, Tn*2006*. The island also encodes genes conferring resistance to tetracyclines, sulphonomides and streptomycin. Both *A. baumannii* strains harbored a small ~8 kb cryptic plasmid which encode putative virulence determinants (TonB-dependent receptor and septicolysin) as well as a XerC/XerD recombination site. Plasmid pAC30b is found only in AC30 but not AC29 and contained the Tn*1548::armA* island that confers resistance to aminoglycosides and macrolides. Interestingly, this island was found in the chromosome of AC29. Both AC30 and AC29 harbored a ~70 kb conjugative plasmid designated pAC30c and pAC29b with pAC29b containing a copy of $bla_{\mathrm{OXA-23}}$. Thus, genomic islands and, to a lesser extent, conjugative plasmids, appeared to play an important role in the dissemination and acquisition of antibiotic resistance determinants in the Terengganu *A. baumannii* strains. Experimental evidence also indicated that polymyxin resistance in AC30 may have developed through a combination of *pmrAB* upregulation and partial impairment of the lipopolysaccharide layer. The bla_{AmpC} variant encoded by both AC29 and AC30 was also shown to be likely an extended-spectrum *Acinetobacter*-derived AmpC (ESAC) conferring resistance to cefepime as well as imipenem. Whole genome sequencing of the two Terengganu *A. baumannii* clinical strains and subsequent experiments enabled a detailed characterization of their genetic repertoire of resistance, thereby giving us an insight into the genetic blueprint of Malaysian isolates of this increasingly important and deadly nosocomial pathogen.

ACKNOWLEDGMENTS

We thank University of Malaya and Universiti Sultan Zainal Abidin for facilities and support. This project was supported by High Impact Research Grant UM.C/625/HIR/MOHE/02 to KLT and UniSZA research grant UDM/09/BR(006) to CCY and ZS. SSL is supported by a University of Malaya Postgraduate Fellowship.

REFERENCES

Adams, M. D., Goglin, K., Molyneaux, N., Hujer, K. M., Lavender, H., Jamison, J. J., et al. (2008). Comparative genome sequence analysis of multidrug-resistant *Acinetobacter baumannii*. *J. Bacteriol.* 190, 8053–8064. doi: 10.1128/JB.00834-08

Adams, M. D., Nickel, G. C., Bajaksouzian, S., Lavender, H., Murthy, A. R., Jacobs, M. R., et al. (2009). Resistance to colistin in *Acinetobacter baumannii* associated with mutations in the PmrAB two-component system. *Antimicrob. Agents Chemother.* 53, 3628–3634. doi: 10.1128/AAC.00284-09

Antunes, L. C. S., Imperi, F., Towner, K. J., and Visca, P. (2011). Genome-assisted identification of putative iron-utilization genes in *Acinetobacter baumannii* and their distribution among a genotypically diverse collection of clinical isolates. *Res. Microbiol.* 162, 279–284. doi: 10.1016/j.resmic.2010.10.010

Arroyo, L. A., Herrera, C. M., Fernandez, L., Hankins, J. V., Trent, M. S., and Hancock, R. E. W. (2011). The *pmrCAB* operon mediates polymyxin resistance in *Acinetobacter baumannii* ATCC 17978 and clinical isolates through phosphoethanolamine modification of lipid A. *Antimicrob. Agents Chemother.* 55, 3743–3751. doi: 10.1128/AAC.00256-11

Ausubel, F. M., Brent, R., Kingston, R. E., Moore, D. D., Seidman, J. D., Smith, J. A., et al. (2002). *Short Protocols in Molecular Biology (5th Edn).* John Wiley & Sons.

Aziz, R. K., Bartels, D., Best, A. A., DeJongh, M., Disz, T., Edwards, R. A., et al. (2008). The RAST Server: rapid annotations using subsystems technology. *BMC Genomics* 9:75. doi: 10.1186/1471-216 4-9-75

Bartual, S. G., Seifert, H., Hippler, C., Rodríguez-valera, F., and Domı, M. A. (2005). Development of a multilocus sequence typing scheme for characterization of clinical isolates of *Acinetobacter baumannii. J. Clin. Microbiol.* 43, 4382–4390. doi: 10.1128/JCM.43.9.4382-4390.2005

Beceiro, A., Llobet, E., Aranda, J., Bengoechea, J. A., Doumith, M., Hornsey, M., et al. (2011). Phosphoethanolamine modification of lipid A in colistin-resistant variants of *Acinetobacter baumannii* mediated by the *pmrAB* two-component regulatory system. *Antimicrob. Agents Chemother.* 55, 3370–3379. doi: 10.1128/AAC.00079-11

Bergogne-Bérézin, E., and Towner, K. J. (1996). Acinetobacter spp. as nosocomial pathogens: microbiological, clinical, and epidemiological features. *Clin. Microbiol. Rev.* 9, 148–165.

Billington, S. J., Jost, B. H., and Songer, J. G. (2000). Thiol-activated cytolysins: structure, function and role in pathogenesis. *FEMS Microbiol. Lett.* 182, 197–205 doi: 10.1111/j.1574-6968.2000.tb08895.x

Bonomo, R. a., and Szabo, D. (2006). Mechanisms of multidrug resistance in *Acinetobacter* species and *Pseudomonas aeruginosa*. *Clin. Infecti. Dis.* 43, 49–56. doi: 10.1086/504477

Bou, G., and Martínez-Beltrán, J. (2000). Cloning, nucleotide sequencing, and analysis of the gene encoding an AmpC beta-lactamase in *Acinetobacter baumannii*. *Antimicrob. Agents Chemother.* 44, 428–432 doi: 10.1128/AAC.44.2.428-432.2000

Brown, B. L., Grigoriu, S., Kim, Y., Arruda, J. M., Davenport, A., Wood, T. K., et al. (2009). Three dimensional structure of the MqsR: MqsA complex: a novel TA pair comprised of a toxin homologous to RelE and an antitoxin with unique properties. *PLoS Pathog.* 5:e1000706. doi: 10.1371/journal.ppat.1000706

Camp, C., and Tatum, O. L. (2010). A review of *Acinetobacter baumannii* as a highly successful pathogen in times of war. *Lab. Med.* 41, 649–657. doi: 10.1309/LM90IJNDDDWRI3RE

Chan, W. T., Moreno-Córdoba, I., Yeo, C. C., and Espinosa, M. (2012). Toxin-antitoxin genes of the Gram-positive pathogen *Streptococcus pneumoniae*: so few and yet so many. *Microbiol. Mol. Bio. Rev.* 76, 773–791. doi: 10.1128/MMBR.00030-12

Chen, C.-C., Lin, Y.-C., Sheng, W.-H., Chen, Y.-C., Chang, S.-C., Hsia, K.-C., et al. (2011). Genome sequence of a dominant, multidrug-resistant *Acinetobacter baumannii* strain, TCDC-AB0715. *J. Bacteriol.* 193, 2361–2372. doi: 10.1128/JB.00244-11

Clinical and Laboratory Standards Institute (2013). *Performance Standards for Antimicrobial Susceptibility Testing; Twenty-Third Informational Supplement.* Wayne, PA: CLSI (Seventeen Informational Supplement M100–S23).

D'Andrea, M. M., Giani, T., D'Arezzo, S., Capone, A., Petrosillo, N., Visca, P., et al. (2009). Characterization of pABVA01, a plasmid encoding the OXA-24 carbapenemase from Italian isolates of *Acinetobacter baumannii*. *Antimicrob. Agents Chemother.* 53, 3528–3533. doi: 10.1128/AAC.00178-09

Darling, A. C. E., Mau, B., Blattner, F. R., and Perna, N. T. (2004). Mauve: multiple alignment of conserved genomic sequence with rearrangements. *Genome Res.* 14, 1394–1403. doi: 10.1101/gr.2289704

Diancourt, L., Passet, V., Nemec, A., Dijkshoorn, L., and Brisse, S. (2010). The population structure of *Acinetobacter baumannii*: expanding multiresistant clones from an ancestral susceptible genetic pool. *PLoS One* 5:e10034. doi: 10.1371/journal.pone.0010034

Dijkshoorn, L., Nemec, A., and Seifert, H. (2007). An increasing threat in hospitals: multidrug-resistant *Acinetobacter baumannii*. *Nat. Rev. Microbiol.* 5, 939–951. doi: 10.1038/nrmicro1789

Doi, Y., Murray, G. L., and Peleg, A. Y. (2015). Acinetobacter baumannii: evolution of antimicrobial resistance – treatment options. *Sem. Respir. Crit. Care Med.* 36, 95–98. doi: 10.1055/s-0034-1398388

Dolejska, M., Villa, L., Poirel, L., Nordmann, P., and Carattoli, A. (2013). Complete sequencing of an IncHI1 plasmid encoding the carbapenemase NDM-1, the ArmA 16S RNA methylase and a resistance-nodulation-cell division/multidrug efflux pump. *J. Antimicrob. Chemother.* 68, 3657–3667 doi: 10.1093/jac/dks357

Dorsey, C. W., Tomaras, A. P., Connerly, P. L., Tolmasky, M. E., Crosa, J. H., and Actis, L. A. (2004). The siderophore-mediated iron acquisition systems of *Acinetobacter baumannii* ATCC 19606 and *Vibrio anguillarum* 775 are structurally and functionally related. *Microbology* 150, 3657–3667. doi: 10.1099/mic.0.27371-0

Eijkelkamp, B. A., Hassan, K. A., Paulsen, I. T., and Brown, M. H. (2011). Investigation of the human pathogen *Acinetobacter baumannii* under iron limiting conditions. *BMC Genomics* 12:126 doi: 10.1186/1471-2164-12-126

Farrugia, D. N., Elbourne, L. D. H., Hassan, K. A., Eijkelkamp, B. A., Tetu, S. G., Brown, M. H., et al. (2013). The complete genome and phenome of a community-acquired *Acinetobacter baumannii*. *PLoS ONE* 8:e58628. doi: 10.1371/journal.pone.0058628

Fournier, P.-E., Vallenet, D., Barbe, V., Audic, S., Ogata, H., Poirel, L., et al. (2006). Comparative genomics of multidrug resistance in *Acinetobacter baumannii*. *PLoS Genet.* 2:e7 doi: 10.1371/journal.pgen.0020007

Garnacho-Montero, J., Ortiz-Leyba, C., Jimenez-Jimenez, F. J., Barrero-Almodovar, A. E., Garcia-Garmendia, J. L., Bernabeu-Wittell, M., et al. (2003). Treatment of multidrug-resistant *Acinetobacter baumannii* ventilator-associated pneumonia (VAP) with intravenous colistin: a comparison with imipenem-susceptible VAP. *Clin. Infect. Dis.* 36, 1111–1118. doi: 10.1086/374337

Grant, J. R., and Stothard, P. (2008). The CGView Server: a comparative genomics tool for circular genomes. *Nucleic Acids Res.* 36, 181–194. doi: 10.1093/nar/gkn179

Hamidian, M., and Hall, R. M. (2014). pACICU2 is a conjugative plasmid of *Acinetobacter* carrying the aminoglycoside resistance transposon Tn*aphA6*. *J. Antimicrob. Chemother.* 69, 1146–1158. doi: 10.1093/jac/dkt488

Hamidian, M., Kenyon, J. J., Holt, K. E., Pickard, D., and Hall, R. M. (2014). A conjugative plasmid carrying the carbapenem resistance gene bla_{OXA-23} in AbaR4 in an extensively resistant GC1 *Acinetobacter baumannii* isolate. *J. Antimicrob. Chemother.* 69, 2625–2628. doi: 10.1093/jac/dku188

Hayes, F., and Kȩdzierska, B. (2014). Regulating toxin-antitoxin expression: controlled detonation of intracellular molecular timebombs. *Toxins* 6, 337–358 doi: 10.3390/toxins6010337

Higgins, P. G., Dammhayn, C., Hackel, M., and Seifert, H. (2010). Global spread of carbapenem-resistant *Acinetobacter baumannii*. *J. Antimicrob. Chemother.* 65, 233–238. doi: 10.1093/jac/dkp428

Hood, M. I., Becker, K. W., Roux, C. M., Dunman, P. M., and Skaar, E. P. (2013). Genetic determinants of intrinsic colistin tolerance in *Acinetobacter baumannii*. *Infect. Immun.* 81, 542–551. doi: 10.1128/IAI.00704-12

Huang, H., Yang, Z.-L., Wu, X.-M., Wang, Y., Liu, Y.-J., Luo, H., et al. (2012). Complete genome sequence of *Acinetobacter baumannii* MDR-TJ and insights into its mechanism of antibiotic resistance. *J. Antimicrob. Chemother.* 67, 2825–2832. doi: 10.1093/jac/dks327

Hujer, K. M., Hamza, N. S., Hujer, A. M., Perez, F., Helfand, M. S., Bethel, C. R., et al. (2005). Identification of a new allelic variant of the *Acinetobacter baumannii* cephalosporinase, ADC-7 beta-lactamase: defining a unique family of class C enzymes. *Antimicrob. Agents Chemother.* 49, 2941–2948. doi: 10.1128/AAC.49.7.2941-2948.2005

Hyatt, D., Chen, G. L., Locascio, P. F., Land, M. L., Larimer, F. W., and Hauser, L. J. (2010). Prodigal: prokaryotic gene recognition and translation initiation site identification. *BMC Bioinformatics* 11:119. doi: 10.1186/1471-2105-11-119

Iacono, M., Villa, L., Fortini, D., Bordoni, R., Imperi, F., Bonnal, R. J. P., et al. (2008). Whole-genome pyrosequencing of an epidemic multidrug-resistant *Acinetobacter baumannii* strain belonging to the European clone II group. *Antimicrob. Agents Chemother.* 52, 2616–2625 doi: 10.1128/AAC.01643-07

Jacoby, G. A. (2009). AmpC beta-lactamases. *Clin. Microbiol. Rev.* 22, 161–182. doi: 10.1128/CMR.00036-08

Jørgensen, M. G., Pandey, D. P., Jaskolska, M., and Gerdes, K. (2009). HicA of *Escherichia coli* defines a novel family of translation-independent mRNA interferases in bacteria and archaea. *J. Bacteriol.* 191, 1191–1199. doi: 10.1128/JB.01013-08

Juhas, M., Crook, D. W., and Hood, D. W. (2008). Type IV secretion systems: tools of bacterial horizontal gene transfer and virulence. *Cell. Microbiol.* 10, 2377–2386 doi: 10.1111/j.1462-5822.2008.01187.x

Jurenaite, M., Markuckas, A., and Suziedeliene, E. (2013). Identification and characterization of type II toxin-antitoxin systems in the opportunistic pathogen *Acinetobacter baumannii*. *J. Bacteriol.* 14, 3165–3172. doi: 10.1128/jb.00237-13

Kim, D. H., Choi, J.-Y., Kim, H. W., Kim, S. H., Chung, D. R., Peck, K. R., et al. (2013). Spread of carbapenem-resistant *Acinetobacter baumannii* global clone 2 in Asia and AbaR-type resistance islands. *Antimicrob. Agents Chemother.* 57, 5239–5246. doi: 10.1128/AAC.00633-13

Lagesen, K., Hallin, P., Rødland, E. A., Staerfeldt, H.-H., Rognes, T., and Ussery, D. W. (2007). RNAmmer: consistent and rapid annotation of ribosomal RNA genes. *Nucleic Acids Res.* 35, 3100–3108. doi: 10.1093/nar/gkm160

Landman, D., Georgescu, C., Martin, D. A., and Quale, J. (2008). Polymyxins revisited. *Clin. Microbiol. Rev.* 21, 449–465. doi: 10.1128/CMR.00006-08

Lean, S.-S., Suhaili, Z., Salwani, I., Nor Iza, A. R., Norlela, O., Fatimah, H. A., et al. (2014). Prevalence and genetic characterization of polymyxin-resistant *Acinetobacter baumannii* isolated from a tertiary hospital in Terengganu, Malaysia. *ISRN Microbiol.* 2014, 1–28 doi: 10.1155/2014/953417

Lean, S.-S., Yeo, C. C., Suhaili, Z., and Thong, K.-L. (2015). Whole-genome analysis of an extensively drug-resistant clinical isolate of *Acinetobacter baumannii* AC12: insights into the mechanisms of resistance of an ST195 clone from Malaysia. *Int. J. Antimicrob. Agents* 45, 178–182. doi: 10.1016/j.ijantimicag.2014.10.015

Lowe, T. M., and Eddy, S. R. (1997). tRNAscan-SE: a program for improved detection of transfer RNA genes in genomic sequence. *Nucleic Acids Res.* 25, 955–964 doi: 10.1093/nar/25.5.0955

Magiorakos, A., Srinivasan, A., Carey, R. B., Carmeli, Y., Falagas, M. E., Giske, C. G., et al. (2011). Multidrug-resistant, extensively drug-resistant and pandrug-resistant bacteria: an international expert proposal for interim standard definitions for acquired resistance. *Clin. Microbiol. Infect. Dis.* 18, 268–281. doi: 10.1111/j.1469-0691.2011.03570.x

Maragakis, L. L., and Perl, T. M. (2008). *Acinetobacter baumannii*: epidemiology, antimicrobial resistance, and treatment options. *Clin. Infect. Dis.* 46, 1254–1263 doi: 10.1086/529198

Merino, M., Acosta, J., Poza, M., Sanz, F., Beceiro, A., Chaves, F., et al. (2010). OXA-24 carbapenemase gene flanked by XerC/XerD-like recombination sites in different plasmids from different *Acinetobacter* species isolated during a nosocomial outbreak. *Antimicrob. Agents Chemother.* 54, 2724–2727. doi: 10.1128/AAC.01674-09

Mihara, K., Tanabe, T., Yamakawa, Y., Funahashi, T., Nakao, H., Narimatsu, S., et al. (2004). Identification and transcriptional organization of a gene cluster involved in biosynthesis and transport of acinetobactin, a siderophore produced by *Acinetobacter baumannii* ATCC 19606. *Microbiology* 50, 2587–2597. doi: 10.1099/mic.0.27141-0

Moffatt, J. H., Harper, M., Adler, B., Nation, R. L., Li, J., and Boyce, J. D. (2011). Insertion sequence IS*Aba11* is involved in colistin resistance and loss of lipopolysaccharide in *Acinetobacter baumannii. Antimicrob. Agents Chemother.* 55, 3022–3024. doi: 10.1128/AAC.01732-10

Moffatt, J. H., Harper, M., Harrison, P., Hale, J. D. F., Vinogradov, E., Seemann, T., et al. (2010). Colistin resistance in *Acinetobacter baumannii* is mediated by complete loss of lipopolysaccharide production. *Antimicrob. Agents Chemother.* 54, 4971–4977. doi: 10.1128/AAC.00834-10

Mosqueda, N., Espinal, P., Cosgaya, C., Viota, S., Plasensia, V., Lerma, F. Á., et al. (2013). Globally expanding carbapenemase finally bursts in Spain: nosocomial outbreak of *Acinetobacter baumannii* producing a plasmid-encoded OXA-23 in Barcelona, Spain. *Antimicrob. Agents Chemother.* 57, 5155–5157. doi: 10.1128/AAC.01486-13

Mosqueda, N., Gato, E., Roca, I., López, M., de Alegría, C. R., Fernández-Cuenca, F., et al. (2014). Characterization of plasmids carrying the blaOXA-24/40 carbapenemase gene and the genes encoding the AbkA/AbkB proteins of a toxin/antitoxin system. *J. Antimicrob. Chemother.* 69, 2629–2633. doi: 10.1093/jac/dku179

Nigro, S. J., and Hall, R. M. (2012). Tn*6167*, an antibiotic resistance island in an Australian carbapenem-resistant *Acinetobacter baumannii* GC2, ST92 isolate. *J. Antimicrob. Chemother.* 67, 1342–1346. doi: 10.1093/jac/dks037

Nordmann, P., and Poirel, L. (2008). *Acinetobacter baumannii* – Basic and Emerging Mechanisms of Resistance. *Euro. Infect. Dis.* 2, 94–97.

Nwugo, C. C., Gaddy, J. A., Zimbler, D. L., and Actis, L. A. (2011). Deciphering the iron response in *Acinetobacter baumannii*: a proteomics approach. *J. Proteomics.* 74, 44–58 doi: 10.1016/j.jprot.2010.07.010

Park, Y. K., Choi, J. Y., Shin, D., and Ko, K. S. (2011). Correlation between overexpression and amino acid substitution of the PmrAB locus and colistin resistance in *Acinetobacter baumannii. Int. J. Antimicrob. Agents.* 37, 525–530. doi: 10.1016/j.ijantimicag.2011.02.008

Pfaffl, M. W. (2001). A new mathematical model for relative quantification in real-time RT–PCR. *Nucl. Acids Res.* 29:e45. doi: 10.1093/nar/29.9.e45

Poirel, L., and Norman, P. (2006). Carbapenem resistance in *Acinetobacter baumannii*: mechanisms and epidemiology. *Clin. Micro. Infect.* 12, 826–836. doi: 10.1111/j.1469-0691.2006.01456.x

Poole, K. (2004). Resistance to beta-lactam antibiotics. *Cell. Mol. Life Sci.* 61, 2200–2223. doi: 10.1007/s00018-004-4060-9

Post, V., White, P. A., and Hall, R. M. (2010). Evolution of AbaR-type genomic resistance islands in multiply antibiotic-resistant *Acinetobacter baumannii. J. Antimicrob. Chemother.* 65, 1162–1170. doi: 10.1093/jac/dkq095

Roca, I., Espinal, P., Villa-Faréz, X., and Villa, J. (2012). The *Acinetobacter baumannii* oxymoron: commensal hospital dweller turned pan-drug-resistant menace. *Front. Microbiol.* 3:148 doi: 10.3389/fmicb.2012.00148

Rodríguez-Martínez, J.-M., Poirel, L., and Nordmann, P. (2010). Genetic and functional variability of AmpC-type β-lactamases from *Acinetobacter baumannii. Antimicrob. Agents Chemother.* 54, 4930–4933. doi: 10.1128/AAC.00427-10

Saranathan, R., Tomar, A., Sudhakar, P., Arunkumar, K. P., and Prashanth, K. (2014). Draft genome sequence of a multidrug-resistant *Acinetobacter baumannii* PKAB07 clinical strain from India belonging to sequence type 195. *Genome Announ.* 2, e00184–e00114. doi: 10.1128/genomeA.00184-14

Segal, H., Garnym, S. M., and Elisha, B. G. (2005). Is IS$_{ABA-1}$ customized for Acinetobacter? *FEMS Microbiol. Lett.* 243, 425–429. doi: 10.1016/j.femsle.2005.01.005

Tian, G.-B., Adams-Haduch, J. M., Taracila, M., Bonomo, R. A., Wang, H.-N., and Doi, Y. (2011). Extended-spectrum AmpC cephalosporinase in *Acinetobacter baumannii*: ADC-56 confers resistance to cefepime. *Antimicrob. Agents Chemother.* 55, 4922–4925. doi: 10.1128/AAC.00704-11

Tian, Q. B., Ohnishi, M., Tabuchi, A., and Terawaki, Y. (1996). A new plasmid-encoded proteic killer gene system: cloning, sequencing, and analyzing hig locus of plasmid Rts1. *Biochem. Biophys. Res. Commun.* 220, 280–284 doi: 10.1006/bbrc.1996.0396

Vila, J., Martí, S., and Sánchez-Céspedes, J. (2007). Porins, efflux pumps and multidrug resistance in *Acinetobacter baumannii. J. Antimicrob. Chemother.* 59, 1210–1215. doi: 10.1093/jac/dkl509

Wieczorek, P., Sacha, P., Hauschild, T., Zórawski, M., Krawczyk, M., and Tryniszewska, E. (2008). Multidrug resistant *Acinetobacter baumannii*–the role of AdeABC (RND family) efflux pump in resistance to antibiotics. *Folia Histochem. Cytobiol.* 46, 257–267. doi: 10.2478/v10042-008-0056-x

Wisplinghoff, H., Decker, M., Haefs, C., Krut, O., Plum, G., and Seifert, H. (2003). Mutations in *gyrA* and *parC* associated with resistance to fluoroquinolones in epidemiologically defined clinical strains of *Acinetobacter baumannii. J. Antimicrob. Chemother.* 51, 177–180. doi: 10.1128/AAC.39.5.1201

Xu, Z., and Hao, B. (2009). CVTree update: a newly designed phylogenetic study platform using composition vectors and whole genomes. *Nucleic Acids Res.* 37, 174–178. doi: 10.1093/nar/gkp278

Zarrilli, R., Pournaras, S., Giannouli, M., and Tsakris, A. (2013). Global evolution of multidrug-resistant *Acinetobacter baumannii* clonal lineages. *Int. J. Antimicrob. Agents* 41, 11 doi: 10.1016/j.ijantimicag.2012.09.008

Zhou, H., Zhang, T., Yu, D., Pi, B., Yang, Q., Zhou, J., et al. (2011). Genomic analysis of the multidrug-resistant *Acinetobacter baumannii* Strain MDR-ZJ06 widely spread in China. *Antimicrob. Agents Chemother.* 55, 4506–4512. doi: 10.1128/AAC.01134-10

Zhu, L., Yan, Z., Zhang, Z., Zhou, Q., Zhou, J., Wakeland, E. K., et al. (2013). Complete genome analysis of three *Acinetobacter baumannii* clinical isolates in China for insight into the diversification of drug resistance elements. *PLoS ONE* 8:e66584. doi: 10.1371/journal.pone.0066584

Zimbler, D. L., Arivett, B. A., Beckett, A. C., Menke, S. M., and Actis, L. A. (2013). Functional features of TonB energy transduction systems of *Acinetobacter baumannii. Infect. Immun.* 81, 3382–3394. doi: 10.1128/IAI.00540-13

14

Comparative Genomic Analysis of the ICE*Sa*2603 Family ICEs and Spread of *erm*(B)- and *tet*(O)-Carrying Transferable 89K-Subtype ICEs in Swine and Bovine Isolates in China

Jinhu Huang[1], *Yuan Liang*[1], *Dawei Guo*[1], *Kexin Shang*[1], *Lin Ge*[1], *Jam Kashif*[2] *and Liping Wang*[1*]

[1] *College of Veterinary Medicine, Nanjing Agricultural University, Nanjing, China,* [2] *Department of Veterinary Pharmacology, Sindh Agricultural University, Tandojam, Pakistan*

Correspondence:
Liping Wang
wlp71@163.com

Integrative and conjugative elements (ICEs) of the ICE*Sa*2603 family have been isolated from several species of *Streptococcus* spp.; however, the comparative genomic and evolutionary analyses of these particular ICEs are currently only at their initial stages. By investigating 13 ICEs of the ICE*Sa*2603 family and two ICE*Sa*2603 family-like ICEs derived from diverse hosts and locations, we have determined that ICEs comprised a backbone of 30 identical syntenic core genes and accessory genes that were restricted to the intergenic sites or the 3′-end of the non-conserved domain of core genes to maintain its function. ICE*Sa*2603 family integrase $\mathrm{Int}_{\mathrm{ICE}Sa2603}$ specifically recognized a 15-bp *att* sequence (TTATTTAAGAGTAAC) at the 3′-end of *rplL,* which was highly conserved in genus *Streptococcus*. Phylogenetic analyses suggest that extensive recombination/insertion and the occurrence of a hybrid/mosaic in the ICE*Sa*2603 family were responsible for the significant increase in ICE diversity, thereby broadening its host range. Approximately 42.5 and 38.1% of the tested *Streptococcus suis* and *Streptococcus agalactiae* clinical isolates respectively contained ICE*Sa*2603 family Type IV secretion system (T4SS) genes, and 80.5 and 62.5% of which also respectively carried $int_{\mathrm{ICE}Sa2603}$, indicating that ICE*Sa*2603 family is widely distributed across these bacteria. Sequencing and conjugation transfer of a novel sequence type ST303 clinical *S. suis* isolate HB1011 demonstrated that the 89K-subtype ICE*Ssu*HB1011 retained its transferrable function, thereby conferring tetracycline and macrolide resistance.

Keywords: integrative and conjugative elements, antimicrobial resistance, horizontal gene transfer, ICE*Sa*2603 family, *erm*(B), *tet*(O), *Streptococcus suis, Streptococcus agalactiae*

INTRODUCTION

Integrative and conjugative elements (ICEs) are self-transmissible mobile genetic elements (MGEs) that primarily reside in the host cell's chromosome, yet have the ability to be transferred between cells by conjugation (Burrus et al., 2002; Burrus and Waldor, 2004). ICEs are considered as mosaic elements with both phage- and plasmid-like features that can integrate into and replicate with the host cell's chromosomes similar to those observed in bacteriophages, as well as transfers, via conjugation in plasmids. The number of identified ICEs continues to increase with the exponential expansion of sequenced complete genomes (Te Poele et al., 2008; Wozniak and Waldor, 2010). Guglielmini et al (Guglielmini et al., 2011) previously described the prevalence and diversity of ICEs by conducting bioinformatics analysis of clustered conjugative apparatus modules in various chromosomal locations. Based on this definition, 18% of sequenced prokaryotic genomes contain at least one ICE, implying that the role of ICEs in horizontal gene transfer is more important than previously conceived.

An ICE typically comprises three distinct modular structures that mediate its integration and excision, conjugation, and regulation. In addition, ICEs contain genes that confer specific phenotypes that are related to its existence in its hosts such as resistance to antibiotics and heavy metals. Therefore, ICEs are important vectors for the horizontal transfer of genetic information, thereby facilitating rapid bacterial evolution (Bi et al., 2011; Rodriguez-Blanco et al., 2012). The SXT/R391 family is the most extensively studied ICE family that was initially identified in *Vibrio cholerae* O139 (Waldor et al., 1996). To date, more than 40 members of this family have been identified in different clinical and environmental *Vibrio* species, as well as related gammaproteobacteria (Mata et al., 2011; Rodriguez-Blanco et al., 2012). The SXT/R391 family shares a core structure containing 52 conserved genes and a common chromosomal integration site at the 5′-end of the *prfC* gene, with other parts all integrating into 3′-ends of distinct element-specific tRNA gene loci (Wozniak et al., 2009; Wozniak and Waldor, 2010). All known elements contain variable DNAs that are inserted into specific positions of the backbone (hot spots) and encoding various beneficial traits such as antibiotic resistant determinants including but not limited to *floR*, *strBA*, *sul2*, *dfrA1*, *dfr18*, *tetA′*, *kanR*, and an emerging extended-spectrum cephalosporin resistance gene bla_{CMY-2} (Harada et al., 2010). SetR, a homolog of the phage λ repressor CI, derepresses the master transcriptional activators required for SXT transfer, including the *int* and *tra* operons (Daccord et al., 2012). Beaber et al. (2004) have shown that DNA-damaging agents mitomycin C and quinolinones which induce the host SOS response, increase the transfer of SXT by several hundred times. The horizontal dissemination of antibiotic resistance genes through ICEs under selective pressure poses new challenges to the use of antimicrobial agents. Extensive knowledge of SXT/R391 family ICEs prompted us to explore the genetic features of other family ICEs.

Comparative analysis of the genomes of different isolates of *Streptococcus agalactiae* has revealed that nearly 2/3 of its regions of diversity in the isolates consist of ICEs or ICE-like elements (Brochet et al., 2008), which include two of the largest families, namely, Tn916 and ICE*Sa*2603. The Tn916 family is extensively distributed across various organisms, including important human pathogens such as *Enterococcus faecalis, Clostridium difficile, Staphylococcus aureus,* and *Streptococcus* spp. (Roberts and Mullany, 2011). Furthermore, its transfer mechanism has been well studied (Roberts and Mullany, 2009). In contrast, the ICE*Sa*2603 family, although identified in various *Streptococcus* species, including *S. agalactiae* and *Streptococcus suis*, has not been comprehensively investigated. In the current study, we first comparatively analyzed 13 ICEs of the ICE*Sa*2603 family and two ICE*Sa*2603 family-like ICEs to perform an in-depth genetic characterization of the core genes that are present in all members of this ICE family. Next, we investigated the potential accessory functions encoded by the variable DNA harbored by these mobile elements as well as the conserved *att* site for integration and the *oriT* site for origination transfer. Phylogenetic analyses of the 30 core genes allowed us to classify the 15 ICEs were into four subgroups. Furthermore, the distribution of these ICEs was determined in *S. suis* and *S. agalactiae*. Finally, an experimental transfer of a swine origin 89K-like subgroup transferable ICE, ICE*Ssu*HB1011, was introduced. It is worth noting that isolates with *erm*(B)- and *tet*(O)-carrying transferable 89K-subtype ICEs were detected in different swine and bovine farms of China, thereby disseminating tetracycline and macrolide resistance genes and contributing to human pathogenesis (Tang et al., 2006).

MATERIALS AND METHODS

Bacterial Strains and Genomic DNA Extraction

A total of 73 *S. suis* isolates from swine (including one *S. suis* strain from ATCC) and 21 *S. agalactiae* isolates from bovine (including one *S. agalactiae* strain from ATCC) were used in the present study (Table S1). Isolates were routinely grown overnight on tryptic soya broth or agar plates supplemented with 5% calf serum at 37°C. The bacterial culture was centrifuged (10,000 g for 5 min at room temperature), and the pellets were harvested and resuspended in TE buffer (10 mM Tris-HCl, 1 mM EDTA, pH 8.0) supplemented with lysozyme (5 mg/mL) and incubated for 30 min at 37°C. Genomic DNA was prepared using a Bacteria DNA kit (Omega, Norcross, GA, USA), following the manufacturer's instructions. The extracted DNA was used as template for PCR and sequencing.

In silico identification of the ICE*Sa*2603 Family

To identify and compare the ICE*Sa*2603 family ICEs, the reference genomes of ICE*Sa*2603 family were obtained from NCBI (**Table 1**). ICE*Sa*2603, ICE*Sde*3396, ICE*Sdy*12394-1, ICE*Spa*43144-1, ICE*Sth*JIM8232-1, ICE*Ssu*32457, ICE*Ssu*BM407-2, and ICE*Ssu*SC84 were previously classified as members of the ICE*Sa*2603 family (Bi et al., 2011; http://db-mml.sjtu.edu.cn/ICEberg/index.php). ICE*Sa*09mas018883, ICE*Ssu*05ZYH33-1 (also designated as 89K according to its size), ICE*Ssu*98HAH33-1, ICE*Ssu*D9, ICE*Ssu*SS12, and ICE*Ssu*HB1011 were grouped into this family because these encode the integrase gene that is closely related to $int_{ICESa2603}$ and share significant sequence alignment and syntenic core structure. ICE*Ssu*T15 and ICE*Slu*van were identified as ICE*Sa*2603 family-like ICEs because these encoded a serine recombinase (SR) family integrase gene instead of the tyrosine family integrase gene, $int_{ICESa2603}$.

TABLE 1 | Properties of the ICES*a*2603 family ICEs.

ICE name	Host strain	Accession	Size (bp)	% GC	Origin	References
ICES*a*2603	*S. agalactiae* 2603V/R	AE009948.1	54349	38	United States	Tettelin et al., 2002
ICES*de*3396	*S. dysgalactiae* subsp. *equisimilis* NS3396	EU142041.1	63668	31	Australia, 2001	Davies et al., 2005, 2009
ICES*dy*12394-1	*S. dysgalactiae* subsp. *equisimilis* ATCC12394	CP002215.1	50568	39	United States, 1939	Suzuki et al., 2011
ICES*pa*43144-1	*S. pasteurianus* ATCC43144	AP012054.1	58476	37	United States	Lin et al., 2011
ICES*th*JIM8232-1	*S. thermophilus* JIM8232	FR875178.1	50501	37	France, 2002	Delorme et al., 2011
ICES*a*09mas018883	*S. agalactiae* 09mas018883	HF952104.1	46903	37	–	Zubair et al., 2013
ICES*su*32457	*S. suis* 32457	FR823304.2	54879	50.5	Italy, 2007	Palmieri et al., 2012
ICES*su*05ZYH33-1	*S. suis* 05ZYH33	CP000407.1	88851	37	China, 2005	Chen et al., 2007
ICES*su*98HAH33-1	*S. suis* 98HAH33	CP000408.1	89154	37	China, 1998	Chen et al., 2007
ICES*su*SC84	*S. suis* SC84	FM252031.1	89165	37	China, 2005	Holden et al., 2009
ICES*su*BM407-2[a]	*S. suis* BM407	FM252032.1	80320	36	Vietnam, 2004	Holden et al., 2009
ICES*su*D9	*S. suis* D9	CP002641.1	55989	39.8	China	Zhang et al., 2011
ICES*su*SS12	*S. suis* SS12	CP002640.1	64284	38.2	China	Zhang et al., 2011
ICES*su*T15[a,b]	*S. suis* T15	CP006246.1	71412	36.3	–	–
ICES*l*uvan[b]	*S. lutetiensis* 5-F9	HE963029.1	94189	44	Netherlands	Bjorkeng et al., 2013
ICES*su*HB1011[c]	*S. suis* HB1011	–	–	–	China, 2010	This study

[a] *The SNF2 family protein was absent (ICESsuBM407-2) or partially absent (ICESsuT15).*
[b] *ICESsuT15 and ICESluvan with a serine recombinase (SR) family integrase were referred as ICESa2603 family-like ICEs.*
[c] *ICESsuHB1011 sequence was not complete.*

Phylogenetic Analyses

The ICES*a*2603 family and family-like ICEs were aligned using ClustalW (Larkin et al., 2007) with default settings. Nucleotide and amino acid conservations were assessed using the appropriate BLAST algorithms. MAUVE 2.4 (Darling et al., 2004) and ACT software (Carver et al., 2005) were used to identify core genes. The core genes of each ICE were further aligned with the corresponding genes of ICES*a*2603 to calculate nucleic acid percentage identity (**Figure 1A**).

To obtain an overview of the relationship of the ICES*a*2603 family and family-like ICEs and its possible evolution, phylogenetic trees were generated from the alignment of whole core genes or each core gene using the neighbor-joining method with bootstrapping ($N = 1000$) parameters by thesoftware, MEGA6 (Tamura et al., 2013) and FigTree v1.4.2.

Evaluation of the *att* and *oriT* Sites

To identify the putative *att* sites, genomic sequences of the ICE-carrying strains were analyzed for the presence of directly repeated sequences flanking the ICEs. The ICES*a*2603 family-like ICES*su*T15, where there are no conserved directly repeated sequences presented, the putative *att* site was predicted by comparing the junction sequence with *S. suis* P1/7.

The predicted *oriT* site of ICES*su*05ZYH33-1 was first identified by Li et al. (2011). We determined that the *oriT* site is conserved in all other ICES*a*2603 family members. MEGA 6 (Tamura et al., 2013) and BioEdit v7 were used to align and mapthe *oriT* site the members of the ICES*a*2603 family.

PCR and Sequencing

To determine the presence of ICES*a*2603 family ICEs in *S. suis* and *S. agalactiae*, primers for the ICES*a*2603 core genes were designed and used in PCR analysis (Table S2). The integrated form and the extrachromosomal circular form of the ICEs of *S. suis* and *S. agalactiae* were detected by using combination primers, P1–P4 (**Figures 1B–D** and Table S2). PCR was also performed to detect the presence of ICE fragments in *S. suis* and *S. agalactiae* by PCR tiling assay as described previously (Sitkiewicz et al., 2011) using specific oligonucleotide primers (**Figure 1E** and Table S2). The products of a representative *S. suis* strain, HB1011, with the ICEs backbone as determined by PCR tiling were sequenced by Genewiz, Inc. (Suzhou, Jiangsu, China).

Antimicrobials and Susceptibility Tests

Tetracycline, erythromycin, streptomycin, and rifampin (Sigma, St. Louis, MO, USA) were used in the present study. MICs were determined according to the guidelines described by the Clinical and Laboratory Standards Institute (CLSI, 2010).

Conjugation Experiments

S. suis BAA-853 and *S. agalactiae* G9, which were susceptible to tetracycline and erythromycin, were used to generate a rifampin- and streptomycin-resistant phenotype by using the stepwise-induced method with sub-MICs of rifampin and streptomycin (Haenni et al., 2010). In the mating experiments, the above two induced strains (tetracycline- and erythromycin-susceptible but rifampin- and streptomycin-resistant) were used as recipients and *S. suis* HB1011 (tetracycline- and erythromycin-resistant but rifampin- and streptomycin- susceptible) with the ICES*su*HB1011 was utilized as donor. Filter mating assays were performed as previously described (Li et al., 2011). The transconjugant was further confirmed by PCR, sequencing, and MLST typing.

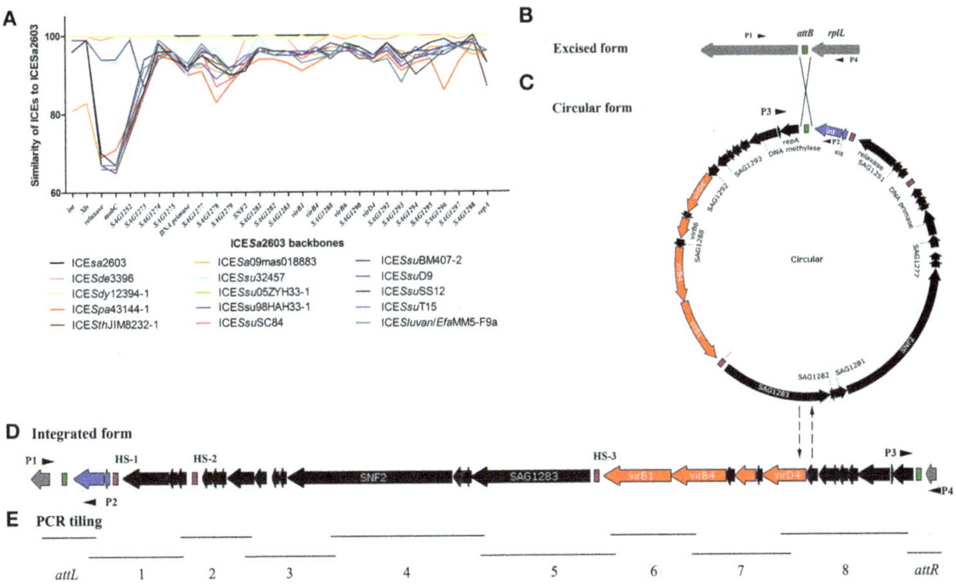

FIGURE 1 | Core genes identity (%) of the ICESa2603 family and schematic diagram of the integration and excision forms. The *att*B **(B)**, *att*ICE **(C)**, *att*L, and *att*R **(D)** sites are shown as green rectangles. The core genes are shown by navy (*int* and *xis*), red (T4SS genes), and black (other core genes). The HS sites are shown as purple rectangles. **(A)**; Core genes identity (%) of each ICE of ICESa2603 family to that of ICESa2603. **(B–D)**; Schematic diagram of the integration and excision forms of the ICESa2603 family. **(E)**; Schematic diagram of PCR tiling to detect the presence of ICEs fragments.

RESULTS

General Features of the ICE*Sa*2603 Family

A list of 13 ICEs of the ICE*Sa*2603 family and two ICE*Sa*2603 family-like ICEs from the NCBI complete genome database were analyzed and compared in this study (**Table 1**). An ICE that encodes an integrase gene closely related to $int_{ICESa2603}$, defined as having >60% gene or protein homology, and has significant sequence alignment (60% nucleic acid identity of core genes) and syntenic core structure were classified as a member of ICE*Sa*2603 family (Bi et al., 2011; **Figure 2A**) (http://db-mml.sjtu.edu.cn/ICEberg/index.php). Although ICE*Ssu*T15 and ICE*Sluvan* contained the backbone sequences of the ICE*Sa*2603 family, its integrase genes belonged to the SR family instead of the tyrosine family site-specific integrase of $int_{ICESa2603}$, and these were referred to as ICE*Sa*2603 family-like ICEs (**Figure 2B**). The strains, which were originally isolated from different countries around the world, belonged to six species of the *Streptococcus* spp. (**Table 1**).

The ICEs of the ICE*Sa*2603 family were first compared by using the MAUVE 2.4 (Darling et al., 2004) and ACT software (Carver et al., 2005) to visualize conserved and variable regions. All ICEs, with sizes ranging from 46,903 bp (ICE*Sa*09mas018883) to 89,165 bp (ICE*Ssu*SC84), shared a common conserved structure and had variable regions (**Table 1** and **Figure 2A**). The syntenic core structure of the ICE*Sa*2603 family backbone (33 kb) contained 30 conserved core genes (**Figures 1A**, **2**). There were three sites within the conserved ICE*Sa*2603 family structure for inserting variable DNAs in all of the ICEs (**Figure 2**). The three sites were named as HS-1, HS-2, and HS-3 (i.e., hotspots) for ICE acquisition of new genes of resistance to antibiotics and heavy metals. The three hotspots were located in the intergenic region of the ICE backbone, suggesting that the acquisition of these variable DNA regions did not interrupted core ICE gene. Additional DNA was inserted into the 3′-end non-conserved region of gene *SAG1277* (Gene ID: 1014084), *SAG1278* (Gene ID: 1014085), *SAG1279* (Gene ID: 1014086), or *SNF2* (Gene ID: 1014087) of some ICEs (**Figure 2**). The four inserted sites were designated as I-1, I-2, I-3, and I-4, respectively. The function of SAG1277, SAG1278, and SAG1279 remains unclear.

The ICE Core Genes

The present study has identified 30 core genes in 13 ICEs, which comprised the recombination module, the conjugation module, and various other genes of unknown functions. Integrases were members of the tyrosine recombinase family that contained a signature R·· H·· RH·· Y active site residue within its C-terminal catalytic domain. ICE*Sa*2603 family integrase $Int_{ICESa2603}$ mediates site-specific recombination between identical *att*ICE and *att*B sites and integration into the 3′-end of 50S ribosomal subunit protein L7/L12 gene (*rplL*) (Haenni et al., 2010), which was determined to be conserved in all *Streptococcus* spp. (Table S3).

The conjugative transfer module of the ICE*Sa*2603 family was grouped into two parts clustered. The first cluster comprised 3 genes (*SAG1250, SAG1251, and SAG1252* in *S. agalactiae* 2603V/R), which were annotated as *Tn5252 orf4 relaxase, Tn5252 orf9 mobC,* and *Tn5252 orf10*, respectively. Protein SAG1250 showed similarities to various prokaryotic DNA relaxases that covalently bind to and nick the DNA at the origin of transfer (*oriT*) with the help of the auxiliary MobC to initiate DNA

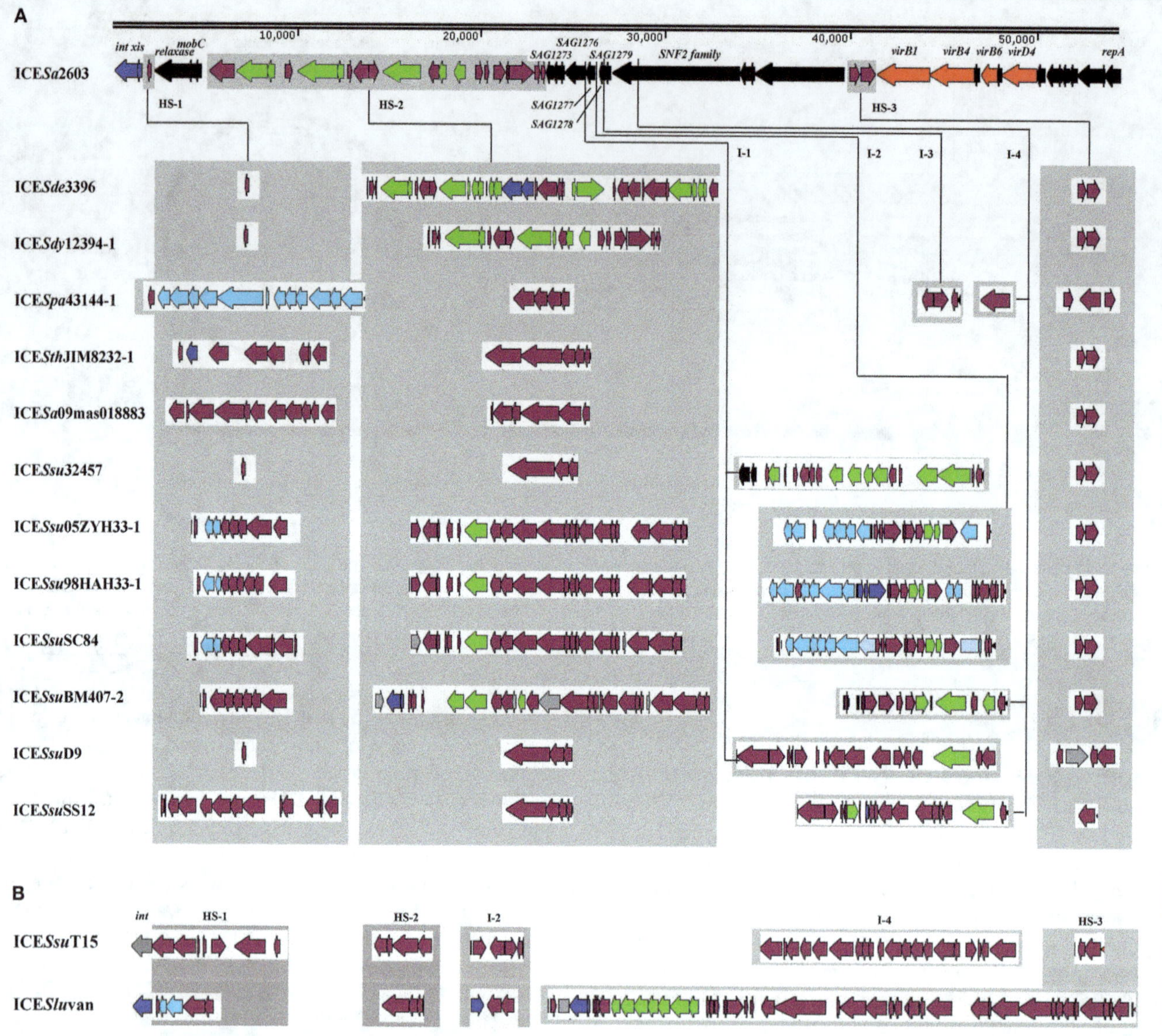

FIGURE 2 | Genomic organization of the ICESa2603 family. HS-1, HS-2, and HS-3 were the hotspots of ICESa2603 family, which were presented in all ICEs. I-1, I-2, I-3, and I-4 could also receive additional genes in part of the ICEs. Genes are shown in different colors: black, ICEs core genes; navy, *int* and *xis*; azure, nisin operon genes; red, T4SS genes; green, antibiotic and heavy metal resistance genes; purple, other cargo genes. **(A)**; ICE*Sa*2603 family ICEs. **(B)**; ICE*Sa*2603 family-like ICEs.

processing (Lanka and Wilkins, 1995), which include ORF4 relaxase (AAC98434.1) of conjugative transposon *Tn5252* (52% identity and 72% similarity) and MobA (AAQ55244.1) of *S. aureus* plasmid pC223 (22% identity and 41% similarity). Protein SAG1251, which is located adjacent to relaxase, shared 57% identity and 79% similarity to *Tn5252* ORF9 MobC (AAC98435.1) and 27% identity and 50% similarity to MobC (AAQ55243.1) of *S. aureus* plasmid pC223. The other part of the conjugation module belonged to a type IV secretion system (T4SS). The VirD4 coupling protein (40% identity to *Agrobacterium tumefaciens* VirD4), which functions as a substrate receptor, binds to the ICE DNA-relaxase complex (relaxosome) and presents it to the mating pair formation (Mpf) channels. Compared to *Agrobacterium tumefaciens* Vir T4SS (Christie, 1997), which are assembled from virB1–VirB11, only VirB1 (transglycosylase, 18% identity), VirB4 (ATPase, 41% identity), and VirB6 (inner-membrane protein, 12% identity) have been identified among the Mpf channels of the ICE*Sa*2603 family (**Figure 2**).

Additional conserved genes were located adjacent to the T4SS genes. Upstream of the *virD4* gene were genes coding for cytosine DNA methylase (*SAG1297*), replication initiation factors (*SAG1299*), and other unknown conserved genes, whereas downstream of the cluster were genes of the *SNF2* family (*SAG1280*), DNA primase (*SAG1276*), and other unknown conserved genes.

Variable DNA Regions

In addition to conserved core genes in all ICEs analyzed, variable DNA regions were also detected in the ICE backbone sequences (**Figure 2**). Most of the variable DNA sequences were inserted at three intergenic hotspots (HS), namely, HS-1, HS-2, and HS-3. The DNA content of various ICEs in HS was highly identical. For example, ICE*Sa*2603, ICE*Sde*3396, ICE*Sdy*12394-1, ICE*Ssu*32457, and ICE*Ssu*D9 had identical contents in HS-1. The DNA contents in HS-2 were more variable (only ICE*Ssu*32457 and ICE*Ssu*D9 shared identical genes), whereas in HS-3, these were more conserved (only ICE*Spa*43144-1, ICE*Ssu*D9, and ICE*Ssu*SS12 varied from the other ICEs). There were other exogenous DNA insertion sites (I), namely, I-1, I-2, I-3, and I-4 in at the 3′-end of the non-conserved region of genes *SAG1277*, *SAG1278*, *SAG1279*, and *SNF2* in some ICEs, respectively. For example, ICE*Ssu*32457 and ICE*Ssu*D9 had exogenous DNAs in I-1, whereas ICE*Ssu*05ZYH33-1, ICE*Ssu*98HAH33-1, ICE*Ssu*SC84, ICE*Ssu*T15, and ICE*Sluvan* had variable DNAs in I-2. Additional DNA inserts were identified at the I-3 site in ICE*Spa*43144-1 and at the I-4 site in ICE*Spa*43144-1, ICE*Ssu*BM407-2, ICE*Ssu*SS12, and ICE*Ssu*T15.

Other notable accessory genes in this family were identified as resistance genes for antibiotics and heavy metals, intact or remnant lantibiotic biosynthesis operon, and other MGEs such as IS elements and Tn916-like elements (**Figure 2**). The antibiotic and heavy metal resistance genes in ICEs are presented in **Table 2**. ICE*Sa*2603, ICE*Sde*3396, and ICE*Sdy*12394-1 carry heavy metal resistance genes, whereas ICEs from *S. suis* mostly contain macrolide resistance gene *erm*(B) and/or tetracycline resistance gene *tet*(M) or *tet*(O). ICE*Spa*43144-1, ICE*Ssu*05ZYH33-1, ICE*Ssu*98HAH33-1, ICE*Ssu*SC84, and ICE*Sluvan* had remnant nisin ORFs, in which a two-component signal transduction system nisK/nisR, is essential for full virulence of highly invasive *S. suis* serotype 2 (Li et al., 2008; Wang et al., 2014). In the first four ICEs, remnant nisin ORFs were presented in two variable regions at the insertion sites of HS-1 and I-2 (**Figure 2A**).

MGEs were also found to be inserted in HS or I sites. For example, IS116 and IS4 family IS elements were inserted in the I-4 region of ICE*Spa*43144-1 and ICE*Ssu*D9, respectively (**Figure 2A**). Moreover, a Tn916-like element was inserted in HS-2 of ICE*Ssu*05ZYH33-1, ICE*Ssu*98HAH33-1, ICE*Ssu*SC84, and ICE*Ssu*BM407-2 (**Figure 2A**). A Tn1549-like element was inserted in I-4 of ICE*Sluvan*, thereby forming a mosaic ICE ((Bjorkeng et al., 2013); **Figure 2B**). The integrase of ICE*Sluvan*, which belongs to theSR family instead of the tyrosine recombinase family of $Int_{ICESa2603}$, resulted in the alteration of the insertion site, thus generating a new ICE (ICE*Sa*2603 family-like ICE).

The *att* Site and *oriT* Site

ICEs of the ICE*Sa*2603 family were found to specially integrate at the 3′-end of the *rplL* gene of *Streptococcus* spp., by recognizing the *att* sequence, TTATTTAAGAGTAAC (**Table 2**). Although different identities of *rplL* gene were observed, the *att* sequence (15 bp from the 3′-end of the *rplL* gene) appeared more conserved in *Streptococcus* spp. (**Table 2** and Table S3), thus

TABLE 2 | Attachment site analysis and resistance profile of the ICE*Sa*2603 family.

ICEs name	% Identity to Int_{2603}	% Identity to $rplL_{2603}$	*att*B (3)	Putative *att* sites	Resistance profile
ICE*Sa*2603	100	100	*rplL*	TTATTTAAGAGTAAC	Hg, Cd, Cu
ICE*Sde*3396	100	–	–	–	Cd, Cu, As
ICE*Sdy*12394-1	100	84	*rplL*	TTATTTAAGAGTGAT	Cd, Cu
ICE*Spa*43144-1	96	84	hypothetical protein	TTATTTAAGAGTAAC	
ICE*Sth*JIM8232-1	96	88	*rplL*	TTATTTAAGAGTAAC	
ICE*Sa*09mas018883	81	99	*rplL*	TTATTTAAGAGTAAC	
ICE*Ssu*32457	99	–[a]	*rplL*	TTATTTAAGAGTAAC	*erm*(B), *aadE*, *aphA*, *tet*(40), *tet*(O/W/32/O)
ICE*Ssu*05ZYH33-1	96	84	*rplL*	TTATTTAAGAGTAAC	*tet*(M), *aadE*
ICE*Ssu*98HAH33-1	96	84	*rplL*	TTATTTAAGAGTAAC	*tet*(M), *aadE*
ICE*Ssu*SC84	96	84	*rplL*	TTATTTAAGAGTAAC	*tet*(M), *aadE*
ICE*Ssu*BM407-2	96	84	*rplL*	TTATTTAAGAGTAAC	*tet*(M), *tet*(L), *tet*(O), *erm*(B), *aadE*, *cat*
ICE*Ssu*D9	99	84	*rplL*	TTATTTAAGAGTAAC	*tet*(O), *erm*(B)
ICE*Ssu*SS12	96	84	*rplL*	TTATTTAAGAGTAAC	*tet*(O), *erm*(B)
ICE*Ssu*T15[b]	–	84	*SSU0877* in *S. suis* P1/7	CCTCTTATGTCAAGTAACTG (*attL*) CATTATTATGACACAATCCC (*attR*)	
ICE*Sluvan*[b]	–	–[a]	*rumA*	CACGTGGAGTGCGTAGTGTT (*attL*) TTCTCAAGGACCAGACAACA (*attR*)	*van* resistance operon
ICE*Ssu*HB1011[c]	98	–	*rplL*	TTATTTAAGAGTAAC	*tet*(O), *erm*(B)

[a]The rplL sequences were not indicated.

[b]The ICESsuT15 and ICESluvan's Int belonging to serine recombinase (SR) family shared low homology with the ICESa2603 tyrosine family site-specific integrase. ICESluvan was inserted in rum, and the att site was experimentally confirmed by Bjorkeng et al. (2013). ICESsuT15 was inserted in the 344 sites of SSU0877 in S. suis P1/7, the putative att site was predicted, and the Int of ICESsuT15 and ICESluvan may recognize CA dinucleotide in certain region of chromosome sites.

[c]ICESsuHB1011 sequence was not complete and the rplL sequences were not known.

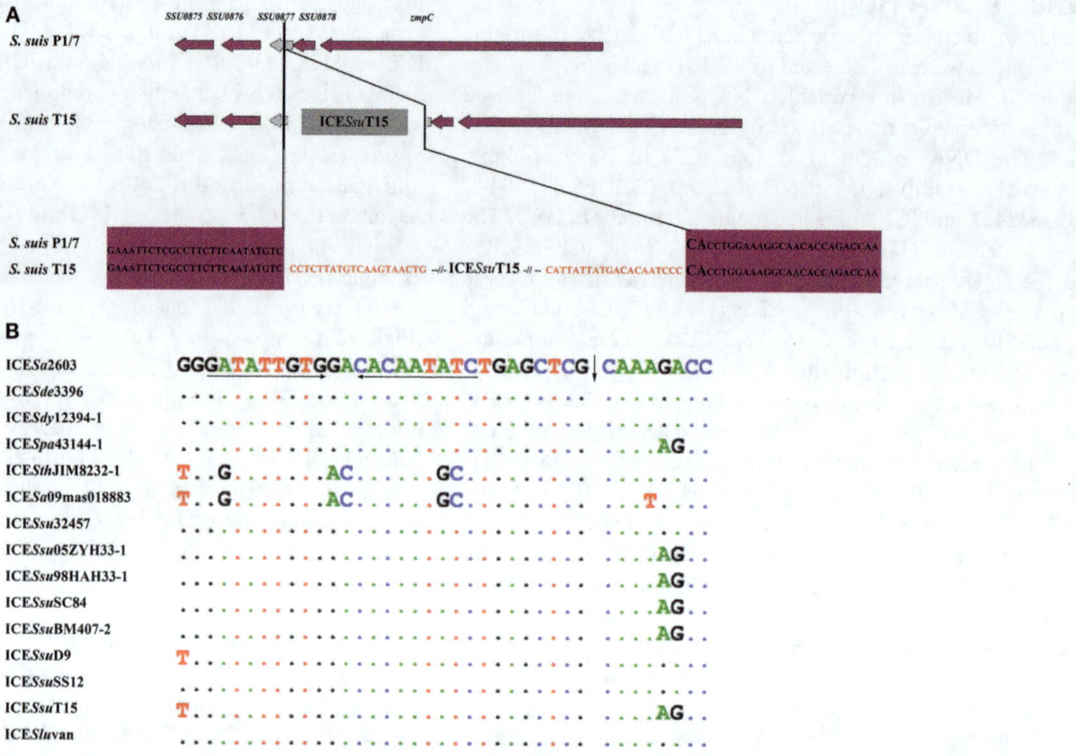

FIGURE 3 | *att* site of ICE*Ssu*T15 (A) and *oriT* site of the ICE*Sa*2603 family (B). (A); ORFs in purple represent the chromosome genes while in gray shows the integraing site of ICE*Ssu*T15, the putative *att* sequence was indicated in red. **(B)**; the *oriT* site of the ICE*Sa*2603 were listed in the top, the bases different from ICE*Sa*2603 were indicated. Arrows under the sequences represent the locations of inverted repeats, and vertical arrows show the nick sites.

providing the sequence basis for its distribution among members of the family. ICE*Ssu*T15 and ICE*Slu*van with Int and belonging to SR instead of tyrosine family integrase of ICE*Sa*2603 family were assumed to have a different *att* site. As shown in **Figure 3A**, there were no consensus insertion sequences have been identified in ICE*Ssu*T15.

A ~200 bp highly conserved region (92–100% identity) highlighted in the visual comparison with Mauve 2.4 (Darling et al., 2004) is considered to be the putative the *oriT* region. The left adjacent region was *SAG1252* (*Tn5252 orf10*, 65–100% identity) gene that showed relatively low identity and the right adjacent region was a variable DNA hotspot site. In all ICEs except for ICE*Sth*JIM8232-1 and ICE*Sa*09mas018883, the *oriT* site contained a 9-nt inverted repeat sequence, GATATTGTG, which served as a binding site for relaxase, and the precise nickase cleavage site was predicted to reside at position 5′-CTCG/CAAA (**Figure 3B**).

Phylogenetic Analyses of the ICE Core Genomes

To analyze the evolution of the ICE core genomes, we first compared the identity of each ICE core genes to the corresponding ICE*Sa*2603 genes (**Figure 1A**). Most ICE core genes exhibited a >90% identity at the nucleotide level. On the other hand, the less conserved genes were located at the junction between core genes and variable region. For example, the DNA processing genes (*SAG1250–SAG1252*) of the conjugative transfer module between HS-1 and HS-2 were much less conserved (65–72%) than other core genes, whereas the *oriT* site remained highly conserved.

Next, we generated phylogenetic trees based on all the 30 core genes (**Figure 4A**) or individual representative genes (**Figure 4B**). The ICEs were then further divided into four subgroups: ICE*Sa*2603 family groups I–III (ICE*Sa*2603-like, 89K-like, and ICE*Sth*JIM8232-1-like) and ICE*Sa*2603 family-like group IV. Interestingly, these subgroups also showed variable features: The ICEs of group IV could be distinguished from that of other groups based on the *int*, only one regulatory gene (*SAG1249*) was found in HS-1 region of group I ICEs (ICE*Sa*2603-like); and a transposon Tn916 was inserted into HS-2 region of group II ICEs, thereby generating mosaic ICEs (**Figures 2, 4A**).

To gain more insights into the evolution of individual core genes, we created phylogenetic trees for each core gene. However, most core genes were determined to be highly conserved, thus making it difficult to generate a reliable tree. Hence, we concentrated on the less conserved genes and selected *int*, *relaxase*, and *mobC* for phylogenetic analyses. **Figure 4B** shows the two subgroups of *int*: the tyrosine recombinase family *int* genes, which were highly conserved and present in all ICE*Sa*2603 families (88–100% identity), and the SR family *int* genes that were present in ICE*Sa*2603 family-like ICEs. According to the tree of

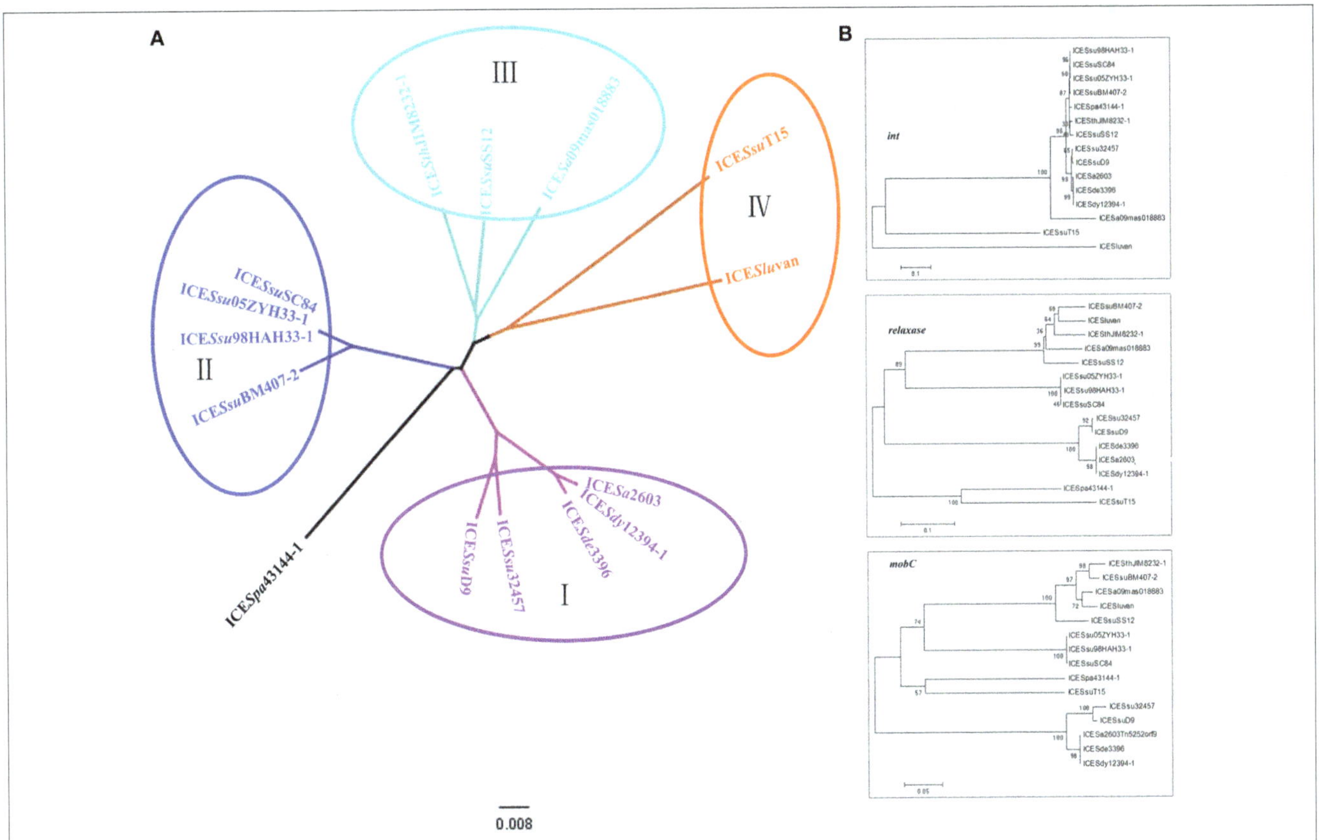

FIGURE 4 | Phylogenetic tree of ICES*a*2603 family. (A) , phylogenetic tree generated from 30 core genes. Four subgroups (I–IV) can be classified; **(B)**, phylogenetic tree of *int*, *relaxase*, and *mobC*.

TABLE 3 | Distribution of ICES*a*2603 family core genes in *S. suis* and *S. agalactiae*.

Strains	$int_{ICESa2603}$	*DNA primase*	T4SS genes				*repA*	ICE at *rplL* site confirmed by PCR[a]
			virB1	*virB4*	*virB6*	*virD4*		
S. suis ($n = 73$)[b]	25	65	32	32	32	31	30	22
S. agalactiae ($n = 21$)[c]	5	19	8	8	8	8	8	5

[a] *The extra chromosomal form and integrating form of ICEs can be seen in* ***Figures 1B–D****.*
[b] *S. suis were isolated from swine in china (Table S1).*
[c] *S. agalactiae were isolated from bovine in china (Table S1).*

relaxase and *mobC*, all ICEs in group I were aligned together and shared ~94% similarity, whereas other ICEs generated distinct branching patterns (with <80% similarity).

Distribution of Members of the ICES*a*2603 Family in *S. suis* and *S. agalactiae*

To evaluate the presence of ICEs in *S. suis* isolates from pigs and *S. agalactiae* isolates from cows in China, several core ICEs genes were detected (**Table 3** and Table S1). Of the 73 *S. suis* isolates screened, 31 were determined to contain ICEs based on the presence of T4SS genes (*virB1*, *virB4*, *virB6*, and *virD4*), in which 80.5% (25/31) also carried $int_{ICESa2603}$. Similarly, 38.1% of the tested *S. agalactiae* isolates harbored T4SS genes, amongst which 62.5% (5/8) were $int_{ICESa2603}$-positive. Different primer pairs P1–P4 were employed and identified the integration/excision form of ICES*a*2603 family in the present study (**Table 3**). **Table 3** shows that 88% (22/25) of the ICES*a*2603 family strains were excised from the chromosomes, thereby forming circular intermediates prior to its conjugative transfer into recipient cells.

To determine whether the ICEs were intact, PCR tiling was performed. Ten overlapping fragments were amplified using the long-PCR method (**Figure 1E** and Table S2). **Table 4** and Table S1 showed that almost all fragments of ICE core regions (fragments 4, 6, 7, and 8) could be amplified in macrolide- and tetracycline-resistant *S. suis*, whereas other fragments (1, 2, 3, 5) linking the variable regions (HS-1–HS-3 and I-1–I-4) were more variable. PCR tiling also detected various ICE

TABLE 4 | Analyses of ICESa2603-family fragments in part of macrolide and tetracycline resistant *S. suis* by PCR tiling assay.

Fragments[a]	attL	1	2	3[b]	4	5	6	7	8	attR
HB1004	+	+	-	-	+	+	+	+	+	+
HB1006	-	-	-	-	+	+	+	+	+	+
HB1011[c]	+	+	+	-	+	+	+	+	+	+
HB1012	+	+	-	-	+	+	+	+	+	+
HB1013	-	-	-	-	+	+	+	-	-	-
NJ4	+	-	-	-	+	+	+	+	-	-
JS07002	+	+	-	-	+	-	+	+	+	+
JS07015	+	+	-	-	+	+	+	+	+	+
SC05017	-	-	-	-	+	+	+	+	+	-
SS2-XY	+	-	-	-	+	+	+	+	+	+
YY060816	-	-	-	-	+	+	+	+	+	-

[a] *The PCR tiling assay and fragments can be seen in* ***Figure 1E****.*
[b] *Failure of amplifying this fragment, probably due to too long additional DNA segments inserted in the 4 regions (I-1 to I-4)*
[c] *The fragment 7 in HB1011 was ~2 kb larger than ICESa2603 core region.*

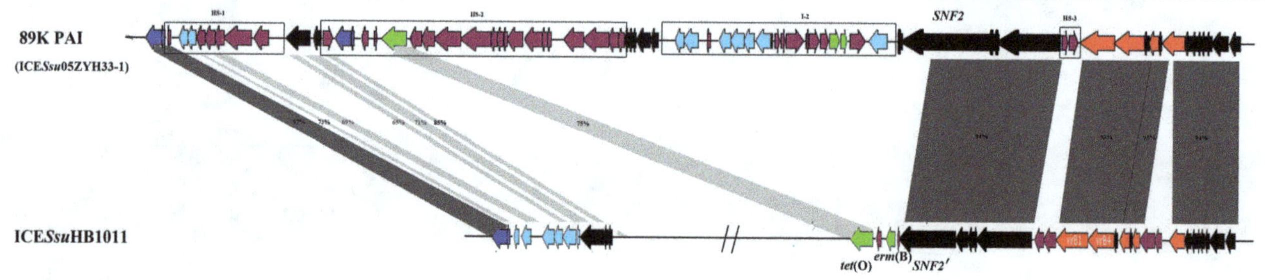

FIGURE 5 | ORF map of ICE*Ssu*HB1011 to the 89K PAI (ICE*Ssu*05ZYH33-1) of *S. suis* 05ZYH33. 89K PAI (ICE*Ssu*05ZYH33-1) was proved to be a pathogenicity island response for the human *S. suis* Outbreak in Sichuan, China in 2005. Genes are shown in different colors: black, ICEs core genes; navy, *int* and *xis*; azure, nisin operon genes; red, T4SS genes; green, antibiotic and heavy metal resistance genes.

subgroups. For example, the strain with fragments 2–8, but without fragments *attL*, 1, and *attR*, belonged to subgroup IV (ICE*Sa*2603 family-like); Furthermore, subgroup I could be distinguished from other subgroups based on the size of fragment 1.

ICE*Ssu*HB1011, a Swine Origin 89K-Like Subgroup Transferable ICEs

Because ICEs play an important role in the horizontal transmission of antibiotic resistance, we selected a swine-derived clinical macrolide- and tetracycline-resistant strain HB1011 (ST303, a new ST type) for subsequent conjugation experiments. The transfer of macrolide and tetracycline resistance from HB1011 to *S.suis* BAA-853 (ST1) was confirmed by sequencing of the PCR products (Figures S1, S2) and MLST typing. The MICs of erythromycin and tetracycline and PCR products of *erm*(B) and *tet*(O) in conjugant JH-1 confirmed that erythromycin and tetracycline resistance genes *erm*(B) and *tet*(O) were indeed transferred (Table S4). We further sequenced the tiling fragments of ICE*Ssu*HB1011, which is an 89K-like ICE (**Figure 5**). The results showed that *erm*(B) and *tet*(O) genes were inserted into the SNF2 gene (I-4, **Figure 1**).

DISCUSSION

ICEs are self-transmissible MGEs that encode the machinery for conjugation, as well as intricate regulatory systems to control excision from chromosomes and its conjugative transfer (Wozniak and Waldor, 2010). ICEs can be considered as mosaic elements that present the combined features of plasmids and bacteriophages. Similar to plasmids, ICEs transfer via conjugation and in parallel to phages, ICEs integrate into and replicate with its host chromosome (Burrus and Waldor, 2004). ICEs often carry accessory genes that encode for antibiotic resistance, virulence factors and various other functions. These are therefore important vectors for the horizontal dissemination of genetic information, thereby facilitating rapid bacterial evolution.

ICE*Sa*2603 is a ~54 kb ICE that was first discovered in *S. agalactiae* 2603V/R (Tettelin et al., 2002). It has the capacity to retain its features of transferability to *S. agalactiae*, *S. dysgalactiae* subsp *dysgalactiae*, and *S. uberis* (Haenni et al., 2010). Subsequently, ICE*Sa*2603-like ICEs have been detected in clinical isolates of *Streptococcus* spp. from both human and animal species from Europe and Asia (**Table 1**). Therefore, it is likely that the members of the ICESa2603 family have been distributed and have the ability to undergo horizontal transmission in *Streptococcus* spp. The number of ICESa2603 family ICEs and its host range may therefore further expand as additional research studies are currently being conducted. However, comparative genomics of ICE*Sa*2603 family ICEs and its evolution have not been reported. Here, we first comparatively analyzed 13 ICEs of the ICE*Sa*2603 family. The results showed that the ICE*Sa*2603 family has a backbone of 30 core genes that encode the core functions of integration/excision, conjugation and regulation.

The integration/excision modules of ICE*Sa*2603 consisted of $\text{Int}_{\text{ICE}Sa2603}$ (SAG1247) and $\text{Xis}_{\text{ICE}Sa2603}$ (SAG1248). $\text{Int}_{\text{ICE}Sa2603}$ belongs to the tyrosine based site-specific recombinases. This family of integrase originated from bacterial phages and conjugate transposons. One member is the integrase from *Bacillus subtilis* conjugative transposon ICE*Bs*1, which mediates integration into the 3′-ends of tRNA-Leu loci (Lee et al., 2007). $\text{Int}_{\text{ICE}Sa2603}$ shares 23% identity with $\text{Int}_{\text{ICE}Bs1}$,but mediate site-specific recombination between identical *att*ICE and *att*B sites and the integration into the 3′-end of the 50S ribosomal subunit protein L7/L12 gene (Haenni et al., 2010). The *attB* sites (TTATTTAAGAGTAAC) of ICE*Sa*2603 family is conserved in *Streptococcus* spp. and even *Enterococcus faecium* and *Enterococcus faecalis*, which provided the sequence basis for the horizontal transfer of this family. To date, members of the ICE*Sa*2603 family has been identified in six species of *Streptococcus* spp. (**Table 2**). The ICE*Sa*2603 family-like ICEs, share 28 of 30 core genes with ICE*Sa*2603 family ICEs, which contain the SR family integrase, although a low level of identity was observed between ICE*Slu*van and ICE*Ssu*T15 (**Figures 1A, 4B**). A study on another SR Int_{Tn5397} has shown that the insertion site contains a central GA dinucleotide; however, no other consensus insertion sequences have been identified (Wang and Mullany, 2000). Based on the experimental transfer of ICE*Slu*van (Bjorkeng et al., 2013) and the analysis of ICE*Ssu*T15 (**Figure 3A**), we believe that *Int* of ICE*Slu*van and ICE*Ssu*T15 recognize the CA dinucleotide in certain regions of chromosomal sites. The ICE*Slu*van is capable of transmission, which provides an illustration that acquisition of a new integration/excision module (most likely via recombination) generates new types of ICEs (here we termed as ICE*Sa*2603 family-like ICEs). Xis, the adjacent recombination directionality factor (RDF), both facilitates excision and inhibits integration. It binds to the ends of the integrated element adjacent to Int binding sites. Deletion of the *xis* gene of ICE*Ssu*05ZYH33-1 demonstrated that xis stimulates, but is not essential for ICE*Ssu*05ZYH33-1 excision from chromosomes (Li et al., 2011).

The conjugation module of ICE*Sa*2603 family is similar with that of plasmids. When ICE*Sa*2603 transfers, a relaxase (SAG1250) first binds to and excises circular ICE DNA at *ori*T sites (Li et al., 2011). Thereafter, with the help of a coupling protein, MobC (SAG1251), the relaxase-DNA complex is presented to type IV coupling protein (T4CP) and the Mpf complex. Finally, the T4CPMpf complex transfers the ICE DNA to the receipt cell. The T4SS complex includes genes encoding for proteins similar to *A. tumefaciens* VirB1, VirB4, VirB6, and VirD4. Zhang et al. reported that these T4SS gene clusters were conserved both in composition and order in the genomes of various *Streptococcal* species and thus collectively described these as a Type–IVC secretion system (Zhang et al., 2012). Knockout of two key components genes (i.e., *virD4* and *virB4*) of the Type–IVC secretion system not only abolished its transferability but also eliminated the lethality of the highly virulent strain and impaired its ability to trigger a host immune response in infected mouse models (Li et al., 2011; Zhao et al., 2011). These results suggest that T4SS has retained its ancestral function as a conjugation module while its capacity to translocate effector protein(s) to the cell surface or into eukaryotic target cells undergoes evolution.

Our understanding of the mechanism underlying the regulation module of the ICE*Sa*2603 family is limited. Comparative analysis of the 30 core genes of ICE*Sa*2603 family ICEs has shown that no common regulators were conserved in all ICEs of this family. Instead, a gene (*SAG1249*) encoding a Cro/CI family transcriptional regulator was detected in all experimentally proven transferable ICEs, despite not being classified as core genes in the present study. This regulator possibly regulates ICE gene expression and ICE transfer (Beaber and Waldor, 2004). Recently, Xu et al. described a two-component system, NisK/NisR, in subgroup II (89K-subgroup, **Figure 4A**) ICEs that contributes to the virulence of *S. suis* serotype 2 (Xu et al., 2014). It is tempting to know whether this two-component system is involved in the regulation of transfer. A bacteriophage abortive infection (Abi) system, AbiE, which assists bacteria from being killed by bacteriophages, has been recently identified as a type IV toxin-antitoxin system in ICE*Sa*2603 (*SAG1284* and *SAG1285*) (Chopin et al., 2005; Dy et al., 2014). Dy et al. (2014) demonstrated that the AbiE system enables plasmid maintenance when these were introduced into pUC19. It is possible that the *abiE* operon and native promoter enable the maintenance of the circular form of ICE*Sa*2603, thereby increasing the frequency for ICE transfer.

Phylogenetic and comparative analyses have shown that most of the core genes of the ICE*Sa*2603 family exhibited 91–100% identity at the nucleotide level (**Figure 1A**). No detectable difference in the degree of conservation of most core genes was observed, which suggests equal selective pressures on core genes of the ICEs. However, some genes (i.e., *Int*, *xis*, *relaxase*, *mobC*, and *SAG1278*) exhibited lower degrees of conservation. The difference in the identity of the *Int* and *xis* genes, as earlier described, was probably generated by recombination. The *relaxase* and *mobC* genes can be divided into four subgroups (**Figure 4B**). The *relaxase* and *mobC* genes exhibited a high level of identity within subgroups (>80%), but a low identity between subgroups (65–80%). These results suggest that the different subgroups of *relaxase* and *mobC* genes (DNA processing module) might have also been generated by recombination of the entire DNA processing module. The relative low level of conservation of *SAG1278*, in which of the 3′-end is often the insertion site (I-2), might have been ascribed to the insertion of accessory genes, as indicated by the lower level of identity of this gene in each ICE. These results also suggest that individual core genes are exposed to different evolutionary pressures.

BLAST analysis of the core conjugation and integration module proteins of the ICE*Sa*2603 family showed that ICE*Sa*2603 shared some similarity with the *Streptococcus pneumoniae* conjugative transposon Tn5252 (Alarcon-Chaidez et al., 1997). Based on limited knowledge, we propose that ICEs were disseminated across a diverse range of hosts, thereby resulting in exposure to different selective stress. To survive stress, cargo genes encoding antibiotic resistance, nitrogen fixation, virulence factors, and various other functions were introduced into the HS and I sites. In the case of the ICE*Sa*2603 family, insertion of variable DNAs might have disrupted the core regulatory module gene (*SAG1249*), but another regulatory module gene could be functioned to compensate and restore stability in the chromosome (Li et al., 2008, 2011). Another benefit is that recombination of the core module increases the adaptability and diversity of ICEs or MGEs. Examples include the ICE*Sa*2603 family-like ICEs, ICE*Sluvan*, in which only the conjugation module int_{Tyr}–*xis* was replaced by SR family int_{SR} yet retained its transferability (Bjorkeng et al., 2013). A very recent study (Marini et al., 2015) has shown that mating between *S. suis* 32457 (donor) and *S. agalactiae* 2603V/R (recipient) yields a hybrid ICE, ICESa2603/ICESsu32457, and interestingly, could be transferred across various *Streptococcus pyogenes* strains. These findings strongly support the concept that recombination plays an important role in horizontal gene transfer and generates novel ICEs or even new types of MGEs. These findings and the above mentioned recombination/insertion/hybrid events further broaden range of ICE diversity. The mosaic patterns of the ICE*Sa*2603 family consisted of genes from groups A, B, and G *Streptococcus* organisms, *Faecalibacterium*, enterococci, and possibly *Listeria innocua* and *Enterococci faecalis*, indicating that recombination and mobilization events are key factors in the assembly of these families. The exchange of integration modules among ICEs or between ICEs and other integrating MGEs (e.g., phages) could potentially create new ICEs with altered insertion site specificities or modified host ranges.

Figure 4A shows that all four subgroups of the ICE*Sa*2603 family were detected in *S. suis*, which is indicative of its important role in the transmission of these types of ICEs of *S. suis*. Screening of ICE*Sa*2603 core genes showed that 42.5% of swine-derived *S. suis* isolates contained ICE T4SS genes and 80.5% of these belong to ICE*Sa*2603 family. These results suggest that *S. suis* isolates harbor ICE*Sa*2603 family ICEs that are extensively distributed in swine in China, which has become a challenge to the hog industry and even a threat to human health.

89K-like subgroup ICEs (ICE*Ssu*98HAH33-1 and ICE*Ssu*05ZYH33-1) were detected in two *S. suis* isolates and have caused two extensive human outbreaks of streptococcal toxic shock–like syndrome (STSLS) in China in 1998 and 2005 (Tang et al., 2006). Further studies (Li et al., 2008; Zhao et al., 2011; Zhong et al., 2014) have demonstrated that 89K is a pathogenicity island (PAI). Li et al. (2008) earlier showed that a two-component signal transduction system (SalK/SalR) is essential for full virulence of highly invasive *S. suis* serotype 2. SalK/SalR, which is also the regulatory module of the nisin lantibiotic operon, was inserted into the HS-1, and possibly acts as a regulator, together with SAG1249. Furthermore, knockout of three key components (*virB1*, *virB4*, and *virD4*) of T4SS significantly reduces but does not abolish the virulence of *S. suis* in a mouse model (Zhao et al., 2011; Zhong et al., 2014). These findings indicate that T4SS harbored by 89K PAI contributes to the development of STSLS and mediates the horizontal transfer of 89K. However, those types of ICEs did not contain genes that encode resistance to macrolides and tetracycline. In the present study, we observed that almost all of the macrolide- and tetracycline-resistant *S. suis* also contained T4SS genes, indicating that ICEs play an important role in spreading macrolide and tetracycline resistance in *S. suis*. The isolated strain HB1011, which was determined to harbor ICE*Ssu*HB1011 that carried *erm*(B) and *tet*(O), which was indicative of its multiple functions, including horizontal transfer resistance and pathogenicity. In addition, other strains of *S. agalactiae* and *S. dysgalactiae subsp. equisimilis* whose ICEs harbored T4SS that was similar to 89K did not cause STSLS. Therefore, it is of great importance to identify factors and mechanisms by which the T4SS of 89K mediates pathogenicity. It has been proposed that ICE*Sa*2603 family T4SS might have retained its ancestral function as a mobile DNA element, but have also evolved its capacity to translocate effector protein(s) to the cell surface or into eukaryotic target cells (Bhatty et al., 2013).

CONCLUSIONS

In summary, the structures of the ICE*Sa*2603 family ICEs in isolates of different *streptococcal* species are clearly illustrated in this study by comparative analyzing 13 ICEs of the ICE*Sa*2603 family and two ICE*Sa*2603 family-like ICEs. Screening of ICE*Sa*2603 core genes shows that ICE*Sa*2603 family ICEs are distributed widely in *S. suis* and *S. agalactiae* in China and mainly belong to the 89K-subtype family ICEs. Furthermore, the spread of the ICEs containing resistance genes *erm*(B) and *tet*(O) among *S. suis* strains by conjugation had a profound effect on the horizontal transmission of antibiotic resistance to macrolides and tetracyclines. Further research is needed to characterize the mechanisms involved in triggering ICEs transfer intra- and inter-species.

AUTHOR CONTRIBUTIONS

JH, DG, and LW developed the concept and designed experiments. JH, YL, KS, and LG performed the experiments and collected the data. JH, YL, and JK conducted all bioinformatics analyses. JH and LW prepared the manuscript. All authors have contributed to, seen and approved the manuscript.

ACKNOWLEDGMENTS

The study was partially supported by the National Natural Science Foundation of China (No. 31572567), Natural Science Foundation of Jiangsu Province of China (No. BK2012771), and a

Project Funded by the Priority Academic Program Development of Jiangsu Higher Education Institutions (PAPD). The authors wish to thank Professor Shile Huang from Louisiana State University, Health Sciences Center in USA for reviewing the manuscript.

REFERENCES

Alarcon-Chaidez, F., Sampath, J., Srinivas, P., and Vijayakumar, M. N. (1997). TN5252: a model for complex streptococcal conjugative transposons. *Adv. Exp. Med. Biol.* 418, 1029–1032. doi: 10.1007/978-1-4899-1825-3_242

Beaber, J. W., Hochhut, B., and Waldor, M. K. (2004). SOS response promotes horizontal dissemination of antibiotic resistance genes. *Nature* 427, 72–74. doi: 10.1038/nature02241

Beaber, J. W., and Waldor, M. K. (2004). Identification of operators and promoters that control SXT conjugative transfer. *J. Bacteriol.* 186, 5945–5949. doi: 10.1128/JB.186.17.5945-5949.2004

Bhatty, M., Laverde Gomez, J. A., and Christie, P. J. (2013). The expanding bacterial type IV secretion lexicon. *Res. Microbiol.* 164, 620–639. doi: 10.1016/j.resmic.2013.03.012

Bi, D., Xu, Z., Harrison, E. M., Tai, C., Wei, Y., He, X., et al. (2011). ICEberg: a web-based resource for integrative and conjugative elements found in Bacteria. *Nucleic Acids Res.* 40, D621–D626. doi: 10.1093/nar/gkr846

Bjorkeng, E. K., Hjerde, E., Pedersen, T., Sundsfjord, A., and Hegstad, K. (2013). ICESluvan, a 94-kilobase mosaic integrative conjugative element conferring interspecies transfer of VanB-type glycopeptide resistance, a novel bacitracin resistance locus, and a toxin-antitoxin stabilization system. *J. Bacteriol.* 195, 5381–5390. doi: 10.1128/JB.02165-12

Brochet, M., Couve, E., Glaser, P., Guedon, G., and Payot, S. (2008). Integrative conjugative elements and related elements are major contributors to the genome diversity of *Streptococcus agalactiae*. *J. Bacteriol.* 190, 6913–6917. doi: 10.1128/JB.00824-08

Burrus, V., Pavlovic, G., Decaris, B., and Guedon, G. (2002). Conjugative transposons: the tip of the iceberg. *Mol. Microbiol.* 46, 601–610. doi: 10.1046/j.1365-2958.2002.03191.x

Burrus, V., and Waldor, M. K. (2004). Shaping bacterial genomes with integrative and conjugative elements. *Res. Microbiol.* 155, 376–386. doi: 10.1016/j.resmic.2004.01.012

Carver, T. J., Rutherford, K. M., Berriman, M., Rajandream, M. A., Barrell, B. G., and Parkhill, J. (2005). ACT: the artemis comparison tool. *Bioinformatics* 21, 3422–3423. doi: 10.1093/bioinformatics/bti553

Chen, C., Tang, J. Q., Dong, W., Wang, C. J., Feng, Y. J., Wang, J., et al. (2007). A glimpse of streptococcal toxic shock syndrome from comparative genomics of S-suis 2 chinese isolates. *PLoS ONE* 2:e315. doi: 10.1371/journal.pone.0000315

Chopin, M. C., Chopin, A., and Bidnenko, E. (2005). Phage abortive infection in lactococci: variations on a theme. *Curr. Opin. Microbiol.* 8, 473–479. doi: 10.1016/j.mib.2005.06.006

Christie, P. J. (1997). Agrobacterium tumefaciens T-complex transport apparatus: a paradigm for a new family of multifunctional transporters in eubacteria. *J. Bacteriol.* 179, 3085–3094.

CLSI (2010). "Performance standards for antimicrobial susceptibility testing; Twentieth informational supplement," in *CLSI Document M100-S20*, ed F. R. Cockerill (Wayne, PA: Clinical and Laboratory Standards Institute).

Daccord, A., Mursell, M., Poulin-Laprade, D., and Burrus, V. (2012). Dynamics of the SetCD-regulated integration and excision of genomic islands mobilized by integrating conjugative elements of the SXT/R391 family. *J. Bacteriol.* 194, 5794–5802. doi: 10.1128/JB.01093-12

Darling, A. C., Mau, B., Blattner, F. R., and Perna, N. T. (2004). Mauve: multiple alignment of conserved genomic sequence with rearrangements. *Genome Res.* 14, 1394–1403. doi: 10.1101/gr.2289704

Davies, M. R., Shera, J., Van Domselaar, G. H., Sriprakash, K. S., and McMillan, D. J. (2009). A novel integrative conjugative element mediates genetic transfer from group G *Streptococcus* to other {beta}-hemolytic Streptococci. *J. Bacteriol.* 191, 2257–2265. doi: 10.1128/JB.01624-08

Davies, M. R., Tran, T. N., McMillan, D. J., Gardiner, D. L., Currie, B. J., and Sriprakash, K. S. (2005). Inter-species genetic movement may blur the epidemiology of streptococcal diseases in endemic regions. *Microbes Infect.* 7, 1128–1138. doi: 10.1016/j.micinf.2005.03.018

Delorme, C., Bartholini, C., Luraschi, M., Pons, N., Loux, V., Almeida, M., et al. (2011). Complete genome sequence of the pigmented *Streptococcus* thermophilus strain JIM8232. *J. Bacteriol.* 193, 5581–5582. doi: 10.1128/JB.05404-11

Dy, R. L., Przybilski, R., Semeijn, K., Salmond, G. P., and Fineran, P. C. (2014). A widespread bacteriophage abortive infection system functions through a Type IV toxin-antitoxin mechanism. *Nucleic Acids Res.* 42, 4590–4605. doi: 10.1093/nar/gkt1419

Guglielmini, J., Quintais, L., Garcillan-Barcia, M. P., De La Cruz, F., and Rocha, E. P. (2011). The repertoire of ICE in prokaryotes underscores the unity, diversity, and ubiquity of conjugation. *PLoS Genet.* 7:e1002222. doi: 10.1371/journal.pgen.1002222

Haenni, M., Saras, E., Bertin, S., Leblond, P., Madec, J. Y., and Payot, S. (2010). Diversity and Mobility of integrative and conjugative elements in bovine isolates of *Streptococcus agalactiae*, *S. dysgalactiae* subsp dysgalactiae, and *S. uberis*. *Appl. Environ. Microbiol.* 76, 7957–7965. doi: 10.1128/AEM.00805-10

Harada, S., Ishii, Y., Saga, T., Tateda, K., and Yamaguchi, K. (2010). Chromosomally encoded blaCMY-2 located on a novel SXT/R391-related integrating conjugative element in a Proteus mirabilis clinical isolate. *Antimicrob. Agents Chemother.* 54, 3545–3550. doi: 10.1128/AAC.00111-10

Holden, M. T. G., Hauser, H., Sanders, M., Thi, H. N., Cherevach, I., Cronin, A., et al. (2009). Rapid evolution of virulence and drug resistance in the emerging zoonotic pathogen *Streptococcus suis*. *PLoS ONE* 4:e6072. doi: 10.1371/journal.pone.0006072

Lanka, E., and Wilkins, B. M. (1995). DNA processing reactions in bacterial conjugation. *Annu. Rev. Biochem.* 64, 141–169. doi: 10.1146/annurev.bi.64.070195.001041

Larkin, M. A., Blackshields, G., Brown, N. P., Chenna, R., McGettigan, P. A., McWilliam, H., et al. (2007). Clustal W and Clustal X version 2.0. *Bioinformatics* 23, 2947–2948. doi: 10.1093/bioinformatics/btm404

Lee, C. A., Auchtung, J. M., Monson, R. E., and Grossman, A. D. (2007). Identification and characterization of int (integrase), xis (excisionase) and chromosomal attachment sites of the integrative and conjugative element ICEBs1 of *Bacillus subtilis*. *Mol. Microbiol.* 66, 1356–1369. doi: 10.1111/j.1365-2958.2007.06000.x

Li, M., Shen, X., Yan, J., Han, H., Zheng, B., Liu, D., et al. (2011). GI-type T4SS-mediated horizontal transfer of the 89K pathogenicity island in epidemic *Streptococcus suis* serotype 2. *Mol. Microbiol.* 79, 1670–1683. doi: 10.1111/j.1365-2958.2011.07553.x

Li, M., Wang, C., Feng, Y., Pan, X., Cheng, G., Wang, J., et al. (2008). SalK/SalR, a two-component signal transduction system, is essential for full virulence of highly invasive *Streptococcus suis* serotype 2. *PLoS ONE* 3:e2080. doi: 10.1371/journal.pone.0002080

Lin, I. H., Liu, T. T., Teng, Y. T., Wu, H. L., Liu, Y. M., Wu, K. M., et al. (2011). Sequencing and comparative genome analysis of two pathogenic *Streptococcus* gallolyticus subspecies: genome plasticity, adaptation and virulence. *PLoS ONE* 6:e20519. doi: 10.1371/journal.pone.0020519

Marini, E., Palmieri, C., Magi, G., and Facinelli, B. (2015). Recombination between *Streptococcus suis* ICESsu32457 and *Streptococcus agalactiae* ICESa2603 yields a hybrid ICE transferable to *Streptococcus* pyogenes. *Vet. Microbiol.* 178, 99–104. doi: 10.1016/j.vetmic.2015.04.013

Mata, C., Navarro, F., Miro, E., Walsh, T. R., Mirelis, B., and Toleman, M. (2011). Prevalence of SXT/R391-like integrative and conjugative elements carrying blaCMY-2 in Proteus mirabilis. *J. Antimicrob. Chemother.* 66, 2266–2270. doi: 10.1093/jac/dkr286

Palmieri, C., Magi, G., Mingoia, M., Bagnarelli, P., Ripa, S., Varaldo, P. E., et al. (2012). Characterization of a *Streptococcus suis* tet(O/W/32/O)-carrying element transferable to major streptococcal pathogens. *Antimicrob. Agents Chemother.* 56, 4697–4702. doi: 10.1128/AAC.00629-12

Roberts, A. P., and Mullany, P. (2009). A modular master on the move: the Tn916 family of mobile genetic elements. *Trends Microbiol.* 17, 251–258. doi: 10.1016/j.tim.2009.03.002

Roberts, A. P., and Mullany, P. (2011). Tn916-like genetic elements: a diverse group of modular mobile elements conferring antibiotic resistance. *FEMS Microbiol. Rev.* 35, 856–871. doi: 10.1111/j.1574-6976.2011.00283.x

Rodriguez-Blanco, A., Lemos, M. L., and Osorio, C. R. (2012). Integrating conjugative elements as vectors of antibiotic, mercury, and quaternary ammonium compound resistance in marine aquaculture environments. *Antimicrob. Agents Chemother.* 56, 2619–2626. doi: 10.1128/AAC.05997-11

Sitkiewicz, I., Green, N. M., Guo, N., Mereghetti, L., and Musser, J. M. (2011). Lateral gene transfer of streptococcal ICE element RD2 (region of difference 2) encoding secreted proteins. *BMC Microbiol.* 11:65. doi: 10.1186/1471-2180-11-65

Suzuki, H., Lefebure, T., Hubisz, M. J., Pavinski Bitar, P., Lang, P., Siepel, A., et al. (2011). Comparative genomic analysis of the *Streptococcus* dysgalactiae species group: gene content, molecular adaptation, and promoter evolution. *Genome Biol. Evol.* 3, 168–185. doi: 10.1093/gbe/evr006

Tamura, K., Stecher, G., Peterson, D., Filipski, A., and Kumar, S. (2013). MEGA6: molecular evolutionary genetics analysis version 6.0. *Mol. Biol. Evol.* 30, 2725–2729. doi: 10.1093/molbev/mst197

Tang, J., Wang, C., Feng, Y., Yang, W., Song, H., Chen, Z., et al. (2006). Streptococcal toxic shock syndrome caused by *Streptococcus suis* serotype 2. *PLoS Med.* 3:e151. doi: 10.1371/journal.pmed.0030151

Te Poele, E. M., Bolhuis, H., and Dijkhuizen, L. (2008). Actinomycete integrative and conjugative elements. *Antonie Van Leeuwenhoek* 94, 127–143. doi: 10.1007/s10482-008-9255-x

Tettelin, H., Masignani, V., Cieslewicz, M. J., Eisen, J. A., Peterson, S., Wessels, M. R., et al. (2002). Complete genome sequence and comparative genomic analysis of an emerging human pathogen, serotype V *Streptococcus agalactiae. Proc. Natl. Acad. Sci. U.S.A.* 99, 12391–12396. doi: 10.1073/pnas.182380799

Waldor, M. K., Tschape, H., and Mekalanos, J. J. (1996). A new type of conjugative transposon encodes resistance to sulfamethoxazole, trimethoprim, and streptomycin in *Vibrio cholerae* O139. *J. Bacteriol.* 178, 4157–4165.

Wang, H., and Mullany, P. (2000). The large resolvase TndX is required and sufficient for integration and excision of derivatives of the novel conjugative transposon Tn5397. *J. Bacteriol.* 182, 6577–6583. doi: 10.1128/JB.182.23.6577-6583.2000

Wang, J., Gao, Y., Teng, K., Zhang, J., Sun, S., and Zhong, J. (2014). Restoration of bioactive lantibiotic suicin from a remnant lan locus of pathogenic *Streptococcus suis* serotype 2. *Appl. Environ. Microbiol.* 80, 1062–1071. doi: 10.1128/AEM.03213-13

Wozniak, R. A., Fouts, D. E., Spagnoletti, M., Colombo, M. M., Ceccarelli, D., Garriss, G., et al. (2009). Comparative ICE genomics: insights into the evolution of the SXT/R391 family of ICEs. *PLoS Genet.* 5:e1000786. doi: 10.1371/journal.pgen.1000786

Wozniak, R. A., and Waldor, M. K. (2010). Integrative and conjugative elements: mosaic mobile genetic elements enabling dynamic lateral gene flow. *Nat. Rev. Microbiol.* 8, 552–563. doi: 10.1038/nrmicro2382

Xu, J., Fu, S., Liu, M., Xu, Q., Bei, W., Chen, H., et al. (2014). The two-component system NisK/NisR contributes to the virulence of *Streptococcus suis* serotype 2. *Microbiol. Res.* 169, 541–546. doi: 10.1016/j.micres.2013.11.002

Zhang, A., Yang, M., Hu, P., Wu, J., Chen, B., Hua, Y., et al. (2011). Comparative genomic analysis of *Streptococcus suis* reveals significant genomic diversity among different serotypes. *BMC Genomics* 12:523. doi: 10.1186/1471-2164-12-523

Zhang, W., Rong, C., Chen, C., and Gao, G. F. (2012). Type-IVC secretion system: a novel subclass of type IV secretion system (T4SS) common existing in gram-positive genus *Streptococcus. PLoS ONE* 7:e46390. doi: 10.1371/journal.pone.0046390

Zhao, Y., Liu, G., Li, S., Wang, M., Song, J., Wang, J., et al. (2011). Role of a type IV-like secretion system of *Streptococcus suis* 2 in the development of streptococcal toxic shock syndrome. *J. Infect. Dis.* 204, 274–281. doi: 10.1093/infdis/jir261

Zhong, Q., Zhao, Y., Chen, T., Yin, S., Yao, X., Wang, J., et al. (2014). A functional peptidoglycan hydrolase characterized from T4SS in 89K pathogenicity island of epidemic *Streptococcus suis* serotype 2. *BMC Microbiol.* 14:73. doi: 10.1186/1471-2180-14-73

Zubair, S., De Villiers, E. P., Fuxelius, H. H., Andersson, G., Johansson, K. E., Bishop, R. P., et al. (2013). Genome sequence of *Streptococcus agalactiae* strain 09mas018883, isolated from a Swedish cow. *Genome Announc.* 1:e00456-13. doi: 10.1128/genomeA.00456-13

15

Tracking Cefoperazone/Sulbactam Resistance Development *In vivo* in *A. baumannii* Isolated from a Patient with Hospital-Acquired Pneumonia by Whole-Genome Sequencing

*Xiaofen Liu[1,2], Huajun Zheng[3], Weipeng Zhang[4], Zhen Shen[1], Miao Zhao[1], Yuancheng Chen[1,5], Li Sun[1], Jun Shi[2] and Jing Zhang[1,5]**

[1] Institute of Antibiotics, Huashan Hospital, Fudan University, Shanghai, China, [2] Roche Innovation Center Shanghai, Shanghai, China, [3] Shanghai-MOST Key Laboratory of Health and Disease Genomics, Chinese National Human Genome Center at Shanghai, Shanghai, China, [4] Division of Life Science, The Hong Kong University of Science and Technology, Hong Kong, China, [5] Key Laboratory of Clinical Pharmacology of Antibiotics, National Population and Family Planning Commission, Shanghai, China

***Correspondence:**
Jing Zhang
zhangj_fudan@aliyun.com

Cefoperazone/sulbactam has been shown to be efficacious for the treatment of infections caused by *Acinetobacter baumannii*; however, the mechanism underlying resistance to this synergistic combination is not well understood. In the present study, two *A. baumannii* isolates, AB1845 and AB2092, were isolated from a patient with hospital-acquired pneumonia before and after 20 days of cefoperazone/sulbactam therapy (2:1, 3 g every 8 h with a 1-h infusion). The minimum inhibitory concentration (MIC) of cefoperazone/sulbactam for AB1845 and AB2092 was 16/8 and 128/64 mg/L, respectively. Blood samples were collected on day 4 of the treatment to determine the concentration of cefoperazone and sulbactam. The pharmacokinetic/pharmacodynamic (PK/PD) indices ($\%T_{>MIC}$) were calculated to evaluate the dosage regimen and resistance development. The results showed that $\%T_{>MIC}$ of cefoperazone and sulbactam was 100% and 34.5% for AB1845, and 0% and 0% for AB2092, respectively. Although there was no available PK/PD target for sulbactam, it was proposed that sulbactam should be administered at higher doses or for prolonged infusion times to achieve better efficacy. To investigate the mechanism of *A. baumannii* resistance to the cefoperazone/sulbactam combination *in vivo*, whole-genome sequencing of these two isolates was further performed. The sequencing results showed that 97.6% of the genome sequences were identical and 33 non-synonymous mutations were detected between AB1845 and AB2092. The only difference of these two isolates was showed in sequencing coverage comparison. There was a 6-kb amplified DNA fragment which was three times higher in AB2092, compared with AB1845. The amplified DNA fragment containing the bla_{OXA-23} gene on transposon *Tn2009*. Further quantitative real-time PCR results demonstrated that gene expression at the mRNA level of bla_{OXA-23} was >5 times higher in AB2092 than in AB1845. These results suggested that the bla_{OXA-23}

gene had higher expression level in AB2092 via gene amplification and following transcription. Because gene amplification plays a critical role in antibiotic resistance in many bacteria, it is very likely that the bla_{OXA-23} amplification results in the development of cefoperazone/sulbactam resistance *in vivo*.

Keywords: cefoperazone/sulbactam, *Acinetobacter baumannii*, whole-genome sequencing, resistance development, PK/PD indices

INTRODUCTION

Acinetobacter baumannii is a non-fermentative, gram-negative opportunistic pathogen that can cause hospital-acquired pneumonia (HAP), blood infections and urinary tract infection, among others (Perez et al., 2007). The rapid spread of the multi-drug-resistant *A. baumannii* has resulted in very limited therapeutic options in the clinic. Sulbactam, a β-lactamase inhibitor, has intrinsic antimicrobial effects against *A. baumannii* by binding to penicillin-binding protein 2 (Peleg et al., 2008). It is commercially available in a combined formulation of ampicillin and cefoperazone. The clinical efficacy of cefoperazone/sulbactam has been shown in previous work (Xin et al., 2013; Sipahi et al., 2014; Xia et al., 2014). This combination has recently been applied to treat critically ill patients receiving continuous venovenous hemofiltration, and 11 of 14 patients survived (Gao et al., 2016). This combination was also administrated to treat neurosurgical patients in a pilot study, with the results showing that cerebrospinal fluid penetration of cefoperazone/sulbactam could be enhanced after neurosurgical impairment of the blood-brain barrier (Wang et al., 2015). In a retrospective review of the outcomes for patients with cefoperazone/sulbactam treated *A. baumannii* bacteremia, 77% of the patients (27/35) presented successful clinical efficacy (Choi et al., 2006). Because the combination has been widely used in the clinic, the resistance rate was monitored and observed to increase from 25.0 to 37.7% from 2004 to 2010 in China (Hu et al., 2016). However, the underlying mechanism is not well understood *in vitro* or *in vivo*, for sulbactam alone or the combination. Sulbactam alone has been reported to cause resistance in *A. baumannii* via PBP3 mutation *in vitro*, despite a lack of information regarding natural pbp3 mutations in clinical isolates (Penwell et al., 2015). The detection and expression of bla_{TEM-1} has been suggested to relate to the minimum inhibitory concentration (MIC) of sulbactam in *A. baumannii* (Waltner-Toews et al., 2011). The combination resistance in *Klebsiella pneumoniae* in an *in vitro* study was reported owing to two different mechanisms, loss of a 39-kDa outer membrane protein and presence of TEM-2 β-lactamase (Rice et al., 1993).

Whole-genome sequencing is powerful and can reveal a vast amount of DNA information from a global perspective. The published complete genome sequences for many bacteria are beneficial and efficient for strain-to-reference sequencing and bioinformatics analysis (Mardis, 2008). They can be used to identify resistance genes as well as minor changes in the genome due to mutation, gene transfer, gene duplication, and amplification. It has been seen as a basis for whole-genome sequencing revealing the mechanism of bacterial pathogen resistance that develops in patients (Snitkin et al., 2011; Wright et al., 2014; Holt et al., 2015). Wright et al. (2016) sequenced 136 strains of *A. baumannii* isolated from patients and showed the genome changes occurred mainly due to the single nucleotide variance in protein coding regions and IS element. Genome sequenced *A. baumannii* showed the colistin resistance was due to the mutations in transcriptional regulatory genes (Cheah et al., 2016). Susceptible *S. aureus* evolved resistance via a 35-point mutation in 31 loci in a patient with a bloodstream infection receiving vancomycin therapy (Mwangi et al., 2007). Tigecycline resistance to *A. baumannii* has been reported to be due to the deletion of three contigs *in vivo* (Hornsey et al., 2011). Additionally, gene amplification was frequently detected due to the antibiotic pressure. The evolution of β-lactamase resistance was studied by exposing *Salmonella typhimurium*, which contain low level β-lactamase resistance, in an environment of progressively increasing concentrations of cephalosporin. The results showed an amplification of bla_{TEM-1} gene copy number followed by acquisition of gene mutations (Sun et al., 2009). A genome-wide assay of *E. coli* strain in 78 different antibiotic environments found that 56 genes were amplified to fit the environment and reproducible increasing in MIC for their corresponding antibiotics was detected (Soo et al., 2011). The amplification of aminoglycoside resistance gene *aphA1* in *A. baumannii* has been considered to be the mechanism underlying the development of tobramycin *in vivo* (McGann et al., 2014).

In the present study, two *A. baumannii* isolates, AB1845 and AB2092, were collected before and after cefoperazone/sulbactam therapy from a patient with HAP. Pharmacokinetics/pharmacodynamics (PK/PD) was applied to evaluate the dosage regimen and development of resistance. Moreover, whole-genome sequencing was employed to investigate the mechanism of cefoperazone/sulbactam resistance of the two isolates developed *in vivo*.

MATERIALS AND METHODS

Medical Records and Treatment

This study was approved by the institutional review board of Huashan Hospital affiliated to Fudan University. Written informed consent was obtained from the patient. A 19-year-old male patient with serious brain trauma was admitted to Huashan Hospital (Shanghai, China). The patient had a high fever with white blood cell counts out of the normal range after being checked into bed for 48 h. Both chest

X-Ray radiophotography and computed tomography showed pneumonia. *A. baumannii* (AB1845) was isolated from the sputum of the patient, revealing bacterial infection. Hence, the patient was treated with meropenem 0.5 g q8h without clinical efficacy. On day 6, the dosage regimen was changed to cefoperazone/sulbactam (2:1, 3 g every 8 h with a 1-h infusion, SULPERAZON®, batch no. 95839116) for 20 days. The treatment was stopped when body temperature and white blood cell counts returned to the normal range; additionally, *A. baumannii* was not detected in the sputum. However, the patient had a fever, and *A. baumannii* (AB2092) was again isolated 4 days after cefoperazone/sulbactam treatment. The cefoperazone/sulbactam treatment was finally considered to have failed both clinically and microbiologically efficacy.

Pharmacokinetics of Cefoperazone/Sulbactam Treatment

Blood samples were collected before the first infusion, immediately after the infusion (0 h), and at 0.5, 1, 2, 4, 6 h afterward on day 4 (i.e., steady state) of the cefoperazone/sulbactam treatment. The drug concentration was determined using the LC-MS/MS method (Zhou et al., 2010), which was validated according to the CFDA guidelines on bioanalytical method validation. The pharmacokinetic parameters of the patients were calculated by Winnonlin (v6.0, phoenix). The patient's pharmacokinetic profiles of cefoperazone and sulbactam were fit into a well-developed population PK model (Chen et al., 2015). The PK/PD indice of cefoperazone and sulbactam was $\%T_{>MIC}$, namely the percentage of the dosing interval for which the plasma concentration of cefoperazone or sulbactam above the MIC. It was calculated based on the drug pharmacokinetic profile and MICs of the combination (Lips et al., 2014).

Bacterial Isolates and Susceptibility Testing

Acinetobacter baumannii AB1845 and AB2092 were isolated from the sputum of the patient before and after cefoperazone/sulbactam therapy. Both isolates were stored at −80°C. The MIC was determined in cation-adjusted Mueller-Hinton broth (Becton-Dickinson, Sparks, MD, USA) using a broth microdilution according to Clinical and Laboratory Standards Institute (CLSI) standards. The MICs of the β-lactam/β-lactamase inhibitor combination, including cefoperazone/sulbactam, carbapenems, colistin, and polymyxin B, were determined.

Genome Sequencing and Analysis

Genomic DNA was extracted using a Genomic DNA Purification Kit (Tiangen, Beijing, China) according to the instruction manual and stored at −80°C before sequencing. A 300-bp paired-end library was constructed for the purified DNA sample following the standard Illumina paired-end protocol. Cluster generation was performed in C-bot, and sequencing was performed on the Illumina Hiseq2500 with 150 cycles. The sequence reads were cleaned using the FASTX toolkit[1]. Genome assembly was performed using Velvet (Ver 1.0.15) (Zerbino and Birney, 2008). Putative protein-coding sequences were determined by combining the prediction results of the glimmer 3.02 (Delcher et al., 2007) and Z-Curve (Guo and Zhang, 2006) program. The phylogenetic tree was constructed using PHYML (Guindon and Gascuel, 2003) by concatenating orthologs based on the protein sequences in the published genome. Functional annotation of CDS was performed by searching the NCBI non-redundant protein database and the KEGG protein database (Kanehisa et al., 2016). Gene annotation was mapped to the *A. baumannii* MDR-ZJ06 genome (from the phylogenetic tree) using Bowtie2 with default settings (Langmead and Salzberg, 2012). Single Nucleotide Polymorphisms (SNPs) from alignments were called using Samtools-0.1.16 (Li et al., 2009), and the output was generated in the pileup format. Gene amplification was detected by comparing the coverage of the reads in AB1845 and Ab2092 using the read depth method (Redon et al., 2006).

Quantitative Real-Time PCR Analysis of *bla*$_{OXA-23}$

The mRNA expression level of *bla*$_{OXA-23}$ was quantified by real-time PCR. AB1845 and AB2092 were cultured in cation-adjusted Mueller-Hinton broth and collected during the early log phase. Total RNA was extracted using a TaKaRa Mini BEST Universal RNA Extraction Kit (TaKaRa Biotechnology, Dalian, China) according to the manufacturer's instruction. Quantitative real-time PCR (qPCR) was performed using a two-step process. RNA was first reverse-transcribed to cDNA (TaKaRa Biotechnology, Dalian, China), and real-time PCR was conducted on an ABI Vii7 (Applied Biosystems, Carlsbad, CA, USA) using a TaKaRa SYBR® FAST qPCR Kit with 40 cycles of denaturation for 5 s at 95°C, annealing for 30 s at 50°C, and extension for 20 s at 72°C. The PCR primers for the *bla*$_{OXA-23}$ gene were F: CCGAGTCAGATTGTTCAAGGA and R: TGTAGAGGCTGGCACATATTC; and those for 16S rRNA were F: GGCGGCTTATTAAGTCGGATG and R: TTCGTACCTCAGCGTCAGTATT. The melting curve analysis was performed immediately after amplification to verify the specificity of the PCR amplification products.

Fluorescence was measured at the end of the annealing-extension phase of each cycle. A threshold value for the fluorescence of all samples was set manually. The reaction cycle at which the PCR product exceeded this fluorescence threshold was identified as the threshold cycle (CT). The relative quantitation was calculated by the $2^{-\Delta\Delta CT}$ method (Schmittgen and Livak, 2008). The student *t*-test was applied to evaluate the significance of the gene expression level of the two isolates.

RESULTS

Antibiotic Susceptibility

Minimum inhibitory concentrations were obtained for cefoperazone/sulbactam combinations, carbapenems, colistin

[1] http://hannonlab.cshl.edu/fastx_toolkit/

and polymyxin B for the pre-therapy isolate AB1845 and the post-therapy isolate AB2092 (**Table 1**). The MICs of the cefoperazone/sulbactam combination were 16/8 mg/L for AB1845; and the MICs of mono cefoperazone and sulbactam were > 128 mg/L and 32 mg/L. The results supported a synergistic effect of the combination (FICI < 0.5, MIC of cefoperazone was set at 128 mg/L for the FICI calculation). However, the MICs of the combination and mono drugs were 128/64, > 128, and 128 mg/L for AB2092. The combination did not display a synergistic effect on AB2092.

Interestingly, AB1845 and AB2092 also developed meropenem resistance with MIC increased from 16 to 64 mg/L, although the bacteria only exposed to meropenem for a short term (0.5 g every 8 h for 4 days). The isolates also developed resistance to imipenem and doripenem. Since both cefoperazone and meropenem are β-lactam antibiotics, the increased MIC in both cefoperazone/sulbactam and carbapenems may imply the similar underlying resistance mechanism. The isolates were only sensitive to colistin and polymyxin B, indicating an alternative choice for the AB2092 treatment. The MICs of the antibiotics are shown in **Table 1**.

TABLE 1 | The minimum inhibitory concentrations (MICs) of different antibiotics for AB1845 and AB2092.

Antibiotics	MIC (mg/L)	
	AB1845	AB2092
Cefoperazone/sulbactam	**16/8***	**128/64**
Cefoperazone	>128	>128
Sulbactam	32	128
Meropenem	**16**	**64**
Imipenem	**16**	**64**
Doripenem	**32**	**128**
Colistin	2	2
Polymyxin B	1	1

**FICI = 0.375; FICI = $MIC_{(A)}/MIC_{(A\ in\ combination)} + MIC_{(B)}/MIC_{(B\ in\ combination)}$. The bold values showed the increasing MIC of AB1845 compared with AB2092.*

Pharmacokinetics and Pharmacodynamics (PK/PD)

Plasma concentrations of cefoperazone and sulbactam at steady state were determined using the validated LC-MS/MS method, and the data were fitted into a developed population PK model (**Figure 1**). The pharmacokinetic parameters of cefoperazone and sulbactam are summarized in **Table 2**. The $T_{1/2}$ of cefoperazone was 4.1 h, which was a bit longer than the reported 1.8 h (Reitberg et al., 1988). This could attribute to liver dysfunction (Boscia et al., 1983) according to the medical records of the patients. The $T_{1/2}$ of sulbactam was 1.3 h, which was comparable to the published data (Reitberg et al., 1988).

Pharmacokinetic/pharmacodynamic indices were calculated to evaluate the drug dosage regimen. The $\%T_{>MIC}$ was the most predictive PK/PD index for cefoperazone and sulbactam (Cooper et al., 2011; Sy et al., 2015), and the MIC of the combination was used for the $\%T_{>MIC}$ calculation. In this study, the $\%T_{>MIC}$ was 100% for cefoperazone and 34.5% for sulbactam for AB1845; and the $\%T_{>MIC}$ was 0% for both cefoperazone and sulbactam for AB2092.

Whole-Genome Sequencing of AB1845 and AB2092

The two isolates were initially typed by PFGE of *ApaI*-digested genomic DNA and shown to be the same strain. Whole-genome

TABLE 2 | Pharmacokinetic parameters of cefoperazone and sulbactam.

Parameters	Units	Cefoperazone	Sulbactam
$T_{1/2}$	h	4.1	1.3
T_{max}	h	1.4	1.3
C_{max}	mg/L	129.6	37.2
AUC_{0-8}	mg h/L	529.1	66.5
AUC_{0-inf}	mg h/L	766.0	68.3
Vd	L	15.5	26.6
CLt	L/h	2.6	14.6
MRT_{0-8}	h	2.6	1.5
MRT_{0-inf}	h	5.7	1.6

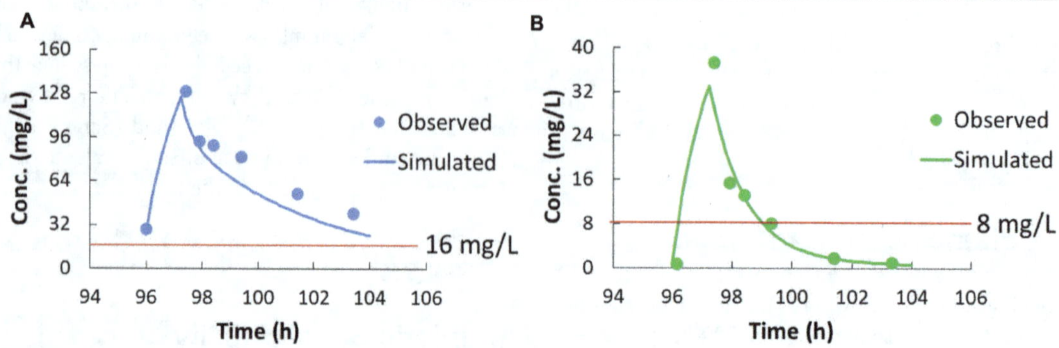

FIGURE 1 | The pharmacokinetic profiles of cefoperazone (A) and sulbactam (B). The dots in **(A,B)** are the concentration determined by LC-MS/MS (before the first infusion on Day 4 and 0.5, 1, 2, 4, 6 h after the infusion), and the curves are the profiles fitted to the population pharmacokinetic model. The red lines were indicated the concentration for the MIC of AB1845.

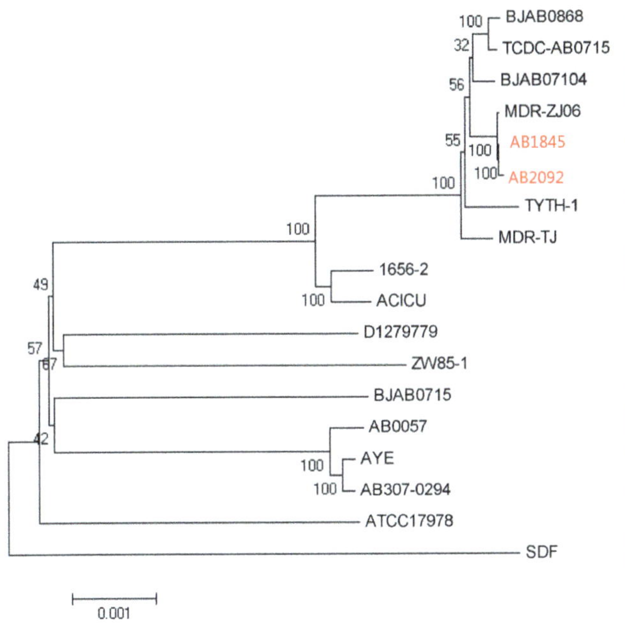

FIGURE 2 | Phylogenetic tree of AB1845 and AB2092 with 16 complete genomes of *A. baumannii* isolates in ftp://ftp.ncbi.nlm.nih.gov/genomes/archive/old_genbank/.

sequencing produced 12,050,346 and 12,191,576 pairs of 150-bp reads for AB1845 and AB2092, respectively. Assembly of the AB1845 and AB2092 genomes resulted in 132 and 131 contigs larger than 500 bp, comprising 4.0 megabases of sequence and representing a median 930-fold coverage. The AB1845 draft assembly has been deposited in GenBank (accession number LVYA00000000), and raw sequence reads for AB1845 and AB2092 have been submitted to NCBI's Sequence Read Archive under the study accession number SRP072783.

Automated gene prediction detected 3,811 and 3,810 putative coding sequences (CDSs) for AB1845 and AB2092, respectively. A phylogenetic tree was constructed, and the genome similarity was compared (**Figure 2**). The 3,718 CDSs were homologous (defined as a BLASTN *e*-value $\leq 1e^{-10}$) to the genome of *A. baumannii* MDR-ZJ06, which was a multi-drug-resistant isolate belonging to International clone II and wide spreading in China (Zhou et al., 2011). MDR-ZJ06 was used as a reference genome for annotating genes in AB1845 and AB2092 (**Figure 3**). The results showed that the two isolates possess 97.6% similarity to MDR-ZJ06, with 921 and 915 SNPs, respectively. Certain genes were missing in both AB1845 and AB2092 compared with MDR-ZJ06. For example, 17 genes had a 15-kb fragment that was missing in the resistant island *Aba*R22, which includes transposition helper proteins, site-specific tyrosine recombinase, transposase protein A and conserved hypothetical proteins, among others.

The CDS of AB1845 was also mapped to the published plasmid sequence. It showed that certain AB1845 genes mapped to plasmid ABKp1 from the *A. baumannii* 1656-2 plasmid with 97.6% similarity.

A comparison of the CDS of AB1845 and AB2092 revealed only 33 SNPs. All of these SNPs were synonymous, indicating that SNP was not the main reason for the development of resistance (Supplementary Table S1).

β-Lactamase and Gene Amplification

A well-known resistance enzyme for β-lactam antibiotics is β-lactamase. The class A, C, and D classes (bla_{OXA-23} and bla_{OXA-51}) β-lactamase were identified in both AB1845 and AB2092 (Supplementary Table S2).

Gene amplification was detected by whole-genome sequencing in combination with the read depth method (Cantsilieris et al., 2012). Approximately 99.9% of the genes had a coverage ratio of AB2092 to AB1845 of approximately 1.0, but six genes (out of 3871 genes) had a coverage ratio over 3.0 (Supplementary Table S3). These six genes were AAA ATPase, DEAD/DEAH box helicases, bla_{OXA-23}, and three hypothetical genes. Among these genes, AAA ATPase and DEAD/DEAH box helicases are involved in the function of DNA replication, repair, transcription, and two uncharacterized genes (Johnson and Mckay, 1999; Iyer et al., 2004). The bla_{OXA-23} was gene encoded class D β-lactamase and conferred resistance to carbapenems and cefoperazone (Pascale and Wright, 2010; Bush, 2013). All six genes were located in a 6-kb sequence fragment in AB1845 designated as *Tn2009*. *Tn2009* was flanked by two *ISAba1* elements that contributed to the duplication of *Tn2009* in AB2902 (Liu et al., 2015). The amplified genes were transcribed in the same orientation (**Figure 4**).

mRNA Level of bla_{OXA-23} in AB1845 and AB2092

Quantitative real-time PCR was conducted to measure the mRNA levels of bla_{OXA-23} in AB1845 and AB2092. Both isolates were collected at the early log phase, and the qPCR results showed that the mRNA level of bla_{OXA-23} was five times higher in AB2092 than in AB1845 (**Figure 5**). This finding was consistent with the gene amplification results.

DISCUSSION

Acinetobacter baumannii commonly colonizes in respiratory tract of the hospitalized patients (Peleg et al., 2008). It is difficult to characterize *A. baumannii* isolated from sputum belongs to the upper airway colonization or the causative pathogen of pneumonia. In this study, the two isolates were considered causative pathogens, as verified by pneumonia testing and symptoms such as cough, fever, and high white blood cell counts. To examine the mechanism of resistance development during cefoperazone/sulbactam treatment, both PK/PD indices and genomic sequencing were employed to understand the problem from a PK/PD and molecular perspective.

Exposure to a sub-optimal antibiotic concentration is the most important factor in the development of bacterial resistance. To

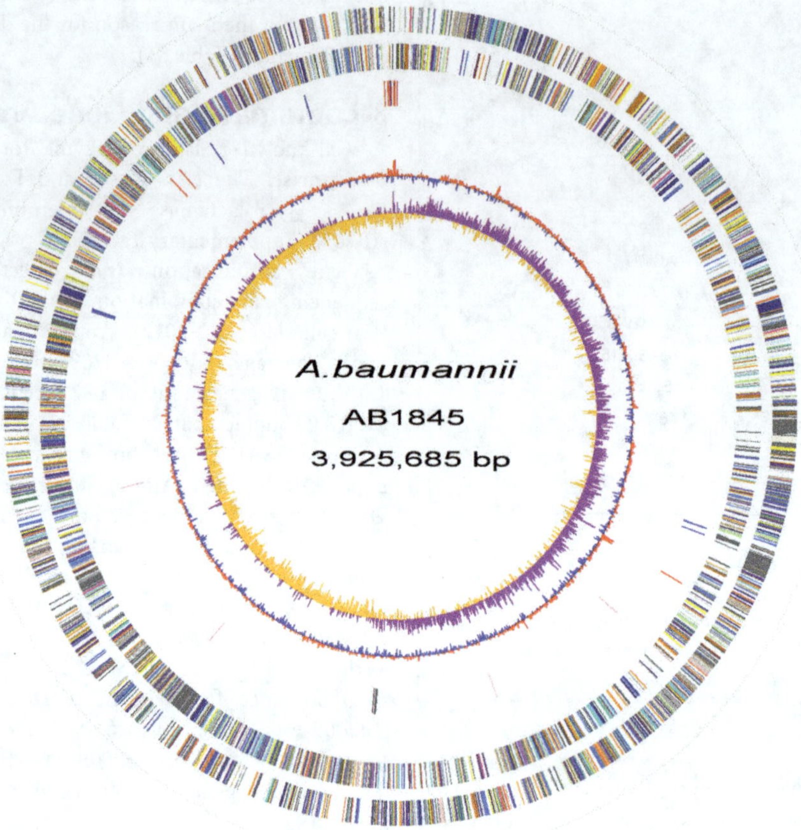

FIGURE 3 | Atlas of the *A. baumannii* AB1845 draft genome. Each concentric circle represents the genomic data for AB1845. The two outer circles illustrate the predicted coding sequences on the plus and minus strands, respectively, colored by functional categories according to the COG classification. The third circle represents the location of the nucleotide substitution between AB1845 and AB2092, with blue representing genes affected by synonymous Single Nucleotide Polymorphisms (SNPs) and red representing SNPs in the intergenic region. The fourth circle displays the loci of the β-lactamase genes (pink) and genes around bla_{OXA-23} (gray). The fifth circle shows the GC content, and the 6th circle (innermost) represents GC skew (G–C)/(G+C) calculated using a 1-kb window.

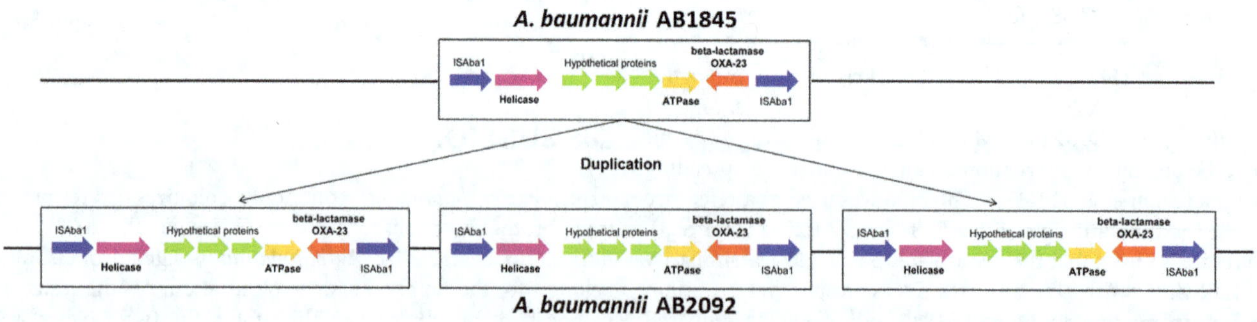

FIGURE 4 | The illustration of the 6-kb DNA fragment of *Tn2009* in AB1845 and AB2092. The fragment was amplified three times in AB2092.

optimize the antibiotic dosage regimen and prevent resistance, the PK/PD which bridges drug exposure and effect, has been widely applied in the clinics (Abdulaziz et al., 2015; Asín-Prieto et al., 2015; Monogue et al., 2015). To date, the PK/PD helps adjust dosage based on the evaluation of steady-state drug concentrations (Martinez et al., 2012). In this study, blood samples were collected at steady state, and the MIC of the bacteria was determined. As the most predictive PK/PD index, the $\%T_{>MIC}$ of cefoperazone and sulbactam was calculated and compared with published PK/PD targets. The published

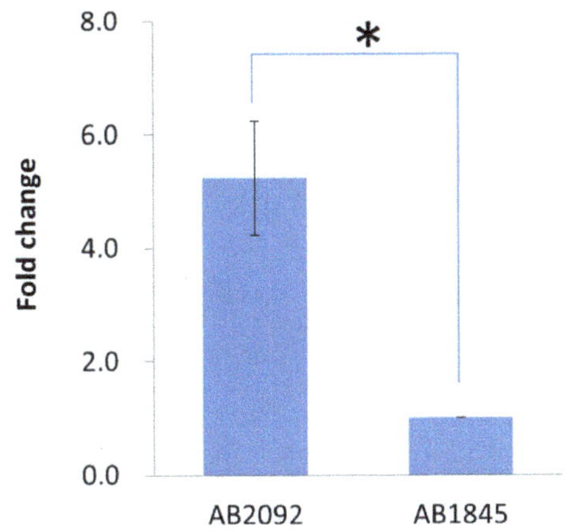

FIGURE 5 | Quantitative real-time PCR results for *bla*$_{OXA-23}$ in AB1845 and AB2092 (*student *t*-test $p < 0.05$).

target $\%T_{>MIC}$ of β-lactam antibiotics was 40–70% (Drusano, 2004); however, there was no published target of sulbactam because of a lack of pharmacodynamic data. In our present dosage regimen (3 g q8h, infusion for 1 h), the target was 100% of cefoperazone and 34.5% of sulbactam for AB1845. Cefoperazone clearly reached its target, whereas there were no criteria for sulbactam. In a study of PK/PD simulation for sulbactam in healthy volunteers, in order for *A. baumannii* with an MIC of 8 mg/L to achieve the target of 40%, the sulbactam had to be administered as a 4-h infusion of 3 g q8h (Jaruratanasirikul et al., 2013). It has even been suggested that the daily dose of sulbactam can be administered up to 6 g for *A. baumannii* infection (Argyres, 2010) to achieve enhanced efficacy. A comparison of these data revealed that sulbactam should be administered at a much higher dose or prolonged infusion time. A suboptimal dose of sulbactam could lead to the development of resistance during treatment. Additional research to identify the PK/PD target of sulbactam specific to *Acinetobacter* is needed.

Whole-genome sequencing was further employed to investigate the mechanism of resistance from a molecular perspective. Comparison of the genome between AB1845 and AB2092 revealed no gene deletions or mutations (only 33 synonymous SNPs). However, the reads depth method showed six amplified genes on *Tn2009*, including *bla*$_{OXA-23}$. The *bla*$_{OXA-23}$ gene encodes the class D β-lactamase which confers resistance to the last resort carbapenem antibiotics. The OXA-23-producing *A. baumannii* is widely disseminated in multi-drug-resistant *A. baumannii*. Although the *bla*$_{OXA-23}$ gene has been mainly identified on plasmids, its chromosomal location has also been reported (Zhou et al., 2011). The transportable elements, *Tn2006*, *Tn2007*, *Tn2008*, and *Tn2009*, play a key role in the transfer of the *bla*$_{OXA-23}$ gene to different locations in one bacterium or to different isolates (Corvec et al., 2007; Liu et al., 2015). *Tn2008* and *Tn2009* have mostly been associated with the transfer of *bla*$_{OXA-23}$ in China (Adams-Haduch et al., 2008; Zhou et al., 2011; Liu et al., 2015). IS elements are associated with the genome gain/loss of genes, especially resistant genes. These two transposons share IS*Aba1* upstream from *bla*$_{OXA-23}$, which belongs to the IS4 family, provides promoter for *bla*$_{OXA-23}$ (Turton et al., 2006; Mugnier et al., 2009), which could explain the gene duplication or amplification in this study. The mRNA level of *bla*$_{OXA-23}$ confirmed the higher expression level of AB2092 and AB1845. There is evidence that gene duplication and amplification in bacteria was directly associated with adaptation to environmental changes, including antibiotic stress (Sandegren and Andersson, 2009). In a study testing the relationships of gene amplification and β-lactams resistance increase showed that 70% colonies of *S. Typhimurium* with increased resistance to cephalosporin had increased *bla*$_{TEM}$ gene copy numbers (Sun et al., 2009). Recent study showed the heteroresistance of *E. Coli* and *S. Typhimurium* was due to the amplification of *pmrD* gene which encoded a protein modifying lipid A. Moreover, the heteroresistance phenotype is associated with different copy numbers of the *pmrD* gene (Hjort et al., 2016). Numerous studies have reported gene duplication and amplification related to bacterial resistance to antibiotics *in vitro* (Bertini et al., 2007). During the macrolide treatment of a *Streptococcus pneumoniae* infection, the bacterium developed resistance due to an 18 bp duplication in the *rplV* gene, leading to the treatment failure. The duplication of *rplV* gene led to the blocking of macrolide binding site on ribosome (Musher et al., 2002). Another example of gene amplification in bacterial isolates from human patients involves the *Streptococcus agalactiae*, a leading cause of neonatal infection. The amplification of a four-fold tandemly amplified 13.5-kbp region and a 92 kbp duplication conferred drug resistance to both sulfonamide and trimethoprim (Brochet et al., 2008).

In this study, the isolates developed resistance to cefoperazone/sulbactam as well as carbapenem antibiotics. The whole-genome sequencing showed the amplified *bla*$_{OXA-23}$ in the isolates is very likely the contribution to the resistance. It is rational to conferred resistance to carbapenem antibiotics by the amplification of *bla*$_{OXA-23}$. However, this is the first time to report the amplification of *bla*$_{OXA-23}$ results in cefoperazone/sulbactam combination resistance. Further studies will be focused on collecting more clinical isolates pairs of *A. baumannii* to support the mechanism proposed.

AUTHOR CONTRIBUTIONS

XL designed the genomic sequencing experiments. XL and HZ performed the data analysis. XL and ZS conducted the qPCR experiments. YC conducted the pharmacokinetic parameter calculation. LS extracted the genomic DNA for sequencing. MZ conducted the MIC test and the PFGE test. JS and JZ conceived of the study and devised the overall study strategy. XL wrote the manuscript with input from all authors.

ACKNOWLEDGMENT

This research is supported by the National Natural Science Foundation of China (Grant No. 8150130235 and 81373493).

REFERENCES

Abdulaziz, M. H., Lipman, J., Mouton, J. W., Hope, W. W., and Roberts, J. A. (2015). Applying pharmacokinetic/pharmacodynamic principles in critically ill patients: optimizing efficacy and reducing resistance development. *Semin. Respir. Crit. Care Med.* 36, 136–153. doi: 10.1055/s-0034-1398490

Adams-Haduch, J. M., Paterson, D. L., Sidjabat, H. E., Pasculle, A. W., Potoski, B. A., Muto, C. A., et al. (2008). Genetic basis of multidrug resistance in *Acinetobacter baumannii* clinical isolates at a tertiary medical center in Pennsylvania. *Antimicrob. Agents Chemother.* 52, 3837–3843. doi: 10.1128/aac.00570-08

Argyres, M. I. (2010). Hospital-acquired infections due to gram-negative bacteria. *New Engl. J. Med.* 363, 1483–1483.

Asín-Prieto, E., Rodríguez-Gascón, A., and Isla, A. (2015). Applications of the pharmacokinetic/pharmacodynamic (PK/PD) analysis of antimicrobial agents. *J. Infect. Chemother.* 21, 319–329. doi: 10.1016/j.jiac.2015.02.001

Bertini, A., Poirel, L., Bernabeu, S., Fortini, D., Villa, L., Nordmann, P., et al. (2007). Multicopy blaOXA-58 gene as a source of high-level resistance to carbapenems in *Acinetobacter baumannii. Antimicrob. Agents Chemother.* 51, 2324–2328. doi: 10.1128/AAC.01502-06

Boscia, J. A., Korzeniowski, O. M., Snepar, R., Kobasa, W. D., Levison, M. E., and Kaye, D. (1983). Cefoperazone pharmacokinetics in normal subjects and patients with cirrhosis. *Antimicrob. Agents Chemother.* 23, 385–389. doi: 10.1128/AAC.23.3.385

Brochet, M., Couve, E., Zouine, M., Poyart, C., and Glaser, P. (2008). A naturally occurring gene amplification leading to sulfonamide and trimethoprim resistance in *Streptococcus agalactiae. J. Bacteriol.* 190, 672–680. doi: 10.1128/jb.01357-07

Bush, K. (2013). The ABCD's of β-lactamase nomenclature. *J. Infect. Chemother.* 19, 549–559. doi: 10.1007/s10156-013-0640-7

Cantsilieris, S., Baird, P. N., and White, S. J. (2012). Molecular methods for genotyping complex copy number polymorphisms. *Genomics* 101, 86–93. doi: 10.1016/j.ygeno.2012.10.004

Cheah, S. E., Johnson, M. D., Zhu, Y., Tsuji, B. T., Forrest, A., Bulitta, J. B., et al. (2016). Polymyxin resistance in *Acinetobacter baumannii*: genetic mutations and transcriptomic changes in response to clinically relevant dosage regimens. *Sci. Rep.* 6, 26233. doi: 10.1038/srep26233

Chen, Y.-C., Zhou, Y.-J., Zhang, J., Wu, J.-F., Guo, B.-N., Wu, X.-J., et al. (2015). "Population pharmacokinetics and pharmacodynamics of cefoperazone/sulbactam (2:1) in adult patients with hospital-acquired pneumoniae," in *Proceedings of the Interscience Conference of Antimicrobial Agents and Chemotherapy*, San Diego, CA.

Choi, J. Y., Kim, C. O., Park, Y. S., Yoon, H. J., Shin, S. Y., Kim, Y. K., et al. (2006). Comparison of efficacy of cefoperazone/sulbactam and imipenem/cilastatin for treatment of *Acinetobacter* bacteremia. *Yonsei Med. J.* 47, 63–69. doi: 10.3349/ymj.2006.47.1.63

Cooper, T. W., Pass, S. E., Brouse, S. D., and Hall, R. G. I. I. (2011). Can pharmacokinetic and pharmacodynamic principles be applied to the treatment of multidrug-resistant *Acinetobacter*? *Ann. Pharmacother.* 45, 229–240. doi: 10.1345/aph.1P187

Corvec, S., Poirel, L., Naas, T., Drugeon, H., and Nordmann, P. (2007). Genetics and expression of the carbapenem-hydrolyzing oxacillinase gene blaOXA-23 in *Acinetobacter baumannii. Antimicrob. Agents Chemother.* 51, 1530–1533. doi: 10.1128/AAC.01132-06

Delcher, A. L., Bratke, K. A., Powers, E. C., and Salzberg, S. L. (2007). Identifying bacterial genes and endosymbiont DNA with Glimmer. *Bioinformatics* 23, 673–679. doi: 10.1093/bioinformatics/btm009

Drusano, G. L. (2004). Antimicrobial pharmacodynamics: critical interactions of 'bug and drug.' *Nat. Rev. Microbiol.* 2, 289–300. doi: 10.1038/nrmicro862

Gao, C., Tong, J., Yu, K., Sun, Z., An, R., and Du, Z. (2016). Pharmacokinetics of cefoperazone/sulbactam in critically ill patients receiving continuous venovenous hemofiltration. *Eur. J. Clin. Pharmacol.* 72, 823–830. doi: 10.1007/s00228-016-2045-x

Guindon, S., and Gascuel, O. (2003). A simple, fast, and accurate algorithm to estimate large phylogenies by maximum likelihood. *Syst. Biol.* 52, 696–704. doi: 10.1080/10635150390235520

Guo, F.-B., and Zhang, C.-T. (2006). ZCURVE_V: a new self-training system for recognizing protein-coding genes in viral and phage genomes. *BMC Bioinformatics* 10:9. doi: 10.1186/1471-2105-7-9

Hjort, K., Nicoloff, H., and Andersson, D. I. (2016). Unstable tandem gene amplification generates heteroresistance (variation in resistance within a population) to colistin in *Salmonella enterica. Mol. Microbiol.* doi: 10.1111/mmi.13459 [Epub ahead of print].

Holt, K. E., Wertheim, H., Zadoks, R. N., Baker, S., Whitehouse, C. A., Dance, D., et al. (2015). Genomic analysis of diversity, population structure, virulence, and antimicrobial resistance in *Klebsiella pneumoniae*, an urgent threat to public health. *Proc. Natl. Acad. Sci. U.S.A.* 112, E3574–E3581. doi: 10.1073/pnas.1501049112

Hornsey, M., Loman, N., Wareham, D. W., Ellington, M. J., Pallen, M. J., Turton, J. F., et al. (2011). Whole-genome comparison of two *Acinetobacter baumannii* isolates from a single patient, where resistance developed during tigecycline therapy. *J. Antimicrob. Chemother.* 66, 1499–1503. doi: 10.1093/jac/dkr168

Hu, F. P., Guo, Y., Zhu, D. M., Wang, F., Jiang, X. F., Xu, Y. C., et al. (2016). Resistance trends among clinical isolates in China reported from CHINET surveillance of bacterial resistance, 2005-2014. *Clin. Microbiol. Infect.* 22(Suppl. 1), S9–S14. doi: 10.1016/j.cmi.2016.01.001

Iyer, L. M., Leipe, D. D., Koonin, E. V., and Aravind, L. (2004). Evolutionary history and higher order classification of AAA+ ATPases. *J. Struct. Biol.* 146, 11–31. doi: 10.1016/j.jsb.2003.10.010

Jaruratanasirikul, S., Wongpoowarak, W., Aeinlang, N., and Jullangkoon, M. (2013). Pharmacodynamics modeling to optimize dosage regimens of sulbactam. *Antimicrob. Agents Chemother.* 57, 3441–3444. doi: 10.1128/AAC.00342-13

Johnson, E. R., and Mckay, D. B. (1999). Crystallographic structure of the amino terminal domain of yeast initiation factor 4A, a representative DEAD-box RNA helicase. *RNA* 5, 1526–1534. doi: 10.1017/S1355838299991410

Kanehisa, M., Sato, Y., Kawashima, M., Furumichi, M., and Tanabe, M. (2016). KEGG as a reference resource for gene and protein annotation. *Nucleic Acids Res.* 44, D457–D462. doi: 10.1093/nar/gkv1070

Langmead, B., and Salzberg, S. L. (2012). Fast gapped-read alignment with Bowtie 2. *Nat. Methods* 9, 357–359. doi: 10.1038/nmeth.1923

Li, H., Handsaker, B., Wysoker, A., Fennell, T., Ruan, J., Homer, N., et al. (2009). The Sequence Alignment/Map format and SAMtools. *Bioinformatics* 25, 2078–2079. doi: 10.1093/bioinformatics/btp352

Lips, M., Siller, M., Strojil, J., Urbanek, K., Balik, M., and Suchankova, H. (2014). Pharmacokinetics of imipenem in critically ill patients during empirical treatment of nosocomial pneumonia: a comparison of 0.5-h and 3-h infusions. *Int. J. Antimicrob. Agents* 44, 358–362. doi: 10.1016/j.ijantimicag.2014.05.011

Liu, L. L., Ji, S. J., Ruan, Z., Fu, Y., Fu, Y. Q., Wang, Y. F., et al. (2015). Dissemination of blaOXA-23 in *Acinetobacter* spp. in China: main roles of conjugative plasmid pAZJ221 and transposon Tn2009. *Antimicrob. Agents Chemother.* 59, 1998–2005. doi: 10.1128/AAC.04574-14

Mardis, E. R. (2008). The impact of next-generation sequencing technology on genetics. *Trends Genet.* 24, 133–141. doi: 10.1016/j.tig.2007.12.007

Martinez, M. N., Papich, M. G., and Drusano, G. L. (2012). Dosing regimen matters: the importance of early intervention and rapid attainment of the pharmacokinetic/pharmacodynamic target. *Antimicrob. Agents Chemother.* 56, 2795–2805. doi: 10.1128/AAC.05360-11

McGann, P., Courvalin, P., Snesrud, E., Clifford, R. J., Yoon, E.-J., Onmus-Leone, F., et al. (2014). Amplification of aminoglycoside resistance gene aphA1 in *Acinetobacter baumannii* results in tobramycin therapy failure. *MBio* 5:e00915-14. doi: 10.1128/mBio.00915-14

Monogue, M. L., Kuti, J. L., and Nicolau, D. P. (2015). Optimizing antibiotic dosing strategies for the treatment of gram-negative infections in the era of resistance. *Expert Rev. Clin. Pharmacol.* 9, 459–476. doi: 10.1586/17512433.2016.1133286

Mugnier, P. D., Poirel, L., and Nordmann, P. (2009). Functional analysis of insertion sequence ISAba1, responsible for genomic plasticity of *Acinetobacter baumannii*. *J. Bacteriol.* 191, 2414–2418. doi: 10.1128/JB.01258-08

Musher, D. M., Dowell, M. E., Shortridge, V. D., Flamm, R. K., Jorgensen, J. H., Le Magueres, P., et al. (2002). Emergence of macrolide resistance during treatment of pneumococcal pneumonia. *N. Engl. J. Med.* 346, 630–631. doi: 10.1056/nejm200202213460820

Mwangi, M. M., Wu, S. W., Zhou, Y., Sieradzki, K., de Lencastre, H., Richardson, P., et al. (2007). Tracking the in vivo evolution of multidrug resistance in *Staphylococcus aureus* by whole-genome sequencing. *Proc. Natl. Acad. Sci. U.S.A.* 104, 9451–9456. doi: 10.1073/pnas.0609839104

Pascale, G. D., and Wright, G. D. (2010). Antibiotic resistance by enzyme inactivation: from mechanisms to solutions. *Chembiochem* 11, 1325–1334. doi: 10.1002/cbic.201000067

Peleg, A. Y., Seifert, H., and Paterson, D. L. (2008). *Acinetobacter baumannii*: emergence of a successful pathogen. *Clin. Microbiol. Rev.* 21, 538–582. doi: 10.1128/cmr.00058-07

Penwell, W. F., Shapiro, A. B., Giacobbe, R. A., Gu, R. F., Gao, N., Thresher, J., et al. (2015). Molecular mechanisms of sulbactam antibacterial activity and resistance determinants in *Acinetobacter baumannii*. *Antimicrob. Agents Chemother.* 59, 1680–1689. doi: 10.1128/aac.04808-14

Perez, F., Hujer, A. M., Hujer, K. M., Decker, B. K., Rather, P. N., and Bonomo, R. A. (2007). Global challenge of multidrug-resistant *Acinetobacter baumannii*. *Antimicrob. Agents Chemother.* 51, 3471–3484. doi: 10.1128/aac.01464-06

Redon, R., Ishikawa, S., Fitch, K. R., Feuk, L., Perry, G. H., Andrews, T. D., et al. (2006). Global variation in copy number in the human genome. *Nature* 444, 444–454. doi: 10.1038/nature05329

Reitberg, D. P., Whall, T. J., Chung, M., Blickens, D., Swarz, H., and Arnold, J. (1988). Multiple-dose pharmacokinetics and toleration of intravenously administered cefoperazone and sulbactam when given as single agents or in combination. *Antimicrob. Agents Chemother.* 32, 42–46. doi: 10.1128/AAC.32.1.42

Rice, L. B., Carias, L. L., Etter, L., and Shlaes, D. M. (1993). Resistance to cefoperazone-sulbactam in *Klebsiella pneumoniae*: evidence for enhanced resistance resulting from the coexistence of two different resistance mechanisms. *Antimicrob. Agents Chemother.* 37, 1061–1064. doi: 10.1128/AAC.37.5.1061

Sandegren, L., and Andersson, D. I. (2009). Bacterial gene amplification: implications for the evolution of antibiotic resistance. *Nat. Rev. Microbiol.* 7, 578–588. doi: 10.1038/nrmicro2174

Schmittgen, T. D., and Livak, K. J. (2008). Analyzing real-time PCR data by the comparative C(T) method. *Nat. Protoc.* 3, 1101–1108. doi: 10.1038/nprot.2008.73

Sipahi, O. R., Arda, B., Nazli-Zeka, A., Pullukcu, H., Tasbakan, M., Yamazhan, T., et al. (2014). Piperacillin/tazobactam vs. cefoperazone/sulbactam in adult low-risk febrile neutropenia cases. *Int. J. Clin. Pract.* 68, 230–235. doi: 10.1111/ijcp.12279

Snitkin, E. S., Zelazny, A. M., Montero, C. I., Stock, F., Mijares, L., Murray, P. R., et al. (2011). Genome-wide recombination drives diversification of epidemic strains of *Acinetobacter baumannii*. *Proc. Natl. Acad. Sci. U.S.A.* 108, 13758–13763. doi: 10.1073/pnas.1104404108

Soo, V. W., Hanson-Manful, P., and Patrick, W. M. (2011). Artificial gene amplification reveals an abundance of promiscuous resistance determinants in *Escherichia coli*. *Proc. Natl. Acad. Sci. U.S.A.* 108, 1484–1489. doi: 10.1073/pnas.1012108108

Sun, S., Berg, O. G., Roth, J. R., and Andersson, D. I. (2009). Contribution of gene amplification to evolution of increased antibiotic resistance in *Salmonella typhimurium*. *Genetics* 182, 1183–1195. doi: 10.1534/genetics.109.103028

Sy, S. K., Zhuang, L., and Derendorf, H. (2015). Pharmacokinetics and pharmacodynamics in antibiotic dose optimization. *Expert Opin. Drug Metab. Toxicol.* 12, 94–114.

Turton, J. F., Ward, M. E., Woodford, N., Kaufmann, M. E., Pike, R., Livermore, D. M., et al. (2006). The role of ISAba1 in expression of OXA carbapenemase genes in *Acinetobacter baumannii*. *FEMS Microbiol. Lett.* 258, 72–77. doi: 10.1111/j.1574-6968.2006.00195.x

Waltner-Toews, R. I., Paterson, D. L., Qureshi, Z. A., Sidjabat, H. E., Adams-Haduch, J. M., Shutt, K. A., et al. (2011). Clinical characteristics of bloodstream infections due to ampicillin-sulbactam-resistant, non-extended-spectrum-beta-lactamase-producing *Escherichia coli* and the role of TEM-1 hyperproduction. *Antimicrob. Agents Chemother.* 55, 495–501. doi: 10.1128/aac.00797-10

Wang, Q., Wu, Y., Chen, B., and Zhou, J. (2015). Drug concentrations in the serum and cerebrospinal fluid of patients treated with cefoperazone/sulbactam after craniotomy. *BMC Anesthesiol.* 15:33. doi: 10.1186/s12871-015-0012-1

Wright, M. S., Haft, D. H., Harkins, D. M., Perez, F., Hujer, K. M., Bajaksouzian, S., et al. (2014). New insights into dissemination and variation of the health care-associated pathogen *Acinetobacter baumannii* from genomic analysis. *MBio* 5:e00963-13. doi: 10.1128/mBio.00963-13

Wright, M. S., Iovleva, A., Jacobs, M. R., Bonomo, R. A., and Adams, M. D. (2016). Genome dynamics of multidrug-resistant *Acinetobacter baumannii* during infection and treatment. *Genome Med.* 8, 26. doi: 10.1186/s13073-016-0279-y

Xia, J., Zhang, D., Xu, Y., Gong, M., Zhou, Y., and Fang, X. (2014). A retrospective analysis of carbapenem-resistant *Acinetobacter baumannii*-mediated nosocomial pneumonia and the in vitro therapeutic benefit of cefoperazone/sulbactam. *Int. J. Infect. Dis.* 23, 90–93. doi: 10.1016/j.ijid.2014.01.017

Xin, X., Jian, L., Xia, X., Jia, B., Huang, W., Li, C., et al. (2013). A multicentre clinical study on the injection of ceftriaxone/sulbactam compared with cefoperazone/sulbactam in the treatment of respiratory and urinary tract infections. *Ann. Clin. Microbiol. Antimicrob.* 12, 38. doi: 10.1186/1476-0711-12-38

Zerbino, D. R., and Birney, E. (2008). Velvet: algorithms for de novo short read assembly using de Bruijn graphs. *Genome Res.* 18, 821–829. doi: 10.1101/gr.074492.107

Zhou, H., Zhang, T., Yu, D., Pi, B., Yang, Q., Zhou, J., et al. (2011). Genomic analysis of the multidrug-resistant *Acinetobacter baumannii* strain MDR-ZJ06 widely spread in China. *Antimicrob. Agents Chemother.* 55, 4506–4512.

Zhou, Y., Zhang, J., Guo, B., Yu, J., Shi, Y., Wang, M., et al. (2010). Liquid chromatography/tandem mass spectrometry assay for the simultaneous determination of cefoperazone and sulbactam in plasma and its application to a pharmacokinetic study. *J. Chromatogr. B Analyt. Technol. Biomed. Life Sci.* 878, 3119–3124. doi: 10.1016/j.jchromb.2010.09.021

16

Wide Distribution of Carbapenem Resistant *Acinetobacter baumannii* in Burns Patients in Iran

Zahra Farshadzadeh*[1], *Farhad B. Hashemi*[1], *Sara Rahimi*[2], *Babak Pourakbari*[3], *Davoud Esmaeili*[4], *Mohammad A. Haghighi*[2], *Ali Majidpour*[5], *Saeed Shojaa*[6], *Maryam Rahmani*[1], *Samira Gharesi*[1], *Masoud Aziemzadeh*[2] and *Abbas Bahador*[1]

[1] Department of Microbiology, School of Medicine, Tehran University of Medical Sciences, Tehran, Iran, [2] Department of Microbiology, School of Medicine, Bushehr University of Medical Sciences, Bushehr, Iran, [3] Pediatrics Infectious Diseases Research Center, School of Medicine, Tehran University of Medical Sciences, Tehran, Iran, [4] Molecular Biology Research Center, Baqiyatallah University of Medical Sciences, Tehran, Iran, [5] Anti-microbial Resistance Research Center, Iran University of Medical Sciences, Tehran, Iran, [6] Department of Microbiology, Faculty of Medicine, Hormozgan University of Medical Sciences, Bandar Abbas, Iran

***Correspondence:**
Abbas Bahador
abahador@tums.ac.ir;
ab.bahador@gmail.com

Antimicrobial resistance in carbapenem non-susceptible *Acinetobacter baumannii* (CNSAb) is a major public health concern globally. This study determined the antibiotic resistance and molecular epidemiology of CNSAb isolates from a referral burn center in Tehran, Iran. Sixty-nine CNSAb isolates were tested for susceptibility to antimicrobial agents using the E test methodology. Multiple locus variable number tandem repeat analysis (MLVA), Multilocus sequence typing (MLST) and multiplex PCR were performed. PCR assays tested for ambler classes A, B, and D β-lactamases. Detection of IS*Aba*1, characterization of integrons, and biofilm formation were investigated. Fifty-three (77%) isolates revealed XDR phenotypes. High prevalence of bla_{OXA-23}-like (88%) and bla_{PER-1} (54%) were detected. IS*Aba*1 was detected upstream of bla_{ADC}, bla_{OXA-23}-like and bla_{OXA51}-like genes in, 97, 42, and 26% of isolates, respectively. Thirty-one (45%) isolates were assigned to international clone (IC) variants. MLVA identified 56 distinct types with six clusters and 53 singleton genotypes. Forty previously known MLST sequence types forming 5 clonal complexes were identified. The Class 1 integron (class 1 integrons) gene was identified in 84% of the isolates. The most prevalent (33%) cassette combination was *aac*A4-*cat*B8-*aad*A1. The IC variants were predominant in the *A. baumannii* lineage with the ability to form strong biofilms. The XDR-CNSAb from burned patients in Iran is resistant to various antimicrobials, including tigecycline. This study shows wide genetic diversity in CNSAb. Integrating the new Iranian *A. baumannii* IC variants into the epidemiologic clonal and susceptibility profile databases can help effective global control measures against the XDR-CNSAb pandemic.

Keywords: antimicrobial resistance, MLST, MLVA, international clone, integron

INTRODUCTION

Acinetobacter baumannii is an important healthcare-associated pathogen that can cause of life threatening infections (Öncül et al., 2014; Girerd-Genessay et al., 2015). *A. baumannii* is the second most common multidrug-resistant cause of nosocomial infection in burn patients in Iran (Alaghehbandan et al., 2012; Azimi et al., 2015). Its infections have become critical challenge to health care systems due to increasing levels of resistance to antimicrobial agents in nosocomial isolates of *A. baumannii* (Cai et al., 2012; Moradi et al., 2015). In developing countries, including Iran, the occurrence of carbapenem non-susceptinle *A. baumannii* (CNSAb) infections is a growing problem in hospitalized burn patients (Mahdian et al., 2015; Zowawi et al., 2015). Inadequate management of the antibiotic therapy of CNSAb infections often leads to the emergence of extensive and pandrug resistant (XDR and PDR) CNSAb strains, which present significant health challenges by prolonging hospitalization, treatment failures, and increased mortality (Liu et al., 2014b). The most important mechanism of carbapenem resistance in *A. baumannii* is the production of carbapenem-hydrolyzing β-lactamases of Ambler classes A, B, and D. In addition to being resistant to all β-lactams available, carbapenemases have a high capacity to spread, since their genes have been commonly found in transferable plasmids containing integrons and insertion sequence (IS) elements (Karah et al., 2012).

Worldwide surveillance has shown that the global population structure of CNSAb isolates is diverse, but a small number of widespread clones, including the international clonal (IC) lineage I–III, also defined based on multilocus sequence typing (MLST) as clonal complexes (CC), may be predominant in healthcare settings (Jeannot et al., 2014). Multilocus variable-number tandem-repeat (VNTR) analysis (MLVA) has been considered as a rapid and cost-effective method for fine-scale typing of *A. baumannii*, enabling inter-laboratory comparison of typing data (Karah et al., 2012). A recent study found the high discriminatory power of MLVA, sufficient to detect diversity among isolates showing identical MLST types (Almeida and Araujo, 2013).

Up-to-date surveillance information regarding genotypic spread, plus local antimicrobial susceptibility patterns of clinical isolates of *A. baumannii*, are necessary for effective drug therapy and control of XDR and PDR *A. baumannii* hospital outbreaks (Fishbain and Peleg, 2010). Considering the points mentioned above, the aim of this study was to determine molecular epidemiology, to identify the dissemination of the most common resistance genes, to investigate the prevalence of integrons and arrangement of integron gene cassettes among CNSAb isolates from a referral burn center in Tehran, Iran.

MATERIALS AND METHODS

Bacterial Isolates

A total of 92 non-repetitive *A. baumannii* strains were collected between January 2012 to May 2013, from the burn wound infections of hospitalized patients in Shahid Motahari hospital, the only referral burn center in Tehran, Iran. Species of these isolates were initially characterized using the API20NE system (bioMérieux, Marcy-l'Etoile, France) and then final identification of isolates were performed by multiplex PCR using *gyrB*-directed primers according to the study of Higgins et al. (2010b).

Antimicrobial Susceptibility Tests and Biofilm Formation Assay

The Clinical and Laboratory Standards Institute (CLSI) guideline (Clinical and Laboratory Standards Institute [CLSI], 2014) for minimum inhibitory concentrations (MICs) using the E test was used to assess the susceptibility of 92 *A. baumannii* isolates to imipenem (Ezy MIC^{TM} strips, Himedia, India). All isolates with MICs of imipenem >4 mg/L were defined as CNSAb isolates and were recruited for further screening (Clinical and Laboratory Standards Institute [CLSI], 2014).The susceptibility of CNSAb isolates to 17 other antimicrobial agents such as amikacin, ampicillin-sulbactam, cefepime, ceftazidime, ciprofloxacin, colistin, gentamicin, levofloxacin, meropenem, minocycline, piperacillin, piperacillin-tazobactam, rifampicin, tetracycline, tigecycline, tobramicin, and trimethoprim-sulfamethoxazole were also carried out using the E test (Ezy MIC^{TM} strips, Himedia, India). The phenotypic presence of Metallo-β-lactamase (MBL) enzymes was detected by imipenem/imipenem + EDTA strips (Ezy MIC^{TM} strips, Himedia, India). The MIC ratio of MBL-E test of $\geq$8 mg/L was interpreted as indicative of MBL activity, according to the manufacturer's instructions. Since there is no breakpoint for tigecycline and rifampicin against *A. baumannii* strains in the CLSI guidelines; therefore, the criteria for interpretation of the MIC values of tigecycline (MIC of $\leq$1 mg/L defined as susceptible and >2 mg/L as resistant) were determined according to the European committee on antimicrobial susceptibility testing (EUCAST) (European Committee on Antimicrobial Susceptibility Testing, 2014) for members of the *Enterobacteriaceae sp.* and CLSI criteria, using breakpoint values suggested for *Staphylococcus aureus* applied to rifampicin (susceptible defined as $\leq$1 mg/L and resistant defined as $\geq$4 mg/L). The MIC_{50} and MIC_{90} of each antibiotic were calculated. *Escherichia coli* ATCC 25922, *Pseudomonas aeruginosa* ATCC 27853, and *E. coli* ATCC 35218 were used as quality control organisms. The phenotype of *A. baumannii* is defined as MDR and XDR according to the International Expert proposal for Interim Standards Guidelines (Magiorakos et al., 2012). The formation assay and quantitative analysis of biofilm were performed according to a previous study (Zhang et al., 2014).

Molecular Detection of β-lactamases-encoding Genes and Characterization of Integrons

A series of PCR amplifications were done for detection of different Ambler class *bla* gene groups: class A; bla_{PER-1}, bla_{VEB-1}, bla_{CTX-M}, bla_{KPC}, bla_{GES}, bla_{TEM} and bla_{SHV-1} (Poirel et al., 2003; Ellington et al., 2007; Bonnin et al., 2011a,b; Potron et al., 2011; Robledo et al., 2011; Climaco et al., 2013); Class B, bla_{IMP-1}, bla_{VIM-2}, bla_{GIM-1}, bla_{SPM-1} bla_{SIM-1} and bla_{NDM-1} (Chen et al., 2011); Class C, bla_{ADC} (Bou and Martinez-Beltran, 2000) and Class D, bla_{OXA-23}-, bla_{OXA-40}-, bla_{OXA-51}-, bla_{OXA-58} (Woodford et al., 2006) and $bla_{OXA-143}$-like (Higgins et al., 2010a) encoding genes were detected. To determine whether IS*Aba1* was present upstream of

bla_{OXA-23}-like, bla_{OXA-51}-like (Turton et al., 2006) and bla_{ADC} (Heritier et al., 2006) genes, PCR mapping experiments were performed. Detection and characterization of class 1, 2, and 3 integrons, as well as gene cassette mapping and sequencing of class 1 integrons were carried out using previously described PCR assay (Dillon et al., 2005).

Genetic Diversity and Population Structure

The IC lineage so-called PCR-based group were identified using three-locus dual assay multiplex PCR as described previously (Turton et al., 2007). Isolates pertaining to the novel variant of the PCR-based group, according to the new combination of amplified products obtained from the two separate multiplex PCRs, did not correspond to those previously defined. Moreover, the CNSAb isolates were genotyped using the MLVA-8 scheme method developed by Pourcel et al. (2011). For clustering analysis, the allele strings were entered into the BioNumerics software v.7.0 (Applied Maths, Sint-Martens-Latem, Belgium) as character values. A cut-off value of 90% similarity was applied to define clusters (Nhu et al., 2014). MLST was performed according to the Bartual method described previously (Bartual et al., 2005). Allele sequences, sequence types (STs), primer sequences and other details are available from the MLST website at http://pubmlst.org. The minimum spanning tree (MSTree) algorithm was used to predict the clonal relationship between all CNSAb isolates. MSTree algorithm was constructed with the BioNumerics software v.7.0 (Applied Maths, Sint-Martens-Latem, Belgium) using a categorical coefficient according to a previous report (Ruan et al., 2013).

Statistical Analysis

Statistical analyses were performed by Student's *t*-test, Fisher's exact test and the chi-square test with the SPSS software package (version 16). *P*-values <0.05 in all experiments were considered as significant.

RESULTS

Antibiotic Susceptibility Testing and Biofilm Assay

In this study, from the 92 *A. baumannii* isolates, 69 isolates were confirmed as CNSAb. The MIC determination of all CNSAb isolates, exhibited non-susceptibility rates $\geq$95% to cefepime, ceftazidime, levofloxacin, meropenem, piperacillin, piperacillin-tazobactam, rifampicin and trimethoprim-sulfamethoxazole (**Table 1**). CNSAb isolates showed high susceptibility to colistin (100%), tigecycline (87%) and minocycline (69%) with MIC_{50} values of 0.01, 0.25, and 5 mg/L, respectively and MIC_{90} values of 1, 3, and 10 mg/L, respectively (**Table 1**). Fifty-three (77%) CNSAb isolates revealed XDR phenotypes. In all XDR isolates following colistin with 100% sensitivity, tigecycline (83%) and minocycline (67%) were the most effective antibiotics. Fifty-six (81%) of all CNSAb isolates formed strong biofilm whereas 13 (19%) of these isolates were considered as weak biofilm forming strains. Our study revealed a significant association of strong biofilm formation with antimicrobial resistance ($P = 0.0001$).

Detection of Antibiotic Resistance Determinants

Phenotypic detection of MBL showed that none of the isolates were positive for MBL. The presence of bla_{OXA51}-like and bla_{ADC} were confirmed in all CNSAb isolates. The high prevalence of bla_{OXA-23}-like (53; 77%) and bla_{PER-1} (37; 54%) were seen in CNSAb isolates. Co-existence of bla_{OXA-23}-like/bla_{OXA-40}-like and bla_{OXA-23}-like/bla_{OXA-58}-like genes were detected in 6 (9%) and 2 (3%) of the isolates, respectively. The PCR results were negative for the other β-lactamase encoding genes including $bla_{OXA-143}$-like, *bla* MBL (bla_{IMP-1}, bla_{VIM-2}, bla_{GIM-1}, bla_{SPM-1} bla_{SIM-1} and bla_{NDM-}), bla_{CTX-M}, bla_{KPC}, bla_{GES} bla_{TEM} and bla_{SHV-1}. IS*Aba1* was detected upstream of bla_{ADC}, bla_{OXA-23}-like, and bla_{OXA51}-like genes in 67 (97%), 29(42%), and 18 (26%) of CNSAb isolates, respectively (**Figure 1**). A significant correlation was observed between the presence of IS*Aba1* in upstream of bla_{OXA-23}-like and increasing MIC of imipenem (MIC $\geq$ 16 mg/L) ($P = 0.042$). In the presence of IS*Aba1* in upstream of bla_{OXA-51}-like and increasing imipenem MICs, there was no significant correlation ($P = 0.347$).

Class 1 integron gene cassettes were identified in 58 (84%) CNSAb isolates while class 2 and 3 integrons were not found. Six different cassette combinations were detected within the

TABLE 1 | The minimum inhibitory concentration (MIC) distribution of 18 antimicrobial agents for the 69 CNSAb isolates as determined by E test.

Antimicrobial agents[a]	MIC (μg/mL)			Non-susceptible (%)
	Range	MIC_{50}	MIC_{90}	
CST	0.016 to 2	0.01	1	0
IPM	≤ 8 *to* ≥ 128	12	32	100
MEM	≤ 8 *to* ≥ 256	24	48	97
PIP	5 *to* ≥ 240	≥240	≥240	99
TZP	0.01 *to* ≥ 240	120	≥240	96
SAM	2/1 *to* ≥ 256/128	32/16	≥256/128	65
CAZ	4 *to* ≥ 256	128	≥256	96
FEP	4 *to* ≥ 256	128	≥256	97
CIP	0.1 *to* ≥ 240	30	60	95
LVX	0.5 *to* ≥ 240	10	60	95
TET	1 *to* 120	5	30	39
MIN	0.1 to 60	5	10	31
TGC	0.023 to 32	0.25	3	13
TOB	1 *to* ≥ 240	30	120	66
GEN	2 *to* ≥ 240	30	120	93
AMK	8 *to* ≥ 256	64	256	86
SXT	5 *to* ≥ 240	30	60	99
RIF	0.1 to 120	10	30	97

[a]*AMK, amikacin; CAZ, ceftazidime; CIP, ciprofloxacin; CST, colisitin; FEP, cefepime; GEN, gentamicin; IPM, imipenem; LVX, levofloxacin; MEM, meropenem; MIN, minocycline; PIP, piperacillin; RIF, rifampicin; SAM, ampicillin-sulbactam; SXT, trimethoprim- sulfamethoxazole; TET, tetracycline; TGC, tigecycline; TOB, tobramicin; TZP, piperacillin-tazobactam.*

Dendrogram, MLVA allelic profiles, MIC ranges of IMP, COL and TGC, and antibiotic resistance determinants of the 69 CNSAb isolates

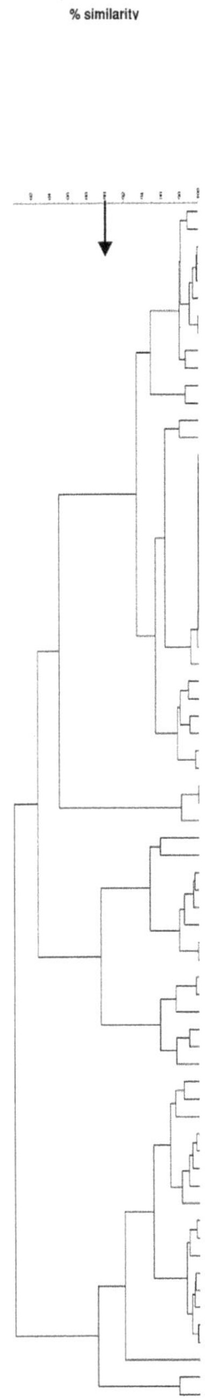

MLVA-8 allelic profile									carbapenemase			IS[a]		Int[b]			MIC[c] (mg/L)					MLST[d]		
MLVA-AB_3530	MLVA-AB_3002	MLVA-AB_2240	MLVA-AB_1988	MLVA-AB_0826	MLVA-AB_0845	MLVA-AB_2396	MLVA-AB_3468	Strain ID	bla_{PER}	bla_{VEB}	OXA Type	ISAba1-bla_{OXA-23}-like	ISAba1-bla_{OXA-51}-like	ISAba1-bla_{ADC}	Integron class I	Gene cassette(s) and order	IMP[e]	COL	TGC	Phenotype[f]	Biofilm[g]	IC[h]	ST[i]	CC[j]
12	N[k]	N	6	17	N	42	38	76			23	*[l]		*	*	5	16	0.016	1	XDR	S	G6	118	92
5	N	N	6	17	N	36	35	75			-			*	*	2	8	2	1	XDR	S	G1	118	92
10	7	4	6	19	N	42	38	349			40			*			32	0.125	3	XDR	S	G1	95	92
10	8	4	6	21	N	42	38	283			-			*	*	6	8	0.016	0.5	MDR	W	G1	95	92
12	7	4	6	21	N	42	38	326	*		40			*	*	6	32	0.19	2	XDR	S	G10	95	92
6	6	N	3	17	N	36	35	74			23			*	*	1	16	1	0.25	XDR	S	G2	118	92
8	8	5	3	17	N	44	43	62	*		23	*	*	*	*	2	128	2	0.5	XDR	S	G9	95	92
8	8	5	3	17	N	44	43	34	*	*	23	*	*	*	*	4	128	2	0.5	XDR	S	G9	669	92
10	6	5	14	17	N	48	48	71	*		23	*		*	*	2	32	0.023	0.75	XDR	S	G1	227	227
5	7	4	10	19	N	42	38	78	*		23,40			*	*	2	16	1	0.75	XDR	M	G2	95	405
13	8	4	15	38	N	46	48	42	*		23			*	*	2	16	0.016	1	XDR	M	G2	102	227
11	6	4	14	32	N	42	53	23	*		23				*	2	8	2	1	MDR	S	G1	404	SG
10	9	4	11	21	N	35	18	84	*	*	23,40			*	*	2	64	1	1	XDR	S	G1	118	92
10	7	4	9	32	N	66	30	1	*		23, 58			*	*	6	32	0.125	0.19	XDR	S	G9	787	92
7	8	4	6	32	N	64	42	21	*		23	*		*	*	1	32	1	0.75	MDR	S	G15	118	92
7	8	4	6	32	N	64	42	63			23			*	*	4	16	2	1	XDR	S	G1	138	92
7	8	4	6	32	N	64	42	82			23			*	*	2	8	0.125	0.25	XDR	S	G17	368	92
7	8	4	6	32	N	64	42	51			23	*		*	*	4	8	0.125	4	XDR	S	G16	512	92
7	8	4	6	32	N	64	42	77			23			*			8	0.016	0.5	XDR	S	G1	801	92
7	8	4	6	32	N	64	42	53	*		23	*		*	*	2	16	1	0.19	XDR	M	G2	546	92
7	8	4	6	32	N	64	42	64	*		23	*		*	*	2	16	0.023	0.38	MDR	S	G1	227	227
7	8	4	6	32	N	64	42	60			23	*		*	*	6	8	0.125	4	XDR	S	G9	245	227
7	8	4	6	32	N	64	42	27	*		23		*	*	*	5	8	0.19	0.38	XDR	S	G9	216	405
7	8	4	6	32	N	64	42	79	*		23		*	*	*	6	16	0.19	0.38	MDR	S	G2	133	225
7	8	4	6	32	N	64	42	73	*		23			*	*	5	16	0.19	0.38	XDR	S	G9	402	225
7	9	4	9	31	N	61	41	22	*		23	*		*	*	1	8	0.016	0.38	XDR	S	G15	118	92
7	9	4	8	19	N	43	30	15			23	*		*	*	2	32	1	0.38	MDR	S	G1	216	405
8	6	4	5	13	N	49	34	16	*	*	23	*	*	*	*	3	64	1	0.19	MDR	S	G16	118	92
8	7	4	4	10	N	65	36	14	*	*	23	*	*	*	*	5	128	0.125	1	XDR	S	G16	345	SG
6	6	4	8	19	N	63	35	13	*	*	23	*		*	*	3	32	0.19	0.75	XDR	S	G16	118	92
11	6	4	8	19	N	49	27	28			23			*			8	1	0.19	XDR	S	G17	800	92
10	7	4	6	10	N	35	18	230			-		*	*			32	0.19	0.75	XDR	S	G1	643	92
13	7	4	7	10	N	49	24	31			23			*			8	0.016	1	MDR	S	G1	227	227
10	6	4	6	30	N	38	16	91	*		23	*		*	*	4	64	0.016	0.5	XDR	S	G16	735	92
10	6	4	6	30	N	38	16	24	*		23	*		*	*	2	64	0.016	0.5	XDR	S	G9	100	227
9	9	3	9	32	N	48	13	19	*	*	23,40	*		*	*	2	32	2	0.5	MDR	S	G1	118	92
2	9	5	11	19	N	45	15	61			23,40			*			8	1	0.19	XDR	S	G2	104	225
10	5	7	7	32	N	66	13	18	*		23			*	*	5	32	0.125	1	XDR	S	G9	133	225
10	8	5	6	10	N	49	16	37		*	23		*	*	*	4	64	1	0.5	XDR	M	G2	225	225
8	7	4	2	10	N	48	13	58	*		23	*			*	5	16	1	1	MDR	S	G16	113	227
11	10	5	6	9	N	46	13	25	*	*	23,58			*	*	4	64	1	1	XDR	M	G2	133	225
5	8	5	9	10	N	61	13	38			23	*	*	*			32	1	4	XDR	M	G2	133	225
10	8	4	10	19	N	61	13	39	*	*	23	*	*	*	*	4	256	2	1	XDR	S	G1	231	227
11	7	5	11	19	N	61	13	54	*		23			*	*	2	16	2	0.38	XDR	S	G1	231	227
11	7	4	17	N	N	61	17	9			23			*	*	5	16	0.19	0.5	MDR	S	G1	788	92
10	9	4	14	N	N	52	13	43			23			*	*	5	16	1	0.5	XDR	S	G1	788	92
12	7	4	6	N	N	42	15	80			23,40			*	*	2	16	1	0.75	XDR	M	G2	670	92
11	8	7	14	N	N	65	33	35	*	*	23	*	*	*	*	2	128	0.19	1	MDR	S	G1	16	405
9	N	6	7	N	N	50	26	8			-			*	*	1	8	0.19	1	XDR	S	G9	365	SG
10	7	6	11	N	N	33	17	17			23			*			32	0.125	0.125	XDR	M	G2	247	227
12	7	5	5	N	N	40	57	336	*		23	*		*	*	1	32	0.25	2	XDR	S	G9	75	92
8	9	4	6	10	N	50	71	49			23			*	*	2	8	0.01	0.19	XDR	S	G1	421	92
12	10	4	3	10	N	65	68	26		*	23	*		*	*	5	128	2	0.75	XDR	S	G16	375	92
10	9	5	11	N	N	47	56	389	*		40		*	*	*	1	32	0.19	16	XDR	S	G9	118	92
11	6	4	10	N	N	49	54	69			23		*	*	*	2	8	1	1	XDR	S	G16	259	227
12	7	5	7	N	N	40	41	68			23			*			16	0.016	0.25	XDR	S	G16	227	227
10	7	6	12	N	N	32	38	351			40			*	*	6	32	0.125	3	XDR	S	G9	246	227
8	8	N	6	N	N	26	27	32		*	23	*	*	*	*	2	128	1	0.125	XDR	M	G2	258	227
10	7	4	11	N	N	48	40	241	*		23		*	*	*	2	32	0.38	0.75	XDR	S	G9	118	92
7	6	4	13	N	N	52	43	44			23	*		*			8	1	1	MDR	W	G1	118	92
12	7	N	11	N	N	42	38	338	*		23	*		*	*	6	32	0.25	2	MDR	S	G1	197	227
10	8	4	9	N	N	51	38	240			23			*	*	2	32	0.25	2	XDR	S	G1	99	227
10	6	4	4	N	N	47	36	36	*	*	23	*	*	*	*	5	128	0.19	0.75	MDR	S	G1	405	227
11	6	5	8	N	N	40	33	20	*		23,40			*	*	4	64	2	0.5	XDR	S	G9	405	227
9	6	4	8	N	N	37	24	7			23		*	*	*	2	8	0.016	0.75	MDR	M	G1	137	92
11	7	5	8	N	N	43	30	50	*		23			*	*	2	16	0.016	0.38	XDR	W	G2	405	227
N	5	6	12	N	N	26	35	30	*		23	*		*	*	4	128	0.016	0.75	XDR	S	G10	225	225
14	8	4	9	19	N	39	56	57			23			*			8	0.016	0.75	XDR	S	G1	95	92
11	8	4	3	10	N	31	57	29	*		23	*		*	*	1	32	1	4	XDR	S	G16	75	92

[a]IS, insertion sequences
[b]Int, Integrons
[c]MIC, minimal inhibitory concentration
[d]MLST, multilocus sequence typing
[e]IPM, imipenem; CST,colistin; TGC, tigecycline
[f]MDR, multidrug- resistant; XDR, extensively drug- resistant
[g]Biofilm formation: S,strong; M, moderate, W, weak
[h]IC, international clonal lineage
[i]ST, sequence type
[j]CC, clonal complex
[k]N, not amplified
[l]*, sign of positive test

FIGURE 1 | Dendogram shows the genetic diversity of 69 carbapenem non-susceptible *Acinetobacter baumannii* isolates by MLVA and MLST; carbapenemase encoding genes; MIC ranges of colisitin, imipenem, and tigecycline; resistance phenotype; biofilm formation and international clonal lineage.

class 1 integrons. The most prevalent cassette combination, *aacA4-catB8-aadA1* (2.3 Kb), were detected in 23 (39%) of class 1 integrons. The other cassette combinations including *arr3-aacA4* (1.3 Kb), *aacC1-orfX-orfX'-aadA1* (2.5 Kb), *orfI-aadA1* (1.5 Kb), *dfrVII* (0.7 Kb) *and dfrXII-orfF-aadA2* (1.7 Kb) were detected in 10 (14.5%), 9 (13%), 7 (10%), 7 (10%), and 2 (3%) CNSAb isolates, respectively. The gene cassette arrangement showed that the class 1 integrons harbor genes encoding resistance to aminoglycosides, chloramphenicol, rifampicin and trimethoprim/sulfamethoxazole. Also, the results of this study show that integron-associated β-lactamases encoding genes were not detected in class 1 integrons-harboring CNSAb isolates.

Genetic Diversity and Population Structure

Three-locus dual assay multiplex PCR showed eight different PCR-based groups (G1, G2, G6, G9, G10, and G15–G17) among the 69 CNSAb isolates. Twenty-five (36%) and 13 (19%) CNSAb isolates belonged to G1 (IC II) and G2 (IC I), respectively. Based on the variations in combination of PCR amplicons of the bla_{OXA-51}-like, *csuE* and *ompA* genes, 31(45%) of the CNSAb isolates belonged to 6 IC variants PCR-based groups (**Table 2**). G9 (14; 45%) was the most common IC variant (**Table 2**).

The MSTree algorithm of the MLST data of this study suggests that 40 previously known STs combined into four CC (**Figure 2**). Thirty-six (52%) and 13 (19%) of all 69 CNSAb isolates belonging to CC-92 and CC-405, respectively. According to the MSTree algorithm, only one singleton ST (ST345) was identified (**Figure 2**). ST118 was the predominant ST, comprising 17.3% of isolates.

By the MLVA typing method, the 69 CNSAb isolates were grouped into 56 distinct MLVA types with 6 clusters and 53 singleton genotypes. Loci MLVA-AB_0845 was absent (N) in all studied isolates. In this study, the VNTR loci MLVA-AB_3002 and MLVA-AB_2240 displayed lower diversity, whereas VNTR loci MLVA-AB_2396 and MLVA-AB_3468 showed higher level of diversity.

Thirty STs corresponded to 30 MLVA types; other STs (118, 95, 75, 227, 405, 231, 788, 225, 227, and 133) included heterogeneous MLVA profiles (**Figure 1**).

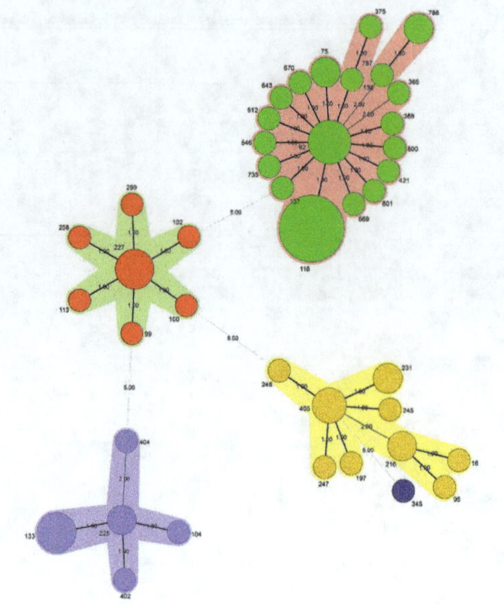

FIGURE 2 | The minimum spanning tree (MStree) was constructed using a categorical coefficient based on multilocus sequence typing (MLST) data. Each circle stands for a different sequence type (ST). The color of a circle and the line clustering the ST with the same color represents an identical ST. Size of the circle corresponds to the number of isolates. The width of the line reflects the genetic distance between ST (heavy and thin lines represent a difference in one allele and two or more alleles, respectively).

In this study, IC variants were classified into 23 STs belonging to CC92, CC405, CC227, CC225 and one singleton (**Figure 1**).

Relationship between IC Lineages with Antibiotic Resistance Determinants

Table 3 shows the frequency of antibiotics resistance in two major epidemic lineages and IC variants. Susceptibility to all antimicrobial agents, except tobramicin, in isolates belonged to IC II was higher than the IC I and IC variants (**Table 3**). It is noteworthy that the IC variants had the highest resistance rate to tigecycline (7; 22.5%) with MIC_{50} and MIC_{90} of 0.25 and 3 mg/L, respectively (**Tables 1** and **3**). Among the 13 carbapenem

TABLE 2 | Combinations of amplicons obtained in the dual multiplex PCRs used to describe 31 *A. baumannii* IC variants.

Variant type (No.)	PCR group 1			PCR group 2		
	csuE 702 bp	bla_{OXA-51}-like 559 bp	*ompA* 355 bp	*csuE* 580 bp	*ompA* 343 bp	bla_{OXA-51}-like 162 bp
G 6[a] (1)	−	−	−	+	−	−
G 9 (14)	−	−	+	+	−	+
G 10 (2)	−	+	−	+	+	−
G 15 (2)	+	−	−	−	−	+
G 16 (10)	−	+	−	+	−	−
G 17 (2)	+	−	+	−	−	+

[a]*G; PCR-based group.*

TABLE 3 | Frequency of antimicrobial non-susceptibility in three major epidemic lineages in the 69 CNSAb isolates.

IC[b] (No.)	% Non-susceptibility to CLSI antimicrobial groups[a]																	
	A								B								O	
	IPM	MEM	CAZ	SAM	AMK	TOB	GEN	CIP	PIP	TZP	FEP	MIN	TET	TGC	LVX	SXT	CST	RIF
IC1 (13)	100	100	100	85	92	62	92	100	100	100	100	48	46	9	100	100	0	100
IC2 (25)	100	92	88	48	84	76	88	88	96	92	96	19	32	5	88	96	0	92
V (31)	100	100	100	61	84	61	100	97	100	97	100	25	39	24	97	100	0	97
Total (69)	100	97	96	65	86	66	93	95	99	96	97	31	39	13	95	99	0	97

[a]*Considerations in the assignment of agents to Groups A, B, and C include clinical efficacy, prevalence of resistance, minimizing emergence of resistance, cost, FDA clinical indications for usage, and current consensus recommendations for first-choice and alternative drugs.* ***Group A*** *are considered appropriate for inclusion in a routine, primary testing panel, as well as for routine reporting of results for the specific organism.* ***Group B*** *comprises agents that may warrant primary testing.* ***Group O*** *(Other) includes agents that have a clinical indication for the organism. Escherichia coli ATCC25922 and Pseudomonas aeruginosa ATCC27853 were used for quality control of antimicrobial susceptibility testing and included in each run.*
[b]*IC, international clonal lineage; V, IC variants; AMK, amikacin; CAZ, ceftazidime; CIP, ciprofloxacin; CST, colisitin; FEP, cefepime; GEN, gentamicin; IPM, imipenem; LVX, levofloxacin; MEM, meropenem; MIN, minocycline; PIP, piperacillin; RIF, rifampicin; SAM, ampicillin-sulbactam; SXT, trimethoprim- sulfamethoxazole; TET, tetracycline; TGC, tigecycline; TOB, tobramicin; TZP, piperacillin-tazobactam.*

non-susceptible IC I, 12 (92%) were XDR while this rate was 14 (56%) and 27 (87%) in the isolates belonging to the IC variants and IC II, respectively. The resistance to beta lactam antibiotics in IC I and IC variant isolates was higher than that in IC II. Thirty-one (100%) IC variants and 22 (88%) of the IC II population formed strong biofilm but only 3 (23%) of IC I had this characteristic. In the present study, resistance to antimicrobials was comparatively higher among IC variant than IC I and IC II CNSAb isolates ($P = 0.0012$ and $P = 0.0001$, respectively). A slightly higher susceptibility to antimicrobials was observed in IC II than IC variants and IC I ($P > 0.05$). Our analysis revealed a significant correlation ($P = 0.003$) between the presence of $bla_{PER1}$and biofilm formation in the CNSAb isolates. The CNSAb isolates belonging to all clones had the bla_{OXA-23}-like gene. All CNSAb isolates classified in IC I, were positive for the bla_{OXA-23}-like gene (**Table 4**). According to **Table 4**, the frequency of IS*Aba1*/bla_{OXA-23}-like and bla_{PER-1} genes in a population belonging to IC variants were significantly higher than IC I and IC II ($P = 0.027$ and $P = 0.003$, respectively). **Table 4** shows that the distribution of different types of class 1 integrons is related to major epidemic lineages with predominance of class 1 integrons belonging to type 2.

DISCUSSION

Nosocomial infections caused by CNSAb are currently among the most difficult to treat. The control of CNSAb infections among hospitalized patients continues to present serious challenges to patient management in developing countries like Iran (Diancourt et al., 2010; Giannouli et al., 2010; Bahador et al., 2013). The main challenge remains choosing the most likely effective antibiotics as determined by *in vitro* testing of antibiotics, based on the local susceptibility patterns (Fishbain and Peleg, 2010). Overall, the findings of this study are consistent with recent studies from Iran that show an alarming trend of CNSAb increase and in MDR *A. baumannii* resistance against a wide spectrum of antimicrobial agents (Moradi et al., 2015). However, in the present study colistin, tigecycline, and tetracyclines exhibited a potent activity against CNSAb isolates. Our data revealed a decrease (10%) of resistance in CNSAb isolates to tigecycline (Bahador et al., 2013), likely due to the implementation of restriction policy on the empirical prescription of tigecycline. Decrease in antibiotic resistance after implementation of restriction policies have been shown in several reports (Altunsoy et al., 2011; Sistanizad et al., 2013). Our finding of positive association in strong biofilm formation and resistance to antimicrobial is consistent with a recent report (Badave and Kulkarni, 2015). This could be due to inadequate penetration of the antimicrobials into the biofilms, and the emergence of persister cells in the biofilms (Penesyan et al., 2015).

The findings of our study on IS*Aba1* elements upstream of bla_{OXA} genes and their role as a promoter to enhance OXA-enzymes expression and MIC of imipenem confirms previous reports (Liu et al., 2014a); however, this finding are in contrast to a recent report by Peymani et al. (2012b) from Iran. Not only did our research find that 42% of CNSAb isolates were the IS*Aba1*-bla_{OXA-23}-like gene; but also, a significant correlation was observed between the presence of IS*Aba1* upstream of bla_{OXA-23}-like gene and increasing imipenem MICs ($\geq$16 mg/L). This dissimilarity might be due to differences in the *A. baumannii* strains available for study. Additionally, our finding of high prevalence of IS*Aba1* sequences upstream of the bla_{ADC} among 97% of CNSAb isolates is consistent with previous report from Iran; and indicate that presence of IS*Aba1* sequences upstream of bla_{ADC} played an important role in increasing resistance to cephalosporins ($MIC_{50} = 128$ mg/L).

However, IC II is the most common strain in the world (Liu et al., 2014a); and majority of the clinical strains (45%) in this study belong to IC variants. The incidence rate of IC variants observed in this study is similar to previously reported rates of 46% from Sweden and 51% from Iran, and is considerably higher than that in other recent reports, including Iran (3%), Latvia (6%), Romania (15%), Italy (23%) and Norway (28%) (Fishbain and Peleg, 2010; Hojabri et al., 2014; Liu et al., 2014a). With regard to a previous study (Fishbain and Peleg, 2010) and the

TABLE 4 | Frequency of antimicrobials resistance determinants in epidemic clonal lineages of the CNSAb isolates.

IC[a] (No.)	Ambler classification No. (%)						ISAba upstream beta-lactamase encoding genes No. (%)			Class 1 integron No. (%)	Class 1 integron types[b] No. (%)					
	Class A		Class D		Co-existence of Class D						1	2	3	4	5	6
	*bla*PER	*bla*VEB	*bla*OXA−23−like	*bla*OXA−40−like	*bla*OXA−23−40−like	*bla*OXA−23−58−like	ISAba1-*bla*OXA−23−like	ISAba1-*bla*OXA−51−like	ISAba1-*bla*ADC							
IC-I (13)	4 (31)	3 (23)	13 (100)	2 (15)	3 (23)	1 (8)	4 (31)	5 (39)	13 (100)	9 (69)	–	6 (46)	–	2 (15)	1 (8)	–
IC-II (25)	10 (40)	4 (16)	17 (68)	2 (8)	1 (4)	0	7 (28)	7 (28)	24 (96)	19 (76)	1 (4)	8 (32)	–	3 (12)	4 (16)	3 (12)
V (31)	24 (77)	6 (19)	24 (77)	5 (16)	2 (6)	1 (3)	18 (58)	6 (19)	30 (97)	30 (97)	6 (19)	9 (29)	2 (6)	4 (13)	5 (16)	4 (13)
Total (69)	48 (69)	13 (19)	54 (78)	9 (13)	6 (9)	2 (3)	29 (42)	18 (26)	67 (97)	58 (84)	7 (10)	23 (33)	2 (6)	9 (13)	10 (14)	7 (10)

[a]*Abbreviations: IC, international clone; V, variants of IC.*
[b]*Type 1, orfI-aadA1; Type 2, aacA4-catB8-aadA1; Type 3, dfrXII-orfF-aadA2; Type 4, aacC1-orfX-orfX'-aadA1; Type 5, arr3-aacA4; Type 6, dfrVII.*

study of Karah et al. (2012), 14 CNSAb isolates showed a new combination of amplicons and were considered in this article as the PCR-based group (G) 15–17. Three of the 6 PCR-based groups in the present study, namely G6, G9, and G10, have been documented in previous studies (Karah et al., 2012; Bahador et al., 2014). Although there are differences among the IC variant types in this study and previous studies in Iran (Peymani et al., 2012b; Rezaee et al., 2013; Hojabri et al., 2014), the IC variants are highly resistant to antibiotics, which suggests that it is conceivable that the MDR phenotype has substantially contributed to their spread in Iran (Bahador et al., 2014).

The results of this study indicate that, in spite of the high rate of antibiotic resistance determinants in the IC I stains in comparison with the IC variant strains, the IC I strains had lower prevalence. This dissimilarity might be due to the higher frequency of strong biofilm formers in the IC variant strains than in the IC I strains. Several studies proved that, cell adhesiveness and ability to form biofilm were higher in the bla_{PER-1} producing isolates (Longo et al., 2014), and this is in agreement with the result of this study. The capability to form biofilm and presence of the bla_{PER} gene were significantly higher in IC variants. In this study, the IC II had slightly higher susceptibility to antimicrobial agents than other clonal lineage population. This might be due to low-frequency of carbapenem-hydrolyzing class D β-lactamase (CHDL) genes and IS*Aba1* elements, as well as low ability of biofilm formation in IC II strain as compared to IC I and IC variant stains.

It has been clearly shown that, the high prevalence of class 1 integrons among MDR *A. baumannii* clinical isolates is responsible for several nosocomial outbreaks globally (Liu et al., 2014b). The resistance rates to all the tested antibiotics, except colistin, tigecycline, minocycline, and tetracycline, of the collected CNSAb isolates were >65%. However, the presence of integron cassette arrays cannot be responsible for all these resistance phenotypes. In Iran, class 1 integrons have been reported previously in >11% of *A. baumannii* isolates (Peymani et al., 2012a; Japoni-Nejad et al., 2013). In the current study, 84% of isolates contain class 1 integrons, but no class 2 and class 3 integrons was detected. This result shows that class 2 and class 3 integrons are not the major resistant determinants in CNSAb isolates in this study, which is consistent with previous studies (Peymani et al., 2012a; Japoni-Nejad et al., 2013).

Only six categories of gene cassettes, including aminoglycoside-modifying enzymes, chloramphenicol-resistant, trimethoprim/sulfamethoxazole-resistant, and rifampicin-resistant were detected in spite of previous studies reporting integrons carrying carbepenemase genes (Japoni-Nejad et al., 2013). This result indicates that acquisitions of carbapenem-hydrolyzing β-lactamase genes plays more important role than the class 1 integrons in carbapenem resistance in CNSAb. The class 1 integrons rate (84%) here defined for a referral burn center in central Iran is higher than the rates found in other geographical regions, including Western Iran (11%) (Salimizand et al., 2014), China (74%) (Wu et al., 2012) and European countries in general (43%) (Castanheira et al., 2014). This result is similar to that of previous studies from the Northwest and center of Iran (Peymani et al., 2012a; Japoni-Nejad et al., 2013).

From literature survey, this is the first study to report carriage of class 1 integrons and associated arrays in *A. baumannii* isolates from burn patients in Iran.

Analysis of the integron cassette arrays shows that most class 1 integrons include the globally distributed *aac*A4-*cat*B8-*aad*A1 array (Liu et al., 2014a), and the cassette array *aacC1-orfX-orfX'-aadA1* has been documented in previous studies in Iran (Japoni-Nejad et al., 2013). Moreover, gene cassettes encoding aminoglycoside- resistance are present in the majority of CNSAb isolates in this study, suggesting a close relationship between high-level resistance rate to aminoglycoside compounds such as gentamicin and amikacin and the *A. baumannii* isolates. The different types of gene cassette array in this study and previous studies in Iran (Japoni-Nejad et al., 2013; Salimizand et al., 2014), suggests that a geographical feature plays a major role in MDR isolates formation, and it could be due to different volumes and patterns of antibiotic consumption in distinct areas. In addition, the results of this study indicate that strains indistinguishable from each other by MLST and MLVA typing can have different cassette arrays, indicating that some of these may be transmitted very efficiently. In this study, amikacin resistance recorded in 10% of the CNSAb isolates was not cassette-encoded. Analysis of the integron cassette arrays show that 50% of these isolates harbor an integron without aminoglycoside resistance genes. In the cases where the class 1 integrons could not be amplified, there were no integrons and alteration of the primer binding site or a large size of the Integron are the most likely explanations for the negative result of PCRs.

According to the findings of this research, MLVA could identify closely epidemiologically related isolates clustered by MLST in some cases (**Figure 1**). In this study, isolates with the same ST, such as ST118, have closely related MLVA type (MTs) (**Figure 1**). This result suggests that the evaluation of MLVA in CNSAb isolates shows its great ability for discrimination of genetically closely related isolates.

In agreement with previous studies (Hojabri et al., 2014), the present results substantiated that CC92 represented the recent widely distributed clone in Iran, accounting for 52.2% of CNSAb isolates examined in this study. This most likely indicates that the isolates belonging to CC92 might benefit with respect to acquiring resistance determinants and surviving in the health care environment, when encountering selection by antibiotics such as carbapenems. The international spread of the CC92 isolates has been reported from all around the world (Karah et al., 2012). This study revealed that the 28 isolates belonging to this lineage were shown to be heterogeneous by MLVA (22 MTs), followed by MLST (13 STs). The 69 CNSAb isolates were shown to be heterogeneous by MLVA (56 MTs) and MLST (40 STs). Taken together, genetic diversity and wide dissemination clones among CNSAb isolates is consistent with recent studies from Iran that show high diversity of STs in *A. baumannii* isolates from burned patients (Hojabri et al., 2014), which are the evidence of dynamic population structure and evolution in progress. The referral of patients from other medical centers of Iran to the hospital in Tehran could be among the possible reasons for this diversity and failure to colonize a specific type. Spread of specific clonally related isolates could have been prevented by the success of infection control programs in the teaching hospital in Iran.

CONCLUSION

Evidence that XDR-CNSAb from burned patients in Iran is rapidly changing toward growing resistance to various antimicrobials, including tigecycline was presented. Despite the increasing resistance to several first-line antimicrobials, all resistant CNSAb isolates remain sensitive to the antimicrobial colistin, a viable agent in controlling XDR-CNSAb outbreaks, especially in developing countries. This is the first study to perform IC analysis and gene cassette mapping of class integrons of *A. baumannii*, obtained from burned patients in Iran. It can be confirmed that the dominant local IC is an IC variant, with the likelihood that G9 is the predominant member. In the current study, the *aac*A4-*cat*B8-*aad*A1 gene cassette array which confers resistance to aminoglycoside and chloramphenicol was identified as the predominant cassette array of class 1 integrons in CNSAb. This study described the widest distribution of $bla_{\mathrm{OXA-23}}$-like-horbering IC ST118 (CC92) and ST405 (CC405) were the principal reason for the rapid increase in the carbapenem resistance rate in burned patients in Iran. The findings highlight the importance of an international drug resistance and molecular epidemiology monitoring network of *A. baumannii* isolates to effect global control measures against XDR-CNSAb.

FUNDING

This research has been supported by Tehran University of Medical Sciences and health Services grant No. 92- 03-30-23186.

ACKNOWLEDGMENT

The authors would like to thank the dedicated laboratory personnel at the Shahid Motahari hospital (Iran University of Medical Sciences Complex), who helped collect specimens for this study.

REFERENCES

Alaghehbandan, R., Azimi, L., and Rastegar Lari, A. (2012). Nosocomial infections among burn patients in Teheran, Iran: a decade later. *Ann. Burns Fire Disasters* 25, 3–7.

Almeida, L. A., and Araujo, R. (2013). Highlights on molecular identification of closely related species. *Infect. Genet. Evol.* 13, 67–75. doi: 10.1016/j.meegid

Altunsoy, A., Aypak, C., Azap, A., Ergönül, Ö, and Balik, I. (2011). The impact of a nationwide antibiotic restriction program on antibiotic usage and resistance against nosocomial pathogens in Turkey. *Int. J. Med. Sci.* 8, 339–344.

Azimi, L., Talebi, M., Pourshafie, M. R., Owlia, P., and Rastegar Lari, A. (2015). Characterization of carbapenemases in extensively drug resistance *Acinetobacter baumannii* in a burn care center in Iran. *Int. J. Mol. Cell Med.* 4, 46–53.

Badave, G. K., and Kulkarni, D. (2015). Biofilm producing multidrug resistant *Acinetobacter baumannii*: an emerging challenge. *J. Clin. Diagn. Res.* 9, 8–10. doi: 10.7860/JCDR/2015/11014.5398

Bahador, A., Raoofian, R., Taheri, M., Pourakbari, B., Hashemizadeh, Z., and Hashemi, F. B. (2014). Multidrug resistance among *Acinetobacter baumannii* isolates from Iran: changes in antimicrobial susceptibility patterns and genotypic profile. *Microb. Drug Resist.* 20, 632–640. doi: 10.1089/mdr.2013.0146

Bahador, A., Taheri, M., Pourakbari, B., Hashemizadeh, Z., Rostami, H., Mansoori, N., et al. (2013). Emergence of rifampicin, tigecycline, and colistin-resistant *Acinetobacter baumannii* in Iran; spreading of MDR strains of novel International Clone variants. *Microb. Drug Resist.* 19, 397–406. doi: 10.1089/mdr.2012.0233

Bartual, S. G., Seifert, H., Hippler, C., Luzon, M. A., Wisplinghoff, H., and Rodríguez-Valera, F. (2005). Development of a multilocus sequence typing scheme for characterization of clinical isolates of *Acinetobacter baumannii*. *J. Clin. Microbiol.* 43, 4382–4390. doi: 10.1128/JCM.43.9.4382-4390.2005

Bonnin, R. A., Nordmann, P., Potron, A., Lecuyer, H., Zahar, J. R., and Poirel, L. (2011a). Carbapenem-hydrolyzing ges-type extended-spectrum β-Lactamase in *Acinetobacter baumannii*. *Antimicrob. Agents Chemother.* 55, 349–354. doi: 10.1128/AAC.00773-10

Bonnin, R. A., Potron, A., Poirel, L., Lecuyer, H., Neri, R., and Nordmann, P. (2011b). PER-7, an extended-spectrum β-lactamase with increased activity toward broad-Spectrum cephalosporins in *Acinetobacter baumannii*. *Antimicrob. Agents Chemother.* 55, 2424–2427. doi: 10.1128/AAC.01795-10

Bou, G., and Martinez-Beltran, J. (2000). Cloning, nucleotide sequencing, and analysis of the gene encoding an AmpC beta-lactamase in *Acinetobacter baumannii*. *Antimicrob. Agents Chemother.* 44, 428–432. doi: 10.1128/AAC.44.2.428-432.2000

Cai, Y., Chai, D., Wang, R., Liang, B., and Bai, N. (2012). Colistin resistance of *Acinetobacter baumannii*: clinical reports, mechanisms and antimicrobial strategies. *J. Antimicrob. Chemother.* 67, 1607–1615. doi: 10.1093/jac/dks084

Castanheira, M., Costello, S. E., Woosley, L. N., Deshpande, L. M., Davies, T. A., and Jones, R. N. (2014). Evaluation of clonality and carbapenem resistance mechanisms among *Acinetobacter baumannii*-Acinetobacter calcoaceticus complex and *Enterobacteriaceae* isolates collected in European and Mediterranean countries and detection of two novel β-lactamases, GES-22 and VIM-35. *Antimicrob. Agents Chemother.* 58, 7358–7366. doi: 10.1128/AAC.03930-14

Chen, Y., Zhou, Z., Jiang, Y., and Yu, Y. (2011). Emergence of NDM-1-producing *Acinetobacter baumannii* in China. *J. Antimicrob. Chemother.* 66, 1255–1259. doi: 10.1093/jac/dkr082

Climaco, E. C., Oliveira, M. L., Pitondo-Silva, A., Oliveira, M. G., Medeiros, M., Lincopan, N., et al. (2013). Clonal complexes 104, 109 and 113 playing a major role in the dissemination of OXA-carbapenemase-producing *Acinetobacter baumannii* in Southeast Brazil. *Infect. Genet. Evol.* 19, 127–133. doi: 10.1016/j.meegid.2013.06.024

Clinical and Laboratory Standards Institute [CLSI] (2014). *Clinical and Laboratory Standards Institute. Performance Standards for Antimicrobial Susceptibility Testing: Twenty fourth Informational Supplement M100-S24CLSI*. Wayne, PA: CLSI.

Diancourt, L., Passet, V., Nemec, A., Dijkshoorn, L., and Brisse, S. (2010). The population structure of *Acinetobacter baumannii*: expanding multiresistant clones from an ancestral susceptible genetic pool. *PLoS ONE* 5:e10034. doi: 10.1371/journal.pone.0010034

Dillon, B., Thomas, L., Mohmand, G., Zelynski, A., and Iredell, J. (2005). Multiplex PCR for screening of integrons in bacterial lysates. *J. Microbiol. Methods* 62, 221–232. doi: 10.1016/j.mimet.2005.02.007

Ellington, M. J., Kistler, J., Livermore, D. M., and Woodford, N. (2007). Multiplex PCR for rapid detection of genes encoding acquired metallo-beta-lactamases. *J. Antimicrob. Chemother.* 59, 321–322. doi: 10.1093/jac/dkl481

European Committee on Antimicrobial Susceptibility Testing. (2014). *Version 4. 0.*

Fishbain, J., and Peleg, A. Y. (2010). Treatment of *Acinetobacter* infections. *Clin. Infect. Dis.* 51, 79–84. doi: 10.1086/653120

Giannouli, M., Cuccurullo, S., Crivaro, V., Di Popolo, A., Bernardo, M., Tomasone, F., et al. (2010). Molecular epidemiology of multidrug-resistant *Acinetobacter baumannii* in a tertiary care hospital in Naples, Italy, shows the emergence of a novel epidemic clone. *J. Clin. Microbiol.* 48, 1223–1230. doi: 10.1128/JCM.02263-09

Girerd-Genessay, I., Bénet, T., and Vanhems, P. (2015). Multidrug-resistant bacterial outbreaks in burn units: a synthesis of the literature according to the orion statement. *J. Burn Care Res.* [Epub ahead of print]

Heritier, C., Poirel, L., and Nordmann, P. (2006). Cephalosporinase over-expression resulting from insertion of ISAba1 in *Acinetobacter baumannii*. *Clin. Microbiol. Infect.* 12, 123–130. doi: 10.1111/j.1469-0691.2005.01320.x

Higgins, P. G., Lehmann, M., and Seifert, H. (2010a). Inclusion of OXA-143 primers in a multiplex polymerase chain reaction (PCR) for genes encoding prevalent OXA carbapenemases in *Acinetobacter* spp. *Int. J. Antimicrob. Agents* 35:305. doi: 10.1016/j.ijantimicag.2009.10.014

Higgins, P. G., Lehmann, M., Wisplinghoff, H., and Seifert, H. (2010b). gyrB Multiplex PCR To differentiate between *Acinetobacter calcoaceticus* and *Acinetobacter* genomic species 3. *J. Clin. Microbiol.* 48, 4592–4594. doi: 10.1128/JCM.01765-10

Hojabri, Z., Pajand, O., Bonura, C., Aleo, A., Giammanco, A., and Mammina, C. (2014). Molecular epidemiology of *Acinetobacter baumannii* in Iran: endemic and epidemic spread of multiresistant isolates. *J. Antimicrob. Chemother.* 69, 2383–2387. doi: 10.1093/jac/dku045

Japoni-Nejad, A., Farshad, S., van Belkum, A., and Ghaznavi-Rad, E. (2013). Novel cassette array in a class 1 integron in clinical isolates of *Acinetobacter baumannii* from central Iran. *Int. J. Med. Microbiol.* 303, 645–650. doi: 10.1016/j.ijmm.2013.09.005

Jeannot, K., Diancourt, L., Vaux, S., Thouverez, M., Ribeiro, A., Coignard, B., et al. (2014). Molecular epidemiology of carbapenem non-susceptible *Acinetobacter baumannii* in France. *PLoS ONE* 9:e115452. doi: 10.1371/journal.pone.0115452

Karah, N., Sundsfjord, A., Towner, K., and Samuelsen, Ø (2012). Insights into the global molecular epidemiology of carbapenem non-susceptible clones of *Acinetobacter baumannii*. *Drug Resist. Updat.* 15, 237–247. doi: 10.1016/j.drup.2012.06.001

Liu, C. C., Tang, C. Y., Chang, K. C., Kuo, H. Y., and Liou, M. L. (2014a). A comparative study of class 1 integrons in *Acinetobacter baumannii*. *Gene* 544, 75–82. doi: 10.1016/j.gene.2014.04.047

Liu, C. P., Shih, S. C., Wang, N. Y., Wu, A. Y., Sun, F. J., Chow, S. F., et al. (2014b). Risk factors of mortality in patients with carbapenem-resistant *Acinetobacter baumannii* bacteremia. *J. Microbiol. Immunol. Infect.* doi: 10.1016/j.jmii.2014.10.006 [Epub ahead of print].

Longo, F., Vuotto, C., and Donelli, G. (2014). Biofilm formation in *Acinetobacter baumannii*. *N. Microbiol.* 37, 119–127.

Magiorakos, A. P., Srinivasan, A., Carey, R., Carmeli, Y., Falagas, M. E., Giske, C. G., et al. (2012). Multidrugresistant, extensively drug-resistant and pandrug-resistant bacteria: an international expert proposal for interim standard definitions for acquired resistance. *Clin. Microbiol. Infect.* 18, 268–281. doi: 10.1111/j.1469-0691

Mahdian, S., Sadeghifard, N., Pakzad, I., Ghanbari, F., Soroush, S., Azimi, L., et al. (2015). *Acinetobacter baumannii* clonal lineages I and II harboring different carbapenem-hydrolyzing-β-lactamase genes are widespread among hospitalized burn patients in Tehran. *J. Infect. Public Health* doi: 10.1016/j.jiph.2015.04.030 [Epub ahead of print].

Moradi, J., Hashemi, F. B., and Bahador, A. (2015). Antibiotic resistance of *Acinetobacter baumannii* in Iran: a systemic review of the published literature. *Osong Public Health Res. Perspect.* 6, 79–86. doi: 10.1016/j.phrp.2014.12.006

Nhu, N. T., Lan, N. P., Campbell, J. I., Parry, C. M., Thompson, C., Tuyen, H. T., et al. (2014). Emergence of carbapenem-resistant *Acinetobacter baumannii* as the major cause of ventilator-associated pneumonia in intensive care unit patients at an infectious disease hospital in southern Vietnam. *J. Med. Microbiol.* 63, 1386–1394. doi: 10.1099/jmm.0.076646-0

Öncül, O., Öksüz, S., Acar, A., Ülkür, E., Turhan, V., Uygur, F., et al. (2014). Nosocomial infection characteristics in a burn intensive care unit: analysis of an eleven-year active surveillance. *Burns* 40, 835–841. doi: 10.1016/j.burns

Penesyan, A., Gillings, M., and Paulsen, I. T. (2015). Antibiotic discovery: combatting bacterial resistance in cells, and in biofilm communities. *Molecules* 20, 5286–5298. doi: 10.3390/molecules20045286

Peymani, A., Farajnia, S., Nahaei, M. R., Sohrabi, N., Abbasi, L., Ansarin, K., et al. (2012a). Prevalence of class 1 integron among multidrug-resistant *Acinetobacter baumannii* in Tabriz, northwest of Iran. *Pol. J. Microbiol.* 61, 57–60.

Peymani, A., Higgins, P. G., Nahaei, M. R., Farajnia, S., and Seifert, H. (2012b). Characterisation and clonal dissemination of OXA-23-producing *Acinetobacter baumannii* in Tabriz, northwest Iran. *Int. J. Antimicrob. Agents* 39, 526–528. doi: 10.1016/j.ijantimicag

Poirel, L., Menuteau, O., Agoli, N., Cattoen, C., and Nordmann, P. (2003). Outbreak of extended-spectrum beta-lactamase VEB-1-producing isolates of *Acinetobacter baumannii* in a French hospital. *J. Clin. Microbiol.* 41, 3542–3547. doi: 10.1128/JCM.41.8.3542-3547.2003

Potron, A., Munoz-Price, L. S., Nordmann, P., Cleary, T., and Poirel, L. (2011). Genetic features of CTX-M-15-producing *Acinetobacter baumannii* from Haiti. *Antimicrob. Agents Chemother.* 55, 5946–5948. doi: 10.1128/AAC.05 124-11

Pourcel, C., Minandri, F., Hauck, Y., D'Arezzo, S., Imperi, F., Vergnaud, G., et al. (2011). Identification of variable-number tandem-repeat (VNTR) sequences in *Acinetobacter baumannii* and interlaboratory validation of an optimized multiple-locus VNTR analysis typing scheme. *J. Clin. Microbiol.* 49, 539–548. doi: 10.1128/JCM.02003-10

Rezaee, M. A., Pajand, O., Nahaei, M. R., Mahdian, R., Aghazadeh, M., Ghojazadeh, M., et al. (2013). Prevalence of Ambler class A β-lactamases and ampC expression in cephalosporin-resistant isolates of *Acinetobacter baumannii. Diagn. Microbiol. Infect. Dis.* 76, 330–334. doi: 10.1016/j.diagmicrobio

Robledo, I. E., Aquino, E. E., and Vazquez, G. J. (2011). Detection of the KPC gene in *Escherichia coli, Klebsiella pneumoniae, Pseudomonas aeruginosa,* and *Acinetobacter baumannii* during a PCR-based nosocomial surveillance study in Puerto Rico. *Antimicrob. Agents Chemother.* 55, 2968–2970. doi: 10.1128/AAC.01633-10

Ruan, Z., Chen, Y., Jiang, Y., Zhou, H., Zhou, Z., Fu, Y., et al. (2013). Wide distribution of CC92 carbapenem-resistant and OXA-23-producing *Acinetobacter baumannii* in multiple provinces of China. *Int. J. Antimicrob. Agents* 42, 322–328. doi: 10.1016/j.ijantimicag

Salimizand, H., Menbari, S., Ramazanzadeh, R., Khonsha, M., and Vahedi, M. S. (2014). DNA fingerprinting and antimicrobial susceptibility pattern of clinical and environmental *Acinetobacter baumannii* isolates: a multicentre study. *J. Chemother.* [Epub ahead of print].

Sistanizad, M., Kouchek, M., Miri, M., Goharani, R., Solouki, M., Ayazkhoo, L., et al. (2013). Carbapenem restriction and its effect on bacterial resistance in an intensive care unit of a teaching hospital. *Iran J. Pharm. Res.* 12, 503–509.

Turton, J. F., Gabriel, S. N., Valderrey, C., Kaufmann, M. E., and Pitt, T. L. (2007). Use of sequence-based typing and multiplex PCR to identify clonal lineages of outbreak strains of *Acinetobacter baumannii. Clin. Microbiol. Infect.* 13, 807–815. doi: 10.1111/j.1469-0691.2007.01759.x

Turton, J. F., Ward, M. E., Woodford, N., Kaufmann, M. E., Pike, R., Livermore, D. M., et al. (2006). The role of ISAba1 in expression of OXA carbapenemase genes in *Acinetobacter baumannii. FEMS Microbiol. Lett.* 258, 72–77. doi: 10.1111/j.1574-6968.2006.00195.x

Woodford, N., Ellington, M. J., Coelho, J. M., Turton, J. F., Ward, M. E., Brown, S., et al. (2006). Multiplex PCR for genes encoding prevalent OXA carbapenemases in *Acinetobacter* spp. *Int. J. Antimicrob. Agents* 27, 351–353. doi: 10.1016/j.ijantimicag.2006.01.004

Wu, K., Wang, F., Sun, J., Wang, Q., Chen, Q., Yu, S., et al. (2012). Class 1 integron gene cassettes in multidrug-resistant Gram-negative bacteria in southern China. *Int. J. Antimicrob. Agents* 40, 264–267. doi: 10.1016/j. ijantimicag

Zhang, D., Xia, J., Xu, Y., Gong, M., Zhou, Y., Xie, L., et al. (2014). Biological features of biofilm-forming ability of *Acinetobacter baumannii* strains derived from 121 elderly patients with hospital-acquired pneumonia. *Clin. Exp. Med.* [Epub ahead of print].

Zowawi, H. M., Sartor, A. L., Sidjabat, H. E., Balkhy, H. H., Walsh, T. R., Al Johani, S. M., et al. (2015). Molecular epidemiology of carbapenem resistant *Acinetobacter baumannii* in the gulf cooperation council states. Dominance of OXA-23-type producers. *J. Clin. Microbiol.* 53, 896–903. doi: 10.1128/JCM.02784-14

17

Dissemination of Novel Antimicrobial Resistance Mechanisms through the Insertion Sequence Mediated Spread of Metabolic Genes

*Leonardo Furi [1,2], Richard Haigh [1], Zaaima J. H. Al Jabri [1], Ian Morrissey [3], Hong-Yu Ou [4], Ricardo León-Sampedro [5,6], Jose L. Martinez [7,8], Teresa M. Coque [5,6,8] and Marco R. Oggioni [1,2]**

[1] Department of Genetics, University of Leicester, Leicester, UK, [2] Dipartimento di Biotecnologie Mediche, Universita di Siena, Siena, Italy, [3] IHMA Europe Sàrl, Epalinges, Switzerland, [4] State Key Laboratory for Microbial Metabolism and School of Life Sciences and Biotechnology, Shanghai Jiaotong University, Shanghai, China, [5] Departamento de Microbiología, Instituto Ramón y Cajal de Investigación Sanitaria, Hospital Universitario Ramón y Cajal, Madrid, Spain, [6] Centro de Investigación Biomédica en Red de Epidemiología y Salud Pública (CIBERESP), Spain, [7] Departamento de Biotecnología Microbiana, Centro Nacional de Biotecnología, Consejo Superior de Investigaciones Científicas, Madrid, Spain, [8] Unidad de Resistencia a Antibióticos y Virulencia Bacteriana (RYC-Consejo Superior de Investigaciones Científicas), Madrid, Spain

***Correspondence:**
Marco R. Oggioni
mro5@leicester.ac.uk

The widely used biocide triclosan selectively targets FabI, the NADH-dependent trans-2-enoyl-acyl carrier protein (ACP) reductase, which is also an important target for the development of narrow spectrum antibiotics. The analysis of triclosan resistant *Staphylococcus aureus* isolates had previously shown that in about half of the strains, the mechanism of triclosan resistance consists on the heterologous duplication of the triclosan target gene due to the acquisition of an additional *fabI* allele derived from *Staphylococcus haemolyticus* (*sh-fabI*). In the current work, the genomic sequencing of 10 of these strains allowed the characterization of two novel composite transposons TnSha1 and TnSha2 involved in the spread of *sh-fabI*. TnSha1 harbors one copy of IS1272, whereas TnSha2 is a 11.7 kb plasmid carrying TnSha1 present either as plasmid or in an integrated form generally flanked by two IS1272 elements. The target and mechanism of integration for IS1272 and TnSha1 are novel and include targeting of DNA secondary structures, generation of blunt-end deletions of the stem-loop and absence of target duplication. Database analyses showed widespread occurrence of these two elements in chromosomes and plasmids, with TnSha1 mainly in *S. aureus* and with TnSha2 mainly in *S. haemolyticus* and *S. epidermidis*. The acquisition of resistance by means of an insertion sequence-based mobilization and consequent duplication of drug-target metabolic genes, as observed here for *sh-fabI*, is highly reminiscent of the situation with the *ileS2* gene conferring mupirocin resistance, and the *dfrA* and *dfrG* genes conferring trimethoprim resistance both of which are mobilized by IS257. These three examples, which show similar mechanisms and levels of spread of metabolic genes linked to IS elements, highlight the importance of this genetic strategy for recruitment and rapid distribution of novel resistance mechanisms in staphylococci.

Keywords: insertion sequence, transposon, antimicrobial drug resistance, metabolism, resistance risk, fabI, IS5 family, IS1182 family

INTRODUCTION

The NADH-dependent *trans*-2-enoyl-acyl carrier protein (ACP) reductase FabI is one of the highly conserved enzymes of the bacterial fatty-acids biosynthesis. The FabI enzyme has been recognized as a novel and promising candidate drug target (Payne et al., 2001; Lu and Tonge, 2008) given the absence of a eukaryotic orthologue and its essential role in the growth of bacterial cells (Heath et al., 2001; Ji et al., 2004); a concept recently challenged by the observation that some bacteria do not require biosynthesis of fatty acids during infection of the host (Brinster et al., 2009). Due to the interest in using FabI as a drug target concerns have been raised about the large scale use of the biocide triclosan which targets the active site of FabI (Schweizer, 2001; Hijazi et al., 2016); these concerns also encompass the wider risk that biocide use *per se* may have on antimicrobial drug resistance (Oggioni et al., 2012, 2013, 2015; Coelho et al., 2013; Maillard et al., 2013; Morrissey et al., 2014). In the specific case considered here the concern is due to the fact that resistance to triclosan is in most bacterial species mediated by mutation of the promoter region or coding sequence of *fabI* (Heath et al., 1999; Slater-Radosti et al., 2001; Ciusa et al., 2012; Oggioni et al., 2012; Grandgirard et al., 2015). In *Staphylococcus aureus* about half of resistant isolates have a novel type of triclosan resistance mechanism which is based on the presence of an alternative copy of *fabI* derived from *Staphylococcus haemolyticus* (*sh-fabI*) (Ciusa et al., 2012). In previous work, we reported that the *sh-fabI* gene appears to be part of a 3022 bp transposable element most probably mobilized by a single copy of the insertion sequence (IS) 1272 (Ciusa et al., 2012). IS1272 was originally identified in *S. haemolyticus* during investigation of homology matches to a truncated IS which is part of the *mec* cassette (Archer et al., 1994, 1996; Tonouchi et al., 1994). IS1272 is part of the IS1182 family of insertion sequences (Siguier et al., 2015) and apart from that truncated version found in the *mec* element is absent from *S. aureus*.

It has long been known that IS elements are able to transpose and/or change the expression of nearby or neighboring genes, that they are significantly involved in plasmid and chromosomal recombination, and that they provide the basic structure of many transposons and mobile resistance elements (Siguier et al., 2014, 2015). Soon after the discovery of ISs, it was observed that IS elements were able to transpose resistance genes between replicons (Hedges and Jacob, 1974; Barth et al., 1976). Most of the attention in the more recent sequence based work on antimicrobial resistance has still focussed on acquired "resistance"-genes which confer resistance either by target modification, inactivation of the antimicrobial compound or efflux; this is in part due to the difficulty of defining housekeeping genes identified in high throughput or metagenomic datasets as true "resistance" genes (Martinez et al., 2015). All of this "resistance"-gene oriented work has therefore overshadowed other mechanisms of resistance including those based on heterodiploidy for a metabolic gene (either resistant or susceptible alleles) which, especially in staphylococci, appears to be a mechanism with low fitness cost (Andersson, 2006; Oggioni et al., 2012). The classic example of such heterodiploidy is the presence of an additional copy of a dihydrofolate reductase (*dhfA*) gene on conjugative elements, which thereby confers resistance to trimethoprim; these include the plasmid located *dhfA* gene which is transposed in *E. coli* by Tn7 (Barth et al., 1976), the Tn4003 transposon of *S. aureus* where the *dfrA* gene is mobilized by IS257 (Needham et al., 1995), and the more recently discovered transposable unit that comprises *dfrG* and IS*256* which is located on the Tn*5801* of different species (León-Sampedro et al., 2016). A more recent example in *S. aureus* involves resistance to mupirocin (pseudomonic acid), a potent inhibitor of the isoleucyl tRNA synthetase; this is conferred by an additional plasmid-encoded *mupA/ileS2* gene, which is again mobilized by IS257 (Gilbart et al., 1993; Woodford et al., 1998). With mupirocin, a disinfectant utilized for skin decontamination of MRSA staphylococci, this increased occurrence of *mupA/ileS* genes in clinical settings where mupirocin was used for decolonization has led to changes in disinfection policies (Hetem and Bonten, 2013). Even if the presence of additional *mupA/ileS* genes were found to confer a fitness defect to *S. aureus*, recent modeling experiments have predicted long-term increases in the prevalence of mupirocin-resistant phenotypes (MupR) given the "universal" use of mupirocin (Deeny et al., 2015).

In order to investigate the nature, sequence conservation, epidemiological distribution, and target site specificity of the 3022 bp transposable element carrying the *sh-fabI* gene we have sequenced the genomes of a series of triclosan resistant *S. aureus* isolates (Ciusa et al., 2012) and compared these data to the vast database of published genomes. The aims of this study were to conduct an in depth characterization of the *sh-fabI* carrying element, and to put this work into context with the other resistance mechanisms in *S. aureus* also based on IS mobilization of metabolic genes, which, in this species, appears to be a highly flexible means for the recruitment and rapid spread of novel resistance traits.

MATERIALS AND METHODS

Bacterial Strains

Sixty-five *S. aureus* strains with reduced susceptibility to triclosan were previously identified by performing standard MIC and MBC assays upon a collection of 1602 clinical isolates (Ciusa et al., 2012; Furi et al., 2013; Grandgirard et al., 2015; Oggioni et al., 2015). Ten out of the 28 isolates carrying the *sh-fabI* gene were selected from this collection for this work (**Table 1**; Ciusa et al., 2012).

Whole Genome Sequencing and Bioinformatic Analysis

The entire genome of 10 *S. aureus* clinical isolates (**Table 1**) with reduced susceptibility to triclosan and positive for *sh-fabI* detection by PCR were sequenced as previously described (Ciusa et al., 2012; **Table 1**). Short reads were assembled using ABySS (British Columbia Cancer Agency, Vancouver, Canada; ver. 1.3.5), improved using the multi-reference based scaffolder MeDuSa (Bosi et al., 2015) and subsequent TnSha1 sequence identification was performed using NCBI's BLAST. 2.3.0+ (Altschul et al., 1997). The sequence of the prototype TnSha1 element corresponds to position 3908 to 887 of GenBank

TABLE 1 | Relevant information of the ten sequenced *S. aureus* isolates (Ciusa et al., 2012).

Strain	TnSha1 insertion site	MLST	Country	Year	MIC (mg/L)	MBC (mg/L)	MRSA
QBR-102278-1091	D	12	Japan	2002	4	32	MSSA
QBR-102278-1107	G	1	Australia	2002	4	32	MSSA
QBR-102278-1203	E	1	France	2002	2	16	MSSA
QBR-102278-1619	A	8	Spain	2002	4	32	MSSA
QBR-102278-2092	A2	8	Canada	2003	4	32	MSSA
QBR-102278-2210	F	83	Mexico	2003	1	32	MSSA
QBR-102278-2351	A2, and B	3	Brazil	2003	8	32	MSSA
QBR-102278-2365	C	8	Brazil	2003	2	32	MSSA
QBR-102278-2376	C	8	Argentina	2003	4	32	MSSA
QBR-102278-2605	A2	8	Japan	2003	32	64	MRSA

accession JQ712986 relative to *S. aureus* strain QBR-102278-1619. No complete sequence of TnSha2 is present in existing complete genomes in GenBank. One of the plasmid versions of TnSha2, which shows a contig break within *fabI*, corresponds to GenBank accession JCAZ01000023 of *S. aureus* strain M0227 (adzpz-supercont1.20.C23). Genbank BLAST searches for TnSha1 and TnSha2 elements were last accessed in December 2015. DNA secondary structures have been predicted by means of the RNAstructure Web Server (Reuter and Mathews, 2010). TnSha1 targets A through G have been confirmed as gene terminator loops by direct visualization of RNA-seq alignment data retrieved from the NCBI Sequence Read Archive (SRA; http://www.ncbi.nlm.nih.gov/Traces/sra/). More precisely Illumina HiSeq data which were previously generated by sequencing of whole RNA (rRNA depleted) extracted from *S. aureus* ATCC51811 (SRA Experiment ID: ERP005246; SRA Sample ID: ERS421566) (Fagerlund et al., 2014) were aligned to genomes of *S. aureus* COL, GenBank accession no. CP000046, and *S. aureus* MW2, GenBank accession no. BA000033, using BWA-MEM (Li, 2014). The bootstrap maximum likelihood tree was obtained using MEGA6 with default parameters. Sequence Types (ST) were defined using the Multilocus Sequence Typing (MLST) web-based service of the Center for Genomic Epidemiology (Lyngby, Denmark; MLST allele sequence and profile data were obtained from: http://pubmlst.org/; Larsen et al., 2012). Contigs containing the TnSha2 element were screened to identify plasmid sequences by means of the PlasmidFinder web-based service of the Center for Genomic Epidemiology (Ver. 1.3; Lyngby, Denmark).

Molecular Analysis of Potential Transfer Intermediates

Genomic DNA was extracted from *S. aureus* strains in the exponential phase of growth using the High Pure PCR template preparation kit (Roche Diagnostics, Germany). Prior to DNA purification the bacterial strains were grown in MHB (Muller Hinton Broth; Beckton Dickinson) in the presence of sub-MIC concentrations of triclosan (2 mg/L for strain QBR-102278-1619 and 4 mg/L for QBR-102278-2351) or following Mitomycin C induction (1 μg/L) (triclosan PHR1338; Sigma-Aldrich) (Ciusa et al., 2012; Oggioni et al., 2012). Negative controls grown in MHB medium only were also used. A FAM labeled TaqMan probe was designed to detect the circular form of the IS1272 and the composite transposon TnSha1 when both were excised from the chromosome (**Table 2**). Detection of strains carrying TnSha1 in either the A or B integration sites via real-time PCR amplification was performed in a LightCycler 480 system (Roche Diagnostics, Germany) using primers annealing between the integration site and the transposon (**Table 2**; Oggioni et al., 2002; Isola et al., 2005; Yesilkaya et al., 2006). Primers LF_30 and LF_31 were used to detect the presence of bacteria with a transposon free integration site A (**Table 2**).

RESULTS

We have recently reported that a 3022 bp chromosomal element composed of the insertion sequence IS1272 and a *fabI* gene of *S. haemolyticus* (*sh-fabI*) confers resistance to triclosan in *S. aureus* (**Figure 1A**; GenBank accession no. JQ712986) (Ciusa et al., 2012). The potential for transposition of this unit is conferred by the presence of an alternative inverted repeat (IRL') in the *S. haemolyticus* chromosome upstream of *sh-fabI* with a high degree of similarity to the inverted repeats of IS1272 (**Figure 1B**). This functional unit, which is composed of the insertion sequence IS1272 and the *sh-fabI* gene, has now being renamed TnSha1 (**Figure 1A**). It should be noted that none of our genomes, nor any deposited sequence, match the originally deposited sequence of IS1272 (Genbank accession U35635), instead all of the hundreds of IS1272 copies show at least six SNPs (four of which are indels) with respect to U35635 (Archer et al., 1996). These differences mean that the updated IS1272 element encodes for a single transposase gene without a stop codon (**Figure 1A**).

In this work the sequence conservation, epidemiological distribution, and target site specificity of TnSha1 were analyzed across a panel of ten sequenced *S. aureus* strains previously reported to carry the *sh-fabI* gene (Ciusa et al., 2012). Sequencing has identified seven different integration sites within the *S. aureus* genomes which have been named A through G (**Figure 2**, **Table 1**). The discovery of seven different TnSha1 integration sites in just 10 strains suggests that there is little or no insertion site preference. Analysis of the intact target sites in the reference strains COL or MW2 showed that TnSha1 integration always

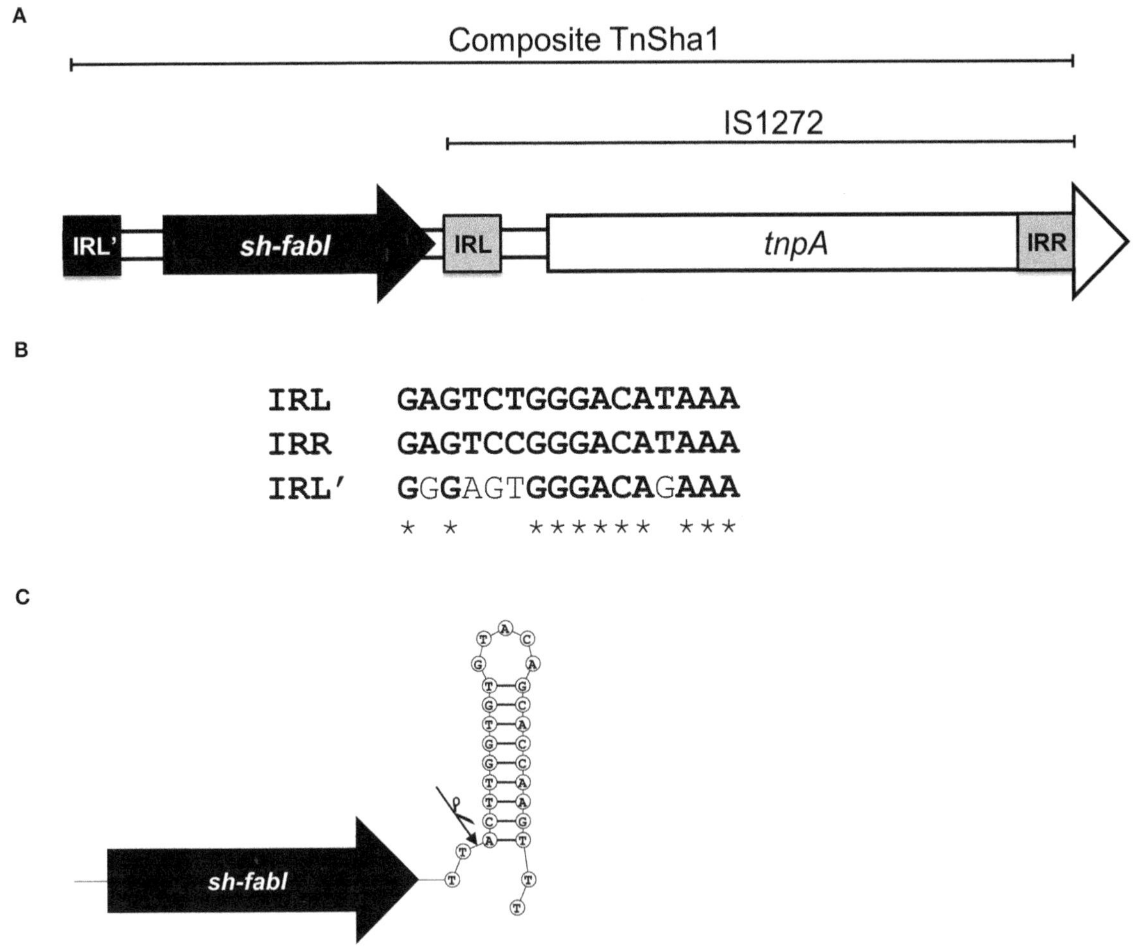

FIGURE 1 | Structure of the TnSha1 transposon. The schematic map of the composite transposon TnSha1 in which the *sh-fabI* gene (black) precedes the IS1272 element is shown in panel **(A)**. The updated IS1272 sequences contains six SNPs (four of which indels) with respect to U35635 and encodes for a single transposase gene without a stop codon within the IS element. The sequence of the prototype TnSha1 element corresponds to position 3908 to 887 of Genbank accession JQ712986 relative to strain QBR-102278-1619. Panel **(B)** shows the inverted repeat sequences of IS1272 (named IRL and IRR) aligned to the IRL' upstream of *sh-fabI* which shows 69% identity. Panel **(C)** indicates the hypothesized integration event of IS1272 downstream of *sh-fabI* to generate the prototype TnSha1 element in *S. haemolyticus*. The arrow indicates the insertion point at the base of the *sh-fabI* terminator hairpin.

TABLE 2 | List of primers.

Name	Position (GenBank ID)	Sequence
LF_26	Taqman probe for TnSha1 (JQ712986)	6FAM-TTCACTTATCCAAGA ACTTTATGTCCCGGA-BHQ-1
LF_27	IRL' of TnSha1 (JQ712986)	ATCCTTGCCGGGGTAAT ACAAC
LF_28	IRL of TnSha1 (JQ712986)	AAAGCGAGCCAACAATACG GAGTA
LF_29	IRR of TnSha1 (JQ712986)	TAGTAGCTCAACGAGCTGA AAATAATC
LF_30	Upstream region flanking the TnSha1 integration site A/A2 (NC_002951)	TTGATTTATTTCCCAGCCTATT CTTTTCA
LF_31	Downstream region flanking the TnSha1 integration site A/A2 (NC_002951)	AGGATGTCGATTTGATTTAT ATTTTTTGTACAT
LF_32	Downstream region flanking the TnSha1 integration site B (NC_002951)	ATCATTTCGTTTATATATAGCA GACATGATAGA

occurs into hairpin structures and, importantly, that the mode of insertion of TnSha1 always produces a partial deletion of the target sequence (**Figure 2B**). This mode of integration into the stem of secondary structures is novel to IS elements (Siguier et al., 2015). For one of the targets sequenced in this work there were two different transposition events which could be detected; these lead to deletions of different sizes in the same target site (A and A2; **Figure 2B**). All seven of the target sites identified by sequencing of our clinical isolates, in addition to other sites subsequently identified in published sequences, were composed of inverted repeat sequences and thus have the potential to form secondary structures with the repeats representing the stems of hairpins (**Figure 2A**). In each case insertion of TnSha1 produced a partial deletion of the hairpin. This type of cleavage of a secondary structure forming sequences appears to indicate that TnSha1 makes blunt and not staggered cuts, which would have resulted in target duplications. Hairpins A through E have been

A

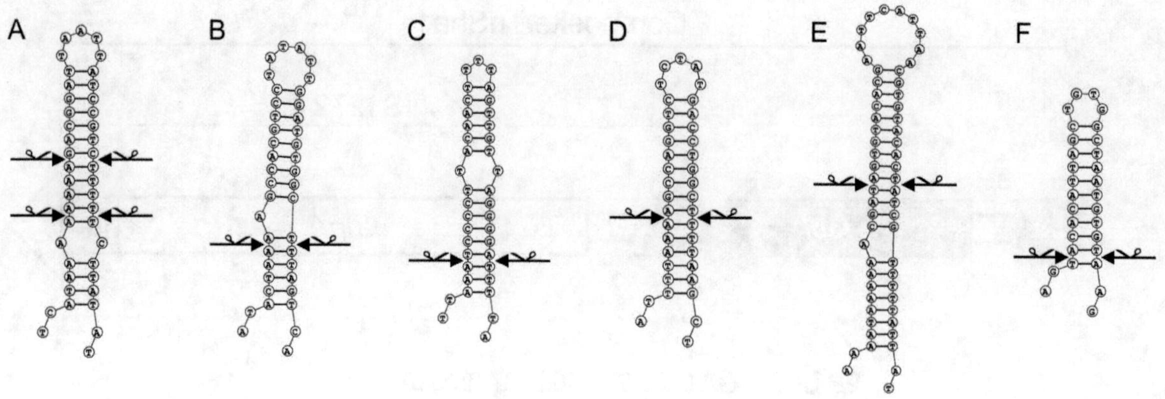

B

```
A)   S. aureus COL          149148 TTTTCATAAAAAAAA------gacggattaattatccgtc-------TTTTTCTTATATCTTT
A2)  S. aureus COL          149148 TTTTCATAAAA------aaaagacggattaattatccgtctttt-------TCTTATATCTTT
B)   S. aureus COL         2545016 CTGGAACGATACTAA--aagccacgtcctatattggatgtggct---TTAGTCATACTTATAT
C)   S. aureus COL         1097865 CTAAATTAATTTAAA-tccccttacaactttttagttgttagggga--TTTTATGCATGTGCTC
D)   S. aureus JKD6159     2563232 TTTAATGATTTTAAA----agccaggtctctatgacctggct-----TTTAAGCTTTTTCCTG
E)   S. aureus MW2         2638638 AAAATAAAAAACGAT-agtgtacacgaattcattaacgtgtacact-ATCGTTTTTATTATGA
F)   S. aureus MW2         1678951 TTCCTATTTTTAAGT----acacattagctgtggctaatgtgt----AAGAACCACTACATAA
G)   S. aureus MW2         2109277 AATAACAAAGAGAATgttattactataacaatagttataacagtaaggaatttttattaaGA

P)   S. haemol. p25-44        5401 TTAGGTTAAAAGAAA---ggagctaacctttagggttagctcc----TTTCTGTGTTCTTTAT
V)   S. aureus pTW20_1        13941 CTTAATAATAAAAAA-gtagaaccatttaaaattaatggttctac--TTTTTAACTATTACTA
W)   S. aureus pWGB745        22280 ATTTTACAAATAAAA---aagaaagtgattatatcactttctt----TCCTTAACTATACGAT
X)   S. aureus pSA1379        20646 TACTTAATAATAAAAaagtagaaccatttaaaattaatggttctactTTTTAACTATTACTAA
Y)   S. epid. pSAP108A        12204 TATTCAAAAAAAAATaatgtattcacgcctaaaggcgtgaatacattATTTTCTCTAAAAAGA
Z)   S. aureus pUSA03          1076 TTTCAAAAAAAAGAG---attatacacctaaaaaggtgtataat---CTTTACGGTTCTGTTG
```

C

```
LPIH6008    TAAATTTTCAAAAAAAAGAGGGGAGTGGGNNNNNNNNCCCGGACTCCTTTACGGTTCTGTTGCAAA
F77919      TAAATTTTCAAAAAAAAGAGGGGAGTGGGNNNNNNNNCCCGGACTCCTTTACGGTTCTGTTGCAAA
W21479      TAAATTTTCAAAAAAAAGAGGGGAGTGGGNNNNNNNNCCCGGACTCCTTTACGGTTCTGTTGCAAA
HOAG6084    TAAATTTTCAAAAAAAAG--GGGAGTGGGNNNNNNNNCCCGGACTC-TTTACGGTTCTGTTGCAAA

F70077      TAAATTTTCAAAAAAAAGA-GAGTCCGGGNNNNNNNNCCCACTCCCCTTTACGGTTCTGTTGCAAA
F12753      TAAATTTTCAAAAAAAAGA-GAGTCCGGGNNNNNNNNCCCACTCCCCTTTACGGTTCTGTTGCAAA
F12753      TAAATTTTCAAAAAAAAGA-GAGTCCGGGNNNNNNNNCCCACTCCC-TTTACGGTTCTGTTGCAAA

target      TAAATTTTCAAAAAAAAGAGATTATACACCTAAAAAGGTGTATAATCTTTACGGTTCTGTTGCAAA
            *****************                             *******************
```

FIGURE 2 | Insertion sites of TnSha1 in bacterial chromosomes and plasmids. The seven target sites found in the 10 *S. aureus* strains sequenced in this work have been named from A to G. Staphylococcal plasmid insertion sites are named P to Z. Panel **(A)** shows the hairpin structures formed by the transposon target sites and the scissors indicate the position of sequence breaks generated upon TnSha1 insertion. Secondary structure predictions were made using RNAstructure web-based software (Reuter and Mathews, 2010). Two independent insertions into target **(A)** have been identified and both cleavage sites are shown which in each case generate blunt-end structures in the stem of the hairpin. The sequences in both panels **(A,B)** refer to TnSha1-free target sites in the chromosomes (COL GenBank accession CP000046 and MW2 GenBank accession BA000033) and plasmids. The target sequences in panel **(B)** report in lower case the sequence deleted upon insertion and the inverted repeats underlined. Panel **(C)** shows evidence for multiple insertions of TnSha1 in target Z in the *repA* terminator of plasmids. The aligned sequences show at least four independent insertions in two opposite orientations of TnSha1 into the *repA* terminator in plasmid pUSA03 (target Z). In this panel TnSha1 sequences are shown in bold italics upper case, the deleted part of the target sequence in lower case, and the *repA* terminator hairpin underlined. Note that in panel **(C)** the upper four TnSha1 sequences are in one orientation while the lower three in the opposite orientation. Note that in panel **(C)** the upper four TnSha1 sequences are in one orientation while the lower three in the opposite orientation. The *S. aureus* strain names are given on the right.

confirmed as transcriptional terminators by mapping of RNAseq data (Fagerlund et al., 2014; **Figure 3**). When analysing the target sites of a further 63 complete TnSha1 elements from published *S. aureus* genomes, we identified an additional 12 target sites confirming a lack of any primary sequence specificity within the target. All of the aforementioned features of TnSha1 are consistent with the known behavior of the IS1272 element alone (Archer et al., 1996); i.e., such as the insertion of IS1272 which occurred into the terminator of *sh-fabI* to generate the prototype TnSha1 (**Figure 1C**), or as in the case of two target sites selected from IS1272 insertions in the genomes of *S. haemolyticus* or *S. warneri* (**Figure 4**). While most sequenced strains contained

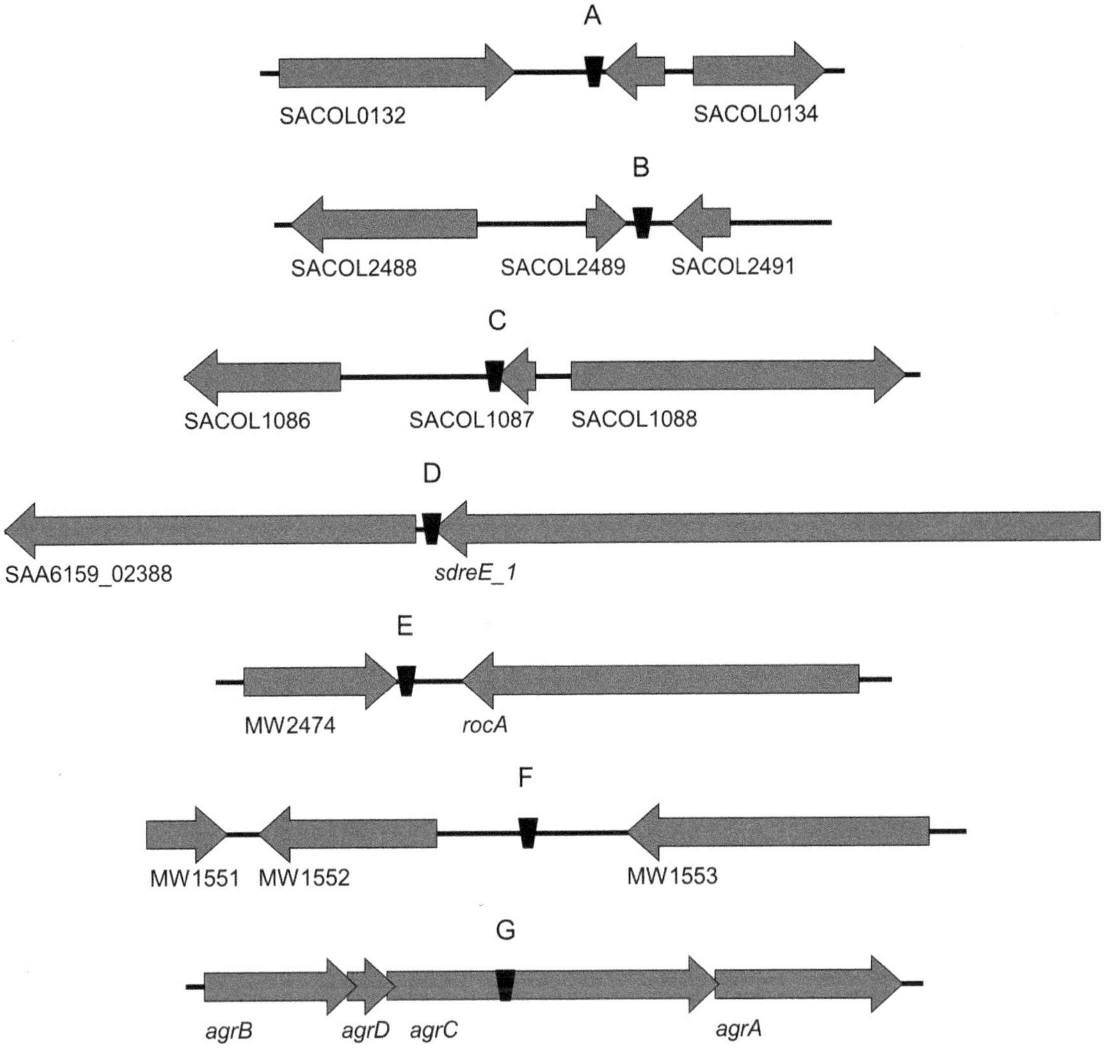

FIGURE 3 | The seven TnSha1 integration sites found in the sequenced *S. aureus* clinical isolates mapped onto annotated genomes. The *S. aureus* reference genomes included COL (GenBank accession no. CP000046; **A–C** insertion sites), JKD6159 (GenBank accession no. CP002114; **D** insertion site), and MW2 (GenBank accession no. BA000033; **E,F** insertion sites). 3/7 TnSha1 elements target predicted terminator hairpin structures **(C–E)**, 1/7 targets the *agrC* coding region **(G)**, and 3/7 target intergenic regions **(A,B,F)**. Hairpins **(A)** through **(E)** have been confirmed as transcriptional terminators by mapping RNAseq data.

just a single copy of TnShal in the chromosome, strain QBR-102278-2351 was an exception and carried two TnShal elements each in two distinct integration sites (**Figure 2**, **Table 1**).

The dynamics and epidemiology of transposition between strains was addressed by searching for other TnShal elements present in Genbank. Database searches identified 133 complete or partial TnShal elements in the genomes of the species *S. aureus*, *S. argenteus*, *S. haemolyticus*, *S. epidermidis*, *S. saprophyticus*, and *S. warneri*, and in a series of staphylococcal plasmids (**Tables 3–5**). Of this latter group the integration between *repA* and IS257 in pUSA03 (target Z in **Table 4**) where TnShal was inserted in two different orientations warrants particular attention (**Figure 2C**). Multi locus sequence typing (MLST) of the strains showed that TnShal was present in *S. aureus* in sequence type 8 ($n = 23$), ST5 ($n = 18$), ST239 ($n = 6$), and ST290 ($n = 6$) strains and in *S. epidermidis* in ST59 ($n = 10$) and ST2 ($n = 6$) strains (**Table 4**, **Figure 5**). Upon determination of the targets of these newly identified TnShal elements it was found that only seven out of the 66 complete TnShal elements were integrated into the same seven target sites, A to G, that we had previously identified (**Figure 2**, **Tables 1**, **4**). Alignment of our previous 10 TnShal elements with the seven new TnShal elements that were integrated at the same target sites showed clustering of the TnShal-SNPs (some elements differed by up to 11 SNPs from the consensus) together with their target sites (**Figure 6**). This pattern is consistent with the insertion of the element into a particular target followed by clonal spread. This scenario appears to be the rule in most cases except for target A2 and Z where the identical TnShal element is present in the same target but in strains of different ST (**Table 4**, **Figure 5**). To test for

A

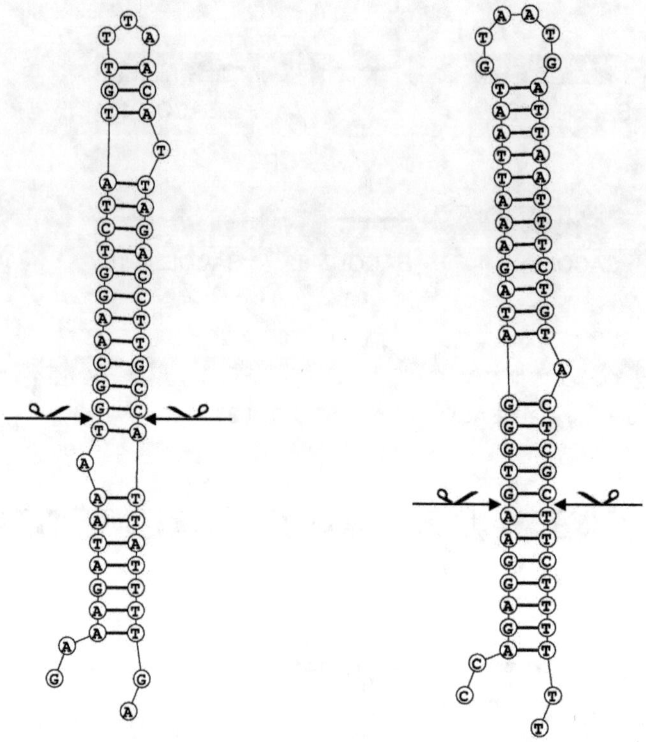

B

S. warneri VCU121 GAAAGATAAAT----ggcaaggtctatgtttaacattagaccttgcc-----ATTATTTTGAC

S. haemolyticus R1P1 TTCCAGAGGAAgtgggatagaaattaatgtaatgattaatttctgtactcgcTTCTTTTTTAT

FIGURE 4 | Analysis of target sites of IS1272 in *S. warneri* and *S. haemolyticus*. Secondary structure predictions of the IS1272 targets in the *S. warneri* strain SG1 (GenBank accession CP003668) are shown in the IS free reference strain VCU121 (GenBank accession NZ_AFEC01000030) (**A** left); and in the *S. haemolyticus* strain Sh29/312/L2 (GenBank accession CP011116) are shown in the IS free reference strain R1P1 (GenBank accession NZ_AJVA00000000) (**A** right). The exact blunt target cleavage sites are indicated by scissors. Panel **(B)** indicates the target sequences showing the inverted repeats (underlined) and deletions (lower case).

TABLE 3 | BLAST search of staphylococcal genomes for TnSha1 and TnSha2.

Species	Element	Chromosomal*	Plasmid**	Total	Total number of genomes	%
S. argenteus	TnSha1	1	1	2	6	33
	TnSha2	–	–	–		–
S. aureus	TnSha1	61	3	64	4127	1.5
	TnSha2	1	–	1		0.02
S. epidermidis	TnSha1	30	1	31	286	10.8
	TnSha2	19	–	19		6.6
S. haemolyticus	TnSha1	3	–	3	141	2.1
	TnSha2	10	–	10		7
S. warneri	TnSha1	2	–	2	7	28.6
	TnSha2	–	–	–		–
S. saprophyticus	TnSha1	1	–	1	5	20
	TnSha2	–	–	–		–
Total		128	5	133	4567	–

**The localization designated as chromosomal may also include plasmids present in genome contigs. **Only includes plasmids deposited as complete plasmids.*

TABLE 4 | List of strains carrying the TnSha1 element on the chromosome.

Species	GenBank ID	Strain name	Insertion site	MLST
S. argenteus	CCEM01000001	Sa_LBSA043	M	–
S. aureus	AIDT01000009	DR10	H	398
S. aureus	AUPV01000018	S100	H	398
S. aureus	CAVU010000033	S1805	E	80
S. aureus	CAWA010000053	S2396	E	1
S. aureus	CFPN01000018	USFL079	L	8
S. aureus	CGGX01000004	USFL145	A	8
S. aureus	CIAK01000002	USFL234	J	8
S. aureus	CIGG01000001	USFL129	C	8
S. aureus	CP003045	71193	H	398
S. aureus	CSDA01000011	USFL101	L	8
S. aureus	CSDP01000002	USFL046	J	8
S. aureus	CSEJ01000001	USFL123	C	8
S. aureus	CSEV01000003	USFL189	A2	828
S. aureus	CSFL01000012	USFL078	L	8
S. aureus	CSHX01000003	USFL190	A2	828
S. aureus	CSJZ01000023	USFL050	nd	8 slv*
S. aureus	CTWU01000023	M705	K	239
S. aureus	CTWW01000088	H211	K	239
S. aureus	CTXO01000001	M1229	K	239
S. aureus	CTXZ01000061	H216	K	239
S. aureus	CTYD01000081	H202	K	239
S. aureus	CVOP01000018	SH3	M	630
S. aureus	CVOU01000018	SH1	M	630
S. aureus	CVRW01000028	M170	K	239
S. aureus	JBFG01000011	KINW6058	I	5
S. aureus	JBGB01000006	FVRH6130	nd	8
S. aureus	JBGS01000014	GGMC6026	I	5
S. aureus	JBLE01000006	SCOA6048	H	8
S. aureus	JBMX01000012	SCOA6012	H	8
S. aureus	JDLS01000019	T78544	I	5
S. aureus	JDOK01000060	F35307	nd	5
S. aureus	JDOU01000017	H27862	I	5
S. aureus	JDOV01000017	H27872	I	5
S. aureus	JDPI01000020	H67656	I	5
S. aureus	JECS01000014	T34011	I	5
S. aureus	JEDV01000017	H64967	I	5
S. aureus	JETJ01000039	T22051	nd	5
S. aureus	JGNE01000030	W41757	nd	5
S. aureus	JHTT01000046	CO-86	L	8
S. aureus	JIXI01000011	PA57	H	398
S. aureus	JIYX01000004	C5086	H	398 slv
S. aureus	JJDE01000001	122	H	398
S. aureus	JURB01000090	84_SAUR	F	5
S. aureus	JUTG01000033	78_SAUR	F	5
S. aureus	JVUC01000150	1315_SAUR	nd	15
S. aureus	JZAL01000002	LHSKBClinical	C	8
S. aureus	LAMP01000014	99-06	L	8
S. aureus	LAMS01000038	99-48	L	8
S. epidermidis	ACHE01000003	BCM-HMP0060	–	59

(Continued)

TABLE 4 | Continued

Species	GenBank ID	Strain name	Insertion site	MLST
S. epidermidis	AGUC01000094	14.1.R1.SE	–	–
S. epidermidis	AHLC01000060	VCU120	–	22
S. epidermidis	AHLF01000017	VCU125	–	384
S. epidermidis	AKGM01000041	NIHLM067	–	333
S. epidermidis	AKGN01000056	NIHLM061	–	332
S. epidermidis	AKGW01000049	NIHLM020	–	7
S. epidermidis	APHT01000038	528m	–	2
S. epidermidis	APHU01000037	41tr	–	2
S. epidermidis	ARWU01000031	UC7032	–	595
S. epidermidis	JUMV01000141	938_SEPI	–	59
S. epidermidis	JUNI01000550	926_SEPI	–	59
S. epidermidis	JUVK01000069	73_SEPI	–	16
S. epidermidis	JUYJ01000034	655_SEPI	–	59
S. epidermidis	JUYK01000190	654_SEPI	–	59
S. epidermidis	JVQK01000079	196_SEPI	–	59
S. epidermidis	JVSC01000077	154_SEPI	–	59
S. epidermidis	JVSC01000077	154_SEPI	–	59
S. epidermidis	JVSZ01000046	134_SEPI	–	59
S. epidermidis	JVTV01000122	1321_SEPI	–	59
S. epidermidis	JWBR01000039	114_SEPI	–	–
S. epidermidis	JWCR01000091	1115_SEPI	–	88
S. epidermidis	JWEH01000024	1063_SEPI	–	2
S. epidermidis	JWFU01000070	1024_SEPI	–	2
S. epidermidis	JWFV01000104	1023_SEPI	–	2
S. epidermidis	JZUK01000044	NGS-ED-1107	–	2
S. epidermidis	JZUL01000024	NGS-ED-1109	–	439
S. haemolyticus	CP011116	Sh29/312/L2	–	–
S. haemolyticus	CUEZ01000014	CN1197	–	–
S. saprophyticus	JXBG01000015	SU8	–	–
S. warneri	CANQ01000015	A487	–	–
S. warneri	JPOW01000002	NGS-ED-1001	–	–

**slv = single locus variant; targets A–G from our genome sequences; targets H–Z from Genbank, nd not determined.*

TnSha1 mobilization we used PCR with a set of divergent primers on TnSha1, another set on IS1272, and a third set of primers targeting the "A" insertion site. Unfortunately, no excision events producing a circular intermediate could be observed when two *S. aureus* strains were tested with or without the presence of triclosan in the culture medium, and after induction with mitomycin C.

Within the multiple entries retrieved from Genbank a subgroup showed identical sequences upstream of TnSha1. Investigation of this genomic region showed that in 30 of the strains the TnSha1 element forms part of a larger element (named TnSha2), which in almost all cases is a plasmid, but which can also be found integrated into the genome where it is generally flanked by a second copy of IS1272 (**Tables 4, 5**). The prototype sequence of TnSha2 (*S. aureus* JCAZ01000023) is an 11.7 kb plasmid in which TnSha1 is inserted into the terminator of *traQ* gene of an 8.7 kb replicon (without fabI and IS1272)

TABLE 5 | List of strains carrying the TnSha1 element on a plasmid.

Species	GenBank ID	Strain name	Plasmid	Insertion site	MLST
S. argenteus	FR821778	MSHR1132	pST75	X2	–
S. aureus	AHVE01000013	CIG1165	Plasmid	X	5
S. aureus	CP012121	USA300_2014.C02	pUSA300_2014.C02	X	8
S. aureus	CP012594	HOU1444-VR	pVR-MSSA_01	W	5
S. aureus	CSBT01000019	USFL338	Plasmid	V	8
S. aureus	JBGG01000015	LAMC6115	Plasmid	Zrc	8
S. aureus	JBJX01000010	SJUD6114	Plasmid	X	8
S. aureus	JBRD01000004	AMMC6015	Plasmid	Z	8
S. aureus	JBSH01000015	HOAG6084	Plasmid	Z	8
S. aureus	JCAZ01000023	M0227	TnSha2	P	5
S. aureus	JDXZ01000007	LPIH6008	Plasmid	Z	8
S. aureus	JEEN01000028	F77919	Plasmid	Z	5
S. aureus	JEOI01000012	W21479	Plasmid	Z	5
S. aureus	JEQM01000030	F70077	Plasmid	Zrc	5
S. aureus	JGFR01000050	F12753	Plasmid	Zrc	5
S. aureus	JGIT01000028	T28653	Plasmid	X	8
S. aureus	JICL01000103	880	pHMPREF1625_3	W	–
S. aureus	JVUX01000076	1299_SAUR	Plasmid	Z	2250
S. epidermidis	GQ900465	SK6536	pSAP110A	Y	–
S. epidermidis	JUPD01000185	890_SEPI	TnSha2	P	2
S. epidermidis	JUUP01000151	749_SEPI	TnSha2	P	2
S. epidermidis	JUYI01000150	656_SEPI	TnSha2	P	2
S. epidermidis	JUZG01000070	634_SEPI	TnSha2	P	2
S. epidermidis	JUZR01000081	623_SEPI	TnSha2	P	2
S. epidermidis	JVBT01000060	568_SEPI	TnSha2	P	2
S. epidermidis	JVHA01000086	439_SEPI	TnSha2	P	2
S. epidermidis	JVHR01000056	422.rep2_SEPI	TnSha2	P	2
S. epidermidis	JVHS01000079	422.rep1_SEPI	TnSha2	P	2
S. epidermidis	JVKK01000090	354_SEPI	TnSha2	P	2
S. epidermidis	JVXB01000116	1249_SEPI	TnSha2	P	2
S. epidermidis	JWBN01000071	1143_SEPI	TnSha2	P	2
S. epidermidis	JWDK01000004	1088_SEPI	TnSha2	P	2
S. epidermidis	JWFS01000043	1026_SEPI	TnSha2	P	2
S. epidermidis	JWFW01000102	1022_SEPI	TnSha2	P	2
S. epidermidis	JWFX01000149	1021.rep2_SEPI	TnSha2	P	2
S. epidermidis	JWFY01000144	1021.rep1_SEPI	TnSha2	P	2
S. epidermidis	JWGH01000004	1013_SEPI	TnSha2	P	2
S. epidermidis	JWGL01000037	101_SEPI	TnSha2	P	2
S. haemolyticus	CUCK01000025	25-38	TnSha2 in chromosome*	P	29
S. haemolyticus	CUCN01000051	25-60	TnSha2	P	–
S. haemolyticus	CUCZ01000051	51-13	TnSha2	P	3
S. haemolyticus	CUDB01000007	51-11	TnSha2 in chromosome	P	3
S. haemolyticus	CUDQ01000054	51-33	TnSha2	P	3
S. haemolyticus	CUDW01000051	51-41	TnSha2	P	2
S. haemolyticus	CUDY01000029	51-43	TnSha2 in chromosome	P	2
S. haemolyticus	CUFB01000040	51-15	TnSha2	P	3
S. haemolyticus	CUGD01000039	127925	TnSha2 in chromosome	P	3
S. haemolyticus	CUGG01000054	113101	TnSha2	P	3

**Chromosomally integrated TnSha2 elements are reported for clarity in this plasmid table.*

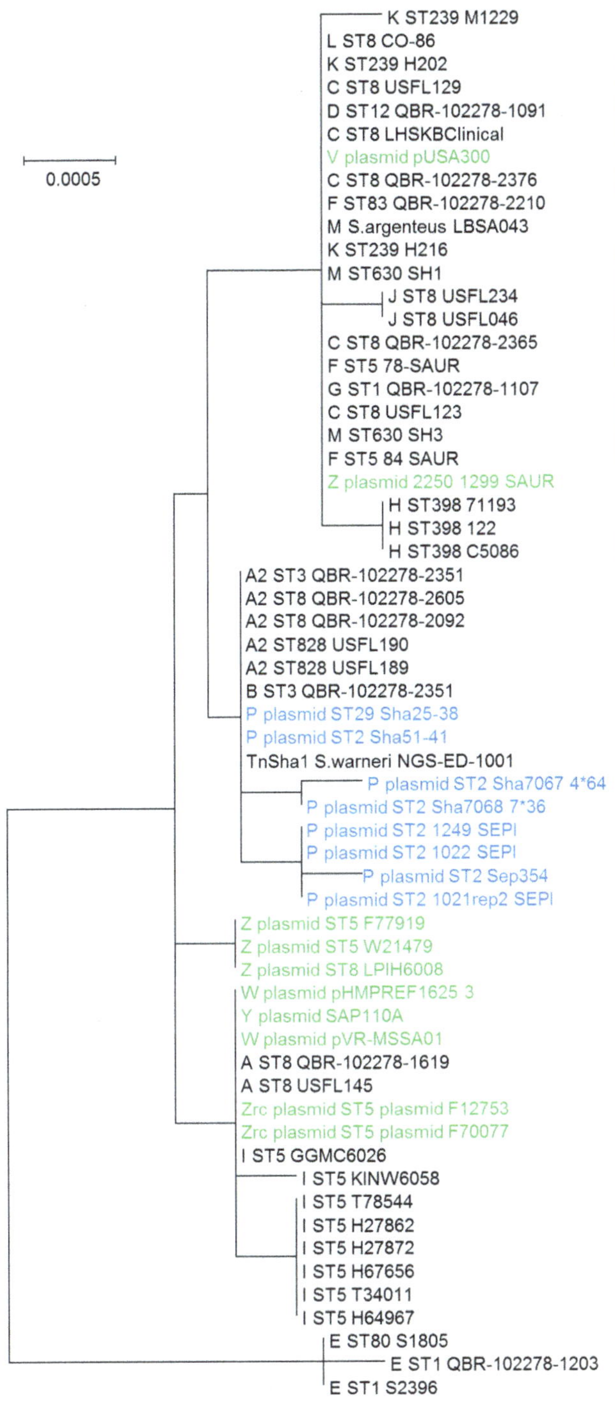

FIGURE 5 | Phylogram of the TnSha1 and TnSha2 elements. The tree contains 57 TnSha1 sequences, 10 of which sequenced in this work (QBR-102278 sequences) and a subset of eight TnSha2 sequences (only the homologous part was used for tree building purposes). Each TnSha1 element is identified by its target **(A–Z)**, followed by the sequence type (ST), and the strain name. TnSha1 elements on *S. aureus* plasmids are colored in green while TnSha2 elements are in blue. For GenBank accession numbers please see **Tables 1, 4**. The bootstrap maximum likelihood tree was obtained using MEGA6 with default parameters. The scale bar is shown in the upper left part. It should be noted that the tree is built based on only few SNPs, which distinguish the different elements.

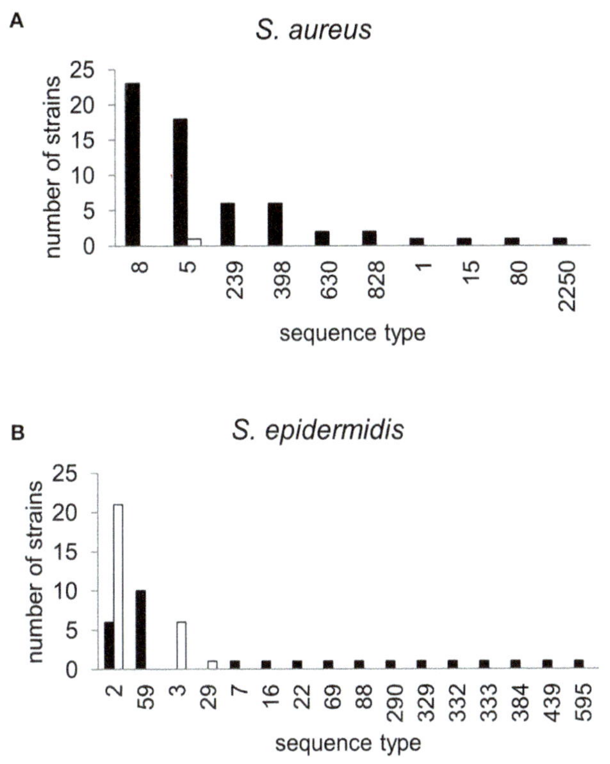

FIGURE 6 | MLST profiles of staphylococcal strains carrying TnSha1 and TnSha2. The sequence type (ST) of 62 *S. aureus* **(A)** and 47 *S. epidermidis* strains **(B)** which carry the *sh-fabI* elements TnSha1 (filled bars) and TnSha2 (open bars) are shown. Data are from staphylococcal genomes accessed in GenBank in December 2015 and sequence types were defined using the MLST web-service of the Center for Genomic Epidemiology (Larsen et al., 2012).

present in *S. haemolyticus* (CUCL01000044), but also in *S. aureus* (JJAQ010000025; **Figure 7**). The *rep* of the plasmid belongs to the Rep_2 family and the DNA polymerase, recombinase and mobilization protein show high sequence identity to those of plasmid p-12228-03 of the *S. epidermidis* reference strain ATCC 12228 (**Figure 7**; Zhang et al., 2003; Lanza et al., 2015). ThSha2 was detected in the genomes of clonally diverse *S. haemolyticus*, and *S. epidermidis* (ST3 $n = 6$, and ST2 $n = 21$, respectively), but only in a single *S. aureus* (ST5) genome, thereby reflecting different dissemination routes for different species (**Table 5**). Upon investigating the origin of deposited genome-strains it was observed that the *S. epidermidis* ST2 strains are of hospital origin (Schoenfelder et al., 2010; Roach et al., 2015) and that all of the *S. haemolyticus* strains with TnSha2 and ST3 that we found in this study are clinical isolates (**Table 5**). Interestingly the database search for IS1272 alone detected no intact elements in *S. aureus*, but did show that about 1/10 of *S. epidermidis* isolates and the majority of *S. haemolyticus* isolates carried IS1272 thereby ensuring that the contigs with the 11.7 kb plasmid in these species are always interrupted in the IS element.

In a phylogenetic tree based on the TnSha1-part of TnSha2, the two elements group into separate branches due to a series of SNPs both in *fabI* and the IS1272 element (**Figure 5**). In

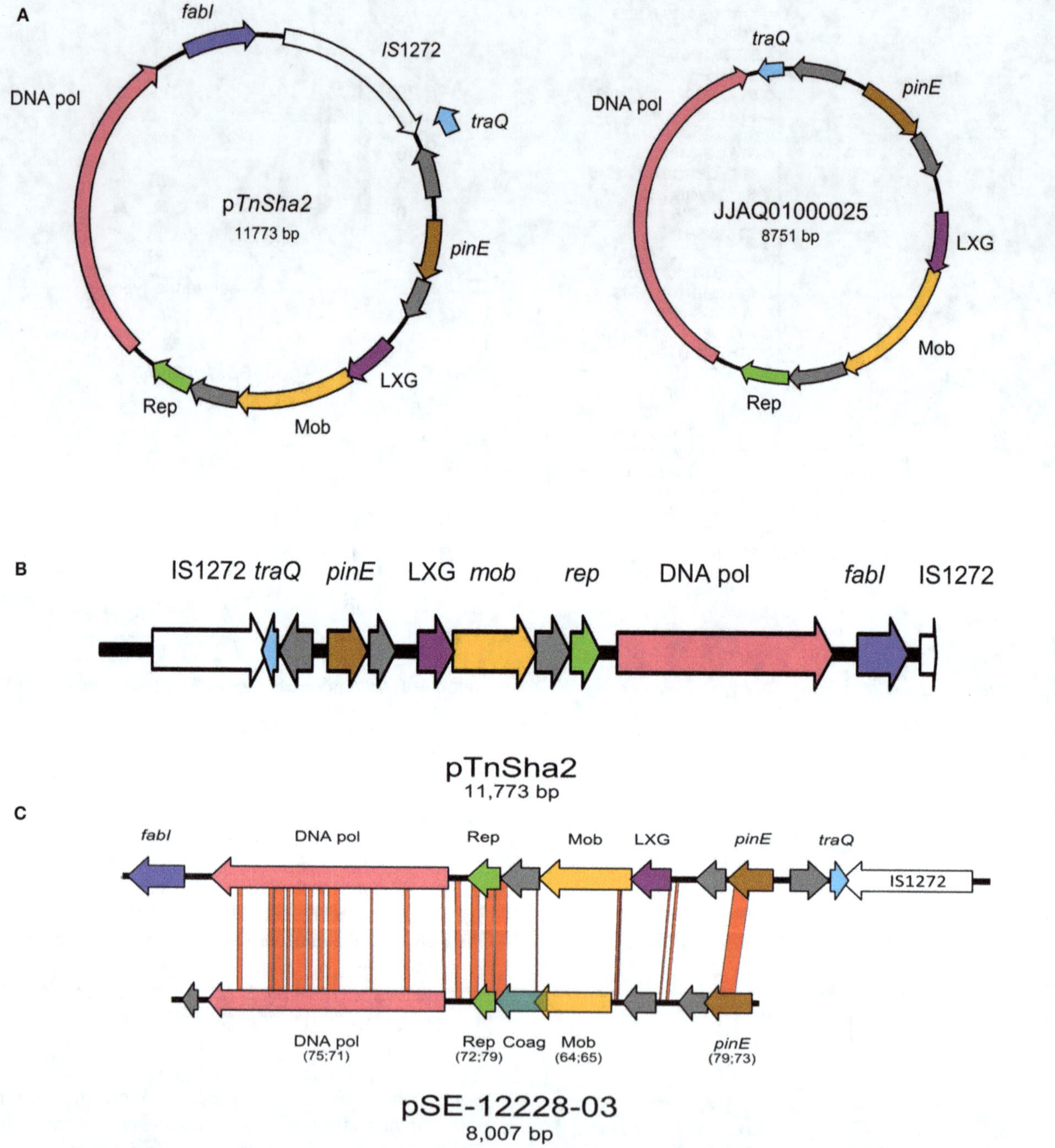

FIGURE 7 | Schematic map of TnSha2 elements. The circular maps of the prototype 11.7 kb plasmid sequence of TnSha2 (*S. aureus* strain M0227, GenBank accession JCAZ01000023) and the target 8.7 kb TnSha1-free plasmid (*S. haemolyticus* CUCL01000044 or *S. aureus* JJAQ010000025) are shown in panel **(A)**. The map of TnSha2 integrated into the *S. haemolyticus* chromosome at a position occupied by a single IS1272 in the reference strain is shown in panel **(B)**. In three occasions the resulting chromosomal island is bordered by two IS1272 elements (CUCK01000025 at position 590129 of strain JCSC1435, CUDY01000029 at 1828271, and CUGD01000039 at 2443315) wile in one case only by the "left" IS1272 element (CUDB01000007 at 1322312). Comparison between pTnSha2 element (*S. aureus* JCAZ01000023) and pSE-12228-03 (*S. epidermidis* NC_005006) is shown in panel **(C)**. Numbers between brackets indicate coverage percentage and identity percentage respectively. The genes encode for: a putative DNA-directed DNA polymerase (DNA pol), a replication protein involved in a theta-type replication mechanism (Rep), a putative mobilization protein (Mob), a domain of a group of polymorphic toxin proteins (LXG), a Site-specific DNA recombinase related to the DNA invertase Pin (PinE), and a conjugal transfer pilin chaperone (TraQ).

four cases the whole 11.7 kb TnSha2 plasmid appears to be integrated into the chromosome of *S. haemolyticus* in the same position that IS1272 elements are found in reference strain JCSC1435 resulting in a chromosomal island flanked in three cases by two IS1272 copies (CUCK01000025 at position 590.150 of JCSC1435, CUDB01000007 at 1.322.311, CUDY01000029 at 1.828.279, and CUGD01000039 at 2.443.339; **Figure 7**; Cavanagh et al., 2014).

In the original publication, which described *sh-fabI* in *S. aureus* QBR-102278-1619 (Ciusa et al., 2012), the TnSha1 element was reported to carry a *sh-fabI* gene with an A577T SNP with respect to the sequence of the core genome *fabI* of *S. haemolyticus* (**Table 6**). Since, this SNP had not previously been associated with triclosan resistance of *sa-fabI* in *S. aureus* the gene was reported to represent a susceptible allele (Ciusa et al., 2012). Whilst analysing the panel of *sh-fabI* genes carried by TnSha1 these were all found to show at least one SNP compared to the core-genome triclosan-susceptible sh-*fabI* of *S. haemolyticus* (**Table 6**). All of these SNPs, except A577T and T578A (see above), have previously been associated with triclosan resistance in the *S. aureus sa-fabI* (**Table 6**; Ciusa et al., 2012; Grandgirard et al., 2015). The only exception is *S. aureus* strain T22051 which carried a wild type *sh-fabI* gene in TnSha1 thus indicating that the second (or third) IS-element mobilized genomic *fabI* copy generally appears to be a resistant allele.

DISCUSSION

A novel mobile chromosomal element based upon IS1272 and containing the *sh-fabI* gene of *S. haemolyticus* has been identified in a large-scale screen for biocide resistance across more than 1600 *S. aureus* isolates (Ciusa et al., 2012). This element, now named TnSha1, had been found to be present in one third of triclosan resistant strains of *S. aureus* while the other triclosan resistant strains of that study had mutations either within or upstream of the enoyl-acyl carrier protein reductase *fabI* gene of the core genome (Ciusa et al., 2012; Grandgirard et al., 2015). A more detailed analysis of these *S. aureus* strains, and of other available staphylococcal genomes, has shown that the situation is actually more complex. The main finding of this was that *sh-fabI* appears to be transferred by two types of elements; TnSha1, which has a single IS1272 element, and the composite TnSha1-carrying plasmid TnSha2 which can also integrate into chromosomal IS1272 copies generating in most cases an IS1272 bordered island (**Figures 1**, **7**). Similar congruent formations where reported for other similar elements as for example IS257 mobilizing the trimethoprim resistance determinant *dfrA* (Leelaporn et al., 1996). As the two elements do not show any obvious species specific characteristics it was somewhat unexpected to observe such a strong divergence in distribution for these elements, with *S. aureus* almost exclusively harboring TnSha1, *S. haemolyticus* prevalently harboring TnSha2, and only *S. epidermidis* which was found to commonly carry both elements. Neither the GC content (TnSha1 is 31.4%, TnSha2 is 31.8%, *S. aureus*, *S. haemolyticus* 33%, and *S. epidermidis* 32%), nor the presence of restriction modification systems nor the origin of the isolates can explain this observation; however we admit that this is as yet based on 4127 *S. aureus* genomes, but upon just 286 *S. epidermidis* and 141 *S. haemolyticus* genomes. Similar dynamics are seen with trimethoprim resistance were the housekeeping *dfrA* gene was acquired from other species and successfully transferred by IS257 across different staphylococcal species (Dale et al., 1995; Leelaporn et al., 1996). Whilst not explaining this difference as such, the observation that the intact IS1272 elements are present in most *S. haemolyticus* isolates, only in a few *S. epidermidis* strains and not at all in *S. aureus* already indicates a species specific "behavior" of the IS element. Future experiments on the transposition of the elements in the three species, will hopefully be more successful than the attempted detection of circular intermediates in *S. aureus* and therefore shed light on the observed differences in the epidemiology of the elements.

The mechanism of TnSha1 and IS1272 target recognition and integration appears to be novel. Both TnSha1 and IS1272 are seen to target hairpin secondary structures with little primary sequence conservation. In addition this insertion into the target appears to produce a blunt-end cut leading to insertion of the element and deletion of part of the target (**Figures 1–4**). This is in line with the original description of IS1272 which reported that the IS did not produce target duplications (Archer et al., 1996). Interestingly the recent review on IS elements by Siguier and colleagues reports that some members of the IS1182 family, of which IS1272 is part, also target palindromic sequences (Siguier et al., 2015).

Albeit that no primary sequence consensus was detected, the hairpin structures targeted by TnSha1 must have some important features in common as we found three targets with evidence for multiple independent insertions i.e., the targets "A" and "X" or the target "Z" where we also observed multiple insertions in both orientations (**Figure 2C**). The absence of insertion into rolling circle replicating plasmids, coupled to the observation of insertions in opposite directions appear to indicate the absence of an orientation bias observed for IS elements inserting into the replication fork (Siguier et al., 2014). In the case of TnSha2 the situation appears to be more varied. The few available data show that TnSha2 can integrate like IS1272. In addition the few and incomplete genomic data we have could indicate that TnSha2 could be mobilized by the hypothetical ORFs present leading to duplications and co-integrate formation (Needham et al., 1995). The capacity to utilize different integration mechanisms has been described also for other transposons like Tn7 (Siguier et al., 2014).

The large number of both TnSha1 and TnSha2 element sequences available in the database has allowed us to investigate the genesis of this recently generated and spreading element. This was hampered in part by the fact that both elements included IS1272 and that these ISs, like most other repeat sequences, tend to generate contig breaks during the assembly of bacterial genomes. Given the absolute conservation of the position of IS1272 in all of the TnSha1 and TnSha2 elements, we favor the hypothesis that the initial steps leading to the formation of these mobile elements were very few or possibly even only a single event. The initial events in the assembly of the element

TABLE 6 | *Sh-fabI* nucleotide sequence of 89 TnSha1 and TnSha2 elements.

Species	Strain	Polymorphic sites in *sh-fabI**						Phenotype**	Comment
		81	106	577	578	593	611		
S. haemolyticus	JCSC1435	T	G	A	T	C	T	S	Reference strain***
S. haemolyticus	0281	.	.	.	.	.	.	S	Triclosan susceptible clinical isolate***
S. aureus	T22051	.	.	.	.	.	.	S	*wt* sequence
S. aureus	KINW6058	C	.	T	.	.	.	R	GTA81GCA V27A; ATT577TTT I193F
S. epidermidis	354_SEPI	.	A	.	.	.	C	R	GTT106ATT V36I; TC611TCC F204S
S. epidermidis	422.rep1_SEPI	.	A	.	.	.	C	R	GTT106ATT V36I; TC611TCC F204S
S. epidermidis	422.rep2_SEPI	.	A	.	.	.	C	R	GTT106ATT V36I; TC611TCC F204S
S. epidermidis	493_SEPI	.	A	.	.	.	C	R	GTT106ATT V36I; TC611TCC F204S
S. epidermidis	568_SEPI	.	A	.	.	.	C	R	GTT106ATT V36I; TC611TCC F204S
S. epidermidis	623_SEPI	.	A	.	.	.	C	R	GTT106ATT V36I; TC611TCC F204S
S. epidermidis	634_SEPI	.	A	.	.	.	C	R	GTT106ATT V36I; TC611TCC F204S
S. epidermidis	656_SEPI	.	A	.	.	.	C	R	GTT106ATT V36I; TC611TCC F204S
S. epidermidis	749_SEPI	.	A	.	.	.	C	R	GTT106ATT V36I; TC611TCC F204S
S. epidermidis	890_SEPI	.	A	.	.	.	C	R	GTT106ATT V36I; TC611TCC F204S
S. epidermidis	1013_SEPI	.	A	.	.	.	C	R	GTT106ATT V36I; TC611TCC F204S
S. epidermidis	1021.rep2_SEPI	.	A	.	.	.	C	R	GTT106ATT V36I; TC611TCC F204S
S. epidermidis	1022_SEPI	.	A	.	.	.	C	R	GTT106ATT V36I; TC611TCC F204S
S. epidermidis	1088_SEPI	.	A	.	.	.	C	R	GTT106ATT V36I; TC611TCC F204S
S. epidermidis	1143_SEPI	.	A	.	.	.	C	R	GTT106ATT V36I; TC611TCC F204S
S. aureus	F12753	.	.	T	.	.	.		ATT577TTT I193F
S. aureus	F70077	.	.	T	.	.	.		ATT577TTT I193F
S. aureus	GGMC6026	.	.	T	.	.	.		ATT577TTT I193F
S. aureus	H27862	.	.	T	.	.	.		ATT577TTT I193F
S. aureus	H27872	.	.	T	.	.	.		ATT577TTT I193F
S. aureus	H64967	.	.	T	.	.	.		ATT577TTT I193F
S. aureus	H67656	.	.	T	.	.	.		ATT577TTT I193F
S. aureus	pHMPREF1625_3	.	.	T	.	.	.		ATT577TTT I193F
S. aureus	pVR_MSSA_01	.	.	T	.	.	.		ATT577TTT I193F
S. aureus	QBR-102278-1619	.	.	T	.	.	.		ATT577TTT I193F
S. aureus	T34011	.	.	T	.	.	.		ATT577TTT I193F
S. aureus	T78544	.	.	T	.	.	.		ATT577TTT I193F
S. aureus	USFL050	.	.	T	.	.	.		ATT577TTT I193F
S. aureus	USFL145	.	.	T	.	.	.		ATT577TTT I193F
S. epidermidis	134_SEPI	.	.	T	.	.	.		ATT577TTT I193F
S. epidermidis	154_SEPI	.	.	T	.	.	.		ATT577TTT I193F
S. epidermidis	196_SEPI	.	.	T	.	.	.		ATT577TTT I193F
S. epidermidis	SAP110A	.	.	T	.	.	.		ATT577TTT I193F
S. aureus	F77919	.	.	.	A	.	.		ATT578AAT I193N
S. aureus	LPIH6008	.	.	.	A	.	.		ATT578AAT I193N
S. aureus	W21479	.	.	.	A	.	.		ATT578AAT I193N
S. aureus	QBR-102278-1203	.	.	.	.	G	.	R	GCT593GGT A198G
S. aureus	S1805	.	.	.	.	G	.	R	GCT593GGT A198G
S. aureus	S2396	.	.	.	.	G	.	R	GCT593GGT A198G
S. argenteus	pST75	.	.	.	.	.	G	R	TTC611TGC F204C
S. argenteus	Sa_LBSA043	.	.	.	.	.	G	R	TTC611TGC F204C
S. aureus	122	.	.	.	.	.	G	R	TTC611TGC F204C
S. aureus	1299_SAUR	.	.	.	.	.	G	R	TTC611TGC F204C

(Continued)

TABLE 6 | Continued

Species	Strain	Polymorphic sites in *sh-fabI**						Phenotype**	Comment
		81	106	577	578	593	611		
S. aureus	1315_SAUR	.	.	.	.	.	G	R	TTC611TGC F204C
S. aureus	71193	.	.	.	.	.	G	R	TTC611TGC F204C
S. aureus	78_SAUR	.	.	.	.	.	G	R	TTC611TGC F204C
S. aureus	84_SAUR	.	.	.	.	.	G	R	TTC611TGC F204C
S. aureus	C5086	.	.	.	.	.	G	R	TTC611TGC F204C
S. aureus	CO-86	.	.	.	.	.	G	R	TTC611TGC F204C
S. aureus	H202	.	.	.	.	.	G	R	TTC611TGC F204C
S. aureus	H216	.	.	.	.	.	G	R	TTC611TGC F204C
S. aureus	LHSKBClinical	.	.	.	.	.	G	R	TTC611TGC F204C
S. aureus	M1229	.	.	.	.	.	G	R	TTC611TGC F204C
S. aureus	QBR-102278-1091	.	.	.	.	.	G	R	TTC611TGC F204C
S. aureus	QBR-102278-1107	.	.	.	.	.	G	R	TTC611TGC F204C
S. aureus	QBR-102278-2210	.	.	.	.	.	G	R	TTC611TGC F204C
S. aureus	QBR-102278-2365	.	.	.	.	.	G	R	TTC611TGC F204C
S. aureus	QBR-102278-2376	.	.	.	.	.	G	R	TTC611TGC F204C
S. aureus	SH1	.	.	.	.	.	G	R	TTC611TGC F204C
S. aureus	SH3	.	.	.	.	.	G	R	TTC611TGC F204C
S. aureus	USA300_2014.C02	.	.	.	.	.	G	R	TTC611TGC F204C
S. aureus	USFL046	.	.	.	.	.	G	R	TTC611TGC F204C
S. aureus	USFL123	.	.	.	.	.	G	R	TTC611TGC F204C
S. aureus	USFL129	.	.	.	.	.	G	R	TTC611TGC F204C
S. aureus	USFL234	.	.	.	.	.	G	R	TTC611TGC F204C
S. epidermidis	1249_SEPI	.	.	.	.	.	G	R	TTC611TGC F204C
S. epidermidis	14.1.R1.SE	.	.	.	.	.	G	R	TTC611TGC F204C
S. epidermidis	NIHLM020	.	.	.	.	.	G	R	TTC611TGC F204C
S. haemolyticus	Sh29/312/L2	.	.	.	.	.	G	R	TTC611TGC F204C
S. aureus	QBR-102278-2092	.	.	.	.	.	C	R	TTC611TCC F204S
S. aureus	QBR-102278-2351	.	.	.	.	.	C	R	TTC611TCC F204S
S. aureus	QBR-102278-2605	.	.	.	.	.	C	R	TTC611TCC F204S
S. aureus	M0277	.	.	.	.	.	C	R	TTC611TCC F204S
S. aureus	USFL189	.	.	.	.	.	C	R	TTC611TCC F204S
S. aureus	USFL190	.	.	.	.	.	C	R	TTC611TCC F204S
S. aureus	W41757	.	.	.	.	.	C	R	TTC611TCC F204S
S. epidermidis	1026_SEPI	.	.	.	.	.	C	R	TTC611TCC F204S
S. haemolyticus	25-38	.	.	.	.	.	C	R	TTC611TCC F204S
S. haemolyticus	25-60	.	.	.	.	.	C	R	TTC611TCC F204S
S. haemolyticus	51-11	.	.	.	.	.	C	R	TTC611TCC F204S
S. haemolyticus	51-13	.	.	.	.	.	C	R	TTC611TCC F204S
S. haemolyticus	51-43	.	.	.	.	.	C	R	TTC611TCC F204S
S. haemolyticus	51-41	.	.	.	.	.	C	R	TTC611TCC F204S
S. haemolyticus	113101	.	.	.	.	.	C	R	TTC611TCC F204S
S. haemolyticus	127925	.	.	.	.	.	C	R	TTC611TCC F204S
S. warneri	NGS-ED-1001	.	.	.	.	.	C	R	TTC611TCC F204S

**Polymorphic sites are indicated with respect to the sh-fabI sequence of S. haemolyticus strain JCSC1435. **Phenotype as defined in references: (Ciusa et al., 2012; Grandgirard et al., 2015). ***These wt S. haemolyticus strains do not carry the TnSha1 element, therefore the fabI sequence reported is the one of the core genome.*

must presumably have happened in *S. haemolyticus* since the *sh-fabI* derives from the core genome of this species and both elements are also present in this species. It is again not easy to hypothesize any rationale for the almost exclusive presence of TnSha1 in *S. aureus,* nor for the preferential presence of TnSha2 in *S. haemolyticus*. The lack of primary target sequence specificity and the peculiarity of targeting hairpin secondary structures make the elements very versatile and would allow for transposition to many locations within the chromosome and on plasmids. While intact IS1272 itself is not present in *S. aureus* the apparent selection for TnSha1 has very efficiently allowed spread of the element in this species. In the case of TnSha2 the presence of mobilization genes could indicate that this plasmid may be horizontally transferable. Unfortunately, as in the case of the *dfr* genes (Dale et al., 1995; Leelaporn et al., 1996), none of these data provide a solid molecular basis to explain the difference in distribution of the two elements among staphylococcal species. The explanation may however come from a closer look at the structure of the elements themselves. Indeed the very limited range of the diversity in the elements, and the SNPs present in them, indicates they have a recent origin. Another important aspect of the dynamics of spread of this element is related to the different triclosan-resistance conferring SNPs which have accumulated in *sh-fabI*. The presence of triclosan susceptible and resistant alleles in the studied population suggests the first event was acquisition of a susceptible *sh-fabI* allele followed by the selection of resistant alleles after the transfer event by a *fabI* targeting agent, most probably triclosan (Ciusa et al., 2012; Oggioni et al., 2012, 2013; Morrissey et al., 2014; Grandgirard et al., 2015). This sequence of events in which the acquisition of a resistance element is followed by the selection of more efficient alleles is also the basis of the diversification of other resistance elements as plasmids-encoded beta-lactamases (Bush, 2013).

Type II fatty acid metabolism, and in particular FabI, are among the most highly investigated targets for development of antimicrobial compounds (Banerjee et al., 1994; Park et al., 2007; Escaich et al., 2011; Kaplan et al., 2012). This remains true even though there is now debate on the essentiality of bacterial fatty metabolism during invasive infection because, at least for group B streptococci, it has been reported that it is possible to forage host derived fatty acids (Brinster et al., 2009; Balemans et al., 2010). In this respect it is therefore of great interest to monitor the spread of genetic elements which could produce reduced susceptibility to new antibiotics whose action is based on targeting and inhibition of FabI. In this regard it is worth noting that our experimental screening of a collection of 1602 world-wide clinical *S. aureus* isolates detected *sh-fabI* in 24/1602 of strains (1.5%; Ciusa et al., 2012), which matched perfectly with the *in silico* screen of a database of microbial genomes that detected *sh-fabI* carrying elements in 65/4127 *S. aureus* strains (1.57%). This exact overlap in prevalence indicates that genome databases of such size can now serve as suitable datasets for epidemiological investigation, even though databases generally lack detailed background information on strains. While 1.5% may not appear to be a high background resistance level in a population, it could still be a worrying presence when introducing a new FabI-targeting agent. Even worse is the detection of *sh-fabI* elements in 14% of *S. epidermidis* isolates; this appears to reach levels at which clinical use, including the use of triclosan as disinfectant for decontamination, could be jeopardized. It remains possible however that the lower sample size, of <300 genomes screened, and the absence of clinical information on these strains may limit the relevance of this observation.

In order to correlate our data with two well-known examples of IS-mobilizable metabolic genes, we have checked the relative occurrence of the *dfrA-thyE* genes (Rouch et al., 1989), conferring trimethoprim resistance, and *ileS2* genes (Needham et al., 1994), conferring mupirocin resistance, in the same 4800 staphylococcal genomes. BLAST searches for *dfrA-thyE* yielded 279 hits in *S. aureus* and 134 in *S. epidermidis*, while *ileS2* yielded 207 hits in *S. aureus* and 18 in *S. epidermidis*. This compares well with the detection rate in our work of 65 TnSha1/2 elements in *S. aureus* and 50 in *S. epidermidis*. These numbers indicate that IS mediated transfer of metabolic resistance genes, in our case *sh-fabI* genes and triclosan resistance, is a highly relevant mechanism for the acquisition and spread of antibiotic resistance. These data show that it is not only plasmids which serve as vectors of IS mediated resistance gene transfers, but that the spread of composite transposons can also be a highly efficient mechanism for such a goal. A similar well-described mechanisms also exist in staphylococci for the antibiotic resistance genes for example in Tn4001, Tn4002, and Tn4003, (Lanza et al., 2015).

In conclusion our data show that IS mediated transposition of metabolic genes represents a vast and growing antimicrobial resistance phenomenon. In addition to the well-described Tn*4003* element, which mobilizes *dfrA* by way of three IS257 thereby conferring trimethoprim resistance (Rouch et al., 1989), or the IS257 mediated mobilization of *ileS2* conferring mupirocin resistance (Needham et al., 1995), this arsenal now includes TnSha1 and TnSha2; these elements utilize IS transposition, IS-targeted integration and plasmid mobilization to allow transfer of the *fabI* gene of *S. haemolyticus* to different staphylococci and thereby contribute to triclosan resistance and, potentially, to resistance for other FabI-targeting drugs. These data show that IS mobilization of metabolic genes is a powerful and highly flexible mechanism that can very rapidly provide resistance phenotypes to vast numbers of strains and species. In this era where thousands of genomes are readily available in public databases the analysis of such IS mediated mobilization of core genome metabolic genes may warrant a more detailed and larger scale investigation.

AUTHOR CONTRIBUTIONS

LF performed the genome sequencing and bioinformatic analysis, wrote the manuscript, and approved the final version. RH participated in bioinformatic analysis, wrote the manuscript, and approved the final version. ZA performed wet lab experiments, participated in bioinformatic analysis, participated in the revision of the manuscript, and approved the final version. HO preformed bioinformatic analysis, discussed results and implication and helped in revision of the manuscript and approved the final version. IM had input in the initial study design, participated in the generation and analysis of data, the revision of the manuscript, and approved the final version. RL preformed bioinformatic analysis, discussed results and implication and helped in revision of the manuscript and approved the final version. JM had input in the initial study design, participated in the analysis of data, participated in the revision of the manuscript, and approved the final version. TC had input in the initial study design, participated in the analysis of data, participated in the revision of the manuscript, and approved the final version. MO designed the study, participated in bioinformatic analysis, wrote the manuscript, and is accountable for all aspects of the work.

FUNDING

The work was in part supported by EC project KBBE-227258 (BIOHYPO). ZA is funded by a fellowship of the Sultan Qaboos University, Oman.

ACKNOWLEDGMENTS

We thank Mike Chandler for helpful discussion. In addition to the authors, Graziella Orefici, Carlo Viti, Ulku Yetis, Hans Joachim Roedger, Ayse Kalkancy, Pilar Visa, Marina Elli, Stephen Leib, Daniel Knight and Diego Mora, participated to the BIOHYPO research project.

REFERENCES

Altschul, S. F., Madden, T. L., Schaffer, A. A., Zhang, J., Zhang, Z., Miller, W., et al. (1997). Gapped BLAST and PSI-BLAST: a new generation of protein database search programs. *Nucleic Acids Res.* 25, 3389–3402. doi: 10.1093/nar/25.17.3389

Andersson, D. I. (2006). The biological cost of mutational antibiotic resistance: any practical conclusions? *Curr. Opin. Microbiol.* 9, 461–465. doi: 10.1016/j.mib.2006.07.002

Archer, G. L., Niemeyer, D. M., Thanassi, J. A., and Pucci, M. J. (1994). Dissemination among staphylococci of DNA sequences associated with methicillin resistance. *Antimicrob. Agents Chemother.* 38, 447–454. doi: 10.1128/AAC.38.3.447

Archer, G. L., Thanassi, J. A., Niemeyer, D. M., and Pucci, M. J. (1996). Characterization of IS1272, an insertion sequence-like element from *Staphylococcus haemolyticus. Antimicrob. Agents Chemother.* 40, 924–929.

Balemans, W., Lounis, N., Gilissen, R., Guillemont, J., Simmen, K., Andries, K., et al. (2010). Essentiality of FASII pathway for *Staphylococcus aureus. Nature* 463, E3. discussion: E4. doi: 10.1038/nature08667

Banerjee, A., Dubnau, E., Quemard, A., Balasubramanian, V., Um, K. S., Wilson, T., et al. (1994). inhA, a gene encoding a target for isoniazid and ethionamide in *Mycobacterium tuberculosis. Science* 263, 227–230.

Barth, P. T., Datta, N., Hedges, R. W., and Grinter, N. J. (1976). Transposition of a deoxyribonucleic acid sequence encoding trimethoprim and streptomycin resistances from R483 to other replicons. *J. Bacteriol.* 125, 800–810.

Bosi, E., Donati, B., Galardini, M., Brunetti, S., Sagot, M. F., Lio, P., et al. (2015). MeDuSa: a multi-draft based scaffolder. *Bioinformatics* 31, 2443–2451. doi: 10.1093/bioinformatics/btv171

Brinster, S., Lamberet, G., Staels, B., Trieu-Cuot, P., Gruss, A., and Poyart, C. (2009). Type II fatty acid synthesis is not a suitable antibiotic target for Gram-positive pathogens. *Nature* 458, 83–86. doi: 10.1038/nature07772

Bush, K. (2013). Proliferation and significance of clinically relevant beta-lactamases. *Ann. N.Y. Acad. Sci.* 1277, 84–90. doi: 10.1111/nyas.12023

Cavanagh, J. P., Hjerde, E., Holden, M. T., Kahlke, T., Klingenberg, C., Flaegstad, T., et al. (2014). Whole-genome sequencing reveals clonal expansion of multiresistant *Staphylococcus haemolyticus* in European hospitals. *J. Antimicrob. Chemother.* 69, 2920–2927. doi: 10.1093/jac/dku271

Ciusa, M. L., Furi, L., Knight, D., Decorosi, F., Fondi, M., Raggi, C., et al. (2012). A novel resistance mechanism to triclosan that suggests horizontal gene transfer and demonstrates a potential selective pressure for reduced biocide susceptibility in clinical strains of *Staphylococcus aureus. Int. J. Antimicrob. Agents* 40, 210–220. doi: 10.1016/j.ijantimicag.2012.04.021

Coelho, J. R., Carrico, J. A., Knight, D., Martinez, J. L., Morrissey, I., Oggioni, M. R., et al. (2013). The use of machine learning methodologies to analyse antibiotic and biocide susceptibility in *Staphylococcus aureus. PLoS ONE* 8:e55582. doi: 10.1371/journal.pone.0055582

Dale, G. E., Broger, C., Hartman, P. G., Langen, H., Page, M. G., Then, R. L., et al. (1995). Characterization of the gene for the chromosomal dihydrofolate reductase (DHFR) of *Staphylococcus epidermidis* ATCC 14990: the origin of the trimethoprim-resistant S1 DHFR from *Staphylococcus aureus*? *J. Bacteriol.* 177, 2965–2970.

Deeny, S. R., Worby, C. J., Tosas Auguet, O., Cooper, B. S., Edgeworth, J., Cookson, B., et al. (2015). Impact of mupirocin resistance on the transmission and control of healthcare-associated MRSA. *J. Antimicrob. Chemother.* 70, 3366–3378. doi: 10.1093/jac/dkv249

Escaich, S., Prouvensier, L., Saccomani, M., Durant, L., Oxoby, M., Gerusz, V., et al. (2011). The MUT056399 inhibitor of FabI is a new antistaphylococcal compound. *Antimicrob. Agents Chemother.* 55, 4692–4697. doi: 10.1128/AAC.01248-10

Fagerlund, A., Granum, P. E., and Havarstein, L. S. (2014). *Staphylococcus aureus* competence genes: mapping of the SigH, ComK1 and ComK2 regulons by transcriptome sequencing. *Mol. Microbiol.* 94, 557–579. doi: 10.1111/mmi.12767

Furi, L., Ciusa, M. L., Knight, D., Di Lorenzo, V., Tocci, N., Cirasola, D., et al. (2013). Evaluation of reduced susceptibility to quaternary ammonium compounds and bisbiguanides in clinical isolates and laboratory-generated mutants of *Staphylococcus aureus. Antimicrob. Agents Chemother.* 57, 3488–3497. doi: 10.1128/AAC.00498-13

Gilbart, J., Perry, C. R., and Slocombe, B. (1993). High-level mupirocin resistance in *Staphylococcus aureus*: evidence for two distinct isoleucyl-tRNA synthetases. *Antimicrob. Agents Chemother.* 37, 32–38. doi: 10.1128/AAC.37.1.32

Grandgirard, D., Furi, L., Ciusa, M. L., Baldassarri, L., Knight, D. R., Morrissey, I., et al. (2015). Mutations upstream of fabI in triclosan resistant *Staphylococcus aureus* strains are associated with elevated fabI gene expression. *BMC Genomics* 16:345 doi: 10.1186/s12864-015-1544-y

Heath, R. J., Rubin, J. R., Holland, D. R., Zhang, E., Snow, M. E., and Rock, C. O. (1999). Mechanism of triclosan inhibition of bacterial fatty acid synthesis. *J. Biol. Chem.* 274, 11110–11114. doi: 10.1074/jbc.274.16.11110

Heath, R. J., White, S. W., and Rock, C. O. (2001). Lipid biosynthesis as a target for antibacterial agents. *Prog. Lipid Res.* 40, 467–497. doi: 10.1016/S0163-7827(01)00012-1

Hedges, R. W., and Jacob, A. E. (1974). Transposition of ampicillin resistance from RP4 to other replicons. *Mol. Gen. Genet.* 132, 31–40. doi: 10.1007/BF00268228

Hetem, D. J., and Bonten, M. J. (2013). Clinical relevance of mupirocin resistance in *Staphylococcus aureus. J. Hosp. Infect.* 85, 249–256. doi: 10.1016/j.jhin.2013.09.006

Hijazi, K., Mukhopadhya, I., Abbott, F., Milne, K., Al-Jabri, Z. J., Oggioni, M. R., et al. (2016). Susceptibility to chlorhexidine amongst multidrug-resistant clinical isolates of *Staphylococcus Epidermidis* from bloodstream infections. *Int. J. Antimicrob. Agents.* doi: 10.1016/j.ijantimicag.2016.04.015. [Epub ahead of print].

Isola, D., Pardini, M., Varaine, F., Niemann, S., Rusch-Gerdes, S., Fattorini, L., et al. (2005). A pyrosequencing assay for rapid recognition of SNPs in Mycobacterium tuberculosis embB306 region. *J. Microbiol. Methods* 62, 113–120. doi: 10.1016/j.mimet.2005.02.004

Ji, Y., Yin, D., Fox, B., Holmes, D. J., Payne, D., and Rosenberg, M. (2004). Validation of antibacterial mechanism of action using regulated antisense RNA expression in *Staphylococcus aureus*. *FEMS Microbiol. Lett.* 231, 177–184. doi: 10.1016/S0378-1097(03)00931-5

Kaplan, N., Albert, M., Awrey, D., Bardouniotis, E., Berman, J., Clarke, T., et al. (2012). Mode of action, *in vitro* activity, and *in vivo* efficacy of AFN-1252, a selective antistaphylococcal FabI inhibitor. *Antimicrob. Agents Chemother.* 56, 5865–5874. doi: 10.1128/AAC.01411-12

Lanza, V. F., Tedim, A. P., Martinez, J. L., Baquero, F., and Coque, T. M. (2015). The plasmidome of firmicutes: impact on the emergence and the spread of resistance to antimicrobials. *Microbiol. Spectr.* 3:PLAS-0039-2014. doi: 10.1128/microbiolspec.plas-0039-2014

Larsen, M. V., Cosentino, S., Rasmussen, S., Friis, C., Hasman, H., Marvig, R. L., et al. (2012). Multilocus sequence typing of total-genome-sequenced bacteria. *J. Clin. Microbiol.* 50, 1355–1361. doi: 10.1128/JCM.06094-11

Leelaporn, A., Firth, N., Paulsen, I. T., and Skurray, R. A. (1996). IS257-mediated cointegration in the evolution of a family of staphylococcal trimethoprim resistance plasmids. *J. Bacteriol.* 178, 6070–6073.

León-Sampedro, R., Novais, C., Peixe, L., Baquero, F., and Coque, T. M. (2016). Diversity and evolution of the *Tn5801-tet*(M)-like integrative and conjugative elements among *Enterococcus, Streptococcus*, and *Staphylococcus*. *Antimicrob. Agents Chemother.* 60, 1736–1746. doi: 10.1128/AAC.01864-15

Li, H. (2014). Toward better understanding of artifacts in variant calling from high-coverage samples. *Bioinformatics* 30, 2843–2851. doi: 10.1093/bioinformatics/btu356

Lu, H., and Tonge, P. J. (2008). Inhibitors of FabI, an enzyme drug target in the bacterial fatty acid biosynthesis pathway. *Acc. Chem. Res.* 41, 11–20. doi: 10.1021/ar700156e

Maillard, J. Y., Bloomfield, S., Coelho, J. R., Collier, P., Cookson, B., Fanning, S., et al. (2013). Does microbicide use in consumer products promote antimicrobial resistance? a critical review and recommendations for a cohesive approach to risk assessment. *Microb. Drug Resist.* 19, 344–354. doi: 10.1089/mdr.2013.0039

Martinez, J. L., Coque, T. M., and Baquero, F. (2015). What is a resistance gene? *Ranking risk in resistomes. Nat. Rev. Microbiol.* 13, 116–123. doi: 10.1038/nrmicro3399

Morrissey, I., Oggioni, M. R., Knight, D., Curiao, T., Coque, T., Kalkanci, A., et al. (2014). Evaluation of epidemiological cut-off values indicates that biocide resistant subpopulations are uncommon in natural isolates of clinically-relevant microorganisms. *PLoS ONE* 9:e86669. doi: 10.1371/journal.pone.0086669

Needham, C., Noble, W. C., and Dyke, K. G. (1995). The staphylococcal insertion sequence IS257 is active. *Plasmid* 34, 198–205. doi: 10.1006/plas.1995.0005

Needham, C., Rahman, M., Dyke, K. G., and Noble, W. C. (1994). An investigation of plasmids from *Staphylococcus aureus* that mediate resistance to mupirocin and tetracycline. *Microbiology* 140 (Pt 10), 2577–2583. doi: 10.1099/00221287-140-10-2577

Oggioni, M. R., Ciusa, M. L., Furi, L., Baldassarri, L., Orefici, G., Cirasola, D., et al. (2012). Lack of evidence for reduced fitness of clinical *Staphylococcus aureus* isolates with reduced susceptibility to triclosan. *Antimicrob. Agents Chemother.* 56, 6068–6069. doi: 10.1128/AAC.01055-12

Oggioni, M. R., Coelho, J. R., Furi, L., Knight, D. R., Viti, C., Orefici, G., et al. (2015). Significant differences characterise the correlation coefficients between biocide and antibiotic susceptibility profiles in *Staphylococcus aureus*. *Curr. Pharm. Des.* 21, 2054–2057. doi: 10.2174/1381612821666150310103238

Oggioni, M. R., Furi, L., Coelho, J. R., Maillard, J. Y., and Martinez, J. L. (2013). Recent advances in the potential interconnection between antimicrobial resistance to biocides and antibiotics. *Expert Rev. Anti Infect. Ther.* 11, 363–366. doi: 10.1586/eri.13.16

Oggioni, M. R., Meacci, F., Carattoli, A., Ciervo, A., Orru, G., Cassone, A., et al. (2002). Protocol for real-time PCR identification of anthrax spores from nasal swabs after broth enrichment. *J. Clin. Microbiol.* 40, 3956–3963. doi: 10.1128/JCM.40.11.3956-3963.2002

Park, H. S., Yoon, Y. M., Jung, S. J., Kim, C. M., Kim, J. M., and Kwak, J. H. (2007). Antistaphylococcal activities of CG400549, a new bacterial enoyl-acyl carrier protein reductase (FabI) inhibitor. *J. Antimicrob. Chemother.* 60, 568–574. doi: 10.1093/jac/dkm236

Payne, D. J., Warren, P. V., Holmes, D. J., Ji. Y., and Lonsdale, J. T. (2001). Bacterial fatty-acid biosynthesis: a genomics-driven target for antibacterial drug discovery. *Drug Discov. Today* 6, 537–544. doi: 10.1016/S1359-6446(01)01774-3

Reuter, J. S., and Mathews, D. H. (2010). RNAstructure: software for RNA secondary structure prediction and analysis. *BMC Bioinformatics* 11:129. doi: 10.1186/1471-2105-11-129

Roach, D. J., Burton, J. N., Lee, C., Stackhouse, B., Butler-Wu, S. M., Cookson, B. T., et al. (2015). A year of infection in the intensive care unit: prospective whole genome sequencing of bacterial clinical isolates reveals cryptic transmissions and novel microbiota. *PLoS Genet.* 11:e1005413. doi: 10.1371/journal.pgen.1005413

Rouch, D. A., Messerotti, L. J., Loo, L. S., Jackson, C. A., and Skurray, R. A. (1989). Trimethoprim resistance transposon Tn4003 from *Staphylococcus aureus* encodes genes for a dihydrofolate reductase and thymidylate synthetase flanked by three copies of IS257. *Mol. Microbiol.* 3, 161–175. doi: 10.1111/j.1365-2958.1989.tb01805.x

Schoenfelder, S. M., Lange, C., Eckart, M., Hennig, S., Kozytska, S., and Ziebuhr, W. (2010). Success through diversity - how *Staphylococcus epidermidis* establishes as a nosocomial pathogen. *Int. J. Med. Microbiol.* 300, 380–386. doi: 10.1016/j.ijmm.2010.04.011

Schweizer, H. P. (2001). Triclosan: a widely used biocide and its link to antibiotics. *FEMS Microbiol. Lett.* 202, 1–7. doi: 10.1111/j.1574-6968.2001.tb10772.x

Siguier, P., Gourbeyre, E., and Chandler, M. (2014). Bacterial insertion sequences: their genomic impact and diversity. *FEMS Microbiol. Rev.* 38, 865–891. doi: 10.1111/1574-6976.12067

Siguier, P., Gourbeyre, E., Varani, A., Ton-Hoang, B., and Chandler, M. (2015). Everyman's guide to bacterial insertion sequences. *Microbiol. Spectr.* 3:MDNA3-0030-2014 doi: 10.1128/microbiolspec.mdna3-0030-2014

Slater-Radosti, C., Van Aller, G., Greenwood, R., Nicholas, R., Keller, P. M., DeWolf, W. E. Jr., et al. (2001). Biochemical and genetic characterization of the action of triclosan on *Staphylococcus aureus*. *J. Antimicrob. Chemother.* 48, 1–6. doi: 10.1093/jac/48.1.1

Tonouchi, N., Tsuchida, T., Yoshinaga, F., Horinouchi, S., and Beppu, T. (1994). A host-vector system for a cellulose-producing Acetobacter strain. *Biosci. Biotechnol. Biochem.* 58, 1899–1901. doi: 10.1271/bbb.58.1899

Woodford, N., Watson, A. P., Patel, S., Jevon, M., Waghorn, D. J., and Cookson, B. D. (1998). Heterogeneous location of the mupA high-level mupirocin resistance gene in *Staphylococcus aureus*. *J. Med. Microbiol.* 47, 829–835. doi: 10.1099/00222615-47-9-829

Yesilkaya, H., Meacci, F., Niemann, S., Hillemann, D., Rusch-Gerdes, S., Group, L. D. S., et al. (2006). Evaluation of molecular-Beacon, TaqMan, and fluorescence resonance energy transfer probes for detection of antibiotic resistance-conferring single nucleotide polymorphisms in mixed *Mycobacterium tuberculosis* DNA extracts. *J. Clin. Microbiol.* 44, 3826–3829. doi: 10.1128/JCM.00225-06

Zhang, Y. Q., Ren, S. X., Li, H. L., Wang, Y. X., Fu, G., Yang, J., et al. (2003). Genome-based analysis of virulence genes in a non-biofilm-forming *Staphylococcus epidermidis* strain (ATCC 12228). *Mol. Microbiol.* 49, 1577–1593. doi: 10.1046/j.1365-2958.2003.03671.x

18

Molecular Epidemiology of Methicillin-Resistant *Staphylococcus aureus* in Africa

Shima M. Abdulgader[1], Adebayo O. Shittu[2], Mark P. Nicol[1,3,4] and Mamadou Kaba[1,3]*

[1] Division of Medical Microbiology, Department of Clinical Laboratory Sciences, Faculty of Health Sciences, University of Cape Town, Cape Town, South Africa, [2] Department of Microbiology, Obafemi Awolowo University, Ile-Ife, Nigeria, [3] Institute of Infectious Disease and Molecular Medicine, Faculty of Health Sciences, University of Cape Town, Cape Town, South Africa, [4] National Health Laboratory Service, Groote Schuur Hospital, Cape Town, South Africa

***Correspondence:**
*Mamadou Kaba,
Division of Medical Microbiology,
Department of Clinical Laboratory Sciences, Faculty of Health Sciences,
University of Cape Town,
Anzio Road, Observatory, 7925,
Cape Town, South Africa
mamadou.kaba@hotmail.com*

Methicillin-resistant *Staphylococcus aureus* (MRSA) infections are a serious global problem, with considerable impact on patients and substantial health care costs. This systematic review provides an overview on the clonal diversity of MRSA, as well as the prevalence of Panton-Valentine leukocidin (PVL)-positive MRSA in Africa. A search on the molecular characterization of MRSA in Africa was conducted by two authors using predefined terms. We screened for articles published in English and French through to October 2014 from five electronic databases. A total of 57 eligible studies were identified. Thirty-four reports from 15 countries provided adequate genotyping data. CC5 is the predominant clonal complex in the healthcare setting in Africa. The hospital-associated MRSA ST239/ST241-III [3A] was identified in nine African countries. This clone was also described with SCC*mec* type IV [2B] in Algeria and Nigeria, and type V [5C] in Niger. In Africa, the European ST80-IV [2B] clone was limited to Algeria, Egypt and Tunisia. The clonal types ST22-IV [2B], ST36-II [2A], and ST612-IV [2B] were only reported in South Africa. No clear distinctions were observed between MRSA responsible for hospital and community infections. The community clones ST8-IV [2B] and ST88-IV [2B] were reported both in the hospital and community settings in Angola, Cameroon, Gabon, Ghana, Madagascar, Nigeria, and São Tomé and Príncipe. The proportion of PVL-positive MRSA carriage and/or infections ranged from 0.3 to 100% in humans. A number of pandemic clones were identified in Africa. Moreover, some MRSA clones are limited to specific countries or regions. We strongly advocate for more surveillance studies on MRSA in Africa.

Keywords: *Staphylococcus aureus*, MRSA, molecular epidemiology, Africa, systematic review

Introduction

Methicillin-resistant *Staphylococcus aureus* (MRSA) is a major public health concern and is responsible for both hospital- and community-associated infections worldwide (De Kraker et al., 2011; CDC, 2013; Falagas et al., 2013; Garza-González and Dowzicky, 2013; Lee et al., 2013; Chen and Huang, 2014). It is estimated that MRSA infections within the health care setting alone affected more than 150,000 patients annually in the European Union, with an additional cost of 380 million Euros (Köck et al., 2010). In the United States of America, 80,461 invasive MRSA infections and 11,285 related deaths occurred in 2011, and an estimated annual burden of between \$1.4 billion and

13.8 billion was attributed to community-acquired MRSA (CDC, 2013; Lee et al., 2013). Besides, MRSA has been established as a pathogen for domestic animals and linked with livestock-associated infections (Verkade and Kluytmans, 2013).

Methicillin resistance is usually due to the *mecA* gene, borne on the staphylococcal cassette chromosome *mec* (SCC*mec*) that codes for a 78-kDa penicillin binding protein (PBP2a), with decreased affinity to methicillin and all beta-lactam antibiotics (Chambers, 1997). To date, eleven SCC*mec* types have been identified (IWG-SCC, 2009). Some cassettes, for example, SCC*mec* II (53 kb) and SCC*mec* III (67 kb), are large and possess mobile genetic elements (MGE), such as integrated plasmids (pUB110, pI258, and pT181) and transposons (e.g., Tn554) (Ito et al., 2001), and are frequently associated with hospital-acquired MRSA (Ma et al., 2002; Ito et al., 2004). In contrast, SCC*mec* IV (21–24 kb) and V (27 kb) are shorter elements, generally susceptible to non-beta-lactam antibiotics, and linked with community MRSA (Chambers and Deleo, 2010). However, the spread of various MRSA clones between the hospital and community settings has made the dichotomous ranking difficult (Deurenberg and Stobberingh, 2008). Recently, a variant *mecA* gene (named *mecC*) which is situated on an SCC*mec* XI element has been described (Shore et al., 2011). It has a higher relative affinity for oxacillin as compared with cefoxitin (Kim et al., 2012), and exhibits only 69% sequence similarity at the nucleotide level and 63% amino-acid identity to *mec*A/PBP2a (Paterson et al., 2014b). Furthermore, based on whole genome sequencing, mutations of the endogenous penicillin-binding proteins (PBP) 1, 2, and 3 in *mecA* and *mecC* negative strains have been postulated as a possible alternative mechanism for beta-lactam resistance in MRSA (Ba et al., 2014).

There is great interest in tracking, identifying and understanding the diversity of MRSA in various settings. Currently, the most widely used molecular techniques include *Staphylococcus* protein A gene typing (*spa*) and multilocus sequence typing (MLST). Studies (particularly using MLST) have provided evidence that a small set of lineages, clonal complex (CC)5, CC8, CC22, CC30, and CC45, are associated with most of the MRSA infections in hospitals (Stefani et al., 2012). Besides, a number of different geographically distinct lineages, CC1, CC8, CC30, and CC80, have also been associated with community MRSA infections (Chatterjee and Otto, 2013), while CC8 and CC30 have been identified as pandemic lineages both in the hospital and community setting (Chatterjee and Otto, 2013). Furthermore, regional clones have been described in Australia (sequence type [ST] 93) (Coombs et al., 2009), India (ST772) (D'Souza et al., 2010; Shambat et al., 2012), South Korea (ST72) (Kim et al., 2007), Taiwan and China (ST59) (Chen and Huang, 2014).

The distribution of MRSA clones in Africa is not well-described. Understanding the molecular epidemiology of MRSA in Africa is important as a recent review indicated that since the year 2000, the prevalence of MRSA appears to be increasing in many African countries and pose a visible threat to the continent (Falagas et al., 2013). Furthermore, there is evidence of the replacement of existing MRSA clones with different and new clonal types in a number of countries (Conceição et al., 2007; Aires-de-Sousa et al., 2008; Albrecht et al., 2011; Espadinha et al., 2013; Lim et al., 2013; Nimmo et al., 2013) but information on this trend is lacking in Africa. The occurrence and changes in clonal identities, and their geographic spread is important to understand the spread and evolution of MRSA.

The Panton-Valentine Leukocidin (PVL) is a two-component pore-forming toxin with cytolytic activity on defined cells of the immune system (neutrophils, macrophages and monocytes) (Löffler et al., 2010; Yoong and Torres, 2013). It is encoded by the *lukS-PV* and *lukF-PV* genes (Boakes et al., 2011), and PVL-producing *S. aureus* exhibit a propensity for causing mainly severe and often recurrent skin and soft tissue infections (Shallcross et al., 2013). In addition, PVL-positive MRSA are associated with community onset-pneumonia (Vandenesch et al., 2003). Although the PVL genes are mainly carried by community-associated MRSA (CA-MRSA) (Vandenesch et al., 2003), data from West and Central Africa showed that at least 40% of clinical methicillin-susceptible *S. aureus* (MSSA) isolates in this region are PVL-positive (Breurec et al., 2011a; Schaumburg et al., 2011; Shittu et al., 2011; Egyir et al., 2014a). Therefore, the acquisition of the *mecA* gene by PVL-positive MSSA and the possible dissemination of PVL-positive CA-MRSA could present a significant challenge in disease management and infection control in resource-limited countries in Africa.

This systematic review examined published literature on the molecular epidemiology of MRSA in Africa. By summarizing currently available data on the continent, our objective was to describe the distribution of MRSA clones, the prevalence of PVL-positive MRSA, and to highlight the need to develop more comprehensive surveillance and reporting systems for multidrug-resistant organisms such as MRSA in Africa.

Methods

This systematic review was conducted in accordance with the preferred reporting items for systematic reviews and meta-analyses (PRISMA) guidelines (Moher et al., 2009).

Literature Search Strategy

The relevant English and French articles available in five electronic databases (MEDLINE, EBSCOhost, ISI Web of knowledge, Scopus, and African Journals Online) were retrieved by two authors using predefined search terms (Table S1). The literature search was conducted until 31 October 2014.

Eligible Article Identification

Figure 1 summarizes the study selection process. All duplicate articles were removed and data on MSSA as well as *in-vitro* studies were also excluded. The eligibility of published reports in this review was based primarily on polymerase chain reaction (PCR) detection of the *mecA* gene, and the use of at least one molecular tool for genotyping of MRSA strains (**Table 1**). In addition, worldwide surveys that covered African countries were also included. An MRSA clone was defined based on the combination of MLST sequence type (ST) and SCC*mec* typing data as previously reported (Okuma et al., 2002). The nomenclature of the SCC*mec* types was as proposed by the International Working

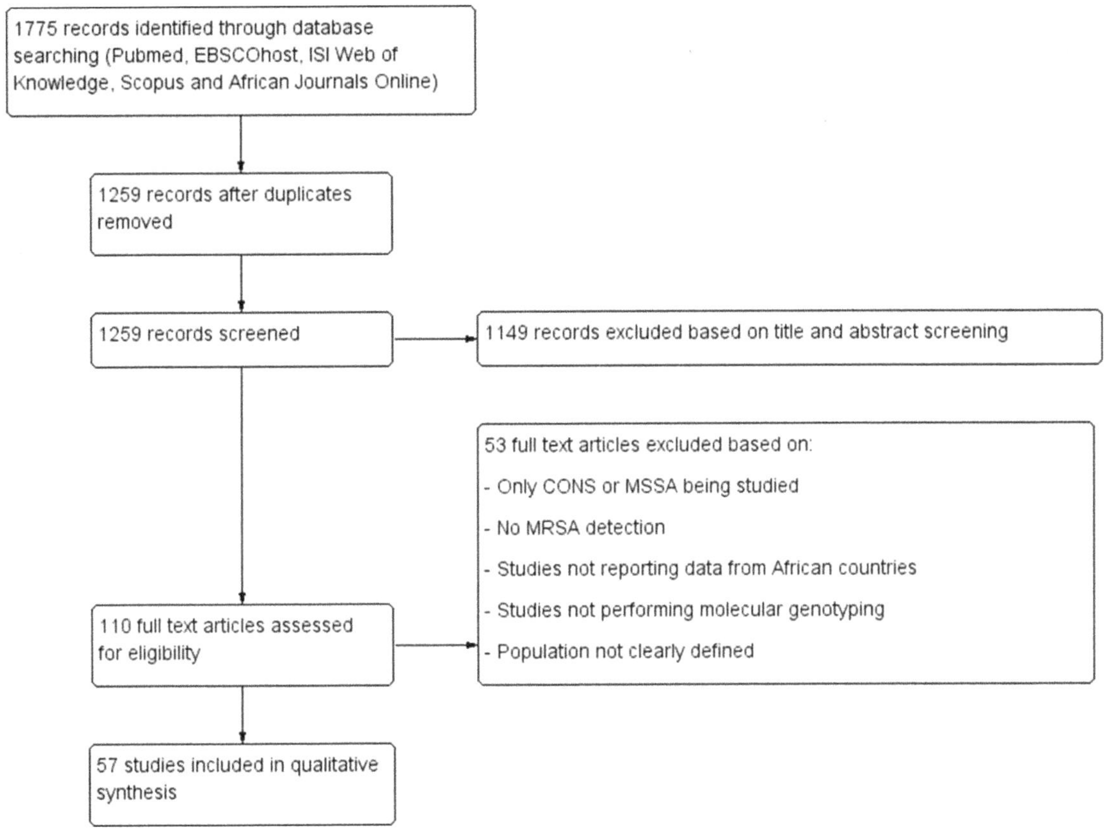

FIGURE 1 | Preferred reporting item for systematic reviews. CONS, coagulase negative staphylococci; MSSA, methicillin susceptible *S. aureus;* MRSA, methicillin resistant *S. aureus.*

Group on the Classification of Staphylococcal Cassette Chromosome Elements (IWG-SCC, 2009). SCC*mec* elements that could not be classified were indicated as non-typeable (SCC*mec*-NT). In this study, we categorized MRSA into various CCs according to the current eBURST scheme, Version 3 (accessed 30 October 2014) (eBURST, www.mlst.net, V3)[1].

Data Extraction and Synthesis

The relevant data were extracted from each of the articles as stated in **Table 1**. Separate articles that analyzed the same *S. aureus* isolates but answered different questions were considered as a single study.

eBURST Analysis

The relationship between the MRSA STs described in this review and other lineages reported world-wide was analyzed using the eBURST scheme. The allelic profiles were downloaded from the MLST website (http://saureus.mlst.net/) which included the African MRSA STs as well as 223 representative and randomly selected STs (from each CC) based on the differences in their allelic profiles. The minimum spanning tree was constructed by the goeBURST algorithm using the Phyloviz software v1.1 (http://www.phyloviz.net/).

[1]eBURST [Online]. Available online at: http://www.mlst.net version 3.

Results

Literature Search

The systematic search of the five electronic databases yielded 1775 articles (**Figure 1**). No additional studies were identified from AJOL. After the removal of duplicate studies and assessment of titles and abstracts, 110 full-text articles were screened, of which 57 studies were considered eligible for the qualitative analysis according to our inclusion criteria.

Characteristics of the Studies Included in the Systematic Review

Most of the data analyzed were obtained from single center studies conducted mainly in five countries; Tunisia ($n = 13$), Nigeria ($n = 9$), South Africa ($n = 7$), Algeria ($n = 5$), and Egypt ($n = 3$) (**Table 1**). Multicenter studies were only reported in two articles (Goering et al., 2008; Breurec et al., 2011b), including a survey which comprised five African countries, Cameroon, Madagascar, Morocco, Niger and Senegal (Breurec et al., 2011b), and an inter-continental multicenter study, which included South Africa (Goering et al., 2008). Only three studies investigated the detection of MRSA in animals (**Table 1**).

In most of the reports included in this study, *S. aureus* was identified by phenotypic and culture characteristics, while molecular identification (16S rRNA, detection of the thermonuclease

TABLE 1 | Characteristics of eligible articles that studied Methicillin resistant *Staphylococcus aureus*.

Country	Study period	Study population (sample type)	No. of *S. aureus* isolates	*S. aureus* molecular identification	No. of MRSA[¶]	Setting (no.)	Genotyping tools						*PVL*	Detection of genes		References
							SCC*mec*	*coa*	*agr*	*spa* typing	MLST	PFGE		Antibiotic resistance	Toxin/ Virulence	
STUDIES CONDUCTED IN HUMANS																
Algeria	2003–2004	Clinical samples from hospitals and community	614	–	204	HA (40)/ CA (21)	✓	–	✓	–	✓	✓	✓	–	–	Ramdani-bouguessa et al., 2006
	2004–2007	Human infections (in-and-out patients)	65	–	23	NR	✓	–	–	✓	✓	–	✓	–	–	Bekkhoucha et al., 2009
	2005–2007	From military hospital (Pus, venous catheter, tracheal aspirates, lumbar puncture fluid, blood culture and urine)	NR	–	64	HA (50)/ CA (14)	✓	–	–	–	–	–	✓	–	✓	Ouchenane et al., 2011, 2013
	2006–2007	Healthy and hospitalized individuals	221^{α} 52^{β}	*gyrA* PCR	99^{α} 23^{β}	HA (65)/ CA (84)	✓	–	✓	✓	✓	NR	✓	–	✓	Antri et al., 2011
	2010–2011	Children and neonates (SSTI, bacteraemia, otitis and bone/joint infections)	129	–	25	HA (15)/ CA (10)	✓	–	–	–	✓	✓	✓	–	–	Djoudi et al., 2013
Angola	2012	Nasal swabs from inpatients and HCW	117	–	68	NR	✓	–	✓	✓	✓	✓	✓	–	✓	Conceição et al., 2014
Egypt	2007–2008	Pus, sputum, wounds, abscess, blood, urine, and discharge	NR	–	21	CA (4)	✓	✓	✓	✓	✓	NR	✓	–	✓	Enany et al., 2010
	NR	SSTI and nasal swabs	38	–	18	CA (18)	✓	–	–	–	–	–	✓	–	–	Sobhy et al., 2012
	NR	Septic wounds, UTI and RTI (nasal swabs)	10	–	7	–	–	✓	–	✓	–	–	–	–	–	El-Jakee et al., 2011*
Gabon	2008–2010	asymptomatic carriers (nares, axillae, inguinal swabs) and patients (abscess, wound, blood and others)	217	*nuc* and 16S rRNA PCR	12	HA (6)/ CA (6)	✓	–	✓	✓	✓	–	✓	–	✓	Schaumburg et al., 2011; Ateba Ngoa et al., 2012
	2010–2013	swabs from *S. aureus* carrier mothers (nasal and mammillary) and their infants (Nasal and pharyngeal)	460	–	9	NR	✓	–	–	✓	✓	–	✓	–	✓	Schaumburg et al., 2014
	NR	Blood culture of one patient	1	–	1	NR	–	–	–	✓	✓	–	✓	–	✓	Huson et al., 2014

(Continued)

TABLE 1 | Continued

Country	Study period	Study population (sample type)	No. of *S. aureus* isolates	*S. aureus* molecular identification	No. of MRSA¶	Setting (no.)	Genotyping tools						*PVL*	Detection of genes		References
							SCC*mec*	*coa*	*agr*	*spa* typing	MLST	PFGE		Antibiotic resistance	Toxin/ Virulence	
Ghana	2011–2012	In-patients and hospital staff	105	*spa* gene PCR	6	NR	✓	–	–	✓	✓	–	✓	–	–	Egyir et al., 2013
	2010–2012	SSTI and blood samples from six hospitals	308	–	9	NR	✓	–	–	✓	✓	–	✓	–	–	Egyir et al., 2014a
	2011–2012	Nasal swabs from apparently healthy carriers	124	–	2	HA (2)	✓	–	–	✓	✓	–	✓	–	–	Egyir et al., 2014b
Kenya	2005–2007	In and out-patients with SSTI boil, abscess, cellulitis and ulcer	84	–	69	NR	✓	–	–	–	–	–	✓	–	–	Maina et al., 2013
	2011	Nasal and axillary skin swabs from hospitalized patients	85	–	6	NR	✓	–	–	✓	✓	–	✓	✓	✓	Aiken et al., 2014
Libya	2009–2010	Nasal swabs from in-patient children, their mothers, out-patient children and HCW	758	–	70	HA (12) CA (6)	–	–	–	–	–	✓	✓	–	–	Al-haddad et al., 2014
Mali	2005	Asymptomatic nasal carriers	88	–	1	CA (1)	–	–	–	–	✓	–	–	–	–	Ruimy et al., 2008
Mozambique	2010–2011	Post-operative, burn wound infections, skin and soft tissue abscesses	99	–	9	HA (8), CA (1)	–	–	–	✓	–	–	✓	✓	–	Van der Meeren et al., 2014
Nigeria	1998–2002	Wounds, aspirate, amniotic fluid	276	–	4	NR	✓	–	–	–	✓	✓	–	–	–	Adesida et al., 2005
	2002–2004	Wound samples, blood cultures, urine, otitis media and ocular related infections	200	–	3	NR	✓	–	–	–	–	✓	–	–	–	Shittu and Lin, 2006b
	2007–2012	Clinical specimen	150	–	12	NR	✓	–	–	–	✓	–	✓	–	✓	Okon et al., 2009, 2014
	2007	Surgical and pediatric patients wound samples, corneal, conjunctival, auricular, genital and nasal swabs	346	–	70	HA (42), CA (28)	✓	–	✓	✓	✓	–	✓	–	✓	Ghebremedhin et al., 2009
	2008–2010	HIV-positive and healthy individuals (nasal swabs)	202	–	26	NR	–	✓	–	✓	✓	–	✓	–	–	Olalekan et al., 2012

(Continued)

TABLE 1 | Continued

Country	Study period	Study population (sample type)	No. of *S. aureus* isolates	*S. aureus* molecular identification	No. of MRSA¶	Setting (no.)	Genotyping tools						*PVL*	Detection of genes		References
							SCC*mec*	*coa*	*agr*	*spa* typing	MLST	PFGE		Antibiotic resistance	Toxin/ Virulence	
	2009	Wound infections, semen, UTI, chronic ulcer, conjunctivitis, throat infections	68	–	11	NR	✓	–	–	✓	✓	–	✓	✓	–	Shittu et al., 2011
	2009–2011	Patients and carriers	62	–	22	NR	✓	–	–	–	✓	–	✓	–	–	Raji et al., 2013
	2010	Clinical samples from patients with burns, septicaemia, wound infections, osteomyelitis, bronchitis and GIT	51	*tuf* gene PCR	15	NR	✓	–	–	✓	✓	–	✓	✓	–	Shittu et al., 2012
	NR	Urine, blood and aspirates, wound, eye and ear, urethral and endocervical swab	116	–	48	HA (40), CA (8)	✓	–	–	–	–	–	✓	–	✓	Terry Alli et al., 2012
South Africa	2001–2003	Wound samples, sputum, otitis media and blood culture	227	*nuc* gene PCR	61	NR	✓	✓	–	–	–	✓	–	✓	–	Shittu and Lin, 2006a; Shittu et al., 2007
	2001–2003	Isolates from 16 laboratories in KZN	241	–	24	NR	✓	–	–	–	✓	–	–	–	–	Essa et al., 2009
	2001–2003	Wounds, sputum, otitis media, urine and blood culture	NR	–	61	NR	✓	–	–	✓	✓	✓	✓	–	–	Shittu et al., 2009
	2005–2006	Bacteraemia, SSTI, urine, catheter tip, cerebrospinal and drainage fluids	NR	–	320	HA	✓	–	–	✓	✓	✓	✓	–	✓	Moodley et al., 2010
	2006–2007	Clinical samples	NR	–	97	HA (79), CA (4)	✓	–	–	✓	–	–	✓	–	–	Makgotlho et al., 2009
	2007–2008 2007–2011	Pus and pus swabs, urine, blood, RTS and CVCT	NR	–	100	CA (10)	✓	–	–	✓	✓	✓	✓	–	✓	Jansen van Rensburg et al., 2011, 2012
	2009–2010	A wide range of clinical specimens mostly SSTI	367	–	56	NR	✓	–	✓	✓	✓	✓	✓	–	–	Oosthuysen et al., 2014
São Tome and Príncipe	2010–2012	Patients and healthy carriers	52	–	14	NR	✓	–	✓	✓	✓	✓	✓	–	✓	Conceição et al., 2013

(Continued)

TABLE 1 | Continued

Country	Study period	Study population (sample type)	No. of *S. aureus* isolates	*S. aureus* molecular identification	No. of MRSA¶	Setting (no.)	Genotying tools						*PVL*	Detection of genes		References
							SCC*mec*	*coa*	*agr*	*spa* typing	MLST	PFGE		Antibiotic resistance	Toxin/ Virulence	
Tanzania	2008	Wound, nasal swab and pus	160	–	24	HA	–	–	–	✓	✓	–	✓	–	–	Moremi et al., 2012
	2010	Apparently healthy children under 5 years (nasal swabs)	114	*nuc* gene PCR	12	CA	–	–	–	–	–	✓	–	–	–	Moyo et al., 2014
Tunisia	1998–2007	Clinical specimens from neutropenic patients	72	*nuc* gene PCR	13	HA (13)	✓	–	–	–	–	✓	–	–	–	Bouchami et al., 2009
	2002	Patients who developed MRSA infections	NR	–	6	HA (6)	–	–	–	–	–	✓	–	–	–	Ben Ayed et al., 2010
	2003–2004	Pus, blood, pleural fluid, venous catheter	NR	–	72	NR	✓	–	✓	–	–	–	✓	–	✓	Ben Nejma et al., 2006
	2003–2004	Pathological samples from different wards	147	–	19	NR	–	–	–	–	–	✓	–	–	–	Ben Saida et al., 2005
	2003–2005	Pus and associated with cutaneous infections	NR	–	64	CA (64)	✓	–	✓	✓	✓	✓	✓	–	✓	Ben Nejma et al., 2013
	2004	Cutaneous pus, blood cultures, urine and puncture fluids	NR	–	34	HA (32), CA (2)	✓	–	–	–	–	–	–	–	–	Ben Jomaa-Jemili et al., 2006
	2004–2005	Cutaneous pus, RTS, urine, blood culture,	475	–	57	NR	–	–	✓	–	–	–	–	–	–	Ben Ayed et al., 2006
	2004–2008	Samples from hospitals and community	NR	–	69	HA (41), CA (28)	✓	–	✓	✓	✓	–	✓	–	–	Ben Jomàa-Jemili et al., 2013
	2006–2008	Children with CA invasive infections bacteraemia and osteomyelitis	36	–	8	CA (8)	✓	–	✓	✓	✓	✓	✓	–	–	Kechrid et al., 2011
	2007	Pus and skin infections	NR	–	11	CA (11)	✓	–	✓	✓	✓	✓	✓	–	–	Ben Nejma et al., 2009
	2008	Pus and blood culture (case report)	2	–	2	NR	✓	–	✓	✓	✓	–	✓	–	✓	Zribi et al., 2011
	2008–2009	Humans in contact with animals	55	–	1	CA (1)	✓	–	✓	✓	✓	–	✓	✓	✓	Ben Slama et al., 2011

(Continued)

TABLE 1 | Continued

Country	Study period	Study population (sample type)	No. of *S. aureus* isolates	*S. aureus* molecular identification	No. of MRSA[¶]	Setting (no.)	Genotyping tools						*PVL*	Detection of genes		References
							SCC*mec*	*coa*	*agr*	*spa* typing	MLST	PFGE		Antibiotic resistance	Toxin/ Virulence	
Uganda	2009–2010	Swabs from patients, HCW and from hospital environment (sinks, door handles, surgical trays, bed and table surfaces)	41	–	41	NR	✓	–	–	–	–	–	✓	✓	✓	Kateete et al., 2011
	2011–2012	SSI	64	*nuc* gene PCR	24	NR	✓	–	–	✓	–	–	✓	–	–	Seni et al., 2013
Multicenter[#]	2007–2008	SSTI, bacteraemia/septicaemia, urine, wounds osteomyelitis and myositis	NR	–	86	CA (9), HA (77)	✓	–	✓	✓	✓	–	✓	–	✓	Breurec et al., 2011b
Multicentre[✖]	2004–2005	Uncomplicated skin infections	292	–	105	HA (3)	✓	–	–	–	✓	✓	✓	–	–	Goering et al., 2008
STUDIES CONDUCTED IN ANIMALS																
Egypt	NR	Cows and buffaloes milk, cattle septic wounds	9	–	5	NR	–	✓	–	✓	–	–	–	–	–	El-Jakee et al., 2011*
Senegal	2009–2011	Pigs (nasal swabs)	73	–	6	NR	✓	–	–	✓	✓	–	✓	–	✓	Fall et al., 2012
Tunisia	2010	Healthy sheep (nasal swabs)	73	–	5	CA (6)	✓	–	✓	✓	✓	✓	–	–	✓	Gharsa et al., 2012

agr, Accessory gene regulator; CA, Community-acquired methicillin-resistant S. aureus; coa, Coagulase gene; CVCT, Central venous catheter tips; GIT, Genital tract infections; gyrA, DNA gyrase gene; HA, Hospital-acquired methicillin resistant S. aureus; HCW, Health care workers; HIV, Human immunodeficiency virus; KZN, KwaZulu-Natal province; MLST, Multilocus locus sequence typing; MRSA, Methicillin-resistant Staphylococcus aureus; No., Number of isolates; NR, Not reported; nuc, Thermonuclease gene; PCR, Polymerase chain reaction; PFGE, Pulsed-field gel electrophoresis; PVL, Panton-Valentine-Leukocidin genes; SCCmec, Staphylococcal chromosomal cassette mec element; RTI, Respiratory tract infections; rRNA, Ribosomal ribonucleic acid; RTS, Respiratory tract specimens; spa, Staphylococcus aureus protein A gene; SSTI, Skin and soft tissue infections; SSI, Surgical site infections; tuf, Elongation factor tu; UTI, Urinary tract infections.

*[¶]MRSA as confirmed by mecA PCR; ✓, Test was conducted; –, Test was not conducted; [α]Hospitalized individuals; [β]Nasal carrier study; [#]African multicenter study which included Cameroon, Madagascar, Morocco, Niger and Senegal; [✖]An international multicenter study which included only South Africa; *Study was conducted in both animal and human host.*

and the elongation factor tu - *nuc, tuf* - genes) was performed in only 12.3% (7/57). The screening for antibiotic resistance and toxin/virulence genes were carried out in seven and 22 studies, respectively (**Table 1**). Furthermore, all the eligible studies analyzed MRSA using at least one genotyping technique, and 59.6% (34/57) provided adequate genotyping data on MRSA clones from 15 African countries (**Tables 1, 2**). Studies included in this systematic review did not investigate on the *mecC* gene.

Community- and Hospital- Acquired MRSA

Overall, 51% (29/57) of the eligible studies provided the potential source (hospital- or community-associated) of the MRSA strains. Only 17.5% (10/57) reported MRSA from community settings (**Table 1**). USA300 (ST8-IV [2B]) and other related sequence types were noted both in health care and community settings in nine African countries (**Tables 1, 2**). The "Brazilian/Hungarian clone" (ST239-III [3A]) was associated with hospital-acquired infections in nine countries (**Tables 1, 2**). Furthermore, the "West Australia MRSA-2" (ST88-IV [2B]) was reported in community- and hospital-acquired infections in several African countries (**Table 2**).

Detection of Panton-Valentine Leukocidin (PVL) Genes

The screening for PVL-associated genes (lukF-PV and lukS-PV) was carried out in 44 studies, and the detection of PVL genes was only reported in 32 studies (**Table 1**). In animals, PVL-positive MRSA (ST5) was described in nasal samples of pigs from Senegal (Fall et al., 2012). In humans, the proportion of PVL-positive MRSA carriage and/or infections ranged from 0.3 to 100%. Studies from Algeria and Tunisia reported higher PVL prevalence while investigations from South Africa reported the lowest prevalence (**Table 3**). Overall, PVL-positive MRSA were more frequently reported with skin and soft tissue infections, and community-associated clones (**Tables 1, 2**). There was no report on the role of PVL in necrotizing pneumonia caused by MRSA in Africa.

MRSA Clones Reported in Africa Using the Current eBURST Scheme

Figures 2, 3 summarize the MRSA clones identified in Africa based on MLST CCs. By the current eBURST scheme, six main CCs were identified: CC5, CC22, CC30, CC45, CC80, and CC88. In addition, a number of diverse *spa* types were identified among the MRSA clones in Africa (**Table 2**), but the distribution of *spa* types t042 and t044 (associated with CC80-IV [2B]) appear to be limited to three North African (Algeria, Egypt and Tunisia) countries (**Table 2**).

Clonal Complex 5

This clonal complex is considered the largest group based on the eBURST scheme (**Figure 3**). It was subdivided into three main clusters and designated as CC5-ST1, ST5, and ST8.

MRSA CC5 with Sequence Type 1

This group was reported in Nigeria (Raji et al., 2013) and Tunisia (Ben Jomàa-Jemili et al., 2013). The clonal type included the PVL-positive ST1-V [5C] isolated from patients in a tertiary hospital in Nigeria (Raji et al., 2013), and the PVL-negative ST1 with a non-typeable SCC*mec* element (*spa* type t035 and *agr* type III) identified in Tunisia (Ben Jomàa-Jemili et al., 2013). In addition, an ST1 related sequence type (ST772-V [5C]), "the Bengal-Bay clone" has been described in Nigeria (Raji et al., 2013).

MRSA CC5 with Sequence Type 5

This clone was documented in 14 studies and consisted of diverse SCC*mec* elements (**Table 2**). The ST5-I [1B]/III [3A] were identified from clinical samples in health care institutions in South Africa (Shittu et al., 2009; Moodley et al., 2010; Jansen van Rensburg et al., 2011; Oosthuysen et al., 2014). ST5-II [2A] has been described in Nigeria (Raji et al., 2013), and Senegal (Breurec et al., 2011b). ST5-IV [2B]-PVL-positive was the dominant clone in hospitalized patients with skin and soft tissue infections in Dakar, Senegal (Breurec et al., 2011b). In addition, ST5-IV [2B] was detected from nasal samples of pigs in the same geographical area (Fall et al., 2012). ST5-IV [2B] has also been identified in Algeria (Ramdani-bouguessa et al., 2006), Gabon (Schaumburg et al., 2011; Ateba Ngoa et al., 2012), Morocco (Breurec et al., 2011b), and South Africa (Essa et al., 2009), while the SCC*mec* IVa [2B] variant was recovered from hospitalized patients in Algeria (Ramdani-bouguessa et al., 2006; Bekkhoucha et al., 2009), Angola (Conceição et al., 2014), and São Tomé and Príncipe (Conceição et al., 2013). Moreover, ST5-IVa [2B] was reported from nasal samples of apparently healthy-hospital workers in Angola (Conceição et al., 2014). Other ST5 and related clones identified are ST5-V [5C] in Angola (Conceição et al., 2014), Cameroon (Breurec et al., 2011b), and Nigeria (Shittu et al., 2011), ST72-SCC*mec*-NT in South Africa (Jansen van Rensburg et al., 2011), ST72-V [5C] in Angola and Ghana (Egyir et al., 2013, 2014b; Conceição et al., 2014), and ST105-II [2A] from a patient in São Tomé and Príncipe (Conceição et al., 2013). Furthermore, ST650-IV [2B] was detected from clinical samples in health care institutions in South Africa (Jansen van Rensburg et al., 2011). Finally, ST2629-V [5C] was described in Angola (Conceição et al., 2014).

MRSA CC5 with Sequence Type 8

MRSA assigned to this clone are widespread and diverse across Africa as evidenced in 27 studies (**Table 2**). The first known early or ancestral clone, ST250-I [1B], was mainly associated with hospital-acquired infections in Ibadan, South-West Nigeria (Ghebremedhin et al., 2009), and recently observed in Ghana (Egyir et al., 2014b). ST8-II [2A] was only described in the KwaZulu-Natal region of South Africa (Essa et al., 2009), while a number of investigations reported ST8-IV [2B] in Angola (Conceição et al., 2014), Cameroon (Breurec et al., 2011b), Gabon (Schaumburg et al., 2011; Ateba Ngoa et al., 2012), Ghana (Egyir et al., 2014b), Madagascar (Breurec et al., 2011b), Nigeria (Adesida et al., 2005; Shittu et al., 2012), São Tomé and Príncipe (Conceição et al., 2013) and South Africa (Essa et al., 2009). The MRSA isolates from Angola possessed the SCC*mec* type IVd element (Conceição et al., 2014). ST612-IV [2B], a double locus variant (dlv) of ST8-IV [2B], and only recently reported as PVL-positive (Oosthuysen et al., 2014), is widespread

TABLE 2 | Methicillin resistant *Staphylococcus aureus* clones reported in 34 eligible studies.

Country	Clonal type ST-SCC*mec*	Clonal complex	*spa* type	PVL status	*agr*	References
Algeria	ST80-IV [2B]	80	ND	+	III	Ramdani-bouguessa et al., 2006
	ST37-IVa [2B]	30	ND	+	III	
	ST239-III [3A]	5	ND	−	I	
	ST239-IVa [2B]	5	ND	−	I	
	ST241-III [3A]	5	ND	−	I	
	ST637-III [3A]	5	ND	−	I	
	ST5-IV, IVa [2B]	5	ND	+, −	II	
	ST635-IV [2B]	80	ND	−	III	
	ST636-NT	22	ND	−	I	
	ST80-IV [2B]	80	t044	+	ND	Bekkhoucha et al., 2009
	ST239-III [3A]	5	t037	−	ND	
	ST5-IVa [2B]	5	t311, t450	−	ND	
	ST88-NT	88	t188, t267	−	ND	
	ST80-IV [2B]	80	t044, t4143	+	III	Antri et al., 2011
	ST241-III [3A]	5	ND	−	III	
	ST8-V [5C]	5	ND	−	I	
	ST80-IVc [2B]	80	ND	+, −	ND	Djoudi et al., 2013
	ST39-II [2A]	30	ND	−	ND	
Angola	ST5-IVa [2B]	5	t105, t311, t11657	−	II	Conceição et al., 2014
	ST8-IVd, V [2B]	5	t104, t1774	−	I	
	ST72-V [5C]	5	t3092	−	I	
	ST88-IVa [2B]	88	t186, t325, t786, t1951, t3869	−	III	
	ST5-V [5C]	5	t6065	−	II	
	ST2629-V [5C]	5	t6065	−	II	
	ST789-V [5C]	7	t091	−	II	
Cameroon	ST5-V [5C]	5	t311	+	II	Breurec et al., 2011b
	ST88-IV [2B]	88	t186	−	III	
	ST8-IV [2B]	5	t024, t121, t451	+	I	
	ST1289-V [5C]	88	t1339	+	III	
Egypt	ST80-IVc [2B]	80	t042, t044, t070, t983	+	III	Enany et al., 2010
	ST30-IVa [2B]	30	t251, t318	+	III	
	ST1010-X[a]	121	t159, t312	+	IV	
Gabon	ST8-IV [2B]	5	t121	+	I	Schaumburg et al., 2011; Ateba Ngoa et al., 2012
	ST88-IV [2B]	88	t186	−	III	
	ST5-IV [2B]	5	t653	−	II	
	ST5-IV [2B]	5	t653	−	ND	Schaumburg et al., 2014
	ST8-NT	5	t112, t121	+	ND	
	ST45-V [5C]	45	t437, t8860	−	ND	
	ST88-IV [2B]	88	t4195	−	ND	
Ghana	ST72-V [5C]	5	t537	−	ND	Egyir et al., 2013
	ST8-V [5C]	5	t064	−	ND	
	ST88-IV [2B]	88	t325, t1951, t2649	−	ND	
	ST72-V [5C]	5	t537	−	ND	Egyir et al., 2014b
	ST8-IV [2B]	5	t121	+	ND	
	ST239-III [3A]	5	t037	−	ND	
	ST250-I [1B]	5	t928	−	ND	
	ST2021-V [5C]	5	t024	−	ND	

(Continued)

TABLE 2 | Continued

Country	Clonal type ST-SCC*mec*	Clonal complex	*spa* type	PVL status	*agr*	References
	ST88-IV [2B]	88	t186	–	ND	
	ST789-IV [2B]	7	t547	+	ND	
	ST508-V [5C]	45	t5132	–	ND	Egyir et al., 2014a
Kenya	ST239-III [3A]	5	t037	–	ND	Aiken et al., 2014
Madagascar	ST8-IV [2B]	5	t121	+	I	Breurec et al., 2011b
	ST30-V [5C]	30	t4686	–	III	
	ST88-IV [2B]	88	t186	–	III	
Morocco	ST239, ST241-III [3A]	5	t037, t138	–	I	
	ST5-IV [2B]	5	t311	+	II	
Niger	ST239, ST241-III [3A]	5	t138	–	I	
	ST239, ST241-V [5C]	5	t037	–	I	
	ST88-IV [2B]	88	t186	–	III	
Nigeria	ST8-IV [2B]	5	ND	–	ND	Adesida et al., 2005
	ST88-IV [2B]	88	t186	+	III	Ghebremedhin et al., 2009
	ST241-IV [2B]	5	t037	–	I	
	ST250-I [1B]	5	t194, t292	–	I	
	ST241-III [3A]	5	t037	–	ND	Shittu et al., 2011
	ST8-V [5C]	5	t064	–	ND	
	ST8-V [5C]	5	t451	–	ND	
	ST94-IV [2B]	5	t008	–	ND	
	ST5-V [5C]	5	t002	–	ND	
	ST241-III [3A]	5	t037	–	ND	Shittu et al., 2012
	ST88-IV [2B]	88	t729, t1603	–	ND	
	ST37-III [3A]	30	t074	–	ND	
	ST39-II [2A]	30	t007	–	ND	
	ST8-V [5C],IV [2B], ST8-NT	5	t064	–	ND	
	ST152-NT	152	t4690	+	ND	
	ST1-V [5C]	5	ND	+	ND	Raji et al., 2013
	ST239-III[3A]$_{mercury}$	5	ND	–	ND	
	ST5-II [2A]	5	ND	–	ND	
	ST8-V [5C]	5	ND	–	ND	
	ST247-I [1B]	5	ND	–	ND	
	ST772-V [5C]	5	ND	+	ND	
	ST88-IV [2B]	88	ND	–	ND	
	ST241-III [3A]	5	ND	–	ND	Okon et al., 2009
Senegal	ST5-IV [2B]*	5	t311	+	ND	Fall et al., 2012
	ST88-IV [2B]*	88	t3489	–	ND	
	ST239, ST241-III [3A]	5	t037, t138	–	I	Breurec et al., 2011b
	ST5-II [2A]	5	t311	+	II	
	ST5-IV [2B]	5	t311	+	II	
	ST88-IV [2B]	88	t168	–	III	
South Africa	ST5-IV [2B]	5	ND	ND	ND	Essa et al., 2009
	ST8-IV [2B]	5	ND	ND	ND	
	ST8-II [2A]	5	ND	ND	ND	

(Continued)

TABLE 2 | Continued

Country	Clonal type ST-SCC*mec*	Clonal complex	*spa* type	PVL status	*agr*	References
	ST239-III [3A]	5	ND	ND	ND	
	ST45-IV [2B]	45	ND	ND	ND	
	ST612-IV [2B]	5	ND	–	ND	Goering et al., 2008
	ST36-II [2A]	30	ND	–	ND	
	ST1173-IV [2B]	5	t064	–	ND	Shittu et al., 2009
	ST1338-IV [2B]	5	t064	–	ND	
	ST239-III [3A]	5	t037	–	ND	
	ST5-III [3A]	5	t045	–	ND	
	ST239-III [3A]	5	t037	–	ND	Moodley et al., 2010
	ST612-IV [2B]	5	t064	–	ND	
	ST5-I [1B]	5	t045	–	ND	
	ST22-IV [2B]	22	t032	–	ND	
	ST22-IV [2B]	22	t891	+	ND	
	ST36-II [2A]	30	t012	–	ND	
	ST239-III [3A]	5	t037	–	ND	Jansen van Rensburg et al., 2011
	ST5-I [1B]	5	t045	ND	ND	
	ST650-IV [2B]	5	t002	ND	ND	
	ST612-IV [2B]	5	t064, t1443, t2196	ND	ND	
	ST72-NT	5	t3092	ND	ND	
	ST22-IV [2B]	22	t032	ND	ND	
	ST36-II [2A]	30	t012, t021	ND	ND	
	ST5-I [1B]	5	t002	–	II	Oosthuysen et al., 2014
	ST8-V [5C]	5	t064	–	I	
	ST612-IV [2B]	5	t064	+	I	
	ST239-III [3A]	5	t021	–	I	
	ST22-V [5C]	22	t891	+	I	
	ST22-IV [2B]	22	t891	–	I	
	ST36-II [2A]	30	t021	–	III	
São Tome and Príncipe	ST5-IVa [2B]	5	t105	–	II	Conceição et al., 2013
	ST105-II [2A]	5	t002	–	II	
	ST8-V [5C]	5	t451	–	I	
	ST8-IV [2B]	5	t451	–	I	
	ST88-IVa [2B]	88	t186, t786	–	III	
Tunisia	ST80-IV [2B]	80	t044	+	III	Ben Nejma et al., 2009
	ST80-IV [2B]	80	t044	+	III	Ben Nejma et al., 2013
	ST728-IVc [2B]	80	t042, t044	+	III	Kechrid et al., 2011
	ST8-IVc [2B]	5	t062	+	II	
	ST80-IVc [2B]	80	t203	+	III	Ben Slama et al., 2011
	ST1-NT	5	t035	–	III	Ben Jomàa-Jemili et al., 2013
	ST247-I [1B]	5	t040	–	I	
	ST239-III [3A]	5	t003	–	I	
	ST241-III [3A]	5	t125	–	I	
	ST97-NT	5	t003	–	I	
	ST1819-I [1B]	5	NS	–	I	
	ST80-IVc [2B]	80	t070	+	III	
	ST2563-IVc [2B]	80	t070	+	III	

(Continued)

TABLE 2 | Continued

Country	Clonal type ST-SCC*mec*	Clonal complex	*spa* type	PVL status	*agr*	References
	ST1440-IVc [2B]	80	t070	+	III	
	ST80-IVc [2B]	80	t1021	–	II	
	ST80-IVc [2B]	80	ND	–	III	
	ST22-NT	22	t998	–	II	
	ST45-NT	45	ND	–	I	
	ST153-NT	80	NST	+	III	
	ST153-NT	80	t044	ND	III	Gharsa et al., 2012

agr, Accessory gene regulator; CC, Clonal complex; NT, Non typeable; ND, Not determined; NST, New spa type; PVL, Panton-Valentine Leukocidin genes; SCCmec, Staphylococcal chromosomal cassette mec element; ST, Sequence type; spa, Staphylococcus aureus protein A gene; X^a*, Unknown SCCmec type other than I, II, III, IV or V; *Clones isolated from pigs; +, PVL positive; –, PVL negative.*

TABLE 3 | Panton-Valentine Leukocidin prevalence as reported by the eligible studies with sample size of 30 or above.

Country	PVL positive (no. positive/total tested)	Prevalence (%)	References
Algeria	46/61	75	Ramdani-bouguessa et al., 2006
	19/64	30	Ouchenane et al., 2011
	94/122	77	Antri et al., 2011
Kenya	14/69	20	Maina et al., 2013
Libya	10/35	29	Al-haddad et al., 2014
Nigeria	33/70	47	Ghebremedhin et al., 2009
South Africa	1/320	0.3	Moodley et al., 2010
	4/97	4	Makgotlho et al., 2009
	5/56	9	Oosthuysen et al., 2014
Tunisia	68/72	94	Ben Nejma et al., 2006
	64/64	100	Ben Nejma et al., 2013
	43/69	62	Ben Jomàa-Jemili et al., 2013
Uganda	30/41	73	Kateete et al., 2011
Multicenter*	20/86	23	Breurec et al., 2011b

**Multicenter study which included Cameroon, Madagascar, Morocco, Niger and Senegal. PVL, Panton-Valentine Leukocidin; no., Number.*

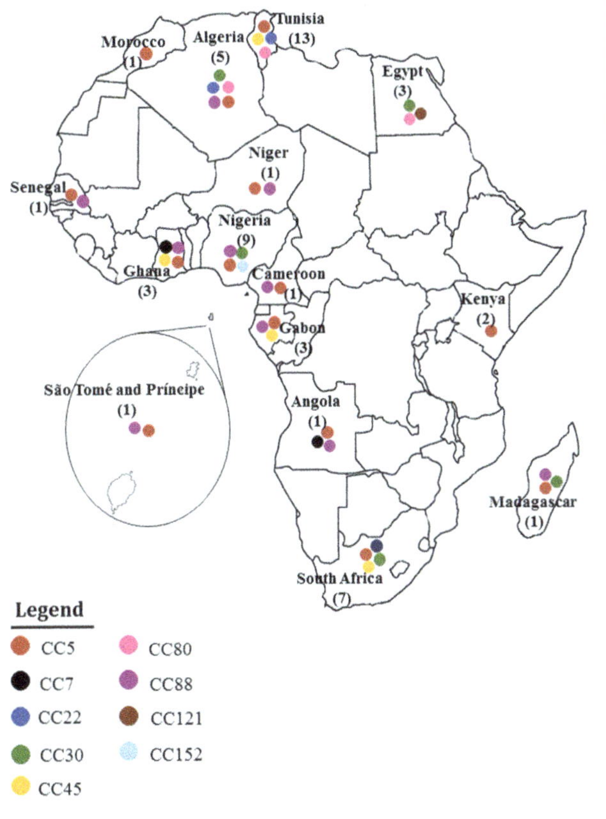

FIGURE 2 | MRSA clones reported in Africa. Each clonal complex is annotated with a colored circle. The number of studies conducted in each country is also indicated.

across South Africa (Goering et al., 2008; Moodley et al., 2010; Jansen van Rensburg et al., 2011; Oosthuysen et al., 2014), alongside other variants such as ST1173/ST1338-IV [2B] (Shittu et al., 2009). The ST8-IV [2B] clone in South Africa was identified from a variety of clinical infections, in particular, bacteraemia, skin and soft tissue and wound infections (Shittu et al., 2009; Moodley et al., 2010; Jansen van Rensburg et al., 2011; Oosthuysen et al., 2014). An ST8-IVc [2B] strain (PVL-positive; *spa* type t062) was identified from a 4 day old male child with community-acquired invasive infection in Tunisia (Kechrid et al., 2011). Furthermore, ST8-V [5C] was described in Algeria (Antri et al., 2011), Angola (Conceição et al., 2014), Ghana (Egyir et al., 2013), Nigeria (Shittu et al., 2011, 2012; Raji et al., 2013), São Tomé and Príncipe (Conceição et al., 2013), and South Africa (Oosthuysen et al., 2014). Other STs observed within the CC5-ST8 cluster include ST8-SCC*mec*-NT in Gabon (Schaumburg et al., 2014) and Nigeria (Shittu et al., 2012), ST94-IV [2B] described in Nigeria (Shittu et al., 2011) and ST97-SCC*mec*-NT in Tunisia (Ben Jomàa-Jemili et al., 2013). In addition, ST247-I [1B] was reported only in Tunisia (Ben Jomàa-Jemili et al., 2013)

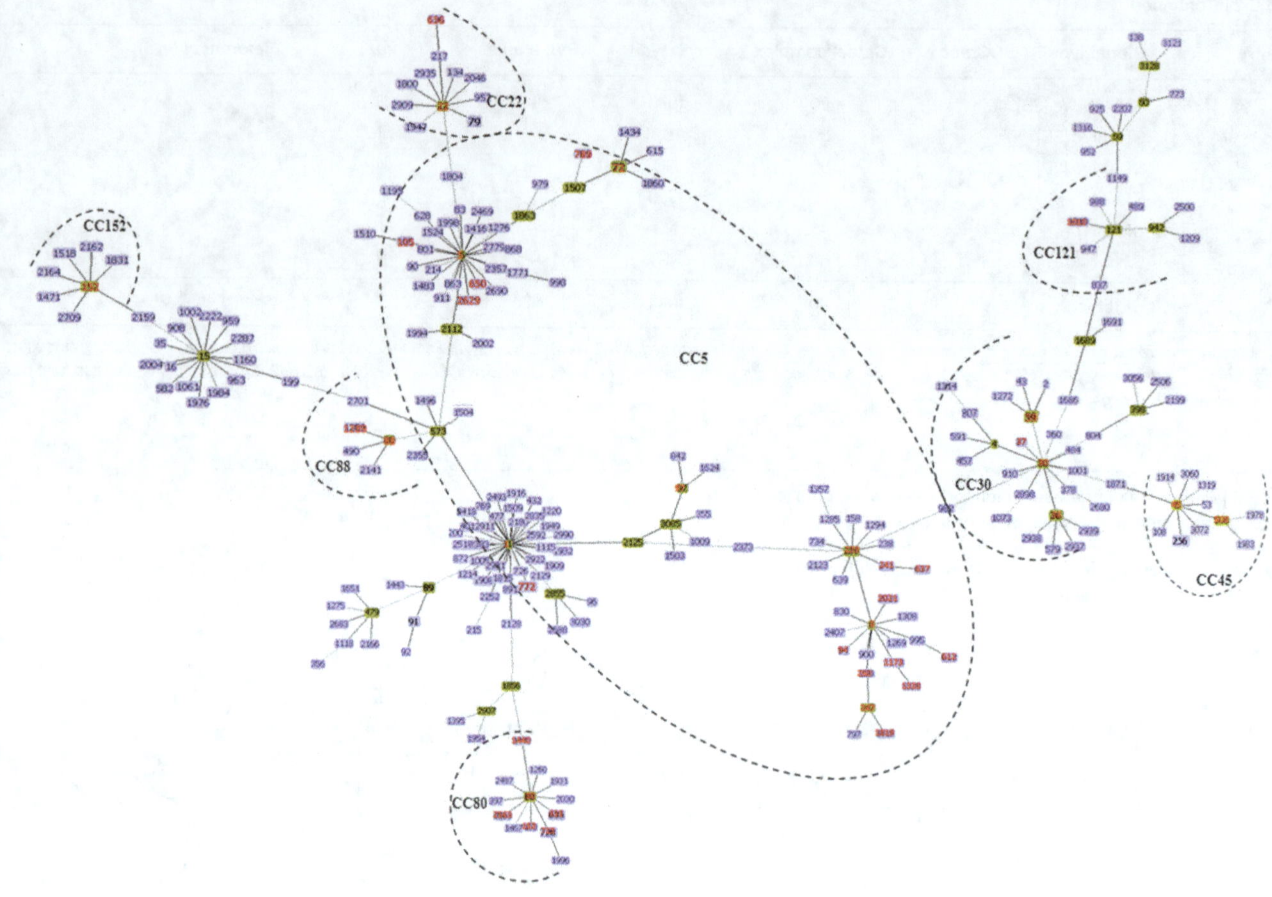

FIGURE 3 | The minimum spanning tree was constructed by the goeBURST algorithm using the Phyloviz software v1.1 (http://www.phyloviz.net/). The allelic profiles were downloaded from the MLST website (http://saureus.mlst.net/) which included the MRSA sequence types (STs) described in this review as well as 223 randomly selected STs (from each CC) based on the differences in their allelic profiles. The Group founder is colored in green and the related STs are in blue. The six main CCs described in this review are indicated by the dotted lines and the STs reported in Africa are indicated by the red color.

and Nigeria (Raji et al., 2013), ST637-III [3A] in Algeria (Ramdani-bouguessa et al., 2006), ST1819-I [1B] in Tunisia (Ben Jomàa-Jemili et al., 2013), and ST2021-V [5C] in Ghana (Egyir et al., 2014b).

The "Brazilian/Hungarian clone" (ST239-III [3A]) is an hybrid of CC30 and CC8 based on a single large chromosomal replacement (Robinson and Enright, 2004), and ST241-III [3A] is a single locus variant (slv) of ST239-III [3A]. These two STs were identified concurrently in Algeria (Ramdani-bouguessa et al., 2006), Morocco, Niger and Senegal (Breurec et al., 2011b), and Tunisia (Ben Jomàa-Jemili et al., 2013). ST239-III [3A] has also been described in Ghana (Egyir et al., 2014b) and Kenya (Aiken et al., 2014), and consistently since 2001 in South Africa (Essa et al., 2009; Shittu et al., 2009; Moodley et al., 2010; Jansen van Rensburg et al., 2011; Oosthuysen et al., 2014). A recent study detected ST239 with the SCC*mec* type $III_{mercury}$ [3A] in a tertiary health care facility in South-West Nigeria (Raji et al., 2013). ST241-III [3A] is the dominant clone in North-East Nigeria (Okon et al., 2009; Shittu et al., 2011, 2012). Interestingly, three SCC*mec* variants, ST239-IVa [2B], ST239/ST241-V [5C], and ST241-IV [2B], and associated with hospital-acquired infections were reported in Algeria (Ramdani-bouguessa et al., 2006), Niger (Breurec et al., 2011b), and Nigeria (Ghebremedhin et al., 2009).

Clonal Complex 22

In Africa, ST22 was identified only in Algeria (Ramdani-bouguessa et al., 2006), South Africa (Moodley et al., 2010; Jansen van Rensburg et al., 2011; Oosthuysen et al., 2014), and Tunisia (Ben Jomàa-Jemili et al., 2013). ST22-IV [2B] was related with hospital-associated infections in the Western Cape and KwaZulu-Natal provinces of South Africa. A variant of ST22 (ST22-V [5C]-PVL-positive) was also reported in an hospital in Western Cape, South Africa (Oosthuysen et al., 2014). The ST22 identified in Tunisia possessed a non-typeable SCC*mec* element (Ben Jomàa-Jemili et al., 2013). Besides, an ST636-SCC*mec*-NT isolate has also been reported in Algeria (Ramdani-bouguessa et al., 2006).

Clonal Complex 30

ST30-IVa [2B]-PVL-positive, also known as "South-West Pacific clone," has been reported in Egypt (Enany et al., 2010), and a multicenter African study identified ST30-V [5C] only in Antananarivo, Madagascar (Breurec et al., 2011b). The hospital associated ST36-II [2A] (UK-EMRSA-16), was described only in South Africa (Goering et al., 2008; Moodley et al., 2010; Jansen van Rensburg et al., 2011; Oosthuysen et al., 2014), while ST39-II [2A] a dlv was identified in an hospital in Algiers, Algeria (Djoudi et al., 2013), and Ile-Ife, South-West Nigeria (Shittu et al., 2012). MRSA assigned with these groups (ST36-II [2A] and ST39-II [2A]) were PVL-negative. Furthermore, two SCC*mec* variants, ST37-IVa [2B] and ST37-III [3A], were reported in Algeria (Ramdani-bouguessa et al., 2006) and Nigeria (Shittu et al., 2012), respectively.

Clonal Complex 45

ST45-IV [2B], the "Berlin clone," was detected in an hospital in the KwaZulu-Natal (South Africa) during a multicenter surveillance study (Essa et al., 2009) and ST45-V [5C] was reported in mother-infant pairs in Gabon (Schaumburg et al., 2014). An MRSA with a non-typeable SCC*mec* associated with community-acquired infections has been identified in Tunisia (Ben Jomàa-Jemili et al., 2013). Finally, ST508-V [5C], a slv to ST45, and also associated with community-acquired infections was described in Ghana (Egyir et al., 2014a).

Clonal Complex 80

The CC80 was limited to three North African countries: Algeria, Egypt, and Tunisia (**Table 2**). The European clone, ST80-IV [2B]-PVL-positive, was first described in Algeria from both hospitalized and outpatients (Ramdani-bouguessa et al., 2006), and has continued to be the leading clone in the country (Ramdani-bouguessa et al., 2006; Bekkhoucha et al., 2009; Antri et al., 2011; Djoudi et al., 2013). ST80-IVc [2B] has been identified in Egypt (Enany et al., 2010), and Tunisia (Ben Slama et al., 2011; Ben Jomàa-Jemili et al., 2013). In addition, sequence types related to ST80 have been recovered from human clinical samples (ST153-SCC*mec*-NT, ST728-IVc [2B], ST1440-IVc [2B], and ST2563-IVc [2B]) (Kechrid et al., 2011; Ben Jomàa-Jemili et al., 2013), and nasal specimen of healthy sheep (ST153-SCC*mec*-NT) (Gharsa et al., 2012) in Tunisia. The afore-mentioned sequence types, ST728, ST1440 and ST2563 belonged to accessory gene regulator (agr) type III and were PVL-positive. Moreover, a PVL-negative ST80-IVc [2B] with *agr* type II has also been detected in Tunisia (Ben Jomàa-Jemili et al., 2013), and a PVL-negative ST635-IV [2B] in Algeria (Ramdani-bouguessa et al., 2006).

Clonal Complex 88

The "West Australia MRSA-2 clone" (WA-MRSA-2), ST88-IV [2B], was reported in both hospital and community settings in eight African countries; Angola (Conceição et al., 2014), Cameroon (Breurec et al., 2011b), Gabon (Schaumburg et al., 2011; Ateba Ngoa et al., 2012), Ghana (Egyir et al., 2013, 2014b), Madagascar (Breurec et al., 2011b), Niger (Breurec et al., 2011b), Nigeria (Ghebremedhin et al., 2009; Shittu et al., 2012; Raji et al., 2013) and Senegal (Breurec et al., 2011b). The MRSA isolates from Angola possessed an SCC*mec* IVa [2B] element (Conceição et al., 2014). PVL-positive ST88-IV [2B] were detected in Nigeria (Ghebremedhin et al., 2009), and an SCC*mec* subtype ST88-IVa [2B] was identified among three health care workers and a patient in São Tomé and Príncipe (Conceição et al., 2013). The ST88-IV [2B] with *spa* type t3489 was also recovered from nasal samples of swine in Senegal (Fall et al., 2012). In addition, an SCC*mec* non-typeable ST88 was described from an out-patient in Algeria (Bekkhoucha et al., 2009), and a strain related to WA-MRSA-2 (ST1289-IV [2B]) was identified in Yaoundé, Cameroon (Breurec et al., 2011b).

Other Clonal Complexes

CC7, CC121, and CC152 have been reported in Africa. The PVL-negative ST789 (assigned with CC7) was identified in Angola (with SCC*mec* IV [2B]) (Conceição et al., 2014). However, in Ghana, ST789 was PVL-positive and carried an SCC*mec* IV element [2B] (Egyir et al., 2014b). An ST1010-PVL-positive (CC121) with non-typeable SCC*mec* element has only been described in Egypt (Enany et al., 2010). Furthermore, PVL-positive MRSA assigned to CC152 (ST152-SCC*mec*-NT) was reported in Nigeria (Shittu et al., 2012).

Discussion

MRSA has been reported in Africa, at least since 1978 (Scragg et al., 1978). This systematic review showed that adequate data on the molecular epidemiology of MRSA are limited, with reports from only 15 of the 54 African countries. No *spa* type was dominant, however, t042 and t044 were the major *spa* types identified in three North African countries (**Table 2**). Moreover, we did not observe a clear distinction between hospital- and community-associated MRSA clones in Africa which is in agreement with other investigations worldwide (Fossum Moen et al., 2013; Pasquale et al., 2013; Sherwood et al., 2013; Tavares et al., 2013). In this systematic review, the use of the current eBURST scheme grouped several African MRSA CCs (CC1, CC5, CC8, and CC7) into a single cluster, (CC5). This raises some concern on a suitable method for discrimination and grouping of *S. aureus* isolates. To overcome the above mentioned issue, whole genome sequencing approach might be the alternative (Dabul and Camargo, 2014).

Although a combination of factors could be responsible for the dissemination of clones between continents, increased movement of human population within or between countries might be one of the potential factors (Rogers et al., 2011). International travel could play a significant role in the transmission of MRSA, particularly the replacement of existing MRSA with fitter and more transmissible clones (Zhou et al., 2014). We observed that the predominant hospital-associated epidemic clones, EMRSA-15 [ST22-IV [2B]) and (EMRSA-16 [ST36-II [2A]), in the United Kingdom (UK) (Johnson et al., 2005) were reported only in South Africa (Goering et al., 2008; Moodley et al., 2010; Jansen van Rensburg et al., 2011; Oosthuysen et al., 2014). Moreover, ST80-IV [2B] (the European clone) has consistently been recognized as the predominant PVL-positive MRSA clone in North Africa

(Ramdani-bouguessa et al., 2006; Bekkhoucha et al., 2009; Ben Nejma et al., 2009, 2013; Enany et al., 2010; Antri et al., 2011; Ben Slama et al., 2011; Ben Jomàa-Jemili et al., 2013; Djoudi et al., 2013). A recent report based on whole genome analysis provided strong evidence that the European ST80-IV [2B] was derived from a PVL-positive MSSA ancestor in sub-Saharan Africa that acquired the SCC*mec* IV element, and clonal spread was enhanced by increased transnational movement (Stegger et al, 2014). However, the factors responsible for the limited spread of the ST80-IV [2B] only in North Africa observed so far are unclear.

The SCC*mec* IV (and its subtypes) and SCC*mec* V were identified in several MRSA clones, and ST5 and ST8 clearly showed more diversity in terms of SCC*mec* types compared with other STs in Africa. The success of these SCC*mec* types (IV and V) could be due to their small sizes and low fitness costs (Enright et al., 2002; Okuma et al., 2002; Monecke et al., 2011). It is also noteworthy that the SCC*mec* types IVa and IVc were identified in genetically unrelated clones, e.g., ST5-IVa [2B] (CC5) in Algeria (Ramdani-bouguessa et al., 2006; Bekkhoucha et al., 2009), São Tomé and Príncipe (Conceição et al., 2013), ST8-IVc [2B] in Tunisia (CC5) (Kechrid et al., 2011), and ST37-IVa [2B] (CC30) in Algeria (Ramdani-bouguessa et al., 2006). This might suggest horizontal gene transfer or independent acquisition (Mašlaòová et al., 2013). Another interesting observation was the detection of the SCC*mec* type IVa and V in the hospital-associated ST239/ST241-III [3A] in Algeria (Ramdani-bouguessa et al., 2006), Nigeria (Ghebremedhin et al., 2009), and Niger (Breurec et al., 2011b). Since ancient MSSA strains for this ST have not been reported (Enright et al., 2002), our observation suggests that acquisition of these SCC*mec* types by MSSA is less likely, and points to the possible replacement of SCC*mec* type III with IV and V on the ST239/241 genome (Li et al., 2013).

Data on the epidemiology of MRSA in animals are limited in Africa (EL Seedy et al., 2012; Fall et al., 2012; Gharsa et al., 2012). Moreover, the genetic relatedness between human and animal MRSA has not been investigated (**Table 2**). It should be noted, however, that human-associated ST5-IV [2B], ST88-IV [2B], and ST153-SCC*mec*-NT have been reported from animals in Tunisia (Gharsa et al., 2012) and Senegal (Fall et al., 2012). Recently, human-associated *S. aureus* lineages were described in captive Chimpanzees in Gabon, Madagascar, Uganda and Zambia (Schaumburg et al., 2012, 2013; Nagel et al., 2013). Notably, a likely case of *S. aureus* transmission from a veterinarian to a chimpanzee from the same sanctuary was demonstrated (Schaumburg et al., 2012). Zoonotic transmission may constitute a major concern in Africa, where there is often substantial exposure to domesticated animals (Fall et al., 2012; Gharsa et al., 2012). Furthermore, animal-adapted clones might undergo further host-adaptive evolutionary changes, which could result in an epidemic spread of new and more virulent strains in the human population (Spoor et al., 2013). Other risk factors for animal to human MRSA transmission, which include contaminated environment (Verkade and Kluytmans, 2013) and meat products (Hamid and Youssef, 2013), have not been investigated in Africa. Livestock-associated MRSA are widespread in Europe, but the transmission of these strains to humans is either rare or limited to people with direct contact with MRSA infected/carrier animals (Verkade and Kluytmans, 2013). Using whole genome sequencing, evidence of zoonotic transmission of MRSA harboring *mecC* was reported in Denmark (Harrison et al., 2013). The *mecC*-positive MRSA, initially known as a livestock MRSA belonging to the CC130, is recognized in both animals and humans in Europe (Paterson et al., 2014a). In addition, this clone has been implicated in severe infections in humans (Paterson et al., 2014b), resulting in one death (García-Garrote et al., 2014). The clinical importance of *mecC*-positive MRSA is not yet clear in Africa as data is unavailable. Therefore, we suggest that surveillance for MRSA should include detection of the *mecC* gene where *mecA* is not detected in resistant isolates.

This systematic review did not seek to provide comprehensive information on the burden of PVL-positive MRSA in Africa. However, it provided some interesting observations on their epidemiology in Africa such as the identification of PVL-positive isolates assigned with CC7 (originally classified with CC152) in Ghana (Egyir et al., 2014a), CC88-IV [2B] in Cameroon (Breurec et al., 2011b) and Nigeria (Ghebremedhin et al., 2009), ST612-IV [2B] in South Africa (Oosthuysen et al., 2014), and CC152 in Nigeria (Shittu et al., 2012). Until now, CC152 was only described in the Balkan region (Francois et al., 2008). The mode of acquisition of the *mecA* gene by ST152 is still unknown, but it might be explained by either its introduction through international travel or the acquisition of the methicillin resistance gene by PVL-positive MSSA, which is prevalent in West and Central Africa (Ruimy et al., 2008; Okon et al., 2009; Breurec et al., 2011a; Schaumburg et al., 2011; Shittu et al., 2011, 2012; Egyir et al., 2014a). These observations highlight the need for further surveillance data (including information on community-acquired necrotizing pneumonia) to understand the epidemiology of PVL-associated *S. aureus* in both hospital and community settings on the African continent.

Conclusion

A number of pandemic MRSA clones were identified in Africa. In contrast, some MRSA clones are limited to specific countries or regions. Although the eBURST snapshot provided a description of the relationship between the MRSA clones reported in Africa and other lineages submitted into the MLST database from other continents, the objective of this review was not to understand the origin of MRSA clones in Africa, as this will require in depth analysis like whole genome sequencing. However, it did show that CC5 is the largest group and predominant in Africa. Nevertheless, the limited data available on MRSA in Africa draw attention to the need for increased surveillance of MRSA and molecular epidemiological studies. We strongly recommend improved co-operation between clinicians and microbiologists in Africa. This synergy could provide an understanding on the local epidemiology of MRSA. In addition, we strongly advocate the establishment of effective diagnostic microbiology facilities that will incorporate high-throughput technologies for monitoring the clonal expansion and dissemination of MRSA. In the meantime, increased networking through collaboration with

S. aureus reference centers could provide support for genotyping services to African countries with limited resources. Finally, population-based surveillances for MRSA are needed to evaluate the situation of community associated MRSA as well as studies on MRSA from animal hosts. To understand the origin of the newly emerged clones, MSSA genotyping needs to be incorporated with MRSA surveillance studies.

Author Contributions

MK, AS, and SMA initiated the project. SMA extracted the data and reviewed the articles with MK. SMA, AS, MN, and MK wrote the manuscript. All the authors reviewed the final version of the manuscript prior to submission for publication

Financial Support

This systematic review was supported by the Organization for Women in Science for the Developing World; Clinical Infectious Diseases Research Initiative, University of Cape Town, South Africa; Bill and Melinda Gates Foundation Global Health Grant (OPP107641), United States of America; Deutscher Akademischer Austausch Dienst, Germany; Carnegie Corporation of New York, United States of America. Any opinions, findings and conclusions, or recommendations expressed in this review are those of the authors, and therefore do not represent the official position of the funders.

Acknowledgments

SMA is supported by the Organization for Women in Science for the developing World (OWSD) and the Drakenstein Child Health Study, Cape Town, South Africa; AS was supported by Deutscher Akademischer Austausch Dienst (DAAD award) Staff Exchange Programme (2013); MK was a recipient of Carnegie Corporation of New York (USA) fellowship and is currently a Wellcome Trust (UK) Fellow.

References

Adesida, S., Boelens, H., Babajide, B., Kehinde, A., Snijders, S., van Leeuwen, W., et al. (2005). Major epidemic clones of *Staphylococcus aureus* in Nigeria. *Microb. Drug. Resist.* 11, 115–121. doi: 10.1089/mdr.2005.11.115

Aiken, A. M., Mutuku, I. M., Sabat, A. J., Akkerboom, V., Mwangi, J., Scott, J. A. G., et al. (2014). Carriage of *Staphylococcus aureus* in Thika Level 5 Hospital, Kenya: a cross-sectional study. *Resist. Infect. Control.* 3:22. doi: 10.1186/2047-2994-3-22

Aires-de-Sousa, M., Correia, B., and de Lencastre, H. (2008). Changing patterns in frequency of recovery of five methicillin-resistant *Staphylococcus aureus* clones in Portuguese hospitals: surveillance over a 16-year period. *J. Clin. Microbiol.* 46, 2912–2917. doi: 10.1128/JCM.00692-08

Albrecht, N., Jatzwauk, L., Slickers, P., Ehricht, R., and Monecke, S. (2011). Clonal replacement of epidemic methicillin-resistant *Staphylococcus aureus* strains in a German university hospital over a period of eleven years. *PLoS ONE* 6:e28189. doi: 10.1371/journal.pone.0028189

Al-haddad, O. H., Zorgani, A., and Ghenghesh, K. S. (2014). Nasal carriage of multi-drug resistant Panton-Valentine leucocidin-positive methicillin-resistant *Staphylococcus aureus* in children in Tripoli-Libya. *Am. J. Trop. Med. Hyg.* 90, 724–727. doi: 10.4269/ajtmh.13-0715

Antri, K., Rouzic, N., Dauwalder, O., Boubekri, I., Bes, M., Lina, G., et al. (2011). High prevalence of methicillin-resistant *Staphylococcus aureus* clone ST80-IV in hospital and community settings in Algiers. *Clin. Microbiol. Infect.* 17, 526–532. doi: 10.1111/j.1469-0691.2010.03273.x

Ateba Ngoa, U., Schaumburg, F., Adegnika, A. A., Kösters, K., Möller, T., Fernandes, J. F., et al. (2012). Epidemiology and population structure of *Staphylococcus aureus* in various population groups from a rural and semi urban area in Gabon, Central Africa. *Acta Tropica* 124, 42–47. doi: 10.1016/j.actatropica.2012.06.005

Ba, X., Harrison, E. M., Edwards, G. F., Holden, M. T. G., Larsen, A. R., Petersen, A., et al. (2014). Novel mutations in penicillin-binding protein genes in clinical *Staphylococcus aureus* isolates that are methicillin resistant on susceptibility testing, but lack the mec gene. *J. Antimicrob. Chemother.* 69, 594–597. doi: 10.1093/jac/dkt418

Bekkhoucha, S. N., Cady, A., Gautier, P., Itim, F., and Donnio, P. Y. (2009). A portrait of *Staphylococcus aureus* from the other side of the Mediterranean Sea: molecular characteristics of isolates from Western Algeria. *Eur. J. Clin. Microbiol. Infect. Dis.* 28, 553–555. doi: 10.1007/s10096-008-0660-x

Ben Ayed, S., Boutiba-Ben Boubaker, I., Boukadida, J., Hammami, S., and Ben Redjeb, S. (2010). Hospital acquired outbreak of methicillin-resistant *Staphylococcus aureus* infection initiated by a health care worker. *Tunis. Med.* 88, 199–202.

Ben Ayed, S., Boutiba-Ben Boubaker, I., Samir, E., and Ben Redjeb, S. (2006). Prevalence of agr specificity groups among methicilin resistant *Staphylococcus aureus* circulating at Charles Nicolle hospital of Tunis. *Pathol. Biol.* 54, 435–438. doi: 10.1016/j.patbio.2006.07.010

Ben Jomaa-Jemili, M., Boutiba-Ben Boubaker, I., and Ben Redjeb, S. (2006). Identification of staphylococcal cassette chromosome mec encoding methicillin resistance in *Staphylococcus aureus* isolates at Charles Nicolle Hospital of Tunis. *Pathol. Biol.* 54, 453–455. doi: 10.1016/j.patbio.2006.07.013

Ben Jomàa-Jemili, M., Ito, T., Zhang, M., Jin, J., Li, S., Ilhem, B.-B. B., et al. (2013). Molecular characterization of methicillin-resistant Panton-valentine leukocidin positive *Staphylococcus aureus* clones disseminating in Tunisian hospitals and in the community. *BMC Microbiol.* 13:2. doi: 10.1186/1471-2180-13-2

Ben Nejma, M. B. E. N., Mastouri, M., Bel, B., Jrad, H., Makhlouf, M., and Nour, M. (2009). Community-acquired methicillin resistant isolated in dermatology department *Staphylococcus aureus*. *Ann. Microbiol.* 59, 823–826. doi: 10.1007/BF03179229

Ben Nejma, M., Mastouri, M., Bel Hadj Jrad, B., and Nour, M. (2013). Characterization of ST80 Panton-Valentine leukocidin-positive community-acquired methicillin-resistant *Staphylococcus aureus* clone in Tunisia. *Diagn. Microbiol. Infect. Dis.* 77, 20–24. doi: 10.1016/j.diagmicrobio.2008.02.010

Ben Nejma, M., Mastouri, M., Frih, S., Sakly, N., Ben Salem, Y., and Nour, M. (2006). Molecular characterization of methicillin-resistant *Staphylococcus aureus* isolated in Tunisia. *Diagn. Microbiol. Infect. Dis.* 55, 21–26. doi: 10.1016/j.diagmicrobio.2005.10.017

Ben Saida, N., Ben Abdallah, H., Hannachi, N., and Boukadida, J. (2005). Multiclonality of methicillin-resistant *Staphylococcus aureus* in a university hospital. *Méd. Mal. Infect.* 35, 363–366. doi: 10.1016/j.medmal.2005.02.014

Ben Slama, K., Gharsa, H., Klibi, N., Jouini, A, Lozano, C., Gómez-Sanz, E., et al. (2011). Nasal carriage of *Staphylococcus aureus* in healthy humans with different levels of contact with animals in Tunisia: genetic lineages, methicillin resistance, and virulence factors. *Eur. J. Clin. Microbiol. Infect. Dis.* 30, 499–508. doi: 10.1007/s10096-010-1109-6

Boakes, E., Kearns, A. M., Ganner, M., Perry, C., Hill, R. L., and Ellington, M. J. (2011). Distinct bacteriophages encoding Panton-Valentine leukocidin

(PVL) among international methicillin-resistant *Staphylococcus aureus* clones harboring PVL. *J. Clin. Microbiol.* 49, 684–692. doi: 10.1128/JCM.01917-10

Bouchami, O., Achour, W., and Ben Hassen, A. (2009). Typing of staphylococcal cassette chromosome mec encoding methicillin resistance in *Staphylococcus aureus* strains isolated at the bone marrow transplant centre of Tunisia. *Curr. Microbiol.* 59, 380–385. doi: 10.1007/s00284-009-9448-1

Breurec, S., Fall, C., Pouillot, R., Boisier, P., Brisse, S., Diene-Sarr, F., et al. (2011a). Epidemiology of methicillin-susceptible *Staphylococcus aureus* lineages in five major African towns: high prevalence of Panton-Valentine leukocidin genes. *Clin. Microbiol. Infect.* 17, 633–639. doi: 10.1111/j.1469-0691.2010.03320.x

Breurec, S., Zriouil, S., Fall, C., Boisier, P., Brisse, S., Djibo, S., et al. (2011b). Epidemiology of methicillin-resistant *Staphylococcus aureus* lineages in five major African towns: emergence and spread of atypical clones. *Clin. Microbiol. Infect.* 17, 160–165. doi: 10.1111/j.1469-0691.2010.03219.x

Centers for Disease Control and prevention. (2013). *Antibiotic Resistance Threats in the United States*, 508. Available online at: http://www.cdc.gov/drugresistance/threat-report-2013/pdf/ar-threats-2013-508.pdf

Chambers, H. F. (1997). Methicillin resistance in staphylococci: molecular and biochemical basis and clinical implications. *Clin. Microbiol. Rev.* 10, 781–791.

Chambers, H. F., and Deleo, F. R. (2010). Waves of resistance: *Staphylococcus aureus* in the antibiotic era. *Nat. Rev. Microbiol.* 7, 629–641. doi: 10.1038/nrmicro2200

Chatterjee, S. S., and Otto, M. (2013). Improved understanding of factors driving epidemic waves. *Clin. Epidemiol.* 5, 205–217. doi: 10.2147/CLEP.S37071

Chen, C. J., and Huang, Y.-C. (2014). New epidemiology of *Staphylococcus aureus* infection in Asia. *Clin. Microbiol. Infect.* 20, 605–623. doi: 10.1111/1469-0691.12705

Conceição, T., Aires-de-Sousa, M., Füzi, M., Tóth, A., Pászti, J., Ungvári, E., et al. (2007). Replacement of methicillin-resistant *Staphylococcus aureus* clones in Hungary over time: a 10-year surveillance study. *Clin. Microbiol. Infect.* 13, 971–979. doi: 10.1111/j.1469-0691.2007.01794.x

Conceição, T., Coelho, C., Santos-Silva, I., de Lencastre, H., and Aires-de-Sousa, M. (2014). Epidemiology of methicillin-resistant and -susceptible *Staphylococcus aureus* in Luanda, Angola: first description of the spread of the MRSA ST5-IVa clone in the African continent. *Microb. Drug Resist.* 20, 441–449. doi: 10.1089/mdr.2014.0007

Conceição, T., Silva, I. S., de Lencastre, H., and Aires-de-Sousa, M. (2013). *Staphylococcus aureus* nasal carriage among patients and health care workers in São Tomé and Príncipe. *Microb. Drug Resist.* 20, 57–66. doi: 10.1089/mdr.2013.0136

Coombs, G., Nimmo, G., Pearson, J., Christiansen, K., Bell, JM, Collignon, PJ, et al. (2009). Prevalence of MRSA strains among *Staphylococcus aureus* isolated from outpatients. *Commun. Dis. Intell. Q. Rep.* 33, 10–20.

Dabul, A. N. G., and Camargo, I. L. B. C. (2014). Clonal complexes of *Staphylococcus aureus*: all mixed and together. *FEMS Microbiol. Lett.* 35, 7–8. doi: 10.1111/1574-6968.12358

De Kraker, M. E. A, Davey, P. G., and Grundmann, H. (2011). Mortality and hospital stay associated with resistant *Staphylococcus aureus* and *Escherichia coli* bacteremia; estimating the burden of antibiotic resistance in Europe. *PLoS Med.* 8:e1001104. doi: 10.1371/journal.pmed.1001104

Deurenberg, R. H., and Stobberingh, E. E. (2008). The evolution of *Staphylococcus aureus*. *Infect. Genet. Evol.* 8, 747–763. doi: 10.1016/j.meegid.2008.07.007

Djoudi, F., Bonura, C., Benallaoua, S., Touati, A., Touati, D., Aleo, A., et al. (2013). Panton-Valentine leukocidin positive sequence type 80 methicillin-resistant *Staphylococcus aureus* carrying a staphylococcal cassette chromosome mec type IVc is dominant in neonates and children in an Algiers hospital. *New Microbiol.* 36, 49–55.

D'Souza, N., Rodrigues, C., and Mehta, A. (2010). Molecular characterization of methicillin-resistant *Staphylococcus aureus* with emergence of epidemic clones of sequence type (ST)22 and ST772 in Mumbai, India. *J. Clin. Microbiol.* 48, 1806–1811. doi: 10.1128/JCM.01867-09

Egyir, B., Guardabassi, L., Esson, J., Nielsen, S. S., Newman, M. J., Addo, K. K., et al. (2014a). Insights into nasal carriage of *Staphylococcus aureus* in an urban and a rural community in Ghana. *PLoS ONE* 9:e96119. doi: 10.1371/journal.pone.0096119

Egyir, B., Guardabassi, L., Nielsen, S. S., Larsen, J., Addo, K. K., Newman, M. J., et al. (2013). Prevalence of nasal carriage and diversity of *Staphylococcus aureus* among inpatients and hospital staff at Korle Bu Teaching Hospital, Ghana. *J. Glob. Antimicrob. Resist.* 1, 189–193. doi: 10.1016/j.jgar.2013.05.006

Egyir, B., Guardabassi, L., Sørum, M., Nielsen, S. S., Kolekang, A., Frimpong, E., et al. (2014b). Molecular epidemiology and antimicrobial susceptibility of clinical *Staphylococcus aureus* from healthcare institutions in Ghana. *PLoS ONE* 9:e89716. doi: 10.1371/journal.pone.0089716

El-Jakee, J. K., Atta, N., Samy, A., Bakry, M. E. A. E., Kandil, M., and Gad El-Said, W. (2011). Antimicrobial resistance in clinical isolates of *Staphylococcus aureus* from bovine and human sources in Egypt. *Glob. Vet.* 7, 581–586.

EL Seedy, F. R., El Hakim, A. S., Syame, S. F., and Osman, N. M. (2012). Advanced techniques used for isolation and characterization of *Staphylococcus aureus* isolated from mastitic buffaloes. *Glob. Vet.* 8, 144–152.

Enany, S., Yaoita, E., Yoshida, Y., Enany, M., and Yamamoto, T. (2010). Molecular characterization of Panton-Valentine leukocidin-positive community-acquired methicillin-resistant *Staphylococcus aureus* isolates in Egypt. *Microbiol. Res.* 165, 152–162. doi: 10.1016/j.micres.2009.03.005

Enright, M. C., Robinson, D. A., Randle, G., Feil, E. J., Grundmann, H., and Spratt, B. G. (2002). The evolutionary history of methicillin-resistant *Staphylococcus aureus* (MRSA). *Proc. Natl. Acad. Sci. U.S.A.* 99, 7687–7692. doi: 10.1073/pnas.122108599

Espadinha, D., Faria, N. A., Miragaia, M., Lito, L. M., Melo-Cristino, J., and de Lencastre, H. (2013). Extensive dissemination of methicillin-resistant *Staphylococcus aureus* (MRSA) between the hospital and the community in a country with a high prevalence of nosocomial MRSA. *PLoS ONE* 8:e59960. doi: 10.1371/journal.pone.0059960

Essa, Z. I., Connolly, C., and Essack, S. Y. (2009). *Staphylococcus aureus* from public hospitals in KwaZulu-Natal, South Africa—infection detection and strain-typing. *South African J. Epidemiol. Infect.* 24, 4–7.

Falagas, M. E., Karageorgopoulos, D. E., Leptidis, J., and Korbila, I. P. (2013). MRSA in Africa: filling the global map of antimicrobial resistance. *PLoS ONE* 8:e68024. doi: 10.1371/journal.pone.0068024

Fall, C., Seck, A., Richard, V., Ndour, M., Sembene, M., Laurent, F., et al. (2012). Epidemiology of *Staphylococcus aureus* in pigs and farmers in the largest farm in Dakar, Senegal. *Foodborne Pathog. Dis.* 9, 962–965. doi: 10.1089/fpd.2012.1197

Fossum Moen, A. E., Tannaes, T. M., and Leegaard, T. M. (2013). USA300 methicillin-resistant *Staphylococcus aureus* in Norway. *APMIS* 121, 1091–1096. doi: 10.1111/apm.12077

Francois, P., Harbarth, S., Huyghe, A., and Renzi, G. (2008). Methicillin-resistant *Staphylococcus aureus*, Geneva, Switzerland, 1993-2005. *Emerg. Infect. Dis.* 14, 304–307. doi: 10.3201/eid1402.070229

García-Garrote, F., Cercenado, E., Marín, M., Bal, M., Trincado, P., Corredoira, J., et al. (2014). Methicillin-resistant *Staphylococcus aureus* carrying the mecC gene: emergence in Spain and report of a fatal case of bacteraemia. *J. Antimicrob. Chemother.* 69, 45–50. doi: 10.1093/jac/dkt327

Garza-González, E., and Dowzicky, M. J. (2013). Changes in *Staphylococcus aureus* susceptibility across Latin America between 2004 and 2010. *Braž. J. Infect. Dis.* 17, 13–19. doi: 10.1016/j.bjid.2012.08.017

Gharsa, H., Ben Slama, K., Lozano, C., Gómez-Sanz, E., Klibi, N., Ben Sallem, R., et al. (2012). Prevalence, antibiotic resistance, virulence traits and genetic lineages of *Staphylococcus aureus* in healthy sheep in Tunisia. *Vet. Microbiol.* 156, 367–373. doi: 10.1016/j.vetmic.2011.11.009

Ghebremedhin, B., Olugbosi, M. O., Raji, A. M., Layer, F., Bakare, R. A., König, B., et al. (2009). Emergence of a community-associated methicillin-resistant *Staphylococcus aureus* strain with a unique resistance profile in Southwest Nigeria. *J. Clin. Microbiol.* 47, 2975–2980. doi: 10.1128/JCM.00648-09

Goering, R. V, Shawar, R. M., Scangarella, N. E., O'Hara, F. P., Amrine-Madsen, H., West, J. M., et al. (2008). Molecular epidemiology of methicillin-resistant and methicillin-susceptible *Staphylococcus aureus* isolates from global clinical trials. *J. Clin. Microbiol.* 46, 2842–2847. doi: 10.1128/JCM.00521-08

Hamid, D., and Youssef, A. (2013). Clinical Features of Methicillin-resistant *Staphylococcus aureus* (MRSA) Infection in Rabbits and its Zoonotic potentials. *Pakistan J. Nutr.* 12, 244–249. doi: 10.3923/pjn.2013.244.249

Harrison, E. M., Paterson, G. K., Holden, M. T. G., Larsen, J., Stegger, M., Larsen, A. R., et al. (2013). Whole genome sequencing identifies zoonotic transmission of MRSA isolates with the novel mecA homologue mecC. *EMBO Mol. Med.* 5, 509–515. doi: 10.1002/emmm.201202413

Huson, M. A. M., Kalkman, R., Remppis, J., Beyeme, J. O., Kraef, C., Schaumburg, F., et al. (2014). Methicillin-resistant *Staphylococcus aureus* as a cause of invasive infections in Central Africa: a case report and review of the literature. *Infection* 42, 451–457. doi: 10.1007/s15010-014-0589-1

Ito, T., Katayama, Y., Asada, K., and Mori, N. (2001). Structural comparison of three types of staphylococcal cassette chromosome mec integrated in the chromosome in methicillin-resistant *Staphylococcus aureus*. *Antimicrob. Agents Chemother.* 45, 1323–1336. doi: 10.1128/AAC.45.5.1323-1336.2001

Ito, T., Ma, X. X., Takeuchi, F., and Okuma, K. (2004). Novel type V staphylococcal cassette chromosome mec driven by a novel cassette chromosome recombinase, ccrC novel type V staphylococcal cassette chromosome mec driven by a novel cassette chromosome recombinase, ccrC. *Antimicrob. Agents Chemother.* 48, 2637–2651. doi: 10.1128/AAC.48.7.2637-2651.2004

IWG-SCC. (2009). Classification of staphylococcal cassette chromosome mec (SCC*mec*): guidelines for reporting novel SCC*mec* elements. *Antimicrob. Agents Chemother.* 53, 4961–4967. doi: 10.1128/AAC.00579-09

Jansen van Rensburg, M. J., Eliya Madikane, V., Whitelaw, A., Chachage, M., Haffejee, S., and Gay Elisha, B. (2011). The dominant methicillin-resistant *Staphylococcus aureus* clone from hospitals in Cape Town has an unusual genotype: ST612. *Clin. Microbiol. Infect.* 17, 785–792. doi: 10.1111/j.1469-0691.2010.03373.x

Jansen van Rensburg, M. J., Whitelaw, A. C., and Elisha, B. G. (2012). Genetic basis of rifampicin resistance in methicillin-resistant *Staphylococcus aureus* suggests clonal expansion in hospitals in Cape Town, South Africa. *BMC Microbiol.* 12:46. doi: 10.1186/1471-2180-12-46

Johnson, A. P., Pearson, A., and Duckworth, G. (2005). Surveillance and epidemiology of MRSA bacteraemia in the UK. *J. Antimicrob. Chemother.* 56, 455–462. doi: 10.1093/jac/dki266

Kateete, D. P., Namazzi, S., Okee, M., Okeng, A., Baluku, H., Musisi, N. L., et al. (2011). High prevalence of methicillin resistant *Staphylococcus aureus* in the surgical units of Mulago hospital in Kampala, Uganda. *BMC Res. Notes* 4:326. doi: 10.1186/1756-0500-4-326

Kechrid, A., Pérez-Vázquez, M., Smaoui, H., Hariga, D., Rodríguez-Baños, M., Vindel, A., et al. (2011). Molecular analysis of community-acquired methicillin-susceptible and resistant *Staphylococcus aureus* isolates recovered from bacteraemic and osteomyelitis infections in children from Tunisia. *Clin. Microbiol. Infect.* 17, 1020–1026. doi: 10.1111/j.1469-0691.2010.03367.x

Kim, C., Milheiriço, C., Gardete, S., Holmes, M. A., Holden, M. T. G., de Lencastre, H., et al. (2012). Properties of a novel PBP2A protein homolog from *Staphylococcus aureus* strain LGA251 and its contribution to the β-lactam-resistant phenotype. *J. Biol. Chem.* 287, 36854–36863. doi: 10.1074/jbc.M112.395962

Kim, E. S., Song, J. S., Lee, H. J., Choe, P. G., Park, K. H., Cho, J. H., et al. (2007). A survey of community-associated methicillin-resistant *Staphylococcus aureus* in Korea. *J. Antimicrob. Chemother.* 60, 1108–1114. doi: 10.1093/jac/dkm309

Köck, R., Becker, K., Cookson, B., van Gemert-Pijnen, J. E., Harbarth, S., Kluytmans, J., et al. (2010). Methicillin-resistant *Staphylococcus aureus* (MRSA): burden of disease and control challenges in Europe. *Euro. Surveill.* 15, 19688.

Lee, B. Y., Singh, A., David, M. Z., Bartsch, S. M., Slayton, R. B., Huang, S. S., et al. (2013). The economic burden of community-associated methicillin-resistant *Staphylococcus aureus* (CA-MRSA). *Clin. Microbiol. Infect.* 19, 528–536. doi: 10.1111/j.1469-0691.2012.03914.x

Li, T., Song, Y., Zhu, Y., Du, X., and Li, M. (2013). Current status of *Staphylococcus aureus* infection in a central teaching hospital in Shanghai, China. *BMC Microbiol.* 13:153. doi: 10.1186/1471-2180-13-153

Lim, K. T., Hanifah, Y. A., Mohd Yusof, M. Y., Ito, T., and Thong, K. L. (2013). Comparison of methicillin-resistant *Staphylococcus aureus* strains isolated in 2003 and 2008 with an emergence of multidrug resistant ST22: SCC*mec* IV clone in a tertiary hospital, Malaysia. *J. Microbiol. Immunol. Infect.* 46, 224–233. doi: 10.1016/j.jmii.2013.02.001

Löffler, B., Hussain, M., Grundmeier, M., Brück, M., Holzinger, D., Varga, G., et al. (2010). *Staphylococcus aureus* Panton-Valentine leukocidin is a very potent cytotoxic factor for human neutrophils. *PLoS Pathog.* 6:e1000715. doi: 10.1371/journal.ppat.1000715

Ma, X. X., Ito, T., Tiensasitorn, C., Chongtrakool, P., Boyle-Vavra, S., Daum, R. S., et al. (2002). Novel type of staphylococcal cassette chromosome mec identified in *Staphylococcus aureus* strains. *Antimicrob. Agents Chemother.* 46, 1147–1152. doi: 10.1128/AAC.46.4.1147-1152.2002

Maina, E. K., Kiiyukia, C., Wamae, C. N., Waiyaki, P. G., and Kariuki, S. (2013). Characterization of methicillin-resistant *Staphylococcus aureus* from skin and soft tissue infections in patients in Nairobi, Kenya. *Int. J. Infect. Dis.* 17, e115–e119. doi: 10.1016/j.ijid.2012.09.006

Makgotlho, P. E., Kock, M. M., Hoosen, A., Lekalakala, R., Omar, S., Dove, M., et al. (2009). Molecular identification and genotyping of MRSA isolates. *FEMS Immunol. Med. Microbiol.* 57, 104–115. doi: 10.1111/j.1574-695X.2009.00585.x

Mašlaňová, I., Doškař, J., Varga, M., Kuntová, L., Mužík, J., Malúšková, D., et al. (2013). Bacteriophages of *Staphylococcus aureus* efficiently package various bacterial genes and mobile genetic elements including SCC*mec* with different frequencies. *Environ. Microbiol. Rep.* 5, 66–73. doi: 10.1111/j.1758-2229.2012.00378.x

Moher, D., Liberati, A., Tetzlaff, J., and Altman, D. G. (2009). Preferred reporting items for systematic reviews and meta-analyses: the PRISMA statement. *PLoS Med.* 6:e1000097. doi: 10.1371/journal.pmed.1000097

Monecke, S., Coombs, G., Shore, A. C., Coleman, D. C., Akpaka, P., Borg, M., et al. (2011). A field guide to pandemic, epidemic and sporadic clones of methicillin-resistant *Staphylococcus aureus*. *PLoS ONE* 6:e17936. doi: 10.1371/journal.pone.0017936

Moodley, A., Oosthuysen, W. F., Dusé, A. G., and Marais, E. (2010). Molecular characterization of clinical methicillin-resistant *Staphylococcus aureus* isolates in South Africa. *J. Clin. Microbiol.* 48, 4608–4611. doi: 10.1128/JCM.01704-10

Moremi, N., Mshana, S. E., Kamugisha, E., Kataraihya, J. B., Tappe, D., Vogel, U., et al. (2012). Predominance of methicillin resistant *Staphylococcus aureus* -ST88 and new ST1797 causing wound infection and abscesses. *J. Infect. Dev. Ctries.* 6, 620–625. doi: 10.3855/jidc.2093

Moyo, S. J., Aboud, S., Blomberg, B., Mkopi, N., Kasubi, M., Manji, K., et al. (2014). High Nasal carriage of methicillin-resistant *Staphylococcus aureus* among healthy tanzanian under-5 children. *Microb. Drug Resist.* 20, 82–88. doi: 10.1089/mdr.2013.0016

Nagel, M., Dischinger, J., Türck, M., Verrier, D., Oedenkoven, M., Ngoubangoye, B., et al. (2013). Human-associated *Staphylococcus aureus* strains within great ape populations in Central Africa (Gabon). *Clin. Microbiol. Infect.* 19, 1072–1077. doi: 10.1111/1469-0691.12119

Nimmo, G. R., Bergh, H., Nakos, J., Whiley, D., Marquess, J., Huygens, F., et al. (2013). Replacement of healthcare-associated MRSA by community-associated MRSA in Queensland: confirmation by genotyping. *J. Infect.* 67, 439–447. doi: 10.1016/j.jinf.2013.07.020

Okon, K. O., Basset, P., Uba, A., Lin, J., Oyawoye, B., Shittu, A. O., et al. (2009). Cooccurrence of predominant Panton-Valentine leukocidin-positive sequence type (ST) 152 and multidrug-resistant ST 241 *Staphylococcus aureus* clones in Nigerian hospitals. *J. Clin. Microbiol.* 47, 3000–3333. doi: 10.1128/JCM.01119-09

Okon, K. O., Shittu, A. O., Kudi, A. A., Umar, H., Becker, K., and Schaumburg, F. (2014). Population dynamics of *Staphylococcus aureus* from Northeastern Nigeria in 2007 and 2012. *Epidemiol. Infect.* 142, 1737–1740. doi: 10.1017/S0950268813003117

Okuma, K., Iwakawa, K., Turnidge, J. D., Warren, B., Bell, J. M., Brien, F. G. O., et al. (2002). Dissemination of new methicillin-resistant *Staphylococcus aureus* clones in the community dissemination of new methicillin-resistant *Staphylococcus aureus* clones in the community. *J. Clin. Microbiol.* 40, 4289–4294. doi: 10.1128/JCM.40.11.4289-4294.2002

Olalekan, A. O., Schaumburg, F., Nurjadi, D., Dike, A. E., Ojurongbe, O., Kolawole, D. O., et al. (2012). Clonal expansion accounts for an excess of antimicrobial resistance in *Staphylococcus aureus* colonising HIV-positive individuals in Lagos, Nigeria. *Int. J. Antimicrob. Agents* 40, 268–272. doi: 10.1016/j.ijantimicag.2012.05.016

Oosthuysen, W. F., Orth, H., Lombard, C. J., Sinha, B., and Wasserman, E. (2014). Population structure analyses of *Staphylococcus aureus* at Tygerberg Hospital, South Africa, reveals a diverse population, high prevalence of Panton-Valentine leukocidin genes and unique local MRSA clones. *Clin. Microbiol. Infect.* 20, 652–659. doi: 10.1111/1469-0691.12452

Ouchenane, Z., Agabou, A., Smati, F., Rolain, J.-M., and Raoult, D. (2013). Staphylococcal cassette chromosome mec characterization of methicillin-resistant *Staphylococcus aureus* strains isolated at the military hospital of Constantine/Algeria. *Pathol. Biol.* 61, 280–281. doi: 10.1016/j.patbio.2013.05.006

Ouchenane, Z., Smati, F., Rolain, J.-M., and Raoult, D. (2011). Molecular characterization of methicillin-resistant *Staphylococcus aureus* isolates in Algeria. *Pathol. Biol.* 59, e129–e132. doi: 10.1016/j.patbio.2009.11.004

Pasquale, T. R., Jabrocki, B., Salstrom, S. J., Wiemken, T. L., Peyrani, P., Haque, N. Z., et al. (2013). Emergence of methicillin-resistant *Staphylococcus aureus* USA300 genotype as a major cause of late-onset nosocomial pneumonia in intensive care patients in the USA. *Int. J. Infect. Dis.* 17, e398–e403. doi: 10.1016/j.ijid.2012.12.013

Paterson, G. K., Harrison, E. M., and Holmes, M. A. (2014a). The emergence of mecC methicillin-resistant *Staphylococcus aureus*. *Trends Microbiol.* 22, 42–47. doi: 10.1016/j.tim.2013.11.003

Paterson, G. K., Morgan, F. J. E., Harrison, E. M., Cartwright, E. J. P., Török, M. E., Zadoks, R. N., et al. (2014b). Prevalence and characterization of human mecC methicillin-resistant *Staphylococcus aureus* isolates in England. *J. Antimicrob. Chemother.* 69, 907–910. doi: 10.1093/jac/dkt462

Raji, A., Ojemhen, O., Umejiburu, U., Ogunleye, A., Blanc, D. S., and Basset, P. (2013). High genetic diversity of *Staphylococcus aureus* in a tertiary care hospital in Southwest Nigeria. *Diagn. Microbiol. Infect. Dis.* 77, 367–369. doi: 10.1016/j.diagmicrobio.2013.08.030

Ramdani-bouguessa, N., Bes, M., Meugnier, H., Forey, F., Reverdy, M., Lina, G., et al. (2006). Detection of methicillin-resistant *Staphylococcus aureus* strains resistant to multiple antibiotics and carrying the panton-valentine leukocidin genes in an Algiers Hospital. *Antimicrob. Agents Chemother.* 50, 1083–1085. doi: 10.1128/AAC.50.3.1083-1085.2006

Robinson, D. A., and Enright, M. C. (2004). Evolution of *Staphylococcus aureus* by large chromosomal replacements. *J. Bacteriol.* 186, 1060–1064. doi: 10.1128/JB.186.4.1060-1064.2004

Rogers, B. A., Aminzadeh, Z., Hayashi, Y., and Paterson, D. L. (2011). Country-to-country transfer of patients and the risk of multi-resistant bacterial infection. *Clin. Infect. Dis.* 53, 49–56. doi: 10.1093/cid/cir273

Ruimy, R., Maiga, A., Armand-Lefevre, L., Maiga, I., Diallo, A., Koumaré, A. K., et al. (2008). The carriage population of *Staphylococcus aureus* from Mali is composed of a combination of pandemic clones and the divergent Panton-Valentine leukocidin-positive genotype ST152. *J. Bacteriol.* 190, 3962–3968. doi: 10.1128/JB.01947-07

Schaumburg, F., Alabi, A. S., Mombo-Ngoma, G., Kaba, H., Zoleko, R. M., Diop, D. A, et al. (2014). Transmission of *Staphylococcus aureus* between mothers and infants in an African setting. *Clin. Microbiol. Infect.* 20, 390–396. doi: 10.1111/1469-0691.12417

Schaumburg, F., Mugisha, L., Kappeller, P., Fichtel, C., Köck, R., Köndgen, S., et al. (2013). Evaluation of non-invasive biological samples to monitor *Staphylococcus aureus* colonization in great apes and lemurs. *PLoS ONE* 8:e78046. doi: 10.1371/journal.pone.0078046

Schaumburg, F., Mugisha, L., Peck, B., Becker, K., Gillespie, T. R., Peters, G., et al. (2012). Drug-resistant human *Staphylococcus aureus* in sanctuary apes pose a threat to endangered wild ape populations. *Am. J. Primatol.* 74, 1071–1075. doi: 10.1002/ajp.22067

Schaumburg, F., Ngoa, U. A., Kösters, K., Köck, R., Adegnika, A. A, Kremsner, P. G., et al. (2011). Virulence factors and genotypes of *Staphylococcus aureus* from infection and carriage in Gabon. *Clin. Microbiol. Infect.* 17, 1507–1513. doi: 10.1111/j.1469-0691.2011.03534.x

Scragg, J. N., Appelbaum, P. C., and Govender, D. A. (1978). The spectrum of infection and sensitivity of organisms isolated from African and Indian children in a Durban hospital. *Trans. R. Soc. Trop. Med. Hyg.* 72, 325–328. doi: 10.1016/0035-9203(78)90118-9

Seni, J., Bwanga, F., Najjuka, C. F., Makobore, P., Okee, M., Mshana, S. E., et al. (2013). Molecular characterization of *Staphylococcus aureus* from patients with surgical site infections at Mulago Hospital in Kampala, Uganda. *PLoS ONE* 8:e66153. doi: 10.1371/journal.pone.0066153

Shallcross, L. J., Fragaszy, E., Johnson, A. M., and Hayward, A. C. (2013). The role of the Panton-Valentine leucocidin toxin in staphylococcal disease: a systematic review and meta-analysis. *Lancet Infect. Dis.* 13, 43–54. doi: 10.1016/S1473-3099(12)70238-4

Shambat, S., Nadig, S., Prabhakara, S., Bes, M., Etienne, J., and Arakere, G. (2012). Clonal complexes and virulence factors of *Staphylococcus aureus* from several cities in India. *BMC Microbiol.* 12:64. doi: 10.1186/1471-2180-12-64

Sherwood, J., Park, M., Robben, P., Whitman, T., and Ellis, M. W. (2013). USA300 methicillin-resistant *Staphylococcus aureus* emerging as a cause of bloodstream infections at military medical centers. *Infect. Control Hosp. Epidemiol.* 34, 393–399. doi: 10.1086/669866

Shittu, A. O., and Lin, J. (2006a). Antimicrobial susceptibility patterns and characterization of clinical isolates of *Staphylococcus aureus* in KwaZulu-Natal province, South Africa. *BMC Infect. Dis.* 6:125. doi: 10.1186/1471-2334-6-125

Shittu, A. O., and Lin, J. (2006b). Antimicrobial susceptibility patterns of *Staphylococcus aureus* and characterization of MRSA in Southwestern Nigeria. *Wounds* 18, 77–84.

Shittu, A. O., Lin, J., and Morrison, D. (2007). Molecular identification and characterization of mannitol-negative methicillin-resistant *Staphylococcus aureus*. *Diagn. Microbiol. Infect. Dis.* 57, 93–95. doi: 10.1016/j.diagmicrobio.2006.05.004

Shittu, A. O., Nübel, U., Udo, E., Lin, J., and Gaogakwe, S. (2009). Characterization of meticillin-resistant *Staphylococcus aureus* isolates from hospitals in KwaZulu-Natal province, Republic of South Africa. *J. Med. Microbiol.* 58, 1219–1226. doi: 10.1099/jmm.0.011452-0

Shittu, A. O., Okon, K., Adesida, S., Oyedara, O., Witte, W., Strommenger, B., et al. (2011). Antibiotic resistance and molecular epidemiology of *Staphylococcus aureus* in Nigeria. *BMC Microbiol.* 11:92. doi: 10.1186/1471-2180-11-92

Shittu, A. O., Oyedara, O., Abegunrin, F., Okon, K., Raji, A., Taiwo, S., et al. (2012). Characterization of methicillin-susceptible and -resistant staphylococci in the clinical setting: a multicentre study in Nigeria. *BMC Infect. Dis.* 12:286. doi: 10.1186/1471-2334-12-286

Shore, A. C., Deasy, E. C., Slickers, P., Brennan, G., O'Connell, B., Monecke, S., et al. (2011). Detection of staphylococcal cassette chromosome mec type XI carrying highly divergent mecA, mecI, mecR1, blaZ, and ccr genes in human clinical isolates of clonal complex 130 methicillin-resistant *Staphylococcus aureus*. *Antimicrob. Agents Chemother.* 55, 3765–3773. doi: 10.1128/AAC.00187-11

Sobhy, N., Aly, F., Abd El Kader, O., Ghazal, A., and Elbaradei, A. (2012). Community-acquired methicillin-resistant *Staphylococcus aureus* from skin and soft tissue infections (in a sample of Egyptian population): analysis of mec gene and staphylococcal cassette chromosome. *Braz. J. Infect. Dis.* 16, 426–431. doi: 10.1016/j.bjid.2012.08.004

Spoor, L. E., McAdam, P. R., and Weinert, L. A. (2013). Livestock origin for a human pandemic clone of Community- Associated Methicillin-Resistant *Staphylococcus aureus*. *MBio* 4, e00356–e00313. doi: 10.1128/mBio.00356-13

Stefani, S., Chung, D. R., Lindsay, J. A., Friedrich, A. W., Kearns, A. M., Westh, H., et al. (2012). Meticillin-resistant *Staphylococcus aureus* (MRSA): global epidemiology and harmonisation of typing methods. *Int. J. Antimicrob. Agents* 39, 273–282. doi: 10.1016/j.ijantimicag.2011.09.030

Stegger, M., Wirth, T., Andersen, P. S., Skov, R. L., Grassi, A., De Simões, M., et al. (2014). Origin and evolution of european methicillin-resistant *Staphylococcus aureus*. *MBio* 5, e01044–e01014. doi: 10.1128/mBio.01044-14

Tavares, A., Miragaia, M., Rolo, J., Coelho, C., and de Lencastre, H. (2013). High prevalence of hospital-associated methicillin-resistant *Staphylococcus aureus* in the community in Portugal: evidence for the blurring of community-hospital boundaries. *Eur. J. Clin. Microbiol. Infec. Dis.* 32, 1269–1283. doi: 10.1007/s10096-013-1872-2

Terry Alli, O. A, Ogbolu, D. O., Mustapha, J. O., Akinbami, R., and Ajayi, A. O. (2012). The non-association of Panton-Valentine leukocidin and mecA genes in the genome of *Staphylococcus aureus* from hospitals in South Western Nigeria. *Indian J. Med. Microbiol.* 30, 159–164. doi: 10.4103/0255-0857.96675

Vandenesch, F., Naimi, T., Enright, M. C., Lina, G., Nimmo, G. R., Heffernan, H., et al. (2003). Community-acquired methicillin-resistant *Staphylococcus aureus* carrying Panton-Valentine leukocidin genes: worldwide emergence. *Emerg. Infect. Dis.* 9, 978–984. doi: 10.3201/eid0908.030089

Van der Meeren, B. T., Millard, P. S., Scacchetti, M., Hermans, M. H., Hilbink, M., Concelho, T. B., et al. (2014). Emergence of methicillin resistance and Panton-Valentine leukocidin positivity in hospital- and community-acquired *Staphylococcus aureus* infections in Beira, Mozambique. *Trop. Med. Int. Health* 19, 169–176. doi: 10.1111/tmi.12221

Verkade, E., and Kluytmans, J. (2013). Livestock-associated *Staphylococcus aureus* CC398: animal reservoirs and human infections. *Infect. Genet. Evol.* 21, 523–530. doi: 10.1016/j.meegid.2013.02.013

Yoong, P., and Torres, V. J. (2013). The effects of *Staphylococcus aureus* leukotoxins on the host: cell lysis and beyond. *Curr. Opin. Microbiol.* 16, 63–69. doi: 10.1016/j.mib.2013.01.012

Zhou, Y. P., Wilder-Smith, A., and Hsu, L.-Y. (2014). The role of international travel in the spread of methicillin-resistant *Staphylococcus aureus*. *J. Trav. Med.* 21, 272–281. doi: 10.1111/jtm.12133

Zribi, M., Etienne, J., El Euch, D., Zribi, H., Bes, M., Meugnier, H., et al. (2011). Detection of the first strain of glycopeptide intermediary *Staphylococcus aureus* in Tunis Rabta hospital. *Path. Biol.* 59, 334–335. doi: 10.1016/j.patbio.2009.10.002

Genotypic and Antimicrobial Susceptibility of Carbapenem-resistant *Acinetobacter baumannii*: Analysis of IS*Aba* Elements and *bla*$_{OXA-23-like}$ Genes Including a New Variant

*Abbas Bahador[1], Reza Raoofian[2,3], Babak Pourakbari[4], Mohammad Taheri[5], Zahra Hashemizadeh[5] and Farhad B. Hashemi[1]**

[1] Department of Microbiology, School of Medicine, Tehran University of Medical Sciences, Tehran, Iran, [2] Legal Medicine Research Center, Legal Medicine Organization, Tehran, Iran, [3] Innovative Medical Research Center, Islamic Azad University, Mashhad, Iran, [4] Pediatrics Infectious Disease Research Center, School of Medicine, Tehran University of Medical Sciences, Tehran, Iran, [5] Department of Microbiology and Virology, Shiraz University of Medical Sciences, Shiraz, Iran

***Correspondence:**
Farhad B. Hashemi
hashemi@fbh.hiv

Carbapenem-resistant *Acinetobacter baumannii* (CR-AB) causes serious nosocomial infections, especially in ICU wards of hospitals, worldwide. Expression of *bla*$_{OXA}$ genes is the chief mechanism of conferring carbapenem resistance among CR-AB. Although some *bla*$_{OXA}$ genes have been studied among CR-AB isolates from Iran, their *bla*$_{OXA-23-like}$ genes have not been investigated. We used a multiplex-PCR to detect Ambler class A, B, and D carbapenemases of 85 isolates, and determined that 34 harbored *bla*$_{OXA-23-like}$ genes. Amplified fragment length polymorphism (AFLP) genotyping, followed by DNA sequencing of *bla*$_{OXA-23-like}$ amplicons of CR-AB from each AFLP group was used to characterize their *bla*$_{OXA-23-like}$ genes. We also assessed the antimicrobial susceptibility pattern of CR-AB isolates, and tested whether they harbored insertion sequences IS*Aba*1 and IS*Aba*4. Sequence comparison with reference strain *A. baumannii* (NCTC12156) revealed five types of mutations in *bla*$_{OXA-23-like}$ genes; including one novel variant and four mutants that were already reported from China and the USA. All of the *bla*$_{OXA-23-like}$ genes mutations were associated with increased minimum inhibitory concentrations (MICs) against imipenem. IS*Aba*1 and IS*Aba*4 sequences were detected upstream of *bla*$_{OXA-23}$ genes in 19 and 7% of isolates, respectively. The isolation of CR-AB with new *bla*$_{OXA-23}$ mutations including some that have been reported from the USA and China highlights CR-AB pervasive distribution, which underscores the importance of concerted national and global efforts to control the spread of CR-AB isolates worldwide.

Keywords: *Acinetobacter baumannii*, *bla*$_{OXA-23-like}$ gene, carbapenemase, novel mutations

INTRODUCTION

Carbapenem-resistant *Acinetobacter baumannii* (CR-AB) can cause severe nosocomial infections particularly among patients in intensive care units (ICUs) around the world (Safari et al., 2013). Inadequate antimicrobial management of CR-AB infections often gives rise to highly resistant strains leading to prolonged hospitalization, treatment failures, and increased mortality (Higgins et al., 2010a). Epidemics of multi-, extensively-, and pandrug-resistant (MDR, XDR, and PDR) CR-AB have been reported from several countries (Kempf and Rolain, 2011; Bahador et al., 2013a; Moradi et al., 2015). In developing countries, such as Iran, challenges in the treatment of CR-AB infections are often exacerbated by widespread nosocomial outbreaks of OXA-type β-lactamase producing MDR-AB (for review see, Moradi et al., 2015). CR-AB are usually resistant to several β-lactams through the expression of chromosomal and plasmid-encoded carbapenemases including Ambler class A (bla_{GES}, and bla_{KPC}), class B (bla_{IMP}, bla_{NDM-1}, bla_{SPM-1}, and bla_{VIM}), and class D ($bla_{OXA-23,40, \text{ and } 58\text{-like}}$; Siroy et al., 2005; Lu et al., 2009; Abbott et al., 2013). While the production of OXA-23 by *A. baumannii* is sufficient to confer resistance to carbapenems, insertion sequence (IS) elements IS*Aba1* and/or IS*Aba4* upstream of $bla_{OXA-23\text{-like}}$ genes enhance the bla_{OXA}-mediated carbapenem resistance of CR-AB (Turton et al., 2006; Lee et al., 2012; Evans and Amyes, 2014). Although there are a few reports from Iran regarding the distribution and/or frequency of the $bla_{OXA-51\text{-like}}$ genes among CR-AB, data about characterization of their bla_{OXA-23} genes and IS*Aba* elements is not available.

In this study, we have characterized bla_{OXA} genes in CR-AB isolates from Iran, and report new variants that harbor novel mutations in their $bla_{OXA-23\text{-like}}$ carbapenemase genes. In addition to analyzing the distribution and frequency of $bla_{OXA-23\text{-like}}$ genes, we have determined the antimicrobial susceptibility patterns of isolates and the presence of IS*Aba1* and IS*Aba4* enhancer elements upstream of their $bla_{OXA-23\text{-like}}$ genes. Characterization of bla_{OXA} genes and assessment of carbapenemase-mediated antibiotic resistance among *A. baumannii* isolates can help efforts to develop databases, which are essential to a comprehensive national surveillance program in Iran, toward the local and global control of CR-AB outbreaks.

MATERIALS AND METHODS

Specimens and Bacterial Isolates and Cultures

A total of 85 non-repetitive clinical specimens were collected during 2011 from the intensive care units (ICUs) of Imam Khomeini Medical Center (IKMC) and Children's Medical Center (CMC) in Tehran, Iran. IKMC and CMC are affiliated with Tehran University of Medical Sciences (TUMS), and both are large referral centers that provide tertiary health care to patients from all over Iran. Specimens were collected from ICUs in surgical (S), internal medicine (M), emergency (E), pediatrics (P), and kidney transplantation (T) wards. Clinical isolates were initially identified as *A. baumannii* using the API20NE system (bioMérieux, Marcy-l'Etoile, France), and were further confirmed by *gyrB* multiplex PCR, as described previously (Higgins et al., 2010b). Specimen sources for *A. baumannii* isolates were as follows: respiratory tract ($n = 51$), urine ($n = 16$), blood ($n = 11$), wound ($n = 5$), and cerebral spinal fluid (CSF; $n = 2$). Twenty six of the *A. baumannii* isolates were part of a previous molecular epidemiologic study (Bahador et al., 2014). Brain heart infusion (BHI) agar plates and Mueller-Hinton broth (MHB; both from Merck, Germany) were used to culture the bacterial isolates.

Antimicrobial Susceptibility Testing

To assess susceptibility of *A. baumannii* clinical isolates, the disk agar diffusion (DAD) method (CLSI, 2015) was carried out according to the Clinical and Laboratory Standards Institute (CLSI) procedures and breakpoint interpretations, using antimicrobial disks containing 19 different antimicrobial agents (Mast Diagnostics, Bootle, UK; **Table 2**). The CLSI guideline for broth microdilution test for minimum inhibitory concentrations (MICs) was used to assess the susceptibility of MDR-AB isolates to colistin (CST), imipenem (IPM), rifampicin (RIF), and tigecycline (TGC). For tigecycline susceptibility tests, the criteria of the European Committee on Antimicrobial Susceptibility Testing (EUCAST) for *Enterobacteriaceae* were used, in which an MIC of $<1\ \mu g/mL$ was defined as susceptible and $>2\ \mu g/mL$ was considered resistant (EUCAST, 2015). Rifampicin susceptibility was interpreted according to CLSI criteria using breakpoint values suggested for *Staphylococcus aureus*, in which susceptible and resistant were defined as $\leq 1\ \mu g/mL$ and $\geq 4\ \mu g/mL$, respectively (CLSI, 2015). *A. baumannii* isolates were defined as MDR, XDR, and PDR according to the definitions provided by Magiorakos (Magiorakos et al., 2012). The MIC geometric mean (MIC_{gm}) of imipenem against $bla_{OXA-23\text{-like}}{}^{+}$ CR-AB isolates were also compared with the MIC_{gm} of non-mutant isolates, and fold-increase calculations were measured against MIC_{gm} of non-mutant strains.

Detection of Carbapenemase Gene IS*Aba1* and IS*Aba4* Insertion Sequences

The overall strategy for the identification of the 34 $bla_{OXA-23\text{-like}}{}^{+}$ CR-AB isolates is shown in **Supplemental Figure 1**. Briefly, we tested all 85 *A. baumannii* isolates for carbapenemase production by the modified Hodge test (Lee et al., 2012), and their chromosomal DNA were tested by two different confirmatory multiplex-PCR assays to identify the most common carbapenemase encoding genes. The criteria to include isolates in this study were the presence of PCR-specific amplicons, confirmed by agarose gel electrophoresis analysis (**Supplemental Figure 1**). A novel in-house multiplex-PCR, referred to as AB-hexaplex-PCR was optimized for the rapid and simultaneous detection of the most common carbapenemase genes, including Ambler class A and B (bla_{KPC}, bla_{GES}, bla_{IMP-1}, bla_{VIM-2}, bla_{NDM-1}, and bla_{SPM-1}) in *A. baumannii* using Primer 3 software (version 4.0; http://primer3.wi.mit.edu/; accessed June 05, 2011). Reference gene sequences were accessed from GenBank [http://www.ncbi.nlm.nih.gov/GenBank (bla_{KPC}:

GQ140348, *bla*$_{\text{GES}}$: GU207844, *bla*$_{\text{IMP-1}}$: EF375699, *bla*$_{\text{VIM-2}}$: GQ288396, *bla*$_{\text{NDM-1}}$: JN794561, and *bla*$_{\text{SPM-1}}$: HM370523; accessed June 04, 2011)], as shown in **Table 1**. The Ambler class D type carbapenemase genes (*bla*$_{\text{OXA-23,24,51,58 like}}$) were detected using the Woodford multiplex PCR assay method (Woodford et al., 2006). Additionally, the AB-hexaplex-PCR distinguished amplicons corresponding to the *bla*$_{\text{IMP-1}}$, *bla*$_{\text{SPM-1}}$, *bla*$_{\text{GES}}$ and *bla*$_{\text{KPC}}$, *bla*$_{\text{NDM-1}}$, and *bla*$_{\text{VIM-2}}$ genes, and isolates that harbored these genes were excluded from our study (representative gel; **Supplemental Figure 2**). The frequency of IS*Aba*1 and IS*Aba*4 elements upstream of *bla*$_{\text{OXA-23-like}}$ and *bla*$_{\text{OXA-51-like}}$ genes were assessed using a series of PCR amplifications. A set of primers, referred to as IS*Aba*1F/OXA-23R, IS*Aba*4F/OXA-23R, and IS*Aba*1F/OXA-51R, is shown in **Table 1**. After our serial

TABLE 1 | Primer sequences and adaptors (and their corresponding reference) utilized in order to identify most common carbapenemase genes in our isolates, and to generate amplicons for AFLP genotyping analysis.

Assay		Primer	Sequence (5′-3′)[†]	Size of amplicon	References
Detection of carbapenemase in the molecular class D	Multiplex PCR	*bla*$_{OXA-51}$likeF	TAATGCTTTGATCGGCCTTG	353	Woodford et al., 2006
		bla$_{OXA-51}$likeR	TGGATTGCACTTCATCTTGG		
		bla$_{OXA-23}$likeF	GATCGGATTGGAGAACCAGA	501	"
		bla$_{OXA-23}$likeR	ATTTCTGACCGCATTTCCAT		
		bla$_{OXA-24}$likeF	GGTTAGTTGGCCCCCTTAAA	240	"
		bla$_{OXA-24}$likeR	AGTTGAGCGAAAAGGGGATT		
		bla$_{OXA-58}$likeF	AAGTATTGGGGCTTGTGCTG	590	"
		bla$_{OXA-58}$likeR	CCCCTCTGCGCTCTACATAC		
Detection of carbapenemase in the molecular classes A and B	hexaplex PCR (h-PCR)	IMP-1F	AACATGGTTTGGTGGTTCTTGT	263	Present study
		IMP-1R	TCCGCTAAATGAATTTGTGGCT		
		VIM-2F	CAATGGTCTCATTGTCCGTGAT	395	"
		VIM-2R	AAATCGCACAACCACCATAGAG		
		NDM-1F	CTGGATCAAGCAGGAGATCAAC	118	"
		NDM-1R	ATTGGCATAAGTCGCAATCCC		
		KPCF	CGCTAAACTCGAACAGGACTTT	640	"
		KPCR	ATAGTCATTTGCCGTGCCATAC		
		blaGESF	GAAAACTTTCATATGGGCCGGA	567	"
		blaGESR	GACCGACAGAGGCAACTAATTC		
		SPM-1F	CCATTGTCTGCAAAAAGTTCGG	439	"
		SPM-1R	AAACATTATCCGCTGGAACAGG		
ISAba1 detection upstream of blaOXA-51	*IsAba-1 F/OXA-51 R*	IsAba-1 F	AAGCATGATGAGCGCAAAG	227	"
		OXA-51 R	GGTGAGCAGGCTGAAATAAAA		
ISAba1 detection upstream of *bla*OXA-23	IsAba-1 F/OXA-23 R	IsAba-1 F	TGAGATGTGTCATAGTATTC	314	"
		OXA-23 R	AGAGCATTACCATATAGATT		
ISAba4 detection upstream of *bla*OXA-23	IsAba-4 F/OXA-23 R	IsAba-4 F	CACAATTTCTGATAAAGATA	327	"
		OXA-23 R	TTTATTAAATTATGCTGAAC		
AFLP	Adaptors	adp MbI	GTAGCGCGACGGCCAGTCGCG	No amplicon	Bahador et al., 2013b
		ADP MbI	GATCCGCGACTGGCCGTCGCGCTAC		
		adp MsI	GTAGCGCGACGGCCAGTCGCGT		"
		ADP MsI	TAACGCGACTGGCCGTCGCGCTAC		
	Pre-amplification	PreAmp Mbo	ACGGCCAGTCGCGGATC	Multiple and variable	"
		PreAmp Mse	CGACGGCCAGTCGCGTTAA		
	Selective primers	Mbo1	PreAmp Mbo + A	Multiple and variable	"
		Mbo2	PreAmp Mbo + T		
		Mbo3	PreAmp Mbo + C		
		Mbo4	PreAmp Mbo + G		
		Mse1	PreAmp Mse + A		
		Mse2	PreAmp Mse + T		
		Mse3	PreAmp Mse + C		
		Mse4	PreAmp Mse + G		

[†] *Nucleotide.*

screening of isolates, 34 isolates were identified that harbored $bla_{\text{OXA-23-like}}$ gene as their sole acquired carbapenemase gene.

AFLP Genomic Fingerprint Analysis

Amplified fragment length polymorphism (AFLP) genotyping of $bla_{\text{OXA-23-like}}{}^{+}$ and $bla_{OXA-51-like}{}^{+}$ isolates was carried out by a modified Vos method (Vos et al., 1995), as described previously (Bahador et al., 2013b). AFLP typing was carried out prior to sequence analysis to ensure thorough examination of the diversity of CR-AB isolates. Briefly, chromosomal DNA was size-verified and double-digested with MboI and MseI (Fermentas, Lithuania). Then DNA fragments were ligated to corresponding adapters using T4 DNA ligase (350 U/μ L, Takara Bio, Japan) followed by the preliminary PCR using PreAmp-Mbo and PreAmp-Mse primers (**Table 1**). Preliminary PCR amplicons served as templates for selective PCR, which generated AFLP genotype profiles upon agarose gel analysis. Initial testing of 36 combinations of primers, including PreAmp Mbo (PreAmp Mbo+A, +T, +C, +G), and PreAmp Mse (PreAmp Mse +A, +T, +C, +G) and *A. baumannii* NCTC12156 DNA as a normalization reference showed that the Mbo4-Mse4 combination generated the clearest AFLP profiles when analyzed using BioNumerics version 5.10 (Applied Maths, Sint-Martens-Latem, Belgium). The similarity between band patterns was calculated using the Dice coefficient, with an optimization of 0.5% and a position tolerance of 1%. The AFLP types were grouped at the 90% similarity cutoff on a dendrogram constructed by the unweighted-pair group method using average linkages (UPGMA).

DNA Sequencing of $bla_{\text{OXA-23-like}}$ Genes

To evaluate an association between changes in the chromosomal carbapenemase gene sequence of isolates and their antimicrobial resistance pattern, a two-step approach was adopted. An initial AFLP assay was carried out on $bla_{\text{OXA-23-like}}{}^{+}$ CR-AB, followed by DNA sequence analysis of the $bla_{\text{OXA-23-like}}$ gene of a representative isolate from each AFLP genotype group.

Briefly, we used a high fidelity Pfu DNA polymerase (Fermentas, Lithuania) to generate $bla_{\text{OXA-23-like}}$ specific amplicons, which were purified using a AccuPrep® PCR Purification Kit (Bioneer, Daejeon, Korea) and cloned into pTZ57R (InsT/A Clone PCR product cloning kit, Fermentas, Vilnius, Lithuania). DNA was then transferred into competent *E. coli* TOP10 cells, which were then isolated using Luria-Bertani (LB) agar supplemented with ampicillin (100 μg /mL). Plasmid DNA was prepared with the AccuPrep Plasmid MiniPrep DNA Extraction Kit, (Bioneer, Daejeon, Korea) and sequenced using an ABI3730 automatic sequencer (Applied Biosystems, CA, USA). The sequences were analyzed using a BLAST algorithm against the NCBI GenBank database [http://www.ncbi.nlm.nih.gov/guide/dna-rna/ (accessed 05.06.11)].

Iodometric Assay of β-lactamase Activity

Bacterial β-lactamase enzymatic activity was determined by an iodometric assay, as described previously (Sawai et al., 1978). Briefly, crude lysates of 16 isolates that represented AFLP groups were extracted using the Saino method (Saino et al., 1982). Briefly, overnight bacterial growth were diluted in MHB broth to a concentration of 10^7 cfu/ml and incubated in a shaker for 2 h at 35°C. As an inducer, imipenem was added at 0.25 of the isolate MIC and incubated for an additional 2 h (Clark, 1996). Bacterial cells were harvested, centrifuged at 4°C, washed twice with 50 mM phosphate buffer saline (PBS; pH 7.0), and re-suspended in 0.1 M PBS (pH 7.0). The suspension was sonicated in an ultrasonic disrupter (Branson Ultrasonics Co., Shanghai, China) at 75 W for 3 min in an ice bath; afterward the disrupted cell suspension was centrifuged at 13,000 × g for 30 min at 4°C. The β-lactamase enzymatic activity of the supernatant fluid (i.e., bacterial lysate) was measured against imipenem using iodometric method described by Doust (Daoust et al., 1973) using reagents prepared, as described previously (Onishi et al., 1974; Sawai et al., 1978; Minami et al., 1980). Briefly, iodine reagent (40 μmol in 0.5 M acetate buffer, pH 4.0) was added to lysate supernatant fluids, after 5 min incubation with imipenem (50 μg/mL) at 30°C. Ten minutes later, samples' absorbance was measured at 620 nm, and imipenem hydrolysis was determined. Activity was reported as the mean of triplicate samples in micromoles of imipenem degraded per minute per milligram of protein in each bacterial extract. Protein concentrations were measured by Bradford assay kit (Pierce™ Coomassie Plus Assay Kit, Thermo Scientific, Ottawa, Canada).

In Silico Analysis and Nucleotide Sequence Accession Numbers

In silico analysis was carried out using GenBank nucleotide database. Predict Protein software (hosted by Rostlab) was also used to predict changes as a result of a frameshift mutation. The nucleotide sequence data were deposited in the GenBank nucleotide database under accession numbers: JQ343842.1, JQ343840.1, JQ343838.1, JQ343836.1, JQ343841.1, JQ343839.1, JQ343837.1, JQ360584.1, JQ360582.1, JQ360580.1, JQ360578.1, JQ360583.1, JQ360581.1, JQ360579.1, JQ360577.1, and JQ061320.1. The novel DNA sequence of $bla_{\text{OXA 23}}$ genes, with "No Full-Match" by GenBank; as well as its corresponding peptide amino acid sequence was submitted to Lahey database (lahey.org/Studies).

RESULTS

Antimicrobial Susceptibility Profiles and AFLP Genomic Fingerprint Analysis

Table 2 shows the susceptibility profiles of all 85 CR-AB isolates against CLSI groups of antimicrobial agents. Overall, CR-AB isolates were most resistant to CLSI group A (51–96%), followed by group B (25–97%) antimicrobials. Overall, up to 96% of isolates were resistant to 12 of the tested antimicrobials; while the rates of resistance to tigecycline, imipenem, and doripenem were 34, 65, and 94%, respectively. The lowest resistance rates among isolates were against colistin (12%), minocycline (25%), and doxycycline (31%). Interestingly, all colistin-resistant isolates were susceptible to tigecycline and/or tobramycin. The frequency of MDR, XDR and PDR isolates were 69, 24, and 0%, respectively; and broadly-resistant CR-AB isolates were most

TABLE 2 | ***In vitro*** **antimicrobial susceptibility results of 85 non-replicate clinical** ***A. baumannii*** **isolates according to CLSI antimicrobial grouping of A, B, and O, as according to the frequency of** ***A. baumannii*** **isolation from the type of ICU ward.**

ICU Ward (No.)	% Resistant to CLSI antimicrobial groups[a]																						
	A									B											O		
	IPM[b]	DOR	SAM	CRO	CAZ	TOB	GEN	CIP	AMK	MIN	DOX	TET	TGC	LVX	PIP	CTX	FEP	TZP	TIM	SXT	CST	NET	RIF
Emergency (12)	8	13	9	14	14	7	12	11	12	5	5	9	8	6	14	14	13	11	14	14	1	6	13
Medical Care (27)	25	30	20	31	32	22	27	22	27	7	7	16	5	19	27	27	33	27	33	32	4	12	27
Pediatrics (8)	6	9	5	9	9	5	9	6	5	1	1	6	4	6	8	7	9	9	8	9	1	5	8
Surgical (31)	22	34	20	34	34	13	29	28	31	11	14	21	13	19	33	33	35	29	33	33	5	14	33
Transplantation (7)	4	8	5	8	7	4	6	7	7	1	4	5	5	5	7	8	7	4	8	8	1	2	7
Total (85)	65	94	59	96	96	51	82	74	81	25	31	58	34	56	89	89	97	80	96	96	12	39	88

[a] *Criteria in the assignment of agents to Groups A, B, and C included clinical efficacy, prevalence of resistance, minimizing emergence of resistance, cost, FDA clinical indications for usage, and current consensus recommendations for first-choice and alternative drugs.* ***Group A*** *are considered appropriate for inclusion in a routine, primary testing panel, as well as for routine reporting of results for the specific organism.* ***Group B*** *comprises agents that may warrant primary testing.* ***Group O (Other)*** *includes agents that have a clinical indication for the organism. Escherichia coli ATCC25922 and Pseudomonas aeruginosa ATCC27853 were used for quality control of antimicrobial susceptibility testing and included in each run.*

[b] *CSF, cerebrospinal fluid; AMK, amikacin; CAZ, ceftazidime; CIP, ciprofloxacin; CRO, ceftriaxone; CST, colistin; CTX, cefotaxime; DOR, doripenem; DOX, doxycycline; FEP, cefepime; GEN, gentamycin; IPM, imipenem; MIN, minocycline; NET, netilmicin; LVX, levofloxacin; PIP, piperacillin; RIF, rifampicin; SAM, ampicillin-sulbactam; SXT, trimethoprim/sulfamethoxazole; TET, tetracycline; TGC, tigecycline; TZP, Piperacillin/tazobactam; TIM, ticarcillin/clavulanic acid; TOB, tobramycin.*

frequently recovered from the surgical and internal medicine ICU wards. However, the frequency of resistant isolates was generally proportional to the number of specimens from each ICU (**Table 2**).

Our analysis revealed that 34 (40%) isolates harbored $bla_{\text{OXA-23-like}}$ genes, and 51 isolates were resistant to carbapenems but did not harbor $bla_{\text{OXA-23-like}}$ genes (**Supplemental Figure 1**). AFLP genotype analysis of $bla_{\text{OXA-23-like}}^{+}$ isolates generated 16 distinct AFLP genotypic groups, labeled genotype A through P. Group C ($n = 6$) was the predominant AFLP type, followed by genotype I ($n = 4$), and genotypes B, K, L, and N ($n = 3$ in each group). While each AFLP group consisted of 1 to 6 isolates, 50% (8/16) of the groups consisted of a single isolate, indicative of a high diversity among CR-AB isolates. Despite this diversity, the antimicrobial susceptibility patterns among 13 (82%) genotypes were similar, with the exception of genotypes A, M, and N (**Figure 1**).

Detection of IS*Aba*1 and IS*Aba*4

Overall, IS*Aba*1 and IS*Aba*4 sequences were present in 61% and 7% of all tested CR-AB isolates, respectively. Among 34 $bla_{\text{OXA-23-like}}^{+}$ isolates, 67% were IS*Aba1*$^{+}$, and 18% were IS*Aba4*$^{+}$ (**Figure 1**); whereas 10 (63%) and 3 (19%) of AFLP genotypes harbored IS*Aba*1 and IS*Aba*4 elements, respectively. IS*Aba*4 was only present among AFLP types G, E, and K isolates, while genotypes A, M, and N has neither IS*Aba*1, nor IS*Aba*4 element (**Table 3**; **Figure 1**). Interestingly, even though the IS*Aba1*$^{+}$ isolates of genotype B ($N = 3$) exhibited an XDR profile, all genotype B isolates remained susceptible to tobramycin and ampicillin–sulbactam.

Among the $bla_{\text{OXA-23-like}}^{+}$ isolates that harbored either IS*Aba*1 or IS*Aba*4 elements, a majority (82%) displayed a distinctive profile of resistance to 9 antimicrobial agents, namely, CAZ, CRO, CTX, DOR, FEP, IPM, PIP, SAM, and TZP (**Figure 1**). The presence of either IS*Aba*1, or IS*Aba*4 imparted resistance to imipenem and doripenem among $bla_{\text{OXA-23-like}}^{+}$ genotypes. However, the genotype A isolate (which was IS*Aba1*$^{-}$ and IS*Aba4*$^{-}$) also showed resistance to imipenem (MIC = 16 μg/mL).

Sequence Analysis of $bla_{\text{OXA-23-like}}$ Genes

Table 3 demonstrates sequence differences between the $bla_{\text{OXA-23-like}}$ specific amplicon among members of AFLP groups vs. the $bla_{\text{OXA-23}}$ sequence of the *A. baumannii* reference strain (referred to as "wild type"). The $bla_{\text{OXA-23}}$ gene sequences of six (37%) AFLP genotypes did not differ from the wild-type strain; however, isolates from 10 (63%) AFLP types had mutations in $bla_{\text{OXA-23}}$ genes. Five different $bla_{\text{OXA-23}}$ gene mutations were detected, and a Genbank search revealed that one of the mutant sequences had been recently reported from the USA, namely $bla_{\text{OXA-366}}$, and three were reported from China (i.e., $bla_{\text{OXA-422, 481, and 482-like}}$ genes). One novel $bla_{\text{OXA-23-like}}$ gene sequence was thus submitted to the Lahey database and assigned as the $bla_{\text{OXA-495-like}}$ gene.

At the points of carbapenemase gene mutations, comparison of a 21-nucleotide sequence of PCR amplicons obtained from isolates in mutant AFLP groups that had >2 members, i.e.,

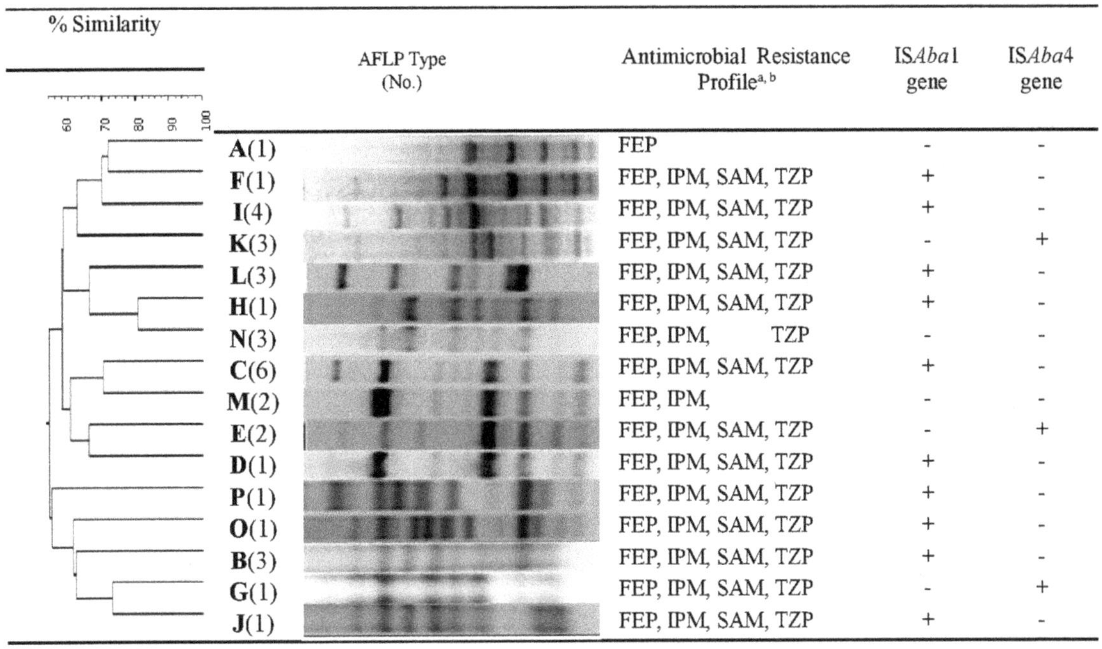

FIGURE 1 | Dendrogram of amplified fragment length polymorphism (AFLP) analysis of genomic DNA extracted from $bla^{(+)}_{OXA\text{-}23\text{-}like}$ CR-AB isolates. Susceptibility to select antimicrobials and IS*Aba* presence is also indicated. Dice coefficient with 0.5% optimization and 1% position tolerance was used. Dendrogram was constructed by unweighted-pair group method using average linkages (UPGMA); and AFLP type identification was defined by groups formed at 90% similarity cutoff. [a]FEP, cefepime; IPM, imipenem; SAM, ampicillin-sulbactam; TZP, Piperacillin/tazobactam. [b]All 16 AFLP types were resistant to the following beta-lactam antimicrobial agents: CAZ, ceftazidime; CRO, ceftriaxone; CTX, cefotaxime; PIP, piperacillin; TIM, ticarcillin/clavulanic acid.

groups L, E, C, and K, showed strong homologies within these AFLP types (**Supplemental Figure 3**). Additionally, **Table 3** displays the AFLP type, imipenem MIC, and IS*Aba* status among various $bla_{OXA\text{-}23}{}^{+}$ isolates according to their specimen source. The absence of mutation(s) in $bla_{OXA\text{-}23}$ genes was associated with lower imipenem MICs as compared to mutant isolates. Imipenem MIC_{gm} for all non-mutant $bla_{OXA\text{-}23}$ gene isolates was 16 μg/mL (range = 8 − 32 μg/mL), whereas the mean MIC for mutants was almost 50 μg/mL (range = 16 − 128 μg/mL). Conversely, alterations of the $bla_{OXA\text{-}23}$ gene were associated with an increased MIC to imipenem among *A. baumannii* isolates. While 16 (47%) of $bla_{OXA\text{-}23\text{-}like}{}^{+}$ isolates harbored IS*Aba1* sequences upstream of their carbapenemase gene, only six (18%) isolates were IS*Aba4*$^{+}$. IS*Aba1* elements were detected upstream of either $bla_{OXA\text{-}23}$ or $bla_{OXA\text{-}51}$ genes, orboth these genes.Since, none of the $bla_{OXA\text{-}23\text{-}like}{}^{+}$ isolates were both IS*Aba1*$^{+}$ and IS*Aba4*$^{+}$, the presence of these elements appeared mutually exclusive among isolates (**Table 3**). The IS*Aba4* sequence was absent among $bla_{OXA\text{-}23\text{-}like}$ genes of non-mutant CR-AB isolates, whereas six of mutant isolates, namely genotypes G, E, and K, were IS*Aba4*$^{+}$ with moderate MICs, ranging from 16 to 64 μg/mL (**Table 3**). Although mutant isolates of genotypes C and D were IS*Aba1*$^{+}$ had the highest imipenem MICs (128 μg/mL), the imipenem MIC of IS*Aba1*$^{+}$ mutants among genotypes O, P, J, and L was 16 μg/mL. Furthermore, all IS*Aba1*$^{-}$ and IS*Aba4*$^{-}$ CR-AB isolates had non-mutant $bla_{OXA\text{-}23}{}^{+}$ genes, namely genotypes A, M, and N isolates, which showed the lowest imipenem MICs (8–16 μg/mL), as well. Surprisingly, three $bla_{OXA\text{-}23\text{-}like}{}^{+}$ isolates (genotypes A and M) had imipenem MICs of 16 μg/mL, but harbored neither IS*Aba1*, nor IS*Aba4* elements upstream of the $bla_{OXA\text{-}23\text{-}like}$ gene, suggestive of other resistance mechanisms in these isolates (**Table 3**).

As shown in **Table 3**, among mutations of the $bla_{OXA\text{-}23\text{-}like}$ genes, insertions/deletions at nucleotide positions 335 and 336 were most frequent (40%), followed by a single substitution at position 771 (30%). Isolates with F, G, O, and P genotypes showed the insertion/deletion mutations at position 335, with imipenem MICs of 16–32 μg/mL, whereas isolates with genotypes E, C, and K had the substitution at position 771 and showed the highest imipenem MICs of 64–128 μg/mL (**Table 3**). Further analysis revealed that three other single substitutions also occurred at positions 376, 625, and 766 among genotypes J and L, but their imipenem MICs were not higher than non-mutants (16 μg/mL). In addition, three isolates with the substitution at position 771 had a 3-4-fold increase in imipenem MIC over that of the non-mutants. A frame-shift mutation at position 355, which corresponds to a change at aa112–118, in genotype D isolate (strain TUMS/BTRF 661) was associated with a four-fold increase in imipenem MIC_{gm} (128 μg/mL) over that of non-mutant isolates. Software prediction showed that the frame-shift mutation may change the subcellular localization of carbapenemase and enhance its secretion rate, which can explain the high MIC of the isolate against carbapenems. *In silico*

TABLE 3 | Comparison of *bla*$_{OXA-23-like}$ gene sequences among CR-AB isolates belonging to various AFLP genotype groups, as compared to *bla*$_{OXA-23}$ gene sequence of *A. baumannii* reference strain, according to the specimen source, MIC against imipenem, β-lactamase activity, and IS*Aba* element status.

No.	Isolates	AFLP type (No.)	Specimen (ICU)[a]	Nucleotide Change(s)	Amino Acid Change(s)	MIC of IMP (mg/L)	Mean of β-lactamase activity[d] (SD)[e]	IS[b] type	
								IS*Aba* 1	IS*Aba* 4
1	*bla*$_{OXA-23}$	N (3)	Urine (S)	None	None	8	ND	–	–
2	"	M (2)	Urine (S)	"	"	16	1.05(0.10)	–	–
3	"	A (1)	Urine (S)	"	"	16	0.87(0.04)	–	–
4	"	H (1)	Blood (P)	"	"	32	2.63(0.15)	+	–
5	"	I (4)	Blood (E)	"	"	16	1.35(0.09)	+	–
6	"	B (3)	CSF (S)	"	"	16	ND	+	–
7	*bla*$_{OXA-482}$	F (1)	Urine (M)	Insertion of A at position 335 and deletion of A at position 336	Ser → Tyr at aa 112	32	0.42(0.04)	+	–
8	"	G (1)	Urine (T)	"	"	16	5.58 (0.91)	–	+
9	"	O (1)	Urine (E)	"	"	16	2.11(0.63)	+	–
10	"	P (1)	Sputum (E)	"	"	16	2.35(0.87)	+	–
11	*bla*$_{OXA-481}$	E (2)	Wound (S)	G → A at position 771	Met → Ile at aa 257	64	1.67(0.09)	–	+
12	"	C (6)	Urine (S)	"	"	128	7.96(0.92)	+	–
13	"	K (3)	Blood (M)	"	"	64	7.75(1.03)	–	+
14	*bla*$_{OXA-495}$	D (1)	Sputum (S)	Frame-shift due to insertion of A at position 335 and deletion at A at position 354	Change in aa112–aa118 motif[c]	128	14.37(1.33)	+	–
15	*bla*$_{OXA-366}$	J (1)	Blood (M)	A → C at position 376, and G → A at position 625	Met → Ile at aa 126, and Glu → Lys at aa 209	16	2.62(0.53)	+	–
16	*bla*$_{OXA-422}$	L (3)	Sputum (S)	G → A at position 766	Glu → Lys at aa 256	16	0.37(0.03)	+	–

[a] *ICUs: E, Emergency; M, Medical Care; P, pediatric; S, surgical; T, transplantation.*
[b] *IS: Insertion sequence.*
[c] *SFTAWE → YIYRLG.*
[d] *μmole imipenem hydrolyzed per min per mg of protein.*
[e] *standard deviation.*
ND, No detectable activity.

comparison of carbapenemase binding domains of wild-type vs. the frame-shift mutant also showed that the binding domain changed from "aa15–24 and aa28–29" to "aa15–22 and aa24–25" motif, which may lead to higher affinity of mutant OXA-23 enzyme for the binding cleft of carbapenems.

Five *bla*$_{OXA-23}$ mutants and one non-mutant isolate showed imipenem MICs of >16 μg /mL. Further *in silico* analysis revealed that the *bla*$_{OXA-23-like}$ gene mutations would lead to up to six amino acid changes in the carbapenemase protein. However, the highest imipenem MICs were associated with a single substitution at position 771 of *bla*$_{OXA-23-like}$ gene, corresponding to aa 257 substitution in carbapenemase among genotypes E, C, and K, which represented 55% of the mutant isolates. All AFLP types with non-mutant *bla*$_{OXA-23}$ genes showed MICs of ≤16 μg/mL for imipenem, except the genotype H isolate (MIC = 32 μg /mL); however, all *bla*$_{OXA-23-like}$ mutants showed high imipenem resistance (MIC = 16–128 μg/mL).

As shown in **Table 3**, we compared the β-lactamase activity of the 16 representative isolates from each AFLP group. Overall, the range of β-lactamase activity of non-mutant isolates was lower (not detectable–2.63 μmoles/min/ mg protein) than the *bla*$_{OXA-23-like}$ mutants (0.37–14.37 μmoles/min/mg protein). Mutant isolates, such as the frame-shift mutant genotype D, showed the highest imipenem MICs (i.e., 64 and 128 μg/mL), as well as the highest β-lactamase activities, i.e., 7.96 and 14.37 μmoles/min/mg protein for genotype C and K, respectively. In contrast, lysates from isolates with no *bla*$_{OXA-23-like}$ gene mutation that had the lowest MICs (e.g., genotypes N and B), and showed no detectable β-lactamase activity. The β-lactamase activity of a imipenem-susceptible (MIC < 4 μg/ml), clinical *A. baumannii* isolate, was also below assay's detection level (data not shown). Among the 4 isolates that had the same mutation at position 335, the β-lactamase activity was between 0.42 and 5.58 μmoles/min/ mg protein, while imipenem their MIC was 16–32 μg/mL. However, with the exception of genotype E, the β-lactamase activity of mutants with position 771 mutation, was increased concomitant with high MICs in these isolates. The majority (70%) of mutant isolates were recovered from two ICU wards; namely, the surgical (n = 4), and the internal medicine (n = 3) ICU; however, no mutant isolates were recovered from the pediatric ICU ward. Among the *bla*$_{\text{OXA-23-like}}$ mutant isolates, eight (80%) were recovered from either urine or sputum specimens (**Table 3**).

Table 4 shows the distribution of IS*Aba*1 or IS*Aba*4 sequences upstream of various *bla*$_{\text{OXA-23-}}$ and *bla*$_{\text{OXA-51-like}}$ genes among the isolates that harbored ≥1 carbapenemase genes; and also their resistance rate against carbapenems. The presence of IS*Aba* upstream of the *bla*$_{\text{OXA-51-like}}$ and*bla*$_{\text{OXA-23-like}}$ genes was associated with high rate of carbapenem resistance. Among all *bla*$_{\text{OXA-51-like}}^{+}$ or *bla*$_{\text{OXA-23-like}}^{+}$ isolates, almost 32% ($n = 27$) lacked either IS*Aba*1, or IS*Aba*4 sequences. CR-AB isolates were consistently more resistant to doripenem than to imipenem, regardless of their *bla*$_{\text{OXA-}}$ genes (**Table 4**). There was no marked difference in carbapenem resistance rates whether the isolates harbored the "*bla*$_{\text{OXA-51-like}}$ gene alone," or "*bla*$_{\text{OXA-51-like}}$ plus *bla*$_{\text{OXA-24-like}}$" genes. Overall, the IS*Aba*1 element was more often associated with *bla*$_{\text{OXA-51-like}}$ gene (20–100%) than with *bla*$_{\text{OXA-23-like}}$ genes; and IS*Aba*$^{+}$ isolates showed high rates of carbapenem resistance, especially against doripenem. Despite this high resistance rate, 13% of *bla*$_{\text{OXA-23}}^{+}$/*bla*$_{\text{OXA-51}}^{+}$ isolates that harbored both IS*Aba*1 and IS*Aba*4 were imipenem susceptible (**Table 4**). Among IS*Aba*$^{+}$ isolates, these elements were upstream of the *bla*$_{\text{OXA-51-like}}$ gene in 60% (31/52) of the isolates, whereas only 31% harbored IS*Aba*1 upstream of the *bla*$_{\text{OXA-23-like}}$ gene. All 13 (15%) CR-AB isolates with *bla*$_{\text{OXA-51-like}}$ gene as their sole carbapenamase gene had IS*Aba*1 elements. Interestingly, even though test isolates showed an overall high resistance rate against carbapenems, 32% of isolates did not harbor either IS*Aba*1 or IS*Aba*4 elements. By and large, there was no marked change in resistance rate among isolates that harbored the *bla*$_{\text{OXA-24-like}}$ gene in combination with other carbapenemase genes (**Table 4**).

DISCUSSION

Infections caused by carbapenem-resistant *A. baumannii* are among the most difficult to treat, especially among ICU patients (Alfandari et al., 2014). In several countries, including Iran, clinicians face serious challenges in choosing an effective combination of antimicrobial agents while treating patients with severe nosocomial CR-AB infections. Efforts to control MDR-AB outbreaks have prompted widespread use of antimicrobials, such as tigecycline and colistin, as therapeutic measures to combat severe infections (Garnacho-Montero et al., 2015). However, appropriate treatment and effective infection control measures require local susceptibility patterns, as well as molecular epidemiologic data, such as the *bla*$_{\text{OXA}}$ gene status of CR-AB isolates. Several surveillance studies have reported widespread nosocomial outbreaks of OXA-type producing *A. baumannii*, and a high prevalence of *bla*$_{\text{OXA}}$ gene-carrying CR-AB in Iran, but data regarding their *bla*$_{\text{OXA-23-like}}$ gene is not available (Moradi et al., 2015).

In the present study, we have genetically evaluated *bla*$_{\text{OXA-23-like}}^{+}$ CR-AB isolates and found high genotypic diversity among the isolates, including variants with new *bla*$_{\text{OXA-23-like}}$ gene mutations. These mutations were associated with up to four-fold increases in MIC levels against imipenem, as compared to non-mutant isolates. Mutations in certain codons associated with a high degree of resistance to imipenem. For instance, substitutions at position 355 of the *bla*$_{\text{OXA-23-like}}$ gene (i.e., the frame-shift mutation) were associated with high-level resistance, whereas position 256 mutations were associated with low-level resistance. Surprisingly, newly-found mutations correspond to regions of the carbapenemase molecule that

TABLE 4 | Frequency of IS*Aba*1 or IS*Aba*4 sequences upstream of various *bla*$_{\text{OXA-}}$ genes among test CR-AB isolates that harbored ≥ 1 carbapenemase genes, and the comparison of percent resistance against carbapenems among CR-AB isolates according to the isolate's *bla*$_{\text{OXA-}}$ gene combination.

No.	Carbapenemase gene(s) of CR-AB isolates (Total=85)	% of isolates with insertion sequences					% Carbapenem Resistance[a]	
		IS*Aba*1 on *bla* $_{\text{OXA-51-like}}$ ($n = 31$)	IS*Aba*1 on *bla* $_{\text{OXA-23-like}}$ ($n = 16$)	IS*Aba*1 on *bla* $_{\text{OXA-51-like}}$ and *bla* $_{\text{OXA-23-like}}$ ($n = 5$)	IS*Aba*4 on *bla* $_{\text{OXA-23-like}}$ ($n = 6$)	Without IS*Aba* ($n = 27$)	DOR[b]	IPM
1	*bla*$_{\text{OXA-51-like}}$ ($n = 13$; 15%)	38	–	–	–	62	77	46
2	*bla*$_{\text{OXA-51-like}}$ and *bla*$_{\text{OXA-23-like}}$ ($n = 54$; 63%)	37	30	9	9	15	100	72
3	*bla*$_{\text{OXA-51-like}}$ and *bla*$_{\text{OXA-24-like}}$($n = 10$; 12%)	30	–	–	–	70	90	40
4	*bla*$_{\text{OXA-51-like}}$, *bla*$_{\text{OXA-23-like}}$ and *bla*$_{\text{OXA-24-like}}$($n = 5$; 6%)	20	0	0	20	60	100	80
5	*bla*$_{\text{OXA-51-like}}$, *bla*$_{\text{OXA-23-like}}$, *bla*$_{\text{OXA-24-like}}$ and *bla*$_{\text{OXA-58-like}}$ ($n = 1$; 1%)	100	0	0	0	0	100	100
6	*bla*$_{\text{OXA-51-like}}$ and *vim-2* ($n = 2$; 3%)	50	0	0	0	50	50	50

[a]*MIC ≤ 8 ug/ml.*

[b]*DOR, doripenem; IPM, imipenem.*

are outside the standard "S-T-F-K, S-X-I, Y-G-N" and "K-S-G" oxacillinase motifs (Couture et al., 1992), suggesting that configurational changes due to novel mutations may also affect oxacillinase activity against carbapenems.

Although a high frequency of MDR (69%) and XDR (24%) *A. baumannii* isolates from this region is consistent with previous reports (D'Arezzo et al., 2011; Potron et al., 2011; Sung et al., 2011), our finding of 26 and 6% increases in resistance to tigecycline and colistin, respectively, (Bahador et al., 2014) is quite worrisome. Fortunately, while all isolates harboring mutations in $bla_{\text{OXA-23-like}}$ genes showed resistance to carbapenem-class antibiotics; they were susceptible to tigecycline and/or tobramycin, which concurs with a recent report that shows potential activity of a number of combinations against MDR *A. baumannii* (Garnacho-Montero et al., 2015). Our data regarding to a high prevalence of $bla_{\text{OXA-23-like}}$ genes among CR-AB from Tehran confirms previous reports (Shahcheraghi et al., 2011), and implies that extra efforts should be focused on controlling the spread of $bla_{\text{OXA-23-like}}{}^{+}$ *A. baumannii* in this area. Clonal outbreaks of OXA-23-producing CR-AB have been reported from several countries (Mugnier et al., 2010). Interestingly, isolates in this study did not harbor any NDM-1 metallo-β-lactamase genes, nor the "bla_{SPM-1} and bla_{GES-1}" genes, which have recently been reported from India and Pakistan (Jones et al., 2014; Sartor et al., 2014), and Tehran (Shahcheraghi et al., 2011), respectively. However, we detected three mutant $bla_{\text{OXA-23-like}}$ genes with identical sequences reported from China.

The data on the presence of IS*Aba*1 sequences upstream from bla_{OXA} genes and enhancement of OXA–enzyme expression confirms previous reports (Sung et al., 2011); however, these findings are in contrast with a report from northwestern Iran, which detected no IS*Aba*1 sequences upstream of the bla_{OXA} gene (Peymani et al., 2012). Together, these results suggest that the CR-AB populations in various parts of Iran are diverse and distinct, which may hint on probable differences in antimicrobial management of *A. baumannii* infections in various regions. Moreover, most of the $bla_{\text{OXA-23}}$ mutant *A. baumannii* isolates were obtained from urine and sputum samples, suggesting that specific infection control protocols regarding urinary catheters and ventilators are possible primary sources of CR-AB transmission.

Although increased imipenem resistance due to a mutation in the $bla_{\text{OXA-23}}$ gene has been reported previously (Lin et al., 2011), to the best of our knowledge, this is the first report of CR-AB $bla_{\text{OXA-23}}$ gene mutants from Iran. It is noteworthy that the TUMS/BTRF661 strain showed the highest MIC (128 μg/mL), implying a greater influence of the frame-shift mutation on carbapenem resistance than any of the substitution mutations. Our future studies will focus on exploring the difference in the MICs of the various mutants (16 vs. 128 μg/mL) and the potential complex interactions between antimicrobial agents and carbapenemase at the molecular level, where the position of the affected motif plays a critical role. While production of carbapenemase remains to be the chief mechanism of carbapenem-resistance in *A. baumannii*, whether additional factors, such as alterations in outer membrane permeability, efflux pumps as with AdeABC (Potron et al., 2015), or OprD porin (Potron et al., 2015), contribute to carbapenem-resistance amongmutants. The variability in the β-lactamase activity and MIC values of variants that share a mutation, suggests that other factors play a role in high MIC levels among mutant CR-AB isolates, and they remain to be explored.

Assuming future confirmation of the correlation of specific mutations with high MICs, carbapenem resistance levels may be predictable by DNA sequence-based detection methods. Also, determination of predominant $bla_{\text{OXA-23}}$ genotype(s) of isolates in various areas and their $bla_{\text{OXA-23-like}}$ gene mutations may serve as a tool for molecular epidemiologic investigations to control the spread of CR-AB infections. While the present study focused on the chromosomal OXA-encoding genes in CR-AB, we also plan to explore the role of mutations in plasmid-encoded $bla_{\text{OXA-23-like}}$ genes among carbapenem-resistant *A. baumannii*, since many of these genes are plasmid-borne.

In conclusion, we report the identification of CR-AB variants that harbor $bla_{\text{OXA-23-like}}$ gene mutations, which are associated with an increased MIC against imipenem. Several $bla_{\text{OXA-23-like}}$ mutant isolates are widespread and have been reported from the USA, and China. The detection of new $bla_{\text{OXA-23}}$ mutant isolates from Iran highlights the importance of concerted efforts, at the national and global levels, toward the control of carbapenem-resistance among *A. baumannii* isolates worldwide.

FUNDING

This research has been supported by Tehran University of Medical Sciences and Health Services grant no. 89. 01-30-10430.

SUPPLEMENTARY MATERIAL

Supplemental Figure 1 | Presentation of study strategy to select $bla_{\text{OXA-23}}{}^{+}$*A. baumannii* clinical isolates for AFLP genotype determination and DNA sequence analysis. Step-wise selection of blaOXA + isolates was carried out using 2 sets of multiplex PCR assay, followed by AFLP genotypic analysis and DNA sequencing of $bla_{\text{OXA-23-like}}$ gene amplicons.

Supplemental Figure 2 | Agarose gel electrophoresis analysis of PCR amplicons specific for Ambler Class A and B carbapenemases. Genomic DNA from clinical *A. baumannii* isolates were analyzed by uniplex and multiplex PCR assay as described (M&M). Lanes 1 through 6; specific bands for (1) NDM1, (2) IMP1, (3) VIM2, (4) SPM1, (5) bla_{GES}, and (6) *KPC* encoding genes. Lane 7; AB-hexaplex (AB-h) PCR products of the above genes. Lane 8; 100 bp DNA markers.

Supplemental Figure 3 | Comparison between DNA sequences of AB0057Ref.seq and PCR amplicon sequences obtained from mutant CR-AB isolates in genotype groups with >2 members showing a 10-nucleotide span in each direction of the point of mutation (total 21 nucleotides). (A) Alignment of amplicons from three isolates of AFLP group L (i.e., TUMS/BTRF443, L2, and L3) as compared to the AB0057Ref.seq sequence, with the mutation at position 766. **(B)** Alignment of AFLP groups E, C, and K amplicons with the AB0057Ref.seq reference sequence, with the mutation at position 771. In both panels (*) indicates sequence identity (or homology), and (–) shows the position of carbapenemase gene mutation.

REFERENCES

Abbott, I., Cerqueira, G. M., Bhuiyan, S., and Peleg, A. Y. (2013). Carbapenem resistance in *Acinetobacter baumannii*: laboratory challenges, mechanistic insights and therapeutic strategies. *Expert Rev. Anti Infect. Ther.* 11, 395–409. doi: 10.1586/eri.13.21

Alfandari, S., Gois, J., Delannoy, P. Y., Georges, H., Boussekey, N., Chiche, A., et al. (2014). Management and control of a carbapenem-resistant *Acinetobacter baumannii* outbreak in an intensive care unit. *Med. Mal. Infect.* 44, 229–231. doi: 10.1016/j.medmal.2014.03.005

Bahador, A., Bazargani, A., Taheri, M., Hashemizadeh, Z., Khaledi, A., Rostami, H., et al. (2013a). Clonal lineages and virulence factors among *Acinetobacter baumannii* isolated from Southwest of Iran. *J. Pure Appl Microbiol.* 7, 1559–1566.

Bahador, A., Taheri, M., Pourakbari, B., Hashemizadeh, Z., Rostami, H., and Mansoori, N. (2013b). Emergence of rifampicin, tigecycline, and colistin-resistant *Acinetobacter baumannii* in Iran; spreading of MDR strains of novel International Clone variants. *Microb. Drug. Resist.* 19, 397–406. doi: 10.1089/mdr.2012.0233

Bahador, A., Raoofian, R., Taheri, M., Pourakbari, B., Hashemizadeh, Z., and Hashemi, F. B. (2014). Multidrug resistance among *Acinetobacter baumannii* isolates from Iran: changes in antimicrobial susceptibility patterns and genotypic profile. *Microb. Drug Resist.* 20, 632–640. doi: 10.1089/mdr.2013.0146

Clark, R. B. (1996). Imipenem resistance among *Acinetobacter baumannii*: association with reduced expression of a 33-36 kDa outer membrane protein. *J. Antimicrob. Chemother.* 38, 245–251.

Clinical and Laboratory Standards Institute (CLSI) (2015). *Performance Standards for Antimicrobial Susceptibility Testing: Twenty fourth Informational Supplement M100-S25.* Wayne, PA: CLSI.

Couture, F., Lachapelle, J., and Levesque, R. C. (1992). Phylogeny of LCR-1 and OXA-5 with class A and class D β-lactamases. *Mol. Microbiol.* 6, 1693–1705. doi: 10.1111/j.1365-2958.1992.tb00894.x

Daoust, D. R., Onishi, H. R., Wallick, H., Hendlin, D., and Stapley, E. O. (1973). Cephamycins, a new family of beta-lactam antibiotics: antibacterial activity and resistance to β -lactamase degradation. *Antimicrob. Agents Chemother.* 3, 254–261. doi: 10.1128/AAC.3.2.254

D'Arezzo, S., Principe, L., Capone, A., Petrosillo, N., Petrucca, A., and Visca, P. (2011). Changing carbapenemase gene pattern in an epidemic multidrug-resistant *Acinetobacter baumannii* lineage causing multiple outbreaks in central Italy. *J. Antimicrob. Chemother.* 66, 54–61. doi: 10.1093/jac/dkq407

Evans, B. A., and Amyes, S. G. (2014). OXA β-lactamases. *Clin. Microbiol. Rev.* 27, 241–263. doi: 10.1128/CMR.00117-13

Garnacho-Montero, J., Amaya-Villar, R., Ferrándiz-Millón, C., Díaz-Martín, A., López-Sánchez, J. M., Gutiérrez-Pizarraya, A., et al. (2015). Optimum treatment strategies for carbapenem-resistant *Acinetobacter baumannii* bacteremia. *Expert. Rev. Anti. Infect. Ther.* 13, 769–777. doi: 10.1586/14787210.2015.1032254

Higgins, P. G., Dammhayn, C., Hackel, M., and Seifert, H. (2010a). Global spread of carbapenem-resistant *Acinetobacter baumannii*. *J. Antimicrob. Chemother.* 65, 233–238. doi: 10.1093/jac/dkp428

Higgins, P. G., Lehmann, M., Wisplinghoff, H., and Seifert, H. (2010b). gyrB multiplex PCR to differentiate between *Acinetobacter calcoaceticus* and *Acinetobacter* genomic species 3. *J. Clin. Microbiol.* 48, 4592–4594. doi: 10.1128/JCM.01765-10

Jones, L. S., Toleman, M. A., Weeks, J. L., Howe, R. A., Walsh, T. R., and Kumarasamy, K. K. (2014). Plasmid carriage of bla NDM-1 in clinical *Acinetobacter baumannii* isolates from India. *Antimicrob Agents Chemother.* 58, 4211–4213. doi: 10.1128/AAC.02500-14

Kempf, M., and Rolain, J. M. (2011).Emergence of resistance to carbapenems in *Acinetobacter baumannii* in Europe: clinical impact and therapeutic options. *Int. J. Antimicrob. Agents* 39, 105–114. doi: 10.1016/j.ijantimicag.2011.10.004

Lee, H. Y., Chang, R. C., Su, L. H., Liu, S. Y., Wu, S. R., Chuang, C. H., et al. (2012). Wide spread of Tn2006 in an AbaR4-type resistance island among carbapenem-resistant *Acinetobacter baumannii* clinical isolates in Taiwan. *Int. J. Antimicrob. Agents* 40, 163–167. doi: 10.1016/j.ijantimicag.2012.04.018

Lin, M. F., Kuo, H. Y., Yeh, H. W., Yang, C. M., Sung, C. H., Tu, C. C., et al. (2011). Emergence and dissemination of blaOXA-23-carrying imipenem-resistant *Acinetobacter* sp in a regional hospital in Taiwan. *J. Microbiol. Immunol. Infect.* 44, 39–44. doi: 10.1016/j.jmii.2011.01.008

Lu, P. L., Doumith, M., Livermore, D. M., Chen, T. P., and Woodford, N. (2009). Diversity of carbapenem resistance mechanisms in *Acinetobacter baumannii* from a Taiwan hospital: spread of plasmid-borne OXA-72 carbapenemase. *J. Antimicrob. Chemother.* 63, 641–647. doi: 10.1093/jac/dkn553

Magiorakos, A. P., Srinivasan, A., Carey, R. B., Carmeli,Y., Falagas, M. E., Giske, C. G., et al. (2012). Multidrug-resistant, extensively drug-resistant and pandrug-resistant bacteria: an international expert proposal for interim standard definitions for acquired resistance. *Clin. Microbiol. Infect.* 18, 268–281. doi: 10.1111/j.1469-0691.2011.03570.x

Minami, S., Inoue, M., and Mitsuhashi, S. (1980). Purification and properties of cephalosporinase in *Escherichia coli*. *Antimicrob. Agents Chemother.* 18, 77–80. doi: 10.1128/AAC.18.1.77

Moradi, J., Hashemi, F. B., and Bahador, A. (2015). Antibiotic Resistance of *Acinetobacter baumannii* in Iran: a systemic review of the published literature. *Osong. Public. Health. Res. Perspect.* 6, 79–86. doi: 10.1016/j.phrp.2014.12.006

Mugnier, P. D., Poirel, L., Naas, T., and Nordmann, P. (2010). Worldwide dissemination of the blaOXA-23 carbapenemase gene of *Acinetobacter baumannii*. *Emerg. Infect. Dis.*16, 35–40. doi: 10.3201/eid1601.090852

Onishi, H. R., Daoust, D. R., Zimmerman, S. B., Hendlin, D., and Stapley, E. O. (1974). Cefoxitin, a semisynthetic cephamycin antibiotic: resistance to β-lactamase inactivation. *Antimicrob. Agents. Chemother.* 5, 38–48. doi: 10.1128/AAC.5.1.38

Peymani, A., Higgins, P. G., Nahaei, M. R., Farajnia, S., and Seifert, H. (2012). Characterisation and clonal dissemination of OXA-23-producing *Acinetobacter baumannii in* Tabriz, northwest Iran. *Int. J. Antimicrob. Agents* 39, 526–528. doi: 10.1016/j.ijantimicag.2012.02.014

Potron, A., Munoz-Price, L. S., Nordmann, P., Cleary, T., and Poirel, L. (2011). Genetic features of CTX-M-15-producing *Acinetobacter baumannii* from Haiti. *Antimicrob. Agents Chemother.* 55, 5946–5948. doi: 10.1128/AAC.05124-11

Potron, A., Poirel, L., and Nordmann, P. (2015). Emerging broad-spectrum resistance in *Pseudomonas aeruginosa* and *Acinetobacter baumannii*: mechanisms and epidemiology. *Int. J. Antimicrob. Agents* 45, 568–585. doi: 10.1016/j.ijantimicag.2015.03.001

Safari, M., Saidijam, M., Bahador, A., Jafari, R., and Alikhani, M. Y. (2013). High prevalence of multidrug resistance and metallo- β-lactamase (MβL) producing *Acinetobacter baumannii* isolated from patients in ICU wards, Hamadan, Iran. *J. Res. Health Sci.* 17, 162–167.

Saino, Y., Kobayashi, F., Inoue, M., and Mitsuhashi, S. (1982). Purification and properties of inducible penicillin β -lactamase isolated from *Pseudomonas maltophilia*. *Antimicrob. Agents Chemother.* 22, 564–570. doi: 10.1128/AAC.22.4.564

Sartor, A. L., Raza, M. W., Abbasi, S. A., Day, K. M., Perry, J. D., Paterson, D. L., et al. (2014). Molecular epidemiology of NDM-1-producing *Enterobacteriaceae* and *Acinetobacter baumannii* isolates from Pakistan. *Antimicrob. Agents Chemother.* 58, 5589–5593. doi: 10.1128/AAC.02425-14

Sawai, T., Takahashi, I., and Yamagishi, S. (1978). Iodometric assay method for β-lactamase with various β-lactam antibiotics as substrates. *Antimicrob. Agents Chemother.* 13, 910–913. doi: 10.1128/AAC.13.6.910

Shahcheraghi, F., Abbasalipour, M., Feizabadi, M., Ebrahimipour,G., and Akbari, N. (2011). Isolation and genetic characterization of metallo-β-lactamase and carbapenamase producing strains of *Acinetobacter baumannii* from patients at Tehran hospitals. *Iran. J. Microbiol.* 3, 68–74.

Siroy, A., Molle, V., Lemaître-Guillier, C., Vallenet, D., Pestel-Caron, M., Cozzone, A. J., et al. (2005). Channel formation by CarO, the carbapenem resistance-associated outer membrane protein of *Acinetobacter baumannii*. *Antimicrob Agents Chemother.* 49, 4876–4883. doi: 10.1128/AAC.49.12.4876-4883.2005

Sung, J. Y., Kwon, K. C., Cho, H. H., and Koo, S. H. (2011). Antimicrobial resistance determinants in imipenem-nonsusceptible *Acinetobacter calcoaceticus-baumannii* complex isolated in Daejeon, Korea. *Korean. J. Lab. Med.* 31, 265–270. doi: 10.3343/kjlm.2011.31.4.265

The European Committee on Antimicrobial Susceptibility Testing (EUCAST) (2015). *Breakpoint Tables for Interpretation of MICs and Zone Diameters Version 5.0.* Available online at: http://www.eucast.org/clinical_breakpoints [Accessed January 01, 2015].

Turton, J. F., Ward, M. E., Woodford, N., Kaufmann, M. E., Pike, R., Livermore, D. M., et al. (2006). The role of ISAba1 in expression of OXA carbapenemase genes in *Acinetobacter baumannii. FEMS. Microbiol. Lett.* 258, 72–77. doi: 10.1111/j.1574-6968.2006.00195.x

Vos, P., Hogers, R., Bleeker, M., Reijans, M., van de Lee, T., Hornes, M., et al. (1995). AFLP: a new technique for DNA fingerprinting. *Nucleic.Acids. Res.* 23, 4407–4414. doi: 10.1093/nar/23.21.4407

Woodford, N., Ellington, M. J., Coelho, J. M., Turton, J. F., Ward, M. E., Brown, S., et al. (2006). Multiplex PCR for genes encoding prevalent OXA carbapenemases in *Acinetobacter* spp. *Int. J. Antimicrob. Agents* 27, 351–353. doi: 10.1016/j.ijantimicag.2006.01.004

Permissions

 Every chapter published in this book has been scrutinized by our experts. Their significance has been extensively debated. The topics covered herein carry significant findings which will fuel the growth of the discipline. They may even be implemented as practical applications or may be referred to as a beginning point for another development.

The contributors of this book come from diverse backgrounds, making this book a truly international effort. This book will bring forth new frontiers with its revolutionizing research information and detailed analysis of the nascent developments around the world.

We would like to thank all the contributing authors for lending their expertise to make the book truly unique. They have played a crucial role in the development of this book. Without their invaluable contributions this book wouldn't have been possible. They have made vital efforts to compile up to date information on the varied aspects of this subject to make this book a valuable addition to the collection of many professionals and students.

This book was conceptualized with the vision of imparting up-to-date information and advanced data in this field. To ensure the same, a matchless editorial board was set up. Every individual on the board went through rigorous rounds of assessment to prove their worth. After which they invested a large part of their time researching and compiling the most relevant data for our readers.

The editorial board has been involved in producing this book since its inception. They have spent rigorous hours researching and exploring the diverse topics which have resulted in the successful publishing of this book. They have passed on their knowledge of decades through this book. To expedite this challenging task, the publisher supported the team at every step. A small team of assistant editors was also appointed to further simplify the editing procedure and attain best results for the readers.

Apart from the editorial board, the designing team has also invested a significant amount of their time in understanding the subject and creating the most relevant covers. They scrutinized every image to scout for the most suitable representation of the subject and create an appropriate cover for the book.

The publishing team has been an ardent support to the editorial, designing and production team. Their endless efforts to recruit the best for this project, has resulted in the accomplishment of this book. They are a veteran in the field of academics and their pool of knowledge is as vast as their experience in printing. Their expertise and guidance has proved useful at every step. Their uncompromising quality standards have made this book an exceptional effort. Their encouragement from time to time has been an inspiration for everyone.

The publisher and the editorial board hope that this book will prove to be a valuable piece of knowledge for researchers, students, practitioners and scholars across the globe.

List of Contributors

Xuan Wang, Gongxiang Chen, Jiachang Cai and Rong Zhang
Second Affiliated Hospital of Zhejiang University, Zhejiang University, Hangzhou, China

Xiaoyan Wu
Second People's Hospital of Jiaxing, Jiaxing, China

Liangping Wang
People's Hospital of Pinghu, Jiaxing, China

Ricarda Sassmannshausen
Institute of Hygiene, University Hospital Münster, Münster, Germany

Ruud H. Deurenberg, John W. A. Rossen and Alexander W. Friedrich
Department of Medical Microbiology, University of Groningen, University Medical Center Groningen, Groningen, Netherlands

Robin Köck
Institute of Hygiene, University Hospital Münster, Münster, Germany
Institute of Medical Microbiology, University Hospital Münster, Münster, Germany

Ron Hendrix
Department of Medical Microbiology, University of Groningen, University Medical Center Groningen, Groningen, Netherlands
CERTE-LvI, Groningen, Netherlands

Annette Jurke
Department of Infectiology and Hygiene, Centre for Health North Rhine-Westphalia, Münster, Germany

Michaela Lipp, Rita Baumert, Bettina Folli, Andrea J. Grisold and Gernot Zarfel
Institute of Hygiene, Microbiology and Environmental Medicine, Medical University Graz, Graz, Austria

Günther Koraimann
Institute of Molecular Biosciences, University of Graz, Graz, Austria

Daniela Toplitsch and Astrid Liebmann
Institute of Hygiene, Microbiology and Environmental Medicine, Medical University Graz, Graz, Austria
Institute of Molecular Biosciences, University of Graz, Graz, Austria

Andreas H. Farnleitner
Interuniversity Cooperation Centre for Water and Health, Vienna, Austria
Research Group Environmental Microbiology and Molecular Ecology, Institute of Chemical Engineering, Vienna University of Technology, Vienna, Austria

Alexander Kirschner
Interuniversity Cooperation Centre for Water and Health, Vienna, Austria
Institute for Hygiene and Applied Immunology, Medical University of Vienna, Vienna, Austria

Carla Novais, Eduarda Silveira and Luísa Peixe
UCIBIO/REQUIMTE, Laboratório de Microbiologia, Departamento de Ciências Biológicas, Faculdade Farmácia, Universidade do Porto, Porto, Portugal

Ana P. Tedim and Val F. Lanza
Servicio de Microbiología, Instituto Ramón y Cajal de Investigación Sanitaria, Madrid, Spain
Consorcio de Investigación Biomédica en Red de Epidemiología y Salud Pública, Barcelona, Spain

Ana R. Freitas
UCIBIO/REQUIMTE, Laboratório de Microbiologia, Departamento de Ciências Biológicas, Faculdade Farmácia, Universidade do Porto, Porto, Portugal
Servicio de Microbiología, Instituto Ramón y Cajal de Investigación Sanitaria, Madrid, Spain

Ricardo Escada
UCIBIO/REQUIMTE, Laboratório de Microbiologia, Departamento de Ciências Biológicas, Faculdade Farmácia, Universidade do Porto, Porto, Portugal
Faculdade de Ciências da Saúde, Universidade Fernando Pessoa, Porto, Portugal

Fernando Baquero and Teresa M. Coque
Servicio de Microbiología, Instituto Ramón y Cajal de Investigación Sanitaria, Madrid, Spain
Consorcio de Investigación Biomédica en Red de Epidemiología y Salud Pública, Barcelona, Spain
Unidad de Resistencia a Antibióticos y Virulencia Bacteriana (RYC-CSIC), Madrid, Spain
Hospital Universitario Ramon y Cajal, Instituto Ramón y Cajal de Investigación Sanitaria, Madrid, Spain
CIBER Epidemiología y Salud Pública, Madrid, Spain
Unidad de Resistencia a Antibióticos y Virulencia Bacteriana Asociada al Consejo Superior de Investigaciones Científicas, Madrid, Spain

Adam P. Roberts and Mohammed Al-Haroni
Division of Microbial Diseases, UCL Eastman Dental Institute, University College London, London, UK

Ana M. Guzman Prieto, Willem van Schaik, Malbert R. C. Rogers and Rob J. L. Willems
Department of Medical Microbiology, University Medical Center Utrecht, Utrecht, Netherlands

Jukka Corander
Department of Mathematics and Statistics, University of Helsinki, Helsinki, Finland

Anirban Sarkar, Goutam Chowdhury and Amit Ghosh
Department of Bacteriology, National Institute of Cholera and Enteric Diseases, Kolkata, India

Gururaja P. Pazhani
Department of Bacteriology, National Institute of Cholera and Enteric Diseases, Kolkata, India
Centre for Drug Discovery and Development, Sathyabama University, Chennai, India

Thandavarayan Ramamurthy
Department of Bacteriology, National Institute of Cholera and Enteric Diseases, Kolkata, India
Center for Human Microbial Ecology, Translational Health Science and Technology Institute, Faridabad, India

Marcus H. Wong, Edward W. Chan and Sheng Chen
Shenzhen Key Laboratory for Food Biological Safety Control, Food Safety and Technology Research Centre, Shenzhen Research Institute, The Hong Kong Polytechnic University, Shenzhen, China
State Key Laboratory of Chirosciences, Department of Applied Biology and Chemical Technology, The Hong Kong Polytechnic University, Hong Kong, China

Yi Li
Laboratory of Clinical Inspection, Henan Provincial People's Hospital, Zhengzhou, China

Michael Iwersen, Marc Drillich and Daniela Klein-Jöbstl
Clinical Unit for Herd Health Management, University Clinic for Ruminants, Department for Farm
Animals and Veterinary Public Health, University of Veterinary Medicine Vienna, Vienna, Austria

Dmitri Sofka and Friederike Hilbert
Institute of Meat Hygiene, Meat Technology and Food Science, Department for Farm Animals and Veterinary Public Health, University of Veterinary Medicine Vienna, Vienna, Austria

Yaowen Chang, Chi Zhang, Wei Zheng and Baodong Ling
Small Molecule Drugs Sichuan Key Laboratory, Institute of Materia Medica, Chengdu Medical College, Chengdu, China

Guangxin Luan, Yanhong Wang, Min Shen, Jingni Yang and Xu Jia
Non-coding RNA and Drug Discovery Laboratory, Chengdu Medical College, Chengdu, China

Ying Xu
Clinical Laboratory, The First Affiliated Hospital, Chengdu Medical College, Chengdu, China

Jinwei Huang
Institute of Antibiotics, The Fifth Affiliated Hospital, Wenzhou Medical University, Lishui, China

Vicenta Cázares-Domínguez, Sara A. Ochoa, Ariadnna Cruz-Córdova, Gerardo E. Rodea, Gerardo Escalona and Juan Xicohtencatl-Cortes
Laboratorio de Investigación en Bacteriología Intestinal, Hospital Infantil de México Federico Gómez, México DF, Mexico

Alma L. Olivares and Norma Velázquez-Guadarrama
Laboratorio de Infectología, Hospital Infantil de México Federico Gómez, México DF, Mexico

José de Jesús Olivares-Trejo
Posgrado en Ciencias Genómicas, Universidad Autónoma de la Ciudad de México, México DF, Mexico

Chang-Ro Lee, Jung Hun Lee, Kwang Seung Park, Byeong Chul Jeong and Sang Hee Lee
National Leading Research Laboratory of Drug Resistance Proteomics, Department of Biological Sciences, Myongji University, Yongin, South Korea

Young Bae Kim
Division of STEM, North Shore Community College, Danvers, MA, USA

Adebayo O. Shittu
Department of Microbiology, Obafemi Awolowo University, Ile-Ife, Nigeria
Institute of Medical Microbiology and Hygiene, Saarland University, Homburg, Germany

Lutz von Müller, Mathias Herrmann and Ulla Ruffing
Institute of Medical Microbiology and Hygiene, Saarland University, Homburg, Germany

Omotayo Oyedara
Department of Biological Sciences, College of Science, Engineering and Technology, Osun State University, Osogbo, Nigeria

Kenneth Okon
Department of Medical Laboratory Services, Federal Medical Centre, Makurdi, Nigeria

Adeola Raji
Department of Microbiology and Immunology, Alfaisal University, Riyadh, Saudi Arabia

Georg Peters and Frieder Schaumburg
Institute of Medical Microbiology, University Hospital Münster, Münster, Germany

Sum Lean and Kwai-Lin Thong
Faulty of Science, Institute of Biological Sciences, Universiti Malaya, Kuala Lumpur, Malaysia

Chew Chieng Yeo
Faculty of Medicine, Biomedical Research Centre, Universiti Sultan Zainal Abidin, Kuala Terengganu, Malaysia

Zarizal Suhaili
Faculty of Bioresources and Food Industry, Universiti Sultan Zainal Abidin, Kuala Terengganu, Malaysia

Jinhu Huang, Yuan Liang, Dawei Guo, Kexin Shang, Lin Ge and Liping Wang
College of Veterinary Medicine, Nanjing Agricultural University, Nanjing, China

Jam Kashif
Department of Veterinary Pharmacology, Sindh Agricultural University, Tandojam, Pakistan

Zhen Shen, Miao Zhao and Li Sun
Institute of Antibiotics, Huashan Hospital, Fudan University, Shanghai, China

Xiaofen Liu
Institute of Antibiotics, Huashan Hospital, Fudan University, Shanghai, China
Roche Innovation Center Shanghai, Shanghai, China

Jun Shi
Roche Innovation Center Shanghai, Shanghai, China

Huajun Zheng
Shanghai-MOST Key Laboratory of Health and Disease Genomics, Chinese National Human Genome Center at Shanghai, Shanghai, China

Yuancheng Chen and Jing Zhang
Institute of Antibiotics, Huashan Hospital, Fudan University, Shanghai, China
Key Laboratory of Clinical Pharmacology of Antibiotics, National Population and Family Planning Commission, Shanghai, China

Weipeng Zhang
Division of Life Science, The Hong Kong University of Science and Technology, Hong Kong, China

Farshadzadeh, Farhad B. Hashemi, Maryam Rahmani, Samira Gharesi and Abbas Bahador
Department of Microbiology, School of Medicine, Tehran University of Medical Sciences, Tehran, Iran

Sara Rahimi, Mohammad A. Haghighi and Masoud Aziemzadeh
Department of Microbiology, School of Medicine, Bushehr University of Medical Sciences, Bushehr, Iran

Babak Pourakbari
Pediatrics Infectious Disease Research Center, School of Medicine, Tehran University of Medical Sciences, Tehran, Iran

Davoud Esmaeili
Molecular Biology Research Center, Baqiyatallah University of Medical Sciences, Tehran, Iran

Ali Majidpour
Anti-microbial Resistance Research Center, Iran University of Medical Sciences, Tehran, Iran

Saeed Shojaa
Department of Microbiology, Faculty of Medicine, Hormozgan University of Medical Sciences, Bandar Abbas, Iran

Richard Haigh and Zaaima J. H. Al Jabri
Department of Genetics, University of Leicester, Leicester, UK

Leonardo Furi and Marco R. Oggioni
Department of Genetics, University of Leicester, Leicester, UK
Dipartimento di Biotecnologie Mediche, Universita di Siena, Siena, Italy

Ian Morrissey
IHMA Europe Sàrl, Epalinges, Switzerland

Hong-Yu Ou
State Key Laboratory for Microbial Metabolism and School of Life Sciences and Biotechnology, Shanghai Jiaotong University, Shanghai, China

Ricardo León-Sampedro
Departamento de Microbiología, Instituto Ramón y Cajal de Investigación Sanitaria, Hospital Universitario Ramón y Cajal, Madrid, Spain
Centro de Investigación Biomédica en Red de Epidemiología y Salud Pública (CIBERESP), Spain

Jose L. Martinez
Departamento de Biotecnología Microbiana, Centro Nacional de Biotecnología, Consejo Superior de Investigaciones Científicas, Madrid, Spain
Unidad de Resistencia a Antibióticos y Virulencia Bacteriana (RYC-Consejo Superior de Investigaciones Científicas), Madrid, Spain

Shima M. Abdulgader
Division of Medical Microbiology, Department of Clinical Laboratory Sciences, Faculty of Health Sciences, University of Cape Town, Cape Town, South Africa

Mamadou Kaba
Division of Medical Microbiology, Department of Clinical Laboratory Sciences, Faculty of Health Sciences, University of Cape Town, Cape Town, South Africa
Institute of Infectious Disease and Molecular Medicine, Faculty of Health Sciences, University of Cape Town, Cape Town, South Africa

Mark P. Nicol
Division of Medical Microbiology, Department of Clinical Laboratory Sciences, Faculty of Health Sciences, University of Cape Town, Cape Town, South Africa
Institute of Infectious Disease and Molecular Medicine, Faculty of Health Sciences, University of Cape Town, Cape Town, South Africa
National Health Laboratory Service, Groote Schuur Hospital, Cape Town, South Africa

Reza Raoofian
Legal Medicine Research Center, Legal Medicine Organization, Tehran, Iran
Innovative Medical Research Center, Islamic Azad University, Mashhad, Iran

Mohammad Taheri and Zahra Hashemizadeh
Department of Microbiology and Virology, Shiraz University of Medical Sciences, Shiraz, Iran

Index

www.ingramcontent.com/pod-product-compliance
Lightning Source LLC
LaVergne TN
LVHW080948210926
848485LV00004B/798